Handbook of Research on Energy-Saving Technologies for Environmentally-Friendly Agricultural Development

Valeriy Kharchenko
Federal Scientific Agroengineering Center VIM, Russia

Pandian Vasant
Universiti Teknologi Petronas, Malaysia

A volume in the Advances in Environmental Engineering and Green Technologies (AEEGT) Book Series

IGI Global
DISSEMINATOR OF KNOWLEDGE

Published in the United States of America by
IGI Global
Engineering Science Reference (an imprint of IGI Global)
701 E. Chocolate Avenue
Hershey PA, USA 17033
Tel: 717-533-8845
Fax: 717-533-8661
E-mail: cust@igi-global.com
Web site: http://www.igi-global.com

Library of Congress Cataloging-in-Publication Data

Names: Kharchenko, Valeriy, 1938- editor. | Vasant, Pandian, editor.
Title: Handbook of research on energy-saving technologies for environmentally-friendly agricultural development / Valeriy Kharchenko and Pandian Vasant, editors.
Description: Hershey, PA : Engineering Science Reference, 2019.
Identifiers: LCCN 2019001912| ISBN 9781522594208 (hardcover) | ISBN 9781522594215 (ebook)
Subjects: LCSH: Agriculture--Energy consumption. | Rural development.
Classification: LCC S494.5.E5 E542 2019 | DDC 338.1--dc23 LC record available at https://lccn.loc.gov/2019001912

This book is published in the IGI Global book series Advances in Environmental Engineering and Green Technologies (AEEGT) (ISSN: 2326-9162; eISSN: 2326-9170)

British Cataloguing in Publication Data
A Cataloguing in Publication record for this book is available from the British Library.

For electronic access to this publication, please contact: eresources@igi-global.com.

Advances in Environmental Engineering and Green Technologies (AEEGT) Book Series

Sang-Bing Tsai
University of Electronic Science and Technology of China Zhongshan Institute, China
Ming-Lang Tseng
Lunghwa University of Science and Technology, Taiwan
Yuchi Wang
University of Electronic Science and Technology of China Zhongshan Institute, China

ISSN:2326-9162
EISSN:2326-9170

Mission

Growing awareness and an increased focus on environmental issues such as climate change, energy use, and loss of non-renewable resources have brought about a greater need for research that provides potential solutions to these problems. Research in environmental science and engineering continues to play a vital role in uncovering new opportunities for a "green" future.

The **Advances in Environmental Engineering and Green Technologies (AEEGT)** book series is a mouthpiece for research in all aspects of environmental science, earth science, and green initiatives. This series supports the ongoing research in this field through publishing books that discuss topics within environmental engineering or that deal with the interdisciplinary field of green technologies.

Coverage

- Industrial Waste Management and Minimization
- Radioactive Waste Treatment
- Renewable Energy
- Policies Involving Green Technologies and Environmental Engineering
- Biofilters and Biofiltration
- Water Supply and Treatment
- Waste Management
- Electric Vehicles
- Pollution Management
- Green Transportation

IGI Global is currently accepting manuscripts for publication within this series. To submit a proposal for a volume in this series, please contact our Acquisition Editors at Acquisitions@igi-global.com or visit: http://www.igi-global.com/publish/.

The Advances in Environmental Engineering and Green Technologies (AEEGT) Book Series (ISSN 2326-9162) is published by IGI Global, 701 E. Chocolate Avenue, Hershey, PA 17033-1240, USA, www.igi-global.com. This series is composed of titles available for purchase individually; each title is edited to be contextually exclusive from any other title within the series. For pricing and ordering information please visit http://www.igi-global.com/book-series/advances-environmental-engineering-green-technologies/73679. Postmaster: Send all address changes to above address.

Titles in this Series

For a list of additional titles in this series, please visit: www.igi-global.com/book-series

Advanced Multi-Criteria Decision Making for Addressing Complex Sustainability Issues
Prasenjit Chatterjee (MCKV Institute of Engineering, India) Morteza Yazdani (Universidad Loyola Andalucía, Spain) Shankar Chakraborty (Jadavpur University, India) Dilbagh Panchal (Dr. B. R. Ambedkar National Institute of Technology (NIT) Jalandhar, India) and Siddhartha Bhattacharyya (RCC Institute of Information Technology Kolkata, India)
Engineering Science Reference • ©2019 • 360pp • H/C (ISBN: 9781522585794) • US $195.00

Amelioration Technology for Soil Sustainability
Ashok K. Rathoure (Biohm Consultare Pvt Ltd, India)
Engineering Science Reference • ©2019 • 280pp • H/C (ISBN: 9781522579403) • US $185.00

Advanced Agro-Engineering Technologies for Rural Business Development
Valeriy Kharchenko (Federal Scientific Agroengineering Center VIM, Russia) and Pandian Vasant (Universiti Teknologi PETRONAS, Malaysia)
Engineering Science Reference • ©2019 • 484pp • H/C (ISBN: 9781522575733) • US $195.00

Spatial Planning in the Big Data Revolution
Angioletta Voghera (Politecnico di Torino, Italy) and Luigi La Riccia (Politecnico di Torino, Italy)
Engineering Science Reference • ©2019 • 359pp • H/C (ISBN: 9781522579274) • US $195.00

Global Initiatives for Waste Reduction and Cutting Food Loss
Aparna B. Gunjal (Asian Agri Food Consultancy Services Ltd, India) Meghmala S. Waghmode (Annasaheb Magar Mahavidyalaya, India) Neha N. Patil (Annasaheb Magar Mahavidyalaya, India) and Pankaj Bhatt (Dolphin (P.G) College of Biomedical and Natural Sciences Dehradun, India)
Engineering Science Reference • ©2019 • 328pp • H/C (ISBN: 9781522577065) • US $195.00

Green Public Procurement Strategies for Environmental Sustainability
Rajesh Kumar Shakya (The World Bank, USA)
Engineering Science Reference • ©2019 • 228pp • H/C (ISBN: 9781522570837) • US $185.00

Climate Change and Its Impact on Ecosystem Services and Biodiversity in Arid and Semi-Arid Zones
Ahmed Karmaoui (Southern Center for Culture & Sciences (SCCS), Morocco)
Engineering Science Reference • ©2019 • 408pp • H/C (ISBN: 9781522573876) • US $225.00

IGI Global
DISSEMINATOR OF KNOWLEDGE

701 East Chocolate Avenue, Hershey, PA 17033, USA
Tel: 717-533-8845 x100 • Fax: 717-533-8661
E-Mail: cust@igi-global.com • www.igi-global.com

Editorial Advisory Board

List of Contributors

Table of Contents

Detailed Table of Contents

This chapter studies the prospects of energy complexes on the basis of renewable energy sources to supply electricity to the stand-alone consumers in different regions of Myanmar. In order to do that, the territory of Myanmar is divided into regions according to their amount of renewable energy sources. The developed methods are for determining the optimum parameters and operation of the energy complex on the basis of renewable energy sources and the cost-effectiveness of those energy complexes was examined. This was for the purpose of a mathematical formulation of the problem of optimization of the energy complex on the basis of renewable energy sources for autonomous rural consumers of the Republic of Myanmar.

At agricultural facilities, the main attention is paid to the formation and maintenance of their microclimate parameters, and mechanization of storage processes. As world experience shows, it is necessary to develop and implement energy-saving systems and the use of renewable energy sources. The authors have developed energy-saving systems based on the heat pump, with upgraded electrical regulators. The developed system (patent 100873), uses thermoelectric elements and a low-potential energy source, to effectively maintain the temperature parameters of the microclimate during long-term storage of potatoes, but it requires a large amount of electricity consumption (30 to 35 kW), so the authors have developed an energy-saving system based on a heat pump (patent 123909). The temperature regime is achieved by using a thermoelectric cooler-heater and an electric heater. The humidifier allows for maintaining the necessary relative air humidity.

Currently, there is an increase in the use of gas turbines. Today they are used in the energy sector: aviation, armed forces, and the navy. The introduction of a new manufacturing technology developed by the authors will make it possible to manufacture cheap and reliable installations and thus ensure an exceptional position on the Russian market for goods and technologies, and taking into account the use of intellectual rights, abroad. The scientific novelty of the sample is the method of calculating small engines with a centrifugal compressor, a centripetal turbine and a combustion chamber with a negative thrust vector of the air flow. It is shown that the developed microgas turbine cogeneration power generator consists of a microturbine engine with a periphery, a free power turbine necessary for the selection of mechanical power, a high-speed electric generator with permanent magnets, an electronic power conversion system, exhaust heat energy recovery system and an automatic control system.

Mono-mode microwave reactors are usually used to heat substances, especially food. This is because heating using a microwave reactor can sustain the flavor, color, and nutrition of the food. Furthermore, this heating technique is cost-effective and time-saving compared to a conventional heating method. The mono-mode reactor is able to determine the absorption of microwave power accurately on the heated substance versus a multimode reactor. In this chapter, a simple and precise mono-mode microwave reactor is designed and developed especially for research laboratories. The advantage of this reactor is to provide a more accurate calibration process, in order to improve the optimum energy use in the heating process, as well as the temperature of the specimen. The reactor can generate output power from 30 watts to 1500 watts, operating at 2.45±0.03 GHz and capable of accommodating a specimen volume of 780 cm3. Pure water is used as a heated specimen to demonstrate the performance and efficiency of this reactor.

Dedicated vertical axial loads on the soil from the wheels of a self-propelled seed drill, the area of the contact patch, the maximum contact pressure for the front and rear wheels and the density of the soil are determined by evaluations and experimental methods. The discrepancy between the theoretical and experimental indicators was: 1.4% and 2.0% for the rear and front wheels in vertical axial loads; 2.8% and 2.2% for the rear and front wheels by the contact area of the tires of the seeder with the soil and the maximum contact pressure; 6.2% – the maximum discrepancy on the values of soil density at a depth of 7.6 cm. Soil hardness was measured in three zones: before the seeder's passage and after each of its passage in a rut behind the front and rear wheels at six different depths, determined by the marks on the soil densimeter tester density. Graphics of dependencies of soil hardness on the depth of measurement were constructed.

A new type of linear cooled photodetectors is considered, of which in the focal region of the optical concentrator mirrors is installed an array of solar cells operating with the low-ratio solar concentration. This work is focused on the theoretical and experimental substantiation of the efficiency increase of photodetectors under conditions of an optimal combination between solar radiation concentration and heat transfer intensity of photovoltaic cells with heat carriers. The problem of obtaining a high temperature liquid due to the limitations of solar cells is solved by organizing the flow of fluid within the thermal collector channels in the focal region of an additional optical concentrator. A mathematical model of engineering characteristic calculation of the Λ - shaped photodetectors and cost calculation of electrical and thermal energy generation is presented. The research results are used in the development of industrial prototypes of photodetectors with a concentration of solar radiation and low production costs.

It is known that in the case of technology use of the supplementary lighting, an irradiation spectral composition heavily influences the effectiveness of the photosynthesis processes, development and productivity of vegetable crops. Hence, the definition of general points at development and projecting of modern phyto-irradiators is one of high-priority tasks in techniques development for plants growing in conditions of protected ground. The research is aimed at reviewing and assessing the effectiveness of existing sources of illumination used in modern systems of supplementary lighting and at deduction of general points of development and projecting of phyto-irradiators based on results of laboratory investigations with the use of modern digital technologies of monitoring and data analysis. The results of the comparative tests of light emitting diodes-based phyto-irradiators showed that the energy consumption per product kilogram is less than in the case of LED-irradiators. Based on the research results, general points were deducted for use at development of modern LED-phyto-irradiators.

The purpose of the research was to establish the mechanism of the magnetic field impact on seeds to determine the most effective mode of pre-sowing treatment of seeds in a magnetic field and design parameters of the device for magnetic treatment of seeds. It is established that under the influence of a magnetic field the rate of chemical reactions occurring in plant cells is accelerated, solubility of salts and acids increases, and permeability of cell membranes accelerates the diffusion of molecules and ions through them. This leads to an increase in the concentration of ions in the cell and oxygen molecules and the growth of water absorption of seeds. Pre-sowing seed treatment promotes increased germination by 25-40%, and germination by 30 - 35%. The most effective pre-sowing treatment of seeds in a magnetic field is a magnetic induction of 0.065 Tl with four reversal magnetization, a pole division of 0.23 m and a seed movement speed of 0.4 m/s. With this mode of treatment, crop yields increase by 20–25%.

An energy audit of the electrical network is required in the process of constructing new electrical networks as well as in justifying the reconstruction need of existing ones. In this chapter, the structure of a mobile measuring complex has been developed to conduct an electrical network survey without disconnecting consumers. The complex can be used to inspect 0.4 kV electrical networks and microgrids of the same voltage class and allows data collection on voltage losses and electric power losses in network elements such as a power lines (electric transmission line), and power transformers. The energy audit is conducted without disconnecting consumers in order to avoid an undersupply of electricity as well as to determine the real operating modes of power networks. Ultimately, the use of the developed measuring complex will increase the reliability of the power supply to consumers and ensure the required quality of the electricity supplied to them.

This chapter discusses the simulation of solar photovoltaic thermal modules of planar and concentrator structures in computer-aided design systems KOMPAS 3D and finite element analysis ANSYS. To create photovoltaic thermal modules, a method for designing their three-dimensional models in the computer-aided design system has been developed. To study the thermal regimes of the created three-dimensional models of modules, a method has been developed for visualizing thermal processes, coolant velocity, and flow lines of a cooling agent in a finite element analysis system. As a result of calculations in the finite element analysis system using the developed method, conclusions can be drawn about the feasibility of the design created with its further editing, visualization of thermal fields, and current lines of the radiator cooling agent. As an illustration of the simulation results, a three-dimensional model of a photovoltaic thermal planar roofing panel and an optimized three-dimensional model of a photodetector of a solar concentrator photovoltaic thermal module are presented.

To study the possibilities of pre-sowing seed activation, irradiation with LEDs emitting in the visible, violet, and near-ultraviolet ranges with a maximum of 405 nm was carried out. As a result of the growing experience, it was found that the height of wheat plants grown from the treated seeds significantly exceeds the control indicators except for the period of 45-55 days. To implement the flow, technology of seed activation with LEDs optoelectronic irradiation unit was developed. The advantages of the given installation are the energy efficiency and of seed treatment efficiency, due to of the optimal radiation spectrum selection and treatment doses.

Use of a variety of electrotechnics is a technologically-efficient and environmentally-friendly technique that increases the productivity of cultivated plants. Stimulation of green plants and vegetable crops in electric field with the intensity of 5-50 kV/m made it clear that the maximum efficiency is observed in the growth period – an increase of up to 30%, compared to the control. Plants have been subjected to stimulation for 3 hours twice a day (in the morning and in the evening). Analysis of studies on the pre-seeding seed stimulation showed that the exposure to pulsed magnetic field improves the dynamics of germination and plant growth at the early stages of development by an average of 10-20%, and more uniform germination helps to ensure high yields.

This chapter is dedicated to establishing characteristics relationships between the induction type impact electromechanical transducer and parameters and quality indicators of electrodynamic effects on the welded joints. The authors developed two-dimensional circle-field mathematical model of transient discharge capacity at the branched electrical circuit with the coil inductance which changes dynamically, allowing by adjusting the parameters of the electromechanical transducer to achieve the necessary technological requirements for the characteristics of electrodynamic processing. Based on mathematical modeling of electrophysical processes in electromechanical transducers, induction type for electrodynamic processing of welded joints, reasonably geometrical parameters massive disk, and the contact area, the necessary conditions are created to reduce residual stresses in the weld joints.

In this chapter, the artificial bee colony (ABC) algorithm was applied to optimize hybrids of wind and solar renewable energy system. The main objective of this research is to minimize the total annual cost of the system by determining appropriate numbers of wind turbine, solar panel, and batteries, so that the desired load can be economically and reliably satisfied based on the given constraints. ABC is a recently proposed meta heuristic algorithm which is inspired by the intelligent behavior of honey bees such as searching for food source and collection and processing of nectar. Instead of gradient and Hessian matrix information, ABC uses stochastic rules to escape local optima and find the global optimal solutions. The proposed methodology was applied to this hybrid system by the help of MATLAB code and the results were discussed. Additionally, it is shown that ABC can be efficiently solve the optimum sizing real-world problems with high convergence rate and reliability. The result was compared with the results of PSO.

Foreword

Handbook of Research on Energy-Saving Technologies for Environmentally-Friendly Agricultural Development is a competent and reckonable research-intensive textbook covering subjects of sustained technical and economic importance worldwide. Due to an increase of carbon emissions contributing to climate change, causing fluctuating impacts on both agricultural plants seeds and biological objects, the need of energy efficient and renewable energy technologies to minimize these effects are being echoed in this book for the environment and sustainable research community. This book supports the contents relevant to multi-disciplinary master's degree in agricultural science and renewable energy engineering.

This book cogitates the timeless principles of energy efficient technologies for agriculture farming system and seeks to demonstrate modern application and case studies. It is perceived that the farms were considered as higher energy consumers and this book investigates it with the development of energy saving technology. It benefits to the performance methods of energy audit for the existing and planned electrical networks without disconnecting agricultural consumers but to improve the reliability of the supply network with microgrids. This would be convenient to incorporate the solar power plants in rural areas and/or farms. In this book, the use of novel algorithm named, Artificial Bee Colony, is discussed to improve the performance of hybrid wind/solar renewable energy system and it minimizes the total annual cost of it with the number of wind mills, solar panels and energy storage batteries calculations, this improves high convergence rate and reliability. This book also presents research findings on a new type of linear cooled photodetectors, theoretically and experimentally investigates the efficiency increment of photodetectors under the conditions of an optimal combination between solar radiation concentration and heat transfer intensity of photovoltaic cells with heat carriers. This book also pertinently studies the influence of solar irradiance onto the effectiveness of the photosynthesis processes and productivity of vegetable crops with phyto-irradiators. Subsequently, this research is aimed at reviewing the evaluation of existing sources of illumination used in modern systems of supplementary lighting and at deduction of general points of development and projection of phyto-irradiators with use of modern digital technologies of monitoring and data analysis. It has led to the evaluation of ozone generators for the increment of vegetables storage life.

In rural farms the substantiation of the economic feasibility of using electrohydraulic water treatment technology is important and is contributed to this book the prospects and scope of electrohydraulic technology with modified design improvement of the treatment process and reduction of the power losses. Energy efficiency in sowing is one of the main objective of this book that contributes to the technological complex of cultivation of cereals for which improvement to traction and coupling properties of the small class tractor for grain crop sowing is studied. The seeds on environmental disturbances were studied in order to layout the principles to follow in the course of pre-sowing treatment and improvement of electro-magnetic stimulation technology for seeds and crops plants. Also, the mechanism of the magnetic field impact on seeds to decide the most effective mode of pre-sowing treatment of seeds in a magnetic field and developed design parameters of the energy saving technology for magnetic treatment of seeds.

The impacts on soil by wheel drives of the self-propelled seed drill were determined by evaluations and experimental methods in order to minimize the impact and the dependencies of soil hardness on the depth of experimental measurement were constructed in this book. The development of environmentally friendly technologies is particularly important, which also contribute to the reduction of resource costs for production and load on the environment.

We extend our respect to all the authors for life-long plausible contributions to this book. We sincerely wish that this book will greatly help to scientists, researchers and students for their enhancement in knowledge to the current state-of-the-art in the energy efficient technologies for agriculture farming system.

Saim Memon
London South Bank University, UK
April 28, 2019

Saim Memon *is a Senior lecturer in Electrical Engineering at London South Bank University, London, UK. He studied BEng(hons) in Electrical Engineering (Mehran UET, Pk), MSc in Mechatronics (Staffordshire University, UK), PhD in Electrical & Electronic Engineering (Loughborough University, UK) and PGCert in Teaching Qualification FE (University of Aberdeen, UK). He is a Chartered Engineer through Engineering Council UK. He is a Fellow of Higher Education Academy and has a Qualified Teacher status by General Teaching Council for Scotland (GTCS). He teaches BEng/MSc modules, supervises BEng/MSc/PhD students. He has multi-disciplinary research/academic experiences in Electrical, Electronic, Solar-Thermal-Vacuum-Systems and Renewable-Energy Engineering. His research experiences are on energy-materials for vacuum-insulated-smart-windows, renewable energy technologies, thermoelectric-materials with vacuum-insulation and heat-storage for the improvement of electric-vehicles charging-efficiency. He has over 35 research publications in the form of high-impact-journals, book-chapter, conferences, book-editor, newsletter and magazines. He is developing and presented his research findings in collaboration with leading scientists in the UK, Europe, USA, Abu Dhabi, Japan, Thailand, Malaysia, Kenya, Peru, Russia and China.*

Foreword

Agriculture is important because it is the primary source of food for humans. Crop production lessens harmful environmental impacts when it encourages the use of renewable resources such as solar energy, biofuels, wind, biomass, geothermal, small-scale hydroelectric, tidal and energy generated by the waves.

Environmentally friendly production technologies are only the beginning of the process and their promotion through modern information technologies means that there are greater possibilities for sharing with producers and consumers thereby affecting the total chain of cultivation from the production, distribution, and storage to the consumption of the product.

In the past little attention has been paid to the need to protect and promote the value of nature in the open landscape. But currently, there are major environmental challenges for the development of future agriculture using the new technological options. The future indicates that agriculture can develop in one of two ways. In the first, it must be cultivated intensively, that is as an industrialized agriculture that generates large quantities of products, forgetting its effects on the environment. In the second, it is an organic-oriented agriculture where production is carried out on a small scale while taking care of the harmful effects on the land and the crops themselves.

Information technology is the fundamental basis of the future of sustained agriculture because it is through scientific knowledge and a strong interaction and communication between producers, researchers and governments that long-term development policies can be created that take into consideration the environment and values related to nature while also, motivating exports that boost an agricultural economy.

This book contains proposals related to the sustainable development of agriculture and livestock using renewable and clean energy that applies innovative knowledge, including modeling and mathematical optimization for the design of efficient devices for the production of energy. The book is another achievement of the Professors Valeriy Kharchenko and Pandian Vasant in their eagerness to spread the latest in agricultural science and technology and the production of crops based on clean energies. Surely, as always, this new book will be a great success for editors and authors who consider it a platform to disseminate the results of their research. Thanks to the editors of IGI-GLOBAL for their efforts to disseminate science in all areas of knowledge, making it accessible to scientists, technologists, engineers and others who live and develop useful knowledge for the welfare of humanity around the globe. Scientists innovate the world!

Gilberto Pérez Lechuga
Autonomous University of the State of Hidalgo, Mexico

Gilberto Pérez Lechuga *(Doctor of Engineering 1993) is Senior Researcher at Applied Mathematics and Operations Research with Industrial applications. He is especially known for his work in the areas of industrial engineering, manufacturing engineering, reliability engineering, financial engineering, logistics engineering and modeling and optimization of agricultural and livestock systems. His areas of interest include stochastic modeling and optimization systems, bio-inspired algorithms (swarm intelligence) and metaheuristics, simulation and the Monte Carlo Method, dynamical emergent systems, stochastic flexible manufacturing systems and mathematical modeling of the supply chain, stochastic finances and stochastic differential equations. He is an industrial consultant with more than 15 years of experience in the analysis, design and implementation of methods and models for the optimization of industrial processes at an international level. He has published more than 70 scientific articles and 7 books and book chapters in his area of specialty. He is currently general director of research at the Autonomous University of the State of Hidalgo, Mexico, and visiting professor at several foreign universities.*

Preface

The growth in the world population and the related increase in the demand for food and other household items set the task of intensifying agricultural production throughout the world. The solution of this problem requires ubiquitous applications of new technologies for cultivating agricultural structures, their primary processing and transportation to places of consumption or deep processing. Some of today's technologies, especially those that are out of date, are characterized by relatively low productivity and, in this connection, high consumption of energy resources. The latter increases the cost of production and, most importantly, leads to an increase in greenhouse gas emissions (i.e., to enhance the anthropogenic impact on the natural environment).

Solving this problem requires the use of new, energy-saving and environmentally less harmful technologies of a variety of different types, and not only in production, but also in the social sphere.

There was such a situation that information of this kind is needed not only directly by the worker in the village. The manufacturers and developers of such technical facilities and technologies and scientific organizations that are engaged in the development of technologies and equipment for the power industry of the village, production processes of various types, as well as the problems of housing and communal solutions, are showing great interest in new efficient means of conducting agricultural business, especially on a small scale, problems of the rural population.

An important issue is the organization of reliable power supply in an autonomous mode. The use of renewable energy sources can play an important role in this issue. And this means that information on the possible technologies of efficient generation of electric and thermal energy, its rational consumption through the use of energy-saving technologies will inevitably be in demand. Of great importance in the organization of intensive activities in rural areas can also be the use of efficient technologies for the cultivation of products and its processing on site with the least expenditure of energy, the amount of which at remote sites may be critically small.

OBJECTIVE OF THE BOOK

The book is intended for agricultural business executives who are going to start and develop production, primarily in agricultural production. It will be useful for practical workers in the agro-industrial sector, small entrepreneurs, local authorities. A useful book will be for scientists, engineers and technicians engaged in research and development of new technologies and new effective technical means to solve the problem of energy consumption decrease in agro industrial sector and in the housing and communal services sector in the countryside.

In the book the information in creation of systems of generation electric and thermal energy with use of renewable energy sources will be displayed. Use of described technologies will allow to create in rural areas efficient and environmentally friendly an energy potential, sufficient for the organization of industrial and social-economic activity in these territories harmless for the surroundings.. The technologies of transfer of the electricity on considerable distances offered in the book will allow to make a choice in favor of more effective way of power supply of territory. I.e. or at the expense of the organization independent RES based power supply, or at the expense of joining to the centralized networks. Useful information on application of innovative power saving up electro technologies in processes of processing of agricultural products, realization of social programs in rural territories will be submitted as well..

The cases of best practice described in the book, specific examples of the implementation of new technologies may provide serious assistance to practitioners in their activities for agricultural production, energy saving technology implementation, intensification processes at the all studies of the rural objects energy supply, effective electricity transmission, electro stimulating technologies, treatment and primary crops proceedings, harvesting machinery, RES based energy systems implementation etc. The wide coverage of the problems envisaged in the preliminary content of the book gives reason to believe that the demand for book from buyers will be quite high.

The book edition provides an information interchange intensification between generators of new technologies and beneficiaries from their use. Such exchange, in turn, undoubtedly, will impulse the further development of the technologies described in the book.

Does not cause the doubts, that fact that for readers will represent a great interest all complex of the stated information and the book will receive undoubtedly great demand at readers.

The book presents the results of studies of various authors from around the world. Some chapters are the result of joint research by various teams, including international ones.

It comprise such topics as advanced technologies in the production of grain with low energy consumption, innovative energy saving technologies in the production of milk and meat, energy-efficient thermal processes in agricultural production, advanced technologies for primary processing agricultural products, intensification of technology processes by electric, magnetic and other influences, PV solar power plants for rural objects energy supply, wind based energy generation installations and systems, cogeneration PV Thermal modules fabrication and application, Rational Use Energy case studies, diagnostic and monitoring methods for energy equipment status in rural conditions, energy efficient electric equipment for heat supply in agriculture, electric pulse cultivation, pre-sowing seed treatment for Improvement of germination and yield growth as well as u se of microwave energy at thermal treatment of grain crops, solutions for diesel, wind and solar power plants joint effective use and so on.

TARGET AUDIENCE.

The book will be useful to a wide range of individuals, such as students of energy and agro-engineering specialties, experts and heads of municipal unions, managers of the ministries and other organizations responsible for development of agricultural production at entrusted with territories. the separate large businessmen beginning rural business and many other public organization for assistance to advancement of new technologies in social sphere.

An extremely useful the book may become for small commodity producers in the countryside, various entrepreneurs of medium and small businesses as well as for inter-regional public organizations for assistance to advancement of new technologies in social sphere in view of environmentally friendly sustainable development. In general the book is intended for representation of the wide public advanced achievements in the field of agro-engineering and related to the rural business energy technical means which will be useful to a wide range of readers and doubtless positive impact on a solution of a problem of a sustainable development of new rural territories.

In addition, the introduction of energy efficient technologies and renewable energy sources into agricultural practice reduces the total cost of primary energy resources, i.e. reduces costs for the production of a unit of agricultural products, which in turn reduces its cost and gives the producer advantages in competing on the agricultural market both within each country, as well as in international agricultural products markets.

In general the book is intended for representation of the wide public advanced achievements in the field of agro-engineering and related to the rural business energy technical means which will be useful to a wide range of readers and doubtless positive impact on a solution of a problem of a sustainable development of new rural territories.

Chapter 1 is dedicated to consideration improvement of traction and coupling properties of the small class tractor for grain crop sowing (By means of the hydropneumatic damping device). Sowing is one of the main operations in technological complex of cultivation of cereals. That's why only with high quality seed distribution along the length and depth of the row can the maximum productivity and yield be achieved. A tractor with seeding machine (seeding machine tractor unit) is subjected to continuously changing external influences that have a negative impact on the performance indicators of the technological operation. Based on the cereal cultivation technology, it is necessary to use tractors with transmissions that can absorb the oscillations and increase the stability of the coulter group of the seeding machine. Since this improves the quality of the operation, reduces the consumption of spent seed and fuel and increases environmental component of the process.

In Chapter 2 evaluation of the effectiveness of the use of programs in the design of power complexes based on renewable energy resources is presented. The chapter aims to described the study the prospects of energy complexes on the basis of renewable energy sources to provide electricity for the stand-alone consumers in different regions of Myanmar. In order to do that the territory of Myanmar is divided into some regions according to their amount of renewable energy sources, developed the methods for determining the optimum parameters and operation of the energy complex on the basis of renewable energy sources and analyzed the cost-effectiveness of those energy complexes in the power sector regional of Myanmar. It is give the mathematical formulation of the problem of optimization of the main parameters and operations of the energy complex on the basis of renewable energy sources for energy supply of numerous autonomous rural consumers of the Republic of Myanmar.

In the Chapter 3 energy saving system (for agriculture) based on heat pump for maintain microclimate of the agricultural objects are considered. At agricultural facilities, the main attention as a rule is paid to the formation and maintenance of their microclimate parameters, and mechanization of storage processes. As world experience shows, it is necessary to develop and implement energy-saving systems and the use of renewable energy sources. The authors have developed energy-saving systems based on the heat pump, with upgraded electrical regulators. The developed system (patent 100873), using thermoelectric elements and a low-potential energy source, allows to effectively maintain the temperature parameters of the microclimate during long-term storage of potatoes, but it requires a large amount of electricity consumption (30...35 kW), so the authors have developed an energy-saving system based on a heat pump (patent 123909). The temperature regime is achieved by using a thermoelectric cooler-heater and electric heater. The designed humidifier allows you to maintain the necessary relative air humidity.

Chapter 4 gives results of a job for creation of gas turbine power plant of low power GTP-10S (Gas turbine power plant). Currently, there is an increase in the use of gas turbines. Nevertheless, today they are used only in the energy sector: aviation, armed forces and navy. The introduction of a new manufacturing technology developed by the author will make it possible to manufacture cheap and reliable installations and thus ensure an exceptional position on the Russian market for goods and technologies, and taking into account the use of intellectual rights, abroad. The scientific novelty of the sample is the method of calculating small engines with a centrifugal compressor, a centripetal turbine and a combustion chamber with a negative thrust vector of the air flow. It is shown that the developed microgas turbine cogeneration power generator consists of a microturbine engine with a periphery, a free power turbine necessary for the selection of mechanical power, a high-speed electric generator with permanent magnets, an electronic power conversion system, exhaust heat energy recovery system and an automatic control system.

Chapter 5 deals with issues related to Milk Pasteurization and Characterization Using Mono-Mode Microwave Reactor and Slotted Coaxial Antenna. Mono-mode microwave reactor is usually used to heat on a substance, especially foodstuffs. This is because heating using the microwave reactor can sustain the flavor, color, and nutrition of the food. Furthermore, this heating technique is cost-effective and time-saving compared to a conventional heating method. The mono-mode reactor is able to determine the absorption of microwave power accurately on the heated substance versus multimode reactor. In this chapter, a simple and precise mono-mode microwave reactor is designed and developed especially for research laboratories. The advantage of this reactor is to provide a more accurate calibration process, in order to improve the optimum energy used in the heating process, as well as the temperature of the specimen, can be determined more accurately. The reactor can generate output power from 30 Watts to 1500 Watts, operating at 2.45±0.03 GHz and capable of accommodating a specimen volume of 780 cm3. Pure water is used as a heated specimen to demonstrate the performance and efficiency of this reactor.

In the Chapter 6 theoretical and experimental evaluation of impact on soil by wheel drives of the self-propelled seeder is considered. It is dedicated to vertical axial loads on the soil from the wheels of a self-propelled seed drill, area of the contact patch, maximum contact pressure for the front and rear wheels and the density of the soil are determined by evaluations and experimental methods. The discrepancy between the theoretical and experimental indicators was: 1.4% and 2.0% for the rear and front wheels in vertical axial loads; 2.8% and 2.2% for the rear and front wheels by the contact area of the tires of the seeder with the soil and the maximum contact pressure; 6.2% - the maximum discrepancy on the values of soil density at a depth of 7.6 cm. Soil hardness was measured in three zones: before the seeder's passage and after each of its passage in a rut behind the front and rear wheels at six different depths, determined by the marks on the soil densimeter tester density. Graphics of dependencies of soil hardness on the depth of measurement were constructed.

In the Chapter 7 readers may find the description of the performance improvement study of linear photovoltaic systems with concentration of solar radiation, submitted by scientist from Kasakhstan. In the chapter a new type of linear cooled photodetectors is considered, on the front walls of which in the focal region of the optical concentrator mirrors is installed an array of solar cells operating with the low-ratio solar concentration. This work is focused on the theoretical and experimental substantiation of the efficiency increase of photodetectors under conditions of an optimal combination between solar radiation concentration and heat transfer intensity of photovoltaic cells with heat carriers. The problem of obtaining a high temperature liquid due to the limitations of solar cells is solved by organizing the flow of fluid within the thermal collector channels in the focal region of an additional optical concentrator. A mathematical model of engineering characteristics calculation of the Λ - shaped photodetectors and cost calculation of electrical and thermal energy generation is presented. The research results are used in the development of industrial prototypes of photodetectors with a concentration of solar radiation and low production costs.

Optimization of spectral composition and energy economy effectiveness of phyto-irradiators with use of digital technologies are considered in the Chapter 8. It is known that in the case of using technology with the supplementary lighting, an irradiation spectral composition influences heavily effectiveness of the photosynthesis processes, development and productivity of vegetable crops. Hence, definition of general points at development and projecting of modern phyto-irradiators is one of high-priority tasks in techniques development for plants growing in conditions of protected ground. The research is aimed at review and effectiveness assessment of existing sources of illumination used in modern systems of supplementary lighting and at deduction of general points of development and projecting of phyto-irradiators based on results of laboratory investigations with use of modern digital technologies of monitoring and data analysis. The results of the comparative tests of light emitting diodes-based phyto-irradiators showed that the energy consumption per a product kilogram is twice less in the case of LED-irradiators. Based on the research results, general points were deducted for use at development of modern LED-phyto-irradiators.

Chapter 9 considers energy-saving technologies for pre-sowing seed treatment in a magnetic field. The purpose of the research was to establish the mechanism of the magnetic field impact on seeds to determine the most effective mode of pre-sowing treatment of seeds in a magnetic field and design parameters of the device for magnetic treatment of seeds. It is established that under the influence of a magnetic field the rate of chemical reactions occurring in plant cells is accelerated, solubility of salts and acids increases, and permeability of cell membranes accelerates the diffusion of molecules and ions through them. This leads to an increase in the concentration of ions in the cell and oxygen molecules and the growth of water absorption of seeds. Pre-sowing treatment of seeds promotes increased germination energy by 25-40%, germination - by 30 - 35%. The most effective pre-sowing treatment of seeds in a magnetic field is a magnetic induction of 0.065 Tl with four reversal magnetization, a pole division of 0.23 m and a seed movement speed of 0.4 m/s. With this mode of treatment, crop yields increase by 20–25%.

Chapter 10 describes mobile measuring complex for conducting mainly rural electric grids survey. An energy audit of the electrical network is required in the process of construction of new electrical networks as well as in justifying the reconstruction need of existing ones. In the paper the structure of a mobile measuring complex has been developed to conduct an electrical network survey without disconnecting consumers. The complex may be used to inspect 0.4 kV electrical networks and microgrids of the same voltage class and allows to obtain data on voltage losses and electric power losses in network elements such as a power line (electric transmission line), a power transformer. The energy audit is proposed to conduct without disconnecting consumers in order to avoid an undersupply of electricity as well as to determine the real operating modes of power networks. Ultimately, the use of the developed measuring complex will increase the reliability of power supply to consumers and ensure the required quality of the electricity supplied to them.

Chapter 11 describes technology of managing reactions of biological objects at anthropogenically transformed territories. Solving the problem of increasing plant resistance, the development of environmentally friendly technologies is particularly important, which also contribute to the reduction of resource costs for production and load on the environment. The research results indicate positive effect of treatment of plants and seeds with electrochemically activated (ECA) water, electric field and impulse pressure (IP). Presowing treatment of seeds with ECA water increases the germination rate and seed germination energy, improves the development of plants, their morphological parameters, etc. The reactions of economically dangerous pests and causative agents of infectious plant diseases to the use of ECA water are individual. The combination of presowing seed treatment with the treatment of vegetative plants provides the highest possible result.

Chapter 12 considers potential evaluation and best practices of solar power plant application in rural areas. Recreational rest and recovery zones require a daily hot water supply, and constant availability of electricity. Therefore, the need for renewable energy sources usage in the Lower Volga region as well as in another ones is obvious and that is why power plants with environmentally friendly component attracts significant attention everywhere. An analysis of the theoretically calculated potential renewable energy makes it possible to assert optimistically that the above mentioned region is promising for the renewable energy autonomous sources implementation. It may be noted that the potential wind and solar energy, in spite of the fact that it is distributed unevenly, is sufficient to provide energy for remote rural communities, and tourist facilities. There is conducted analysis of availability of actinometrical data required for heliotechnical calculations and are identified areas of applicability and accuracy of the information received from various meteorological information sources.

In the Chapter 13 development and research of PVT modules in computer-aided design and finite element analysis systems are considered. The chapter discusses the simulation of solar photovoltaic thermal modules of planar and concentrator structures in the computer-aided design systems KOMPAS 3D and finite element analysis ANSYS. To create photovoltaic thermal modules, a method for designing their three-dimensional models in the computer-aided design system has been developed. To study the thermal regimes of the created three-dimensional models of modules, a method has been developed for visualizing thermal processes, coolant velocity, and flow lines of a cooling agent in a finite element analysis system. As a result of calculations in the finite element analysis system using the developed method, conclusions can be drawn about the feasibility of the design created with its further editing, visualization of thermal fields and current lines of the radiator cooling agent. As an illustration of the simulation results, a three-dimensional model of a photovoltaic thermal planar roofing panel and an optimized three-dimensional model of a photodetector of a solar concentrator photovoltaic thermal module are presented.

In the Chapter 14 seed pre-activation study by means of LEDs radiation is given. To study the possibilities of seeds presowing activation, irradiation with LEDs emitting in the visible, violet and near ultraviolet ranges with a maximum of 405 nm was carried out. As a result of the growing experience, it was found that the height of wheat plants grown from the treated seeds significantly exceeds the control indicators except for the period of 45...55 days. To implement the flow technology of seed activation with LEDs optoelectronic irradiation unit was developed. The advantages of the given installation is the energy efficiency and of seed treatment efficiency, due to of the optimal radiation spectrum selection and treatment doses.

In the Chapter 15 improvement of technology of electrical and magnetic stimulation of seeds and crop plants is considered. Use of a variety of electrotechnics is a technologically efficient and environmentally friendly technique, which increases the productivity of cultivated plants. Stimulation of green plants and vegetable crops in electric field with the intensity of 5-50 kV/m made it clear that the maximum efficiency is observed in the growth period - an increase of up to 30%, compared to the control.

Plants have been subjected to stimulation for 3 hours twice a day (in the morning and in the evening). Analysis of studies on the preseeding seed stimulation showed that the exposure to pulsed magnetic field improves the dynamics of germination and plant growth at the early stages of development by an average of 10....20%, and more uniform germination helps to ensure high yields.

Chapter 16 describes linear electromechanical transducer in the systems of welded joints of electrodynamic processing developed by Ukrainian scientists. Research is dedicated to establishing characteristics relationships between the induction type impact electromechanical transducer and parameters and quality indicators of electrodynamic effects on the welded joints. Developed two-dimensional circle-field mathematical model of transient discharge capacity at the branched electrical circuit with the coil inductance which changes dynamically, allowing by adjusting the parameters of the electromechanical transducer to achieve the necessary technological requirements for the characteristics of electrodynamic processing. Based on mathematical modeling of electrophysical processes in electromechanical transducers induction type for electrodynamic processing of welded joints, reasonably geometrical parameters massive disk and the contact area in which the necessary conditions are created to reduce residual stresses in the weld joints.

Chapter 17 presents artificial bee colony based optimization of hybrid wind and solar renewable energy system above artificial bee colony (ABC) algorithm was applied to optimize hybrids of wind and solar renewable energy system. The main objective of this research is to minimize the total annual cost of the system by determining appropriate numbers of wind turbine, solar panel and batteries, so that the desired load can be economically and reliably satisfied based on the given constraints. ABC is a recently proposed meta heuristic algorithm which is inspired by the intelligent behavior of honey bees such as searching for food source, collection and processing of nectar. Instead of Gradient and Hessian matrix information, ABC uses stochastic rules to escape local optima and nd the global optimal solutions. The proposed methodology was applied to this hybrid system by the help of MATLAB code and the results were discussed. Additionally, it is shown that ABC can be efficiently solve the optimum sizing real world problems with high convergence rate and reliability. The result was compared with the results of PSO.

In the Chapter 18 the study of luminescence spectra of seeds of crop species for diagnostic quality is described. Optical luminescent biological objects diagnostics methods and biotissues are considered. According to the previously developed method, excitation and photoluminescence spectra agricultural plants seeds, including cereals, legumes, fodder, technical and vegetable, were measured. The typical excitation spectrum lies in the range of 355...500 nm and has two maxima: the main one at 424 nm and the side one at 485 nm. The luminescence spectrum lies in the range of 420 ... 650 nm and has a maximum in the region of 500...520 nm. The maximum luminescence is less pronounced than in the excitation spectrum. The measured spectral luminescence characteristics forage plants seeds by scarification. Due to the scarification forage plants seeds spectral characteristics increase. In Galega seeds with multiple scarification observed qualitative changes in the excitation spectrum associated with the appearance of a new maximum at a wavelength of 423 nm. Similarly for clover seeds. The obtained results can be used to create seed diagnostics devices.

In the Chapter 19 researches in the field of technology electrohydraulic effect impact on water are described. The purpose of the study is to investigate the technology and technical means of electrohydraulic action on water. The authors justify the relevance of research. The design of the original negative electrode tip is being developed to increase the density of the electromagnetic field and reduce power loss. The design parameters of the electrohydraulic installation are shown. Modeling of factors influencing the process of electrohydraulic treatment of water according to the Plackett-Berman plan and the random balance method is carried out; significant and insignificant factors are identified. The operation modes of the electrohydraulic installation are determined and optimized experimentally. The substantiation of the economic feasibility of using electrohydraulic water treatment technology in farms is being conducted. The prospects and scope of electrohydraulic technology are determined.

The book will be useful to a wide range of persons, such as students of power and agroengineering specialities, experts and heads of municipal unions, managers of the ministries and other organisations responsible for development of agrarian industry, the businessmen dealing with such a business and many others as for organisations like big consortium, inter-regional public organisation for assistance to advancement of new technologies in social sphere "sustainable development."

Valeriy Kharchenko
Federal Scientific Agroengineering Center VIM, Russia

Pandian Vasant
Universiti Teknologi PETRONAS, Malaysia

Acknowledgment

First of all the editors would like to express their sincere gratitude to all the people which took part in the process of preparing and editing and publishing of the book.

This is undoubtedly all the authors of the presented chapters, and to an even greater degree, those of them who, apart from preparing their manuscripts, devoted a lot of time and effort to review the manuscripts of other authors. Their efforts certainly contributed to ensuring the high quality of the book. We would like to personally highlight such authors as Bolshev V., Budnikov D., Samarin G., Vasilyev A., who with great care and very conscientiously reacted to this important process of preparing the manuscript of the book.

It would be unfair not to gratefully acknowledge the significant role of the editorial board, the majority of whose members have made a worthy contribution to the formation of a high-level book.

It is strictly necessary to mention the terrific role that IGI Global publishing house has played in order for the information, included in to the book, to become published.

Finally, the editors would like to express many thanks to the entire staff of IGI Global publishing houses who helped the editors and authors in preparing the manuscript of this book professionally and extremely friendly, and have made a lot to make this book the edition of premium quality.

The editors are confident that this book will certainly be useful to a wide circle of readers, from starting student to the specialists of high ranks.

Valeriy Kharchenko
Federal Scientific Agroengineering Center VIM, Russia

Pandian Vasant
Universiti Teknologi PETRONAS, Malaysia

Acknowledgment

Chapter 1

Improvement of Traction and Coupling Properties of the Small Class Tractor for Grain Crop Sowing by Means of the Hydropneumatic Damping Device

Sergey Evgenevich Senkevich
https://orcid.org/0000-0001-6354-7220
Federal Scientific Agroengineering Center VIM, Russia

Pavel Vladimirovich Lavrukhin
Don State Agrarian University, Russia

Anna Aleksandrovna Senkevich
Don State Agrarian University, Russia

Pavel Aleksandrovich Ivanov
Don State Agrarian University, Russia

Nikolay Viktorovich Sergeev
Don State Agrarian University, Russia

ABSTRACT

Sowing is one of the main operations in the technological complex of cultivation of cereals. Only with high quality seed distribution along the length and depth of the row can the maximum productivity and yield be achieved. A tractor with a seeding machine is subjected to continuously changing external influences that have a negative impact on the performance indicators of the technological operation. Based on the cereal cultivation technology, it is necessary to use tractors with transmissions that can absorb the oscillations and increase the stability of the coulter group of the seeding machine. Since this improves the quality of the operation, reduces the consumption of spent seed and fuel and increases environmental component of the process.

DOI: 10.4018/978-1-5225-9420-8.ch001

INTRODUCTION

The growth of the world population and the related increase in demand for food poses new challenges for agricultural production. One of these challenges is the intensification of agricultural production worldwide. The solution to this problem requires the use of new techniques and new technologies for the cultivation of crops.

Sowing is one of the main operations in the technological complex of measures for the cultivation of grain. Only at high quality of distribution of seeds on length and depth of a row the maximum productivity can be received. The tractor with the seeder (sowing machine-tractor unit) is exposed to continuously changing external influences, which have a negative impact on the performance of the technological operation.

Based on the technology of grain cultivation, for sowing it is necessary to use a tractor with transmissions that can absorb fluctuations and increase the stability of the coulters group. This improves the quality of the operation, reduces the consumption of spent seed, and fuel, improves the environmental component of the process.

The efficient functioning of the sowing machine and tractor unit you can improve due to installation in the transmission of the tractor is a hydro-pneumatic damping device with a variable moment of inertia. This device reduces the fluctuations of the external traction load transmitted to the engine, stabilizes the work of the coulter group of the trailed seeder, and reduces the vibration load of the power transmission.

Studies conducted on an experimental sample of the tractor showed that to improve the efficiency of the technological operation-sowing, it is possible by increasing the efficiency of the sowing machine-tractor unit on the basis of a tractor of class 1.4, by introducing a hydro-pneumatic damper into the transmission. The analysis shows that the use of a hydro-pneumatic damping device in the transmission of a tractor of small traction class is an effective constructive measure that improves the output of the unit and improves the sowing operation.

BACKGROUND

Effect of Transmission Elasticity on the Functioning of Machine-Tractor Units (MTU)

In mobile agricultural units, the variability of external factors in the interaction of working bodies of machines with the processed environment (soil, plants) and movers with the field surface determines the complex nature of the movement of individual points, which characterizes to a large extent the quality of a number of operations (plowing, sowing, etc.) (Lurie, 1981).).

The continuous and random nature of changes in external influences causes fluctuations in the load regime, both by tractor traction and torque on the motor shaft (Ageev, 1978). As noted by L. E. Ageev, due to the random nature of the effects, the output parameters and performance of the units can be considered as random functions or random sequences. As a result, it is necessary to apply empirical models based on the theory of experiment planning. The author gives the model and laws of distribution of operational and technological input parameters of various agricultural units. However, we are only interested in the sowing unit.

Tests of the sowing unit consisting of a tractor of class 1.4 and a grain seeder (figure 1) showed that the increase in operating speeds from 0.71 to 3.68 m/s significantly affects the parameters of the distribution of traction resistance. It is found that the mathematical expectation $\overline{R}_a$ increases from 4460 to 5700 N, that is, by 27.8%, and the measure of the tractive resistance scattering – the mean square deviation σ_R – varies within 340...1360 N. The coefficient of load variation ν_R and the degree of its unevenness δ_R varied from 7.6 to 23.8% and from 0.46 to 1.43, respectively. The oscillations of the tractive resistance of the sowing unit in the marked speed range are close to the law of normal distribution.

Figure 2 (b) shows, as an example, the laws of distribution of the depth of seeding of winter rye. It was established experimentally, that on the increased transfers of the sowing unit there is an increase in the measure of dispersion of the depth of seeding and a decrease in its mathematical expectation from 64.9 to 55.5 mm, that is, by 14.5%. The mean square deviation of the embedding depth in the range of operating speeds of 0.71 ... 3.68 m / s increased from 5.7 to 10.3 mm, and the coefficient of variation (variability) from 8.8 to 18.6%. The probability of matching the empirical and theoretical frequencies of the seed depth distribution was in the range of 0.35-0.73(Ageev, 1978). Also, studies have found that the depth of seeding of some cops at different speeds of the sewing unit are variable.

N. G. Kuznetsov in their work, leads of different spectral density traction resistance when working the ploughing, cultivating and sowing aggregates. Tractors of these units are equipped with a constant-power engine and a pneumatic-hydraulic linkage. Since we are interested only in the sowing unit, the

Figure 1. The law of distribution of traction resistance of sowing unit (tractor of class 1.4 with grain seeder): 1 – 0.71 m/s; 2 – 1.52 m/s; 3 – 1.85 m/s; 4 – of 2.17 m/s; 5 – 2.55 m/s; 6 – 3.10 m/s; 7 – 3.68 m/s. (Ageev, 1978)

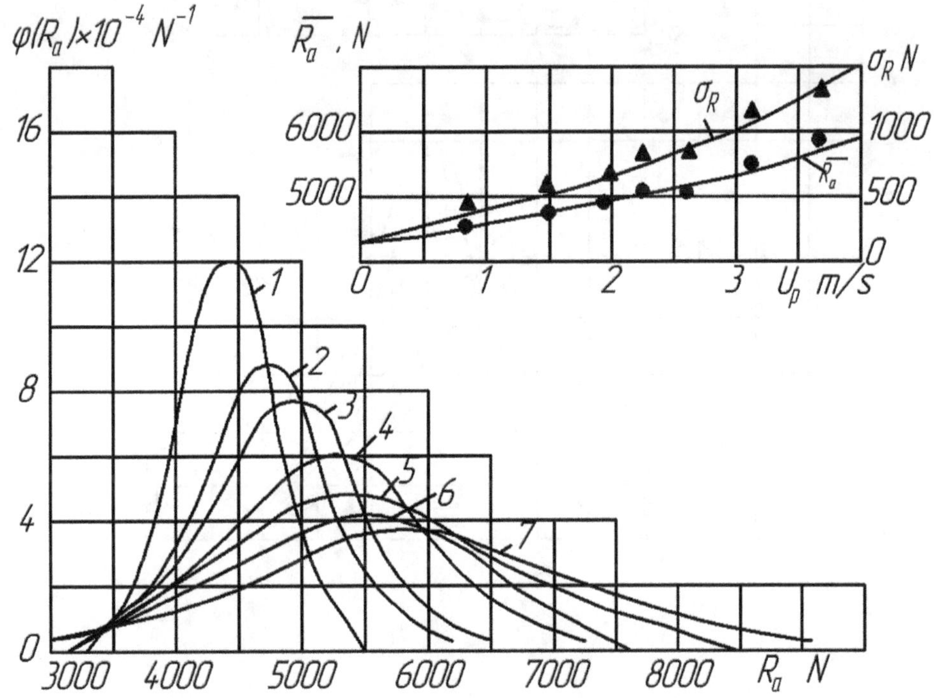

Figure 2. Laws of the resistance moment distribution M_c on the motor shaft (a) and depth of seeding h (b) if the work of the sowing unit: 1. 0.71 m/s; 2. 1.52 m/s; 3. 1.85 m/s; 4. of 2.17 m/s; 5. 2.55 m/s; 6. 3.10 m/s; 7. 3.68 m/s.
(Ageev, 1978)

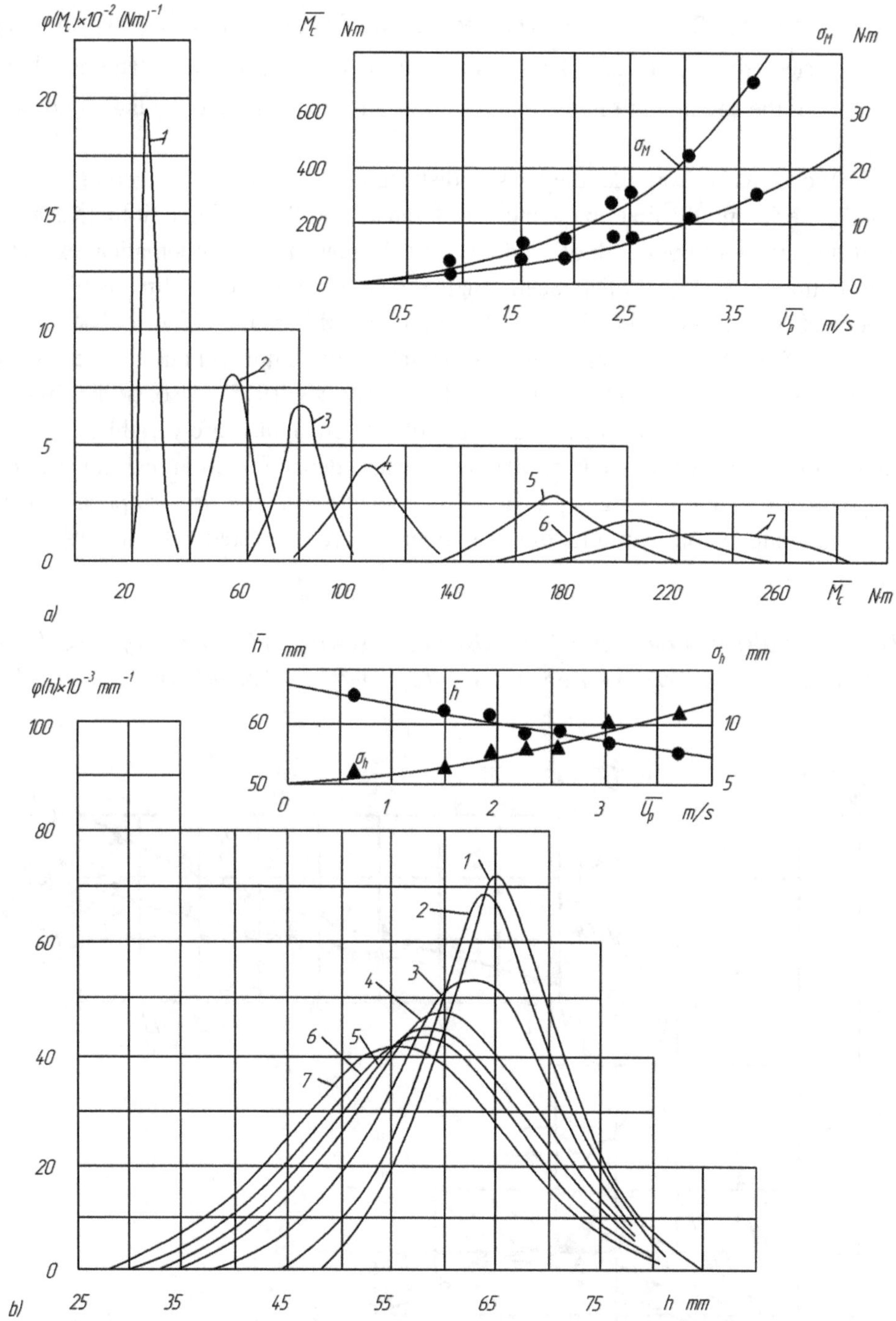

Figure 3. Spectral densities of the horizontal component of the tractive effort on sowing
SE with SL (Solid line) – series engine with a serial linkage
SE with HPL (Dashed line) – series engine with a hydro-pneumatic linkage
CPE with SL (Dash-dotted line with two points) - constant power engine with serial linkage
CPE with HPL (Dash-dotted line with one point) - constant power engine with a hydro-pneumatic linkage
(Kuznetsov, 2006)

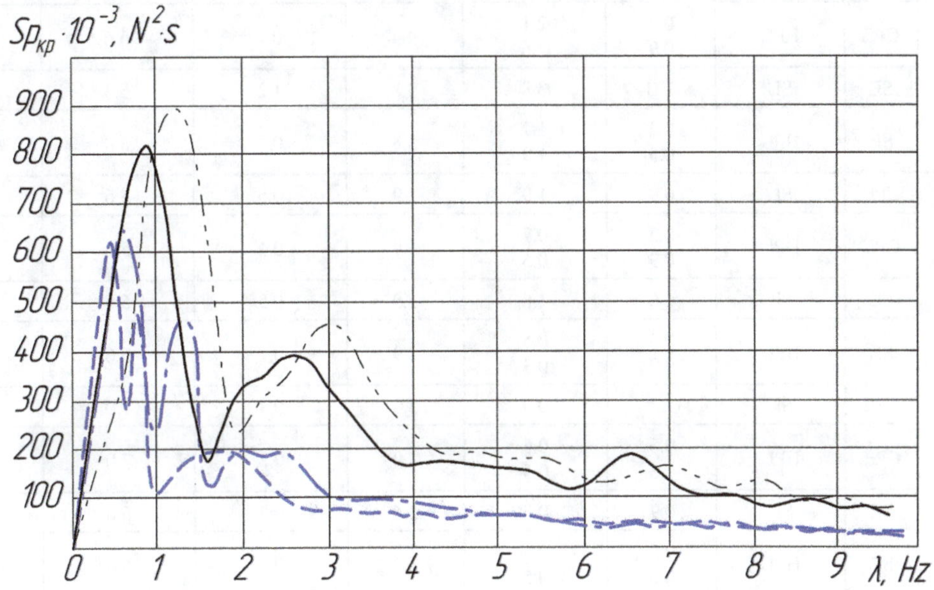

spectral densities of the horizontal component of the pulling force on the sowing are shown in figure 3. The spectral densities of the traction resistances $S_{P_{KP}}$ of different MTU with a constant power engine (CPE) and a serial engine when using a serial linkage (SL) have three pronounced frequencies at which a surge is observed in the range from 0 to 8 Hz (Fig.3)) (Kuznetsov, 2006).

Figure 3 shows the data for tractors with pneumatic-hydraulic linkage (HPL) for the corresponding MTU, the range of which is placed in the frequency band from 0 to 3 Hz. In table 1 the author gives a specified frequency and the values of the spectral densities for a variety of MTU.

The maximum amplitude of the spectrum $S_{1P_{PK}}$ is located in a narrow band λ from 0.5 to 1.2 Hz with a slight shift in its higher frequencies in the transition from plowing to cultivation and sowing. This effect is explained by the increase in the speed of MTU. The third component $S_{3P_{PK}}$ of the spectral density of the traction force on cultivation is practically absent.

The first component is formed by the interaction of the working bodies of the agricultural machine with the soil, the heterogeneity of its physical and mechanical properties of the macrorelief. The second and third components are caused by longitudinal-angular oscillations of the tractor (Kuznetsov, 2006). Thus N. D. Kuznetsov proved the effectiveness of the use of elastic elements in the machine-tractor unit using methods of statistical dynamics. However, when analyzing the operation of the sowing unit, the emphasis was placed only on the performance indicators (traction force, engine resistance moment, etc.), and the technological performance of the sowing unit (depth of incorporation), unfortunately, were not given.

Table 1. Frequency composition of tractive effort with different MTU and linkage mechanisms

Operation	MTU	Linkage	λ_1, Hz	$S_{1P_{PK}} \cdot 10^{-6}$, N²•s	λ_2, Hz	$S_{2P_{PK}} \cdot 10^{-6}$, N²•s	λ_3, Hz	$S_{3P_{PK}} \cdot 10^{-6}$, N²•s
Plowing	CPE	SL	0.63	5.3	4.4	1.4	6.6	0.6
	CPE	HPL	0.3 0.9	2.1 2.6	4.4	0.1	6.6	0.08
	SE	SL	0.31	3.8	3.8	1.2	6.3	0.9
	SE	HPL	0.2 0.5	1.7 1.9	3.8	0.2	6.3	0.09
Cultivation	CPE	SL	0.9	1.9	1.9	0.5	5.6	0.1
	CPE	HPL	0.3 0.9	0.7 0.5	1.9	0.1	5.6	0.02
	SE	SL	0.6	1.8	1.9	0.5	5.6	0.1
	SE	HPL	0.3 0.9	0.5 0.3	1.9	0.1	5.6	0.02
Sowing	CPE	SL	1.3	0.9	3.0	0.4	7.0	0.1
	CPE	HPL	0.6 1.3	0.6 0.5	1.9	0.2	7.0	0.04
	SE	SL	0.9	0.8	2.5	0.4	6.6	0.2
	SE	HPL	0.5 0.8	0.6 0.5	1.7	0.2	6.6	0.05

Also, the effectiveness of the use of elastic vibration dampers on sowing units, proves in his work N. M. Bespamyatnova (2002). (Bespamyatnova, 2002). She indicates that the vibration damper should have a mobile frequency, that is, to be a self-adjusting mechanism.

The search for ways to improve the quality of vibration dampers in industry led to the emergence of non-metallic elements for damping vibrations: rubber, air, liquid. However, these elements (except hydraulic) have not yet been widely used in agricultural machinery. Hydraulic shock absorbers, reducing the amplitude of the oscillations, do not change the frequency regime, what is necessary in this problem (Feuerhuber, Lindert, &Schlacher, 2013; Polifke, 2016;). When damping with air in the low-frequency resonance zone decreases and the frequency and amplitude of the oscillations. However, when using a pneumatic damper, it is necessary to use powerful shock absorbers, since low friction leads to a decrease in the damping decrement, so such elastic elements increase the duration of the system oscillations. Hence followers the need to use combined dampers that combine sufficient vibration damping with low mass and volume, that is, to introduce viscous and elastic elements into the scheme (Shekhovtsov, Sokolov-Dobrev, &Potapov, 2016; Zhengatal, 2016; Polivaev, Gorban, Vorohobin &Vedrinsky, 2018).

After analyzing the experimental data, the author (Bespamyatnova, 2010) concludes that the problem of uniform seeding at a given depth is solved as a problem of minimizing the amplitude of one of the inertial elements of the coulter group system, for this, the principle of dynamic vibration damping is used. To do this, it is necessary to introduce a link with a non-oscillating characteristic (elastic-viscous mechanism) that increases the system stability into the scheme.

There are many design solutions aimed at reducing the dynamic loads in the transmission and improving the operating conditions of the engines. Most fully the above problems are solved by setting the transmission torque Converter or fluid coupling (Li, Sun, Hu &Liu, 2019; Deboli, Calvo & Preti, 2017; Nekhoroshev, 2014; Nekhoroshev, & Nekhoroshev, 2015). Despite the undeniable advantages, torque converters have disadvantages that constrain their widespread use in agricultural tractors. Theoretical and experimental studies (Nekhoroshev, 2014; Nekhoroshev & Nekhoroshev, 2015; Senkevich at al., 2019) it is established that the improvement of the dynamic performance of the tractor can be achieved by the installation of elastic elements in the transmission of the tractor.

The positive effect of elastic elements on the dynamic performance of tractors are noted in all these studies, but their design provides a change in the stiffness of the transmission in a small strictly limited range. The flexibility of elastic elements during operation is not regulated. To reduce torsional vibrations and reduce dynamic loads in the transmission, the most widely used are special elastic-damping elements installed in the driven clutch discs, or elastic drive in the driving wheels (Polivaev, Gorban, Vorohobin & Vedrinsky, 2018; Polivaev, Zhidenko, Maksimov, & Levkin, 2017; Polivaev, & Vedrinsky, 2014; Polivaev, & Vedrinsky 2013;).

The elastic drive in the wheels and the dampfer in the clutch do not sufficiently dampen vibrations during steady-state driving. Compensation of resistance moment oscillations is the most rational way to optimize the loading conditions of the engine of the machine-tractor unit. Numerous researchers have proved the effectiveness of the introduction of various elastic bonds in the elements and links of the machine-tractor unit (Polivaev, Gorban, Vorohobin & Vedrinsky, 2018; Zoz, & Grisso, 2003.).

V. P. Goryachkin in the study of the spring placed between the engine and the agricultural tool noted that in the presence of an elastic bond, the duration of the impact acting on the agricultural tool is stretched over time, so that the average pulling force will be less than in an inelastic impact (Goryachkin, 1965).

The source of fluctuations in the speed of the crankshaft of the engine can be not only the vibrations of the traction resistance, as studies have shown, but also the fluctuations of the transmission shafts, so to reduce the dynamic loads on the engine, it is necessary to increase the flexibility of the transmission (Shekhovtsov et al., 2016; Polivaev, 2016). Note, among the researchers there is no consensus on the installation location in the transmission of elastic elements.

In the works of professor Polivaev the influence of elastic drive of the driving wheels on some performance indicators of the MTU is considered. The installation of elastic elements on the semi axes of the driving wheels reduces the vibrations of moments by 30 to 40%, and the installation of elastic damping elements by 40 to 70%, with a sudden change in the load, the peak torque values on the semi axis are reduced by 40 to 50%, and on the motor shaft by 25 to 36%.

The installation of an elastic drive of the driving wheels also leads to a reduction in energy consumption for the movement of the machine-tractor unit per unit path by 10...15% and an increase in the tractor traction efficiency, which increases the productivity of the MTU.

The reduction of energy consumption in the elastic drive is due to more stable loading of the tractor, increased smoothness, reduced slipping (up to 13...16%) and fluctuations in speed. The increase in traction efficiency of the tractor is due to a decrease in the unevenness of the load and the speed of rotation of the power transmission elements.

Along with the positive aspects, it should be noted that the elastic-damping elements installed on the wheel transmit significant torques, so such devices are bulky and have insufficient reliability. In addition, they cannot protect the engine from torsional vibrations arising in the transmission of the tractor, and have a linear characteristic.

To reduce the levels of torsional vibrations in the power transmissions of energy-saturated tractors, elastic and friction dampers (springs, etc.) have been installed in the driven disks (Polifke, 2016; Polivaev, Vedrinsky, & Derkanosova, 2016). The influence of the main parameters of such elastic-friction dampers on the torsional vibrations of the tractor power transmission has been studied by many scientists (Polifke, 2017; Polifke, 2017).

As noted by research Kravchenko and Senkevich (Senkevich, Kravchenko et al., 2019; Kravchenko, 2010; Senkevich A., 2008;Patent 2398147; Patent 2222440) due to significant fluctuations in the external load, the achievement of the potential performance of the machine and tractor unit perhaps by installing an elastic damping mechanism with a nonlinear characteristic in the transmission of the tractor (figure 4). The elastic damping mechanism (EDM) is designed for smooth running of the MTU during acceleration, reducing dynamic loads in the transmission, as well as to perform protective functions against external load fluctuations due to variable stiffness and automatic change of the gear ratio.

The EDM consists of the following components and systems (RU 2299 135):

- Planetary gear
- Drive of oil pumps of the gearbox, hydraulic system and tractor control
- Oil pump device with dosage and protection system
- Hydropneumatic accumulator with a system of filling and protection

The crown gear 2 of the planetary gear 1 is connected to the flywheel of the engine 17 and transmits the torque to the shaft 14 of the drive of the primary shaft of the gearbox through satellites connected to the driver and the solar gear 3. The driver has a rigid connection with the drive shaft of the gearbox.

The Central gear 3 of the reducer through the drive 4 rotates the pliable link of the elastic element of the pump 5. In order to reduce the pressure of the working fluid during operation, a pump and a motor are included in the installation, which receive a drive from one gear. The pump and motor have common suction and discharge manifolds. The suction manifold through the pipeline with the shut-off valve 19 is connected to the oil valve 20. The discharge manifold connects the hydraulic pumps to the oil cavity of the pneumatic hydraulic accumulator 28 (cavity I). The battery cavity (cavity III) separated from cavity I by a piston with seals (piston with cavity II) is filled with compressed air. Filling is carried out through the crane 35 from the receiver of the pneumatic system of the tractor (0.7 - 0.9 MPa). The law of the oil supply to the pneumatic hydraulic accumulator can be changed by an adjustable throttle 18. The maximum pressure in the discharge cavity of the pumps is limited by a safety valve 19. The oil is discharged into the tank 26 when the valve 19 is actuated and the valve 20 is open. The drive shaft of the gearbox and the drive shaft of the pumps are driven through the planetary gearbox 1.

Based on the above material, we assume that the efficiency of the seeding machine-tractor unit (MTU) can be improved by installing an elastic-damping mechanism in the tractor's transmission. This mechanism will reduce the oscillations of the external traction load transmitted to the engine and to the coulter group of the trailed seeder.

The purpose of the research is:

1. Experimentally confirm that by reducing the oscillations of the traction force and the speed of movement (due to the use of an elastic-damping mechanism), the uniformity of seeding can be improved

Figure 4. Scheme of the elastic damping mechanism (EDM) in transmission of tractor: 1. planetary gear; 2. crown gear; 3. reactive link; 4. oil pump drive gear; 5. oil pump; 6. hydraulic motor drive gear; 7. hydraulic motor; 8. overrunning clutch of the pump; 9. overrunning clutch of the motor; 10. oil pump suction line; 11. oil pump pressure line; 12. pressure line of the hydraulic motor; 13. suction line of the hydraulic motor; 14. the shaft drove (associated with the primary shaft of the gearbox); 15. transmission; 16. clutch; 17. engine; 18. two-stage adjustable throttle; 19.safety valve; 20. control valve; 21.the control highway; 22.hydraulic distributor; 23. valve spool; 24 and 25. left and right return springs; 26. hydraulic tank; 27.the hydropneumatic accumulator; I. the discharge cavity; II. cavity control (volume); III. the cavity of the pneumatic (air); 28. hydraulic cylinder of pneumatic hydraulic accumulator; 29. free piston; 30. piston with rod; 31. there is a piston position regulator; 32. oil wire; 33. oil channel; 34. control pneumatic line; 35. air tap.
(RU 2299 135 C1)

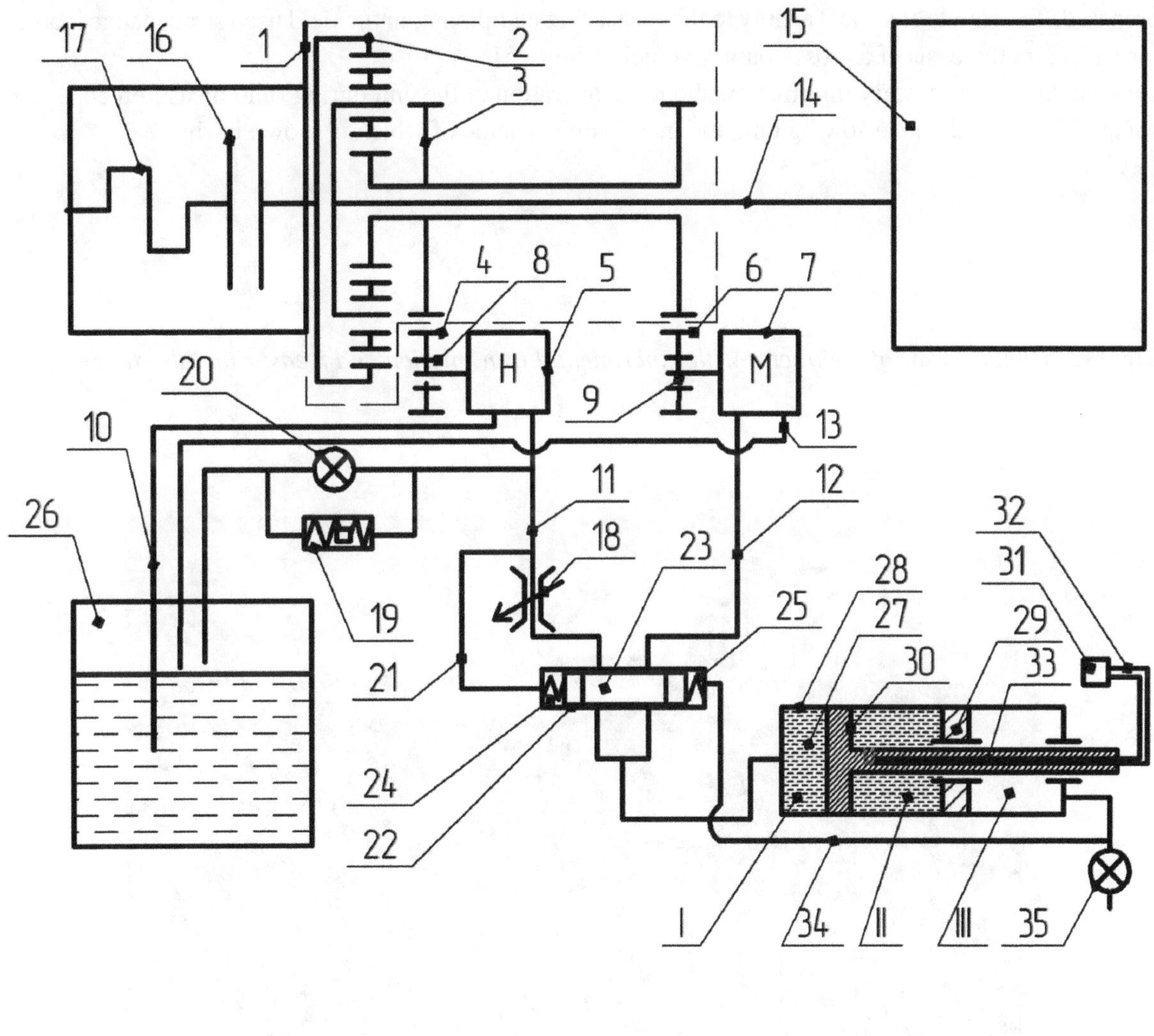

2. Experimentally confirm that the use of an elastic-damping mechanism reduces the slipping of the drive wheels of the tractor (this reduces the abrasion of the fertile topsoil and improves the environmental friendliness of the technological operation)
3. Experimentally confirm that the use of an elastic damping mechanism reduces the load on the transmission, fluctuations in the angular velocity of the engine crankshaft, and fuel consumption. And improves the performance of the machine-tractor unit

THE RESULTS OF EXPERIMENTAL STUDIES

For carrying out field experimental researches the machine-tractor unit as a part of the wheel tractor of class 14 kN (with experimental transmission) and the seeder grain capture width of 5.4 meters was prepared. To provide electrical energy to the measuring equipment, in the field used strain gauge laboratory TL-2 on the basis of a cross-country vehicle (figure 5).

To obtain the necessary information about the operation of the unit during field measurements, sensors were installed on the sowing unit, the installation scheme of which is shown in figure 6.

Figure 5. General view of the tractor with experimental transmission and measuring laboratory

Figure 6. Scheme of installation of sensors on the investigated sowing MTU: 1. rpm sensor of the crankshaft of the engine; 2, 5. the speed sensors and leading pure mango wheels, respectively; 3. to a torque sensor; 4. the traction link; 6. oil pressure switch; 7. speed sensor gear drive oil pump; 8. flow meter fuel

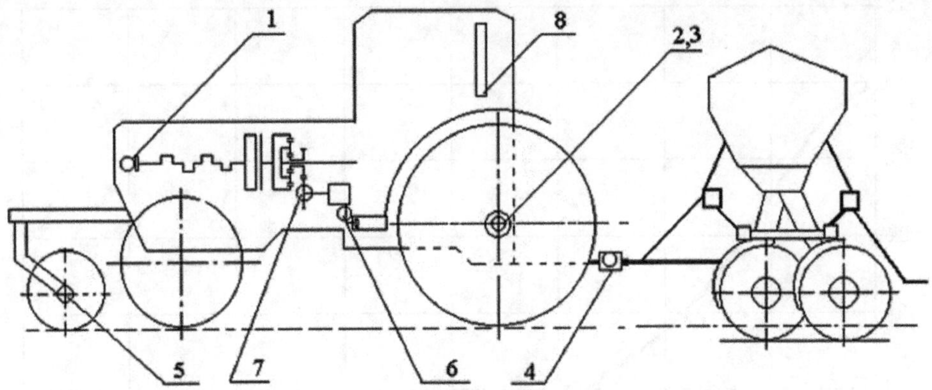

THE RESULTS OF THE STUDY OF WHITE SUPPORT BASE

For sowing is very important to prepare the soil surface. Therefore, studies have been conducted microprofile support base (field surface). Processing of the received data was carried out on the personal computer by the method proposed by A. A. Silayev (Silaev, 1963). Statistical indicators characterizing the microprofile field, is presented in table 2.

To the field microprofile characterize normalized autocorrelation function and spectral density effects of microprofile (figure 7 and 8, respectively) having been used. The calculation was made according to the program of spectral analysis.

The normalized autocorrelation function of the field microprofile action is a decreasing function. The graph of this function is constructed for the velocity V=1 m/s. To determine the autocorrelation function of the microprofile effect for any other speed, it is sufficient for each value of the autocorrelation function to divide the argument value by the value of this speed in m/s, leaving the value of the desired function unchanged (Lurie, 1969; Lurie, 1981; Bespamyatnova, 2010).

Graphs of the spectral density of the white exposure are also decreasing functions. They show that the predominant frequencies of exposure to the micro profile are in the low frequency range. Higher frequencies are filtered due to the smoothing effect of the pneumatic tire when rolling on irregularities.

Table 2. Microprofile indicators

No.	Indicators	Values
1	The maximum height of the irregularities, m	0.018
2	The minimum height of the irregularities, m	-0.021
3	Dispersion, m^2	0.00009456
4	Standard deviation, m	0.009724

Figure 7. Normalized autocorrelation function effects of the support base microprofile when driving at a speed of 1 m/s

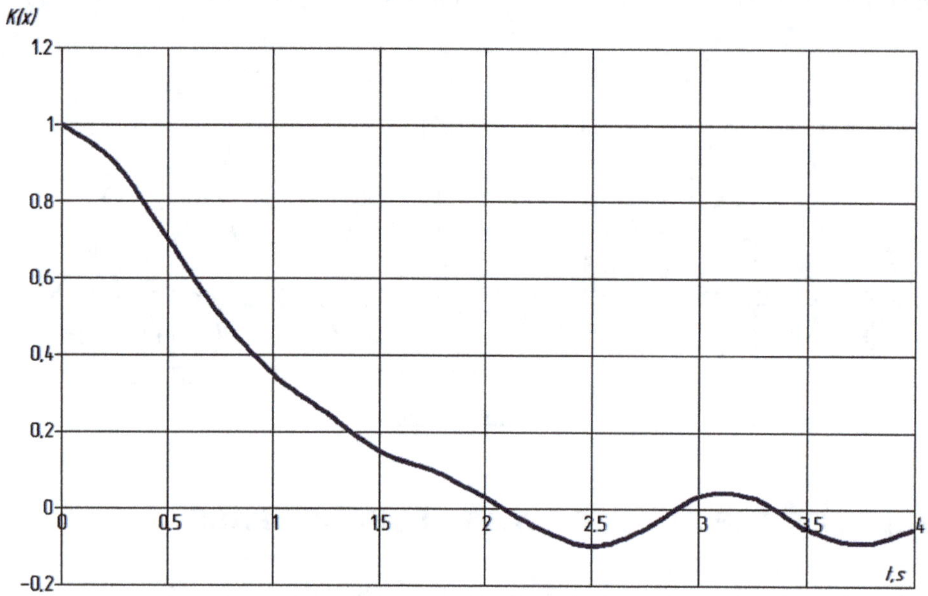

Figure 8. The spectral density of the action of the microprofile of the support base at different speeds.

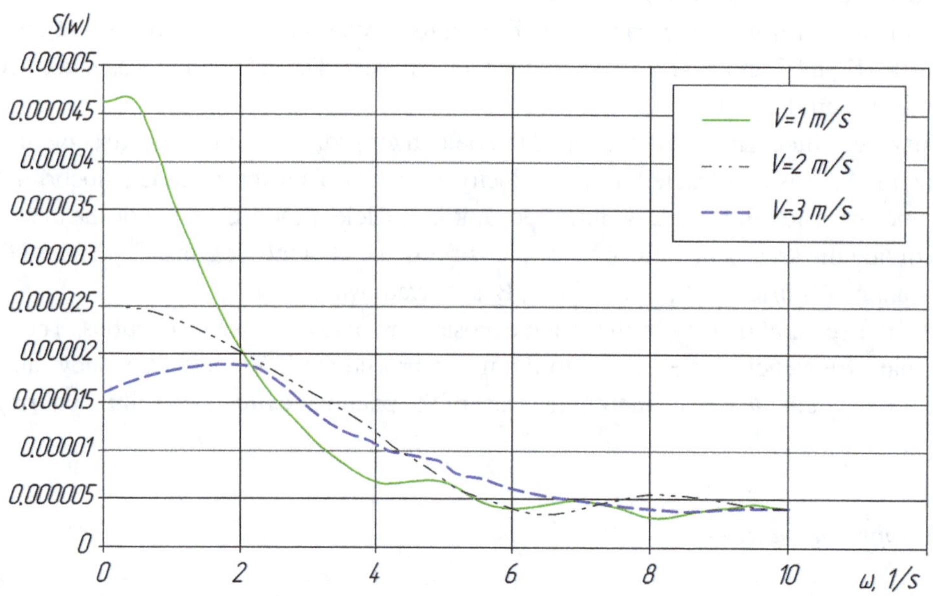

DETERMINATION OF THE SEEDING DEPTH

To seeding depth determine on the etiolated plants part on the day of sowing on one of the repetitions of the experiment on two adjacent passages, rows outside the track of the wheels of the unit are noted, using the randomization method. After germination (with the appearance of 2-3 leaves, after 2 weeks), in certain areas from the every, the plants are cut above ground part. The remaining part of the plants in the soil, together with the seed, is dug out and measured with an error of no more than ±1 mm. The distance from the seed to the cut is an indicator of the depth of seeding. Plants are dug at intervals of 10-15 cm in a row.

For each row of coulters from two adjacent passes of the seeder, the depth of seeding should be determined:

- Not less than 100 plants with a three-row arrangement of coulters on the seeder;
- Not less than 200 plants with two-row arrangement of coulters on the seeder;
- Not less than 300 plants with a single row arrangement of coulters on the seeder.

The number of plants to be dug on each row is determined by the formula:

$$n_p = \frac{200}{n_c^p \cdot n_c^c} \quad (1)$$

where n_c^p is the number of considered coulters in a row, pcs.; n_c^c is the number of rows in the drill coulters, pieces; 200 is total number of plants to be measured.

In addition, the depth of sowing is determined for at least two coulters following the tractor wheels, using the randomization method. The number of measurements must be at least 100.

The results winter wheat seeds quality determining used in the experiment are presented in table 3.

The quality of seeding is presented in table 4.

Table 3. Quality of winter wheat seeds

Indicator	Value
Crop suitability	0.87
Weight of 1000 seeds, g	40.5
Number of seeds sown, PCs/ PM	86
The number who rise seeds, pieces/PM	81
Conditional weight of seeds sown at 1 PM, kg	0.004
Conditional weight of one seed, g	0.0405
Relative field germination	0.94
the number of considered coulters, pieces	3
the number of rows in the drill coulters, pieces	2
Number of plants to be dug on each row	33

Table 4. Depth of seeding

Indicator	Options		Improvement, %
	Without EDM	With EDM	
Set depth of seeding, sm	5.0	5.0	–
Mathematical expectation, sm	5.36	5.12	4.5
Dispersion, sm^2	0.656	0.372	43.3
Standard deviation, sm	0.81	0.61	24.7
Coefficient of variation, %	15.11	11.96	–

Comparison of the obtained data shows that the experimental unit more steadily holds a given depth of seeding.

The installation of an elastic damping element (EDM) in the tractor transmission, as can be seen from table 4, improves the uniformity depth of sowing of coulters, improves the standard deviation characterizing the stability of the process. Thus, a quantitative assessment of the depth of seeding indicates that the experimental unit performs better sowing, due to the fact that the work of the Coulter group is stabilized. For further analysis of the seeding unit, a correlation-spectral analysis of the depth of seeding into the soil was carried out. According to the analysis results, normalized autocorrelation functions and spectral densities of changes in the depth of seeding, which are mutual mathematical transformers through the cosine of the Fourier transform, are obtained.

The correlation functions characterizing the random process in the time domain for the analysis of the depth of seeding have no physical meaning, but they are necessary for the construction of normalized spectral densities. In this regard, there is no need to provide graphs of correlation functions of the depth of seeding.

The graphs of normalized spectral densities of seed depth are presented in figure 9.

The normalized spectral density of the implementation of the depth of seeding into the soil by the sowing unit (figure 9), built on the basis of the normalized autocorrelation function shows that in the unit with the elastic damping mechanism the maximum of dispersions falls on the frequency of 1.0 s^{-1}.

Analysis of the normalized spectral density obtained without an EDM (figure 9) shows that in the range 0-1.5 s^{-1} there are four peaks at the following frequencies: 0.6 s^{-1}, 0.9 s^{-1}, 1.3 s^{-1}. There is a predominant frequency of 1.6 s^{-1}, as well as there are peaks in the high frequency zone of 2.5 s^{-1}, 2.9 s^{-1}.

All this indicates that the distribution of seeds in depth, sowing unit with elastic damping mechanism in the tractor transmission (figure 9), occurs with one frequency (in other words, at one depth).

On the other hand, the distribution of seeds in depth, sowing unit without elastic damping mechanism in the transmission of the tractor, occurs at different frequencies, and the dispersion values are approximately the same and are located throughout the cut (the seeds lie at different depths).

RESULTS OF ENERGY EVALUATION OF SEEDING MACHINE-TRACTOR UNIT

In accordance with the program of experimental studies, the energy assessment of the seeding unit equipped with an elastic damping mechanism in the transmission of the tractor was carried out. Energy indicators are presented in table 5.

Figure 9. Normalized spectral density of the embedding depth seed to soil

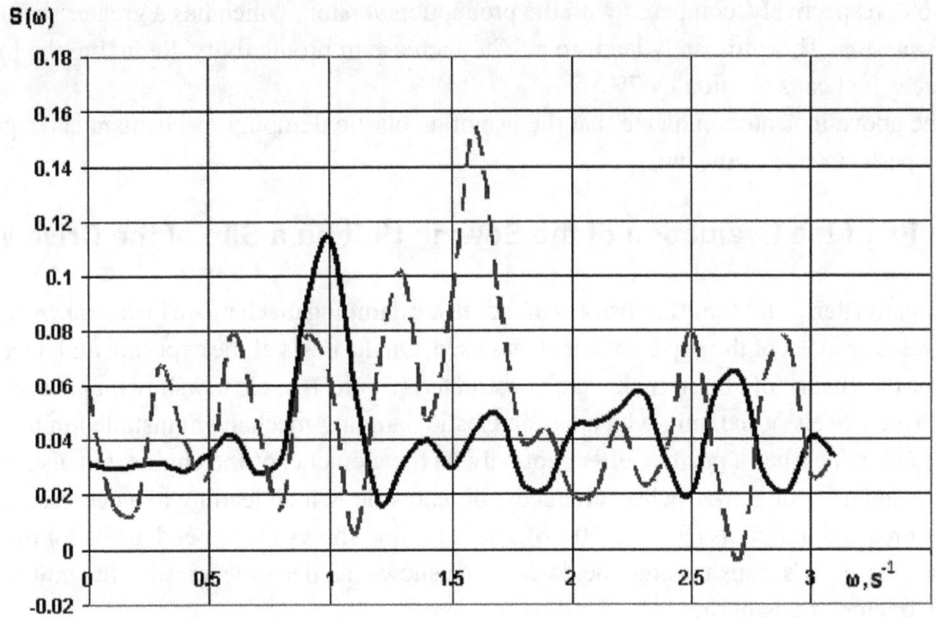

Table 5. Energy performance of sowing MTU

Nº	Energy Performance	Unit Without EDM in Transmission	Unit With EDM in Transmission
1	Traction resistance P_{kp}, N	5713.79	5079.52
2	The speed of movement of the unit V, m/s (km/h)	2.34 (8.42)	2.48 (8.93)
3	The frequency of rotation of the crankshaft ω_1 rad/s	239.32	246.49
4	Hour fuel consumption G kg/h	9.85	9.16
5	The propellers slipping δ, %	14.31	13.23
6	Unit capacity W, ha/h	4.55	4.81
7	Per hectare fuel consumption G_{ha}, kg / ha	2.17	1.91

The data of the table show that the seeding unit MTU with EDM in the transmission has better energy performance (speed, performance and per hectare fuel consumption) than the same MTU serial execution.

From table 5, it can be seen that the drag force is reduced by 11.1%. The value of this force is determined by the external influence of the soil on the working bodies of the seeder and the interaction of links through elastic and deforming connections between them. The installation of an EDM in the transmission of the sowing machine-tractor unit creates more favorable conditions for the formation of the nature of the load of the tractor units, both under unsteady driving conditions and under steady-state ones. The consequence of this is a reduction in the load on the engine and drive wheels of the tractor, causing the angular velocity of the engine crankshaft and the translational speed of the tractor increase by 3% and 6%, respectively, compared with the production version, which has a greater stiffness of the power transmission. This ultimately leads to a 5.7% increase in productivity. Reducing the load on the engine reduces fuel consumption by 7%.

All of the above indicators indicate that the use of an elastic damping mechanism can significantly improve the performance of the unit.

The Results of the Evaluation of the Sowing Unit on a Slip of the Drive Wheels

As an additional criterion for the effectiveness of the elastic damping mechanism in the transmission of the tractor, the relative index of the slip coefficient was used. On the basis of the experimental data obtained, the graphs of the change of slip in time were constructed (figure 10), clearly showing a decrease in the slip of the drive wheels of the unit, which has an elastic damping mechanism installed in the transmission of the tractor. This has a positive effect not only on the elements of the engine, but also on the soil. Reducing the intensity of "jerks" wheels reduces soil abrasion, which leading to a decrease in fertility.

In the figures and diagrams (Figures 10-16), the solid line shows the dependencies for the unit with the EDM in the tractor's transmission, the dashed line shows the dependencies for the unit without the EDM in the tractor's transmission.

The need for the use of instantaneous slip coefficient was needed to identify the frequency characteristics of the process of slipping, and to find out the parameters of the variation series, allowing a better understanding of the processes. The instantaneous value of the slip coefficient was determined by a well-known formula, and the data for the calculation were obtained with the help of measuring equipment and statistically processed. Statistical indicators on slipping δ sowing unit are given in table 6.

Figure 10. Change of slipping in sowing unit time

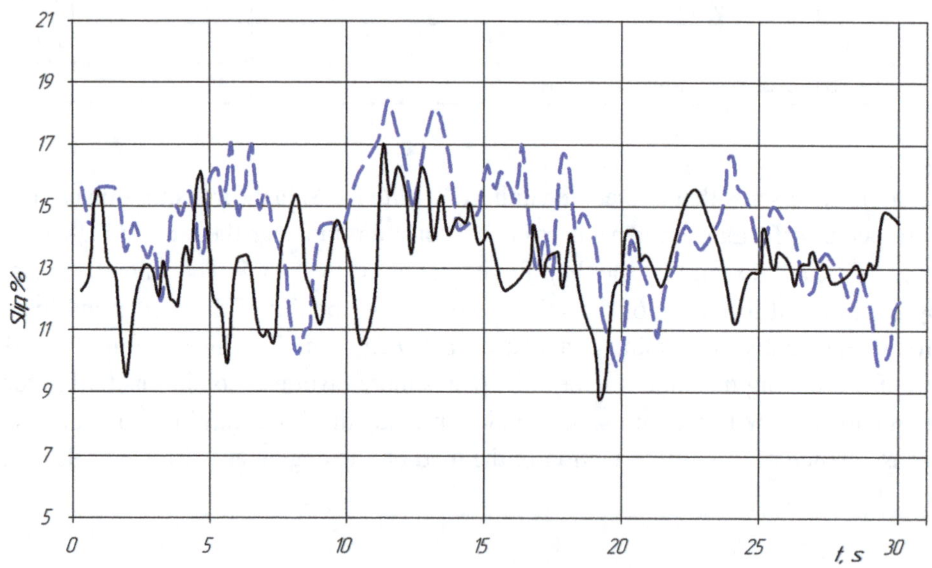

Figure 11. The correlation function of the change of slipping through time sowing unit

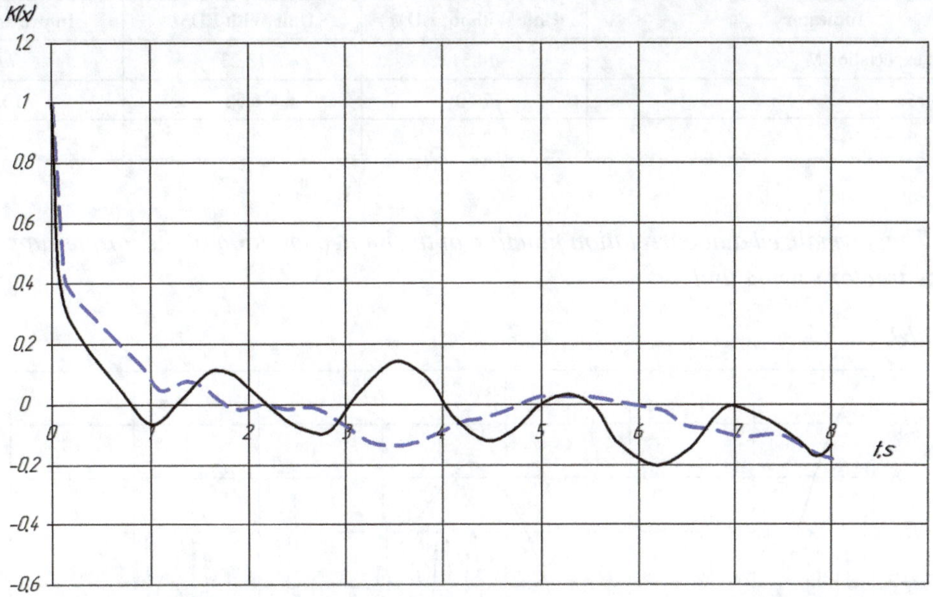

Figure 12. Seeder spectral density of slip change in time

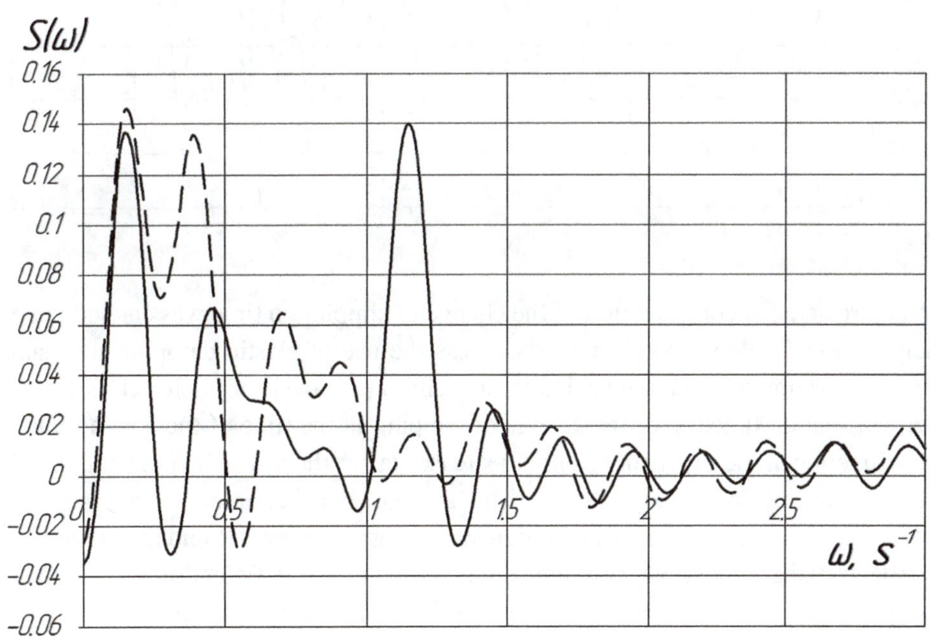

Table 6. Statistical indicators on the slip δ sowing unit

Indicators	Unit Without EDM	Unit With EDM	Improvement %
Mathematical expectation M_{δ}, %	14.31	13.23	7.6
Dispersion D_{δ}, %	7.40	6.62	10.5
Standard deviation σ_{δ}, %	2.72	2.57	5.5

Figure 13. The normalized autocorrelation function of the implementation of the torque supplied by the motor of the tractor sowing unit

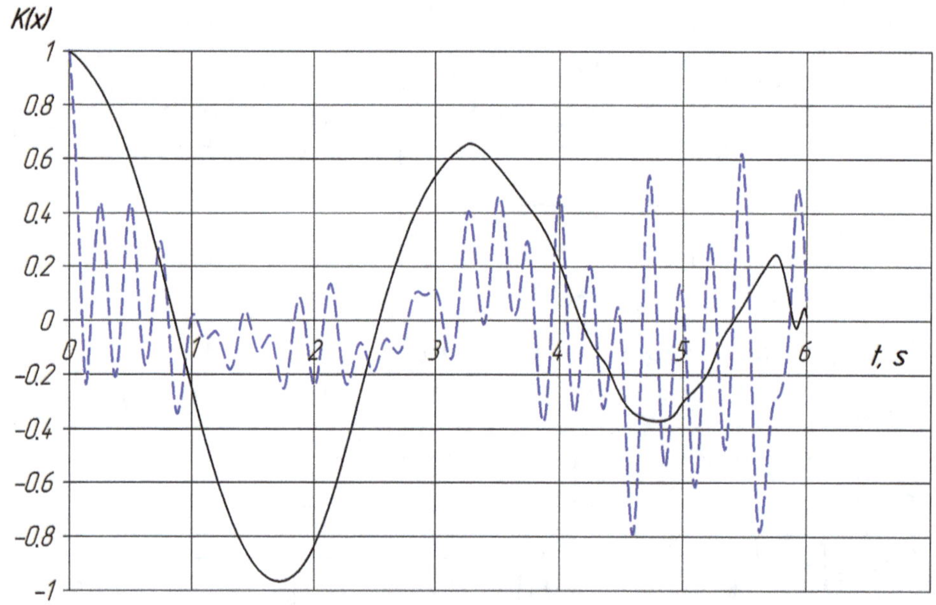

Also for a more detailed consideration of the change of slipping in time was carried out correlation-spectral analysis, which allows to qualitatively assess the use of elastic damping mechanism for the purpose of modernization of sowing machine-tractor unit on the basis of tractor class 1.4.

The given correlation function of the change of slipping in the time of the sowing unit (figure 11) shows that the function of the experimental unit decreases faster-the time of decreasing 0.7 - 0.9 s, in the serial version, the time of decreasing 1.6-1.8 s. This is a positive phenomenon, as it indicates the rapid adaptation of the engine of the experimental unit to changing conditions. Rapid adaptation of the engine to changing effects makes it possible to assume an increase in the coefficient of adhesion to the support surface, however, additional studies have not been conducted to confirm or refute the above assumption.

On the graph (figure 12) of the spectral densities S(ω), giving an idea of the frequency composition of the process of slipping, the prevailing frequencies are seen. The cross-section of frequencies S (ω) for the experimental and serial variants is approximately the same and is about 1.5 s^{-1}.

Table 7. Results of torque quantification developed by the engine of the tractor of the sowing unit

Statistical Indicator	Torque, Nm					
	Measurement Area by Options					
	Without EDM			With EDM		
	1	2	3	1	2	3
Mathematical expectation	425.5	421.4	412.2	422.3	417.5	404.5
Dispersion	349.65	299.25	256.67	183.80	60.92	202.07
Standard deviation	18.70	17.30	16.02	13.56	7.81	14.22
Confidence interval	23.97	22.18	20.54	17.38	10.01	18.23

Table 8. Results of quantitative evaluation of the pull force on the hook

Statistical Indicator	Pull Force on Hook, N					
	Measurement Area by Options					
	Unit Without EDM			Unit With EDM		
	1	2	3	1	2	3
Mathematical expectation	5772.5	5746.5	5548.7	4890.6	5157.6	5026.4
Dispersion	5169.8	103839.4	64946.2	1827.7	94291.9	134065.6
Standard deviation	71.9	322.2	254.8	42.8	307.1	366.2
Confidence interval	92.2	413.1	326.7	54.9	393.7	469.5

Figure 14. Normalized spectral density of realization torque of the sowing unit developed by the tractor engine

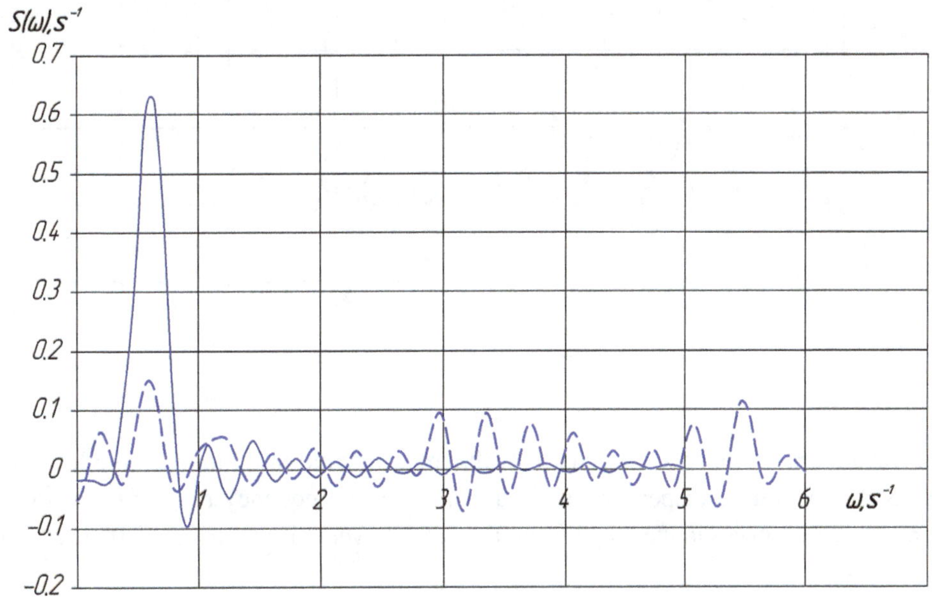

Figure 15. The normalized autocorrelation function of realization of traction force on a hook of a tractor of the sowing unit

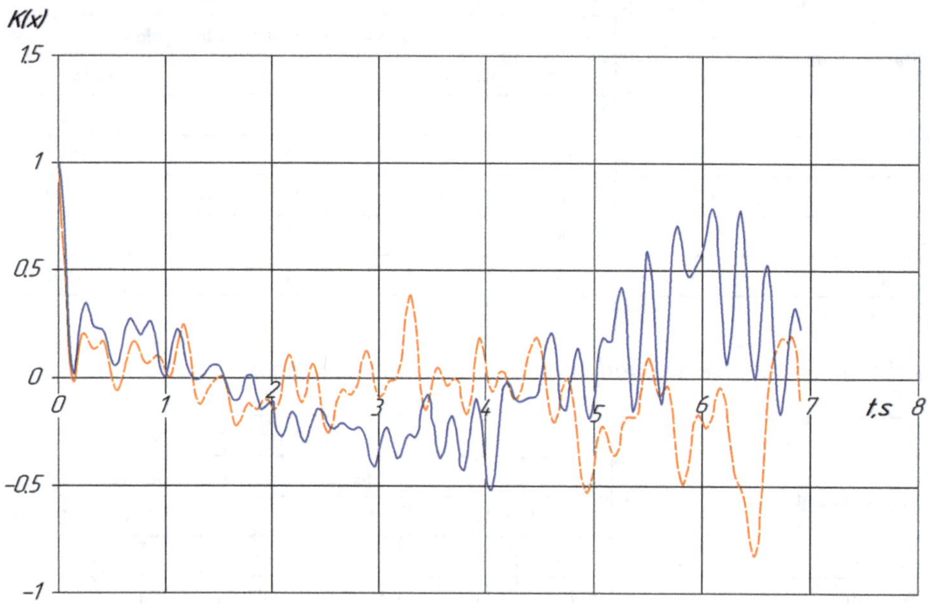

Figure 16. Normalized spectral density of the thrust force realization on the tractor hook of the sowing unit

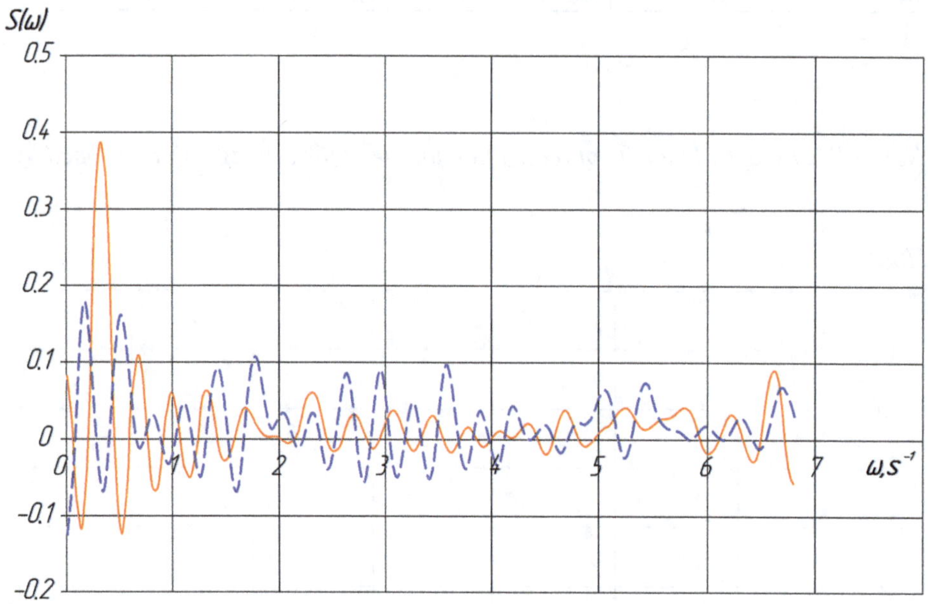

However, the nature of the spectral density of the experimental unit indicates the low frequency of the process, as the maximum dispersions shifted to the lower frequency range from 0 to 1.0 s^{-1}. The process of changing the curves in the range from 0 to 1.0 s^{-1} is very similar to each other, which indicates the identity of the process of changing the slip in time. The spectrum of change of slipping in time of the serial unit from 0 to 1.5 s^{-1} has four dominant peaks at 0.2 s^{-1}, 0.4 s^{-1}, 0.7 s^{-1} and 0.9 s^{-1}, and in the experimental version there are only two of them is 0.2 s^{-1} and 0.45 s^{-1}.

Also, the dispersion values of the experimental unit are less important than the serial version. This is a positive phenomenon, as it indicates that the drive wheels of the experimental unit less slipping. The maximum spectral density of the experimental unit falls on the low-frequency range in the area of 0.2 s^{-1}

However, in the high-frequency region at a frequency of 1.23 s^{-1}, there is vertex of dispersion graphic of the of the change in the slipping of the experimental unit, indicating that the maximum value of the change in the slipping of the engine falls on certain marks characterizing the change in the type of gear-box (change in the gear ratio in the planetary mechanism), while the serial unit is continuously slipping in the range from 0 to 1.5 s^{-1}.

Estimation of Quantitative Indicators of Sowing Unit Working Processes

Also, for a more detailed study of sowing process unit operation, it was necessary to assess the qualitative and quantitative indicators of the working process of the sowing unit in two versions:

1. With elastic damping mechanism in tractor transmission;
2. With locked elastic damping mechanism (hereinafter referred to as "serial").

When the unit was working in the field, its main operating parameters were recorded.

The time of experience and the distance traveled were recorded using the system of automatic accumulation and Processing of metrological information (SANO). The results of the experiment were processed using a special software package. Three sections of the data array were obtained. Each site was statistically processed. The results of the statistical processing, namely the quantitative assessment, were presented in tables 7-8.

The confidence interval is defined by the formula

$$\Delta = t \cdot \sigma \tag{2}$$

where t is the student coefficient (Spiridonov, 1981). σ is the mean square deviation.

Analyzing table 7, we observe a decrease in the mathematical expectation of the engine torque at all three sites. In the first and second sections of the mathematical expectation decreased by 1%, in the third section there is a decrease of 1.9%. In turn, the standard deviation is reduced in the first section by 27.5%, in the second section by 54.9% and in the third section by 11.2% compared to the option without EDM. Thus, judging by the experimental data, the engine of the tractor, which has an elastic-damping mechanism, is less loaded (as evidenced by the mathematical expectation) and more protected from changing loads (this is evidenced by the standard deviation).

Analyzing table 8, we observe a decrease in the mathematical expectation of the traction force on the tractor hook. In the first section, the mathematical expectation is reduced by 15.3%, in the second section there is a decrease of 10.2%, and in the third section there is a decrease of 9.4% compared to the option without an elastic damping mechanism.

In turn, the standard deviation decreases by 40.5% in the first area, by 4.7% in the second area, and there is no decrease in the third area, as the standard deviation increases by 43.7%. The increase in the standard deviation may be due to the fact that the unit with an elastic damping mechanism fell on a harder layer of soil, which led to an increase in the uneven operation of the unit. In other words, there were no comparable conditions in the third section.

Thus, judging by the experimental data, the transmission of the tractor, which has an elastic-damping mechanism, is less loaded (as evidenced by the mathematical expectation) and more protected from changing loads (this is evidenced by the standard deviation).

Since the elastic-damping mechanism allows to stretch the shock load in time, and smoothes the vibrations, as a result of this, there is a decrease in the mathematical expectation of the traction load on the hook, and, consequently, the work of the Coulter group of the seeder is improved because the standard deviation of the depth of incorporation is reduced.

Qualitative Assessment of the Sowing Unit: Description of Frequency Characteristics

To describe the nature of the random stationary process, which is the process of movement of the seeding unit, a correlation-spectral analysis of the experimental data was carried out. According to the results of the analysis, normalized autocorrelation functions and spectral densities of changes in the parameters of the unit are obtained, the graphs of which are shown in figures 13-16. Normalized autocorrelation functions and normalized spectral densities of processes are constructed for the velocity of 2.4 m/s.

To carry out correlation and spectral analysis, the process of operation of the machine-tractor unit in serial and experimental versions in three-fold repetition of each option was recorded.

For the three plots obtained normalized autocorrelation function and the spectral density of realizations of the parameters of the sowing unit tractor. In this Chapter, we present graphs for only one plot.

The normalized autocorrelation function of realization of the torque developed by the engine of the tractor of the sowing unit (figure 13) shows that time of decline of function at the serial unit makes 0,1 - 0,2 s, unlike the unit with the elastic damping mechanism in transmission at which time of decline of correlation function 0,8-0,9 s.

The time of recession characterizes the speed of the system response to changes in the environment. In our case, the engine without EDM reacts to changes more strongly, indicating its vulnerability to fluctuations in external traction load. Conversely, the engine of the unit, which has an elastic damping mechanism, is more protected from external load fluctuations, since its crankshaft works smoothly, in comparison with the unit without EDM.

The normalized spectral density of torque realization (figure 13) developed by the engine of the sowing unit tractor obtained on the basis of the normalized autocorrelation function shows that in the unit with EDM the maximum of dispersions falls on the frequency of 0.6-0.7 s^{-1}. This indicates that the system reacts to the appearance of the external environment and changes the processed environment in some way. In the present case, the autonomy of the engine operation is manifested. That is, it operates at one dominant frequency.

In the unit without EDM there is a stretched spectrum of dispersions (figure 14), which indicates the operation of the engine in different frequency ranges. There are three peaks in the low-frequency range from 0 to 1.3 s^{-1}, four peaks in the high-frequency range from 2.9 to 4.2 $s^{-1}$and two peaks in the frequency range from 5 to 5.5 s^{-1}. This indicates the unstable operation of the engine unit without an elastic damping mechanism in the transmission of the tractor. That in turn leads to fuel overruns, the emergence of fluctuations in the transmission, transmitted to the Coulter group seeders.

The cutoff frequency, which determines the upper limit of the frequency spectrum of the random process also indicates the narrow-band nature of the engine unit with EDM in the tractor transmission ($\omega_c = 2\ s^{-1}$). The unit without EDM cutoff frequency is $\omega_c = 6\ s^{-1}$.

The normalized autocorrelation function (figure 15) of the implementation of the traction force on the hook of the tractor of the sowing unit shows that the decay time of the unit function with an elastic damping mechanism and without an elastic damping mechanism in the transmission is approximately the same and is 0.1-0.2 s.in the unit with an elastic damping mechanism, there is an increase in the correlation in the range of 5.0-7.0 s. This indicates that the process of the operation of the sowing unit with EDM in the tractor transmission is more stable.

The normalized spectral density of realization of the traction force on the hook of the tractor of the sowing unit (figure 16), constructed on the basis of the normalized autocorrelation function shows that in the unit with the elastic damping mechanism the maximum of dispersions falls on the frequency of 0.3 s^{-1}.

The cutoff frequency at the site 2 (figure 16), of the reaction force implementation on the hook of the tractor sowing unit with EDM in the tractor transmission is $\omega_c = 2{,}7\ s^{-1}$.

In the unit without EDM there is a stretched spectrum of dispersions (figure 16), which indicates the operation of the unit in different frequency ranges and the worst adaptation to changing conditions (the system does not respond to changing conditions). There are two peaks in the low-frequency region at 0.2 s^{-1} and 0.6 s^{-1}, two peaks in the frequency range 1.4 s^{-1} and 1.7 s^{-1}, four peaks in the frequency range from 2.5 s^{-1} to 3.7 s^{-1}, as well as in the frequency range from 5.0 $s^{-1}$to 5.7 s^{-1}.

The cutoff frequency (figure 15) of the traction force on the hook of the tractor of the sowing unit without EDM in the tractor transmission is $\omega_c = 6.3\ s^{-1}$.

CONCLUSION AND FURTHER DIRECTIONS OF RESEARCH

Research Findings

In general, the section "results of experimental studies" can draw the following conclusions:

1. The data obtained show that the use of EDM can reduce the resistance of the coulter group of seeders transmitted to the engine by an average of 15% in the field.
2. As a result of the use of EDM in the transmission assembly productiveness is creased by 5.7% (by reducing the slippage of the engine), which is 1.82 hectares per shift, and the improvement of agricultural performance in sowing (uniformity of the depth of seeding) is 32.8% compared to the production version.
3. The angular velocity of the engine crankshaft using an elastic element is increased by 3%, which indicates an improvement in engine performance.
4. Due to the more stable operation of the engine hour consumption and per hectare fuel consumption decreased by 7.5% and 13.6%, respectively.
5. Analysis of the engine torque values showed a decrease in the mathematical expectation of the engine torque at all three sites by an average of 10.1%. Thus, judging by the experimental data,

the engine of the tractor, which has an elastic-damping mechanism, is less loaded (as evidenced by the mathematical expectation) and more protected from changing loads (this is evidenced by the standard deviation).

6. Analysis of the values of the traction force on the tractor hook showed a decrease in the mathematical expectation of the traction force on the tractor hook by an average of 11.6%. Consequently, the transmission of the tractor, which has an elastic damping mechanism, is less loaded and more protected from changing loads.
7. Analysis of the depth of seeding seed sowing unit, showed that the installation of an elastic damping mechanism in the transmission of the machine-tractor unit contributes to an increase in the uniformity of the coulters by 4.5%, and 24.7% improves the standard deviation characterizing the stability of the process performed by the sowing machine-tractor unit.

The analysis shows that the installation of an elastic damping mechanism in the transmission of the tractor of the sowing machine-tractor unit is a very effective constructive measure that improves the output of the unit in real operating conditions.

FURTHER DIRECTIONS OF RESEARCH

Future research and further practical application of EDM are aimed at studying its impact on the work of the MTU during such technological operations as plowing and transporting.

REFERENCES

Ageev, L. E. (1978). *Osnovy rascheta optimal'nykh i dopustimykh rezhimov raboty mashinno-traktornykh agregatov* [Basics of calculating the optimal and allowable modes of operation of machine-tractor units]. Leningrad, Russia: Kolos.

Bespamyatnova, N. M. (2002). Nauchno-metodicheskiye osnovy adaptatsii pochvoobrabatyvayushchikh i posevnykh mashin [Scientific and methodical bases of adaptation of tillage and sowing machines]. – Rostov n/D: OOO .

Bespamyatnova, N. M. (2010). Vibratsii v tekhnologicheskikh protsessakh [Vibrations in technological processes]. Zernograd: VNIPTIMESKh.

Deboli, R., Calvo, A., & Preti, C. (2017). Whole-body vibration: Measurement of horizontal and vertical transmissibility of an agricultural tractor seat. *International Journal of Industrial Ergonomics, 58*, 69–78. doi:10.1016/j.ergon.2017.02.002

Feuerhuber, K., Lindert, S. O., & Schlacher, K. (2013). Vibration attenuation by semi-active dampers. *Paper presented at 14th International Conference on Computer Aided Systems Theory (EUROCAST)*, Las Palmas, Spain.

Goryachkin, V. P. (1965) Zemledel'cheskaya mekhanika. Sobraniye sochineniy [Agricultural mechanics. Moscow, Russia: Kolos.

Kravchenko, V. A. (2010). *Povishenie dinamicheskih i ekspluatatsionnyh pokazateley selskohozyastvennyh mashinno-traktornyh agregatov* [Increase of dynamic and operational parameters of agricultural machine and tractor units] [PhD dissertation]. FSEE HE ABSSAEA, Zernograd Russia.

Kravchenko, V. A., Senkevich, S. E., Senkevich, A. A., Goncharov, D. A., & Duryagina, V. V. (2008). RU Patent 2398147 F16H47/04.

Kravchenko, V. A., Senkevich, S. E., Senkevich, A. A., Yarovoy, V. G., & Tolstoukhov, Yu. S. (2004). RU Patent 2222440, B60K 17/10

Kuznetsov, N. G. (2006). *Stabilizatsiya rezhimov raboty skorostnyh mashinno-traktornyh agregatov* [Stabilization of the operating modes for high-speed machine-tractor units]. Volgograd, Russia: Volgograd State Agricultural Academy.

Li, B., Sun, D., Hu, M., & Liu, J. (2019). Automatic starting control of tractor with a novel power-shift transmission. *Mechanism and Machine Theory*, *131*, 75–91. doi:10.1016/j.mechmachtheory.2018.09.012

Lurie, A. B. (1969). *Dinamika regulirovaniya navesnykh selskokhozyaystvennykh agregatov* [The dynamics of regulation of mounted agricultural units]. Leningrad, Russia: Mechanical Engineering.

Lurie, A. B. (1981). *Statisticheskaya dinamika sel'skokhozyaystvennykh agregatov* [Statistical dynamics of agricultural units]. Moscow, Russia: Kolos.

Nekhoroshev, D. A. (2014). *Stabilizatsiya rezhimov MTA izpolzovaniem pnevmogidravlicheskoy mufty stsepleniya* [Stabilization of MTU modes using pneumatic hydraulic clutch] [PhD dissertation]. Volgograd State Agrarian University, Volgograd, Russia.

Nekhoroshev, D. D., & Nekhoroshev, D. A. (2015). Energeticheskoe sredstvo v transmissii kolesnogo traktora [Power means in the transmission of the wheeled tractor]. *Unique Research of the XXI Century*, *12*(12), 40–41.

Polifke, G. (2016). Hydrodamp - hydraulic torsional vibration damper for tractors. *Paper presented at Conference on Agricultural Engineering*, Koln, Germany.

Polifke, G. (2017). Hydrodamp - hydraulic torsional vibration damper for tractors and construction machines design for a race-sensitive drive train. Paper presented at *VDI Conference on Couplings and Clutch Systems in Drives 2017 / 2nd VDI Conference on Vibration Reduction in Mobile Systems*, Ettlingen, Germany.

Polivaev, O. I., Gorban, L. K., Vorohobin, A. V., & Vedrinsky, O. S. (2018). Decrease of dynamic loads in mobile energy means. *Paper presented at Conference Series: Materials Science and Engineering (IOP)*, Voronezh, Russia. 10.1088/1757-899X/327/4/042083

Polivaev, O. I., & Vedrinsky, O. S. (2013). *Vliyanie uprugodempfiruyushchego privoda vedushchikh koles traktora na dinamicheskie nagruzki v transmissii pri razgone MTA* [Influence of the elastic-damping drive of the tractor's driving wheels on the dynamic loads in the transmission when accelerating the MTU]. (pp. 80–83). Innovative Technologies and Technical Means for Agroindustrial Complex.

Polivaev, O. I., & Vedrinsky, O. S. (2014). Analiz vliyaniya uprugodempfiruyushchego privoda koles na dinamicheskie nagruzki v transmissii traktora pri razgone [Analysis of the effect of the elastic-damping drive of the wheels on the dynamic loads in the tractor's transmission during acceleration]. *The Bulletin of the Don Agrarian Science*, *28*, 5–9.

Polivaev, O. I., Vedrinsky, O. S., & Derkanosova, N. M. (2016). Povishenie dolgovechnosti stsepleniya traktorov za schet uprugofriktsionnogo dempfera [Increasing the durability of tractor traction due to elastic-friction damper]. *Science and Education in Modern Conditions*, *12*, 226–230.

Polivaev, O. I., Zhidenko, K. S., Maksimov, I. I., & Levkin, I. G. (2017). Uprugo-dempfiruyushchiy privod vedushchih koles traktora [Elastic-damping drive of the tractor's driving wheels]. *Paper presented at Modern Trends in the Development of Technology and Technical Means in Agriculture*, Voronezh, Russia.

Senkevich, A. A. (2008). Povysheniye effektivnosti funktsionirovaniya posevnogo mashinno-traktornogo agregata putem ustanovki v transmissiyu traktora klassa 1,4, uprugodempfiruyushchego mekhanizma [Increase the efficiency of the sowing machine-tractor unit by installing a class 1.4 tractor, an elastic-damping mechanism in the transmission] [PhD dissertation]. FSEE HE ABSSAEA, Zernograd Russia. 142 p.

Senkevich, S., Kravchenko, V., Duriagina, V., Senkevich, A., & Vasilev, E. (2019). Optimization of the Parameters of the Elastic Damping Mechanism in Class 1.4 Tractor Transmission for Work in the Main Agricultural Operations. In P. Vasant, I. Zelinka, & G. W. Weber (Eds.), *Intelligent Computing & Optimization. ICO 2018*. Cham: Springer. doi:10.1007/978-3-030-00979-3_17

Shekhovtsov, V. V., Sokolov-Dobrev, N. S., & Potapov, P. V. (2016). Decreasing of the Dynamic Loading of Tractor Transmission by means of Change of the Reactive Element Torsional Stiffness. *Procedia Engineering*, *150*, 1239–1244. doi:10.1016/j.proeng.2016.07.129

Shekhovtsov, V. V., Sokolov-Dobrev, N. S., & Potapov, P. V. (2016). Decreasing of the Dynamic Loading of Tractor Transmission by means of Change of the Reactive Element Torsional Stiffness. *Procedia Engineering*, *150*, 1239–1244. doi:10.1016/j.proeng.2016.07.129

Silaev, A. A. (1963). *Spektral'naya teoriya podressorivaniya transportnykh mashin* [Spectral theory of the suspension of transport vehicles]. Moscow, Russia: Mashgiz.

Spiridonov, A. A. (1981). *Planirovanie eksperimenta pri issledovanii tekhnologicheskikh protcessov* [Planning an Experiment in the Study of Technological Processes]. Moscow, Russia: Mashinostroenie.

Spiridonov, A. A. (1981). *Planirovanie eksperimenta pri issledovanii tehnologicheskikh protsessov* [Planning an experiment in the study of technological processes]. Moscow, Russia: Mechanical Engineering.

Zheng, E. L., Fan, Y. D., Zhu, R., Zhu, Y., & Xian, J. Y. (2016). Prediction of the vibration characteristics for wheeled tractor with suspended driver seat including air spring and MR damper. *Journal of Mechanical Science and Technology*, *30*(9), 4143–4156. doi:10.100712206-016-0826-x

Zoz, F., & Grisso, R. (2003). Traction and Tractor Performance. *Paper presented at Agricultural Equipment Technology Conference*, Louisville, KT.

KEY TERMS AND DEFINITIONS

Autocorrelation Function: Is the dependence of the relationship between the function (signal) and its shifted copy on the magnitude of the time shift.

Correlation Function: Is a function of time and spatial coordinates, which determines the correlation in systems with random processes.

Elastic-Damping Mechanism (EDM): Is a mechanism that damps vibrations and dissipates energy inside the system under the effect of cyclic loads. In this case, the mechanical energy of the oscillations is converted into the thermal energy.

Etiolated Part of the Plant: Is the part of the plant that grew in the dark (in the soil) and is characterized by the absence of chlorophyll, the abnormal shape of the organs and the weak development of the cell membranes.

Gearbox (GB): A change speed gearbox for changing the torque and for transfer it from engine to wheels.

Machine-Tractor Unit (MTU): Is a tractor connected to a working machine or machines. A tractor in the unit is a source of mechanical energy which helps the machine to move around the field, for example when a tractor pulls a cultivator.

Mobile Agricultural Unit: Is a source of mechanical energy. With it, the machine moves around the field, for example, a tractor pulls a plow or a cultivator behind it. In addition, the tractor can actuate the working bodies of another machine, such as a silage cutter. Sometimes he sets in motion both the machine and its working organs, like those of a potato, flax, beet harvester, harvester, and mower.

Slipping Tractor Thrusters: Is the interaction of a tractor wheel or track with a supporting surface, accompanied by a reduction in the rate of translational movement, a change in traction characteristics due to different characteristics of the supporting surface and traction resistance.

Spectral Density Function (Or Power Spectrum): Refers to the characteristics by which the basic properties of stationary random processes are analyzed. The spectral density function characterizes the harmonic composition of the studied process and determines the energy spectrum.

Chapter 2
Evaluation of the Effectiveness of the Use of Programs in the Design of Power Complexes Based on Renewable Energy Resources

Thu Yein Min
National Research University "Moscow Power Engineering Institute," Russia

Michael G. Tyagunov
National Research University "Moscow Power Engineering Institute," Russia

Galina M. Deriugina
National Research University "Moscow Power Engineering Institute," Russia

ABSTRACT

This chapter studies the prospects of energy complexes on the basis of renewable energy sources to supply electricity to the stand-alone consumers in different regions of Myanmar. In order to do that, the territory of Myanmar is divided into regions according to their amount of renewable energy sources. The developed methods are for determining the optimum parameters and operation of the energy complex on the basis of renewable energy sources and the cost-effectiveness of those energy complexes was examined. This was for the purpose of a mathematical formulation of the problem of optimization of the energy complex on the basis of renewable energy sources for autonomous rural consumers of the Republic of Myanmar.

DOI: 10.4018/978-1-5225-9420-8.ch002

INTRODUCTION

With the developments of technological progress, the requirements for energy efficiency increases with every year. Practically in all developed countries RES development projects are being developed and implemented. The basic principle of the energy use based on renewable energy is to extract it from the processes constantly occurring in the environment and provide them for technical use. Technical progress has now reached such a level, with which energy generation is determined not only economic expediency, as well as a number of other factors, the most significant of which are environmental, social, and that are associated with human development and energy security. RES has significant advantages in terms of ecology and social significance. RES developed significantly in countries with limited resource base whose energy security is directly depends on the supply of energy (primarily oil and gas). A very complex, but urgent task is the system of optimization of parameters and HES (Hybrid Energy Systems) modes based on renewable energy sources. To create the HES requires special informational, mathematical and software for the feasibility study of projected the HES in the conditions of countries where market relations are only at the stage of their formation. In this regard, for efficient use of the RES resources, selection of the composition of generating installations based on renewable energy sources and their optimal parameters have reliable data on the flow of resources and optimize the energy parameters and operating modes of the HES. According to world experience, the use of only one type of RES in systems power supply of autonomous consumers does not always allow to provide reliable and uninterrupted power supply due to physical features of RES. In this regard, the autonomous consumer of power supply for Myanmar is most efficiently organized by sharing energy sources based on renewable energy sources, in particular, solar, wind and water. For efficient operation of HES as part of power plants based on renewable energy, it is necessary to create software that will allow optimizing the design parameters and operating modes of all HES components according to a given optimality criterion.

BACKGROUND

Currently, one of the biggest problems of the Republic of Myanmar, which is a developing country, is the need to raise the social standard of living of a large rural population, which is largely determined by the level of consumer supply of cheap electricity. In 2017, Myanmar produced 18 billion kWh of electricity, and the demand for electricity amounted to 20 billion kWh (Aung & Shestopalova, 2016). The specific energy consumption per person was only 200 kWh / year (Aung & Shestopalova, 2016). The level of energy consumption in Myanmar is the lowest in comparison with neighboring developing countries. The installed capacity of power plants operating in the Unified Energy System (UES) of Myanmar was 5,390 MW in 2017.

In Myanmar, there is a shortage of electricity in all sectors, including the municipal sector (Hla, 2015). Currently, Myanmar's national electricity grid does not cover the entire territory of the country.

The National Grid (NG) covers only 38% of the country's population. NG does not cover mountain areas due to the high cost of transmission lines. Mountain and remote regions have only local networks of autonomous power supply.

This means that 62% of the population of Myanmar live in a decentralized area with an unguaranteed energy supply. Of the 64907 rural settlements, only 7% are connected to the National Grid. Most autonomous consumers in rural areas use diesel or gasoline generators. The cost of electricity for a centralized consumer is 3.5-7.5 US cents / kWh. The cost of electricity in the villages is much higher (about 50-90 US cents / kWh) (Aung & Shestopalova, 2016; Aung, 2015; Aung, Malinin & Shestopalova, 2016).

Thus, the most pressing problem of Myanmar is the lack of power supply to the population. Another important issue is the heating and air conditioning of residential buildings.

Myanmar has a monsoon climate with three main seasons: the rainy season is from late May to late October, the cooler season is from late October to mid-February, and the hot season is from mid-February to late May. In the hot season, the average monthly temperature in many parts of Myanmar exceeds +30 ° C. In the mountainous areas and northern areas of the country it is much cooler (in the cold season the temperature sometimes reaches -5 °C), in the valleys the temperature does not exceed + 20 °C (Aung & Shestopalova, 2016; Hadden, 2008). Temperature fluctuations during the year are large, which urgently requires solving the above problem.

At present, Myanmar is trying to reduce the deficit in power generation in the country. For this purpose, new gas thermal and hydroelectric power plants are being built. But these measures improve power supply only in cities where there is a connection with the National Energy System.

For remote areas where it is impossible to connect to the national grid, you need to use local energy resources. These primarily include renewable energy sources.

Global use of renewable energy in 2014 amounted to 18.4% (IRENA,2016). In 2016, IRENA offers a roadmap for the future of renewable energy (REmap). This map is a plan to double the use of renewable energy in global energy consumption by 2030 (Zay & Tyagunov, 2016). Now 40 countries of the world are parts of Remap, but only two ASEAN countries - Indonesia and Malaysia. (Kyaw, 2012).

The population of ASEAN countries will increase to 715 million by 2025. The economy will grow by more than 5% per year, as a result, the demand for energy will grow rapidly. Until 2025, in ASEAN countries there will be an annual growth in energy demand by 4%, and in 2025 energy consumption will be 60% higher than in 2016 (Irena, 2016).

Myanmar is one of the developing countries of ASEAN. Due to its geographical position, Myanmar has a large renewable energy resource (Melia, 2012). Currently, the most common types of renewable energy used to provide electricity to autonomous consumers in the developing world with a large share of the rural population, which is very typical for Myanmar, are biomass energy, solar energy and wind energy, as well as small hydropower (Lin, 2011).

However, some power plants based on renewable energy sources do not provide a guaranteed power supply to consumers. One of the possible solutions in this case is the creation of energy complexes comprising solar (photovoltaic), heat pump, hydropower and other power plants, as well as energy accumulators of various types - electrochemical, thermal, mechanical, etc. This is a topic that is considered in this chapter.

The justification for the feasibility of constructing a HES based on renewable energy sources, carried out at the early stages of their design, depends on the methods used to calculate the parameters of HES elements. Most of the known mathematical models for calculating wind, solar and hydropower resources are based on empirical relationships and coefficients that are valid only for a specific area. Thus, it can be argued that in the early stages of design, the reliability of data on the hydro, wind and climatic characteristics of the terrain of the proposed HES construction is extremely low. Partial compensation for this is the use of techniques based on calculations based on simulated rows of observed parameters of water flow, wind speed and direction, weather, precipitation, etc. over the past periods of observations with subsequent statistical processing of the results (Bezrukikh, 2010).

Similar programs for the feasibility study of the parameters and composition of the HES have been developed earlier, one of which is the HOMER program, which, as shown below, has a number of deficiencies and limitations (Homer).

MAIN FOCUS OF THE CHAPTER

System Properties of Small Distributed Power Supply Systems

Energy systems include in their composition of consumers who form the demand for electrical, thermal, mechanical and other types of energy consumed, producers (generators) of the required types of energy, means of delivering energy to the consumer, a flexible backup system of energy suppliers and effective management of all elements of the system.

Important features of small power systems are:

1. A relatively small distance from the place of production to the place of energy consumption, which significantly reduces the amount of losses along the length of the delivery lines (electrical, mechanical, hydraulic, thermal, etc.)
2. A relatively small number of energy-consuming installations, the unit capacity of which is comparable to the installed capacity of the entire power grid
3. A relatively small number of energy-generating installations, the unit capacity of which is comparable to the installed capacity of the entire power grid
4. Close interdependence of operating modes of all elements of the power system, affecting the stability and reliability of its work

The target installation of energy systems is usually formulated as "a complete and uninterrupted energy supply of energy consumers in accordance with their requirements for the quantity and quality of energy, the energy consumption schedule in a given time interval". In other words, the parameters of the generating part of the system are selected based on the needs of consumers, the structure and parameters of which do not depend on their energy supply (Dobermann, 2016).

General Provisions of the Design of the Power Complex

Designing an energy complex on the basis of the Sun, wind and DGEN (Diesel) is a multistage, iterative one. Iterations are performed on time variable (in calculations of solar radiation and wind energy arrival), slope angles of the receiving site (optimization of PV (solar photovoltaic module) production) and layout of wind turbines on the site to minimize aerodynamic shading losses. All design stages are considered separately for each type of PV and wind turbines.

Optimality criteria of HES:

1. Minimum capital and operating costs;
2. Minimum space required for equipment placement;
3. Minimum power consumption from the network/backup power source (diesel).

Criteria for selection of HES site:

1. High intensity of PV and wind energy inflow
2. The simplicity of the terrain (the minimum shading of the PV, the minimum shading of wind turbines, the possibility of self-cleaning PV, ease of installation of wind turbines and PV);
3. Soil infertility and deep underground water (unfit for agricultural use of the land is relatively cheap for rent);
4. Location near an electrical substation with sufficient capacity for connection of HES;
5. Availability of infrastructure (transport, roads to reduce construction costs).

In this paper, economic indicators were taken as the main criterion for optimality in order to ensure a minimum of discounted costs over the calculated period of time (T = 20 years),

$$E = \sum_{j=1}^{T_{calc}} E_j \left(\frac{1}{1+\varepsilon} \right)^j \tag{1}$$

where T_{calc} - settlement period; j is the ordinal number of the year; E - the total cost for the billing period; ε - the discount rate adopted is 0.1 (10%).

The task of optimizing the parameters and modes of HES based on renewable energy sources requires special information, mathematical and software tools for its feasibility study.

In this regard, for efficient use of renewable energy resources, selection of generator sets based on renewable energy sources and their optimal parameters, it is necessary to have reliable data on the flow of resources and optimize the energy parameters and operating modes of the HES. For optimization, the standard Microsoft Excel program is used - integer linear programming.

HES is a technical system that combines in a single technological process generators of electrical, thermal, and other types of energy, energy accumulators, switching and transmission of energy, means of converting primary energy into a type suitable for use by consumers. The virtual (structural and functional) model of the HES includes all possible elements, some of which are shown in Figure 1. In the future, the virtual model will be transformed into the technical models of HES, which include the types and sources of energy that are considered in their socio-economic conditions efficiency for the region.

The results of calculations developed by the author of the program in EXCEL are compared with the results of the HOMER program.

Figure 1. Block diagram of a typical HES for rural energy supply using a model based on Microsoft excel
Where PV – solar photovoltaic installation, WIND – wind generators, DGEN– diesel power plant, AB – battery, k-charge – charge controller, inv – Inverter, load – consumers of electricity.

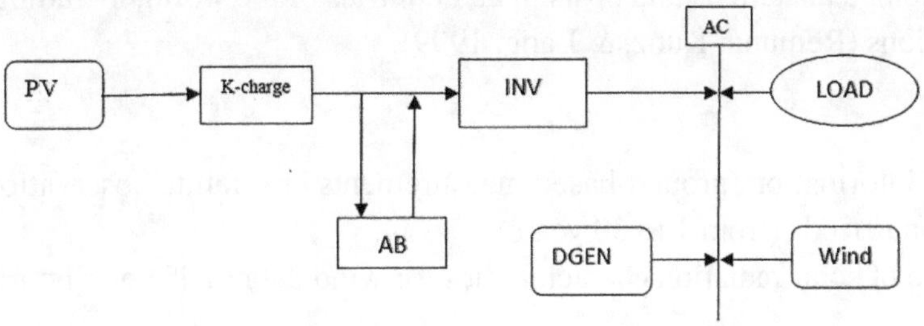

Information Support of Solar and Wind Energy Calculations and Specialized Databases on Solar Radiation (SR) and Wind Characteristics for the Conditions of Myanmar and All Over the World

Information sources are the most accessible and relevant for use as evidence. Experimental measurements of solar radiation and wind speed in some areas of Myanmar are as follows:

1. Myanmar Database
2. Database "schedule Russia"
3. Database, "NASA"
4. Database "METEONORM"
5. Database "RETScreen"
6. IRENA Database (international renewable energy Agency)
7. Database "SWERA" Solar and Wind Energy Resource Assessment (assessment of solar and wind energy resources) (Russian meteorological site)

The main sources of data on the arrival of solar radiation and wind resources are:

1. Ground weather stations
2. Satellite data
3. Computer generated interpolations, some using terrestrial and satellite data sources (for example, the "Meteonorm" software)

Various studies conducted in the world have revealed that the most reliable data sources are data obtained from terrestrial weather stations. The use of satellite observations has limitations and inaccuracies and suggests the presence of different models for their recalculation on the ground. For example,

1. Sensors do not distinguish between clouds and snow cover
2. The error is much higher in the highlands and near large ponds
3. Measurements are carried out in the upper atmosphere and models are required to recalculate the SR (solar Radiation) and wind speed on Earth

Tables 1 and 2 present the main information sources on SR for the conditions of Myanmar and the whole world. Today in the world there are many different databases on various meteorological data provided by meteorological stations and posts in all countries of the world, including on actinometrical meteorological stations (Remund, Kunz, & Lang, 1999).

They differ:

1. By sources of information (ground-based measurements or satellite observations)
2. Data collection periods (from 1 to 30 years)
3. Representation of solar radiation characteristics for wind data (half-hour, hourly, monthly, annual values)
4. Spatial interpolation capabilities.

Table 1. Basic information sources on SR for the conditions of myanmar and the world

Database	Region	Source	Observation Period	Availability	Data
Meteonorm	All over the world	8325 points: AMS, 5 geostationary satellites, models	1991-2010	Software (http://meteonorm.com/ (paid)	Monthly time clock is modeled
NASA	All over the world	1195 points: satellite data models	1983-1993	Free (http://eosweb/larc.nasa.gov/cgi-bin/ sse/sse.cgi.)	Monthly
World Radiation Data Center BND (WRDC)	All over the world	1195, 29 of them on the territory of the Russian Federation	1964	Free (http://wrdc.mgo.rssi.ru)	Daily, monthly sentries
RET Screen	All over the world	Various sources, including WRDC and NASA	1961-1990	Software (free)	Monthly
3TIER	All over the world	satellite data	1991-2008	Paid	Monthly
HPC	Russia and the territory Soviet Union	166 weather stations, data Applied Handbook on Soviet Union Climate (St. Petersburg: hydrometeoizdat, 1992)	-	Developer Department of Hydropower and Renewable Energy Sources, MPEI	Monthly

Table 2. Basic information sources on wind characteristics for the conditions of Myanmar and the world

Name (Developer)	A Brief Description of	Access Mode on the Internet
SBD "Weather vane" (developer SIC)	Averaged data from a number of observations for 1965-1976 and reduced to the conditions of an open area at an altitude of 10m according to 2337 masses of Russia	Department of Hydropower and Renewable Energy Sources National Research University "Moscow Power Engineering Institute"
SBD «NASA»	Averaged data from a number of observations since 1983. till 1995, a small amount of MS	http://eosweb.larc.nasa.gov/cgi-bin/sse/ sse.cgi
SBD «Weather Russia»	Series of observations since 1999. according to nv 1341 MS	http://info.space.ru

In different databases, information on solar radiation and wind characteristics (Table 1 and Table 2) is given averaged over different time intervals. Since this work considers the local energy complex, the choice of the optimal composition of which will be based on the power balance, and there should be averaged data for a time interval of at least 1 hour, therefore, for further calculations, Meteororm on CP and Weather Schedule were chosen as the main sources wind (Deriugina, Zay, & Tyagunov, 2017).

Calculating Wind Potential in the Conditions of Myanmar

Calculations on the assessment of wind resources of Myanmar were first performed according to the data of the Meteoronm program (Zay & Vissarionov, 2013) which, as shown in (Bezrukikh, 2010), is not applicable to the assessment of wind resources. The paper presents the refined results of Myanmar's wind energy potential based on multiyear data (from 1995 to 2017) of 49 ground Weather Stations (WS) at a height of 10 m from the ground from the Weather site. At the sites of all 49 MS, the calculation of the main average annual characteristics of the wind was carried out according to the method described in (Bezrukikh, 2010), at different heights: 10 m, 30 m, 50 m, 80 m and 100 m. The average annual wind speed was raised to different heights according to the power dependence with the exponent "0,2". According to the results of the calculation, maps of the distribution of average annual wind speed and specific power at different heights were constructed and the gross potential of Myanmar's wind power at heights was estimated approximately: 10 m - 3128 TWh per year, 30 m - 6047 TWh per year, 50 m - 8215 TWh per year, 80 m - 10,892 TWh per year, 100 m - 12,452 TWh per year.

According to the results of the calculations, it can be said that most of the territory of Myanmar (about 93% of the territory) belongs to areas with weak wind energy potential, where at an altitude of 100 m the average long-term wind speed is less than 4.2 m / s (Figure 2 and 4) and the specific power wind current less than 109 W / m^2, i.e. in these areas, the use of wind energy is unpromising even for network wind power plants (wind turbines). The greatest wind activity is observed in the western part of the country.

Technique allowed to clarify the value of the gross potential of wind energy at the territory of Myanmar according to the data of SBD "RP" according to an algorithm corresponding to described in the clause 2.3 (see Figure 3).

When calculating the potential, the territory of Myanmar was divided into zones without taking into account the complex and diverse terrain, which did not allow for the division of the territory of the country into zones on a grid with a step of 1^0 x1^0, as is customary in methods of NASA.

The wind potential was also calculated at different heights. Modeling (recalculation) of average annual wind speed to a height h for the conditions of land-based WS of Myanmar and coastal (sea) territories was carried out according to a power dependence of the vertical wind profile:

$$V_0^h = V_0^{10} \mathrm{u} \left(\frac{h}{10}\right)^{m_0} \qquad (2)$$

where h is the height from the surface of the earth; V_0^h, V_0^{10} – average wind speeds at the sites of MC at a height of h and 10 m; m_0 is the average exponent (in foreign sources, the Hellman coefficient), which for the conditions of ground-based MS was adopted "0.2" in accordance with IEC 61400-1 and the maritime territory - "0.14" in accordance with IEC 61400-3.

Figure 2. Map of the distribution of mean annual wind speed in the territory of Myanmar at an altitude of 100 m (W/m²)

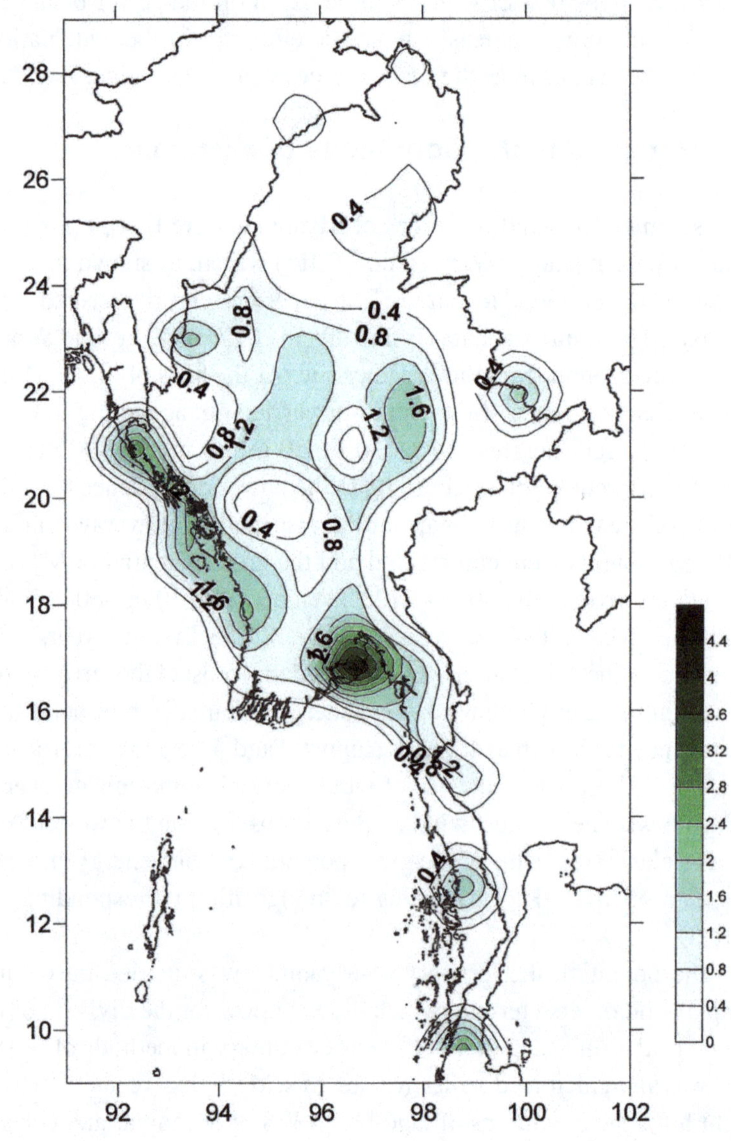

Calculation of theoretical wind potential E^{gross} was held for the average year according to the formula:

$$E^{gross} = \sum_{i=1}^{n} E_i^{gross} = \sum_{i=1}^{n} \frac{N_{specific}.F_i.T}{20} \tag{3}$$

where n is the number of zones into which the territory is divided, with the same physiographic conditions (geographic location, landscape conditions, the presence and type of water surfaces, etc.); $N_{specific}$

Figure 3. Distribution of gross wind energy potential over the territory Myanmar at a height of 10m, W/m²

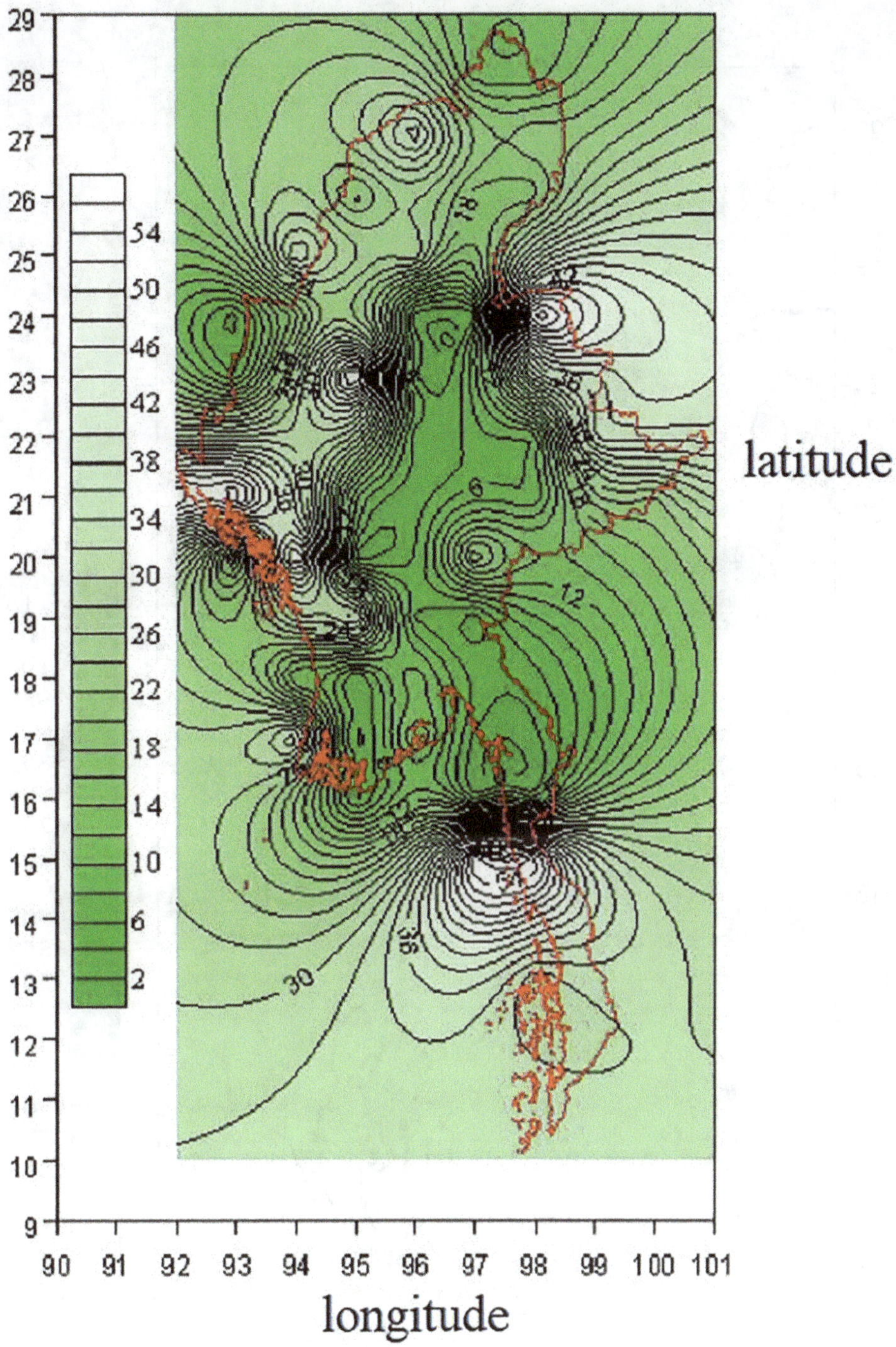

Figure 4. Map of the distribution of specific wind power (m/s) in the territory of Myanmar at an altitude of 100 m (W/m²)

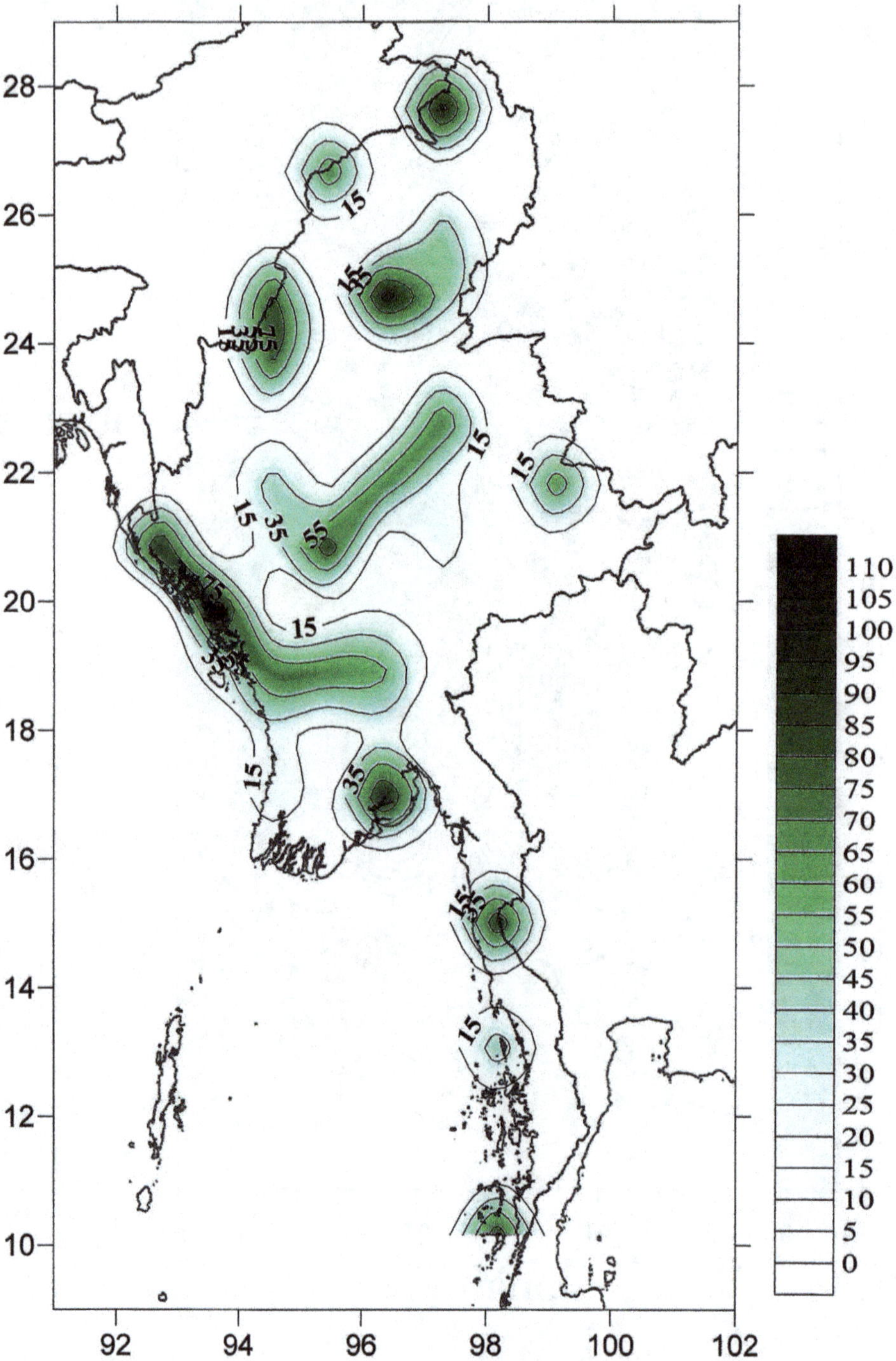

Figure 5. The territory of the location of the marine wind in the Andaman Sea

- Area for placement of marine wind farms

……

– power density of the wind flow in the ith zone, kW / m^2; F_i is the area of the ith zone, the total area of all zones is the area of the territory of Myanmar - 678,500 km^2; T = 8760 h - the number of hours per year.

Thus, Myanmar's wind potential is at altitude: 6047 TWh per year at 30 m; at 50 meters - 8215 TWh per year; at 80 m - 10892 TWh per year; 100 m - 12452 TWh per year. The refined gross wind energy potential of Myanmar at a height of 10 m from the earth's surface is 3128 TWh per year, which is 1.7 times less than 5230 TWh / year, and 1.7 times more than, which is 1820 TWh / year.

Evaluation of the theoretical wind potential of the marine areas of Myanmar was carried out only for the Andaman Sea on the western coast of Myanmar, whose coastline is about 2,224 km. Since they consider profitable marine wind farms, remote from the coast not further than 50 km and not closer 1-1.5 km and at a depth of no more than 50 meters, a preliminary monitoring of the marine areas of the west coast of Myanmar was carried out and the area for the placement of marine wind farms was calculated to be 80,400.9 km^2 (in Figure 5 it is highlighted in red).

In the selected area for the placement of the offshore wind, five points were selected in its different parts, for which data on wind speed and wind directions were obtained from the Blended Coastal Wind bank of the Windpro program at a height of 10 m, as on the Weather Schedule site Information is provided only on ground-based MS. The main perennial energy characteristics of the wind were calculated under the conditions of 5 points of the Andaman Sea and it was revealed that they did not change significantly in the chosen territory: average long-term wind speeds from 5.1 m / s to 5.3 m/s; power densities of wind flow from 163 W/m^2 to 177 W/m^2. In this regard, it was decided to assess the theoretical potential of the Andaman Sea on the western coast of Myanmar without dividing into zones according to only one point A (94'15 ", 18'00") (Figure 5), where the energy characteristics of the wind correspond average values of the entire territory: average annual wind speed - 5.2 m/s; power density of the wind flow 170 W/m^2(Zay & Tyagunov, 2015a).

Figures 6-9 show the main energy characteristics of the wind at point A at a height of 10 m.

The calculation of the main average annual characteristics of the wind at different heights (from 10 m to 130 m) under the conditions of point A was carried out according to formula (2) with the exponent "0.14." The calculation of the theoretical wind potential of the selected area for the placement of offshore wind farms was carried out according to the formula (3) when it was divided into 2 zones: remote from the coast at a distance of 1.5 km to 30 km, the area of which was about 58,999.89 km^2; remote at

Table 3. Average multi-year energy characteristics of the wind and theoretical wind potential at different heights of the sea area of the western coast of Myanmar

Height Above Sea Level, m	The average. Speed, m/s	Average beats. Power, kW/m^2	E_{val}, TWh per Year at a Distance From the Coast L, km		
			1,5 - 30	30 - 50	1,5 -50
10	5,18	0,17	4422,02	1604	6026,01
30	6,04	0,27	7014,73	2544,45	9559,19
50	6,49	0,34	8693,36	3153,34	11846,7
80	6,93	0,41	10590,54	3841,5	14432,04
100	7,15	0,45	11631,09	4218,94	15850,03
130	7,42	0,50	12986,03	4710,42	17696,45

Figure 6. Long-term recurrence of wind speed in the Andaman Sea at 10 m

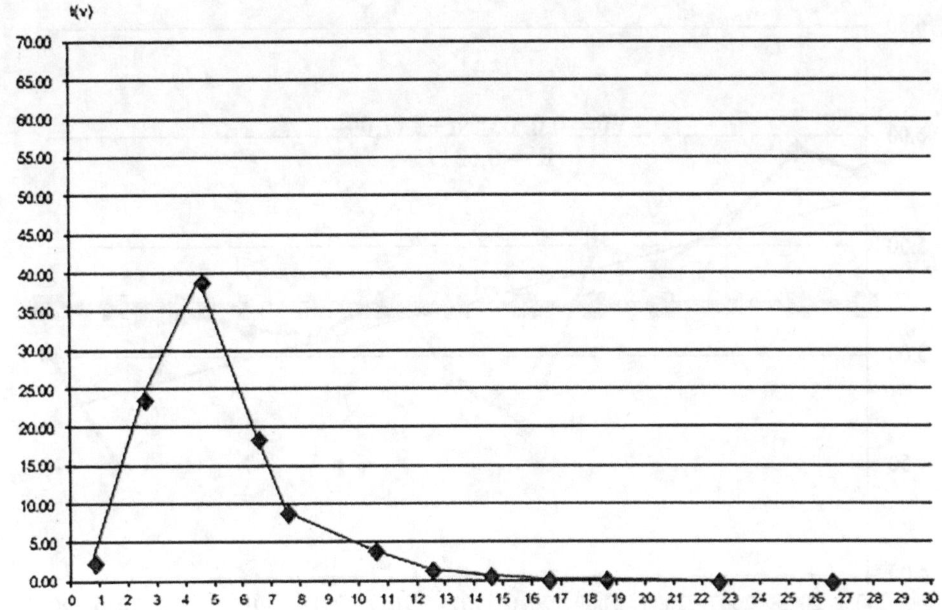

Figure 7. Long-term recurrence of wind directions in the Andaman Sea

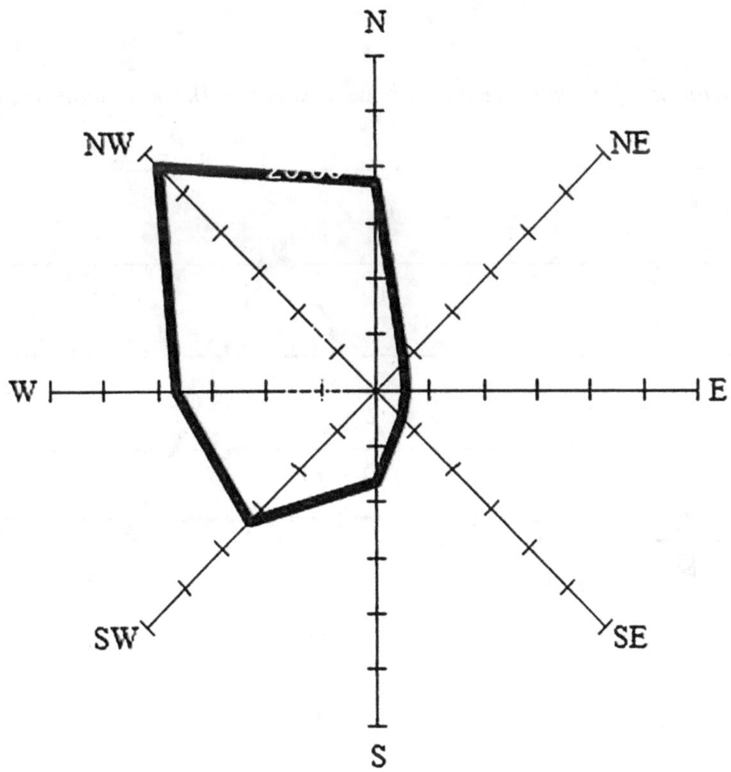

Figure 8. Long-term variation of average annual wind speed in the Andaman Sea at a height of 10 m

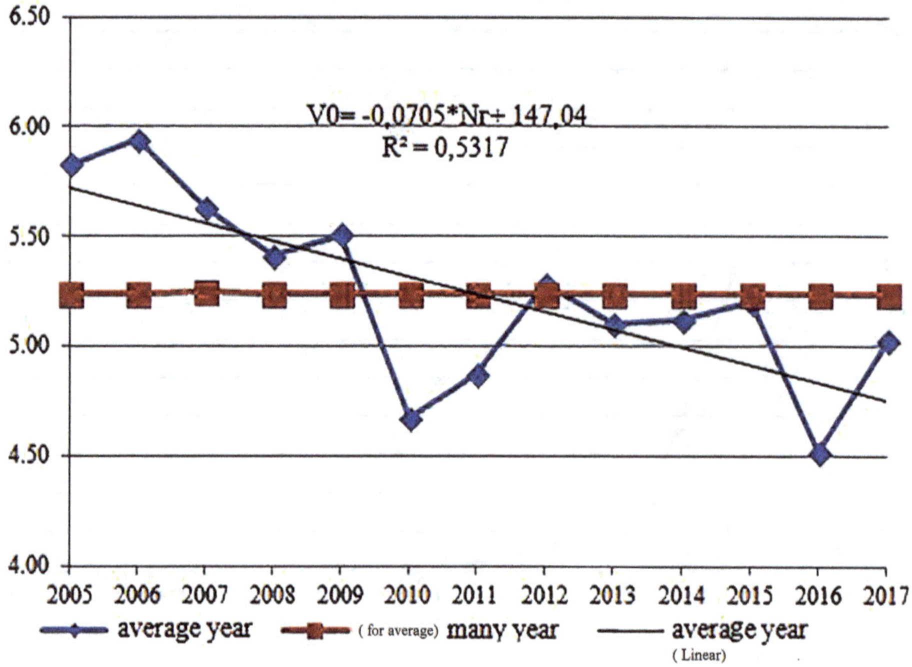

Figure 9. Annual variation of the average monthly wind speed in the Andaman Sea at a height of 10 m

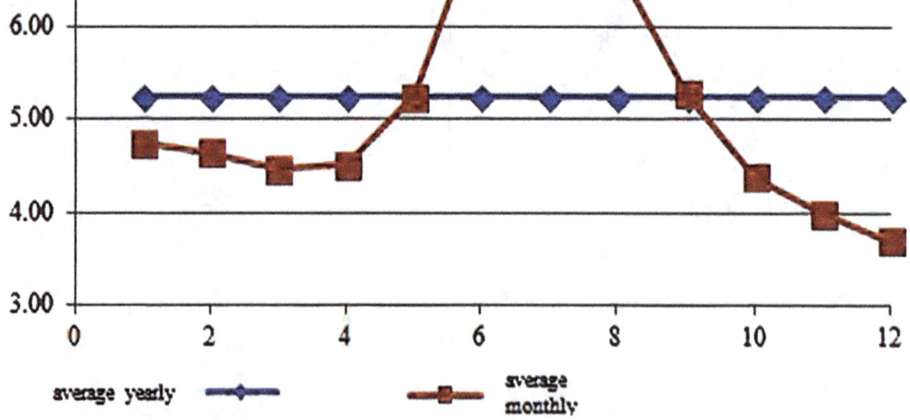

a distance of 30 km to 50 km from the coast, the area of which is about 21401.01 km^2. The calculation results are presented in the table 3.

Analysis of the results of the calculations revealed that the wind potential at all altitudes (from 10 m to 100 m) of only the marine area, remote from the coast from 1.5 km to 30 km and an area of about 8% of the entire territory of Myanmar, is comparable to the wind potential of the entire territory Myanmar (Remund, Kunz, & Lang, 1999). Thus, there can conclude about the feasibility and prospects of building marine wind farms in the Adaman Sea along the west coast of Myanmar. The final conclusion can be made only after a feasibility study (Deriugina, Zay & Tyagunov,2017).

Myanmar's total technical solar energy resources are 2814.6 TWh / year. The technical potential of wind energy using FD 4.0 installations (3 kW) is 34.5 TWh / year, using Hummer 6.4 installations (5 kW) - 70.9 TWh / year and Hummern 8.0 (10 kW) - 97.5 TWh / year (Zay & Tyagunov, 2017).

Features of Selecting Solar Photovoltaic (PV) Modules in the Conditions of Myanmar

Solar energy is the largest energy resource in Myanmar and can be considered as a source for solving energy problems. In (Deriugina, Zay & Tyagunov, 2017), the gross potential of solar energy resources throughout Myanmar was determined, which amounted to 1.15 million TWh per year. Figure 10 shows a map of the distribution of the average annual arrival of solar radiation on a horizontal receiving site in Myanmar.

Myanmar's total technical solar energy resources are 2814.6 TWh / year (Zay & Tyagunov, 2017).

Given the hot climate throughout Myanmar (the average annual temperature is not lower than 24 °C), the choice of the type of solar photovoltaic modules (PV) should be made not only by price indicators. First of all, it is necessary to decide which solar cell technology - thin-film or crystalline - is better to use in a particular place, since its choice is determined by the choice of PV location, which also depends on taking into account many factors (Remund, Kunz, & Lang, 1999). The following indicators should be considered when choosing the type of PV:

1. Energy-efficiency

In the absence of an area limit, more often choose an PV with lower energy-efficiency values (5-8%), but at the same time, to obtain the same power as for FEM PV with an efficiency > 13%, more costs will be required for different wires and ground.

2. Easy installation (Less complicated installation)

Supplying PV with pre-drilled mounting holes is generally preferred over frameless modules, although some frameless modules, such as self-adhesive laminates (peel-and-stick laminates), are even easier to install than framed frames;

3. Availability
4. Weight
5. Warranty. Most guarantees apply for 1-3 years after manufacturing and 20-25 years of operation

Figure 10. Myanmar map showing the average annual solar arrival radiation horizontal reception area in Myanmar (kW h / m² per day)

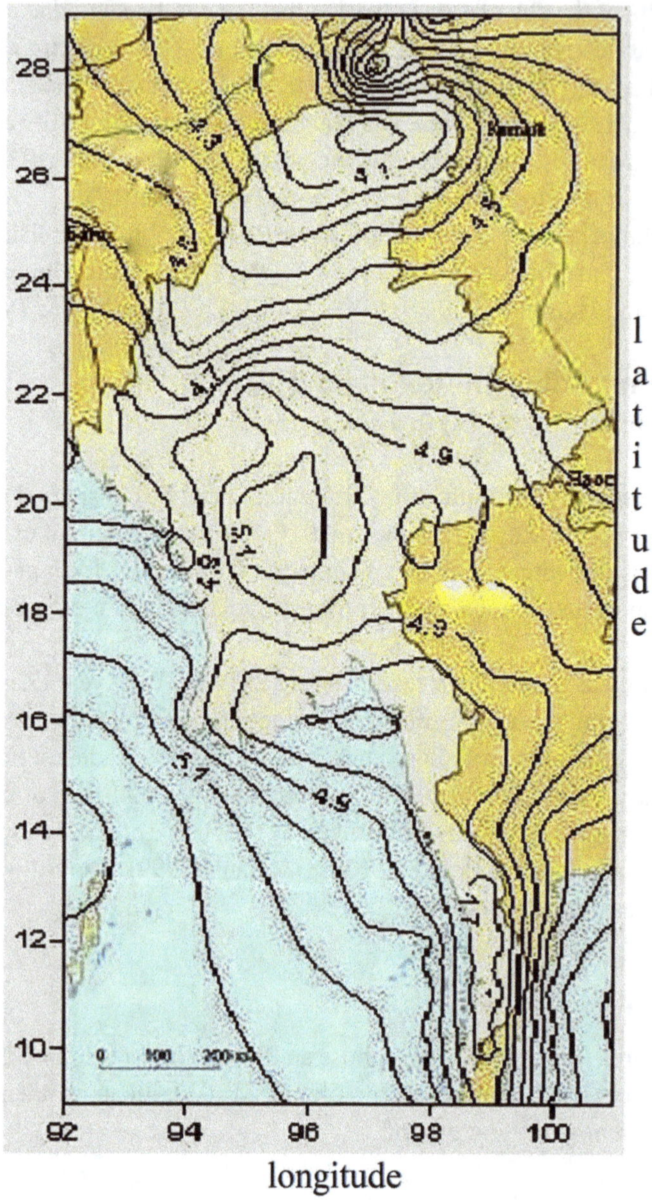

6. Degradation. Thin-film PVs degrade faster than PV based on crystalline silicon
7. Current, voltage and power parameters. Each of these must be compatible with the characteristics (direct and alternating current) of the converter and the limitations of the connection with alternating current
8. Aesthetics, especially in the construction of new facilities, where PV will be integrated
9. The history of the manufacturer

Currently, manufactured by manufacturers of FEM KM may have deviations on passport data ± 2%. As studies have shown for Myanmar, the use of thin-film modules is most promising, since at high temperatures thin-film PVs show greater efficiency (Aung, 2013).

For example, Table 4 presents the main technical and economic data on 15 selected PV options based on silicon of various technologies: single-crystal, polycrystalline and amorphous specific power in the range from 100 W to 300 W, for which energy efficiency indicators were calculated from and without taking into account the influence temperature on efficiency in the conditions of the central part of Myanmar (Mandalay).

The results of the calculations are presented in table 5.

The calculations of the main indicators of energy efficiency were made according to the formulas:

- Installed capacity utilization $CAPEX_{PV}^{uti.}$, equal to the ratio of the energy produced by the PV for the year to the energy that the PV would generate when operating for a year with an installed capacity,

$$CAPEX_{PV}^{uti} = \frac{E_{PV}^{year}}{N_{PV}^{max} \cdot 8760} \tag{4}$$

where E_{PV}^{year} - annual energy production PV.

Table 4. The main technical and economic data on previously selected versions of the SPS

No.	Brand	Type of PV Modules	PV Model	β_t	F_{PV}, m²	η_{MAX}	P_{peak}, W	Price, $/ Piece	Specific Price $/W
1	Germany	Monocrystalline silicon	axitec 270м	0.006	1.63	0.159	270	370	1.25
2	China		fsm 270m	0.006	1.63	0.16	270	290	1
3	China		fsm 160m	0.006	0.96	0.165	160	176	1.11
4	China		fsm 210m	0.006	1.31	0.19	210	232	1.2
5	China		fsm 200m	0.006	1.28	0.149	200	211	1.05
6	Germany	Polycrystalline silicon	axitec 300п	0.006	1.94	0.155	300	344	1.14
7	Germany		axitec 260п	0.006	1.63	0.159	260	299	1.15
8	China		fsm 200п	0.006	1.31	0.149	200	179	0.9
9	China		fsm 250п	0.006	1.63	0.153	250	248	0.99
10	China		fsm 150п	0.006	0.98	0.153	150	147	0.98
11	Switzerland	Amorphous silicon	pramac mcph p7	0.0025	1.43	0.089	125	100	0.81
12	Germany		bekar	0.0025	1.54	0.065	105	187	1.78
13	China		tw-tf-120w	0.0025	1.43	0.084	120	122	1.02
14	Russia		p7125lpramac (hewel)	0.0025	1.3	0.089	125	123	0.98
15	China		tw-tf 120	0.0025	1.43	0.084	120	140	1.16

Table 5. Indicators of energy efficiency of pre-selected variants of SPS

No.	Brand	Type of PV Modules	Model FSM	Efficiency With T			Efficiency Without T			Δδ,%	T_c^{max}, ^0C
				E_{PV}^{year}, kWh per Year	K_{PV}^{uti}	$E_{PV}^{sp.}$, kWh / m²	E_{PV}^{year}, kWh per Year	K_{PV}^{uti}	$E_{PV}^{sp.}$, kWh / m²		
1	Germany	Mono-crystalline silicon	axitec 70м	451.99	0.191	277.29	525.27	0.222	322.25	16.21	69.01
2	China		fsm 270m	449.52	0.190	275.78	528.57	0.223	324.28	17.59	71.65
3	China		fsm 160m	273.02	0.195	284.40	321.03	0.229	334.41	17.59	71.65
4	China		fsm 210m	429.01	0.233	327.49	504.45	0.274	385.08	17.59	71.65
5	China		fsm 200m	328.73	0.188	256.82	386.54	0.221	301.98	17.59	71.65
6	Germany	Poly-crystalline silicon	axitec300p	524.41	0.200	270.32	609.44	0.232	314.14	16.21	69.01
7	Germany		axitec 60p	451.99	0.198	277.29	525.27	0.231	322.25	16.21	69.01
8	China		fsm 200p	336.43	0.192	256.82	395.60	0.226	301.98	17.59	71.65
9	China		fsm 250p	429.85	0.196	263.71	505.45	0.231	310.09	17.59	71.65
10	China		fsm 150p	258.44	0.197	263.71	303.89	0.231	310.09	17.59	71.65
11	Sweden	Amorp-hous silicon	mcph p7	241.87	0.221	169.14	257.94	0.236	180.38	6.65	71.65
12	Germany		bekar	190.23	0.207	123.53	202.88	0.221	131.74	6.65	71.65
13	China		tw-tf-20w	228.28	0.217	159.64	243.45	0.232	170.25	6.65	71.65
14	Russia		p7125lpramac ewel)	219.88	0.201	169.14	234.49	0.214	180.38	6.65	71.65
15	China		tw-tf 120	228.28	0.217	159.64	243.45	0.232	170.25	6.65	71.65

- Indicator of specific energy per unit area of PV:

$$E_{PV}^{sp.} = \frac{E_{PV}^{year}}{F_{PV}} \quad (5)$$

where F - the solar area in PV (m²).

The calculation of the annual production of SPS in the work was carried out taking into account the influence of the ambient temperature (T) on the efficiency of the Ecological management system (EMS) and without EMS. The calculation of the instantaneous power PV taking into account the ambient temperature is made according to the formula:

$$P_{PV} = \mu_t . F_{PV} . E_{\Sigma}^{\beta\gamma} \quad (6)$$

where η_t - instantaneous efficiency (efficiency) of the PV at the actual ambient temperature Ta; F_{PV} - area of one solar energy in PV (m²); $E_{\Sigma}^{\beta\gamma}$ – the total arrival of solar radiation on the receiving site inclined at an angle β.

When calculating instant performance η_t various types of losses (connection loss, transmission loss, etc.) are assumed to be zero.

The instantaneous efficiency of the PV at the actual ambient temperature is determined by the formula (Habibet et al., 1999):

$$n_t = n_r . n_{pt} \left[1 - \beta_t . \left(T_c - T_r\right)\right] \tag{7}$$

where η_r – maximum efficiency of PV under standard conditions indicated in the passport data by the manufacturer; η_{pt} - is assumed to be 1, if the FMS operates at the maximum power point, i.e. with an MRI device; T_c – temperature inside the solar cells PV (°C), T_r – standard temperature (25 °C) ; β_t - temperature coefficient of efficiency, depending on the material of the PV, varies for silicon SPM in the range from 0.003 to 0.006 per degree Celsius (1 °C) and is usually set in the passport data by the manufacturer. In the absence of information on β_t in this paper, it is assumed: for amorphous silicon (thin film technology) - 0.003; for crystalline silicon - 0.006. If the temperature inside the PV is less than 250 C, then the efficiency of the solar module η_t is taken to be η_r.

The temperature inside the solar cells PV is determined on the basis of the energy balance proposed by Duffie in 1991, according to the formula:

$$T_c = T_a + R \cdot \left(\frac{\tau\alpha}{U_l}\right) \tag{8}$$

where T_a - ambient temperature (°C); U_L - heat loss coefficient (W /m^2 on 1 °C); τ and α represent, respectively, the transmittance and the absorption coefficient of the solar energy . Total heat loss coefficient $(\tau\alpha/U_L)$, can be estimated at standard temperature (NOCT) using the formula (Chedid et al., 1996; Duffie et al., 1991):

$$\frac{\tau\alpha}{U_l} = \frac{NOCT - 20}{800} \tag{9}$$

Therefore, the instantaneous efficiency of the PV can be determined by the formula:

$$\eta_t = \eta_r \cdot \eta_{pt} \left\{1 - \beta_t \left(T_c - T_r\right) - \beta_t \cdot R \left(\frac{NOCT - 20}{800}\right)\left(1 - \eta_r \cdot \eta_{pt}\right)\right\} \tag{10}$$

Parameters η_{pt}, β_t, $NOCT$, F_{PV} are parameters that depend on the type module and must be specified by the manufacturer of the modules. If the manufacturer does not specify the NOCT parameter, then it is assumed to be 470 °C.

An analysis of the results showed that not taking the ambient temperature into account when calculating the annual output of various PV models from Mono-crystalline and polycrystalline silicon leads to an overestimate of its value by 16.21% and 17.59%, respectively, and for PV models from amorphous silicon it leads to an overestimate values of 6.65%, which affects the share of PV in the autonomous consumer's load schedule, and, accordingly, the annual volume of fuel savings per DGEN.

The final choice of the PV brand and their quantity in the HES is determined on the basis of a feasibility study.

Calculating the Financial and Economic Efficiency of the Energy Complex

In the calculations of financial and economic efficiency of HES, to determine the most profitable of the options considered (PV+ Wind + DGEN + HP), the economic efficiency of the construction and operation of the energy complex as a whole is calculated, i.e. The revenues of the project are funds from saving diesel fuel, and expenditures - investments in equipment (PV, Wind, inverter, HP, DGEN and AB) with a factor, which takes into account additional unaccounted expenses (connecting wires, etc.) and fuel costs of operating DGEN. In the calculations, the diesel fuel saving on DGEN is determined by the replacement PV, as well as by the heat pump unit, since in the absence of the heat pump unit, the necessary heat energy for heating, air conditioning and hot water supply would be derived from electrical energy from the DGEN (Aung, 2013).

A number of modern criteria can be applied to assess the economic efficiency of the power complex. Criteria can be both integral (characterizing an object for the entire period of its construction and operation), and simple (calculated for a specific year of operation of the object). The integral criteria include: net present value (NPV). Simple criteria include: gross income, payback period.

These calculations are carried out to determine the economic efficiency of the energy complex as a whole (economic efficiency), regardless of the method of financing its construction. The above performance criteria are calculated based on the analysis of revenues from saved diesel fuel and expenses in the form of operating costs and capital investments in equipment (PV, Wind, inverter, HP, DGEN and AB), as well as the cost of fuel and fuel and lubricants to ensure a guaranteed power supply .Comparison of various options for the composition of HES equipment and the selection of the best of them is performed using the following indicators: net present value (NPV);

Net present value when calculating in current (basic) prices and at a constant discount rate is determined by the formula:

$$NPV = \sum_{t=1}^{Tpro.year} E_{t\ cost}^{fuel} \cdot \left(1 + \frac{\varepsilon_d}{100\%}\right)^{-t} - \sum_{t=1}^{Tpro.year} \left(CAPEX_t^{HES} + C_{t\ HES}^{opt.}\right) \cdot \left(1 + \frac{\varepsilon_d}{100\%}\right)^{-t} \quad (11)$$

where $T_{pro.year}$, years - an estimated period of 20 years; $CAPEX_t^{HES}$, $ - investments in the energy complex depend on the composition of the equipment (PV, DGEN, HP, AB) in the tth year; $C_{t_{HES}}^{opt}$, $ - general operating expenses of HES in t[th] year; $E_{t_{cost}}^{fuel}$, $ - the cost of the purchase and delivery of fuel, which will be saved in the tth year due to PV and HP, are determined by the amount of fuel saved $\Delta B_t^{HES.fuel}$ and its value V_t^{fuel}, ε_d, % – the value of the discount rate:

$$E_{t\ cost}^{fuel} = \Delta B_t^{ecopt} . V_t^{fuel} \quad (12)$$

$$V_t^{fuel} = V_c^{fuel}.\left(1+\varepsilon_{inf}\ /100\%\right)^{(t-expected)} \tag{13}$$

where $V_c^{fuel.}$, \$/t - the cost of fuel at current prices, $1+\varepsilon_{inf}/100\%$ – constant expected annual rate (percentage) of inflation (12.17%).

Costs associated with the operation of HES, in general, $C_{t_{HES}}^{opt.}$ consist of the costs of capital and current repairs and other costs:

$$C_{t\ HES}^{opt.} = C_t^{rep} + C_t^{cur.rep} + C_t^{other} + C_t^{fuel} \tag{14}$$

Where $C_t^{rep.}$, \$ - the cost of capital repairs per year t accepted 2%; $C_t^{cur.rep.}$, \$ - costs for current repairs per year t; $C_t^{other.}$, \$ - other costs per year t; C_t^{fuel}, \$ - costs of fuel cost per year t.

The costs of capital and current repairs are determined as a percentage of the total investment:

$$!_t^{rep} = CAPEX.\frac{\mu_{rep}}{100}\% \tag{15}$$

where $\mu_{rep.}$, % – rate of deductions for major repairs (0.5%):

$$C_t^{cur.rep} = CAPEX.\frac{\mu_{cur.rep}}{100}\% \tag{16}$$

where $\mu_{cur.rep.}$, % – the rate of deductions for maintenance (0.1%).

Other costs are usually taken in the range of 0.5% - 5% of the sum of all other operating costs, in this work 0.5% were taken:

$$C_t^{other} = 0.005.\left(C_t^{rep} + C_t^{cur.rep}\right) \tag{17}$$

If the NPV of the investment project for the settlement period is positive, then the project is effective and the question of its implementation can be considered. At the same time - the higher the level of NPV, the more effective the project (Burmistrov, Vissarionov, Deryugin et al., 2007).

The second indicator is the investment payback period. $T_{payback}^{disc}$ – determined by the time interval (from the beginning of the project), beyond which the integral effect becomes positive. In other words, this is the period (years) during which the initial investment on the investment project is covered by total income. It is determined from the condition:

The discounted payback period is determined from the following equality:

$$\sum_{t=1}^{T_{pay\ back}^{disc}} E_{t\ cost}^{fuel} \cdot \left(1+\frac{\varepsilon_d}{100\%}\right)^{-t} - \sum_{t=1}^{T_{pay\ back}^{disc}} \left(CAPEX_t^{HES} + C_{t\ HES}^{opt}\right) \cdot \left(1+\frac{\varepsilon_d}{100\%}\right)^{-t} = 0 \tag{18}$$

Simple criteria: income, simple payback period are determined, similar to discounted:

$$D = \sum_{t=1}^{T@} E_{t\ cost}^{fuel} - \sum_{t=1}^{T@} \left(CAPEX_{t}^{HES} + C_{t\ HES}^{opt.} \right) \tag{19}$$

$$\sum_{t=1}^{T_{pay\ back}^{disc}} E_{t\ cost}^{fuel} - \sum_{t=1}^{T_{pay\ back}^{disc}} \left(CAPEX_{t}^{HES} + C_{t\ HES}^{opt.} \right) = 0 \tag{20}$$

The calculation of the financial performance of various types of HES is carried out under the following conditions:

1. Discount rate $\varepsilon_d = 10\%$
2. Settlement period $T_p = 20$ years
3. Inflation index $\varepsilon_{inf} = 12.17\%$

Investments in DGEN and PV are made in the first year of construction of the facility. Capital investments in a HP, inverter and AB are also made in one year, but after 10 years the costs in inverter and AB are again made. To select the optimal composition of the equipment of the energy complex, approach to the choice of the optimal composition of the equipment of the power complex. The choice of the optimal composition of the equipment of the power complex to provide an autonomous consumer is made on the basis of a feasibility study from 6 options:

1. Option 1 – Solar PV, DGEN, HP
2. Option 2 – Solar PV, DGEN
3. Option 3 – Solar PV, HP, DGEN, AB
4. Option 4 – Solar PV, DGEN, AB
5. Option 5 – Solar PV, Wind, DGEN, AB
6. Option 6 – Wind, DGEN and AB

When choosing the optimal composition and amount of equipment for all variants of the composition of equipment, preselected models of DGEN (Gesan L 6 key -Japan) are considered with consideration of HP and (Robin-Subaru ED 6.0 / 230-SLE) without consideration of HP, HP (NIBE F2040-8kW - Switzerland) and AB (Delta GX 12-100 - China). But only the number of ABs in options 3, 4, 5 and 6 is specified. The choice of the PV brand is based on a feasibility study of the equipment composition in option 1. For the other three options for the equipment composition, only the number of PV is specified. The final optimal composition and quantity of HES equipment is selected on the basis of a business case.

For the economic substantiation of the composition and quantity of equipment of the energy complex, it is necessary to preliminarily determine the annual output of the PV and DGEN, as well as fuel economy on the DGEN, which, if there are HP in the power complex, is made up of fuel savings due to the PV and HP, in the absence of a HP, only from fuel economy due to PV (options 2 and 4). This information is obtained on the basis of carrying out calculations on the power balance.

For the composition of the power complex equipment, options 1 and 2 when choosing the number of PVs are limited to the condition: the share of free energy from the PV should not exceed 5% of the annual production of the PV, since with a larger share of free energy it is necessary to consider the ways of its realization. For the composition of the equipment options 3 and 4 free energy from PV is accumulated in AB, the number of which is chosen in such a way that, on the one hand, there is maximum displacement of the DGEN from the load diagram, and on the other hand, full use of free energy from the PV.

The balance of power of the power complex is determined by the composition of the generating plants and the presence of AB. Below, as an example, the power balance is given for the first four variants of the HES composition, since, as was shown above, the use of wind power in small power complexes is inefficient throughout almost the entire territory of Myanmar (Burmistrov, Vissarionov, Deryugina, Kuznetsova, Kunakin, Malinin, 2009).

Power Balance For The Composition of The HES 1 and 2 Options

Method of power balance of the energy complex of the equipment composition options 1 (PV + DGEN + HP) and 2 (PV + DGEN). The project of the energy complex as part of the DGEN + PV + HP, designed to supply an autonomous consumer, is being considered. DGEN is considered as a guaranteed source of energy supply, since PV have zero guaranteed capacity. PV acts as part of the power complex as a backup power and is designed to reduce the production of DGEN and save fossil fuels. When adding a heat pump unit to an energy complex, the consumption of electrical energy for heating, conditioning and hot water supply is reduced 4 times. The power consumption of a HP is taken into account in the consumption load schedule. Also, due to the addition of a heat pump unit to the power complex, there is an additional fuel saving on a DGEN.

The calculation is carried out for one year T = 1 year. Δt = 1 hour was chosen as the calculated time interval.

In the case of the possibility of providing electricity to the consumer only at the expense of the PV, the DEA is taken out of operation completely and power is supplied only by the PV. During periods of lack of solar energy (at night), the consumer is provided with energy at the expense of DGEN. In the remaining time intervals, the operating mode of the DGEN and PV is optimized according to the criterion of the most crowding out of DGEN generation and taking into account the restrictions on the allowable minimum and maximum load of the DGEN.

Baseline information for determining the energy efficiency of PV, which is part of the decentralized energy complex (HES). The annual number of average hourly values of the electrical power of the consumer (load graph) P_{load} (t), where t = 1, 2 ... 8760 is the number of hours per year, W or kW, depending on the composition of the equipment:

Option 1 - $P_{load}(t) = P_{app}(t) + P_{heat}(t) \,/\, CAPEX_{HP}$

Option 2 - $P_{load}(t) = P_{app}(t) + P_{heat}(t)$.

Where $CAPEX_{HP}$ - Coefficient of performance (COP) HP.

1. Annual range of average hourly values of the total arrival of SR (solar Radiation) on an inclined platform $R_{\Sigma}^{\beta}(t)$, where t = 1, 2 … 8760 – number of hours per year, W /m^2.
2. **Parameters of PV:** PV area A_{PV}, Efficiency PV η_{PV}, quantity PVZ $_{PV}$.
3. Annual range of average hourly temperatures $T_c(t)$ the environment.
4. **Parameters of DGEN:** Minimal $N_{DGEN}^{\min}$ and maximum $N_{DGEN}^{\max}$ permissible operating power, specific fuel consumption DGEN $B_{specific}^{DGEN} = 1,4l / h$ taking into account HP and $B_{specific}^{DGEN} = 1,7l / h$ excluding HP.

By simulating the operating mode of the HES for the period T =1 year based on the consumer power balance for each calculated time interval Δt = 1 hour (i = 1, 2, ..., 8760), for different brands of PV (three types), and the number and model of DGEN are determined by:

1. Hourly working capacity PV $N_{PV}(t)$ and DGEN $N_{DGEN}(t)$
2. Fuel economy on DGEN by calculating PV $\Delta B^{year}(t)$
3. Average hourly free (excess) power PV $N_{free}^{PV}(t)$ and DGEN $N_{free}^{DGEN}(t)$

The final indicators of energy efficiency of PV and DGEN are:

4. Annual values of PV $\frac{\delta_m}{\lambda_m} \approx 0.25 \cdot 10^{-3}$ and DGEN $E_{year}^{DGEN}(t)$ electricity in kWh%
5. Annual fuel consumption for DGEN in tons $B^{year}(t)$
6. Annual free energy due to PV $N_{free}^{PV}(t)$ and DGEN $N_{free}^{DGEN}(t)$ in kWh%
7. Utilization of installed power PV $CAPEX_{PV}^{uti}$
8. The number of working hours of PV and DGEN (t = 1, 2, 3, …, 8760)
9. Annual fuel economy at DGEN $\Delta B^{year}(t)$ at the expense of PV and HP (option 1) and only at the expense of PV (option 2).

The algorithm for calculating the operating mode of HES is based on the power balance equation, that is, the equality of the generated and consumed power at any time t:

$$P_{load}\left(t\right) = N_{PV}\left(t\right) - \Delta N_{PV}^{load.}\left(t\right) + N_{DGEN}\left(t\right) - \Delta N_{DGEN}^{load}\left(t\right) \tag{21}$$

where t, h - the number of hours per year; $P_{load}(t)$, kW – average consumer capacity per hour; $N_{PV}(t)$ and $N_{DGEN}(t)$, kW – average hourly working capacity of PV and DGEN; $\Delta N_{PV}^{load}(t)$ and $\Delta N_{DGEN}^{load}(t)$, kW – average hourly power loses of PV and DGEN.

The power loss of HES in the first approximation can be determined by the following relationship:

$$\Delta N_{EC}^{load}\left(t\right) = \sum N_{loss}^{DGEN}\left(t\right) + \Delta N_{loss}^{PV}\left(t\right) + \Delta N_{loss}^{power}\left(t\right) \tag{22}$$

where $\Delta N_{loss}^{DGEN}(t) = 5\%$ - losses for own needs depending on the operating capacity of DGEN

$\Delta N_{loss}^{PV}(t) = 8\%$ - losses for own needs depending on the installed capacity of PV
$\Delta N_{loss}^{power}(t) = 10\%$ - power transmission loss

Depending on the ratio of power PV, DGEN and customer needs, the following options are possible:

1. $P_{load}(t) \leq [N_{PV}(t) - \Delta N_{loss}^{PV}(t)]$ - the potential total PV capacity (minus auxiliary energy losses and transmission losses) exceeds the energy demand of the consumer at a given time, then: the DGEN is idling, covering only its own needs; PV capacity is limited to the level required by the consumer $N_{PV}\left(t\right) = P_{load}\left(t\right)$ and appears $N_{free}^{PV}(t) = N_{PV}(t) - P_{load}(t)$. In this algorithm, it can accept the limitation: the annual share of free energy due to PV from the total annual produced PV when choosing the number of PV should not exceed 5%.
2. 2. $P_{load}(t) \geq [N_{PV}(t) - \Delta N_{loss}^{PV}(t)]$ - PV can't provide all the needs of the consumer. Accordingly, it is necessary to "add" power from DGEN, the power of which (without taking into account losses) in the first approximation is determined as follows:

$$N_{DGEN}\left(t\right) - \Delta N_{loss}^{DGEN}\left(t\right) = P_{load}\left(t\right) - \left[N_{PV}\left(t\right) - \Delta N_{loss}^{PV}\left(t\right)\right] \tag{23}$$

Since there is a limit on the minimum operating capacity of a DGEN, the following options are possible:

1. $N_{DGEN}(t) - \Delta N_{loss}^{DGEN}(t) \geq N_{DGEN}^{\min}(t) - DGEN$ operates with a power exceeding the minimum allowable. Power DEA determined. Free energy is zero in this case $N_{free}^{DGEN}(t) = 0$.
2. $N_{DGEN}(t) - \Delta N_{loss}^{DGEN}(t) < N_{DGEN}^{\min}(t) - DGEN$ should work with a power less than the minimum allowable. It is necessary to increase the power of the DGEN to the minimum allowable power and there is an excess free power generated by the DGEN: $N_{free}^{DGEN}(t) = N_{DGEN}^{\min}(t) - N_{DGEN}(t)$.

Determination of the annual values of the production of DGEN, PV and free energy produced by the formulas:

$$E_{year}^{DGEN} = \sum_{i=1}^{8760} N_{DGEN}\left(t\right).\Delta t_i \tag{24}$$

$$E_{year}^{PV} = \sum_{i=1}^{8760} N_{PV}\left(t\right).\Delta t_i \tag{25}$$

$$E_{year}^{free\ DGEN} = \sum_{i=1}^{8760} N_{free\ DGEN}\left(t\right).\Delta t_i \tag{26}$$

$$E_{year}^{free\,PV} = \sum_{i=1}^{8760} N_{free\,PV}(t).\Delta t_i \quad (27)$$

Annual fuel consumption for DGEN (ΔB) is determined by the formula:

$$E^{year} = \sum_{i=1}^{8760} N_{DGEN}(t).\Delta t_i.\beta_{sp.fual.con} \quad (28)$$

Where $\beta_{sp.fual.con}$ - specific fuel consumption (Liter/kW);

Fuel economy at DGEN due to PV (ΔB) is determined by the formula:

$$E^{year} = \sum_{i=1}^{8760} N_{PV}(t).\Delta t_i.\beta_{sp.fual.con} \quad (29)$$

Fuel economy at DGEN due to PV (ΔB) is determined by the formula:

$$\Delta B^{year} = \sum_{i=1}^{8760} N_{PV}(t).\Delta t_i.\beta_{sp.fual.con} \quad (30)$$

Power Balance For The Composition of The HES 3 and 4 Options

The power balance equation takes the form:

$$P_{load}(t) = P_{PV}(t) + P_{DGEN}(t) + P_{AB}^{recharge}(t) - P_{PV}^{free}(t) - P_{DGEN}^{free}(t) - P_{AB}^{charge}(t) \quad (31)$$

Where $P_{AB}^{recharge}(t)$ and $P_{AB}^{charge}(t)$ – The average hourly working powers of charge and discharge of the battery. Free energy from PV is redistributed by AB.

DESCRIPTION OF THE PROGRAM HOMER AND DEVELOPED BY THE AUTHOR

The HOMER program is a computer program developed by the American National Renewable Energy Laboratory (NREL). The HOMER program is designed to calculate low power systems and compare various energy production methods. The "HOMER" program simulates the physical behavior of the power system and its cost over the period of operation, including installation cost and cost of further operation, allows the designer to compare various options for the composition of the power system equipment and determine its technical and economic advantages, helps to determine the risks associated with weather variability. With the help of the GOMER complex, it is possible to calculate both an autonomous network and a grid-connected network that generates thermal and electrical energy, which contains various combinations of energy sources (PV, WIND, micro-hydro, biogas plants, DGEN, micro turbines, fuel cells, accumulator and hydrogen sources, etc.).

The program structure contains 3 main modules: modeling, optimization and sensitivity analysis. The HOMER program presents the design results in the form of tables and graphs, which makes it easier to compare different configurations and to determine the economic and technical advantages of various systems. The general view of the program window is shown in Figure 10. The “HOMER” program file contains all the information about the technological parameters, costs and the availability of resources that must be analyzed in the project. The result of the HOMER program is the cost of energy ($ / kWh) and the current cost of the power system for various types of low-power hybrid power system designs (Homer).

To achieve the goal of the work you need the following basic data:

1. Design location
2. Electrical and thermal load
3. Hybrid power system (PV, WIND, DGEN, AB, etc.)
4. Cost elements (PV, WIND, DGEN, Converter, Battery, gasoline, etc.)
5. Connection or without connection to the power supply network

When creating a new project by using HOMER program, generates a list of elements calculated in this work, the photovoltaic system (PV, Wind, DGEN, AB, Converter, and electrical load) (shown on Figure 11).

Figure 11. HES based on PV+WIND+DGEN+AB for power supply of the village using HOMER program

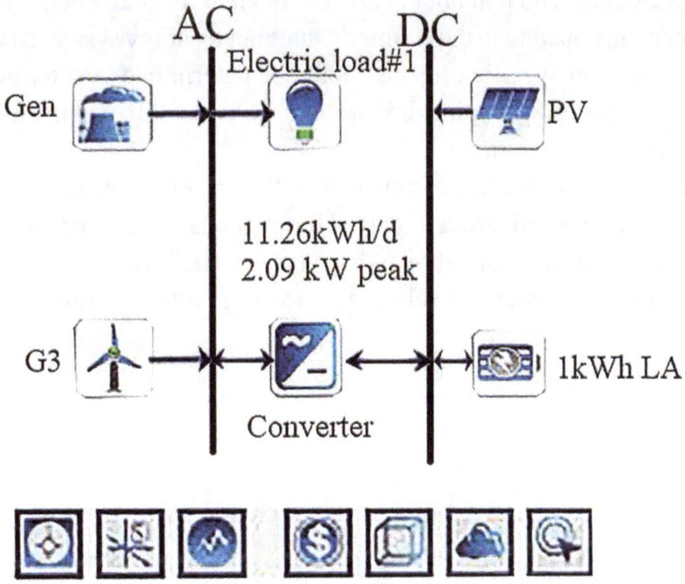

The author has developed a program in the Microsoft Excel software environment that selects the most energy-efficient version of the HES equipment composition in accordance with the consumer load schedule by the criterion of max electricity generation from power plants based on renewable energy sources (solar, wind, heat pumps).

CONCLUSION

Criteria and data validation methods for estimating gross wind potential obtained from various sources of information. The effectiveness of the proposed technique is shown by the example of the applicability of four SBD to determine the total wind potential of the energy of Myanmar.

1. Specified Myanmar's gross wind potential at different heights: 10 m - 3128 TWh per year, 30 m - 6047 TWh per year, 50 m - 8215 TWh per year, 80 m - 10,892 TWh per year, 100 m - 12,452 TWh in year. It was revealed that most of the territory of Myanmar (about 93% of the territory) belongs to areas with weak wind energy potential, where at an altitude of 100 m the average long-term wind speed is less than 4.2 m / s, and the specific wind power current is less 109 W / m^2, i.e., in these areas, the use of wind energy is unpromising even for network wind power plants (wind turbines). Gross wind potential at all altitudes (from 10 m to 100 m) only in the marine zone, remote from the coast from 1.5 km to 30 km and covering about 8% of the entire territory of Myanmar, is comparable to the wind potential of the entire territory of Myanmar.
2. Analysis of hydropower resources has shown that the regions with three major rivers (the Irrawaddy, the Chindwin and the Thanlwin), the northern, western and eastern highlands of Myanmar have the greatest potential. The potential of the coastal areas of Myanmar is small, and the construction of small hydropower plants in these areas is impractical.
3. Analysis of tidal energy resources showed that the potential of the Andaman Sea is greater than that of the Bay of Bengal. The potential is especially great for tidal energy in southern Myanmar, where tidal power plants operate in the country's integrated energy system as well as in local energy systems. The feasibility of specific projects should be determined by a feasibility study.
4. It is shown that when choosing PV, it is important to consider the manufacturing technology of an PV, since Myanmar has a hot climate.
5. Experimental verification of typical parameters of PV power plants based on PV, wind turbines, AB and DGEN for the standard power supply of a rural consumer in Myanmar using the GIS tools developed by RES Myanmar. It was shown that a typical HES based on renewable energy sources for a typical Myanmar consumer in a selected region pays off in 6 years.

REFERENCES

Aung, H. M. (2015). Renewable Energy Development in Myanmar. In The Energy Dialogue between Russia and ASEAN in the Field of Renewable Energy and Clean Energy Technologies.

Aung, Ko., Malinin, N. K., & Shestopalova, T. A. (2016). *Statement of the problem of optimizing the composition of the energy complex on the basis of renewable energy sources for the integrated power supply to the rural consumer of Myanmar.*

Aung, Ko., & Shestopalova, T. A. (2016). Problems of creating a comprehensive system of electricity and heat supply of the off-line consumer of Myanmar. In *Proceedings of the Twenty-second international scientific and technical conference of students and graduate students "Radio electronics, Electronics and Energy"* (p. 316). Moscow: Nauka.

Aung, T.P. (2013, June). Introduction of MES and Current status of Energy Efficiency in Myanmar. Energy and Renewable Energy Committee. University of Technology Thonburi Campus, Bangkok, Thailand.

Bezrukikh, P. P. (2010). Wind power.

Burmistrov, A.A., Vissarionov, V.I., & Deryugin, G.V. (2007). Methods for calculating renewable energy resources: a tutorial. MPEI.

Burmistrov, A. A., Vissarionov, V. I., Deryugina, G. V., Kuznetsova, V. A., Kunakin, D. N., & Malinin, N. K. (2009). *Methods for calculating the resources of renewable energy sources.* MPEI.

Dapice, D. (n.d.). *Electricity Demand and Supply in Myanmar.*

Deriugina, G.V., Zay, Y.L., & Tyagunov, M.G. (2017). Data verification for use in the regional renewable energy sources geographic information system. Energetik, 5, 36-40.

Dobermann, T. (2016, March). Energy in Myanmar. International Growth Centre.

Hadden, R. L. (2008). The geology of Burma (Myanmar): An annotated bibliography of Burma's geology, geography and earth science. Geospatial Information Library (GIL).

Hanoi, M. (2012, November). Renewable Energy and Rural Development in Myanmar.

IRENA. (2016). *REmap: Roadmap for a Renewable Energy Future, 2016 Edition. International Renewable Energy Agency*. Abu Dhabi: IRENA.

IRENA. (2016). *Renewable Energy Outlook for ASEAN: a REmap Analysis. International Renewable Energy Agency (IRENA). Abu Dhabi and ASEAN Centre for Energy*. Jakarta: ACE.

Kyaw, S.S.N. (2012, March). Myanmar Electricity Outlook with reference to Demand Scenario.

Lin, U. (2011, June). ASEAN Countries' Presentation on Renewable Energy Projects and Business Opportunities (Myanmar), Myanmar Engineering Society (MES). In ASEAN Countries Presentation on Renewable Energy, BITEC, Bangkok.

Remund, J., Kunz, S., & Lang, R. (1999). METEONORM: Global meteorological database for solar energy and applied climatology. In *Solar Engineering Handbook (version 4.0)*. Bern, Germany: Meteotest.

Zay, L. & Tyagunov, M.G. (2015a). Prototype of the pilot GIS Regional Renewable Energy Sources. In *International Congress REENCON-XXI Renewable Energy XXI Century: Energy and Economic Efficiency*, Moscow, (pp. 169-172). Moscow: Nauka [Science].

Zay, L. & Tyagunov, M.G. (2016, July).Creating regional geographic information system to determining optimal placements of power generation based on renewable energy resources. *Papers of the SGEM Conference*, Republic of Bulgaria.

Zay, L. & Tyagunov, M.G. (2016). Evaluation of the technical potential of solar, wind, tidal and small electric power industry of Myanmar. In *Proceedings of the Twenty-second International Scientific and Technical Conference of Students and Postgraduates "Radio Electronics, Electrical Engineering and Energy,"* Moscow, MPEI, (pp. 321). Moscow: Nauka [Science].

Zay, Y.M. & Vissarionov, V.I. (2013). Study of the energy characteristics of regional wind energy in the Republic of the Union of Myanmar [Ph.D. thesis]. Moscow Energy Institute.

Zay, Y.L. & Tyagunov, M.G. (2017).Research on the efficiency of the use of solar-wind-hydraulic energy complex in the republic of Myanmar union.

ADDITIONAL READING

Aung, Ko., & Shestopalova, T. A. (2019). *Research efficiency of use energy complex on solar and heat-pumping installations in regional Myanmar energy.*

KEY TERMS AND DEFINITIONS

Criteria: Are generally categorized into the three groups: energy, environmental and economic criteria. Determination of optimal energy mix comes down to determination of the percentage share of each component of renewable energy supply in defined boundary of the observed problem.

Geographic information system (GIS): Is a system designed to capture, store, manipulate, analyze, manage, and present all types of geographical data. The key word to this technology is Geography – this means that some portion of the data is spatial.

Optimization Parameter: (Or a decision variable, in the terms of optimization) is a model parameter to be optimized. For example, the number of nurses to employ during the morning shift in an emergency room may be an optimization parameter in a model of a hospital. The OptQuest Engine searches through possible values of optimization parameters to find optimal parameters. It is possible to have more than one optimization parameter.

Renewable Energy: Is energy that is collected from renewable resources, which are naturally replenished on a human timescale, such as sunlight, wind, rain, tides, waves, and geothermal heat.

Small Hydro: Is the development of hydroelectric generation facilities on a scale corresponding to river discharge and potential, and which is suitable for local community and industry, or to contribute to distributed generation in a regional electricity grid.

Solar Energy: Is radiant light and heat from the Sun that is harnessed using a range of ever-evolving technologies such as solar heating, photovoltaics, solar thermal energy, solar architecture, molten salt power plants and artificial photosynthesis.

Wind Energy: Wind energy (or wind power) describes the process by which wind is used to generate electricity. Wind turbines convert the kinetic energy in the wind into mechanical power. A generator can convert mechanical power into electricity.

Chapter 3
Energy Saving System Based on Heat Pump for Maintain Microclimate of the Agricultural Objects:
Energy Saving System for Agriculture

Giedrius Ge
Lithuanian Energy Institute, Lithuania

Irina Georgievna Ershova
https://orcid.org/0000-0003-1126-3837
Federal Scientific Agroengineering Center VIM, Russia

Alexey N. Vasilyev
https://orcid.org/0000-0002-7988-2338
Federal Agricultural Research Centre VIM, Russia

Dmitry Tikhomirov
Federal Scientific Agroengineering Center VIM, Russia

Gennady Nikolaevich Samarin
https://orcid.org/0000-0002-4972-8647
Federal Scientific Agroengineering Center VIM, Russia

Dmitrii Poruchikov
Federal Scientific Agroengineering Center VIM, Russia

Mikchail Arkadievich Ershov
Pharmasintez, Russia

ABSTRACT

At agricultural facilities, the main attention is paid to the formation and maintenance of their microclimate parameters, and mechanization of storage processes. As world experience shows, it is necessary to develop and implement energy-saving systems and the use of renewable energy sources. The authors have developed energy-saving systems based on the heat pump, with upgraded electrical regulators. The developed system (patent 100873), uses thermoelectric elements and a low-potential energy source, to effectively maintain the temperature parameters of the microclimate during long-term storage of potatoes, but it requires a large amount of electricity consumption (30 to 35 kW), so the authors have developed an energy-saving system based on a heat pump (patent 123909). The temperature regime is achieved by using a thermoelectric cooler-heater and an electric heater. The humidifier allows for maintaining the necessary relative air humidity.

DOI: 10.4018/978-1-5225-9420-8.ch003

INTRODUCTION

It is known that the loss of vegetables is up to 40%, which reduces to 5-10% while maintaining the necessary parameters of humidity and air temperature with an active ventilation system (Sircov, 2009), and mechanization of vegetable storage processes. As the world experience shows, the development and implementation of energy-saving systems (Besekersky, Popov, 2004) and the use of renewable energy sources (Rodionova, Borovkov, & Ershov, 2012) are effective in this situation, which also sharply reduces the consumption of energy resources (Vasilyeva, Novikova, Timofeev, 2011).

The search for new energy-saving technologies to maintain the air temperature in the potato storage is an important task (Bulatov, Krukov, Chan, 2014). In this regard, the use of a heat pump, effectively functioning with the use of modernized high-speed electric regulators, and supporting the temperature regime of potato storage, is relevant (Kolchin, Fomin, 2006).

In this regard, the scientific work is aimed at maintaining the parameters of the microclimate in the vegetable store using a modernized heat pump on a low-potential energy source (LPES).

The objective of this research work is the development of parameters and operating modes power saving system based on heat pump for the formation of the climate for agriculture on the example of careful analysis. Theoretical studies were conducted using the theory of machines and mechanisms, differential and integral calculus, and analytical methods. The program KOMPAS-3D V16 performed two-dimensional simulation of energy-saving systems. The use of modern computer technology in this work provides the possibility of simultaneous solution of the equation of dynamics of the electric regulator with a solid filler and an electric heater, represented by the aperiodic link of the first order.

BACKGROUND

Currently, thermoelectric heat pumps (HP) and HP on a low-potential energy source are widely used (Gorshkov, 2004). Thermoelectric HP operating on Peltier and Thomson effects have low efficiency and high cost (Ray & McMichael, 1982). However, the company TERMIONA (Moscow) developed thermoelectric elements that increase the cooling capacity and efficiency of generation by 15-30%, which will significantly increase the energy efficiency of devices (Naer & Garachuk, 1982). Currently, on the instructions of the Ministry of industry and energy of the Russian Federation a program for the development of non-traditional energy in Russia, including HP is being developed (Kalnin & Savitsky, 2000).

Therefore, the relevance of the use of low-potential energy is now objectively justified (Vasiliev & Krundyshev, 2002).

Production of HP in each country is focused, first of all, on satisfaction of requirements of the domestic market. Abroad, the active introduction of heat pumps contributes to the International Energy Agency. The European Heat Pump Association presented data on sales of heat pump equipment in 2015, while the heat pump market grew by 10% (Berzan, Robu, & Sheet, 2011).

In the USA, Japan and some other countries, air-air reversible HP intended for heating and summer air conditioning were the most widespread, while in Europe water-water and water-air HP were more widespread (Kireev, Lazeev, & Stepanenko, 2003). In Sweden and other Scandinavian countries, the availability of cheap electricity has led to the development of large HP (Vasilyev, 2006). Thus, 50% of the population of Stockholm city use heating, using a low-potential energy source (Baltic sea water). In the Netherlands, Denmark and other countries in the region, gas is the most affordable fuel and, therefore,

HP is rapidly developing with a gas engine. For Germany, it is extremely important to replace imported fuel from oil and reduce environmental pollution, so the development of HP with electric drive, as well as from gas diesel engines (Levin, 1999). The German heat pump stock is totaling out at 880 000 units (Sabel & Weinhold, 2019). "In 2020, we will have one million heat pumps that will produce about 8 billion kWh of renewable energy, equivalent to 15% of EU-requested renewable energy in Finland," says Jussi Hirvonen, executive director of the Finnish Heat Pump Association SULPU.

The potential use of non-traditional renewable energy sources in Russia is large (Briganti, 2001), but their effective use requires the ability to calculate the possibility of generating energy from such sources anywhere in the area, as well as to predict the economic feasibility of renewable energy compared to fossil fuels (Syromyatnikov, 1989).

The search for patent analogues - patent N° 2354897 (Nakoryakov, Elistratov, 2007), N° 2285872, N° 2152567. Considered heat pumps are not designed to regulate the temperature of vegetable stores. It can be seen from the above schemes that for the effective functioning of the HP, it is necessary to use regulators with automatic control, especially in cases where it is required to change the parameters of the temperature regime of the air in the storage during the transition from the cooling mode to the heating mode, and Vice versa.

Analyzing expediency of use of this or that option of the scheme of the choice of type of energy saving system for formation of a microclimate on objects of agriculture, it is possible to note that the most perspective from the point of view of energy saving is use of LPSE of soil.

MAIN FOCUS OF THE CHAPTER

The developed block diagram of thermal regulation of storage energy-saving system based on a heat pump is designed to form a microclimate in vegetable stores and ensure long-term storage of vegetables with minimal energy, heat, cold (Fig. 1).

The main elements for supplying the necessary energy are the ground heat exchanger 2 (Vasilyeva, 2012) and the Converter LPSE 3 with its heat pump and regulators (Bulatov et al., 2014).

In the scheme of the system, an aqueous solution of calcium chloride with a concentration of 3.83% (not freezing at a temperature of -2 °C) was selected, circulating in the ground heat exchanger.

The main element of the Converter 3 (Fig. 1) is a heat pump with electric regulators, which turns the LPSE into heat energy and cold source to maintain the microclimate in the vegetable store.

In 1824, Carnot first used a thermodynamic cycle to describe the process, and this cycle remains the fundamental basis for comparing and evaluating the efficiency of HP (Vasilyeva et al., 2011). According to the principle of operation, heat pumps are classified into absorption and vapor compression pumps (Atkinson, 2000).

It is known that heat pumps are designed to increase (decrease) the value of heat q_1 (most often extracted from the environment) due to the cost of mechanical energy l_κ and subsequent use of the resulting heat $q_1 = q + |l_\kappa|$ consumer (Gomaa, 2015). The efficiency of the heat pump is estimated by the coefficient of heat transformation-the ratio of the cooling capacity of the heat pump to the power consumed by the compressor or the ratio of the received heat $q_2 = q_1 + |l_\kappa|$ to the spent mechanical energy $|l_\kappa|$:

Figure 1. Block diagram of temperature control of greenhouses: 1 – the vegetable store, 2 – soil heat exchanger, 3 – energy Converter (heat pump), 4 – administrative building (object of heating (cooling)), 5 – channel LPSE, 6 – channel vent duct, 7 – channel of the transformed energy, 8 – channel of exhaust LPSE, 9 – channel of waste energy

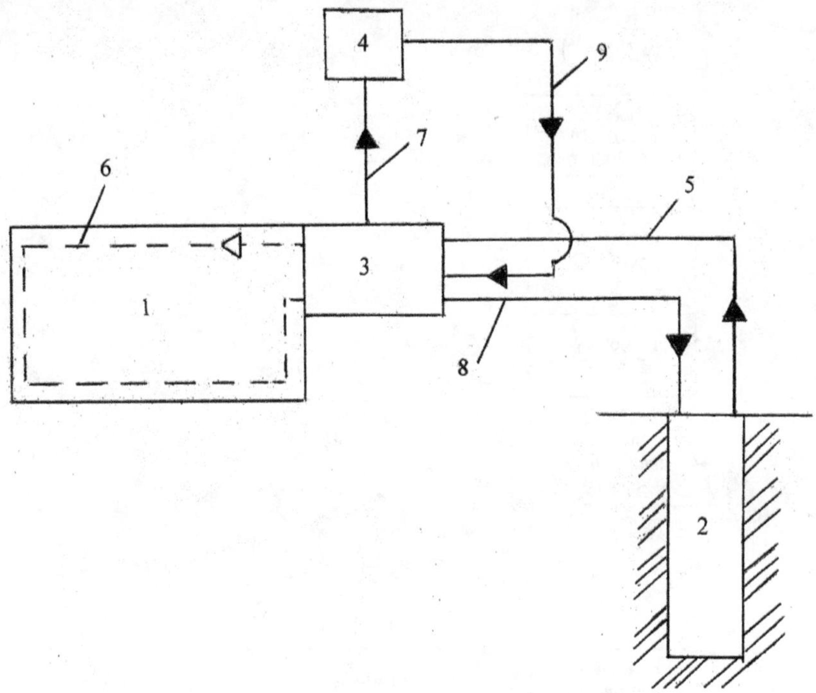

$$\psi = \frac{q_2}{|l|} = \frac{q_1}{|l|} + 1 = \varepsilon + 1 \quad (1)$$

where ε is the heating coefficient.

Flow control diagram of the energy carrier includes an electric regulator containing a thermoelectric module or an electric heater that acts on a solid filler that drives a rod with valves.

An algorithm for matching the modes of operation of the heat pump with controlled and controlled parameters of the microclimate of the vegetable store (Fig. 2) allows you to determine the operating modes of the heat pump controller.

If the temperature of the vegetable store does not correspond to the set value, then using the electric regulators of the heat pump in the Converter, the required temperature is set, necessary for supplying air to the vegetable store.

In General, the temperature control system of the vegetable store consists of a setpoint programmer (P) 1; a regulator (R) 2 and a control object (CO) 3 (Fig. 3).

On the diagram “I” denotes the set of external information which is supplied on the programmer (Chugayeva, 2012). The task of constructing an optimal system is to synthesize a controller and a programmer for a given object, which in a certain sense best solve the problem of control. Mathemati-

Figure 2. The algorithm of matching the modes of operation of the heat pump with controlled and controlled parameters of the microclimate of the vegetable store: Topt-is optimal storage temperature (°C); t-storage time (min.); Δt – increment of time (min.); tkon-storage time-10 months (min.); HP, CP - power of heat and coolant (W); TH, TC - temperature of heat and coolant (°C), TV - temperature of vegetable store (°C)

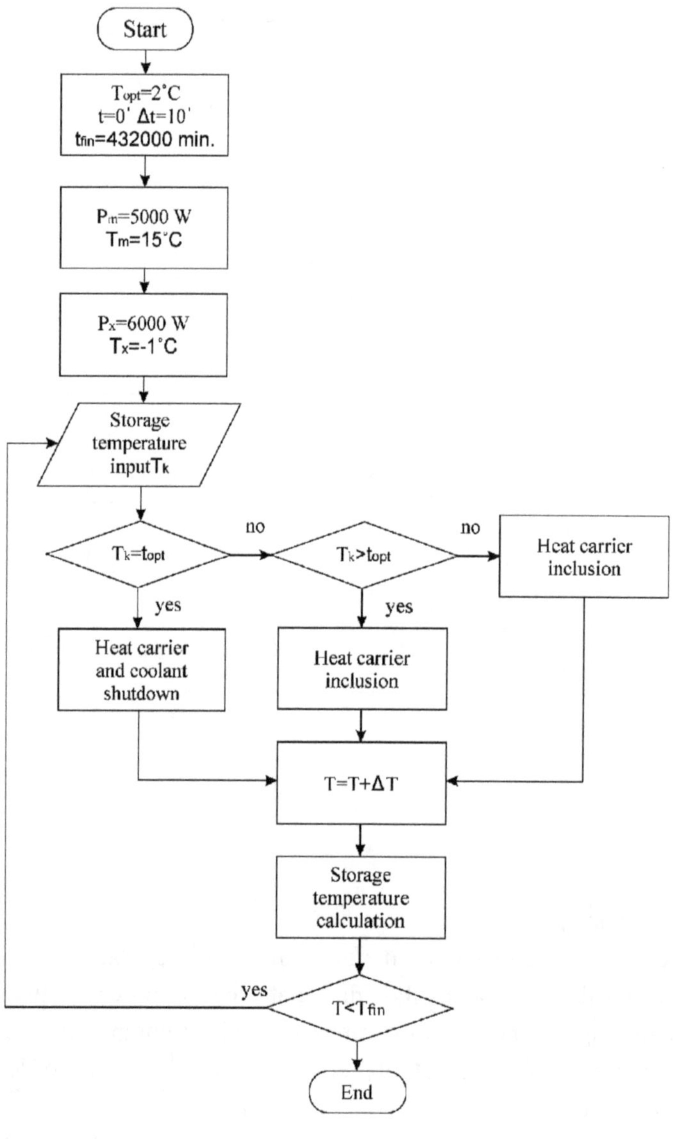

cally, these problems can be formulated uniformly and solved by the same methods, but at the same time these problems have specific features that make it appropriate at a certain stage of their separate consideration. Features are caused by the fact that the solution of the first problem is connected, as a rule, with definition of program management, and the solution of the second task – with definition of management with feedback.

Figure 3. Automatic control system: 1-setpoint programmer; 2-regulator; 3-control object

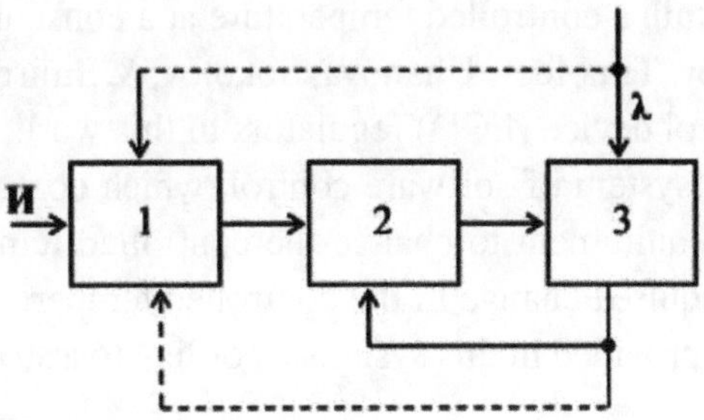

Thus, software control is a control in the form of a function from time, feedback control is a control in the form of a function from phase coordinates (Sage & White, 1982). The temperature control system of the vegetable store can be structurally considered as a single-circuit, presented in the form of conditional images of links and connections between them. In General, the system consists of two enlarged structural units: the object (vegetable store) and the regulator.

The connection from the object to the controller is called the main feedback. The figure shows that the temperature control system is a closed system of directional links.

Vegetable store (object) is characterized by the coordinates: the temperature of the inlet T_i, outlet temperature T_O and load coordinate T_e. On the lines of communication provided the transmission ratios relations K_{c1}, K_{c2}, K_{c3}.

With the help of special setting bodies of the task, which are equipped with regulators, you can set a particular value of the temperature of the task ZD; in accordance with this parameter, the value of the controlled parameter, which will be supported by the regulator, is determined.

With the help of load (disturbance) the object changes the specified mode. Disturbance effects are commonly referred to as effects that seek to disrupt the required functional relationship between the driving force and the controlled temperature. In this case, the driving force is understood to be the effect on the system that determines the required law of change in the controlled temperature.

As a result of the regulatory impact of the regulator on the vegetable store, the latter returns to the equilibrium mode while ensuring the desired value of the parameter. The regulator acts on the vegetable store through the RO, which converts the change obtained at the outlet of the regulator of the temperature of the energy carrier entering the vegetable store. Requirements for the accuracy of maintaining a controlled temperature in the statics and dynamics are allowed here the use of simple designs of static regulators and regulation only by the deviation of the controlled temperature (Ershova, Ershov, & Poruchikov, 2017).

The regulator, as well as the vegetable store with its system, are components of the vegetable store air SART. The deviation of the controlled air temperature from the set value depends, on the one hand, on the properties of the regulator and, on the other hand, on the properties of the vegetable store and its cooling-heating system.

Thus, the operation of temperature control of vegetable storage includes five main stages: measurement of the controlled air temperature of vegetable storage, comparison of the measured value with the value set by the temperature, processing of this difference in the regulator, processing of the control signal into the control action; return of the controlled temperature to the set value under the regulation.

In the existing SART on all modes of the vegetable store operation, the algorithm of operation contains an instruction to maintain a controlled temperature at a constant value, i.e. it is a stabilization system (Vasiliev, Ershova, Belov, Timofeev, Uhanova, Sokolov, & Smirnov, 2018). Using the electrical elements in the Executive control device (ECD) regulators in this work, in addition to the stabilization system, we propose to create a system of software control, which contains an algorithm for the functioning of the SART, i.e. the requirement to change the controlled temperature in accordance with a predetermined function. The required change in the controlled temperature by the load and during the storage of the vegetable store is provided in this system according to a strictly defined program (Kalinin, & Kudryavtsev, 2007).

Structural diagram of the SART greenhouses (Fig. 4) explains the principle of operation of the system.

The sensing element of the temperature sensor 2 registers the temperature in the potato area. The setting device 4, depending on the mode of operation of the vegetable store, determines the required mode of changing the temperature of the object. The plug of the three-way valve has a cutout, which, when rotated by an electric motor, allows you to open (close) the branch pipe, close (open) the branch pipe or simultaneously open (close) the branch pipes.

The temperature of the object is converted by the sensor 2 into a signal $U_d(T)$, which is compared in the comparison scheme 3 with the load signals $U_H(P)$ and the master(software) device $U_3(T)$. The misalignment signal, which determines the deviation of the object temperature from the required value, is fed to the control unit 6, which is then converted into a powerful effect $U_C('')$, which controls the regulator. The regulator 5 directs the flow of energy to the vegetable store or to the heating (cooling) object and then the regulating effect $\mu(T)$ from the regulator is fed to the vegetable store, changing its temperature and reducing the misalignment to the minimum possible value for this system.

Thus, the basis of the considered system of automatic temperature control is a closed loop of influences, which are internal influences. The system $U_4(''), U_=(\ ), U_7(''), U_C(''), \mu('')$ is influenced by external influences, which include: the specified adjustable parameter (the required temperature of the object depending on the load) $U_H(P)$, disturbing effects λ (fluctuations in the power of heat in the object; changes in ambient temperature, etc.) and other secondary effects.

For the practical implementation of temperature control in accordance with the organization chart should be a concept, i.e. to choose the principle of operation of each block of the system (e.g., thermos-receiver - a chromel-kopel thermocouple, control unit – microprocessor controller – electric actuator that controls the operation of regulator valves in the trunk of the coolant). However, regardless of the specification of individual blocks of the scheme and the principle of their operation, the General principle of operation and construction of the block diagram of a closed cooling system remains unchanged.

Vegetable store is considered as an adjustable object heat exchanger. The object itself is the cooling-heating system of the vegetable store, which is dynamically equivalent to the aperiodic link. The input signal is the difference between the current values of the energy carrier temperature, load and the temperature setting value, the output signal is the control signal at the output of the control unit. The inflow of heat into the system can vary from a minimum to a certain maximum, corresponding to the full opening (closing) of the main valve on the vegetable store. Change of the regulating influence is carried out by means of the Executive mechanism which speed in both parties is limited.

If the task consists in the fastest closing (opening) of branch pipes of regulating body, then at first it is necessary to open (close) branch pipes as quickly as possible. If the initial temperature deviation from the set value is large, it may happen that in the process of regulating the nozzle for bypass will be completely

Figure 4. Block diagram of the cooling system-heating vegetable store: 1 - the object of regulation; 2-measuring temperature Converter (temperature sensor); 3-comparison circuit; 4-master device; 5-regulator; 6-control unit; 7-load sensor; 8-heating object

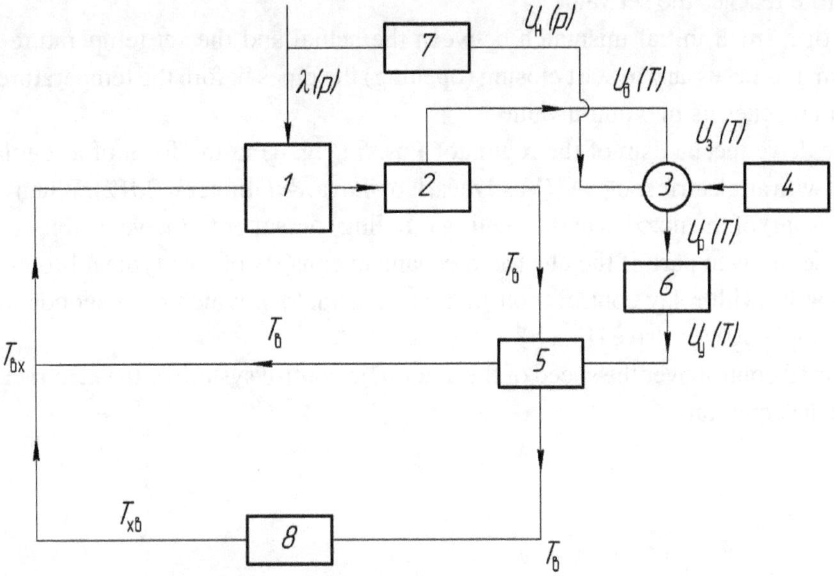

Поз. 5

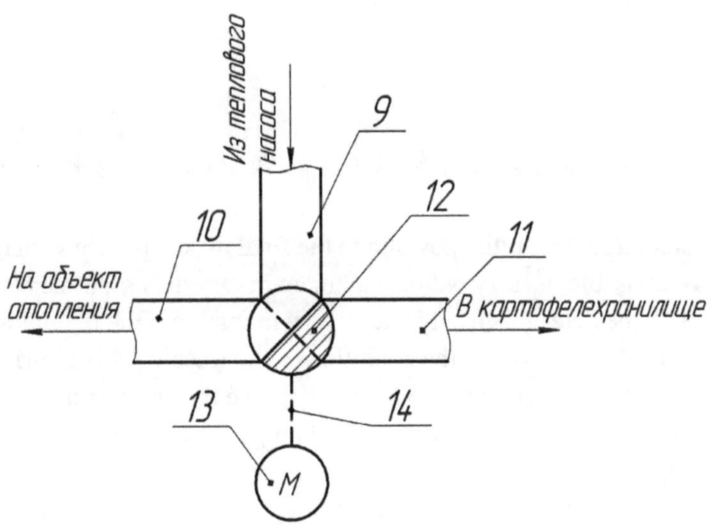

open (closed). At some point in time, when the temperature of the energy carrier in the cooling-heating system is close to the set, it is necessary to begin to close (open) the nozzles. This is necessary in order to equal the arrival of the temperature of the cooling system (heating) to a predetermined value of heat (cold), transmitted to the cooling system (heating) and withdrawn from it were the same. The start time of the closure shall be chosen so that the specified state is reached at the maximum speed of movement of the control plug. If you start to close (open) the pipe a little later, due to the limited speed of movement of the actuator, the set temperature will be reached earlier, and it will be necessary to reduce the closing speed – otherwise the heat exchange balance will come before the temperature at the outlet of the vegetable store reaches the set value.

In the case of a small initial mismatch between the actual and the set temperature of the cooling (heating) system, it is necessary to start closing (opening) the pipes before the temperature of the cooling (heating) system reaches its maximum value.

We draw the drive mechanism of the regulator Pos. 5 (Fig. 4) in the form of a regulatory body – a three-way crane with an electric motor M (Vasilyeva, Novikova, & Timofeev, 2012). When the tube rotates, the opening (closing) of the nozzles to the heating (cooling) object or to the vegetable store takes place.

Let us consider a given part of the electric mechanism consists of two typical links – aperiodic and integrating actuators with relay control (constant speed actuator), which corresponds to the structural scheme of the simplest servo drive (Fig. 5).

Find the optimal control over the speed of the automatic control system of the electric actuator, which is described by the equations:

$$\frac{dX}{dt} = Y\,; \quad \frac{dY}{dt} = U - Y \tag{2}$$

a control action is limited by the condition:

$$\left|U\right| \leq 1 \tag{3}$$

In the initial position of the coordinate system $\%_0 < 0$, $Y_0 = 0$; at the end of the control process $X_к = Y_к = 0$.

Time $t_к$ of moving the system from the initial position to the final in accordance with the task should be minimal. In this case, it is advisable to carry out management in two stages. At the same time, at the first stage, during the time t1 (to be determined), it is necessary to maintain the regulatory effect at the maximum level, i.e. in this case U = +1, so that the condition of the tracking drive during acceleration is the greatest. At the second stage, it is necessary to vigorously brake the drive using the anti-switching mode, for which it is necessary to change the sign of the regulating effect, leaving it maximum in absolute value, i.e. U = -1 (Fig. 6).

In this case, the motor control system M should be optimal for performance. In this scheme, the electric motor through the gearbox is connected to the regulator and produces energy supply, depending on the temperature in the vegetable store (potato storage) or on the heating object. Taking into account

Figure 5. Block diagram of the electric actuator

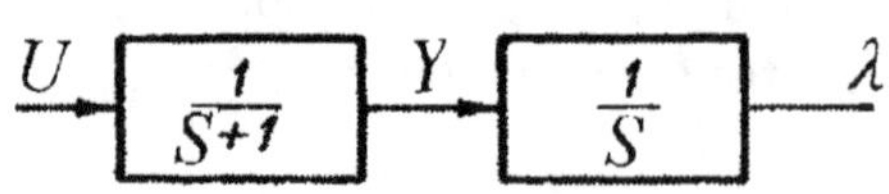

Figure 6. Structural diagram of the electrical mechanism: PO – three-way stopcock – regulatory body; Ω – the rotation speed of the engine; M_c – static torque; r_p – resistor; U_1 – is applied to a motor voltage; i_1 – the current in the circuit motor armature

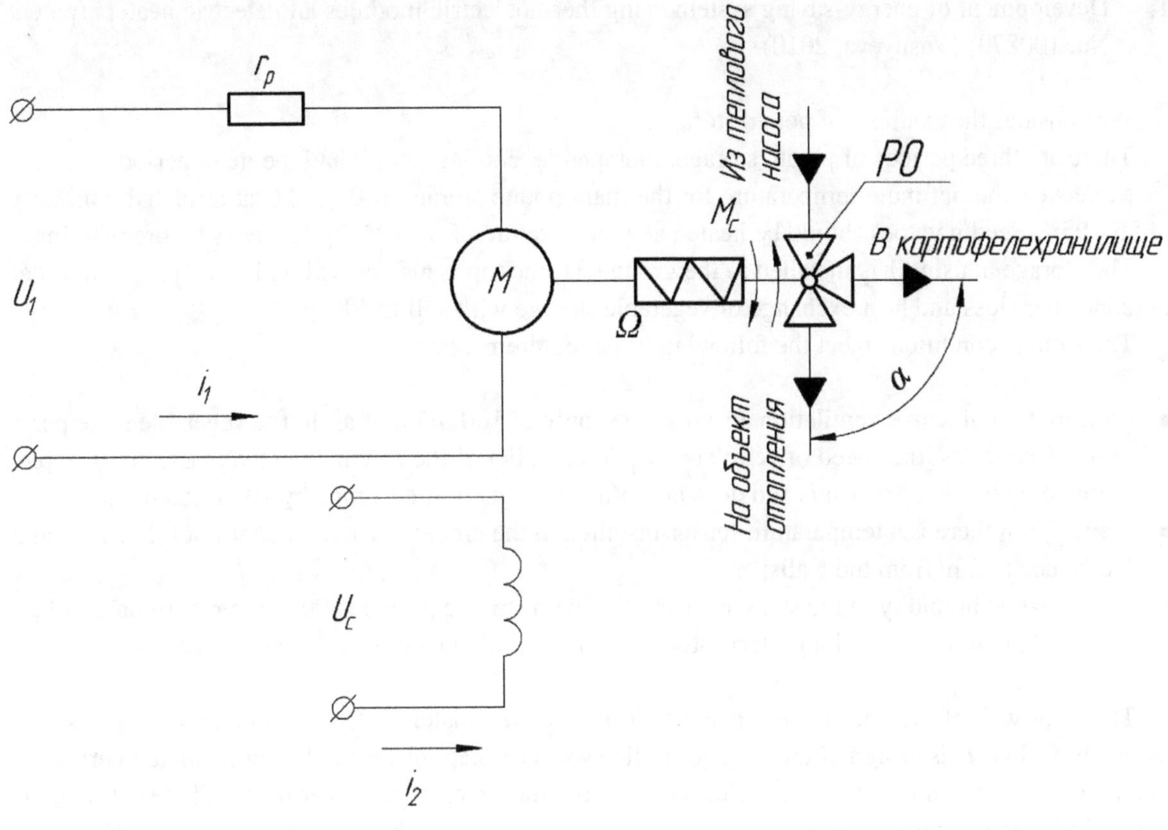

the actual operating conditions of the regulatory body, a requirement may be made that the developed executive mechanism ensure that the RO plug moves from the initial position to the final position in a minimum of time.

Using the technique of three-factor active planning experiment type 2^3 and the program "Statistic V5.0" empirical expressions are obtained.

Empirical expression patterns of duration of the transient process electrical controller depending on the capacity of EP and the temperature of the solid filler at a constant temperature energy source equal to 60°C:

$$t = 40{,}42 + 1{,}986\, T_{н} – 0{,}425\, N_{эл} – 0{,}011\, T_{н}^{2} - 0{,}005\, T_{н}N_{эл} - 0{,}003\, N_{эл}^{2},\ с \quad (4)$$

Empirical expression of the model of the amount of heat of the energy carrier depending on the power of EP and the temperature of the solid filler at a constant temperature of the energy carrier equal to 60 °C:

$$Q = - 5901{,}256 + 159{,}581 \bullet T_{н} + 117{,}6879 \bullet N_{эл} - 0{,}824 \bullet T_{н}^{2} - 0{,}625\, T_{н}N_{эл} - 0{,}114\, N_{эл}^{2},\ Вт \quad (5)$$

Energy-saving systems based on a heat pump, with upgraded electrical regulators, to form a microclimate in vegetable stores are developed.

1. Development of energy-saving system using thermoelectric modules and electric heaters (patent No. 100873) (Vasilyeva, 2010).

We consider the example of potato storage.

There are three periods of potato storage: therapeutic, cooling period and the main period.

Moreover, the optimum temperature for the main potato storage is 2 – 4 °C at a relative humidity of 85 - 95%. Seed potatoes should be heated at a temperature of 8 - 10 °C two weeks before planting.

The storage housing 1 is installed in the ground 31, the top is also closed with soil (Fig. 7) in order to reduce heat loss and heat exchange of vegetable storage with soil LPSE.

The storage conditions meet the following basic requirements:

- the system of active ventilation ensures the supply of 100 m3/h of air to the vegetable store per 1 m3 of products; the speed of ventilated air at the outlet of the channels to the mass, for example, potato tubers, is up to 1 m/s and does not differ throughout the storage by 10...Fifteen %;
- every 5 m there is a temperature sensor installed in the embankment to a depth of 0.5-0.7 m; at a distance of 2 m from the walls;
- the relative humidity in the storage during the main storage period is 90%; the rate of natural loss of seed potatoes during long - term storage without artificial cooling does not exceed 4-6%.

The front wall of the housing 1 is made of heat-insulated material. Door 4 and a window 5 is hermetically sealed. This design of the housing 1 allows you to keep the heat at low ambient temperatures. The required room temperature 3 is achieved from the thermoelectric cooler-heater (TCH) 10 operating in the "heating" mode, made as a heat pump, and the electric heater 11 (Patent 31637) (Timofeev, Chekmarev, & Galkina et al., 2003).

In case 1, there are 7 containers of potatoes in 2 - 3 layers. Remote control 2 is an enlarged view of the extension element A. it contains: a closed loop cooling system, electronic controls, TCH 8, 10; the electric heater 9. The closed circuit of the cooling system serves to remove heat from the "hot", "cold" junctions TCH 8, 10 and includes: electric pump 15, expansion tank 19, heat exchanger 18. The heat exchanger 18 receives cold water, where the heat exchange of cold water with a closed circuit of the cooling system takes place, while the heated cold water goes through the channel 38 to the object of consumption (Patent 100873) (Vasilieva, &Timofeev, 2011).

Energy-saving system of microclimate formation in the vegetable store works as follows. Prepared for storage in containers potatoes forklift 32 lay in storage 1. After that begins the healing period of potatoes. To do this, the set point 22 in the storage 1 is set to a predetermined temperature, for example,18 °C In this case, the temperature sensor 16 gives a signal to the comparison unit 21, here is also a signal from the set point 22. In the comparison unit 21 there is a comparison of these signals, the mismatch signal enters the control unit 20. From the control unit 20, the processed signal is fed to the voltage reverse unit 23, where the polarity of the voltage is reversed and the electricity is supplied to the TCH 8. TONE 8 begins to work as a" heater", i.e. as a heat pump. In addition, electricity is supplied to the electric heater 9 and as a result of the operation of the TCH 8 and the electric heater 9, the storage 1 is heated to a predetermined temperature. The control unit 20 starts the electric pump 15, which begins

Figure 7. Schematic diagram of energy-saving system of microclimate formation in vegetable store: 1-storage building; 2-control station; 3-warm room; 4-entrance door; 5-storage window; 6-humidifier; 7 - containers with potatoes; 8, 10-thermoelectric cooler-heater (TONE); 9,11-electric heater; 12,13-electric fans; 14-electric drive; 15-electric pump; 16,17-temperature sensors; 18-heat exchanger; 19-expansion tank; 20 control unit; 21 – block of comparison; 22- knobs; 23 – unit reverse voltage; 24 – timer; 25 – control unit with timer; 26 – remote control heater space; 27 – switch; 28 – unit heater control facilities; 29 – block comparison; 30 – reference; 31 – ground; 32 – forklift; 33,34,35,36,37,38 – coolant "hot", "cold" junctions of TCH; 39,40,41,42,43,44,45,46,47,48 – channels of supply of electricity; 49,50,51,52,53,54,55,56,57 – channels of signal; 58 – channel for exhaust air into the atmosphere; 59 – ozonator

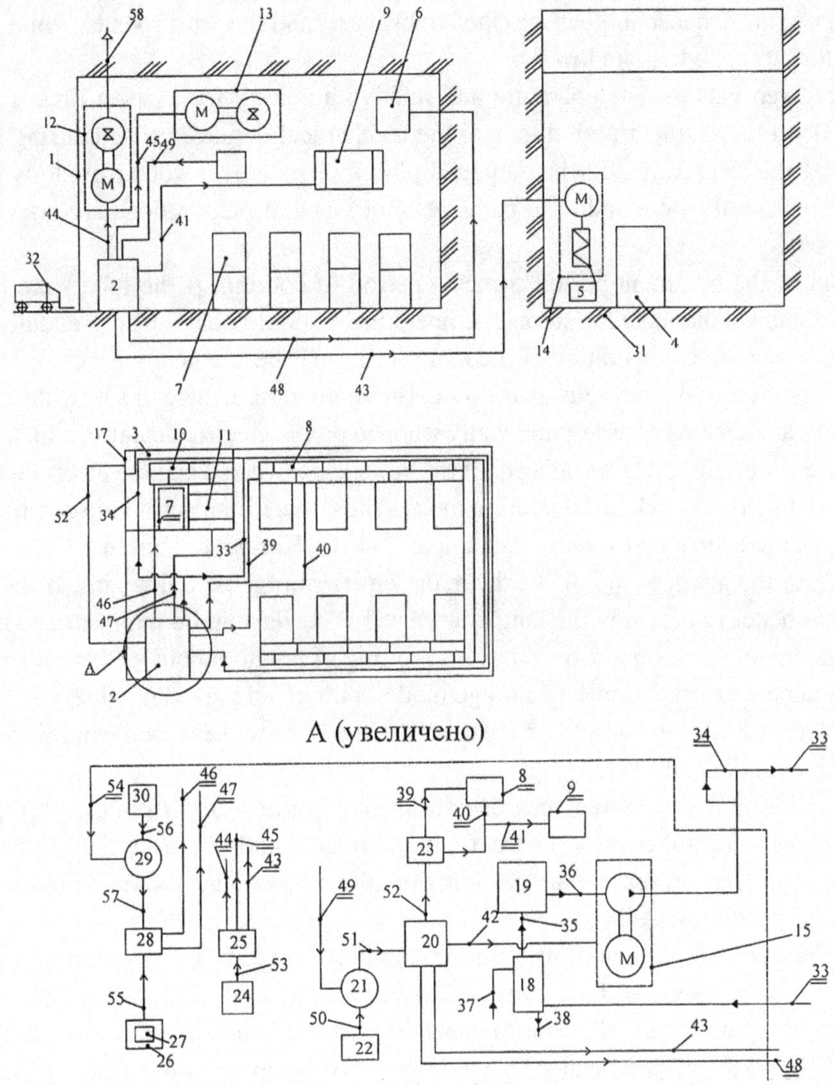

to move the cooling water along the closed circuit, while removing heat from the cold junctions of the TCH 8 (Patent 2350850) (Rymkevich, Kostyrya, & Kachkin, 2009).

The control unit 20 supplies electricity to the air humidifier 6 (patent 117256) (Vasilyeva, 2012), as a result of which the specified humidity, for example, 90%, is achieved in the vegetable store 1. At the same time, the ozonator 59 begins to work according to the given program (Pshechenkov, & Stockings, 2008). Thus, the necessary temperature of potatoes during the treatment period of storage is provided; and the potatoes are in the dark.

For air exchange timer 24 set the time of ventilation. After this time, the timer sends a signal to the control unit 25, which starts the electric fans 12, 13 and the electric drive 14, which opens the window 5, as a result of which the storage 1 is ventilated. In the vegetable store with active ventilation check the tightness of the system, which should provide air supply to the potatoes at least 50 m^3 per 1 ton of tubers. In this case, the exhaust air goes into the atmosphere, and the fan 13 vents, while fresh air enters the vegetable store through the window 5.

Thus, at the given values of temperature and relative humidity for a given time, for example, 20 days, tubers undergo a treatment period to heal the mechanical damage caused during harvesting and transportation (at the same time, on the damaged places of potatoes, wound tissue is formed, which protects the tuber from the penetration of pathogens into it) and preparations are made for long-term storage of potatoes.

After the end of the treatment period comes a period of cooling. If the tubers are healthy, with a minimum of mechanical damage, the storage temperature 1 should be reduced gradually by 0,5 °C per day for 20 to 30 days to the temperature of the main storage (Pshechenkov & Stockings, 2008).

Reduction of storage temperature is as follows. The temperature value is set by the set point 22, in this case, if necessary, electricity is automatically stopped on the electric heater 9, which ceases to heat, and in the reverse control unit 23, the polarity of the voltage is reversed and the electricity is supplied to the TCH 8, which begins to work in the cooling mode. Thus, there is an automatic control of the storage temperature 1 to a predetermined value, for example, 2-4 °C (Kurgan & Pankin, 2016).

After cooling of the potatoes to 2-4 °C, the main winter storage period begins. If the quality of potatoes is low, it is better to maintain the temperature at 1-3 °C during the main storage period in order to slow down the life of microorganisms that cause rotting of potato tubers.

Support of temperature and humidity storage mode in the main period is achieved by the operation of the TCH 8; electric heater 9; humidifier 6; fans 12, 13 and due to the sealed window 5 and as a result of heat exchange with the soil LPSE.

The spring period of potato storage begins with the release of tubers from the period of deep rest: in the early varieties – from the end of February, in the middle and late-from the beginning of March. Starting from this time, to prevent potato germination, the temperature should be reduced to 1.5 - 2.0 °C (Kuznetsov & Spiridonov, 2009).

Before shipping potatoes to the consumer, the temperature should be increased to 12 °C to avoid black spots and damage to the potatoes. To do this, the loader 32 container of potatoes 7 from the storage 1 moves to the warm room 3, while the control panel 26 switch 27 signals the control unit 28, which includes the TCH 10 and the electric heater 11, while the TCH 10 operates in the heating mode, the room temperature is controlled by the comparison unit 29 using the temperature sensor 17 and the set point 30. When the temperature of potatoes reaches 12 °C, the container with potatoes 7 is sent to the consumer.

The developed system, using thermoelectric elements and LPSE (soil and groundwater), allows to effectively maintain the temperature parameters of the microclimate during long-term storage of potatoes, but it requires a large amount of electricity consumption (30...35 kW), so we have developed a second energy-saving system.

2. Development of an energy-saving system based on a heat pump with modernized electric regulators for the formation of a microclimate in vegetable stores (patent No. 123909).

In the developed system, the LPSE as a result of heat exchange gives energy to the salt solution with a concentration of 3.83%; a temperature of 6 - 8 °C, which is fed through the channel 10 to the evaporator 1 (Fig. 8).

Inside the heat pump, the refrigerant passes through the evaporator and transfers the heat collected from the ground to the internal circuit of the heat pump. The internal circuit of the heat pump is filled with a refrigerant, for example R-134a (Tetrafluoroethane CH_2F-CF_3 with a pressure of 40,603 bar at a critical point that does not affect the ozone layer).

The refrigerant, having a low boiling point, passes through the evaporator while passing from the liquid state to the gaseous state. From the evaporator, the gaseous refrigerant, having a low temperature, enters the compressor 2, where it is compressed, while the temperature and pressure increase. Further, the

Figure 8. Structural diagram of the heat pump of vegetable store: 1-evaporator, 2-compressor, 3-condenser, 4, 7-electrical regulators, 5-vegetable store, 6-pump, 8-choke, 9 - additional evaporator; 10, 11-LPSE channels

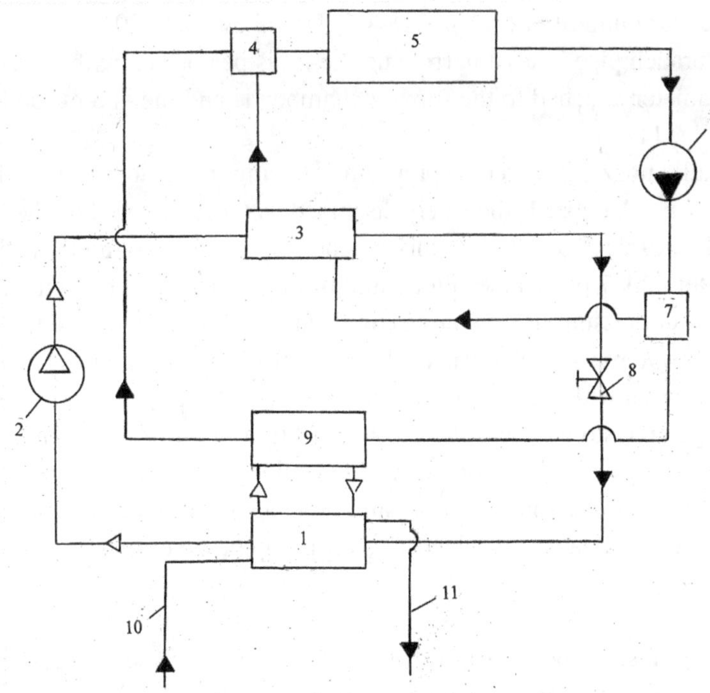

compressed and hot gas enters the condenser 3, where heat exchange occurs between the hot gas and the coolant from the heating pipeline of the vegetable store 5 (Patent 123909) (Vasilieva & Timofeev, 2013).

When the refrigerant is supplied to the additional evaporator 9 in the summer, the heat pump produces cold, which is fed to the vegetable store 5, while the coolant supply is regulated by electric regulators 4, 7, i.e., stopping the supply of coolant from the condenser 3 to the vegetable store 1.

3. Energy-saving system based on a heat pump using an electric regulator with a solid filler and an electric heater to form a microclimate in vegetable stores.

In the device for regulating the temperature regime of potato storage, a source of thermal energy 15 and a source of cold 16 of any design are provided, what energy carriers through electromagnetic valves 17, 18 and the main electric regulator 1 provide the necessary energy flow to the vegetable store 13 (Fig. 9).

4. Energy-saving system based on a heat pump for the formation of a microclimate in vegetable stores.

Schematic diagram of energy-saving system (Fig. 10) contains four conditional circulation circuits.

The I circuit includes: the supply channels 1 and the discharge channels 2 of the low-potential energy source (the remaining elements of the first circuit are not shown in the figure).

The II circuit includes: the main evaporator 6; the electric compressor 3, the condenser 4, the throttle valve 5, the electric regulator 8, the evaporator 7.

The third cooling circuit includes: capacitor 4, the electric regulators 12, 13; heating the object 14; a vegetable store 9; the electric pump 10.

The IV circuit of the coolant include: an additional evaporator 7; vegetable store 9; an electric regulator 13; an electric pump 11.

For the formation of microclimate in the greenhouses and it the desired relative humidity is achieved by use of the developed air humidifier (Patent 118406, Timofeev et al., 2012).

In this device, the branch pipe 3 has a narrowing, i.e., it is performed with a certain taper, with the end of the reduced diameter attached to the mixing chamber 1, and the end of the large diameter – to the duct channel 17 (Fig. 11).

In the electrohydraulic nozzle 2, water is sprayed by flowing it at a higher speed through a series of nozzle holes. In the mixing chamber 1, the finely dispersed water is captured by the flow of air flowing at a high speed from the nozzle 3, resulting in mixing and the necessary saturation of the air with water.

Water from the water tank 7 passes the filter 9 and through the channel 15 enters the electric water pump 6, in which the water pressure rises to the required value and is fed to the electrohydraulic nozzle 2, which fires and delivers water in a finely dispersed form to the mixing chamber 1. The mixing chamber 1 is mixing and air saturation with water. The obtained water-air mixture expands through the nozzle 4 is supplied into the vegetable store, where there is a saturation of air of a greenhouse with a water-air mixture as long as the humidity reaches a predetermined value. When the relative humidity of the vegetable store air is 95%, the comparison unit stops supplying a signal to the control unit 11, which stops supplying energy to the electric fan 5, the electric water pump 6, the electrohydraulic nozzle 2 and the device stops working.

6. Development of the reserve power supply of the control unit of the energy-saving system of the vegetable store by direct conversion of thermal energy into electrical energy.

Figure 9. Schematic diagram of an energy-saving system based on a heat pump using an electric regulator with a solid filler and an electric heater for the formation of a microclimate in vegetable stores: 1-the main electric regulator with three nozzles:2-branch pipe for supply of energy carrier; 3-branch pipe for removal of energy carrier in vegetable store; 4-branch pipe for removal of energy carrier; 5-rod; 6,7-valves; 8,9-saddles; 10 - the case of the filler; 11-solid filler; 12-electric heater; 13-vegetable store; 14-temperature sensor; 15-source of thermal energy; 16-source of cold; 17,18-electromagnetic valves; 19, 20-additional electric regulators; 21-object of heating; 22-consumer of cold; 23-control unit; 24-comparison unit; 25-setpoint; 26, 27, 28, 29, 30, 31, 32, 33, 34, 35, 36, 37, 38 – energy channels; 39, 40, 41, 42, 43 – electric power supply channels; 44, 45, 46 – electric signal supply channels

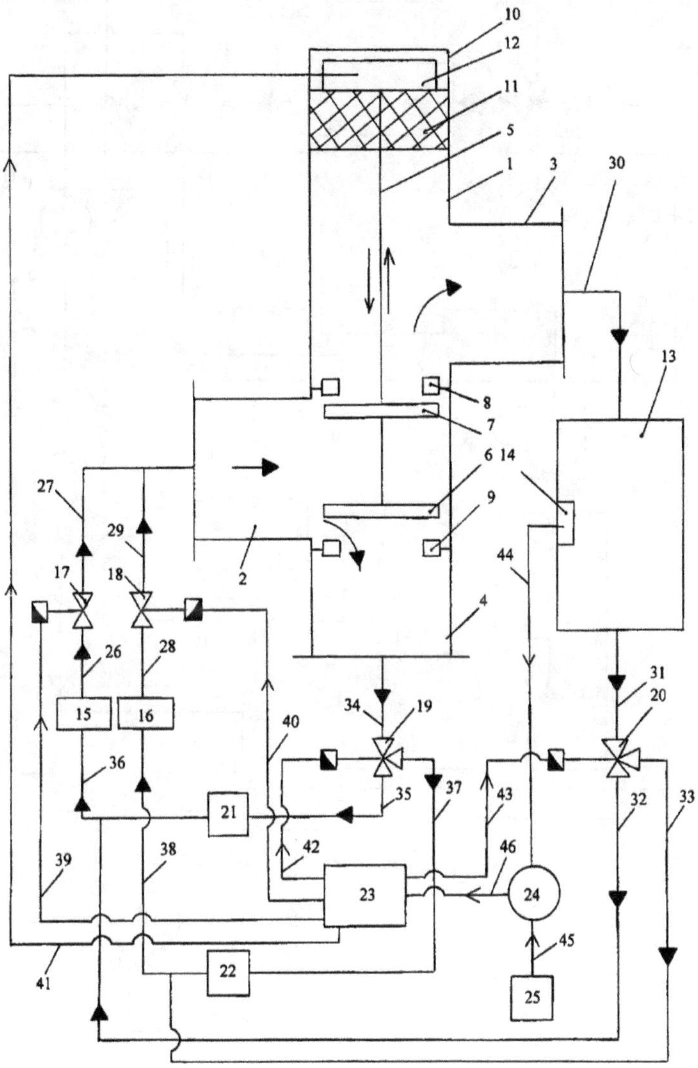

Figure 10. Energy saving system of formation of microclimate in the vegetable store at the base of the heat pump with updated electrical controls: 1, 2 – channels, and inlet and outlet of the low-potential power source; 3 – electric compressor; 4 – condenser; 5 – throttle valve; 6 – the main evaporator, 7 – additional evaporator; 8 – electric controller with a solid filler and EP ; 9 – vegetable store; 10, 11 – electric pump; 12 – relay switching regulator; 13 – electric controller with a solid filler and TM; 14 – the object of heating; 15 - temperature sensor; 16-control unit; 17 - comparison unit; 18 - setpoint; 19 -power supply; 20 - switch; I - LPSE circuit; II - refrigerant circuit; III - coolant circuit; IV - coolant circuit

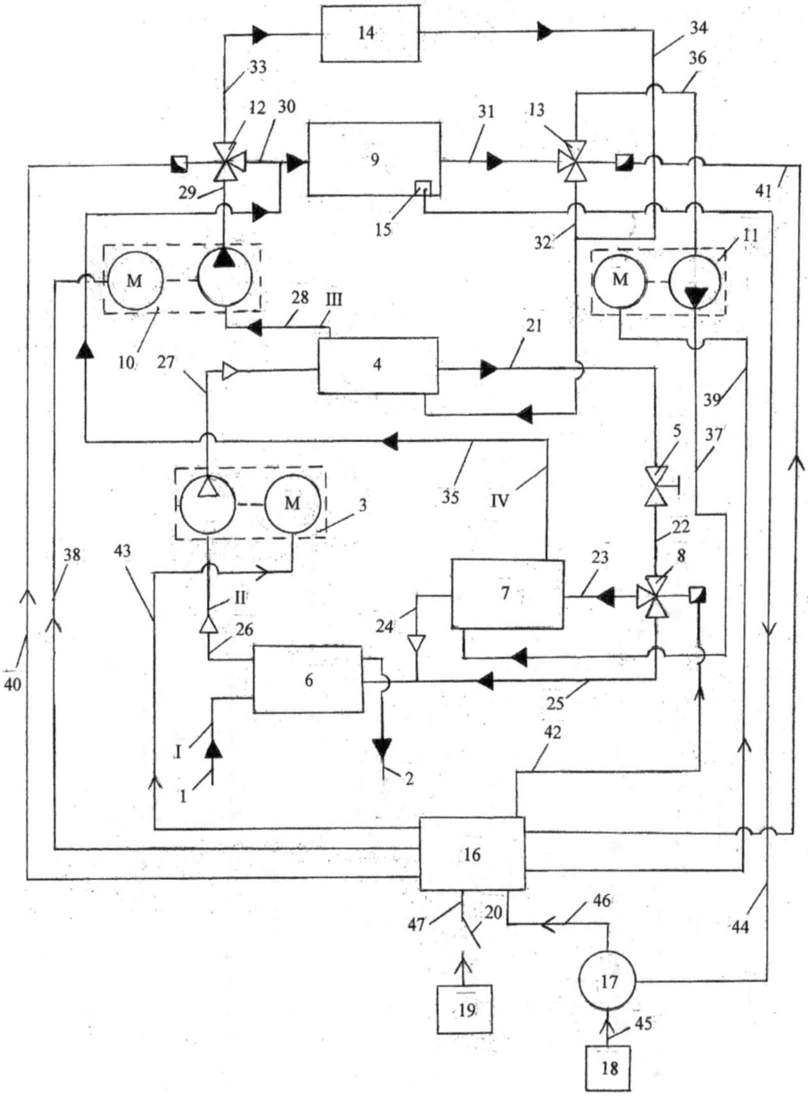

For emergencies the device for direct conversion of thermal energy in electric with use of the electric regulator with a firm filler and an electric heater is developed.

The main elements of this device are a gas heating boiler, thermoelectric generator modules (TM), heating circuit, electric regulator, consumer, battery and automation elements (Fig. 12).

Figure 11. Schematic diagram of the device for regulating the relative humidity in the vegetable store: 1-mixing chamber, 2-electro-hydraulic nozzle; 3, 4-nozzles; 5-electric fan; 6-electric water pump; 7-water tank; 8-float valve; 9-filter; 10-relative humidity sensor; 11-control unit; 12-setpoint; 13-comparison unit

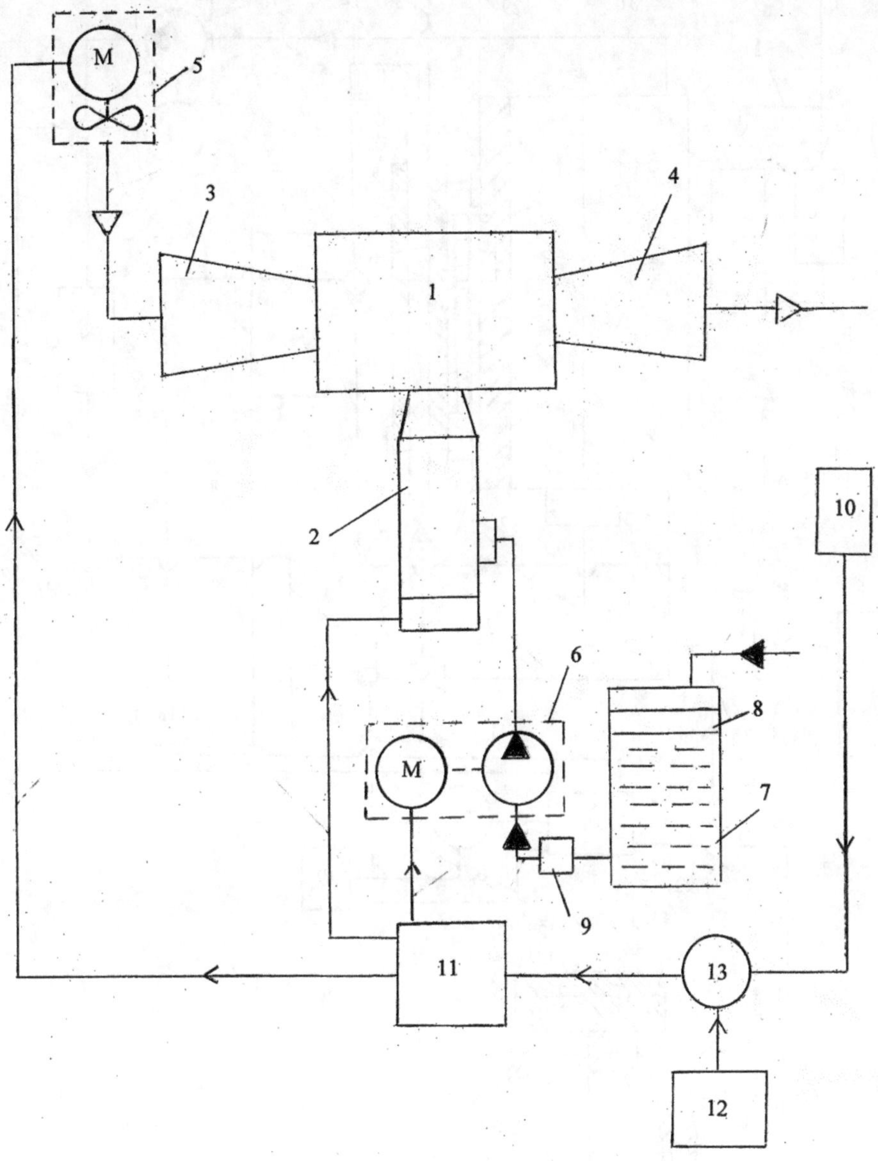

The body of the heating boiler is a hot heat exchanger, and is used to install TM and cold heat exchanger. TM are installed between cold and hot heat exchangers. The cold heat exchanger is cooled by the heat carrier of the heating circuit, and the hot heat exchanger is heated from the body of the heating boiler. When this device is turned on, the gas burner starts to work, while the hot heat exchanger and the heating circuit coolant are heated (Rezgo, & Nikolaeva, 2010).

Figure 12. Schematic diagram of the device for direct conversion of thermal energy into electrical energy

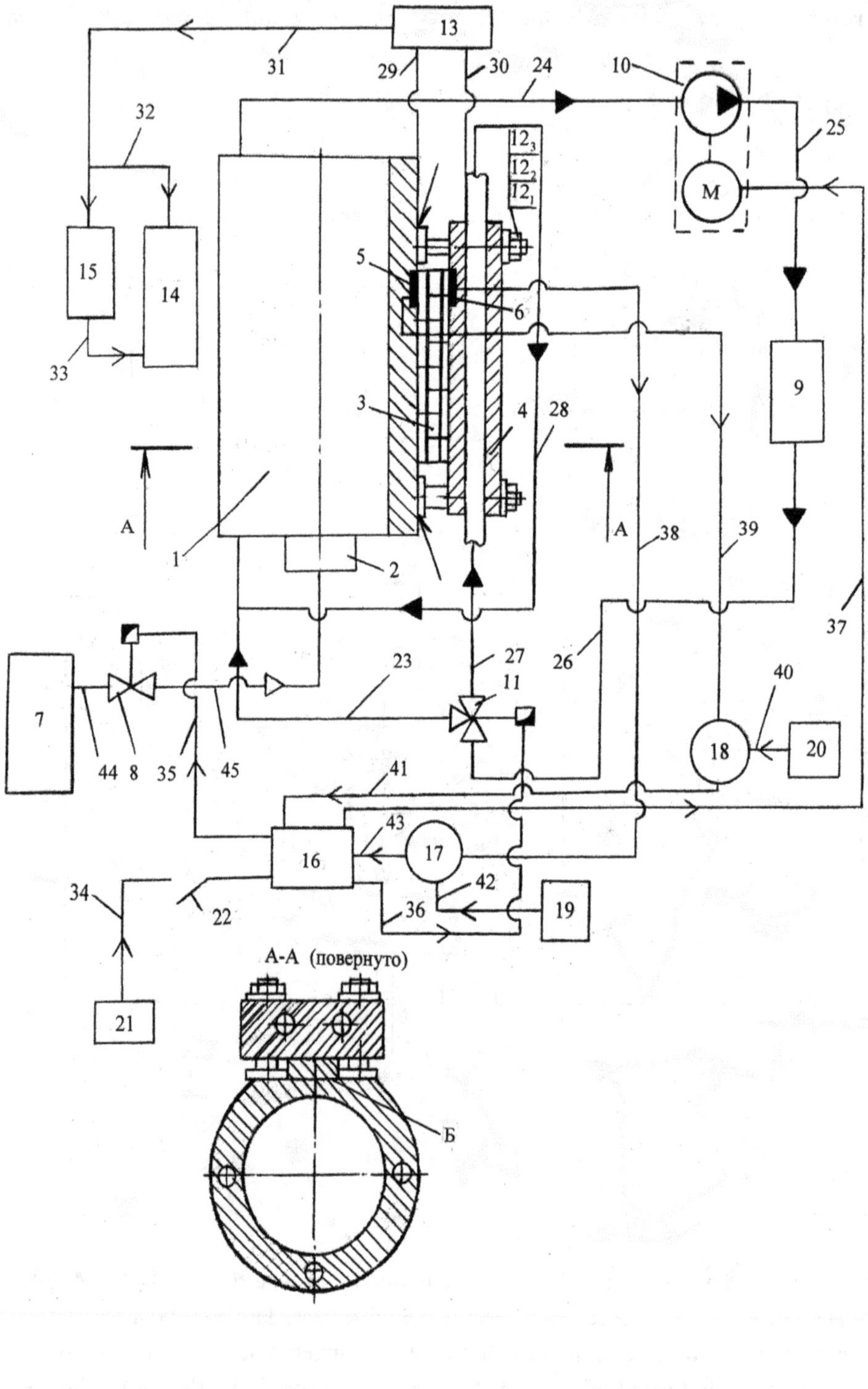

The hot heat exchanger as a result of heat exchange heats the hot junctions of the thermoelectric generator module (TEGM). The coolant of the heating circuit heats the heating object, cools the cold heat exchanger and cold junctions of the TEGM.

On the outer surface of the hot heat exchanger, a flat ground is installed, which is designed to install the TEGM and the tight connection of the heat exchangers. This design allows you to provide a clamp of TGEM to the heat exchanger with the required surface pressure, saving metal hot heat exchanger to the plug connection, consisting of a hot heat exchanger, TEGM, a cold heat exchanger.

The hot junctions of TGEM heat from the hot exchanger and the cold junctions TGEM cooled in the heat exchange with the cold heat exchanger. The temperature of the cold heat exchanger is controlled by the heating circuit coolant.

The cold heat exchanger 4 must have a high thermal conductivity, so it is recommended to manufacture it from aluminum. Depending on the temperature of the cold junctions, the electric controller regulates the flow of the heating circuit coolant.

The solenoid valve serves to supply the necessary gas to the burner 2. The battery 15 is used to accumulate electricity during operation of this device and its use by the consumer in case of non-working condition of the device.

The developed device for direct conversion of thermal energy into electrical energy works as follows.

The solenoid valve is actuated and the gas enters the burner which starts to work. The heat carrier of a heating circuit by means of the pump begins to circulate, and the electric regulator controls giving of the required quantity of the heat carrier in heat exchangers.

In this case, the coolant of the heating circuit as a result of heat exchange cools the cold heat exchanger 4, the cold junctions of the TEGM 3 and enters the electric controller 11 and the circulation process of the coolant of the heating circuit resumes.

The temperature of hot and cold junctions of the TAGM is controlled by temperature sensors and automation elements.

On the hot junctions of the TEGM, heat is absorbed from the hot heat exchanger, and on the cold side the heat is removed by the heat carrier of the heating circuit minus the electricity received on the external load.

On the external load in the voltage Converter 13, a voltage equal to the thermoelectric power station is created, minus the voltage drop and internal resistance, electricity is supplied to the consumer 14 and to the battery 15 and electricity is accumulated. The received electricity in the battery 15 is used by the consumer 14 when the device is not working.

The amount of electricity received depends on the number of TAGS, the temperature difference between the junctions. In this regard, the consumer himself determines the required number of TEGM.

SOLUTIONS AND RECOMMENDATIONS

It is known that thermoelectric cooler-heater has significant advantages:

- The possibility of obtaining cold and heat, using the Peltier effect, in the absence of moving parts and the refrigerant (Patent 52468) (Selivanov, Timofeev, Chekmarev et al., 2006)
- Versatility, that is, the ability to transfer the thermoelectric device from the cooling mode to the heating mode by reversing the DC
- Reliability and long service life

These properties make it possible to create a high-speed temperature control device that provides an increase in the efficiency of automatic control of a given temperature when storing potatoes in the autumn-winter-spring time.

Thanks to the developed design of the system, it is possible to maintain the temperature regime of storage of vegetables, which leads to a decrease in their losses during long-term storage.

Thus, the system of maintaining the temperature regime of the vegetable store with the use of a heat pump with modernized electric regulators produces heat, and through the use of an additional evaporator – cold, which is used in the warm season, as a result of which the optimal temperature in the vegetable store is maintained throughout the year.

The following advantages can be distinguished in the upgraded heat pump:

- Cooling and heating of the vegetable store is carried out to maintain the temperature during long-term storage
- Economy – the cost of production of 1 kW / h of thermal energy is 3-5 times lower than other sources
- Energy saving-the system allows to produce heat and cold using renewable energy-low-potential source of soil energy, while not using traditional energy sources
- Safe and eco-friendly – uses no toxic, polluting, flammable and explosive cooling and coolants and refrigerants
- Reliability-the upgraded heat pump has a minimum of moving parts with a high service life

The developed energy-saving systems based on the heat pump (Patent 109507, 2011 and Patent 52468, 2006) provide vegetable store, which supports the main optimal parameters of the microclimate: temperature, relative humidity.

Thus, the developed device for regulating the relative humidity in the vegetable store will maintain the necessary parameters of humidity in the vegetable store and ensure long-term storage of products with minimal losses until a new crop.

Thus, the device for direct conversion of thermal energy into electrical energy can be used in boilers for heating vegetable stores in the autumn and winter and for electricity.

FUTURE RESEARCH DIRECTIONS

It is known that the heat pump evaporator may freeze during operation. The authors estimate that the required amount of heat energy for heating an object of 100 m^2 is 12.76 kW. In order to produce 1 kWh of energy it is necessary to freeze 10.75 kg of water. It was found that when water with a mass of 1 kg freezes, 334.94 kJ of energy is released. The resulting energy of the water-ice phase transition is planned to be used in agriculture for heating and cooling, and this task is a promising topic for future research. Also, we are planning to use electro hydraulic shock (Belov, 2018).

CONCLUSION

The developed energy-saving systems based on the heat pump with modernized electric regulators allow to maintain the temperature regime of the vegetable store by using solar energy, low-potential and artificial energy sources and an absorption refrigeration machine, throughout the year. Absorption heat converter has a simple design, and electricity is used only to drive pumps and ensure the functioning of the automation elements. The temperature regime is achieved by using a thermoelectric cooler-heater and electric heater.

Designed humidifier allows you to maintain the necessary relative humidity (85-95%) in the room for long-term storage of potatoes.

Developed with the use of thermoelectric generator of the electric regulator, the direct converter of thermal energy into electrical energy allows to obtain electricity in emergency situations and use it in the converter of LPSE.

REFERENCES

Abdalla, G. (2015). Performance Characteristics of Automotive Air Conditioning System with Refrigerant R134a and its alternatives. *International Journal of Energy and Power Engineering*, *4*(3), 168–177.

Atkinson, L. K. (2000). *Single pressure absorption heat pump analysis* [dissertation]. Georgia Institute of Technology.

Belov, A. A. (2018). Modeling the assessment of factors influencing the process of electro-hydraulic water treatment. *Vestnik Ngiei*, *11*, 103–112.

Berzan, V. P. (2011). Aspects of the problem of stimulating the introduction of heat pumps. Problems of regional energy, 1, 91–94.

Besekersky, V. A., & Popov, E. P. (2004). *Theory of automatic control systems*. *Profession*, *4*, 752.

Briganti, A. A. (2001). *Residential Heat Pumps*. Moscow. *ABOK*, *5*, 24–30.

Bulatov, U. N., Krukov, A. V., & Hung, C.Z. (2014). Automatic controllers for distributed generation plants. Bratsk: Bratsk State University.

Chugayeva, V. I. (2008). Patent 2187888. Russian Federation.

Ershova, I.G., Ershov M.A., Poruchikov D.V. (2017). Justification of the regulation of the transmission of a low-potential energy source in a system based on a heat pump. *Journal Fundamental foundations of mechanics*, *2*, 32-33.

Gorshkov, V. G. (2004). Heat pumps. Analytical review. In Handbook of industrial equipment.

Kalinin, M. I., & Kudryavtsev, E. P. (2006). Patent 2292000. Russian Federation.

Kalnin I.M., Savitsky I.K. (2000). Heat pumps: yesterday, today, tomorrow. *Refrigeration equipment*, *10*, 2-6.

Kireev, V. V., Lazeev, N. A., & Stepanenko, P. P. (2003). Saving energy resources through the use of natural cold. Storage and processing of agricultural resources, 10, 10-13.

Kolchin, N. N., & Fomin, S. L. (2006). *Potato storage: State and development prospects.* Potatoes and vegetables. *KARTO and OV*, *1*, 28–31.

Kurgan, V. P., & Pankin, A. A. (2016). Optimal speed control electromechanical positioning system. Bulletin of the Samara State Technical University, 1(49), 116-121.

Kuznetsov, A. I., & Spiridonov, V. T. (2009). The state and prospects for the development of potato in Chuvashia. In *Proceedings of the scientific-practical conference "Prospects for the innovative development of potato."* Cheboksary: PMC CR "Agro-innovations.".

Levin, M. I. (1999). Current state of the problem of diesel automatics in foreign practice and domestic experience. *Dvigatelestroyeniye*, *4*, 28–31.

Martin, S., & Katja, W. (2019). BWP Market Figures 2018: Sustainable growth with upward air, a clear signal for politics. Retrieved from https://www.waermepumpe.de/fileadmin/user_upload/2019-01-28_BWP_Absatzzahlen_2018_fin.pdf

Naer, V. A., & Garachuk, V. K. (1982). Theoretical bases of thermoelectric cooling. *Energy*, 120.

Nakoryakov, V. E., & Elistratov, S. L. (2007). Ecological aspects of the use of vapor compression heat pumps. *Energy*, *4*, 76–83.

Pshechenkov, K. A., & Stockings, B. A. (2008). *Tuber ozonation reduces storage loss, improves product quality and increases yield. KARTO and OV*, *3*, 7.

Ray D., McMichael D. (1982). Heat pumps. *Energoizdat*, *224*.

Rezgo, G. Y., & Nikolaeva, M. A. (2010). *Introduction of innovative storage technologies as a way to solve the problem of ensuring food security. Food Industries*, *4*, 35–37.

Rodionova, A., V., Borovkov M., S., Ershov, M., A. (2012). Justification of the selected frequency of electromagnetic radiation in physical prophylaxis of harbols. *Niva Povolz'A*, *1*, 108-110.

Rymkevich, A. A., Kostyrya, A. M., & Kachkin, A. A. (2009). Patent 2350850.

Sage E.P., & White C.S. (1982). Optimal control systems. *Radio and communication*, *392*.

Selivanov, M. I., Timofeev, V. N., & Chekmarev, G. E. (2006). Patent 52468. Russian Federation.

Sircov, D. V. (2009). Chuvashia, the prospects for the development of potato. *Agrotrade*, *1*, 8–11.

Syromyatnikov, V.F. (1989). Adjustment of automation of ship power plants. *Shipbuilding*, *352*.

Timofeev, N., Timofeev, A. V., & Timofeev, D. V. (2006). Patent 118406. Russian Federation.

Timofeev, V. N., Chekmarev, G. E., & Galkina, N. A. (2002). Patent 31637. Russian Federation.

Vasiliev, A., Ershova, I., Belov, A., Timofeev, V., Uhanova, V., Sokolov, A., & Smirnov, A. (2018). Energy-saving system development based on heat pump. *Amazonia Investiga*, *17*, 219–227.

Vasiliev, G.P. (2002). Economically reasonable level of thermal protection of buildings. *Energy saving*, *5*, 54-57.

Vasiliev, G. P., & Krundyshev, N. S. (2002). Energy-efficient rural school in the Yaroslavl region. *ABOK*, *5*, 34–38.

Vasilieva, I. G., & Timofeev, V. N. (2006). Patent 100873. Russian Federation.

Vasilieva, I. G., & Timofeev, V. N. (2013). Patent 123909. Russian Federation.

Vasilyev, G. P. (2006). *Heat and cool supply of buildings and structures using low-potential thermal energy of the surface layers of the Earth*. Moscow: Red Star.

Vasilyeva, I. G. (2006). Patent 117256. Russian Federation.

Vasilyeva, I. G. (2006). Patent 109507. Russian Federation.

Vasilyeva, I. G. (2010). Improving the efficiency of storage of agricultural products at catering facilities. *Food Industries*, *8*, 19–21.

Vasilyeva, I. G. (2012). Analysis of ground heat exchangers to maintain the temperature regime in vegetable storehouses. In *Proceedings of the international scientific-practical conference "Actual issues of improving the technology of production and processing of agricultural products: Mosolovskie readings"* (pp. 137-138).

Vasilyeva, I.G., Novikova G.V., & Timofeev V.N. (2011). Mechanization of potato storage processes. *Agrarian Science for Agriculture*, *2*, 90-94.

Vasilyeva, I.G., Novikova G.V., Timofeev V.N. (2011). Improving the efficiency of storage of potatoes at catering facilities. *Bulletin of the International Academy of Refrigeration*, *4*, 27- 29.

Vasilyeva, I. G., Novikova, G. V., & Timofeev, V. N. (2012). Improvement of the electronic three-way valve in the heat pump installation of the potato storage. In *Proceedings of the VIII All-Russian Scientific and Practical Conference of Young Scientists, Postgraduates and Students "Youth and Innovation"* (pp. 205-208).

ADDITIONAL READING

Aksenov, V. I., Bubnov, N. G., Klinova, G. I., Iospa, A. V., & Gevorkyan, S. G. (2010). Phase transformations of water in frozen soils under the influence of cryopegs. *Science*, *1*, 40–51.

Doronin, Y. P. (1997). Growth and melting of sea ice. Sea ice. Collection and analysis of observational data, physical properties and forecasting of ice conditions: a reference guide. Gidrometeoizdat, 107-125.

Frolova, I. Y., & Gavrilo, V. P. (1997). Sea ice Collection and analysis of observational data, physical properties and forecasting of ice conditions.

Johnson J.B., Metzner R.C. (1990). Thermal expansion coefficients for sea ice.

Makshtas, A. P., Andreas, E. L., Svyashchennikov, P. N., & Timachev, V. F. (1998). *Accounting for clouds in sea ice models*. U.S. *Army CRREL Rep*ort.

Nazintsev, Y. L., & Panov, V. V. (2000). *Phase composition and thermophysical characteristics of sea ice*. St. Petersburg: Gidrometeoizdat.

Permyakova P.P., & Romanov P.G. (2000). *Heat and salt transfer in frozen unsaturated soils*. Yakursk: YF Publishers.

Chapter 4
Gas Turbine Power Plant of Low Power GTP-10S

Valentin Gusarov
Federal Scientific Agroengineering Center VIM, Russia

Leonid Yuferev
Federal Scientific Agroengineering Center VIM, Russia

Zahid Godzhaev
Federal Scientific Agroengineering Center VIM, Russia

Aleksandr Parachnich
Federal Scientific Agroengineering Center VIM, Russia

ABSTRACT

Currently, there is an increase in the use of gas turbines. Today they are used in the energy sector: aviation, armed forces, and the navy. The introduction of a new manufacturing technology developed by the authors will make it possible to manufacture cheap and reliable installations and thus ensure an exceptional position on the Russian market for goods and technologies, and taking into account the use of intellectual rights, abroad. The scientific novelty of the sample is the method of calculating small engines with a centrifugal compressor, a centripetal turbine and a combustion chamber with a negative thrust vector of the air flow. It is shown that the developed microgas turbine cogeneration power generator consists of a microturbine engine with a periphery, a free power turbine necessary for the selection of mechanical power, a high-speed electric generator with permanent magnets, an electronic power conversion system, exhaust heat energy recovery system and an automatic control system.

DOI: 10.4018/978-1-5225-9420-8.ch004

INTRODUCTION

For the harmonious development of the world, economy requires a balanced development of energy. World energy must be reliable with a focus on promising developments.

The modern development of scientific and technological progress, directly related to the development of the energy complex, allows a modern person to use his usual civilization achievements, which provide a high and comfortable standard of living for people living in countries with developed infrastructure.

The organization of trouble proof power supply of agricultural facilities in remote areas is fraught with great difficulties, due to significant investments in the construction and maintenance of low-voltage distribution networks.

As a result of the low payback of low voltage distribution networks, it is proposed to organize the power supply of agricultural consumers using the technology of building micro grids based on renewable energy sources (RES). For reliable operation of micro networks, it is necessary to use generating equipment that is capable, if necessary, of ensuring uninterrupted long-term power supply to agricultural facilities. This is an urgent need for the power supply of autonomous consumers.

Literature Review

At present, geographically distributed small power plants are being created in Europe, America, Australia and even Africa in non-electrified territories in the immediate vicinity of the consumer, which can be grouped together in a network and serve several consumers who live in isolation. The benefit of using them is associated not only with reduced fuel losses in plants operating in low-efficiency modes, but also with low maintenance costs. As a result of increasing the accuracy of energy consumption metering devices, it increases the accuracy of automated monitoring of power plant parameters, it becomes possible to use modern energy complexes capable of operating in an automatic mode, with a more reliable security system with less investment (Gusarov, 2004).

The practice and experience of many advanced countries shows that for reliable power supply in an autonomous electric network there must be both generators working on renewable energy sources and generators working on traditional hydrocarbon fuel. The main sources of electricity at autonomous facilities should be gasoline and diesel generators, micro gas-turbine generators operating mainly on hydrocarbon fuels, as well as small hydropower plants, biofuel generators, etc. Generators operating on renewable energy sources reduce the cost of electricity and reduce the need for continuous use of fuel generators, but are not the main and guaranteed sources of energy (Gusarov, 2016).

Of the traditional fuel generators on the world market, the most preferred is a micro gas turbine unit. Which can be used as a fuel, both liquid and gaseous type of fuel. It also has the greatest time to failure, and can work for a long time without shutting down in automatic mode.

The possibility of obtaining a large amount of thermal energy from gas turbine power plants implies a faster payback of the project (Adamavichus, 2016).

Gas turbine power plants used as heat-electric equipment for high-power thermal power stations and district heat-electric stations are economically justified. Today, gas-fired power plants have a low unit cost of fuel and equipment, and the relatively low cost of construction and installation and, subsequently, low operating costs, most favorably affect consumer interest in purchasing such a power plant (Gusarov, 2017).

Excessive heat energy makes it possible, with the use of absorption chillers, without the cost of electricity, to adjust the air conditioning of residential and industrial premises. Coolant cooled in this way can be used in various industrial processes. The production of electricity, heat and cold from one installation is called three generation.

High efficiency of use of gas turbine installations provides a wide operating range of changes in electrical loads from 1 to 115%.

The main advantage of using gas turbines is the low content of harmful emissions in the range of 9-25 ppm, which allows them to be placed directly in places where people live. This parameter is better than that of piston power plants - their closest competitors (Kharchenko, 2013).

The use of gas turbine power plants, as compared with piston gas, allows to reduce capital expenditures on the use of exhaust gas cleaning systems.

Gas turbine power plants have minor vibrations and noise in the range of 65–75 dB (which corresponds to the noise level of a household vacuum cleaner). Special sound insulation to reduce noise is generally not applicable (Gusarova, 2013).

In today's Russian energy market, microturbines occupy a more than modest niche. From a small number of offers, the overwhelming majority comes from foreign manufacturers (Pilavachi, 2000).

The company "MTT Micro Turbine Technology BV" (Netherlands) created the micro-gas-turbine installation "Ener Twin", which has 3 kW of electrical and 15 kW of thermal power. According to its performance, this is a micro CHP station; it is designed for use as a heating boiler and a home power plant in households and in small enterprises. The design of the installation focused on low cost, reliability, low noise and low operating costs (Mago, 2012).

The heart of EnerTwin is a modified turbo-compressor installation from a car in combination with a high-frequency electric generator. The turbine operates at 180-240 thousand revolutions per minute, and at the same time produces on demand of the system 1-3 kW of electrical power and 6-15 kW of thermal energy with a total efficiency of up to 90% for at least 30,000 hours (Pilavachi, 2002).

For external control, EnerTwin has Open Therm, RS-485, 0-10V and Modbus interfaces, with which you can remotely change the system parameters, as well as there are internal automatic control systems that automatically adjust to peak loads. The noise level of Ener Twin is 55 dB (a), which is usually not louder than similar generators with an internal combustion engine. Fuel is currently used only natural gas. Now, together with the Scientific Research Institute of Oil and Heat (IWO), the issue of using fuel oil is being worked out, as well as the version for liquefied gas (LPG).

With the successful passing of tests, it is expected to enter the market at the cost of Ener Twin system at 12500-13000 Euro (which is not a large amount for European consumers), taking into account the fact that the lifetime of the installation will be at least 30 thousand hours.

Table 1 presents the technical characteristics of micro CHP.

Capstone microturbines occupy a dominant position in the Russian market of microgas turbines. The firm "Capstone turbine corporation" produces micro gas-turbine installations of models Capstone C200, Capstone C65, Capstone C30 and Capstone C15, with an electrical capacity of 200, 65, 30 and 15 kW, respectively.

The design of these microturbines allows working on various types of fuel: natural gas of high and low pressure, biogas, kerosene, diesel fuel, biodiesel fuel, liquefied gas, low-calorific gases, fuels with high sulfur content.

Table 1. Technical characteristics of the micro-CHP "EnerTwin"

N°	Parameter	Value
1	Electric power (max / min)	3,0 /1,0 kWt
2	Heat output (max / min)	14,4 /5,0 kWt
3	Electric efficiency (max / min)	15 /10%
4	Maximum total efficiency (depends on the parameters of the heating system, for example, the temperature of the return pipeline)	87%
5	Rotor rotation speed (max / min)	240 / 180 thousand rpm
6	Gas consumption	1,87 /0,84 nm^3/ hour
7	Fuel	natural gas
8	Parameters of the heating system (supply / return pipe)	80/60 °C
9	Noise, at a distance of 1 m	55 dB
10	Geometrical dimensions	970 x 610 x 1120mm
11	Station weight	225 kg
12	Chimney diameter	100 mm
13	Output voltage	230 V/50 Hz

A microturbine engine consists of only one moving part - a rotating shaft, on which an electric generator, a compressor and the turbine itself are located. The unit does not use gearboxes or other mechanical drives. A unique design feature of the engine is the use of air bearings, due to which a record speed of rotation of the shaft is achieved - 96,000 revolutions per minute. This innovation makes it possible to abandon the use of oil, the high consumption of which in other types of equipment is a significant part of operating costs. Low operating temperatures reduce the level of emissions of nitrogen oxides, so that the level of emissions of CO and NOx does not exceed 9 ppm, which allows microturbines to be attributed to one of the most environmentally friendly sources of energy generation. Another unique feature of the Capstone turbines is the layout of the main assembly units. The compact case contains a compressor, a combustion chamber, a heat exchanger, a turbine itself and permanent magnets of an electric generator. The generator is cooled by the oncoming air flow, which eliminates the need to organize a liquid cooling system and increases the reliability and efficiency of the equipment during operation. Thanks to the use of an air-to-air heat exchanger (heat exchanger) in the design of a turbo engine, microturbines have a high electrical efficiency for turbine generators - up to 35%. The recuperator uses the heat energy of the exhaust to preheat the air in the combustion chamber, which reduces the amount of fuel consumed almost twice (Iacoboni, 2012).

Table 2 presents the main technical characteristics of microturbines.

When comparing the technical characteristics of the Capstone C30 and Capstone C15, it can be seen that the geometrical dimensions and weight of the installations are almost equal, with the electric power different by half. This indicates that the engineering solutions for reducing the geometric parameters of the engine of the Capstone C15 model, while maintaining the specified efficiency, were not found. Essentially, the model Capstone C15 is the same model Capstone C30, with the same engine, generator, frame, only with a limited electrical output of 15 kW.

The market value of a single Capstone C30 or Capstone C15 in the Russian Federation is $ 80,000.

Table 2. Capstone microturbine specifications.

Microturbine Parameters	Capstone C30	Capstone C65	Capstone C200	Capstone C1000
Electric power, kW	30	65	200	1000
Electricity efficiency, %*	26±2	29±2	33±2	33±2
Total efficiency when using heat, %	80	80	66-80	66-80
Operating voltage range, B	380-480	380-480	380-480	380-480
Maximum current in phase, A	46	100	275-290	1450 (400 V)
Current frequency, Hz	50	50	50	50

At the Russian production enterprise JSC SKB Turbina, a model range of MSTI power plants with a capacity from 60 to 200 kW is being produced [4]. These plants can be used as cogeneration, by connecting additional heat exchange equipment. The fuel is natural gas. Table 3 presents the technical characteristics of MGTU-100

On the Russian market today, foreign companies are represented, such as: Alstom, General Electric, GE, Kawasaki, MANTURBO AG, Mitsubishi Heavy Industries, Rolls-Royce, SIEMENS, Solar Turbines, Turbomach, Microturbine Power Stations —Capstone - Calnetix.

Table 3. Technical characteristics of MGTI-100

Options	Value
Equivalent power, kW	not less than 383.5
Electric output power, kW	100,0
Electric power at generator terminals, kW	118,5
Electric efficiency, %	not less than 25,0
Specific power, kW / kg	not less than 0,17
Nominal output voltage, V	380 (± 10%)
Nominal frequency of alternating current, Hz	50 (± 0,4%)
Rotor speed, n, rpm	65000
Air consumption, kg / s	1,039
Fuel consumption, g / s	9,035
Degree of pressure increase, $p_к$	5,0
Turbine expansion ratio, $p_т$	4,1
Temperature in front of the turbine, °C	897(±50)
MSTU exhaust temperature, °C	358,5(±35)
Hourly fuel consumption, kg / h	32,5
Specific fuel consumption, m^3 / (kWh)	0,155
Assigned resource hours	60000
Overall dimensions (by container), mm (Length × Width × Height)	3056x1574x1736
Weight, not more, kg	2500

Gas turbine units of the same capacity as piston gas, diesel or gasoline, have a much lower specific weight. In accordance with building codes, it is allowed to place low-power gas turbine power plants on roofs or technical floors of residential and public buildings. This technical capability is an important financial factor in urban development, because it saves costly square meters and helps solve the problem of locating an autonomous heat and power plant.

Gas turbine power plants have high technical reliability and ease of maintenance and operation. Some gas turbine power plants have been continuously operating for 5-7 years (Kulagin, 2012).

Manufacturers of modern gas turbines repair plants without transportation to the manufacturer, as they bring a replacement turbine or combustion chamber in advance, which significantly reduces the time to complete the overhaul to 4-6 working days and reduces repair costs.

The advantage of gas turbine installations is their long service life, up to 200,000 hours, overhauls are carried out after 30000-60000 hours. In the operating mode, gas turbine installations consume engine oil significantly less than piston gas.

Gas turbine power plants can operate in any climatic conditions. Autonomous power supply in remote areas is economically justified by eliminating the costly construction of power lines. In places with developing infrastructure, gas turbine power plants reduce the shortage of electrical and thermal energy.

The use of gas turbine power plants in block-modular systems allows for a smooth increase in power plant capacity. The creation of modular gas turbine plants consisting of unified power units and a common control system allows for a short period of time to increase the electrical power with the lowest capital and time costs (Gusarova, 2013).

The high level of automation of the control system of the gas turbine power plant, allow operation without the direct presence of maintenance personnel. The operation of a gas turbine power plant can be monitored remotely, via various television and radio communication channels. The design provides for the presence of integrated systems of automatic protection and fire suppression, necessary for emergency situations.

Overwhelmingly, the technology and design of gas turbines are based on the technology of aircraft engines. As in aircraft engines, gas turbine power plants contain a multi-stage compressor that compresses atmospheric air and supplies it under high pressure to the combustion chamber, where it mixes with fuel. During the combustion of the fuel-air mixture, thermal and kinetic energy is released. Kinetic energy is converted into mechanical work, due to the supply of a jet of an expanding working fluid with a guide vane onto the wheels of a free turbine.

Part of the kinetic energy is spent on the compressor. The rest of the energy is transferred to the rotation of the generator. This work is a useful work gas turbine power plant. After the turbine, the exhaust gas enters the utilizer for the selection of thermal energy (Gusarova, 2013).

Gas turbine engines have the highest power density per unit of weight among all heat engines and can reach 6 kW / kg (Smirnov, 2016).

Currently, there is an increase in the use of gas turbines in Russia and abroad. Nevertheless, today they are used only in the energy sector: aviation, armed forces and the navy. In FSSI FSAC VIM there is a development of a micro gas-turbine power plant (MGTPP) for the power supply of industrial and agricultural facilities. The main criterion for its development is the use of existing components on the market produced by the Russian industry. The use of micro gas-turbine power plants in the Russian industry is a new step towards improving the efficiency of the country's production complex. However, there is no single method for calculating engines of small gas turbines, where, for example, turbo compressors of an internal combustion engine are used. This development proposes a new technology for manufacturing micro gas-turbine power plants of low power (Gusarov, 2013).

Micro gas-turbine stationary power plants with a capacity of up to 30 kW have never been manufactured in Russia and are not manufactured to this day, Russian manufacturers have no experience in manufacturing such plants. The introduction of a new manufacturing technology developed by the author will make it possible to manufacture cheap and reliable installations and thereby ensure an exceptional position on the Russian market of goods and technologies, and taking into account the use of intellectual rights, also abroad. The developed methods and technologies for the production of micro gas-turbine power plants provide for technical inspection at least once a year. Installations provide multifunctional use and thereby increase the efficiency of application of process equipment in operation. Therefore, the organization of production according to the developed technologies and the development of innovative components and assemblies for the domestic production of micro gas-turbine power plants is an important task. The author has developed and manufactured several samples of micro gas-turbine power plants with a capacity of up to 10 kW, for which a declaration of compliance with the EAEU has been obtained (Valitskas, 2013).

BACKGROUND

The scientific novelty of the sample is the method of calculating small engines with a centrifugal compressor, a centripetal turbine and a combustion chamber with a negative thrust vector of the air flow.

In this paper, an analysis of the existing design solutions of gas turbine power plants (GTPP) and MGTPP was carried out, as a result of which a radial centripetal turbine has been found to be more suitable for the following reasons:

- A turbocompressor of this type is particularly suitable for small gas turbine engines, since by constructively connecting the turbine to the impeller of a compressor with the same outer diameter, the design can be made more compact; - the rotor of the radial centripetal turbine, unlike the axial-turbine rotor, consisting of a disk and individual blades, can be made more cheaply from forging or by precision casting
    - A radial turbine of small size can theoretically be even more efficient than the corresponding axial turbine due to the significantly smaller influence of the Reynolds number on its characteristics and, therefore, scale;
    - The radial turbine has a higher strength and reliability in operation compared to an axial turbine;
    - The blades of the radial turbine are practically insensitive to the action of small solid particles trapped in gaseous products of combustion, while the ingress of solid particles on the blades of the axial turbine can cause serious erosion of the blades;
    - The radial turbine has higher differential pressure in the stage than the axial-type turbine. For differential pressure in the stage, about 3 or 4, can be used 2 or 3-speed radial turbine;

- A radial turbine with adjustable nozzle vanes can maintain its calculated (maximum) efficiency in a relatively wide power range and thus have significantly better performance at partial loads than an axial-type turbine;
- Like the axial, a radial-type turbine, used as free or power, has the same characteristics of maximum torque when starting at a low speed.

It is shown that the developed micro gas turbine cogeneration power generator consists of a micro turbine engine with peripherals, a free power turbine necessary for the selection of mechanical power, a high-speed electric generator with permanent magnets, an electronic power conversion system, exhaust heat energy recovery system and an automatic control system (Gusarov, 2012).

The micro turbine engine was developed with a unique approach to maximize the use of unified assemblies and parts manufactured by Russian industry. So, as a turbine and compressor, a serial turbocharger engine was used, for which a special unique combustion chamber was developed. This development was associated with major problems that were overcome. The operation of the turbocharger is ensured by the lubrication system on the engine oil, which includes: an oil tank, an oil pump, an oil filter and an oil cooling radiator. Oil pressure of 2.5 - 5.0 kg / cm^2 is provided by a standard oil pump used in the Russian industry, the working temperature is 80 - 95 ºC and maintained with an oil radiator. It is recommended to change the oil at least every 8700 hours of operation.

On the (Figure 1) shows the kinematic block diagram of the developed MGTPP

The fuel of a micro gas turbine engine is biogas, propane-butane, methane, mine gas, synthesis gas, etc. The unique combustor developed allows you to adjust the quality of the gas-air mixture depending on the gas used or its quality. This allows you to empirically find the optimal ratio of gas to air, providing the best technical characteristics of the operating cycle of the engine (Smirnov, 2016).

A positive result for this solution can be achieved by changing the area of tangential holes in the mixer of the combustion chamber. The engine compressor provides a pressure in the combustion chamber of 0.8 - 1.6 kg / cm^2. The temperature of the working fluid at the turbine inlet is 520 - 560 °C. The increased coefficient of excess air provides the level of NO_2 emissions at a level of 0.4–0.6 ppm.

For efficient operation of a combustion chamber with a low temperature of its surface, of the order of 40 - 45 °C, at a propane-butane mixture combustion temperature of 1950 °C, basic gas-dynamic relationships have been developed that help determine the main geometric characteristics of a combustion chamber with a vortex air flow converter and its power characteristics . An engineering methodology has been developed for calculating these characteristics on the basis of fundamental hydrodynamic aerodynamics, which describes the movement of a swirling quasi-potential laminarized air flow in special combuster channels and criteria for using gas energy in the form of a whirlwind flow.

The low temperature of the working fluid at the outlet of the turbine makes it possible to use cheap low-temperature materials for the production of the casing of the combustion chamber, gas pipelines, impellers of the free power turbine, plate heat exchanger. The nominal rotor speed of the engine is 60 - 98 thousand rpm. The efficiency of the turbocharger is 70 - 73% (Gusarov, 2012).

On the (Figure 2) shows a gas turbine engine of equivalent power of 70 kW, weight 11.3 kg.

To convert the kinetic energy of the gas flow into the energy of rotation, a multi-stage radial free power turbine has been developed.

To determine the effectiveness of the free turbine, a method for calculating multi-stage radial turbines has been developed. At present, the production of MGTPP with an estimated electrical power of 30 kW, based on the TKR-NE221 turbo-compressor, has begun.

Figure 1. Kinematic scheme of MSTU. 1– starting and cooling system; 2 - compressor; 3 - combustion chamber; 4 - turbine; 5 - free power turbine; 6 - shaft cooling radiator; 7 - electric generator; 8 - hydraulic station (oil); 9 –oil cleaning filter; 10 - oil radiator; 11 - emergency gas solenoid valve; 12 - gas tap supply level; 13 - gas valve servo; 14 - ignition system, 15 - cold air duct, 16 - vortex recuperative heat exchanger, 17 - heated air duct, 18 - hot gas duct.

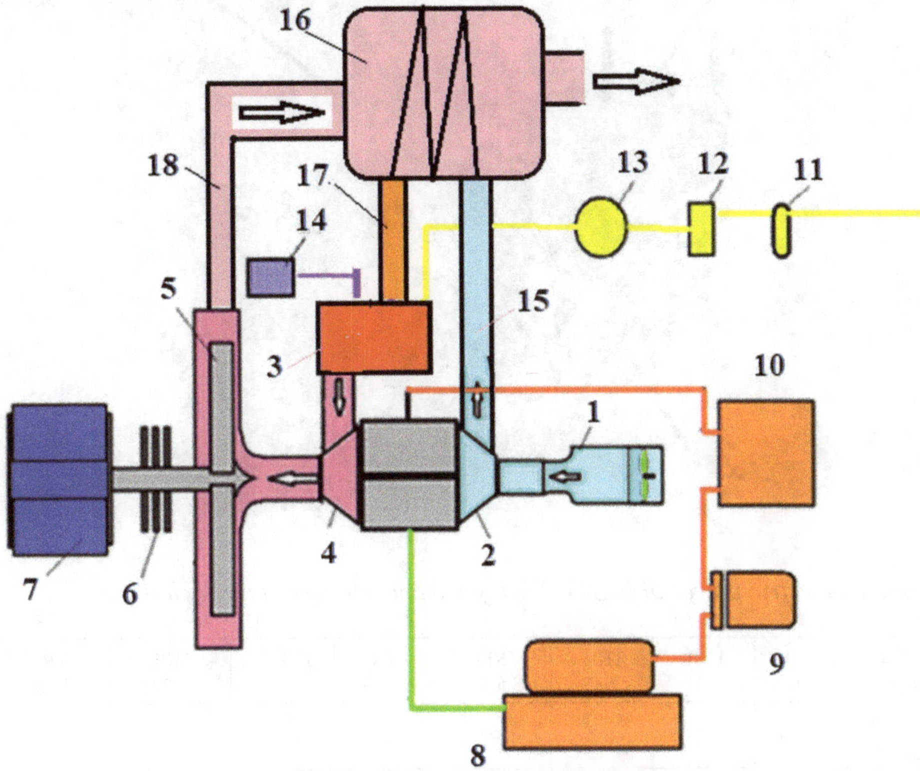

The micro gas turbine engine is designed to drive a free power turbine and an electric generator.

MGTPP consisting of a turbocharger and a free turbine is not kinematically connected with the turbocharger, and the rotation is transmitted by the gas-dynamic method.

The turbocharger is a single-stage centrifugal compressor rotated by a single-stage gas turbine. Counterclockwise rotation from the exit side of the turbine.

The paper describes the experimental studies carried out at the stage of testing the theory of using a turbocharger engine as the main unit (compressor and turbine) of the GTE for long-term operation at a speed of 96–98 thousand rev^{-1}, showed that the engine developed is stable and technical specifications are given.

Experimental studies were also carried out, when working on propane-butane, with other, more powerful turbo-compressors: TKR 7N6, TKR-NE221, and others with air flow rates from 0.14 to 0.45 kg / s. All of them showed good performance and high reliability. When working on methane, the electrical power increases by 1.3 times.

The calculated electrical power of the engines based on turbo compressors of the internal combustion engine using propane-butane mixture is given in table 4.

Figure 2. Gas turbine engine. 1 - combustion chamber, 2 - compressor; 3 - turbine; 4 - coupling.

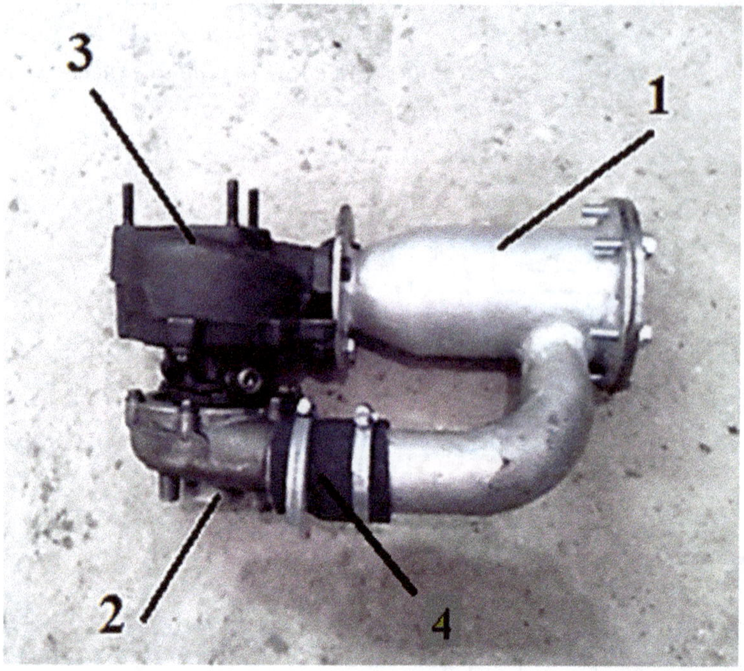

Table 4. Estimated electric power of MGTPP using various models of turbochargers.

The Name of the Turbocharger	TKR-6	TKR-7H1	TKR-7H6	TKR-8.5H1	TKR-8.5H3	TKR-9-12	TKR-NE221	TKR-11N2	TKR-K36
Productivity of the Compressor, kg / s	0.14	0.27	0.3	0.22	0.2	0.25	0.45	0.27	0.42
Max. Email Power of Installation, kW	7.3	14.2	15.8	11,6	10,5	13,1	23,7	14.2	22,1

For normal engine start, a starting and cooling system is provided, consisting of an inlet air filter, a pneumatic starter and an air valve. The air flow created by the pneumatic starter has soft contact with the engine compressor and provides the required amount of air into the combustion chamber, which is a guarantee for the engine to go to normal operation. An innovative solution is that after the engine is finished, the pneumatic starter automatically turns on, allowing the rotor to rotate to cool to the required temperature and thereby preventing the formation of carbon deposits in the sliding bearing, which in turn increases engine life (Gusarov, 2017).

The main problem of starting a gas turbine engine, made on the basis of a turbocharger engine, is the supply of air into the combustion chamber with simultaneous rotation of the rotor of the engine. Since it is technically impossible to mechanically connect to the rotor due to accurate balancing and high rotational speed, it is necessary to supply 60 g / s of air for a successful launch, and the rotor speed must be at least 16,000 rpm^{-1}, and therefore special starting device. (Figure 3) shows the designed and manufactured starting device.

Figure 3. Developed starting device of a micro gas turbine engine.

The starting and cooling system of a micro gas turbine engine with a starting compressor with an air valve involves starting the gas turbine engine by supplying compressed air from the double air intake side to the compressor. Start produced by air from the starting compressor. After the shutdown of the micro gas turbine engine, the starting compressor is re-started without fuel supply and the combustion chamber, the turbine and the rotor bearings are cooled. A device for starting and cooling a micro gas turbine engine contains a starting compressor with an air valve connected to a compressor of a micro gas turbine engine by a transition coupling. The starting compressor and the air valve are located in a double air intake, which ensures smooth transmission of torque to the engine rotor, forced cooling of the combustion chamber, turbine and rotor bearings.

The developed device is illustrated in (Figure 4), which shows the general scheme of the device to ensure the launch.

The task of the developed design is to create an engine starting system without a gearbox using a starting compressor and the possibility of cooling it after switching off.

The tasks for the GTE combustion chamber are solved in the proposed design of the launch scheme in that the air intake has two nozzles. On one is the starting turbocharger, on the other - the air valve. The starting turbocharger is used when starting and stopping the GTPP to provide the starting rotor speed of the micro gas turbine engine and cooling it after work is finished. Promotion of the turbocharger takes

Figure 4. Diagram of the start of the micro gas turbine engine

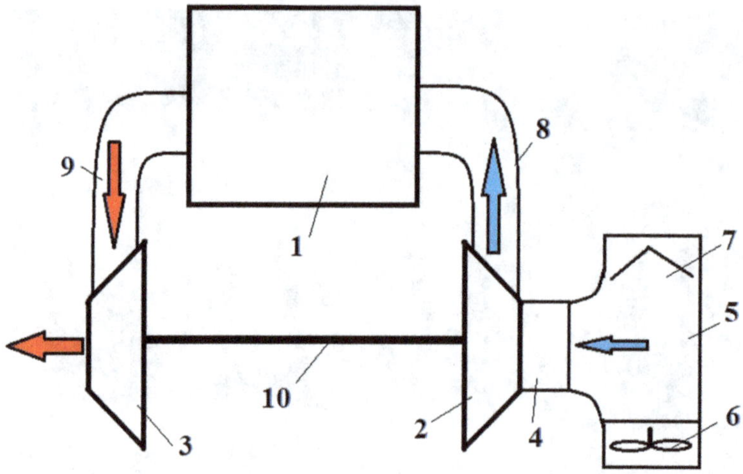

place by the air flow coming from the starting compressor to the blades of the engine compressor. The air valve allows you to achieve accurate air metering for engine start-up, automatically closes when the micro gas turbine engine starts and stops and opens during operation, due to the thrust generated by the compressor. At the moment when the rotor of the engine starts to operate, the air valve opens and the starting compressor is turned off. Next, the air supply is promoted compressor engine.

The device for starting and cooling a micro gas turbine engine with a starting compressor with an air valve consists of combustion chamber 1, compressor 2, turbine 3, transition coupling 4, double air intake 5, start compressor 6, air valve 7, cold air duct 8, hot air duct 9, rotor shaft ten.

The device works as follows. The rotor is started due to the pressure of the air mass generated by the starting compressor to the engine compressor through a double air intake and transitional coupling, with the air valve closed (Balakshin, 2015).

The micro gas turbine engine is cooled by supplying air mass through the engine compressor, a cold air duct, a combustion chamber, a turbine, a hot air duct, through a double air intake and a transition sleeve created by the starting compressor and ensuring the rotor rotates when there is no fuel supply before it cools. The absence of a rigid connection of the mechanisms rotating with high frequency of the transmitting torque is ensured by the use of a starting compressor. Forced cooling of a micro gas turbine engine is carried out without burning the fuel, by turning on the starting compressor with a gradual decrease in its rotational speed and turning it off after reaching an engine temperature of 60-80 °C, which distinguishes it from existing analogues, therefore, it meets the "novelty" criterion.

Comparison of the claimed invention with other known solutions in the field of technology has shown that identical signs on the features distinguishing the claimed invention from analogues have not been identified, and therefore it meets the criterion of "inventive step".

To carry out experimental studies of a developed and manufactured micro gas turbine engine having a high rotor speed, the problem arose of determining a sufficiently accurate rotor speed. It is impossible to connect counting and reading devices, contact and contactless with magnets, to the rotor of the turbocharger, since in any case the balancing of the rotor is disturbed. To solve this problem, a unique device was developed, allowing to determine the rotor speed of the engine.

The novelty of the development lies in the fact that the rotor of the turbocompressor, namely the fastening nut of the compressor turbine, is applied a reflective coating illuminated by a laser light emitter, the reflection of which is reflected in the light receiving device, the signal from which is fed to the reading electronic conversion device. The scheme of such a device is shown in (Figure 5).

The rotor speed measurement system shown in (Figure 6) consists of: an input nozzle of a turbocharger 1, a compressor turbine 2, a nut 3, a coupling 4, a metal tube 5, an optical fiber cable 6, a laser light emitter 7 with a light beam 8, an electron-transforming device 9, electrical line 10, on-Board power source 11, the reflector 12, the light absorbent 13.

GTE Rotor Rotation

The working engine rotor speed reaches 60-100 thousand rev / min. The compressor nut is divided in diameter and painted as follows: one half is covered with reflective paint and is a reflector, the second half is covered with light absorbing paint and is a light absorber. The metal tube containing the fiber optic cable is brought to the nut at a distance of 5-8 mm and positioned so that the emitted light from the laser light emitter hits the mounting nut and, reflecting from the reflector located on it, into which fiber optic cable is placed, through which the signal enters the electron-converting device. When you turn on the

Figure 5. TKR compressor nut with reflective and light-absorbing coating

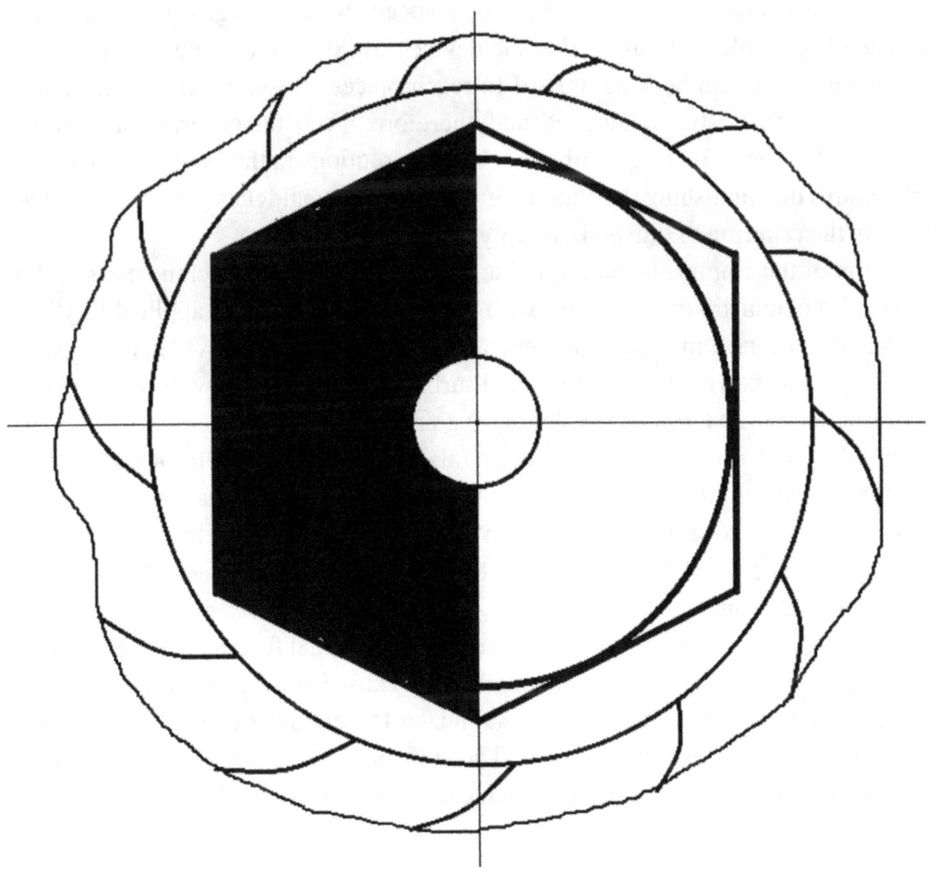

Figure 6. Diagram of the device frequency measurement system

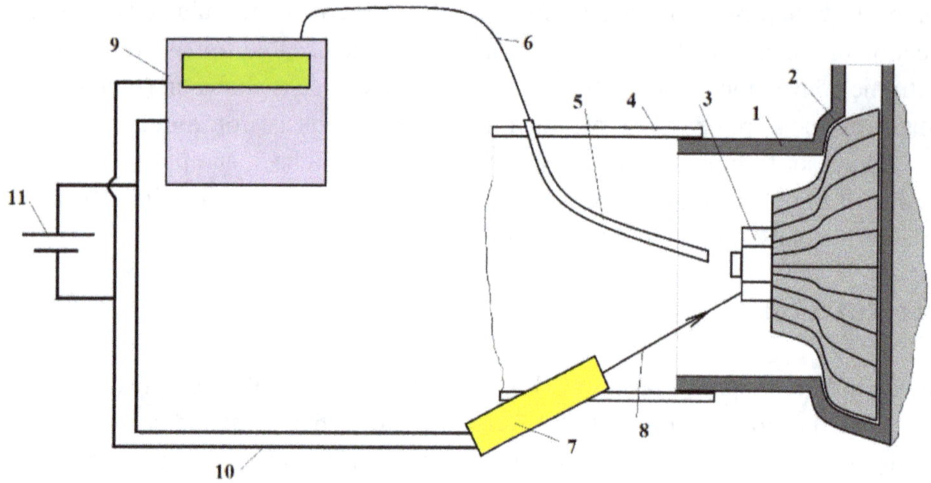

laser light emitter, the light beam falls on a rotating nut, on the light spot of which the reflector or light absorber alternately hits a rotating spot. A reflector in pulsed mode transmits reflected light through a fiber-optic cable to an electronic conversion device. A laser light emitter and an electronic conversion device are connected to an onboard power source via an electrical line.

Studies of instruments for measuring the rotational speed above the high-speed machine parts on the market have made it possible to establish that the device developed and manufactured differs from the known analogues in that it can be used in hard-to-reach places, closed cavities, ducts, gearboxes, etc. This distinguishes it from existing analogues, and, therefore, meets the criterion of "novelty."

Comparison of the proposed design with other known solutions in the field of technology has shown that identical features distinguishing the claimed invention from analogues have not been identified, and therefore it meets the criterion of "inventive step".

The application of the claimed invention in hard-to-reach places of machine parts and mechanisms of various areas of the industry provides him with the criterion "industrial applicability".

Along with axial turbines, in which the general direction of steam flow closely enough coincides with the direction of the axis of the turbine, radial turbines, where the flow of the working gas occurs in the plane of rotation of the turbine impeller, have gained some acceptance.

Along with radial centrifugal turbines, there are also radial centripetal turbines in which the working gas flows from the center to the periphery. If the flow of the working gas at the exit from the radial stage then goes to other stages and has a direction along the axis of the turbine, then such turbines are called radial-axial turbines. Unlike the Yungstrom turbines, the developed radial turbine stage with such a scheme has fixed guide vanes.

However, unlike the axial stages, where the action of centrifugal forces on the working blades leads to their stretching, in the radial stage the centrifugal forces cause bending stresses in the working blades. Thus, the working blades of the radial foot are subjected to bending stresses, both from the dynamic effect of the gas jet and from centrifugal forces. The rims of the blades of one disk are included in the gaps between the rims of the other, thanks to which each crown of the blades is simultaneously working for its own disk and guides for another. The bending moment created by centrifugal forces is determined by the peripheral speed and size of the blades.

The free turbine (FT) is designed to convert the energy of the gas produced by the gas generator into power on the drive shaft.

A free turbine has been developed - three-stage, radial, reactive.

The free turbine is designed as a separate unit (module) and, if necessary, can be replaced without replacing the gas generator.

The free turbine (Figure. 1) consists of the following main units:

- Nozzle apparatus
- Turbine rotor
- Support of the turbine and generator

The gas supply from the gas generator to the free turbine is carried out through an adapter that is the body part of the nozzle apparatus.

The main power element of the free turbine is the support of the turbine. The rotor shaft ST relies on two ball bearings (front and rear) located in the generator covers. The axial forces acting on the turbine rotor are perceived by a ball bearing installed in the generator housing of the rear bearing.

The free turbine drives the rotor of the free turbine, and it in turn rotates the generator. The gas energy on the blades of the nozzle apparatus and the impeller is converted into torque on the rotor shaft. The torque is transmitted to the rotor located at the end of the generator rotor.

The main elements of the nozzle apparatus of a free turbine are:

- Outer casing
- Nozzle vanes
- Outer ring
- Splitter
- Seal segments
- Inspection hatches plugs
- Floating rings

The nozzle on the output side of the flange of the outer ring is rigidly connected to the support of the free turbine, on the input side has a movable telescopic connection with the support of the gas generator along the outer and inner flanges.

Structurally, the main elements of the nozzle apparatus are as follows:

The outer casing is a prefab assembly consisting of a front flange, a cone and a rear flange. On the case there are removable side poluobichayki for inspection of the input edges of the working and nozzle blades of the free turbine.

Nozzle vanes - cut from sheet steel with a thickness of 3 mm. The nozzle vanes in the outer casing are centered with rings made on the outer shelves of the blades, and in the inner case with rings made on the lower shelf of the scapula.

Sealing segments with screwed-on rings are installed in both the nozzle unit and the turbine rotor. Nodes with sealing elements provide a minimum radial clearance between the rotor and the stator of the free turbine.

Docking apparatus with the flanges of the gas generator is carried out as follows:

Floating rings are installed at the rear flange of the housing, providing a telescopic connection with the generator supports.

The concentricity of the rotor is ensured by longitudinal-transverse adjustment relative to the generator supports.

The rotor of the free turbine includes:

- Turbine wheel, blades, outer rings of blades, sealing plates
- Shaft, labyrinths, fasteners
- Coupling bolts
- Half-browns
- Distance ring; reflector; labyrinth sleeve

Figure 7 shows a nozzle and a three-stage radial turbine, 1 - nozzle; 2 – turbine

A three-stage free turbine with a guide vane has a rotation opposite to the turbocharger.

MAIN FOCUS OF THE CHAPTER

The parameters of the gas-air path are supplied with indices determined by the corresponding sections in the turbocharger circuit.

Initial data:

1. Turbocharger speed n1 = 90000 rpm
2. Turns of the output shaft of the free turbine n2B = 8000 rpm.
3. Fuel consumption G_T = 4.5 g / s = 0.0045 kg / s.
4. Air flow through the compressor G_B = 260 g / s = 0.26 kg / s
5. Gas flow through the nozzle apparatus $G_{rca} = G_T + G_B$ = 0, 2628 kg / s
6. Gas consumption through the turbine G_G = 0.27 kg / s

Figure 7. Free three-stage turbine

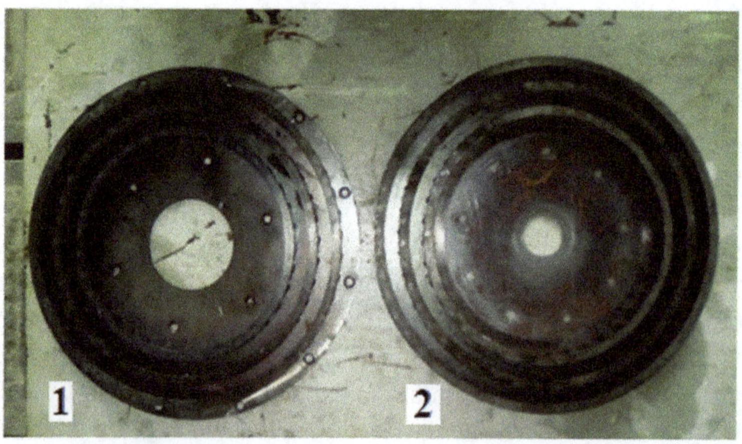

In accordance with the calculation, the design documentation was developed and a centrifugal three-stage radial free power turbine was built, supplying the gas flow to the vortex plate recuperative heat exchanger.

The task of the vortex recuperative heat exchanger (VRHE) is to increase the efficiency of the fuel turbine generator by using the heat of exhaust gases emitted into the atmosphere (HEG ~ 350 ° C) for heating the air entering the combustion chamber of the installation from a pneumatic compressor (HG ~ 18 ° C).

In gas turbine installations, useful work is carried out due to the kinetic energy of a gas moving at high speed. The working body in these installations are the products of combustion generated during the combustion of fuel. The gas flow is created by the flow of gas from the turbine nozzles. The fuel turbogenerator operates on a cycle with the supply of heat at p = Const. The theoretical cycle of a gas turbine unit with fuel combustion at p = Const and adiabatic compression of air in the compressor is presented in (Figure 8).

The thermal efficiency of the installation is determined by:

$$\eta t = -(T_1 / T_2).$$

With an increase in temperature T_2 to T'_{2}, the maximum temperature of the cycle T_3 to T'_3 increases, hence the expansion work increases and the effective efficiency of the installation increases. One of the options for increasing the temperatures T_2 and T_3 is to increase the temperature of the air stream entering the combustion chamber through the use of a regenerative heat exchanger.

Figure 8. Theoretical cycle of a gas turbine unit with fuel combustion at p = Const.

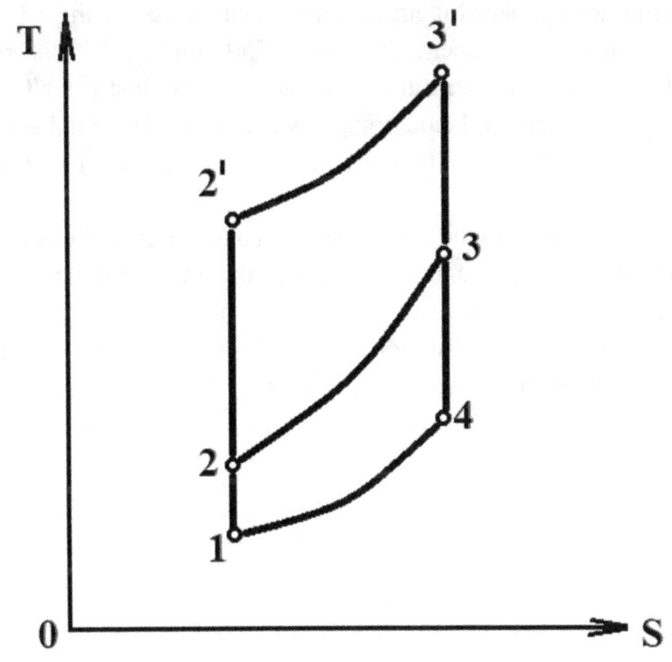

The task of calculating the heat exchanger is to determine the size of the heating surface necessary to transfer a given amount of heat. Further, the amount of heat transferred and the measurement of the temperature of the working fluid are determined from the known dimensions of the heating surface.

In order to reduce the size and mass of heat exchangers or increase their heat output, various methods of heat exchange intensification are widely used. All these methods pursue one goal - at a given speed and dimensions of the heat exchanger channel, increase the heat transfer coefficient. For this purpose, various kinds of turbulizers, spiral ribs, screw inserts, twisted ribbons, corrugations, etc. are used. Nevertheless, these methods of heat exchange intensification inevitably cause an increase in hydraulic resistance. Usually, intensifiers (turbulizers) increase the heat exchange coefficient by 30–40%, but the hydroresistance of the working medium in the channels of the heat exchanger increases by 40–60%.

When developing a turbine vortex recuperator, the vortex method of heat exchange intensification, which includes the use of spherical holes on the heat exchange surface, was used, is one of the most effective, since its implementation causes an accelerated increase in the heat transfer coefficient compared to an increase in the hydraulic resistance coefficient;

- Heat transfer coefficient should increase to 60%
- The efficiency of the heat exchanger will increase ~ 10-11%
- The heat exchanger envelope decreases by 15-20%

The use of heat transfer areas formed by special relief from hemispherical depressions as an intensifier significantly intensifies heat transfer and reduces aero-hydrodynamic resistance (in energy-exchange channels) due to the formation of vortex formations above the hemispherical depressions creating a steady swirling flow perpendicular to the direction of the main flow mass flow.

ON the (Figure 9 and Figure 10) show the duct assembly and the heat exchanger plate.

ON the (Figure 11) shows the appearance of a micro gas-turbine power plant with an installed free three-stage power single-disk centrifugal turbine and a vortex recuperative heat exchanger.

Tests to determine emissions of harmful emissions were carried out using a Testo T350S instrument at a temperature of 20 °C and an atmospheric pressure of 750 mm Hg.

The results showed the content of harmful emissions in the exhaust gas NO_2 is 0.4 ppm when the unit is in mode with the temperature of the outgoing gas to 560 °C. The obtained values are lower than those of modern Russian and foreign installations and fully meet the latest sanitary and environmental requirements.

The developed micro gas turbine engine can be used as a basic source of electrical energy as a part of the micropowergrid (Adomavicius, 2013). For this purpose the inverter transforming tension developed by the gas turbine in the high-stabilized quality voltage can be used (Kharchenko, 2019). Creation of microgrid on the basis of a resonant transmission system of electrical energy can be other option of use of the electrical energy developed by the gas turbine (Yuferev, 2018).

Figure 9. Duct assembly of an experimental vortex recuperative heat exchanger

Figure 10. Heat exchanger plates A, B, C of the experimental vortex recuperative heat exchanger

Figure 11. Appearance of a micro gas turbine installation, where a) right side view, b) left side view

FINDINGS

1. Turbocompressors of the internal combustion engine are suitable for small gas-turbine engines, since the structural connections of the turbine with the compressor impeller fully correspond to the parameters of the MGTPP necessary for operation.
2. The developed techniques for selecting the turbocharger of the internal combustion engine make it possible to manufacture a gas turbine engine of various power. In addition to those described in the work of turbochargers, there are more productive inflator assemblies installed on diesel engines of railway locomotives and marine engines.
3. A three-stage free power turbine was designed and manufactured on one disc, having labyrinth seals with lower cost and lower labor costs relative to the existing analogues at each stage.
4. Parts and assemblies of the vortex recuperative heat exchanger were manufactured, providing the estimated fuel economy of a microgas turbine unit up to 40%.
5. A technique has been developed for experimental studies of a recuperative heat exchanger for an experimental microgas turbine unit GTP-10C.
6. Microgas turbine drives with a capacity of up to 300 kW, suitable for use in automotive manufacturing have never been produced, the development of production in this direction will significantly increase the intellectual level of engineering in Russia.

REFRENCES

Adamavichus, V.B., Gusarov, V.A., Valitskas, I.Y., & Kharchenko V.V. (2016). Arguments and new opportunities to increase the share of energy supply of agriculture through renewable energy. *Innovations in agriculture*, *5*, 391-396.

13. Adomavicius, V., Kharchenko, V., Valickas, J., & Gusarov, V. (2013). Res-based microgrids for environmentally friendly energy supply in agriculture. In *Conference Proceeding - 5th International Conference, TAE 2013: Trends in Agricultural Engineering 2013*, Prague, Czech Republic, September 2-3 (pp. 51-55).

Adomavicius, V. B., Kharchenko, V. V., Gusarov, V. A., & Valitskas, I. Yu. (2013). Sources of controlled power in micro networks. *Alternative Energy and Ecology*, *7*, 54–59.

Balakshin, OB, Kukharenko, B.G. (2015). The emergence and development of flutter rotor blades axial turbocharger. *Problems of mechanical engineering and reliability of machines*, *3*, 24-29.

Ganiev, R.F., Balakshin, OB, Kukharenko, B.G. (2015). Turbulence flow and flutter rotor blades in an axial turbocharger. *Problems of mechanical engineering and reliability of machines*, *6*, 3-10.

Gusarov, V.A. (2014). Improving the efficiency of electronic devices for autonomous power supply of rural consumers. *Innovations in agriculture*, *2*, 39-48.

Gusarov, V.A. (2017). Microgas turbine power plants for autonomous power supply. *Gas turbine technologies*, *6*, 8 - 12.

Gusarov, V. A., Gusarova, O. F., & Kulagin, Ya. V. (2013). The use of gas turbine technologies in agricultural production. *Alternative Energy and Ecology*, *2*, 76–79.

Gusarov, V. A., Zadde, V. V., Nikitin, B. A., & Kargiev, V. M. (2004). Automatic complex uninterrupted power supply. In *Proceedings of the International Scientific and Technical Conference* (Vol. 4, pp. 263 – 268).

Gusavor, V. A. (2016). Prospects for distributed energy. *Bulletin of the VIESH*, *3*, 77–83.

Iacoboni, M., Chow, J., & Wheless, E. (2012). Calabasas Landfill Microturbine Power Generation Project: Lessons Learned after One-Year of Operation. Retrieved from https://www.capstoneturbine.com/

Kharchenko, V., Gusarov, V., & Bolshev, V. (2019). Reliable Electricity Generation in RES-Based Microgrids. In Handbook of Research on Smart Power System Operation and Control (pp. 162-187).

Kharchenko, V. V., & Gusarov, V. A. (2015). The principles of the formation of a generating complex of microgrids based on renewable energy sources. *Bulletin of Agrarian Science Don*, *32*, 71–83.

Mago, P., & Luck, R. (2012). Evaluation of the potential use of a combined micro-turbine organic Rankine cycle for different geographic locations. *Applied Energy*, *102*, 1324–1333. doi:10.1016/j.apenergy.2012.07.002

Pilavachi, P. A. (2000). Power generation with gas turbine systems and combined heat and power. *Applied Thermal Engineering*, *20*(15-16), 1421–1429. doi:10.1016/S1359-4311(00)00016-8

Pilavachi, P. A. (2002). Mini - and micro-gas turbines for combined heat and power. *Applied Thermal Engineering*, *22*(18), 2003–2014. doi:10.1016/S1359-4311(02)00132-1

Smirnov, A.V., Kosmynin, A.V., Khvostikov, A.S., Shchetinin, B.C., Ivanova, N.A. (2016). Problems of operation of turbochargers ICE and ways to improve their reliability. *Problems of mechanical engineering and machine reliability*, *2*, 67 - 71.

Yuferev, L. (2018). The Resonant Power Transmission System. In Handbook of Research on Renewable Energy and Electric Resources for Sustainable Rural Development (pp. 534-560). doi:10.4018/978-1-5225-3867-7.ch022

Chapter 5
Milk Pasteurization and Characterization Using Mono-Mode Microwave Reactor and Slotted Coaxial Antenna

Suhail Abdullah
https://orcid.org/0000-0003-4897-9396
Unversiti Teknologi Malaysia, Malaysia

Kok Yeow You
https://orcid.org/0000-0001-5214-7571
Universiti Teknologi Malaysia, Malaysia

Cheong Yew Chong
Tunku Abdul Rahman University College, Malaysia

Mohamed Sultan Mohamed Ali
Universiti Teknologi Malaysia, Malaysia

ABSTRACT

Mono-mode microwave reactors are usually used to heat substances, especially food. This is because heating using a microwave reactor can sustain the flavor, color, and nutrition of the food. Furthermore, this heating technique is cost-effective and time-saving compared to a conventional heating method. The mono-mode reactor is able to determine the absorption of microwave power accurately on the heated substance versus a multimode reactor. In this chapter, a simple and precise mono-mode microwave reactor is designed and developed especially for research laboratories. The advantage of this reactor is to provide a more accurate calibration process, in order to improve the optimum energy use in the heating process, as well as the temperature of the specimen. The reactor can generate output power from 30 watts to 1500 watts, operating at 2.45±0.03 GHz and capable of accommodating a specimen volume of 780 cm³. Pure water is used as a heated specimen to demonstrate the performance and efficiency of this reactor.

DOI: 10.4018/978-1-5225-9420-8.ch005

INTRODUCTION

Microwave Reactors

In 1945, the heating effect of microwave energy was accidentally discovered by Percy Spencer when he noticed the melted candy bar while he was objected to the active radar. He decided to test it with popcorn and egg and the food was cooked using microwave. In 1947, the first commercial microwave oven so-called "Radarange" was developed by Raytheon with weight of 340 kg and height of 1.8 meter (Osepchuk, 1984). In 1967, the first domestic (home model) microwave oven was commercialized byAmana (a division of Raytheon).

Microwave heating is characterized by conventional heating being penetrative heating, non-contact which reduces of overheating of material surfaces, short thermal gradients, volumetric heating, energy saving, environmentally friendly, and fast which increases the production rate (Bogdal, 2005; Groisman and Gedanken, 2008).

Recently, the use of microwave reactor is growing rapidly in the process of heating in food stuff factories, such as food dehydration/frozen drying and food pasteurization/sterilization for storage control. The reason is due to the tendency of water to absorb microwave energy and generate the heat within the food. When the raw food is exposed to the microwave, the water molecules in the food will be induced to rotate and produce heat as shown in Figure 1 (You, 2017). Thus, the rate of water removal and the effective used energy are higher than hot-air drying method. Besides, the microwave heating is capable of maintaining original texture structure, nutrition, flavor and color of the food at specify heat temperature compared to conventional oven drying techniques.

The comparison of the efficiency of used energy between various heating techniques is listed in Table 1 (Constellation, 2016).

There are two microwave frequencies allocated by the US Federal Communications Commission (FCC) for industrial, scientific and medical (ISM) use, which are 915 MHz and 2.45 GHz. Normally, most of the microwave heating applications are devoted to 2.45 GHz, since it provides a suitable compromise between power deposition and penetration depth as well as it is an unlicensed operating frequency. Villamiel et al. (1996; 1998) have demonstrated the effectiveness of microwave heating over conventional heating proved that volumetric heating at 2.45GHz produces lower denaturation levels of whey proteins and β-lactoglobulin as compared with conventional pasteurization. In addition, Coronel et al. (2003) used 915 MHz microwave applicator based on continuous-flow and it is shown that milk is heated rapid

Table 1. Comparison of energy efficiency for various heating techniques (Constellation, 2016)

Heating Techniques	Temperature (°C)	Heating Time	Used Energy (kWh)
Electric Oven	350	1 Hour	2.0
Electric Convection Oven	325	45 Minutes	1.39
Gas Oven	350	1 Hour	0.90
Electric Frying Pan	420	1 Hour	0.90
Toaster Oven	425	50 Minutes	0.95
Electric Crockpot	200	7 Hours	0.70
Microwave Oven	High	15 Minutes	0.36

Figure 1. Microwave heating mechanisms: Water molecules are oriented when exposed to microwave (You, 2017)

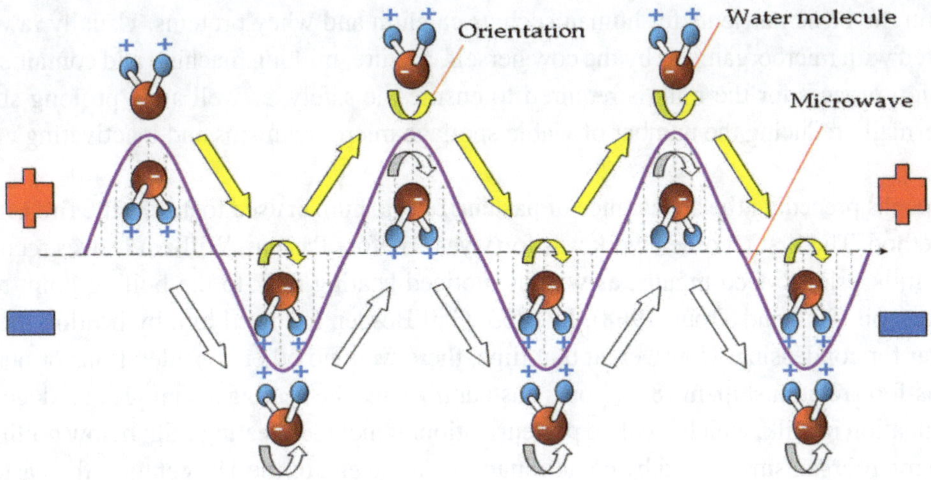

than water. While, three parameters were used by Clare et al. (2005) to evaluate the microwave heating as compared with UHT sensory, microbiological, and biochemical; the experiments exhibit similar biological and biochemical effects, while the sensory changes exhibited in UTH. The studies indicate the effectiveness of using dielectric heating to produce pasteurized milk that satisfying the biological, biochemical and sensory requirements avoiding an extensive heat damage.

Although, microwave reactors have been widely used, however, the efficiency of use and precision control of heating process are less concerned by users and reactor builders, especially applications in the laboratories. Microwave reactors can be categories in two types, namely multi-mode and mono-mode microwave reactor. The multi-mode reactor, such as domestic microwave oven, is capable of accommodating a high volume of food heating. However, energy efficiency in heating is relatively low and food temperature monitoring is less accurate due to the existence of multi-reflection of microwave energy in the heating cavity and the non-uniform heating area. On the other hand, efficient coupling between microwave energy and food is relatively high in mono-mode reactor cavity and it is ability to precisely control food heating (heating time and power level control), as well as the thermal properties of the food can be accurately characterized (specify heating temperature for certain application). In fact, the increase in temperature in food during heating is depended on the dielectric properties of the food (to be discussed in Section 2).

In this chapter, simple and precise mono-mode microwave reactor is designed, fabricated, and tested, especially for food thermal characterization in laboratories. Several scientific analysis of heating process is described. This chapter presents two proposed systems for milk pasteurization application, the first is conventional designed mono-mode reactor and the second is by proposing slotted coaxial antenna as pasteurizer.

Brief History of Milk Pasteurization

Milk contain the basic nutrients for human such as calcium and whey proteins. Usually raw milk gets contaminated with microorganisms by the cow herself, pasture, milking machine and containers. Hence, pasteurization process for the milk is required to ensure the safety, as well as to prolong shelf-life of milk by thermally reducing the number of viable spoilage microorganisms and inactivating undesirable enzymes.

In the epochs preceding the appearance of pasteurization, human used to utilize the fire as food preservation method. The first recorded work was forty years before Pasteur, William Dewes recommended to heat the milk prior to feed infants, as well as, noticed heating milk to the boiling point reduces its tendency to spoil (Hall and Trout, 1968). In 1853, Gail Borden followed him by heating the milk and adding sugar for condensing. However, at that time, there were no other considerations or observations about germs-food relationship. In 1860, Louis Pasteur invented the heating technique to reduce the bacterial contamination in milk, which is called pasteurization. It includes heating milk below boiling point to destroy the microorganisms caused by contamination. At the end of the 18th century, the pasteurization entered to the commercial sector, where firstly applied as a common practice in the mid-1880s by Denmark followed by USA on 1893. After that, several milk-borne diseases were recognized before 1900.

In the early 19th century, there were many fluctuations in the reliability of milk as pasteurization, with many fears of the possibility of reducing the nutritional value of milk when Pasteurization will be adopted. Despite many concerns, however, many factories use pasteurization secretly to maintain their shelf life. In 1899, Smith's research declared that heating milk at 60 °C for 15 min will kill *Mycobacterium Tuberculosis*. On the other hand, Russel and Hasting research indicates that heating milk at 60 °C for 10 min will kill the *Mycobacterium Tuberculosis*. From 1890 to 1927, there was a recommendation by 31 studies on heating milk at 60 °C for 20 min as margin to ensure complete destruction of MT. Although there were 26 reports on thermal death time of *Mycobacterium Tuberculosis* from 50 °C to 100 °C for 1 min to 6 hours within the year of 1883 to 1906, but public health organizations did not have serious attention to the former studies.

In 1924, North and Park confirmed the earlier reports of *Mycobacterium Tuberculosis* and related thermal death time. Consequently, Public Health Reports issued the standards for milk pasteurization which compelling dairy industry to pasteurize the milk with temperature not less than 61.1 °C for 30 min in approved equipment. Despite the popularity and spreading of using "Holder" method in pasteurization, but it is lowering the production rate. Hence, new equipment was underway and being initialized for milk pasteurization, such as heat exchanger with concept of High Temperature Short Time (HTST) in order to increase the rate of production. It includes heating milk at 71.7 °C for 15 seconds. In 1933, Public Health Reports adopted the HTST and add it to the standards of the pasteurization. Nevertheless, the thermal death time of microorganism for HTST is still not recognized and 25% of food-borne diseases were caused by milk.

This condition encouraged Workman in 1941 to examine the microbial destruction at HTST temperature and period and it presented complete destruction of microorganism. In 1949, the presence of Coxiella burnetii (the major cause of *Q*-fever) noticed in pasteurized milk which indicates high thermal resistivity. Therefore, Public Health Report updated new standard for the batch pasteurization to be 62.8 °C for 30 min. Based on what preceded, the quality indication of pasteurization relies on thermal destruction of Coxiella Burnetii (Westhoff, 1978).

Recently, HTST is a common pasteurization method accredited in many countries. HTST provides shelf-life variations from 3 to 21 days at storage temperature of 7 °C after treatment at 72 °C for 15 seconds (Rossitto et al., 2012; Meunier and Sandra, 2016). Although pasteurization inactivate microorganisms, spores, and thermoduric bacteria are difficult to destroy in HTST pasteurization. Thus, it is preferred to keep the pasteurized milk in proper uniform refrigeration below 4 °C, which is difficult to achieve. Ultra-High Temperature (UHT) treatment has been used to overcome inadequate refrigeration storage problem by heating milk 135–140 °C for a few seconds. Therefore, UHT prolonged the shelf-life to be 3 to 6 months at ambient temperature. However, extreme thermal treatment of milk causes degradations in nutrition's quality and changing the sensory properties of the milk. Although the HTST is usually preferred on the UHT due to such reasons, there is still a problem encounters the conventional heating which is the fouling of heat exchangers which in turn causes insufficient heating causes reduction of food quality and flavor changes (Kudra et al., 1991; Boxler et al., 2013). Hence, free-fouling technique has an increase in demand; therefore, the emerging microwave heating technology came up to solve such problems.

Milk Pathogens

Besides *Mycobacterium Tuberculosis*, cow milk also contains other pathogens or bacteria as listed in Table 2. In fact, the percentage of pathogens in raw milk mainly depends on the weather or season and the process of producing milk from the cow.

THEORY

Dielectric Properties

The ability of the biological/food specimens to absorb microwave energy and transfer the energy into heat, in which can be evaluated using relative complex permittivity, ε_r parameter as:

$$\varepsilon_r = \varepsilon_r' - j\varepsilon_r'' \tag{1}$$

where ε_r' is represented the specimen's ability to store energy and ε_r'' denotes the ability of specimens to transform the stored energy into heat. In fact, the dielectric properties of polar dielectric specimens, such as water, can be modelled using Debye/Cole-Cole relaxation as:

$$\varepsilon_r = \varepsilon_\infty + \frac{\varepsilon_s - \varepsilon_\infty}{1 + \left(j\omega\tau\right)^{1-\alpha}} \tag{2}$$

where ε_∞ is the permittivity at the high frequency limit and ε_s is the static or very low frequency permittivity. On the other hand, τ is the characteristic relaxation time of the specimen. The exponent parameter α (value between 0 and 1) allows to describe different spectral shapes.

Table 2. Previous study of percentage of pathogens in raw milk sample

Pathogen	Pathogen Properties	Percentage of Pathogen Affected in Raw Milk Sample	Reference
Bacillus cereus	Gram-positive, rod-shaped, aerobic, facultative anaerobic, motile, beta-hemolytic.	113 out of 206 raw milk samples	Christiansson et al., 1999
		9 out of 100 raw milk samples	Ahmed et al., 2016
		50 out of 50 raw milk samples	Gundogan et al., 2014
		32 out of 119 raw milk samples	Yobouet et al., 2014
		46, 50, and 21 out of 192, 190, and 148 raw milk	Moussa-Boudjemaa et al., 2004
		2 out of 298 raw milk samples	Muckly et al., 2016
		257, 97, 98, and 26 out of 485, 172, 174, and 112 raw milk samples	Donovan et al., 1959
Brucella	Small, Gram-negative, nonmotile, non-spore-forming, rod-shaped (coccobacilli).	16 out of 99 raw milk samples	Kaynak-Onurdag et al., 2016
		10 out of 50 raw milk samples	Rock et al., 2016
		17, 0, 4, and 57 out of 112, 100, 72, and 1629 raw milk samples	Kessy et al., 2015
Campylobacter Jejune	Gram-negative, rods curved or spiral, motile, non-spore-forming.	2 and 4 out of 50 and 38	El-Sharoud et al., 2009
		52, 8, 5, 22, 0, 18, 0, 0, 7, 2, 41, 1, 2, 1, and 6 out of 1528, 1382, 594, 10720, 191, 828, 293, 1274, 150, 21, 409, 108, 130, 237,120 raw milk samples, respectively	Mioni et al., 2015
		8, 20, and 1 out of 1050, 450, and 6 raw milk samples	Barakat et al., 2015
Clostridium	Anaerobic, Gram-positive, spore-forming rod.	34, 22, and 15 out of 90, 55, 41 raw milk samples	Barakat et al., 2015
		2 out of 135 raw milk samples	Montaz et al., 2015
		40 out of 50 raw milk samples	Naguib et al., 1972
		78 out of 79 raw milk samples	Capelli et al., 2014
Coxiella Brunetti	Small, Gram-negative, coccobacillary.	8 out of 58 raw milk samples	Tonucci et al., 2013
		20 out of 130 raw milk samples	Loftis et al., 2010
		9, 6, and 16 out of 31, 133, and 31 raw milk samples	Caudron et al., 2007
		36 and 59 out of 214 and 344 raw milk samples	Ho et al., 1995
		81 and 71 out of 139 and 117 raw milk samples	Caudron et al., 2007
		4, 3, and 0 out of 26, 5 and 7 raw milk samples	Hirai et al., 2012
		83, 46, and 32 out of 152, 113, and 135 raw milk	Magnino et al., 2009
		4, 6, 5, and 4 out of 10, 7, 10 and 10 raw milk samples	Natale et al., 2009
		9 out of 21 raw milk samples	De Bruin et al., 2012
		18, 23, 58, 56, 40, and 103 out of 20, 24, 61, 60, 43, and 108 raw milk samples	Loftis et al., 2010
Escherichia coli (E. coli)	Gram-negative, facultative anaerobic, rod-shaped, coli-form.	57, 9, 10, 12, 13, 11, 8, and 12 out of 100, 20, 20, 20, 20, 20, 20, and 20, respectively	Soomro et al., 2002
		55 out of 72 raw milk samples	Ombarak et al., 2016
		68 out of 75 raw milk samples	Lubote et al., 2014
		48 out of 50 raw milk samples	Paneto et al., 2007
		0 and 3 out of 10 and 5 raw milk samples	Razzaq et al., 2016
		35 out of 153 raw milk samples	Vivegnis et al., 1999
		7 out of 30 raw milk samples	Elbagory et al., 2016
		37 out of 50 raw milk samples	Gundogan et al., 2014
		8 out of 125 raw milk samples	Mansouri-Najand et al., 2007
		38 out of 100 raw milk samples	Thaker et al., 2012
		13, 11, and 28 out of 30, 22, and 31 raw milk samples	De Souza et al., 2016
		13 out of 50 raw milk samples	Bali et al., 2013

continued on following page

Table 2 Continued

Pathogen	Pathogen Properties	Percentage of Pathogen Affected in Raw Milk Sample	Reference
Listeria Monocytogenes	Gram-positive, non-spore-forming, motile, facultative anaerobic, rod-shaped.	5 out of 100 raw milk samples	Kevenk et al., 2016
		2, 1, 1, and 4 out of 60, 60, 80, and 80 raw milk samples	Mahm et al., 2010
		0, 4, 3, and 5 out of 47, 250, 100, and 80 raw milk	Aygun et al., 2006
		21 and 6 out of 63 and 113 raw milk samples	D'agostino et al., 2004
		2 out of 153 raw milk samples	Magnani et al., 2004
		1 out of 141 raw milk samples	Waak et al., 2002
		28, 145, and 10 out of 774, 8716, and 145 raw milk	Daminelli et al., 2015
		65 out of 860 raw milk samples	Karns et al., 2010
		1, 4, and 8 out of 36, 36 and 426 raw milk samples	Fedio et al., 2013
		1 out of 59 raw milk samples	Husu et al., 1990
		3 out of 72 raw milk samples	Bacci et al., 2014
		5 out of 100 raw milk samples	Usman et al., 2014
		7 and 13 out of 343 and 65 raw milk samples	Seyoum et al., 2015
		8 out of 696 raw milk samples	Marnissi et al., 2014
Mycobacterium	Aerobic, rod-shaped, non-endospores-forming, possess capsules.	5 and 0 out 50 and 10 raw milk samples	Eftekhari et al., 2016
		111 out of 345 raw milk samples	Slana et al., 2008
		4, 3, and 10 out of 143, 100, and 100 raw milk samples	Stephan et al., 2007
		789 out of 2934 and 90 out of 789 raw milk samples	Ricchi et al., 2016
		10 and 2 out of 145 both of raw milk samples	Usman et al., 2016
		13 out of 32 raw milk and 2 out 20 pasteurized milk	Wacker et al., 2011
		4 out of 83 raw milk samples	Ben Kahla et al., 2011
		5 out of 102 raw milk samples	Cadmus et al., 2010
		5 out of 400 raw milk samples	Smith et al., 2007
		19, 67, 2, 10 out of 244, 567, 244, and 567 raw milk	Anderson et al., 2004
		50, 1, 35, and 0 out of 389, 389, 357 and 357 raw milk	Kazwala et al., 1998
		32 out of 805 raw milk samples	Van Kessel et al., 2016
Salmonella	Rod-shaped, Gram-negative, non-spore-forming.	24 and 45 out of 200 and 200 raw milk samples	O' Donnell et al., 1995
		6 out of 173 raw milk samples	Haitsma et al., 2009
		129 out of 244 raw milk samples	Vaillant et al., 2008
		3 and 2 out of 29 and 29 raw milk samples	Yamazi et al., 2009
		0, 0, and 0 out of 36, 18 and 210 raw milk samples	Gwida et al., 2014
		42, 24, and 4 out of 200, 200, and 50 raw milk samples	Ahmed et al., 2014
Shigella	Gram-negative, facultative aerobic, non-spore-forming, nonmotile, rod-shaped.	3 out of 240 raw milk samples	Rahman et al., 2015
		11 and 2 out of 100 and 50 raw milk samples	Nazari et al., 2014

continued on following page

Table 2 Continued

Pathogen	Pathogen Properties	Percentage of Pathogen Affected in Raw Milk Sample	Reference
Staphylococcus aureus	Gram-positive, round-shaped, nonmotile, non-spore-forming.	52 out of 246 raw milk samples	Luzzana et al., 2007 Giezendanner et al., 2009
		23 out of 23 raw milk samples	Loncarevic et al., 2005
		65 out of 65 raw milk samples	Bartolomeoli et al., 2009
		18 out of 34 raw milk samples	Jørgensen et al., 2005
		13, 157, 165, and 31 out of 51, 963, 220, and 82 raw milk samples	Hunt et al., 2012
		81 out of 117 raw milk samples	Fagundes et al., 2010
		18 out of 245 raw milk samples	Uyanik et al., 2012
		3 out of 10 raw milk samples	Pexara et al., 2016
		45 out of 60 raw milk samples	Rola et al., 2016
		122 out of 244 raw milk samples	Rall et al., 2008
		38 and 8 out of 44 and 11 raw milk samples	SunYoung et al., 2010
		34 out of 44 raw milk samples	Jakobsen et al., 2011
		35 out of 73 raw milk samples	Bartolomeoli et al., 2009
		31 out of 44 raw milk samples	D'amico et al., 2011

For instance, pure water is used as a specimen under test, since the physical properties of the water is well known as listed in Table 3. The value of C_p for pure water is 4.186 J/(kg °C). Generally, the dielectric properties (dielectric loss factor, ε_r'') of the most biological/chemical specimens are depended on temperature, T, such as pure water, thus a real-time microwave power tuning is necessary during the heating process. The typical relationship between the loss factor, ε_r'', mechanism of water molecules and operating frequency is shown in Figure 2. In addition, the mechanisms of water molecules in food specimens are different when exposed to different ranges of operating frequency, f as shown in Figure 3. The parameter values, ε_∞, ε_s, τ, and α of the water in (2) as a function of temperature, T are extracted from are listed in Table 4.

Absorption Microwave Power

The microwave power, P (in W/m^3) that is dissipated or absorbed by the specimen during heating process is given as:

$$P = 2\pi f \varepsilon_o \varepsilon_r'' \left|E\right|^2 \tag{3}$$

where f, ε_o, ε_r'', and E are the operating frequency (typical: 915 MHz or 2.45 GHz), permittivity in vacuum ($\approx$ 8.854188x10−12 F·m^{-1}), relative loss factor of the specimen, and a supplied microwave energy within the specimen (TE$_{10}$ mode of incident electrical field, in V/m), respectively. The loss factor, ε_r'' is temperature-dependent parameter and its influences the energy absorption or attenuation of the specimen.

From equation (3), the specific absorption rate (SAR) (in W/kg) of the specimen can be defined as:

Table 3. Physical properties of pure water.

T (°C)	ρ (kg/m³)	ε_r''	
		915 MHz	2.45 GHz
20	0.99821	4.254	10.901
25	0.99705	3.643	9.382
30	0.99565	3.154	8.152
40	0.99222	2.430	6.310
50	0.98805	1.908	4.969
60	0.98319	1.489	3.887
70	0.97774	1.138	2.976
80	0.97176	0.871	2.280
90	0.96530	0.746	1.953
100	0.95837	0.849	2.221

Figure 2. Relationship between ε_r'', mechanism of water molecules and frequency, f.

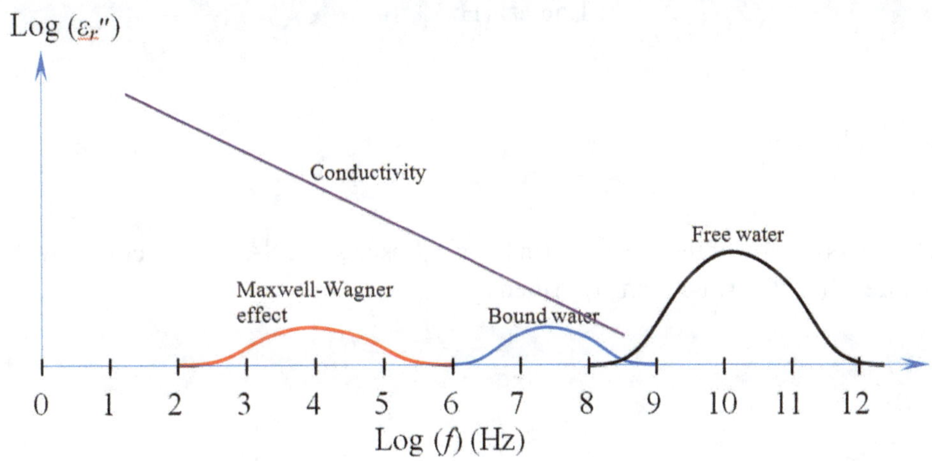

Table 4. The Cole-Cole parameters for water as the function of temperature.

Cole-Cole Parameters for Water Valid for (DC to 20 GHz) and (0 °C to 100°C)	
$\varepsilon_s = 87.9144 - 0.404399T + 9.58726\times10^{-4}\,T^2 - 1.32892\times10^{-6}\,T^3 < \pm\, 0.0055$	(Hamelin et al., 1998)
$\varepsilon_\infty = 4.5$	(Grant et al., 1957)
$\tau = 17.68703 - 0.62240692T + 0.01263T^2 - 0.1354\times10^{-3}T^3 + 0.005631\times10^{-4}T^4$ (ps), $\pm$ 0.0372 ps	(Grant et al., 1957)
$\alpha = 0.02$	(Grant et al., 1957)

Figure 3. The mechanisms of water molecules when exposed to different range of operating frequency.

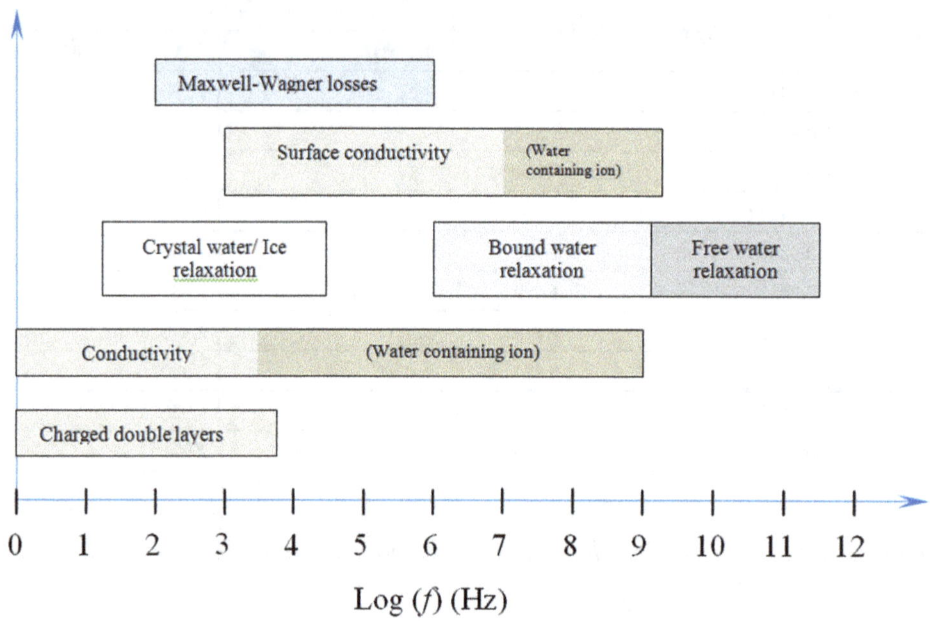

$$\text{SAR} = \frac{2\pi f \varepsilon_o \varepsilon_r''}{\rho} |E|^2 = \frac{P}{\rho} \tag{4}$$

The ρ is the mass density of the specimen (in kg/m^3). Besides, the SAR ties in closely with the specimen temperature, T (in °C), which can be written as:

$$\begin{aligned} \text{SAR} &= C_p \frac{\partial T}{\partial t} \\ &\approx C_p \frac{\Delta T}{\Delta t} \end{aligned} \tag{5}$$

where C_p is the specific heat capacity of the specimen [in J/(kg °C)] and t is the time (s). From equations (4) and (5), the value of absorbed P by the specimen can be estimated as:

$$P \approx C_p \rho \frac{\Delta T}{\Delta t} \tag{6}$$

Once $\Delta T/\Delta t$, ρ and ε_r'' are known, the absorbed P can be calculated from equation (6). The calculated P using equation (3) can be compared and used to validate the measured P results obtained from (6). As the value of P increases, the heating rate will also rise due to the increasing of temperature, T in short period of time.

Penetration Depth of Microwave

The penetration depth, D (in unit meter) of microwave is played an important role in the determination of maximum thickness of food specimens which can be heated directly by microwaves (Sun et al., 2016). The formulation D in terms of ε_r' and ε_r'' can be simplified as:

$$D \approx \frac{\lambda_o \sqrt{\varepsilon_r'}}{2\pi\varepsilon_r''} \quad (7)$$

Penetration depth, D of microwave is greater and more effective at lower frequencies due to the microwave has larger values of wavelength, λ_o. Hence, normally, domestic microwave heating and some industrial applications are located at operating frequency of 2.45 GHz. On the other hand, 915 MHz is preferred for industrial/commercial microwave heating. Besides, the higher values of ε_r'' reduce the depth of penetration, D and lead to surfacial heating, whereas lower values of ε_r'' will lead to uniform volumetric heating (Singh et al., 2015). Table 5 shows the properties of penetration depth, D for milk at 2.45 GHz. The values of ε_r' and ε_r'' are calculated from the equations (10) and (11). It is found that D for milk is nearly 1.1 cm and its value is increased with temperature, T rise.

Governing Equations

Electromagnetic Wave Equation

For finite cylindrical symmetry cases, such as insulated monopole antenna, based on the principal of TEM mode, the non-vanish components are H_ϕ, E_ρ and E_z. Therefore, the wave equation can be simplified to a scalar equation for H_ϕ, are yielded as,

Table 5. Physical properties of raw milk at 2.45 GHz.

T (°C)	ρ (kg/m³)	ε_r'	ε_r''	λ_o (m) (for Free-Space)	D (m)
25.0	1.0248	60.7	13.3	0.122	0.0114
30.0	1.0220	60.2	13.1	0.122	0.0115
35.0	1.0188	59.7	12.9	0.122	0.0117
40.0	1.0151	59.1	12.6	0.122	0.0119
45.0	1.0108	58.6	12.4	0.122	0.0120
50.0	1.0058	58.1	12.2	0.122	0.0122
55.0	1.0002	57.6	11.9	0.122	0.0124
60.0	0.9937	57.0	11.7	0.122	0.0126
65.0	0.9863	56.5	11.5	0.122	0.0127
70.0	0.9780	56.0	11.3	0.122	0.0129
75.0	0.9687	55.5	11.0	0.122	0.0132

$$\left(\frac{\partial^2}{\partial\rho^2}+\frac{1}{\rho}\frac{\partial}{\partial\rho}-\frac{1}{\rho^2}+\frac{\partial^2}{\partial z^2}+k^2\right)H_\phi=0 \tag{8}$$

The differential equation of (8) which governs the spatial of radiation and it can be solved by numerical routine, such as Finite Element Method (FEM) and Moment Method (MoM) (You, 2015).

Bio-Heat Transfer Equation

In this study, the heat diffusion in the insulated monopole is applied on the milk pasteurization. The advantage of a monopole compared to other applicators is that the size of the monopole is small. A bio-heat equation has been implemented to predict the heat distribution of the insulated slotted monopole which is surrounded by a milk specimen including the obtained specific absorption rate (SAR) distribution using the wave equation (8). The simplified bio-heat transfer equation is given as:

$$pC_p\frac{\partial T}{\partial t}=\kappa\left(\frac{\partial^2 T}{\partial\rho^2}+\frac{1}{\rho}\frac{\partial T}{\partial\rho}+\frac{\partial^2 T}{\partial z^2}\right)+P_{ext} \tag{9}$$

where p = 1.03 kg/m^3, C_p = 3.93×10^{3}J/(kg °C), and κ = 0.56 J/(kg °C) in (9) are the density, specific heat capacity, and thermal conductivity of the milk specimen, respectively. Symbol P_{ext} (= P) is the external heat sources which is related to electric field, E, which is given in equation (3) in Section 2.2. Once the values of P_{ext} and ε_r'' are obtained from simulation [using Equation (8)] and measurement, equation (9) can be solved using the FEM-based simulator, namely COMSOL.

METHODOLOGY

Dielectric Measurements

The dielectric measurements were achieved using Keysight 85070E dielectric probe connected to a Keysight E5071C Vector Network Analyzer via coaxial cable and controlled by PC-based software as shown in Figure 4. Before the dielectric measurement of milk was done, the calibration process was performed at the probe aperture using three standard references, namely half-space air, metal short terminator, and de-ionized water (at ~25 °C), respectively.

After calibration procedure was completed, the probe was immersed into the milk with a slant and covering the entire area of the probe to ensure there are no bubbles attached into the surface of the coaxial probe. The milk samples were thermally processed using Fisher Scientific Isotherm Heated Magnetic Stirrer. The dielectric measurements of milk were repeated at each temperature step from 25 °C to 75 °C starting the measurements at ±0.5 °C as a probe cooling margin for accurate monitoring. The temperature, T of the raw milk samples were controlled by a water path surrounding the sample's container. The temperature, T readings were monitored by thermocouple connected to Fluke 287 multimeter as shown in Figure 4 (a).

Figure 4. (a) Dielectric measurement set-up. (b) Keysight 85070E dielectric probe

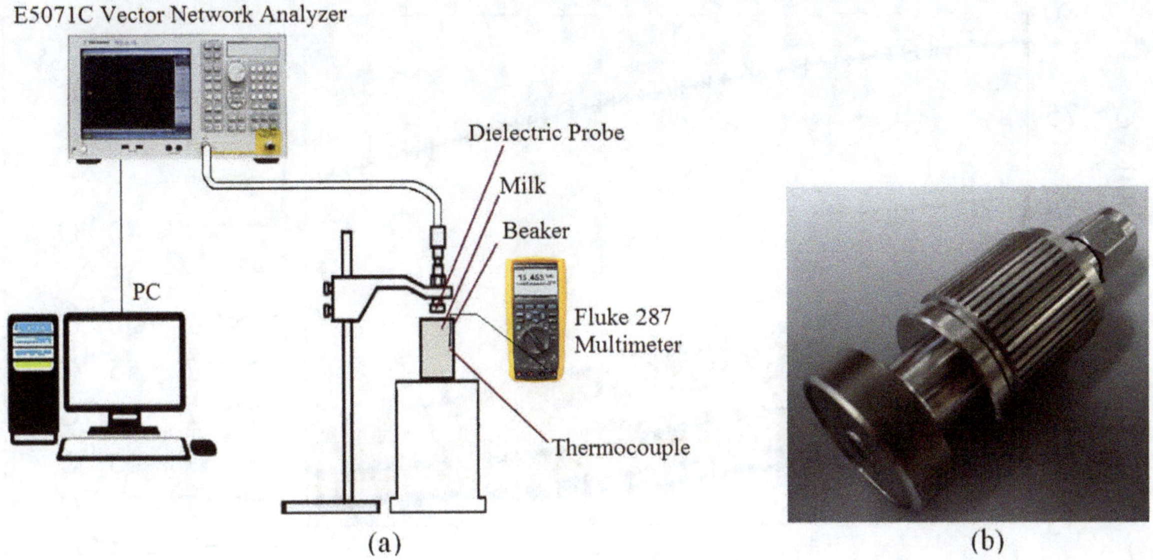

Since the main chemical constituent of the milk is water (82% to 87% water content), thus the milk is classified as polar solvent material and its value of ε_r can be modeled using Debye/Cole-Cole polar model (As mentioned in Section 2.2). An accurate monitoring of dielectric properties of the milk at various temperature degrees is required for accurate heat distribution simulation and pasteurization efficiency. Figures 5 (a) and (b) depict the measured dielectric constant, ε_r' and loss factor, ε_r'' of cow's raw milk at wide range of batch-based pasteurization temperatures from 25 °C to 75 °C, as well as for deionized water at 25 °C.

The ε_r' of milk is linearly decreased with temperature as well as the frequencies. Increasing in the high temperature, T causes increasing the intermolecular vibrations; Consequently, the hydrophobic interactions of milk's protein will be increased which in turn causes an interruption to the ordered water molecule arrangements that causes reduction in ε_r' (Vasbinder and De Kruif, 2003; Qian et al., 2017; Zhu et al., 2015; Herve et al., 1998; Al-Holy et al., 2005). Figure 5 (b) portrays a new pattern in cow's milk dielectric loss factor, ε_r'' which has not covered clearly in the literatures. The value of ε_r'' at low frequencies (< 1.9 GHz) is higher as the temperature, T goes high and decreasing exponentially over such frequency range. This dipole losses are caused by water dipole rotation.

While, the higher temperature causes decreasing in ε_r'' at higher frequencies from 1.9 GHz to 6 GHz. This is due ionic losses that cause migration of ions.The ionic conductivity, σ isstrongly affected by the temperature, especially at lower frequencies and physically it depends on the presence of salts in the milk. The variation in dielectric measurements is substantially affected by milk compositions and the cleanliness of milking and storage temperature. Figure 6 shows the measured ε_r' and ε_r'' versus temperature, T at operating frequency of 2.45 GHz, since the microwave frequency used for milk pasteurization is 2.45 GHz. The linear correlation between ε_r' and T at 2.45 GHz is given as:

$$\varepsilon_r' = 63.355 - 0.10537 \times T \text{ for } 25\,^\circ\text{C} \le T \le 75\,^\circ\text{C} \tag{10}$$

Figure 5. Measured (a) dielectric constant, ε_r' and (b) loss factor, ε_r'' versus operating frequency, f.

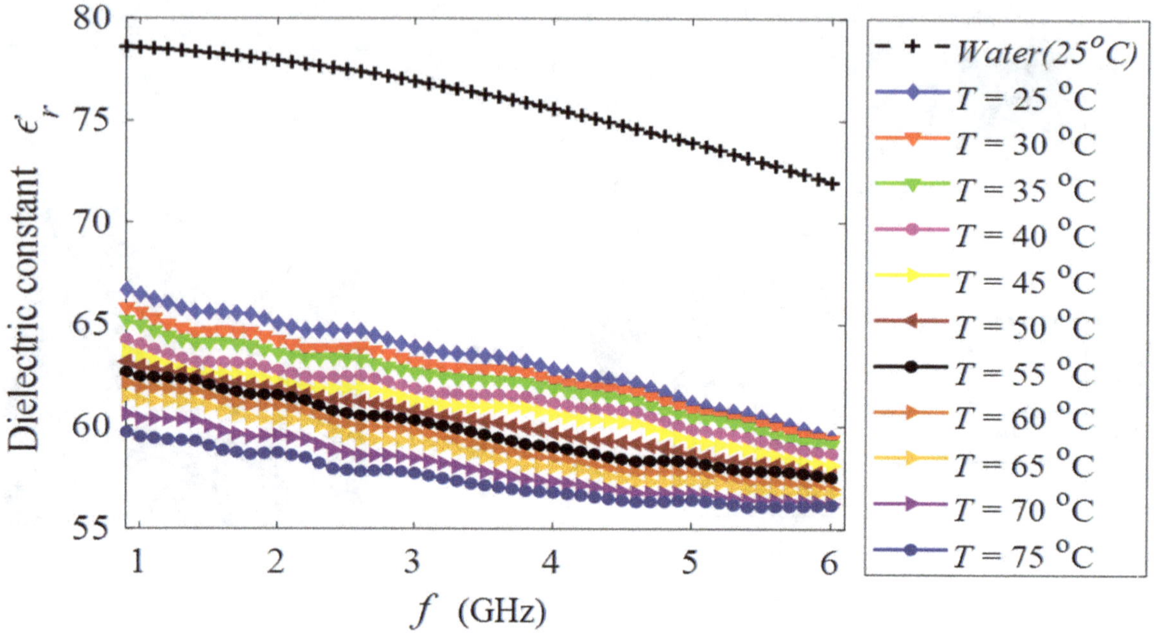

(a)

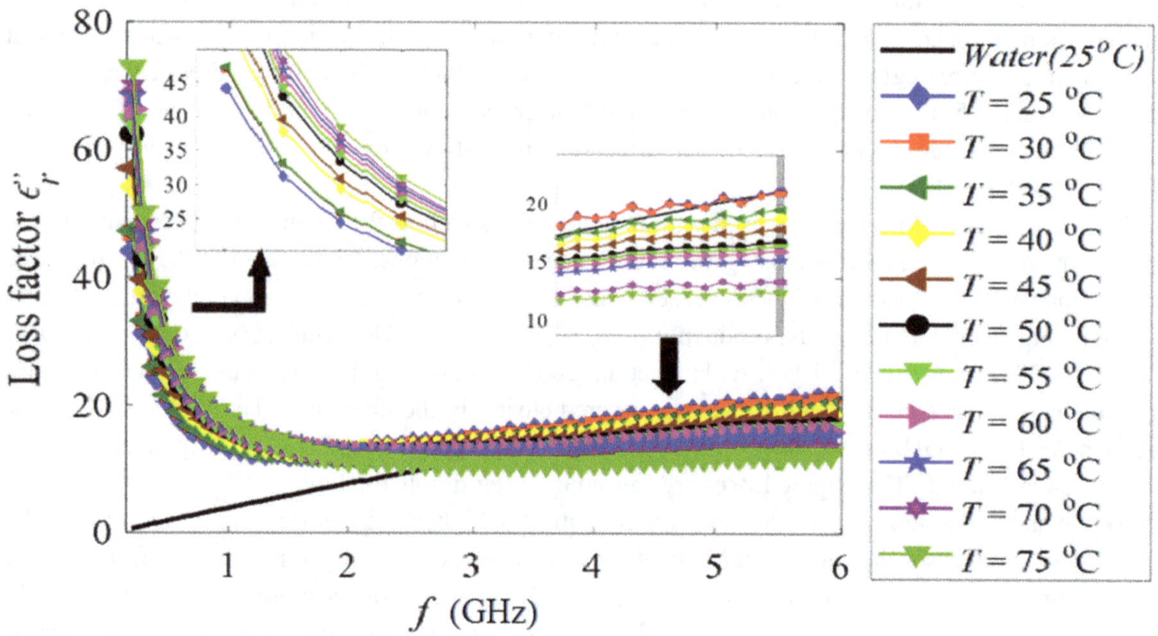

(b)

Figure 6. Measured (a) dielectric constant, ε_r' and (b) loss factor, ε_r'' versus temperature, T at 2.45 GHz.

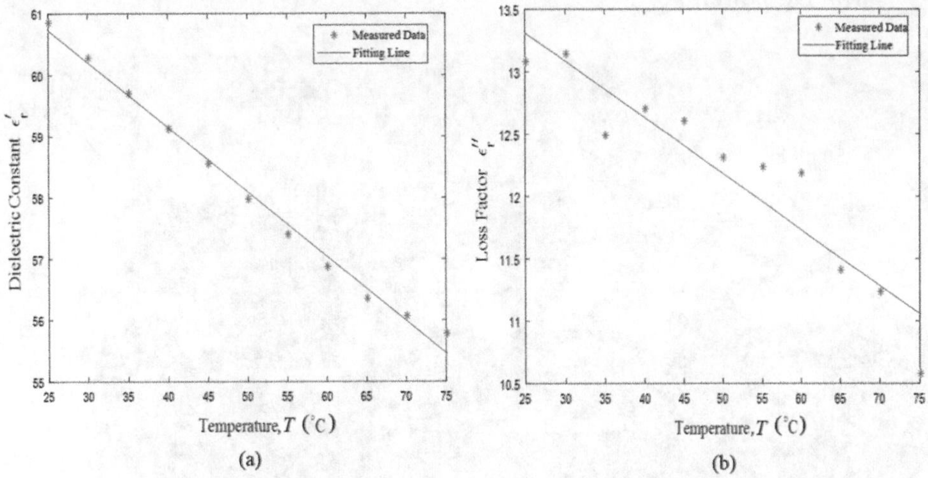

On the other hand, the related linear equation between ε_r'' and T can be written as:

$$\varepsilon_r'' = 14.46 - 0.04578 \times T \text{ for } 25\text{°C} \leq T \leq 75\text{°C} \quad (11)$$

Conventional Mono-Mode Microwave Reactor

Prototype Specifications

The prototype reactor uses a standard aperture size of S-band rectangular waveguide (WR430 waveguide) as a heating cavity where it allows mono mode (TE_{10} mode) of microwave to propagate in the cavity at frequency of 2.45 GHz (located on the industrial, scientific and medical (ISM) band). The reactor is mainly consists of DC power adjustable microwave generator (Samsung magnetron), customize applicator, waveguide isolator (up to 3 KW), three-screw waveguide tuner, and match load terminator (up to 3 KW). It should be noted that the isolator part (To prevent reflected wave from damaging the magnetron) is combination of circulator and match terminator.

In fact, the microwave output power, $P_{microwave}$ (in Watt) from the magnetron is controlled by the DC power supply, P_{input} (in Watt). The relationship between $P_{microwave}$ and P_{input} are measured using Mini-Circuits PWR-SEN-6G+ power sensor and 40-dB dual-loop waveguide coupler as shown in Figure 7. The DC power, P_{input} is increased gradually and the corresponding value of $P_{microwave}$ is measured as shown in Figure 7 (b) and their correlation is given as equation (12). From equation (12), it is clear that the generator started to generate microwave power requires DC power above 86 Watts.

$$P_{microwave} = 0.76066739 \times P_{input} - 86.00062282\,,\ r^2 = 0.9688 \quad (12)$$

Figure 7. (a) Microwave output power, $P_{microwave}$ measurement set-up. (b) Variations in microwave output power, $P_{mircowave}$ with DC power, P_{input}.

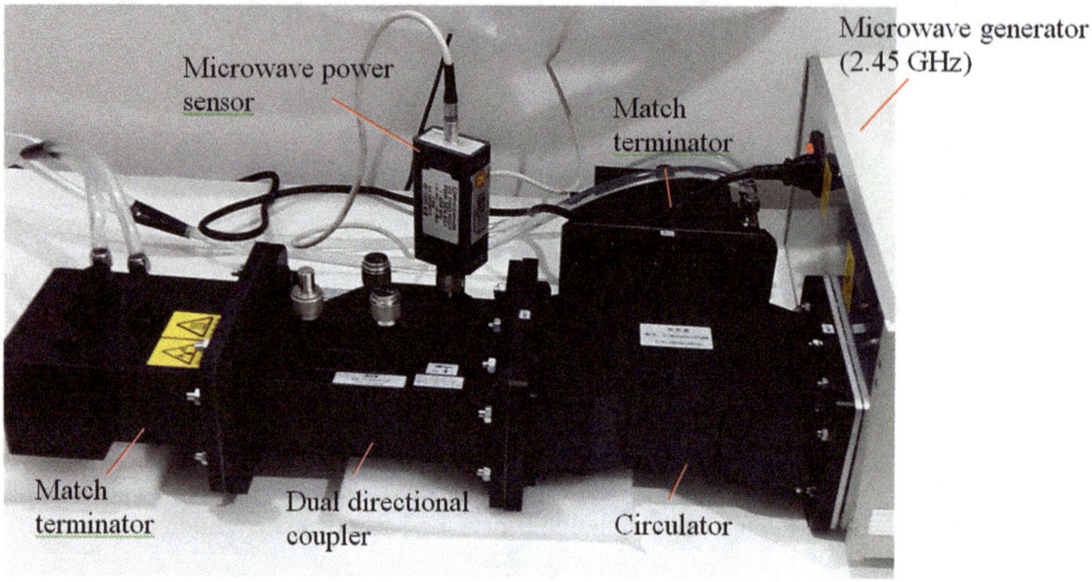

(a)

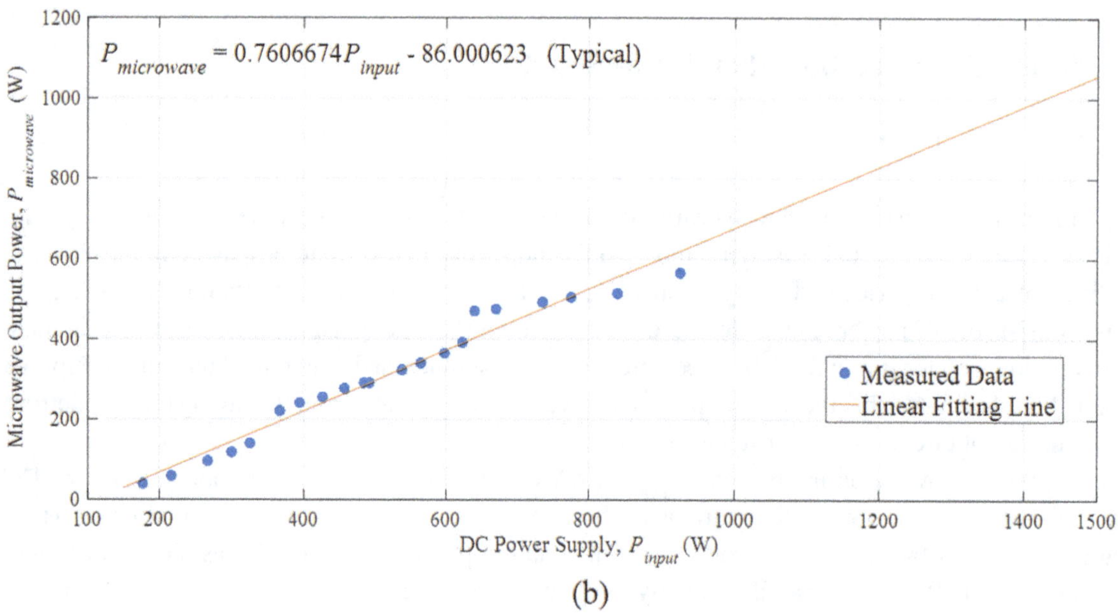

(b)

Customize Applicator

Figures 8 (a), (b) shows the custom applicator and custom quartz glass specimen holder. The customized cavity applicator is made of aluminum and top side of the cavity's wall is allowed to open to insert a sample into the cavity. The top side wall is designed to have three holes in order to release the heating pressure inside the cavity. Surrounding the hole was enclosed with ferrite rods to reduce microwave leakage, since ferrite is a microwave absorbing material as shown in Figure 8(a). Additionally, sensory devices can be placed into the specimen through the holes for real-time properties measurement of the sample during heating, such as monitoring the temperature of the sample. Besides, 3 mm thickness of quartz glass is customized as a specimen holder in which it has applicable heating temperature for long hours up to 1200 °C and capable of accommodating a maximum specimen volume of 780 cm^3.

Cavity Coupling Calibration

Before conducting a heating process, a calibration procedure is required in order to achieve the internal cavity coupling matching when the waveguide cavity is filled with milk specimen. The calibration is capable of optimizing absorption microwave power of specimen which may save the used power and heating time. The calibration setup consists of a match load waveguide, applicator filled with specimen, three-screw waveguide tuner, waveguide-coaxial adaptor, and Mini-Tiny vector network analyzer (VNA) as shown in Figure 9 (a). The cavity coupling matching can be tuned with three-screw waveguide tuner. The screws on waveguide tuner are tuned and the reflected signal in the cavity is monitoring by VNA until resonance achieve (minimum reflected power) at frequencies near 2.45 GHz as shown in Figure 9 (b).Here should be noted that when cavity matching is done, the first screw is usually first tuned and followed by a third screw, then the second one. A complete mono-mode microwave reactor system is shown in Figure 10, where the sample temperature can be monitored by entering the thermocouple into the applicator through the hole on top side wall of the applicator.

Modified Mono-Mode Microwave Reactor (Slotted Coaxial Applicator)

A slotted coaxial applicator is designed to match the complex impedance of the particular milk specimen at the operating frequency of 2.45 GHz to the impedance of the radiating applicator in order to ensure

Figure 8. (a) Custom waveguide applicator. (b) Custom quartz glass used as milk specimen holder.

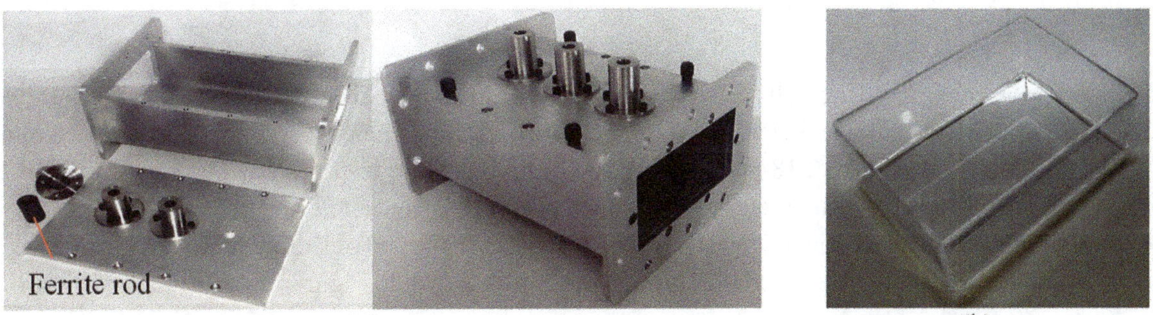

(a) (b)

Figure 9. (a) Cavity coupling calibration set-up. (b)VNA user interface guide (GUI); Cavity achieved resonance at frequencies near 2.45 GHz.

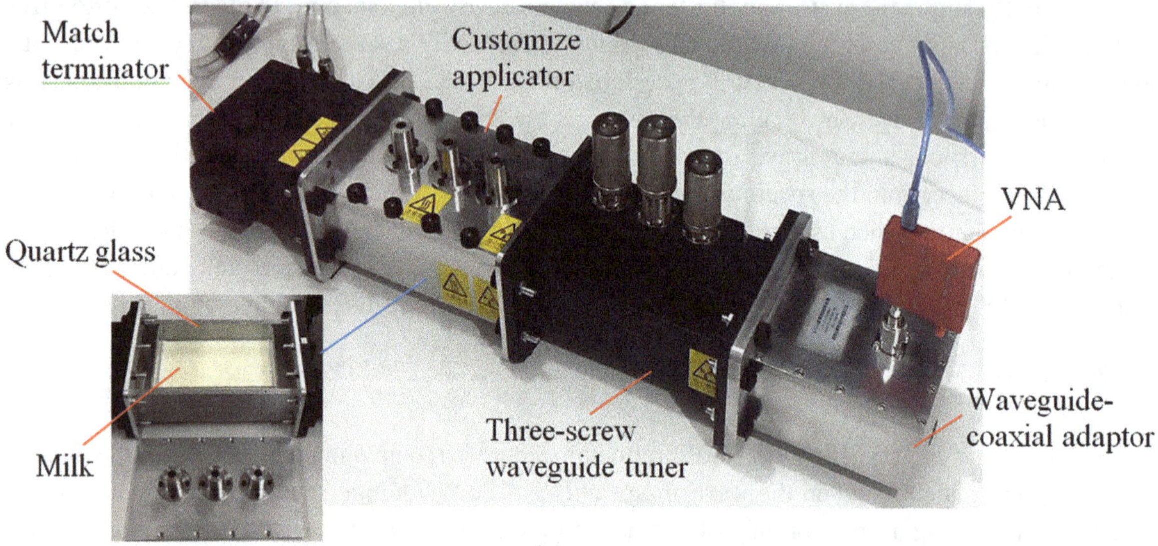

(a)

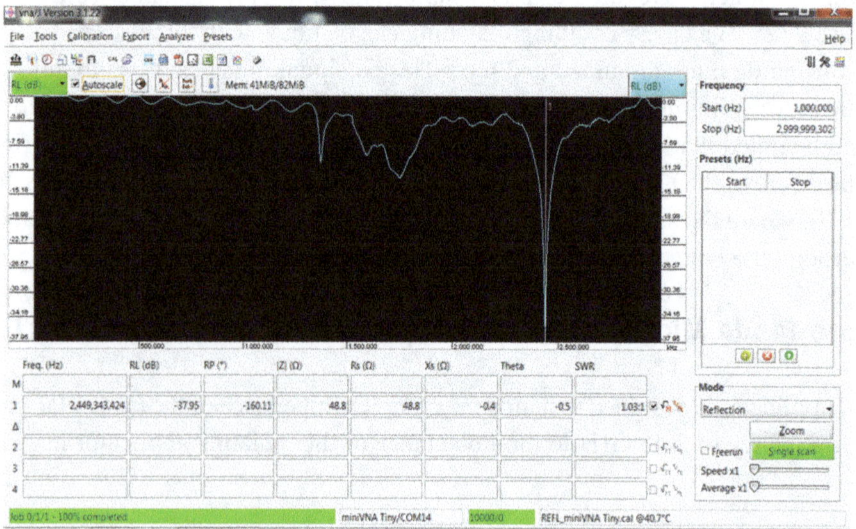

(b)

the microwave energy is efficiently delivered into the milk specimen. The slotted coaxial applicator is fabricated from a RG405/U semi-rigid coaxial cable with 0.51 mm of inner conductor diameter, 1.68 mm of filled Teflon diameter, and 2.18 mm of outer conductor diameter, respectively. The end of the cable has been soldered to become short circuit and one other end has been connected to the SMA connector. A small portion of the outer conductor (at a location adjacent to the end of short circuit) has been moved to allow the microwave to be emitted out through the gap as shown in Figure 11 (a).

Figure 10. Complete mono-mode microwave heating Prototype.

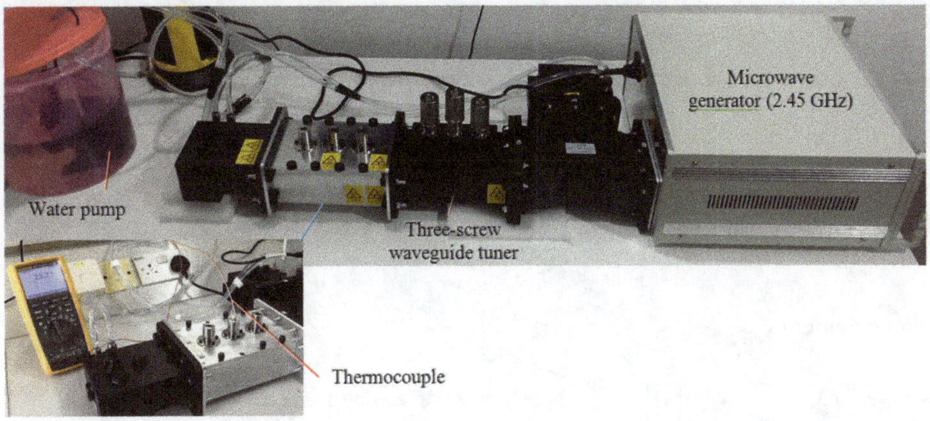

Figure 11. (a) The monopole has one slot sections in order to concentrate the heat distribution around the slot area. (b) The slotted monopole is shielded by Teflon.

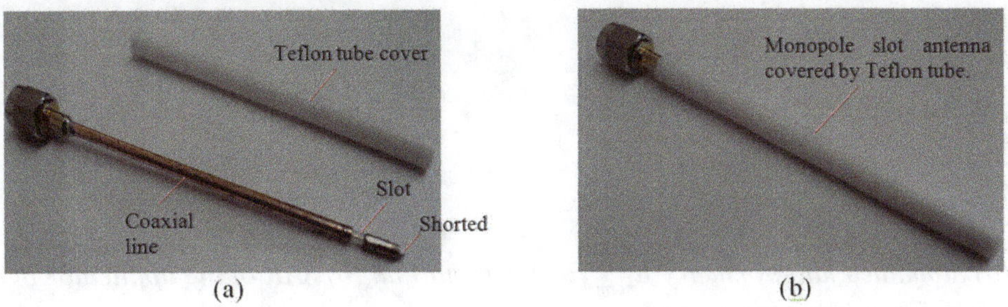

Figure 12. Simulated and measured $|S_{11}|$ of the applicator versus operating frequency.

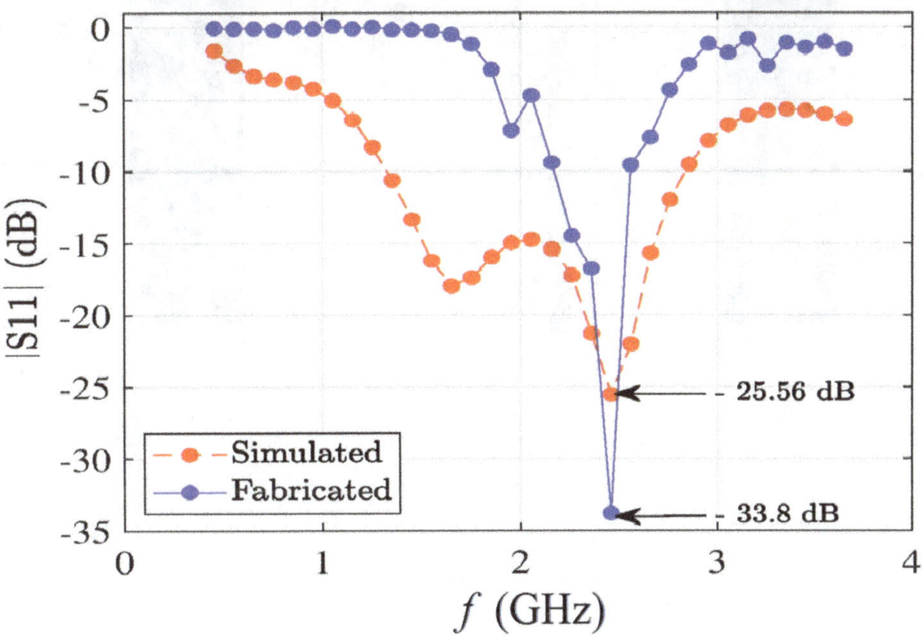

Figure 13. Modified mono-mode microwave reactor system

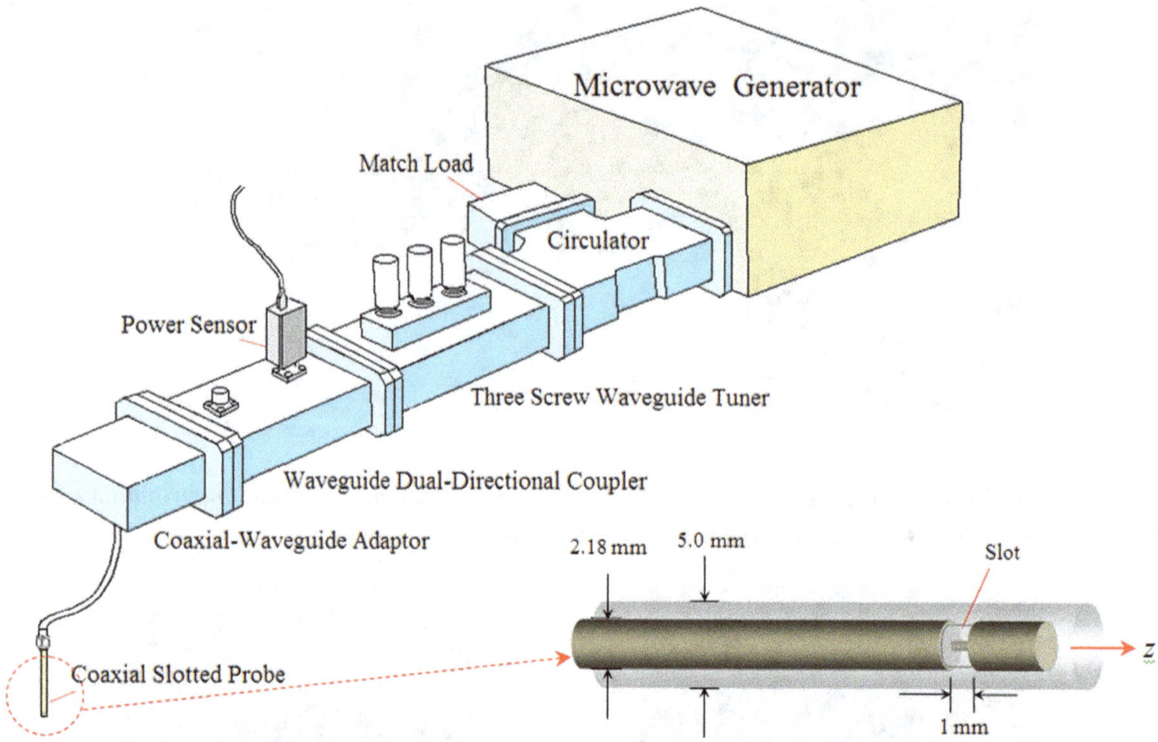

Figure 14. (a) Simulated half symmetry E_{norm} field contour and (b) SAR of the applicator in milk after 10 minutes excited by microwave power

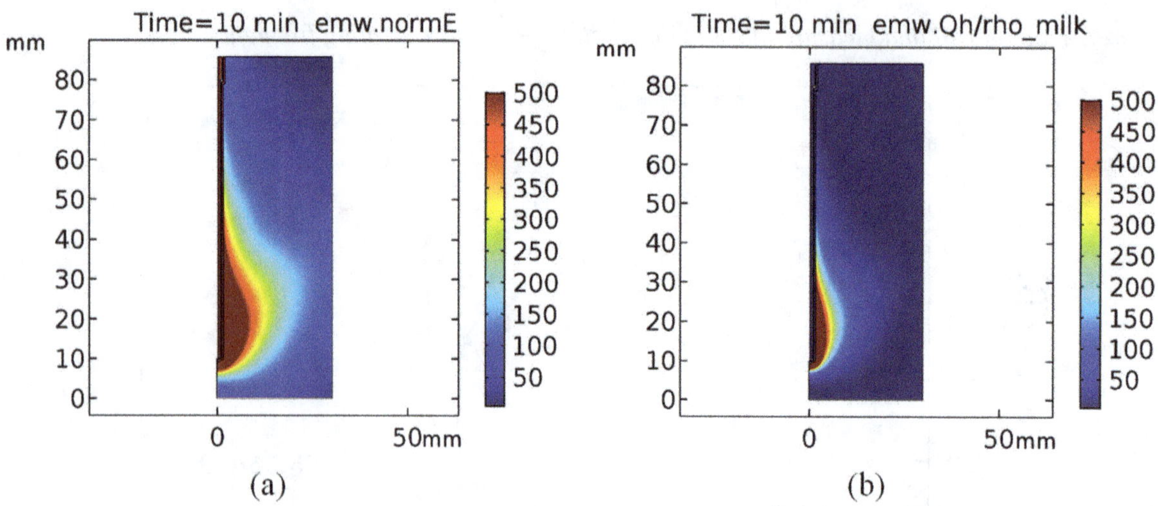

Figure 15. (a) Simulated half symmetry E_{norm} field contour and (b) SAR of the applicator in milk after 20 minutes excited by microwave power

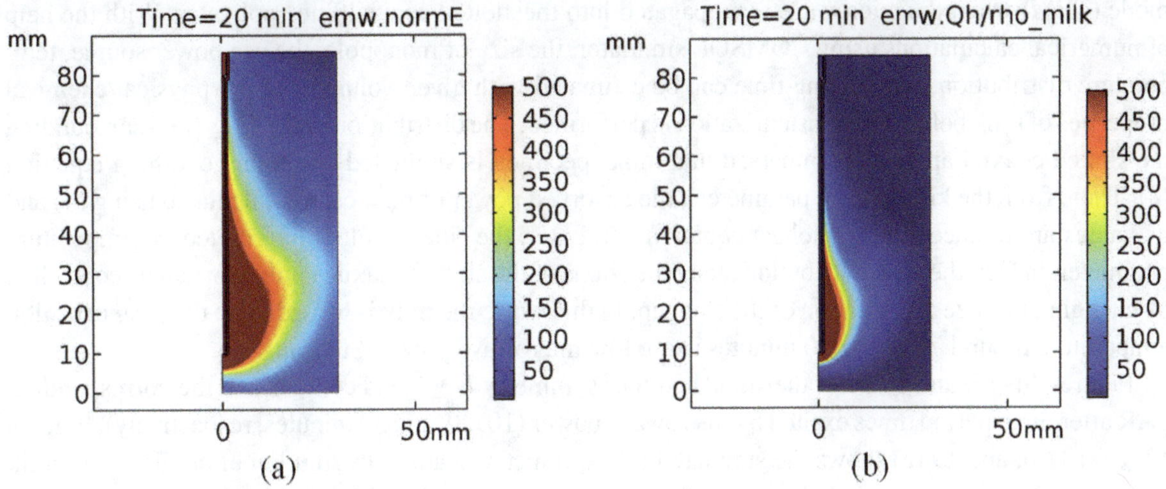

Figure 16. (a) E_{norm} of simulated antenna at 30 min. (b) SAR of simulated antenna after 30 minutes excited by microwave power

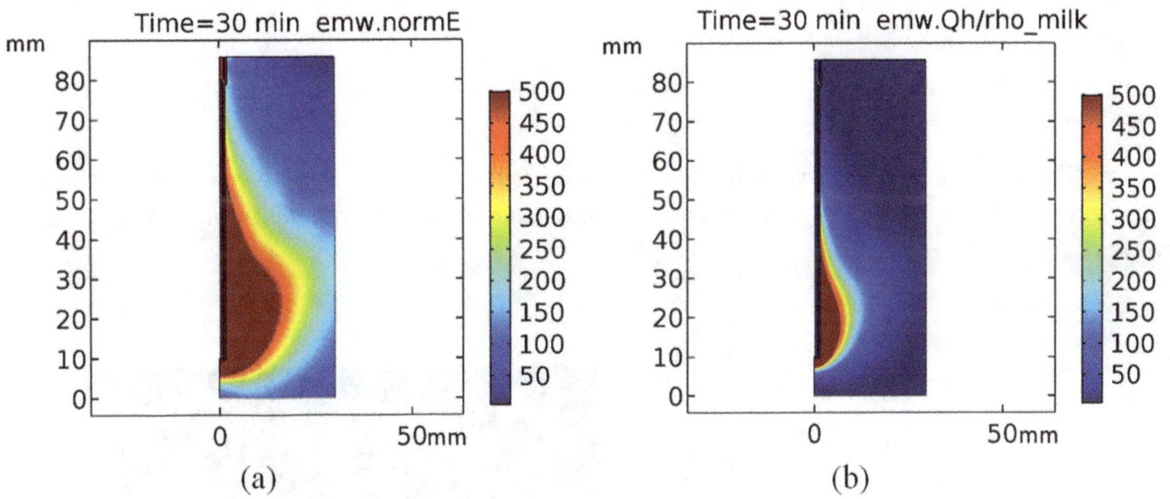

Finally, the slotted coaxial applicator is shielded by Teflon tube ($\varepsilon_r = 2.05$) to avoid the contact between the outer conductor metal and the milk specimen, as well as prevent the leakage of electric charges from the conductor into the milk as shown in Figure 11 (b). In other word, the Teflon tube is used as a hygienic protector when the slotted coaxial applicator is contacted directly with the milk specimen.

The fabricated slotted coaxial applicator is tested using Keysight E5071C Vector Network Analyzer. The return loss, $|S_{11}|$ of the applicator immersed in raw milk is measured and compared with simulated result as shown in Figure 12. Obviously, the matching impedance ($|S_{11}| < -20$ dB) of the applicator is existed at 2.45 GHz, thus optimum microwave power can be released from applicator into milk specimen.

The modified mono-mode microwave reactor system is shown in Figure 13. The dominant transverse electric, TE_{10} mode propagates in the rectangular waveguide is converted to transverse electromagnetic mode (TEM) when the microwave is propagated into the slotted coaxial line applicator. With the help of numerical calculations using COMSOL simulator, the size of monopole, the use power source, temperature distribution, and heating time can be estimated with given volume and the physical/chemical properties of milk before milk pasteurization is performed. The distribution of the E_{norm} field surrounding the slotted coaxial applicator immersed in a milk specimen is simulated based on governing equation (8). Then, from the known E_{norm} parameters, the absorbed power of milk can be calculated using (4) and its values are replaced in the bioheat equation (9). Later, the final result is interpreted in temperature parameter. In fact, the heat distribution along the z-axis of the slotted coaxial applicator can be controlled by changing the size and location of the slot gap. In this work, the milk is heated up to the pasteurization temperature around 72 °C for 30 minutes using low microwave power (30 Watt).

Figures 14, 15, and 16 show the simulated half symmetry E_{norm} field contour and the corresponding SAR after three period times excited by microwave power (10, 20, and 30 minutes, respectively). Figures 17 (a), 18 (a), and 19 (a) shows the simulated half symmetry heating distribution of applicator in milk specimen. On the other hand, the corresponding measured heating distribution using VarioCAM infrared thermal imaging camera (manufactured by InfraTec.) after 10, 20, and 30 minutes excited by microwave power via applicator are shown in Figures 17 (b), 18 (b), and 19 (b), respectively.

Figure 17. (a) Simulated axisymmetric distribution of temperatures in a half-space. (b) Measured temperature distribution in milk using thermal imaging camera after 10 minutes excited by microwave power at 2.45 GHz

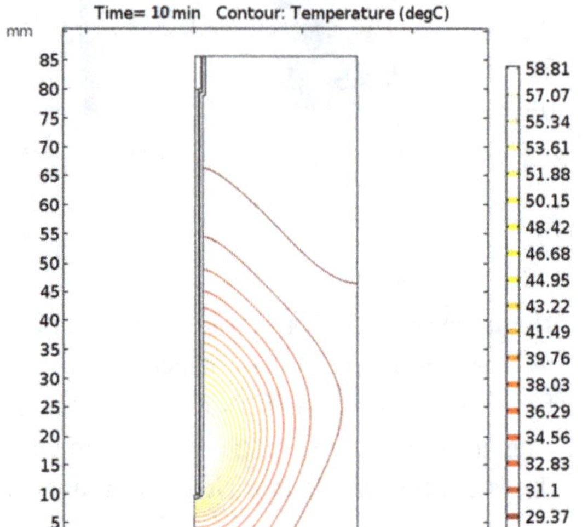

(a)

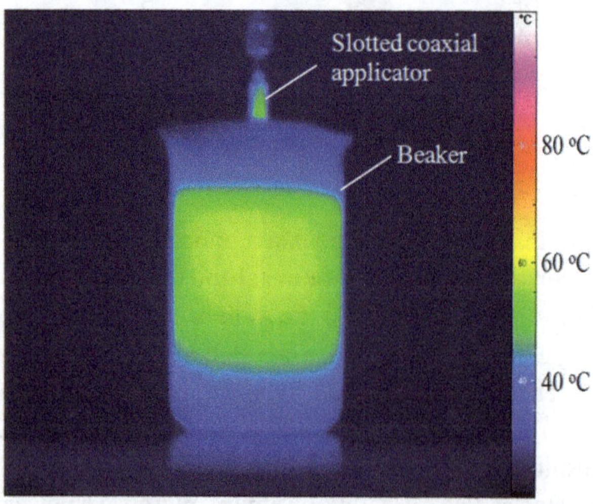

(b)

Figure 18. (a) Simulated axisymmetric distribution of temperatures in a half-space. (b) Measured temperature distribution in milk using thermal imaging camera after 20 minutes excited by microwave power at 2.45 GHz

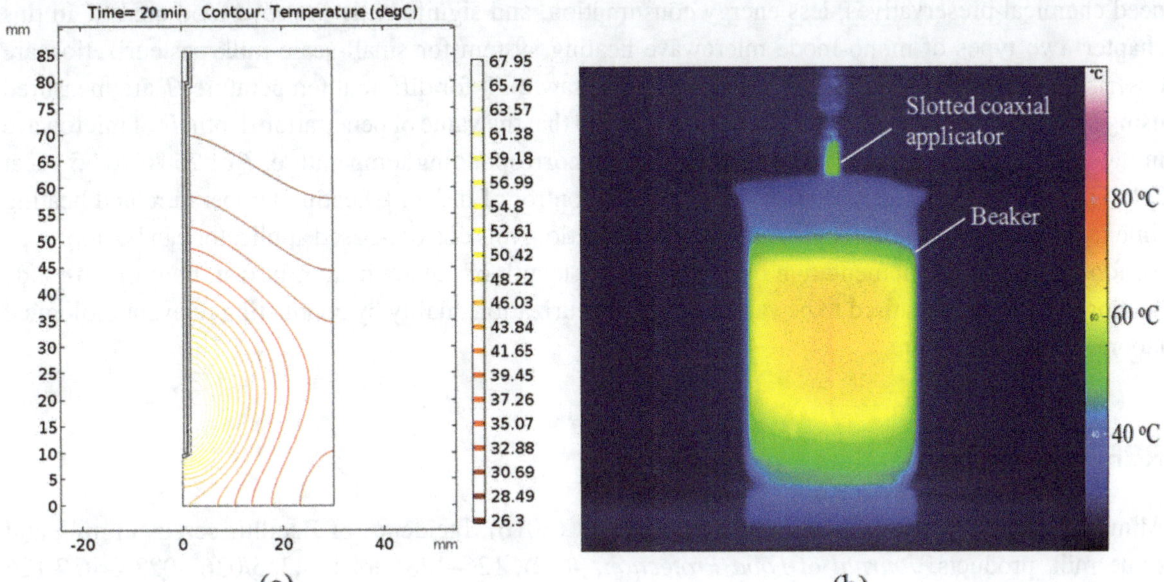

(a) (b)

Figure 19. (a) Simulated axisymmetric distribution of temperatures in a half-space. (b) Measured temperature distribution in milk using thermal imaging camera after 30 minutes excited by microwave power at 2.45 GHz

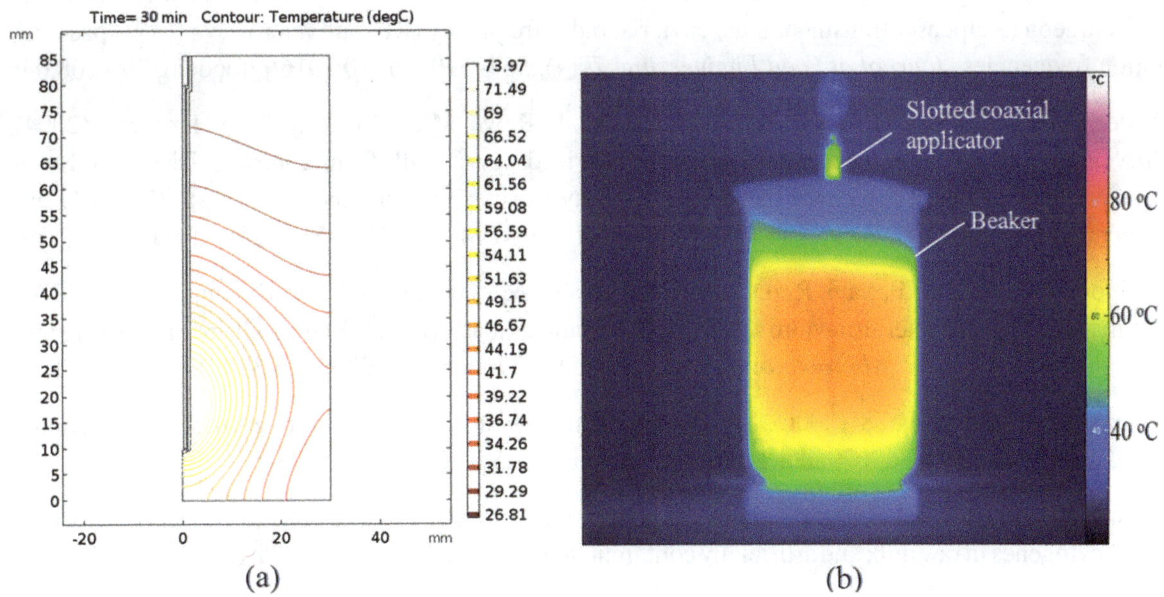

(a) (b)

CONCLUSION

Food pasteurization by microwave is one kind of green food processing technique, which it does not need chemical preservatives, less energy consumption, and significantly improve food quality. In this chapter, two types of mono-mode microwave heating system for small-scale milk pasteurization are described. In this study, dielectric properties of the raw milk for different temperature, T are measured using dielectric probe. On the other hand, it was found that the value of penetration depth, D of microwave in raw milk is varied from 1.1 cm to 1.3 cm for the corresponding temperature, T of 25 °C to 75 °C at 2.45 GHz. This study is expected to provide good control of the milk heating temperature and heating time in future microwave milk pasteurization. In addition, this coaxial-based applicator can be improved for food industrial implementation by extending the length of the applicator. In fact, both mono-mode heating systems are required to be validated for pasteurization quality by conducting relevant biological laboratory tests in future.

REFERENCES

Ahmed, A. A. H., Moustafa, M. K., & Marth, E. H. (2016). Incidence of Bacillus cereus in milk and some milk products. *Journal of Food Protection*, *46*(2), 126–128. doi:10.4315/0362-028X-46.2.126 PMID:30913599

Ahmed, A. M., & Shimamoto, T. (2014). Isolation and molecular characterization of Salmonella enterica, Escherichia coli O157: H7 and Shigella spp. from meat and dairy products in Egypt. *International Journal of Food Microbiology*, *168–169*, 57–62. doi:10.1016/j.ijfoodmicro.2013.10.014 PMID:24239976

Al-Holy, M., Wang, Y., Tang, J., & Rasco, B. (2005). Dielectric properties of salmon (Oncorhynchus keta) and sturgeon (Acipenser transmontanus) caviar at radio frequency (RF) and microwave (MW) pasteurization frequencies. *Journal of Food Engineering*, *70*(4), 564–570. doi:10.1016/j.jfoodeng.2004.08.046

Anderson, W., O'Reilly, C. E., Grant, I. R., Donaghy, J., Rowe, M., O'Mahony, P., & Harvey, P. (2004). Surveillance of bulk raw and commercially pasteurized cows' milk from approved Irish liquid-milk pasteurization plants to determine the incidence of mycobacterium paratuberculosis. *Applied and Environmental Microbiology*, *70*(9), 5138–5144. doi:10.1128/AEM.70.9.5138-5144.2004 PMID:15345392

Arrigoni, N., Cammi, G., Savi, R., De Cicco, C., Ricchi, M., Garbarino, C. A., & Pongolini, S. (2016). Estimation of Mycobacterium avium subsp. paratuberculosis load in raw bulk tank milk in Emilia-Romagna Region (Italy) by qPCR. *MicrobiologyOpen*, *5*(4), 551–559. doi:10.1002/mbo3.350 PMID:26991108

Aygun, O., & Pehlivanlar, S. (2006). Listeria spp. in the raw milk and dairy products in Antakya, Turkey. *Food Control*, *17*(8), 676–679. doi:10.1016/j.foodcont.2005.09.014

Bacci, C., Lanzoni, E., Alpigiani, I., Boni, E., Vismarra, A., Bonardi, S., & Brindani, F. (2014). Listeria monocytogenes in raw milk and artificially contaminated aliquots. *Large Animal Review*, *20*(4), 175–180.

Bali, O. S., Lajnef, R., Felfoul, I., Attia, H., & Ayadi, M. A. (2013). Original article detection of escherichia coli in unpasteurized raw Milk. *International Journal of Agricultural and Food Science*, *3*(2), 53–55.

Barakat, A. M. A., Sobhy, M. M., El Fadaly, H. A. A., Rabie, N. S., Khalifa, N. O., Hassan, E. R., & Zaki, M. S. (2015). Zoonotic hazards of campylobacteriosis in some areas in Egypt. *Life Science Journal*, *12*(7), 9–14.

Bartolomeoli, I., Maifreni, M., Frigo, F., Urli, G., & Marino, M. (2009a). Occurrence and characterization of Staphylococcus aureus isolated from raw milk for cheesemaking. *International Journal of Dairy Technology*, *62*(3), 366–371. doi:10.1111/j.1471-0307.2009.00498.x

Bartolomeoli, I., Maifreni, M., Frigo, F., Urli, G., & Marino, M. (2009b). Occurrence and characterization of Staphylococcus aureus isolated from raw milk for cheesemaking. *International Journal of Dairy Technology*, *62*(3), 366–371. doi:10.1111/j.1471-0307.2009.00498.x

Ben Kahla, I., Boschiroli, M. L., Souissi, F., Cherif, N., Benzarti, M., Boukadida, J., & Hammami, S. (2011). Isolation and molecular characterisation of Mycobacterium bovis from raw milk in Tunisia. *African Health Sciences*, *11*(3 Suppl. 1), S2–S5. doi:10.4314/ahs.v11i3.70032 PMID:22135638

Bogdal, D. (2005). Interaction of Microwaves with Different Materials. In D. Bogdal (Ed.), Microwave-Assisted Organic Synthesis: One Hundred Reaction Procedures. The Netherlands: Elsevier. doi:10.1016/S1460-1567(05)80014-5

Boxler, C., Augustin, W., & Scholl, S. (2013). Fouling of milk components on DLC coated surfaces at pasteurization and UHT temperatures. *Food and Bioproducts Processing*, *91*(4), 336–347. doi:10.1016/j.fbp.2012.11.012

Capelli, E., Sacchi, R., Ghitti, M., Feligini, M., Bonacina, C., Panelli, S., & Brambati, E. (2014). One-year investigation of Clostridium spp. occurrence in raw milk and curd of Grana Padano cheese by the automated ribosomal intergenic spacer analysis. *Food Control*, *42*, 71–77. doi:10.1016/j.foodcont.2014.02.002

Caudron, C., Héchard, C., Natorp, J. C., Berri, M., Souriau, A., Rodolakis, A., & Camuset, P. (2007). Comparison of Coxiella burnetii shedding in milk of dairy Bovine, Caprine, and Ovine Herds. *Journal of Dairy Science*, *90*(12), 5352–5360. doi:10.3168/jds.2006-815 PMID:18024725

Chandrasekaran, S., Ramanathan, S., & Basak, T. (2012). Microwave material processing-a review. *AIChE Journal. American Institute of Chemical Engineers*, *58*(2), 330–363. doi:10.1002/aic.12766

Chong, C. M., & Datta, A. K. (1999). Thawing of foods in a microwave oven: II. Effect of load geometry and dielectric properties. *The Journal of Microwave Power and Electromagnetic Energy*, *34*(1), 22–32. doi:10.1080/08327823.1999.11688385 PMID:10355128

Christiansson, A., Bertilsson, J., & Svensson, B. (1999). Bacillus cereus spores in raw milk: Factors affecting the contamination of milk during the grazing period. *Journal of Dairy Science*, *82*(2), 305–314. doi:10.3168/jds.S0022-0302(99)75237-9 PMID:10068952

Clare, D. A., Bang, W. S., Cartwright, G., Drake, M. A., Coronel, P., & Simunovic, J. (2005). Comparison of sensory, microbiological, and biochemical parameters of microwave versus indirect UHT fluid skim milk during storage. *Journal of Dairy Science*, *88*(12), 4172–4182. doi:10.3168/jds.S0022-0302(05)73103-9 PMID:16291608

Constellation. (2016). Which is more energy efficient? Microwave vs toaster oven vs oven. Retrieved from https://blog.constellation.com/2016/05/20/toaster-oven-vs-microwave-energy-efficiency/

Coronel, P., Simunovic, J., & Sandeep, K. P. (2003). Temperature profiles within milk after heating in a continuous-flow tubular microwave system operating at 915 MHz. *Journal of Food Science*, *68*(6), 1976–1981. doi:10.1111/j.1365-2621.2003.tb07004.x

D'agostino, M., Wagner, M., Vazquez-boland, J. A., Kuchta, T., Karpiskova, R., Hoorfar, J., & Cook, N. (2016). A validated PCR-based method to detect Listeria monocytogenes using raw milk as a food model—Towards an international standard. *Journal of Food Protection*, *67*(8), 1646–1655. doi:10.4315/0362-028X-67.8.1646 PMID:15330529

D'amico, D. J., & Donnelly, C. W. (2011). Characterization of Staphylococcus aureus strains isolated from raw milk utilized in small-scale Artisan cheese production. *Journal of Food Protection*, *74*(8), 1353–1358. doi:10.4315/0362-028X.JFP-10-533 PMID:21819666

Daminelli, P., Bertasi, B., Dalzini, E., Losio, M. N., Bernini, V., & Varisco, G. (2015). Survey of prevalence and seasonal variability of Listeria monocytogenes in raw cow milk from Northern Italy. *Food Control*, *60*, 466–470.

de Bruin, A., van der Plaats, R. Q. J., de Heer, L., Paauwe, R., Schimmer, B., Vellema, P., & van Duynhoven, Y. T. H. P. (2012). Detection of Coxiella burnetii DNA on small-ruminant farms during a Q fever outbreak in the Netherlands. *Applied and Environmental Microbiology*, *78*(6), 1652–1657. doi:10.1128/AEM.07323-11 PMID:22247143

de Souza, V., Ribeiro, L. F., Oliveira, M. C., Borges, L. A., de Medeiros, M. I. M., Maluta, R. P., & Barbosa, M. M. C. (2016). Antimicrobial resistance and virulence factors of Escherichia coli in cheese made from unpasteurized milk in three cities in Brazil. *Foodborne Pathogens and Disease*, *13*(9), 469–476. doi:10.1089/fpd.2015.2106 PMID:27258947

Donovan, K. O. (1959). The occurrence of Bacillus Cereus in milk and on dairy equipment. *The Journal of Applied Bacteriology*, *22*(1), 131–137. doi:10.1111/j.1365-2672.1959.tb04617.x

Eftekhari, M., & Mosavari, N. (2016). Isolation and molecular identification of Mycobacterium from commercially available pasteurized milk and raw milk samples collected from two infected cattle farms in Alborz Province, Iran. *International Journal of Mycobacteriology*, *5*, S222–S223. doi:10.1016/j.ijmyco.2016.10.005 PMID:28043568

El-Sharoud, W. M. (2009). Prevalence and survival of Campylobacter in Egyptian dairy products. *Food Research International*, *42*(5–6), 622–626. doi:10.1016/j.foodres.2009.01.009

Ellison, W. J. (2007). Permittivity of pure water at standard atmospheric pressure, over the frequency range 0-25 THz and the temperature range 0-100 °C. *Journal of Physical and Chemical Reference Data*, *36*(1), 1–18. doi:10.1063/1.2360986

Fagundes, H., Barchesi, L., Filho, A. N., Ferreira, L. M., & Oliveira, C. A. F. (2010). Occurrence of staphylococcus aureus in raw milk produced in dairy farms in São Paulo State, Brazil. *Brazilian Journal of Microbiology*, *41*(2), 376–380. doi:10.1590/S1517-83822010000200018 PMID:24031507

Fang, Y., Cheng, J., Roy, R., Roy, D. M., & Agrawal, D. K. (1997). Enhancing densification of zirconia-containing ceramic-matrix composites by microwave processing. *Journal of Materials Science*, *32*(18), 4925–4930. doi:10.1023/A:1018624223909

Fedio, W. M., & Jackson, H. (2013). Incidence of Listeria monocytogenes in raw bulk milk in Alberta. *Canadian Institute of Food Science and Technology Journal*, *23*(4–5), 236–238.

Giezendanner, N., Meyer, B., Gort, M., Müller, P., & Zweifel, C. (2009). Rohmikh-Assoziierte Staphylococcus Aureus Intoxikation Bei Kindern. *Schweizer Archiv fur Tierheilkunde*, *151*(7), 329–331. doi:10.1024/0036-7281.151.7.329 PMID:19565455

Grant, E. H., Buchanan, T. J., & Cook, H. F. (1957). Dielectric behavior of water at microwave frequencies. *The Journal of Chemical Physics*, *26*(1), 156–161. doi:10.1063/1.1743242

Groisman, Y., & Gedanken, A. (2008). Continuous flow, circulating microwave system and its application in nanoparticle fabrication and biodiesel synthesis. *The Journal of Physical Chemistry C*, *112*(24), 8802–8808. doi:10.1021/jp801409t

Gude, V. G. (2017). Microwave–Mediated Biofuel Production. In Microwave–Mediated Biofuel Production. doi:10.1201/9781315151892

Gundogan, N., & Avci, E. (2014). Occurrence and antibiotic resistance of Escherichia coli, Staphylococcus aureus and Bacillus cereus in raw milk and dairy products in Turkey. *International Journal of Dairy Technology*, *67*(4), 562–569. doi:10.1111/1471-0307.12149

Gwida, M. M., & AL-Ashmawy, M. A. M. (2014). AL-Ashmawy, M. A. M. (2014). Culture versus PCR for Salmonella species identification in some dairy products and dairy handlers with special concern to its Zoonotic importance. *Veterinary Medicine International*, 1–5. doi:10.1155/2014/502370

Haitsma, O., De Jonge, R., Besselse, M., Mulder, B., Van Pelt, W., Van Steenbergen, J., & Van Duynhoven, Y. T. H. P. (2009). A prolonged outbreak of Salmonella Typhimurium infection related to an uncommon vehicle: Hard cheese made from raw milk. *Epidemiology and Infection*, *137*(11), 1548–1557. doi:10.1017/S0950268809002337 PMID:19296867

Hall, C. W., & Trout, G. M. (1968). *Milk Pasteurization*. Westport, Connecticut: AVI Publishing Company, Inc.

Hamelin, J., Mehl, J. B., & Moldover, M. R. (1998). The static dielectric constant of liquid water between 274 and 418 K near the saturated vapor pressure. *International Journal of Thermophysics*, *19*(5), 1359–1380. doi:10.1023/A:1021979401680

Herve, A. G., Tang, J., Luedecke, L., & Feng, H. (1998). Dielectric properties of cottage cheese and surface treatment using microwaves. *Journal of Food Engineering*, *37*(4), 389–410. doi:10.1016/S0260-8774(98)00102-2

Ho, T., Htwe, K. K., Yamasaki, N., Zhang, G. Q., Ogawa, M., Yamaguchi, T., & Hirai, K. (1995). Isolation of Coxiella burnetii from dairy cattle and ticks, and some characteristics of the isolates in Japan. *Microbiology and Immunology*, *39*(9), 663–671. doi:10.1111/j.1348-0421.1995.tb03254.x PMID:8577279

Hunt, K., Schelin, J., Rådström, P., Butler, F., & Jordan, K. (2012). Classical enterotoxins of coagulase-positive Staphylococcus aureus isolates from raw milk and products for raw milk cheese production in Ireland. *Dairy Science & Technology*, *92*(5), 487–499. doi:10.100713594-012-0067-4

Husu, J. R., Seppänen, J. T., Sivelä, S. K., & Rauramaa, A. L. (1990). Contamination of Raw Milk by Listeria monocytogenes on Dairy Farms. *Journal of Veterinary Medicine, Series B*, *37*(1–10), 268–275.

Karns, J. S., Gorski, L., McCluskey, B. J., Perdue, M. L., & Van Kessel, J. S. (2010). Prevalence of Salmonellae, Listeria monocytogenes, and Fecal Coliforms in bulk tank milk on US dairies. *Journal of Dairy Science*, *87*(9), 2822–2830. PMID:15375040

Kaynak-Onurdag, F., Okten, S., & Sen, B. (2016). Screening Brucella spp. in bovine raw milk by real-time quantitative PCR and conventional methods in a pilot region of vaccination, Edirne, Turkey. *Journal of Dairy Science*, *99*(5), 3351–3357. doi:10.3168/jds.2015-10637 PMID:26971148

Kazwala, R. R., Daborn, C. J., Kusiluka, L. J. M., Jiwa, S. F. H., Sharp, J. M., & Kambarage, D. M. (1998). Isolation of Mycobacterium species from raw milk of pastoral cattle of the Southern Highlands of Tanzania. *Tropical Animal Health and Production*, *30*(4), 233–239. doi:10.1023/A:1005075112393 PMID:9760715

Kessy, B. M., John, J., Rajashekara, G., Kassem, I. I., Gebreyes, W., Kazwala, R. R., & Kashoma, I. P. (2015). Prevalence and antimicrobial resistance of Campylobacter isolated from dressed beef Carcasses and raw milk in Tanzania. *Microbial Drug Resistance*, *22*(1), 40–52. PMID:26153978

Kevenk, T. O., & Terzi Gulel, G. (2016). Prevalence, antimicrobial resistance and Serotype distribution of Listeria monocytogenes isolated from raw milk and dairy products. *Journal of Food Safety*, *36*(1), 11–18. doi:10.1111/jfs.12208

Kudra, T., & Van, F. R. (1991). Heating characteristics of milk constituents in a microwave pasteurization system. *Journal of Food Science*, *56*(4), 931–934. doi:10.1111/j.1365-2621.1991.tb14608.x

Loftis, A. D., Priestley, R. A., & Massung, R. F. (2010). Detection of Coxiella burnetii in commercially available raw milk from the United States. *Foodborne Pathogens and Disease*, *7*(12), 1453–1456. doi:10.1089/fpd.2010.0579 PMID:20704507

Loncarevic, S., Jørgensen, H. J., Løvseth, A., Mathisen, T., & Rørvik, L. M. (2005). Diversity of Staphylococcus aureus enterotoxin types within single samples of raw milk and raw milk products. *Journal of Applied Microbiology*, *98*(2), 344–350. doi:10.1111/j.1365-2672.2004.02467.x PMID:15659189

Lubote, R., Shahada, F., & Matemu, A. (2014). Prevalence of Salmonella spp. and Escherichia coli in raw milk value chain in Arusha, Tanzania. *American Journal of Research Communication*, *2*(9), 1–13.

Luzzana, M., Morandi, S., Cremonesi, P., Pisoni, G., Moroni, P., Brasca, M., & Castiglioni, B. (2007). Detection of enterotoxigenic Staphylococcus aureus isolates in raw milk cheese. *Letters in Applied Microbiology*, *45*(6), 586–591. doi:10.1111/j.1472-765X.2007.02231.x PMID:17916131

Magnani, M., Brandi, G., Casiere, A., Amagliani, G., Omiccioli, E., & Bruce, I. (2004). Direct detection of Listeria monocytogenes from milk by magnetic based DNA isolation and PCR. *Food Microbiology*, *21*(5), 597–603. doi:10.1016/j.fm.2003.10.008

Magnino, S., Vicari, N., Boldini, M., Rosignoli, C., Nigrelli, A., Andreoli, G., & Fabbi, M. (2009). Rilevamento di Coxiella burnetii nel latte di massa di alcune aziende bovine Lombarde. *Large Animal Review*, *15*(1), 3–6.

Mahmoodi, M. M. (2010). Occurrence of Listeria monocytogenes in raw milk and dairy products in Noorabad, Iran. *Journal of Animal and Veterinary Advances*, *9*(1), 16–19. doi:10.3923/javaa.2010.16.19

Mansouri-Najand, L., & Khalili, M. (2007). Detection of shiga-like toxigenic Escherichia coli from raw milk cheeses produced in Kerman-Iran. *Veterinarski Arhiv*, *77*(6), 515–522.

Marnissi, E. (2014). Presence of Listeria monocytogenes in raw milk and traditional dairy products marketed in the north-central region of Morocco. *African Journal of Food Science*, *7*(5), 87–91. doi:10.5897/AJFS2013.0992

Marshall, J. C., Soboleva, T. K., Jamieson, P., & French, N. P. (2016). Estimating bacterial pathogen levels in New Zealand bulk tank milk. *Journal of Food Protection*, *79*(5), 771–780. doi:10.4315/0362-028X.JFP-15-230 PMID:27296424

Masateru, N., & Masato, M. (2013). Single-mode microwave reactor used for continuous flow reactions under elevated pressure. *Industrial & Engineering Chemistry Research*, *52*(12), 4683–4687. doi:10.1021/ie400199r

Meunier-Goddik L., Sandra, S. (2016). Liquid Milk Products: Pasteurized Milk. *Reference Module in Food Science*.

Mioni, R., Giacometti, F., Cascone, G., Bianchi, M., Piva, S., Serraino, A., ... Merialdi, G. (2015). Human campylobacteriosis related to the consumption of raw milk sold by vending machines in Italy: Quantitative risk assessment based on official controls over four years. *Preventive Veterinary Medicine*, *121*(1–2), 151–158. PMID:26142145

Moussa-Boudjemaa, B., Kihal, M., Lopez, M., & Gonzalez, J. (2004). The incidence of Bacillus cereus spores in Algerian raw milk: A study of the chief sources of contamination. *Archiv fur Lebensmittelhygiene*, *55*(4), 94–96.

Naguib, K., & Shouman, M. T. (1972). Identification and typing of Clostridia in raw milk in Egypt. *The Journal of Applied Bacteriology*, *35*(4), 525–530. doi:10.1111/j.1365-2672.1972.tb03733.x PMID:4346755

Natale, A., Busani, L., Comin, A., De Rui, S., Buffon, L., Nardelli, S., & Ceglie, L. (2009). First report of bovine Q-fever in north-eastern Italy: Preliminary results. *Clinical Microbiology and Infection*, *15*, 144–145. doi:10.1111/j.1469-0691.2008.02154.x PMID:19438637

Nazari, R., Godarzi, H., Rahimi Baghi, F., & Moeinrad, M. (2014). Enterotoxin gene profiles among Staphylococcus aureus isolated from raw milk. *Majallah-i Tahqiqat-i Dampizishki-i Iran*, *15*(4), 409–412. PMID:27175141

O'Donnell, E. T. (1995). The incidence of Salmonella and Listeria in raw milk from farm bulk tanks in England and Wales. *International Journal of Dairy Technology*, *48*(1), 25–29. doi:10.1111/j.1471-0307.1995.tb02433.x

Ortolani, M. B. T., Yamazi, A. K., Moraes, P. M., Viçosa, G. N., & Nero, L. A. (2009). Microbiological quality and safety of raw milk and soft cheese and detection of Autochthonous Lactic acid bacteria with Antagonistic activity against Listeria monocytogenes, Salmonella Spp., and Staphylococcus aureus. *Foodborne Pathogens and Disease*, *7*(2), 175–180. doi:10.1089/fpd.2009.0390 PMID:19839761

Osepchuk, J. M. (1984). A history of microwave heating applications. *IEEE Transactions on Microwave Theory and Techniques*, *32*(9), 1200–1224. doi:10.1109/TMTT.1984.1132831

Paneto, B. R., Schocken-Iturrino, R. P., Macedo, C., Santo, E., & Marin, J. M. (2007). Occurrence of toxigenic Escherichia coli in raw milk cheese in Brazil. *Arquivo Brasileiro de Medicina Veterinária e Zootecnia*, *59*(2), 508–512. doi:10.1590/S0102-09352007000200035

Porto, A. F., Sadicoff, B. L., Amorim, M. C. V., & De Mattos, M. C. S. (2002). Microwave-assisted free radical bulk-polyaddition reactions in a domestic microwave oven. *Polymer Testing*, *21*(2), 145–148. doi:10.1016/S0142-9418(01)00061-7

Qian, F., Sun, J., Cao, D., Tuo, Y., Jiang, S., & Mu, G. (2017). Experimental and modelling study of the denaturation of milk protein by heat treatment. *Han-gug Chugsan Sigpum Hag-hoeji*, *37*(1), 44–51. doi:10.5851/kosfa.2017.37.1.44 PMID:28316470

Rahman, T., Akon, T., Sheuli, I. N., & Hoque, N. (2015). Microbiological analysis of raw milk, pasteurized milk and yogurt samples collected from different areas of Dhaka city, Bangladesh. *Journal of Bangladesh Academy of Sciences*, *39*(1), 31–36. doi:10.3329/jbas.v39i1.23655

Rall, V. L. M., Vieira, F. P., Rall, R., Vieitis, R. L., Fernandes, A., Candeias, J. M. G., & Araújo, J. P. (2008). PCR detection of staphylococcal enterotoxin genes in Staphylococcus aureus strains isolated from raw and pasteurized milk. *Veterinary Microbiology*, *132*(3–4), 408–413. doi:10.1016/j.vetmic.2008.05.011 PMID:18572331

Rana, K. K., & Rana, S. (2014). Microwave reactors: A brief review on its fundamental aspects and applications. *Open Access Library Journal*, *1*, 1–20.

Ratanadecho, P., Aoki, K., & Akahori, M. (2002). The characteristics of microwave melting of frozen packed beds using a rectangular waveguide. *IEEE Transactions on Microwave Theory and Techniques*, *50*(6), 1495–1502. doi:10.1109/TMTT.2002.1006410

Razzaq, A. (2016). Occurrence of Shiga toxin producing E. coli from raw milk. *Pure and Applied Biology*, *5*(2), 270–276. doi:10.19045/bspab.2016.50035

Ricchi, M., Savi, R., Bolzoni, L., Pongolini, S., Grant, I. R., De Cicco, C., & Arrigoni, N. (2016). Estimation of *Mycobacterium avium* subsp. *paratuberculosis* load in raw bulk tank milk in Emilia-Romagna Region (Italy) by qPCR. *MicrobiologyOpen*, *5*(4), 551–559. doi:10.1002/mbo3.350 PMID:26991108

Rock, K. T., Mugizi, D. R., Ståhl, K., Magnusson, U., & Boqvist, S. (2016). The milk delivery chain and presence of Brucella spp. antibodies in bulk milk in Uganda. *Tropical Animal Health and Production*, *48*(5), 985–994. doi:10.100711250-016-1052-3 PMID:27026231

Rola, J. G., Czubkowska, A., Korpysa-Dzirba, W., & Osek, J. (2016). Occurrence of Staphylococcus aureus on farms with small scale production of raw milk cheeses in Poland. *Toxins*, *8*(3), 62. doi:10.3390/toxins8030062 PMID:26950152

Rossitto, P. V., Cullor, J. S., Crook, J., Parko, J., Sechi, P., & Cenci-Goga, B. T. (2012). Effects of UV Irradiation in a continuous turbulent flow UV reactor on microbiological and sensory characteristics of cow's milk. *Journal of Food Protection*, *75*(12), 2197–2207. doi:10.4315/0362-028X.JFP-12-036 PMID:23212017

Seyoum, E. T., Woldetsadik, D. A., Mekonen, T. K., Gezahegn, H. A., & Gebreyes, W. A. (2015). Prevalence of listeria monocytogenes in raw bovine milk and milk products from central highlands of Ethiopia. *Journal of Infection in Developing Countries*, *9*(11), 1204–1209. doi:10.3855/jidc.6211 PMID:26623629

Short, A. L. (1955). The temperature coefficient of expansion of raw milk. *The Journal of Dairy Research*, *22*(1), 69–73. doi:10.1017/S0022029900007561

Singh, S., Gupta, D., Jain, V., & Sharma, A. K. (2015). Microwave processing of materials and applications in manufacturing industries: A review. *Materials and Manufacturing Processes*, *30*(1), 1–29. doi:10.1080/10426914.2014.952028

Smith, W. L., Ruzante, J. M., Cullor, J. S., Gardner, I. A., & Thornton, C. G. (2007). Modified culture protocol for isolation of Mycobacterium avium subsp. paratuberculosis from raw milk. *Foodborne Pathogens and Disease*, *3*(4), 457–460. PMID:17199528

Soomro, A. H., Arain, M. A., Khaskheli, M., & Bhutto, B. (2009). Isolation of Escherichia Coli from raw milk and milk products in relation to public health sold under market conditions at Tandojam, Pakistan. *Pakistan Journal of Nutrition*, *1*(3), 151–152.

Stephan, R., Schumacher, S., Tasara, T., & Grant, I. R. (2007). Prevalence of Mycobacterium avium subspecies paratuberculosis in Swiss raw milk cheeses collected at the retail level. *Journal of Dairy Science*, *90*(8), 3590–3595. doi:10.3168/jds.2007-0015 PMID:17638968

Sun, J., Wang, W. L., & Yue, Q. Y. (2016). Review on microwave-matter interaction fundamentals and efficient microwave-associated heating strategies. *Materials (Basel)*, *9*(4), 1–25. doi:10.3390/ma9040231 PMID:28773355

Usman, A. M., Kwagga, J., Junaid, K., & Abdulkadir, I. (2016). Detection of mycobacteria in raw milk and assessment of risk factors among Fulani herdsmen in Bwari Area Council, Abuja, Nigeria. *International Journal of Infectious Diseases*, *45*, 248. doi:10.1016/j.ijid.2016.02.554

Usman, M., & Mukhtar, N. (2014). Prevalence of Listeria Monocytogenese in raw milk in Iran. *Ncbi. Nlm.Nih. Gov*, *7*(12), 8–11.

Vasbinder, A. J., & De Kruif, C. G. (2003). Casein-whey protein interactions in heated milk: The influence of pH. *International Dairy Journal*, *13*(8), 669–677. doi:10.1016/S0958-6946(03)00120-1

Villamiel, M., López-Fandiño, R., Corzo, N., Martínez-Castro, I., & Olano, A. (1996). Effects of continuous flow microwave treatment on chemical and microbiological characteristics of milk. *European Food Research and Technology*, *202*(1), 15–18. PMID:8717091

Villamiel, M., Munoz, M. M., Hernandez, A., & Corzo, N. (1998). β-lactoglobulin denaturation and furosine formation during continuous microwave treatment of milk at temperatures of 90-120 °C. *Milch-wissenschaft. Milk Science International*, *53*(8), 434–436.

Waak, E., Tham, W., & Danielsson-Tham, M. L. (2002). Prevalence and fingerprinting of Listeria monocytogenes strains isolated from raw whole milk in farm bulk tanks and in dairy plant receiving tanks. *Applied and Environmental Microbiology*, *68*(7), 3366–3370. doi:10.1128/AEM.68.7.3366-3370.2002 PMID:12089016

Wacker, R., Augustin, J. C., Boulais, C., Ben Cheikh, M. H., & Peladan, F. (2011). Modeling the occurrence of Mycobacterium avium subsp. paratuberculosis in bulk raw milk and the impact of management options for exposure mitigation. *Journal of Food Protection*, *74*(7), 1126–1136. doi:10.4315/0362-028X.JFP-11-005 PMID:21740715

Westhoff, D. C. (1978). Heating milk for microbial destruction: A historical outline and update. *Journal of Food Protection*, *41*(2), 122–130. doi:10.4315/0362-028X-41.2.122 PMID:30795181

Yamasaki, S., Awasthi, S. P., Hinenoya, A., Shima, A., Elbagory, A. R. M., Iguchi, A., & Ombarak, R. A. (2016). Prevalence and pathogenic potential of Escherichia coli isolates from raw milk and raw milk cheese in Egypt. *International Journal of Food Microbiology*, *221*, 69–76. doi:10.1016/j.ijfoodmicro.2016.01.009 PMID:26824810

Yobouet, B. A., Kouamé-Sina, S. M., Dadié, A., Makita, K., Grace, D., Djè, K. M., & Bonfoh, B. (2014). Contamination of raw milk with Bacillus cereus from farm to retail in Abidjan, Côte d'Ivoire and possible health implications. *Dairy Science & Technology*, *94*(1), 51–60. doi:10.100713594-013-0140-7

You, K. Y. (2015). *RF Coaxial Slot Radiators: Modeling, Measurements, and Applications*. U.S.: Artech House.

You, K. Y. (2017). Materials Characterization Using Microwave Waveguide System. In S. K. Goudos (Ed.), *Microwave Systems and Applications* (pp. 341–358). U.S.: InTech. doi:10.5772/66230

You, K. Y., You, L. L., Yue, C. S., Mun, H. K., & Lee, C. Y. (2017). Physical and Chemical Characterization of Rice Using Microwave and Laboratory Methods. In Rice - Technology and Production (pp. 81 - 99). U.S.: InTech. doi:10.5772/66001

Zhu, X., Guo, W., Jia, Y., & Kang, F. (2015). Dielectric properties of raw milk as functions of protein content and temperature. *Food and Bioprocess Technology*, *8*(3), 670–680. doi:10.100711947-014-1440-5

Chapter 6
Theoretical and Experimental Evaluation of Impact on Soil by Wheel Drives of the Self-Propelled Seeder

Alexandr Vladimirovich Lavrov
Federal Scientific Agroengineering Center VIM, Russia

Maksim Nikolaevich Moskovskiy
https://orcid.org/0000-0001-5727-8706
Federal Scientific Agroengineering Center VIM, Russia

Natalia Sergeevna Kryukovskaya
Federal Scientific Agroengineering Center VIM, Russia

ABSTRACT

Dedicated vertical axial loads on the soil from the wheels of a self-propelled seed drill, the area of the contact patch, the maximum contact pressure for the front and rear wheels and the density of the soil are determined by evaluations and experimental methods. The discrepancy between the theoretical and experimental indicators was: 1.4% and 2.0% for the rear and front wheels in vertical axial loads; 2.8% and 2.2% for the rear and front wheels by the contact area of the tires of the seeder with the soil and the maximum contact pressure; 6.2% – the maximum discrepancy on the values of soil density at a depth of 7.6 cm. Soil hardness was measured in three zones: before the seeder's passage and after each of its passage in a rut behind the front and rear wheels at six different depths, determined by the marks on the soil densimeter tester density. Graphics of dependencies of soil hardness on the depth of measurement were constructed.

DOI: 10.4018/978-1-5225-9420-8.ch006

INTRODUCTION

The experience of the development of many countries of the world, including Russia, shows that grain production is a basic function of the agro-industrial complex. It determines the level of the country's food security and the development of many sectors of the national economy. It is accounted one fifth of all the costs of agricultural production and more than 60% of its profits on the grain production. Therefore, the grain market is considered the most important strategic area of interest of any developed country.

The demand for seed grain production in Russia for 2018 is 11.3 million tons per year in the region. For the future 5 years, according the tasks of import substitution, it is planned to increase the production area and, as a result, the share of demand for grain seeds will increase by at least 25...30%. By 2026, according to the federal scientific and technical program of agricultural development for 2017-2025, it will plane to introduce and to use seed production technologies of the highest categories (original and elite) and to reduce the level of dependence on imports seeds by at least 30 percent (Russian Ministry of Agriculture, 2017, 2018).

The quantity and quality of the harvest depends from the method of sowing, the technical means for its implementation, the periods and seeding rates. To increase production volumes and provide agricultural producers with high quality sowing material, it is necessary to equip selection and seed farms with modern universal sowing machines adapted for sowing in various soil and climatic conditions (Lavrov, 2018).

Now days selection machinery and equipment are equipped with standard kinematic transmission systems and drives. The share of imported machinery and equipment ranges from 60 to 80% (Kryajkov, Godzhaev, Shevtcov & Lavrov, 2015). It is used manual or mechanical seed supply system during sowing in the design of selection seeders. This factor contributes to the growth of seed losses, greater energy intensity and labor costs of the process.

Federal Scientific Agro-engineering Center VIM began to solve this problem in 2017. In 2018 It was designed self-propelled selection seeder on the base of the self-propelled chassis Agromash 30 SC with an intelligent system of sowing seed.

According to the conducted research, it has been established that practically all models of tractors and self-propelled machines create maximum contact pressure by means of movers on the soil above permissible values (Hetz, 2001; Ziyaee & Roshani, 2012). Thus, in view of the extreme actuality of the saving soil fertility preservation when evaluating the agrotechnical indicators of a self-propelled seeder, it is necessary first of all to conduct research to determine the harmful effects of propulsion on the soil (Lavrov, Kryukovskaya & Petrischev, 2018).

The soil is compacted as a result of the movement of agricultural machinery, its particles are compressed, which contributes reduction of space for water and air. Soil compaction depends on many factors. So, the soil containing one-dimensional particles is compacted better than the soil of particles of different sizes. Increasing the moisture content also contributes to increased soil compaction. And the presence of organic impurities, on the contrary, leads to less compaction. Optimal for growing plants is soil containing 50% of soil particles and 25% of air and water each (Sergeev, 2017).

Deformation of the soil accumulate with each passage of agricultural machinery in the fields. American scientists have found an increase in soil density by 20% over 40 years of exposure to agricultural equipment, and the damage caused in this regard is estimated at 1.18 billion dollars. Russian scientists have found that during the period of cultivation and harvesting of crops, various agricultural equipment passes across the field up to 17 times. At the same time, driving systems affect the soil from 6 to 70 times in 10-12% of the treatment area, from 1 to 6 times in 65-80% of the field, and only 10-15% of the area is not affected. The depth of soil compaction is 30-80 cm, especially the upper fertile layer is disturbed. It is known that an increasing of soil density by 0.1 g/cm^3 leads to a decrease in grain yield by up to 600 kg/ha and to 1500-2500 kg/ha potato crop (Pogodin, Kuchko, Barsukevich & Shatilo, 2008; Polivaev & Voicshev, 2013; Adams, Slake, Martin & Boelter, 1960).

The most compacted areas of the field are the tracks along which agricultural machinery moves and the headlands at the edges of the field with the greatest compaction in the upper soil layer (Balbuen, Terminiello, Claverie, Casado & Marlats, 2000), and the area of the plots of the field compacted by agricultural equipment and the depth of compaction depend on the type of crop cultivation technology used (Tullberg, Hunter, Paull & Smith, 1990).

For field operations in early spring and during sowing, the permissible specific pressure of agricultural machinery thrusters on the soil for most of its types is pressure 0.4-0.6 kg/cm^2, 0.8 kg/cm^2 on the plowed field and 1-1.5 kg/cm^2 on field transport works. But basically, most modern machines do not meet these requirements (Locshenko, 2016).

With an increase of soil density, its resistivity increases, crumbling decreases, lumpiness forms, physical, chemical and agrophysical properties deteriorate, and erosion processes are developed. Under such conditions roots of plants will not develop normally and the yield is significantly reduced (Polivaev & Voicshev, 2013).

The aim of the research is evaluation of the impact on the soil of the undercarriage systems of a self-propelled seed seeder, determination experimentally and theoretically values of soil compaction at different depths before and after the passage of the drill and their analysis.

METHODS AND RESULTS

In the FSAC VIM we developed the self-propelled selection seeder, the general view is shown on Figure 1. The technical characteristics of the self-propelled seeder are presented in Table 1.

The seeder is designed for single-grain sowing of cereals, leguminous plants and groats on the plots of the second stage of selection operations. When the gear ratio of the gears of the gearbox changes, the seeding rate is adjusted. To set the parameters of the plots and the sowing apparatus, the global seeding control system (GSC) is used. This system controls the power supply of the cassette feeding mechanism's electric motors and notifies with a sound signal when empty cassettes are detected during ordinary seeding.

Laboratory and field researches were carried out in the ISA-branch of the Federal Scientific Agro-engineering Center VIM (Ryazan region, village Podvyaze) during the cultivation of wheat, in accordance with generally accepted methods. It was performed operational and technological tests to determine the agrotechnical properties, and laboratory and field tests to determine the main parameters and characteristics of the self-propelled selection seeder.

Figure 2 shows a block-scheme of theoretical and experimental studies of the impact on the soil of a self-propelled selection seed drill.

Figure 1. General view of self-propelled seeder

THEORETICAL RESEARCHES OF THE IMPACT ON THE SOIL SEEDER

Determination of Vertical Axial Loads

The design scheme for determining the vertical axial loads of a self-propelled seed drill on the soil is shown in Figure 3.

For determine the seeder weight on the front and rear wheels, it is necessary to evaluate the system of equations:

$$\begin{cases} \sum M_{GC} = G_{FW} \cdot \left(L_{base} - L_{GC}\right) - G_{RW} \cdot L_{GC} = 0, \\ G_{FW} + G_{RW} = G_{GC}, \end{cases} \tag{1}$$

where:

G_{FW} is weight of the seeder falling on the front wheels, kg;
G_{RW} is weight of the seeder falling on the rear wheels, kg;
L_{base} is wheelbase, m;
L_{GC} is the distance from the center of the rear wheel to the center of gravity of the seeder (horizontally in the longitudinal direction), m.

The result is:

Table 1. Technical characteristics of self-propelled seeder

No.	Parameter	Value
1	Type	Self-propelled, universal wheel mobile energy mean
2	Weight, kg	2860
3	Engine	D-120 (diesel four-stroke two-cylinder with air cooling system and two-row vertical arrangement of cylinders)
4	The operating power of the engine, kW / HP	22.1 / 30
5	Specific fuel consumption, g/kWh (g/HPh)	245 (180)
6	- number of gears forward running - number of reverse gears	6 6
7	Velocity, km/h	5 - 25
8	Differential lock	+
9	- Track width of front wheels, mm - Track width of rear wheels, mm	1324 – 1424 1314 - 1484
10	Wheelbase, mm	2500
11	Agrotechnical clearance, mm	450
12	Seeding mechanism drive	From the wheel telemetry sensor
13	Width of capture, mm	1250
14	Number of rows	3 – 6
15	A separate funnel	One feed hopper per row
16	Rotary dispenser	2 – 6 rows, the seed is distributed on all rows
17	Central dispenser sowing material	The sowing material is distributed into several rows
18	Cassette table	The seed is poured in 4 or 6-row magazines and sowed automatically
19	The system of distribution of seed	6 conical dispensers with a diameter of 120 mm or 4 with a diameter of 195 mm
20	System of control	System of Global Seed Control (GSC)
21	Anchor coulters single disk coulters Double disc coulters	ISARIA, NODET ACCORD CX NODET

$$G_{FW} = 1075\,\text{kg},\ G_{RW} = 1785\,\text{kg}.$$

Accordingly, it can be calculated that one front wheel carries a load of 537.5 kg, and one rear wheel carries a load 892.5 kg from the seeder weight.

Determination of the Contact Patch Area of the Seeder Tires With Soil

The following types of wheels are mounted on the seeder. The rear wheels are YaF-394 (12.40 R28), the front wheels are Ya-387-1 (6.50-16). For calculate the area of the contact patch with the soil for the rear and front wheels, we used evaluating scheme of wheel deformation (Figure 4) (Izmailov, Shevtsov, Lavrov, Godzhaev & Pryadkin, 2015; Shevtsov, Soloveychik, Rusanov & Lavrov, 2014; Soane, 1970).

Figure 2. Block diagram of theoretical and experimental studies of the impact on the soil of self-propelled seeder

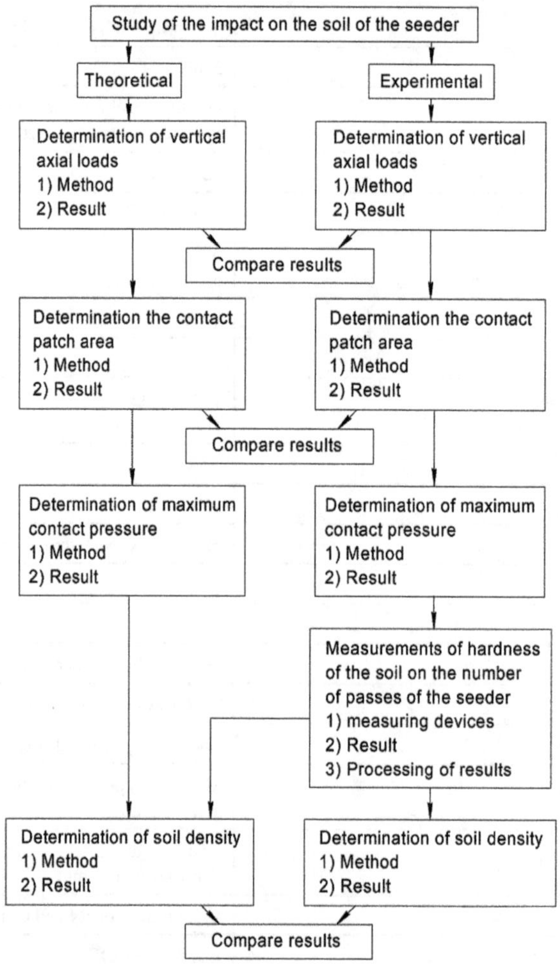

The contour area of the contact patch is determined by the formula:

$$A_k = a_k b_k \pi / 4 \quad (2)$$

in which:

$$a_k = C_3 \sqrt{f(D - f)} \quad (3)$$

$$b_k = 2\sqrt{f(2R_{Eq} - f)} \quad (4)$$

Figure 3. Evaluating scheme for determining the vertical axial loads on the soil of a self-propelled selection seeder

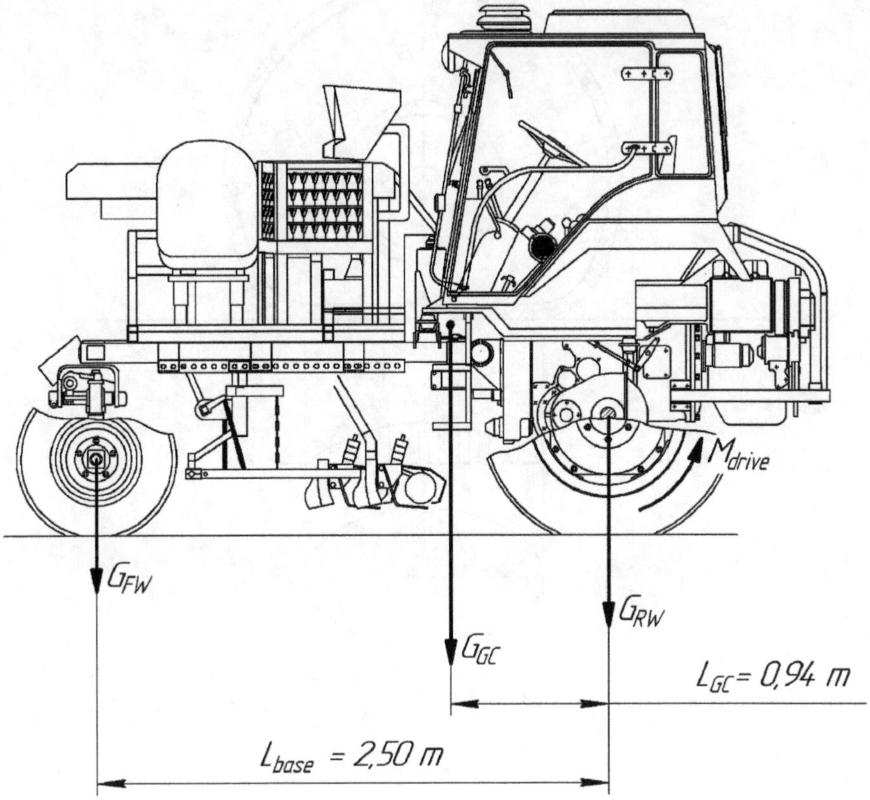

$$R_{Eq} = \left(B + H\right) / 2.5 \quad (5)$$

$$H = \left(D - d\right) / 2; \quad (6)$$

$$C_3 = \frac{20.5}{11.9 + \frac{D}{B} - \frac{\left|n - 9\right|}{2} - 3} \quad (7)$$

where:

D, B is the outer diameter and section width of the tire, m;
d is the diameter of the rim, m;
H is the height of the tire profile, m;
n is ply rating;

Figure 4. Evaluating scheme of wheel deformation

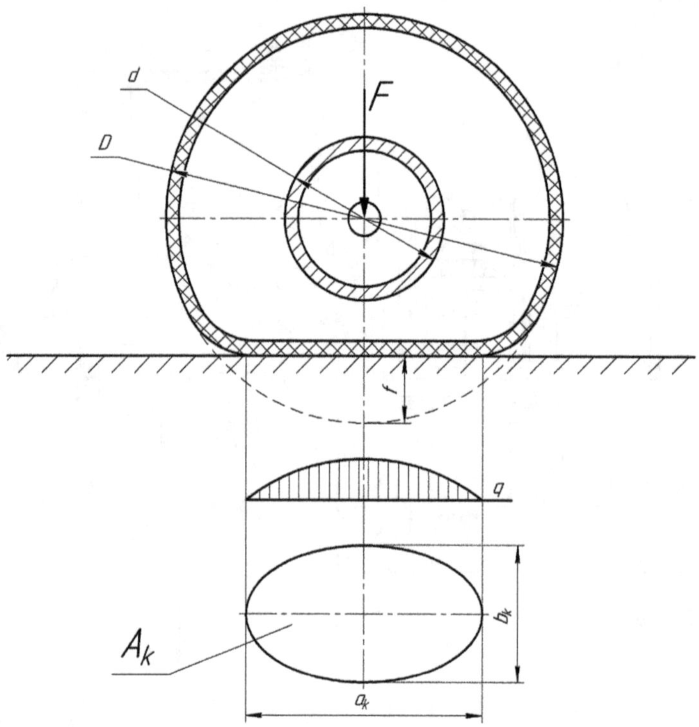

f is deflection, m;
C_3 is coefficient depending on the size and layering of the tire.

The value of the permissible static deflection of the tire is calculated by the formula:

$$\left|f\right| = R_{fr} - R_{st} \quad (8)$$

where:

R_{fr} is free radius of the wheel, m;
R_{st} is static wheel radius, m.

Tire deflection is calculated by the formula:

$$f = \frac{C_2 F}{2 \cdot \left(p_w + p_0\right)} + \sqrt{\left[\frac{C_2 F}{2 \cdot \left(p_w + p_0\right)}\right]^2 + C_1 F} \quad (9)$$

where:

C_1, C_2, p_0 are constant for a given tire, the coefficients;
p_w is the internal air pressure in the tire, kPa;
F is vertical wheel load, kN.

$$p_0 = 16.7 \cdot n \cdot \left(\sqrt{D / B} - 1.4\right) - 28 \geq 0 \tag{10}$$

If $p_0 < 0$ in the calculations is taken to be zero.

Load rates and tyre pressures are shown in Table 2 and Figure 5.

The relationship between vertical load, tire deflection and tire pressure are described by the equation (Godzhaev, Shevtsov, Lavrov & Rusanov, 2015):

$$F = \frac{\left|f\right|^2}{C_1 + C_2 \cdot \left|f\right| / \left(p_w + p_0\right)} \tag{11}$$

From which it turns out:

$$\frac{\left|f\right|^2}{F} = C_1 + C_2 \cdot \frac{\left|f\right|}{\left(p_w + p_0\right)} \tag{12}$$

Universal characteristics of tires are presented in the Table 3.

Table 2. Standards of loads and tire pressure to determine the mode of operation

The Tyre Designation	Ply Rating	Bus Load, kg, at an Internal Pressure of, kPa, at Speeds Up to 30 km/h							
		80	**90**	**100**	**110**	**120**	**130**	**140**	**150**
6,50-16 (Ya-387-1)	6	-	-	-	-	-	-	390	400
12,4 R28 (YaF-394)	8	970	1040	1110	1170	1240	1285	1350	1405
		160	**170**	**180**	**190**	**200**	**210**	**220**	**230**
6,50-16 (Ya-387-1)	6	415	430	445	460	475	490	505	510
12,4 R28 (YaF-394)	8	1450	-	-	-	-	-	-	-
		240	**250**	**260**	**270**	**280**	**290**	**300**	**310**
6,50-16 (Ya-387-1)	6	525	540	555	565	580	590	600	615
12,4 R28 (YaF-394)	8	-	-	-	-	-	-	-	-

Figure 5. Load standards and tire pressure to determine the mode of operation

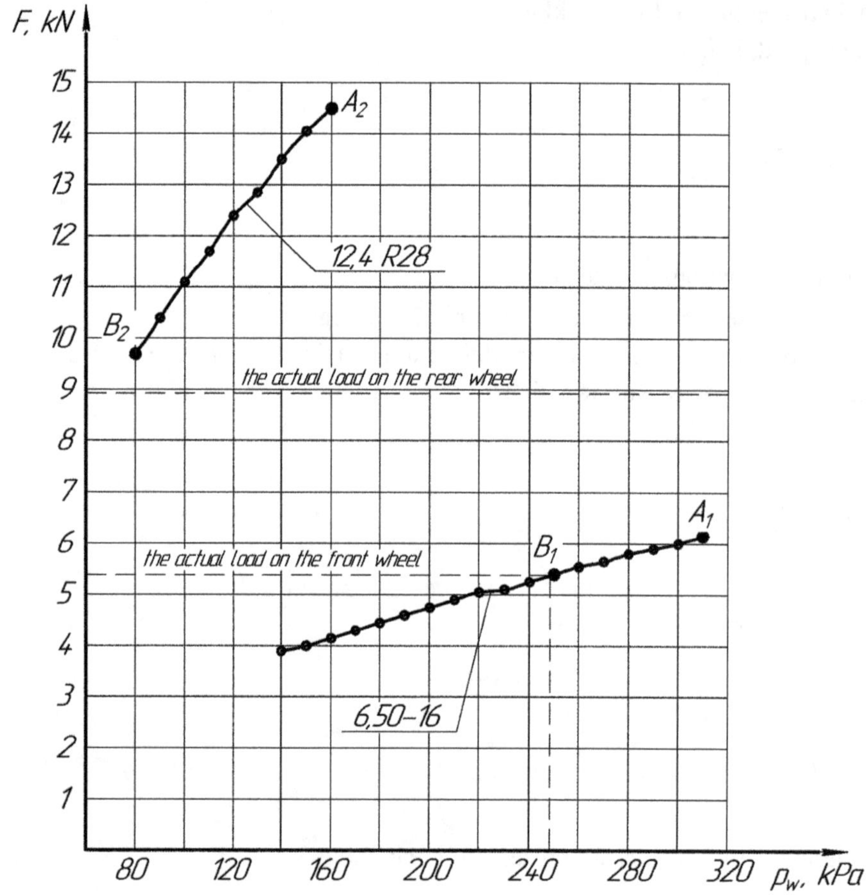

Table 3. Universal tire characteristics 6.50-16 (the points A_1 and B_1) and 12.4 R28 (the points A_2 and B_2)

Indicator	Dimension	Type of Tyre			
		6.50-16		12.4 R28	
		A_1	B_1	A_2	B_2
$\frac{\lvert f\rvert^2}{F}\cdot 10^4$	m^2/kN	1.46	1.67	2.91	4.36
$\frac{\lvert f\rvert}{p_w + p_0}$	m/kPa	0.86	1.03	3.08	4.96

Substituting the dates from table 3 into the resulting equation (12), we obtain two systems of equations, the solution of which allows us to determine the coefficients C_1 and C_2 for the tires of the front and rear wheels.

For the front wheels:

$$\begin{cases} 1.46 = 10^4 \cdot C_1 + 0.86 \cdot C_2 \\ 1.67 = 10^4 \cdot C_1 + 1.03 \cdot C_2 \end{cases} \Rightarrow \begin{cases} C_1 = 0.48 \cdot 10^{-4} \\ C_2 = 1.15 \end{cases}$$

For the rear wheels:

$$\begin{cases} 2.91 = 10^4 \cdot C_1 + 4.36 \cdot C_2 \\ 3.08 = 10^4 \cdot C_1 + 4.96 \cdot C_2 \end{cases} \Rightarrow \begin{cases} C_1 = 0.55 \cdot 10^{-4} \\ C_2 = 0.767 \end{cases}$$

Using the coefficients, the deflection and the contact area of the tires of the front and rear wheels are found. It is known that the pressure in the tires of the front wheels was 280 kPa, and that in the tires of the rear wheels was 120 kPa (controlled by a manometer). The source data and the obtained values according to the results of calculations for the front and rear wheels of the planter for convenience are combined in Table 4.

Determination of the Maximum Contact Pressure of the Wheel Propulsion Seed Drill on the Soil

The maximum contact pressure is calculated by the formula:

$$q_{\max} = \frac{K_2 F}{K_1 A_k} \tag{13}$$

where:

K_2 =1.5 is the coefficient of the longitudinal unevenness distribution of pressure;
K_1 is the coefficient of reduction of the contact area of the tire wheels to the working conditions on the soil base, depending on the diameter of the wheel.

The results of the calculations are also presented in Table 4.

Analyzing the obtained results, we can conclude that the construction of a self-propelled seeder does not meet the standards for allowable propulsion pressure of agricultural equipment on the soil, which is 0.4-0.6 kg/cm^2 for sowing (Loschenko, 2016). The pressure on the soil of the front wheels exceeds the standards by 2.9 times, and the rear - by 2.1 times.

Table 4. Calculation results in determining the contour area of the contact patch of the front and rear wheels of the planter with the soil and their maximum contact pressure on the soil

N°	Parameter	Designation	Front Wheel	Rear Wheel
1	The outer diameter of the wheel, m	D	0.760	1.250
2	Section width, m	B	0.175	0.315
3	The diameter of the rim, m	d	0.406	0.711
4	Ply rating	n	6	8
5	The height of the tire profile, m	H	0.177	0.269
6	Equivalent radius, m	R_{Eq}	0.141	0.234
7	Coefficient depending on the size and layering of the tire	C_3	1.75	1.66
8	Free radius of the wheel, m	R_{fr}	0.390	0.640
9	Static wheel radius, m	R_{st}	0.360	0.575
10	Permissible static deflection of the tyre, m	$\|f\|$	0.030	0.065
11	Coefficient, kPa	p_0	40.53	51.10
12	Coefficient, m²/kN	C_1	$0.48 \cdot 10^{-4}$	$0.55 \cdot 10^{-4}$
13	Coefficient, 1/m	C_2	1.15	0.767
14	Tyre pressure, kPa	p_{wf}	280	120
15	Deflection, m	f	0.028	0.050
16	The length of the ellipse of contact patch, m	a_k	0.251	0.405
17	The width of the ellipse of contact patch, m	b_k	0.169	0.289
18	Contour area of the contact patch, m²	A_k	0.0333	0.0919
19	Coefficient of reduction of the contact area of the wheel tire to the working conditions on the soil base	K_1	1.4	1.15
20	Maximum contact pressure, kPa	q_{max}	172.1	126.6

EXPERIMENTAL RESEARCH OF THE IMPACT BY THE SEEDER

Determination of Vertical Axial Loads

For the practical determination of the vertical axial loads from the impact of the wheels of the seeder on the soil, we used mobile platform scales VSP4-B-3000 (2000x1000) with a measurement error of 2 kg. The seeder was installed by the front wheels on the supporting surface, and the rear wheels on the scales (Figure 6). Taking dates of the scales, it turned out that the weight of the seeder per rear wheel is 1810 kg. Then the seeder was moved, so that the rear wheels were on the supporting surface, and the front on the scales. The weight of the seeder per front wheel, according to the indications of platform scales, was 1053 kg.

It is clear that the total weight does not correspond to reality and differs by 3 kg. This is due to the error in measuring custom weights.

Figure 6. Determination of the weight of the seeder falling on the rear wheels

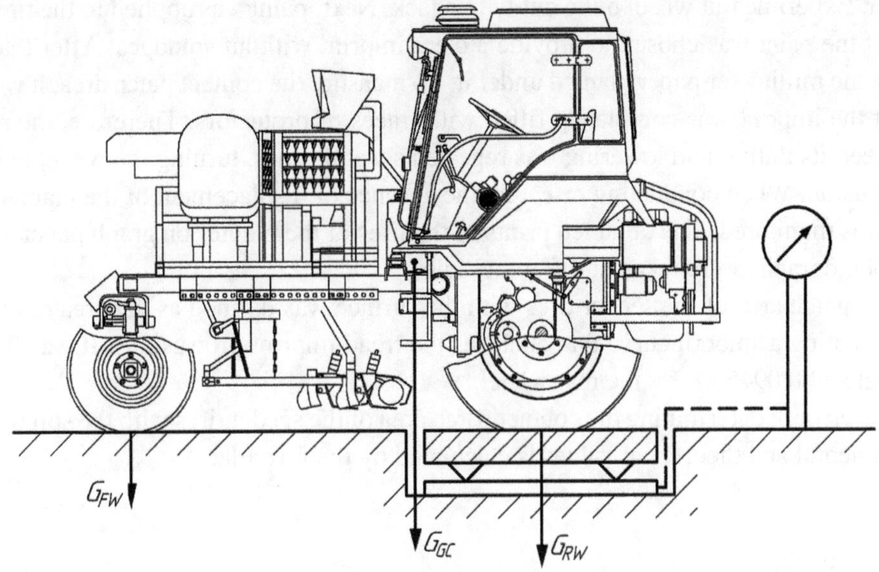

The relative error of the discrepancy in the theoretical data for the rear wheels is calculated by the formula:

$$\delta = \frac{\left|G_{RW} - G_{RWE}\right|}{G_{RW}} \cdot 100\% \tag{14}$$

where:

G_{RW} is theoretical weight of the seeder falling on the rear wheels, kg;
G_{RWE} is experimental weight of the seeder falling on the rear wheels, kg.

Substituting values, it turns out 1.4%.
For the front wheels, the relative error is calculated by analogy and corresponds to 2.0%.

Determination of the Contact Patch Area of the Seeder Tires With Soil

Experimental determination of the contour area of the contact patch of the front and rear wheels of the planter was carried out indoors with a concrete base, the deviation from the flatness of the surface of which within the dimensions of the planter did not exceed 5 mm. The contact area of the tires of the seeder with the base was measured first under the front wheel, then under the rear wheel. A steel sheet with a thickness of 20 mm was placed under the test wheel to provide the necessary values of surface irregularities (no more than 1 mm) within the designated area. On top of it was attached graph paper. The position of the wheels of the planter corresponded to the rectilinear movement.

The pressure in the tires of the front wheels was 280 kPa, and in the rear 120 kPa and was controlled by manometer. Experimental wheel hung out with a jack. Next, paint was applied to the tire treads. The consistency of the paint was chosen to provide a clear imprint without smudges. After that, the wheel went down on the millimeter paper planted under it. To measure the contact patch area, it was necessary to ensure that the imprint was completely filled with traces of protectors. Therefore, the procedure of lifting the wheel, its tinting and lowering was repeated several times, turning the wheel to the width of the tread protrusion. When conducting research, the absence of displacement of the planter in the horizontal plane was monitored. The obtained prints of the tires of the planter on graph paper are presented in Figure 7 (photographs were taken at different scales).

The contact patch area of the seeder tires with the surface was defined as the area of the figure obtained by delineating a smooth curve that encloses the tread imprint protrusions. It was 0.0340 m^2 for the front wheel and 0.0945 m^2 for the rear wheel.

The relative error of determining the contact patch area of the seeder tires with the soil when comparing the experimental and theoretical values is calculated by the formula:

$$\delta = \frac{\left|A_k - A_{ke}\right|}{A_k} \cdot 100\% \qquad (15)$$

where:

A_{ke} is experimentally measured contact patch area of the seeder tires with soil, m^2.

Accordingly, the relative error of experimental determination of the contact patch area for the front wheels is 2.2%, and for the rear wheels it is 2.8%.

Figure 7. Imprints of front (left) and rear (right) seeder tires to determine the contact patch area

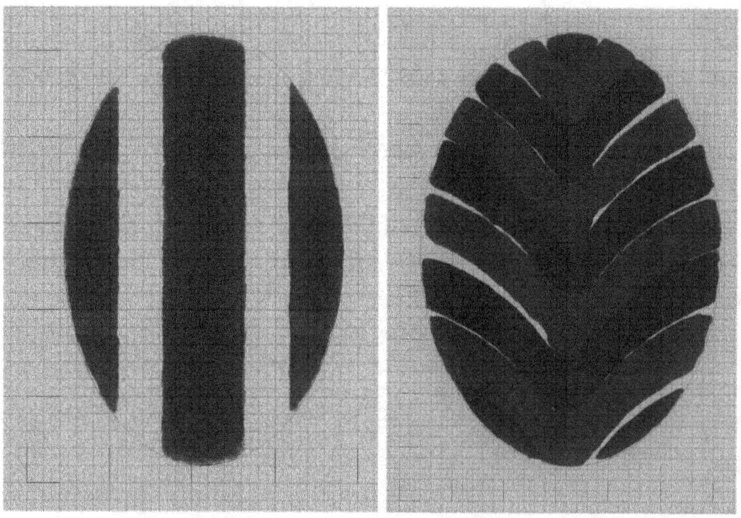

By using experimental values of the contact area of the seeder wheels, we calculated the maximum contact pressure of the wheels on the ground. It is 168.6 kPa for the front wheels and It is 123.1 kPa for the rear wheels.

Measurements of Soil Hardness by the Multiplicity of Passes

The research of the change in soil hardness after the seeders was carried out on a field with a chernozem-podzolic soil type, the hardness was measured with a soil densitometer of Wile company Farmcomp. The magnitude of the hardness was measured on the field in three zones: before the passage of the seeder and after its each passage in a rut behind the front and rear wheels at six different depths, determined by the marks on the densitometer. In total, three seeder passes were performed. For each zone and depth, 20 measurements were made, at least 40 cm apart from each other. In total, 840 measurements were taken.

The marks on the densitometer correspond to the depth of the soil hardness measurement in inches, and the watch type meter indicates the soil hardness in psi. Therefore, for convenience of displaying information about measurements and calculating the average values of soil hardness, data in these values are used.

The average value of hardness for each zone and depth of measurement is determined by the formula:

$$x_{med} = \frac{\sum_{i=1}^{n} x_i}{n} \tag{16}$$

where:

x_i is soil hardness of the i-th measurement, lb/in²;
n is amount of measurements.

For an absolute evaluation of the measure of the spread of hardness values for each zone and depth of measurement, the standard deviation is calculated, which determined by the formula:

$$\sigma = \sqrt{\frac{\sum_{i=1}^{n} \left(x_i - x_{med}\right)^2}{n}} \tag{17}$$

To evaluate the measure of the spread of hardness values relative to the values themselves, the coefficient of variation is calculated:

$$V_\sigma = \frac{\sigma}{x_{med}} \cdot 100\% \tag{18}$$

The results of measurements and calculated average values of soil hardness are presented in Tables 5-11. The obtained average values of soil hardness mean square deviations and coefficient of variation

Table 5. The results of measurements of soil hardness before the passage of the seeder at different depths and their average values, pound/inch²

Measurement Depth, Inch	The Number of Measurements																				x_{med}
	1	2	3	4	5	6	7	8	9	10	11	12	13	14	15	16	17	18	19	20	
3	40	50	50	50	50	60	50	50	40	50	50	50	50	60	50	50	50	50	50	50	50
6	100	100	110	110	90	100	90	100	100	100	100	110	90	90	90	100	110	100	110	100	100
9	110	110	110	120	120	120	120	120	120	130	130	120	120	130	130	120	110	120	120	120	120
12	180	170	160	170	170	170	180	180	170	160	170	160	160	160	170	170	180	170	180	170	170
15	170	170	170	160	160	170	180	180	160	170	170	160	170	170	180	180	180	170	160	170	170
18	170	170	180	170	170	170	170	170	160	170	170	160	180	180	180	170	160	160	170	170	170

Table 6. The results of measurements of soil hardness after the first pass of the seeder behind the front wheel at different depths and their average values, pound/inch²

Measurement Depth, Inch	The Number of Measurements																				x_{med}
	1	2	3	4	5	6	7	8	9	10	11	12	13	14	15	16	17	18	19	20	
3	120	120	130	130	130	130	130	140	130	140	130	130	130	120	130	130	140	130	130	130	130
6	130	130	140	140	140	150	140	150	150	140	140	130	140	140	140	140	130	140	150	140	140
9	150	150	150	150	160	160	150	140	150	150	140	150	150	140	150	150	150	150	160	150	150
12	180	180	180	180	180	180	170	180	180	180	170	180	180	180	180	180	180	190	190	180	180
15	180	180	180	180	190	190	180	180	180	180	180	170	170	180	170	180	180	180	190	180	180
18	190	180	180	180	180	180	170	180	170	180	190	180	180	180	180	170	180	180	180	190	180

Table 7. The results of measurements of soil hardness after the first pass of the seeder behind the rear wheel at different depths and their average values, pound/inch²

Measurement Depth, Inch	The Number of Measurements																				x_{med}
	1	2	3	4	5	6	7	8	9	10	11	12	13	14	15	16	17	18	19	20	
3	150	150	150	140	150	150	150	160	150	150	160	150	150	140	150	150	150	150	150	150	150
6	150	150	160	160	160	170	160	160	160	160	170	160	160	160	170	160	160	150	160	160	160
9	160	170	170	160	170	180	170	170	160	170	180	170	170	170	180	170	170	170	170	170	170
12	190	190	180	190	190	200	190	180	190	180	200	190	190	190	200	190	190	190	190	190	190
15	190	190	190	200	190	190	200	180	190	190	190	180	180	190	190	200	190	190	190	190	190
18	200	190	190	190	200	190	190	190	180	190	180	190	180	190	200	190	190	190	190	190	190

depending on the depth of measurement and the number of passes of the seeder are grouped in Table 12. In this case, the units of measurement for the values there are centimeters for the depth of measurement and kgf/ cm² for hardness indicators, according to SI.

Analyzing the data, we can conclude that the measured values of soil hardness are reliable, since all the calculated coefficients of variation are less than 10%.

Table 8. The results of measurements of soil hardness after the second pass of the seeder behind the front wheel at different depths and their average values, pound/inch²

Measurement Depth, Inch	The Number of measurements																				x_{med}
	1	2	3	4	5	6	7	8	9	10	11	12	13	14	15	16	17	18	19	20	
3	160	160	160	170	160	160	160	160	150	160	150	160	160	160	170	160	160	160	160	160	160
6	160	170	170	180	170	170	160	170	160	170	170	170	170	170	180	170	170	180	170	170	170
9	170	180	180	190	180	190	180	170	180	180	180	190	180	170	170	180	180	190	180	180	180
12	190	190	190	210	190	210	200	190	200	210	210	210	200	200	190	200	200	210	200	200	200
15	190	210	200	210	190	210	190	200	190	200	200	210	200	210	190	210	190	200	210	190	200
18	190	210	210	210	200	210	190	190	190	200	200	200	190	210	190	210	190	190	210	210	200

Table 9. The results of measurements of soil hardness after the second pass of the seeder behind the rear wheel at different depths and their average values, pound/inch²

Measurement Depth, Inch	The Number of Measurement																				x_{med}
	1	2	3	4	5	6	7	8	9	10	11	12	13	14	15	16	17	18	19	20	
3	160	170	180	160	170	160	180	160	180	170	170	170	180	160	160	180	170	180	180	160	170
6	170	170	190	170	180	180	180	180	190	180	180	180	170	180	180	180	180	190	190	180	180
9	180	190	180	180	180	200	180	190	200	190	200	190	190	190	200	190	190	190	190	200	190
12	190	200	200	190	200	210	200	210	210	200	220	210	200	210	220	210	200	200	210	210	205
15	190	190	210	200	190	220	200	210	200	210	200	220	210	210	220	220	200	190	200	210	205
18	190	190	210	200	200	220	200	200	200	210	200	220	210	200	210	210	200	200	210	210	205

Table 10. The results of measurements of soil hardness after the third pass of the seeder behind the front wheel at different depths and their average values, pound/inch²

Measurement Depth, Inch	The Number of Measurements																				x_{med}
	1	2	3	4	5	6	7	8	9	10	11	12	13	14	15	16	17	18	19	20	
3	170	180	180	170	170	180	190	190	170	180	180	180	190	180	190	180	180	170	180	190	180
6	180	190	180	190	190	200	200	190	190	200	200	180	180	190	200	190	190	180	180	200	190
9	190	210	190	200	200	210	200	190	210	200	200	190	200	190	210	210	210	190	190	210	200
12	200	210	200	210	210	220	210	210	220	210	210	210	220	200	210	210	220	200	200	220	210
15	200	200	200	220	210	220	220	220	210	220	200	200	210	210	210	200	210	210	220	210	210
18	220	210	210	220	200	210	210	210	200	220	200	200	200	220	220	200	220	210	220	200	210

For greater clarity, the measurement results are presented in the form of graphs of the dependences of the average indicators of soil hardness on the depth of measurement with a soil density meter (Figure 8). It was revealed that the soil is noticeably compacted after the first pass and during subsequent passes the compaction value changes slightly. Moreover, with an increase in the depth of measurement, this trend is observed especially clearly.

Table 11. The results of measurements of soil hardness after the third pass of the seeder behind the rear wheel at different depths and their average values, pound/inch²

Measurement Depth, Inch	The Number of Measurements																				x_{med}
	1	2	3	4	5	6	7	8	9	10	11	12	13	14	15	16	17	18	19	20	
3	190	180	180	200	200	200	200	190	190	180	200	190	180	180	190	180	190	180	200	200	190
6	190	190	200	210	210	210	190	190	190	200	210	200	200	200	210	200	210	190	190	210	200
9	200	210	210	220	210	210	210	200	200	210	210	210	220	200	220	200	200	200	200	200	210
12	210	220	220	210	210	210	200	220	210	220	220	210	230	220	230	220	210	210	220	200	215
15	220	210	210	200	210	200	210	220	220	230	210	220	220	230	230	210	200	220	220	210	215
18	230	220	220	200	200	200	220	210	230	230	210	220	210	230	230	200	200	220	210	210	215

Table 12a. The average values of soil hardness, mean square deviations and coefficient of variation before and after the passage of the seeder

Measurement Depth, cm	Position of Measurement of Soil Hardness											
	Before the Passage of the Seeder			1st Pass of Front Wheel			1st Pass of Both Wheels			2nd Pass of Front Wheel		
	The Hardness of the Soil, x_{med}, kgf/cm²	σ, kgf/cm²	$V_σ$, %	The Hardness of the Soil, x_{med}, kgf/cm²	σ, kgf/cm²	$V_σ$, %	The Hardness of the Soil, x_{med}, kgf/cm²	σ, kgf/cm²	$V_σ$, %	The Hardness of the Soil, x_{med}, kgf/cm²	σ, kgf/cm²	$V_σ$, %
7.6	3.52	4.5	8.9	9.14	5.5	4.2	10.55	4.5	3.0	11.25	4.5	2.8
15.2	7.03	7.1	7.1	9.84	6.3	4.5	11.25	5.5	3.4	11.95	5.5	3.2
22.9	8.44	6.3	5.3	10.55	5.5	3.7	11.95	5.5	3.2	12.66	6.3	3.5
30.5	11.95	7.1	4.2	12.66	4.5	2.5	13.36	5.5	2.9	14.06	7.7	3.9
38.1	11.95	7.1	4.2	12.66	5.5	3.0	13.36	5.5	2.9	14.06	7.7	3.9
45.7	11.95	6.3	3.7	12.66	5.5	3.0	13.36	5.5	2.9	14.06	8.4	4.2

Table 12b. The average values of soil hardness, mean square deviations and coefficient of variation before and after the passage of the seeder

Measurement Depth, cm	2nd Pass of Both Wheels			3rd Pass of Front Wheel			3rd Pass of Both Wheels		
	The Hardness of the Soil, x_{med}, kgf/cm²	σ, kgf/cm²	$V_σ$, %	The Hardness of the Soil, x_{med}, kgf/cm²	σ, kgf/cm²	$V_σ$, %	The Hardness of the Soil, x_{med}, kgf/cm²	σ, kgf/cm²	$V_σ$, %
7.6	11.95	7.7	4.6	12.66	7.1	3.9	13.36	8.4	4.4
15.2	12.66	6.3	3.5	13.36	7.7	4.1	14.06	8.4	4.2
22.9	13.36	7.1	3.7	14.06	8.4	4.2	14.76	7.7	3.7
30.5	14.41	8.1	3.9	14.76	7.1	3.4	15.12	8.1	3.7
38.1	14.41	10.2	5.0	14.76	7.7	3.7	15.12	9.2	4.3
45.7	14.41	8.1	3.9	14.76	8.4	4.0	15.12	11.2	5.2

Figure 8. Graphs of the dependence of soil hardness on the depth of measurement

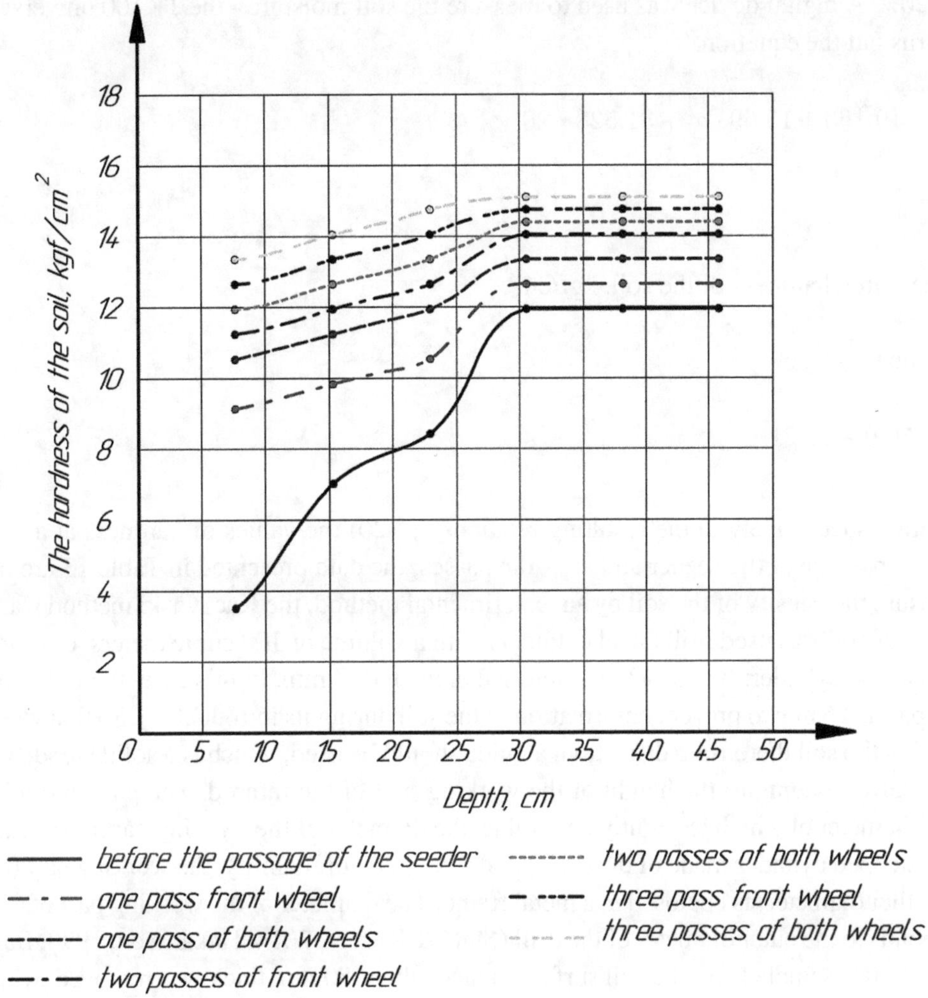

Determination of Soil Density in a Theoretical and Practical Methods

To determine the density of the soil by the theoretical method, the dependence is used (Rusanov, 1998):

$$H_s = -19.163 + 15.69 \cdot \rho_s + 31.528 / W_s \tag{19}$$

where:

H_s is the hardness of the soil, MPa,
ρ_s is soil density, g/cm³,
Ws is soil moisture, %.

During testing a self-propelled seeder, soil moisture was measured at a depth of 3 inches (7.6 cm) and it was 20%. A digital device was used to measure the soil moisture - the TK100 universal moisture meter. It turns out the equation:

$$H_1 \cdot 0.1 = -19.163 + 15.69 \cdot \rho_s + 31.528 / 20$$

where:

H_1 is the measured hardness of the soil, kgf/cm^2.

From which:

$$\rho_s = H_s / 156.9 + 1.121 \tag{20}$$

Substituting successively in the resulting relationship (20) the values of hardness at a depth of 7.6 cm before the passage of the seeder and after the passes, the data presented in Table 13 are obtained.

For measure the density of the soil by an experimental method, the Kaczynski method was used. For the sampling of soil was used drill (steel cylinder) with a volume of 100 cubic meters. cm. Geometrical dimensions of the cylinder: height 40 mm, internal diameter 57 mm. In this case, the inner diameter of the cutting part is 56 mm to prevent deformation of the soil during its introduction. So that when the drill is immersed in the soil there is no distortion, a guide should be used, which is a solid wooden plate with a thickness corresponding to the height of the working part of the ramrod. The plate has a hole in the middle, the diameter of which is slightly larger than the diameter of the working part of the ramrod. The ramrod is a stepped cylinder made of durable wood. For greater reliability, the working part (cylindrical part of a smaller diameter) is enclosed in a metal frame. The diameter of the working part of the cleaning rod corresponds to the outer diameter of the drill (Malý & Kučera, 2014; Hakansson, 1990; Bajla, 1998).

Before taking a sample from the soil surface, a layer about 50 mm thick was removed, while providing a flat area and avoiding soil deformation. The walls of the drill were slightly greased with petroleum jelly, after which the drill was inserted into the guide hole. From above the ramrod was attached to the borax and, pressing on the ramrod by hand, the drill pressed into the soil. This process lasted until the

Table 13. The density of the soil before and after the passage of the seeder at a depth of 7.6 cm

Position of Measurement of Soil Hardness	The Hardness of the Soil, kgf/cm^2	Soil Density, g/cm^3
Before the passage of the seeder	3.52	1.143
1st pass of front wheel	9.14	1.179
1st pass of both wheels	10.55	1.188
2nd pass of front wheel	11.25	1.193
2nd pass of both wheels	11.95	1.197
3rd pass of front wheel	12.66	1.202
3rd pass of both wheels	13.36	1.206

ramrod entered the guide hole to a cylindrical part of a larger diameter. Then the ramrod and the guide were carefully removed, a PCB plate was superimposed on the upper part of the drill (for this purpose, any solid and smooth plate is suitable) and the drill submerged in the soil was dug with a spatula. Next, the soil under the drill cut with a small margin. Without removing the textolite plate, the drill rose, turned over and the excess soil protruding beyond the edges of the drill was cut off with a sharp knife (Figure 9). The side walls of the drill were cleaned from adhering soil particles. The obtained soil sample was poured into a dry plastic bag, on which was attached a label with information about the place of sampling for laboratory tests.

Soil samples were taken before the seeder's passage, after the first, second and third passes in a rut behind the front and rear wheels, according to the procedure described above. In addition, for each case, three samples were taken to exclude random deviations in soil density indices. Plastic bags with soil samples to protect against heating and wetting were folded in a bucket and covered with a towel on top.

In the laboratory, the samples were weighed on a VK 3000.1 electronic scale with a measurement error of 0.1 gram. Knowing the volume of the samples, we calculated the density of the soil using the formula:

$$\rho = \frac{m}{V} \tag{21}$$

where:

m is the weight of the soil sample, g;
V is the volume of the soil sample, cm^3.

Figure 9. Obtaining a soil sample to determine its density by the method of Kaczynski

The relative error of practical and theoretical results of determining the density of the soil is calculated by the formula:

$$\delta = \frac{\left|\rho_E - \rho_T\right|}{\rho_E} \cdot 100\% \quad (22)$$

where:

ρ_E is density value obtained by the experimental method, g/cm^3;
ρ_T is density value obtained by calculation method, g/cm^3.

The obtained values of the mass, density of soil samples and the relative error of practical and theoretical results of determining the density of the soil are presented in Table 14.

The obtained data shows (Table 14) that the density of the soil increases from 1.080 g/cm^3 to 1.177 g/cm^3 after a three-time pass of the seeder, and the discrepancy between the results obtained by theoretical and experimental methods are insignificant (not more than 6.2%).

Table 14. The obtained values of the mass and density of soil samples and the relative error of theoretical and practical results

Position of Measurement of Soil Hardness	The Number of Measurements	Soil Sample Weight, g	The Average Weight of the Soil for Measurement Position, g	Soil Density, g/cm^3	Relative Error, %
Before the passage of the seeder	1	107.6	108.0	1.080	5.83
	2	108.3			
	3	108.1			
1st pass of front wheel	1	110.9	111.0	1.110	6.20
	2	111.5			
	3	110.6			
1st pass of both wheels	1	114.1	113.8	1.138	4.40
	2	113.3			
	3	114.0			
2nd pass of front wheel	1	115.0	115.2	1.152	3.56
	2	115.6			
	3	115.0			
2nd pass of both wheels	1	116.5	116.1	1.161	3.10
	2	116.0			
	3	115.8			
3rd pass of front wheel	1	116.5	116.8	1.168	2.91
	2	117.0			
	3	116.9			
3rd pass of both wheels	1	117.5	117.7	1.177	2.46
	2	117.2			
	3	118.4			

CONCLUSION

It was evaluating impact on the soil of the driving systems of a self-propelled seeder:

1. Vertical axial loads acting on the soil from the wheels of a self-propelled seeder are determined by calculation and experimental methods. The discrepancy between the obtained values is 1.4% for the rear wheels and 2.0% for the front wheels.
2. The size of the contact patch of the tires of the front and rear wheels of the seeder with the soil was determined by calculation and experimental methods. At the same time, theoretical and practical data do not differ significantly from each other. The relative error of experimental determination of the contact patch area for the front wheels is 2.2%, and for the rear wheels it is 2.8%. It was revealed that the pressure of the wheels on the soil does not comply with the permissible standards for sowing operations and exceeds them by 2.9 and 2.1 times for the front and rear wheels, respectively.
3. It was determined the maximum contact pressure of the wheel propulsion of the seeder on the ground for theoretical and practical data.
4. During field tests of the seeder, the soil hardness was measured in three zones: before the seeder passage and after each pass in the track, behind the front and rear wheels at six different depths, indicated by the marks on the soil density meter. Graphs of dependencies of soil hardness on the depth of measurement were constructed. It was revealed that the soil is noticeably compacted after the first pass and during subsequent passes the compaction value changes slightly. Moreover, with an increase in the depth of measurement, this trend is observed especially clearly.
5. The density of the soil is determined by calculation and experimental methods in three zones: before the seeder passes and after each pass in the track behind the front and rear wheels at a depth of 7.6 cm. The results obtained differ from each other by a maximum of 6.2%.
6. In order to reduce the harmful effects of self-propelled selection movers on the soil, it is necessary to conduct research aimed to reducing the maximum contact pressure of propulsion on the soil by using ultra-low-pressure tires and the number of passages of the seeder by combining operations in the cultivation of selection crops.

REFERENCES

Adams, E. P., Slake, G. R., Martin, W. P., & Boelter, D. H. (1960). Influence of soil compaction on crop growth and development. *Trans. 7-th Intern. Congr. Soil Science*, *1*, 171–178.

Bajla, J. (1998). *Penetrometrické merania pôdnych vlastností* [Penetrometrical Measurements of the Soil Properties]. Nitra: SUA in Nitra.

Balbuen, R. H., Terminiello, A. M., Claverie, J. A., Casado, J. P., & Marlats, R. (2000). Soil compaction by forestry harvester operation, evolution of physical properties. *Revista Brasileira de Engenharia Agrícola e Ambiental*, *4*, 453–459.

Godzhaev, Z.A., Shevtsov, V.G., Lavrov, A.V., & Rusanov, A.V. (2015). Metodika rascheta maksimalnogo kontaktnogo davleniya kolesnogo dvizhetelya na pochvu s ispolzovaniem universalnoj kharakteristiki shiny [The method of calculation of maximum contact pressure wheel on soil mover with the use of universal bus characteristics]. *Alternative Energy Sources in the Transport and Technological Complex: Problems and Prospects for Rational introduction*, *2*(1), 83-89.

Hakansson, I. (1990). A method for characterizing the state of compactness of the plough layer. *Soil & Tillage Research*, *16*(1-2), 105–120. doi:10.1016/0167-1987(90)90024-8

Hetz, E. J. (2001). Soil compaction potential of tractors and other heavy agricultural machines used in Chile. *Agricultural Mechanization in Asia. Africa and Latin America*, *32*, 38–42.

Izmailov, A., Shevtsov, V., Lavrov, A., Godzhaev, Z., & Pryadkin, V. (2015). Application of the Universal Tire Characteristic for Estimating the Maximum Pressure of a Pneumatic Tractor Wheel on the Ground. *SAE*. doi:10.4271/2015-01-2760

Kryajkov, V. M., Godzhaev, Z. A., Shevtcov, V. G., & Lavrov, A. V. (2015). Park traktorov: Sostoyanie i napravlenie razvitiya [Tractor fleet: the status and direction of development]. *Rural Mechanic*, *9*, 3–5.

Lavrov, A., Smirnov, I., & Litvinov, M. (2018). Justification of the construction of a self-propelled selection seeder with an intelligent seeding system. In *MATEC Web of Conferences*. 10.1051/matecconf/201822405011

Lavrov, A. V. (2018). K voprosu razrabotki intellektualnoj tekhnologii vyseva semennogo materiala [To the question of the development of intellectual technology of seeding seed material]. *Engineering Bulletin of the Don*, *4*. Retrieved from http://ivdon.ru/ru/magazine/archive/n4y2018/5329

Lavrov, A. V., Kryukovskaya, N. S., & Petrischev, N. A. (2018). Otsenka vozdejstviya na pochvu kolesnykh dvizhitelej samokhodnoj selektsionnoj seyalki [Evaluation of impact on soil wheel drivers of self-propelled selection seeder]. *Agrimachinery and Energy*, *4*(21), 95–106.

Loschenko, A. V. (2016). *Snizhenie uplotneniya pochvy i povrezhdeniya kornevoj sistemy dvizhitelyami mobilnykh ehnergeticheskikh sredstv* [Reducing of soil compaction and root damage by movers of mobile energy tools]. In V.I. Orobinskiy & V.G. Kozlova (Eds.), *Proceedings of the international scientific-practical conference: Modern scientific and practical solutions of the XXI century* (pp. 231-234). Voronezh, Russia: Voronezh State Agrarian University.

Malý, V., & Kučera, M. (2014). Determination of mechanical properties of soil under laboratory conditions. *Research in Agricultural Engineering*, *60*(Special Issue), 66–69. doi:10.17221/37/2013-RAE

Ministry of agriculture of the Russian Federation (2017). *Federalnaya nauchno-tekhnicheskaya programma razvitiya selskogo khozyajstva na 2017 - 2025 gody [Federal scientific and technical program of agricultural development for 2017-2025]*. Moscow, Russia: Printing office of the Ministry of agriculture of the Russian Federation.

Ministry of agriculture of the Russian Federation. (2018). *Itogi raboty otrasli rastenievodstva v 2017 godu i zadachi na 2018 god [The results of the work of the crop industry in 2017 and objectives for 2018]*. Moscow: Printing office of the Ministry of agriculture of the Russian Federation.

Pogodin, N. N., Kuchko, V. V., Barsukevich, F. A., & Shatilo, S. V. (2008). Uplotnenie pochv selskokhozyajstvennoj tekhnikoj [Soil compaction by agricultural machinery]. *Land Improvement*, *1*(59), 70–74.

Polivaev, O. I., & Voicshev, V. S. (2013). Snizhenie uplotneniya pochvy dvizhitelyami mobilnykh ehnergeticheskikh sredstv [Reduction of soil compaction by movers of mobile energy resources]. *Bulletin of Voronezh State Agrarian University*, *1*(36), 57–59.

Rusanov, V.A (1998). *Problema pereuplotneniya pochv dvizhitelyami i ehffektivnye puti ee resheniya [The problem of soil compaction movers and effective ways to solve it]*. Moscow, Russia: all-Russian Institute of mechanization.

Sergeev, A. G. (2017). *Ehkologicheskaya problema – uplotnenie pochvy* [Ecological problem – soil compaction]. In *Materials of the IV all-Russian scientific-practical conference in Saratov state agrarian University named after N.I. Vavilova: Technogenic and natural safety* (pp. 338-340). Saratov, Russia: Amirit.

Shevtsov, V. G., Soloveychik, A. A., Rusanov, A. V., & Lavrov, A. V. (2014). The use of universal characteristics of a tire in determining maximum pressure of a wheel running on soil. In *Proceedings of the International Research & Practice Video Conference: Topical research trends in the twenty-first century: Theory and practice* (pp. 169-173). Voronezh, Russia: Voronezh Public Forest Engineering Academy.

Soane, B. D. (1970). The ground pressure of wheels and traks. *Power Farm*, *44*(4), 40–44.

Tullberg, J. N., Hunter, M. N., Paull, C. J., & Smith, G. D. (1990). Why control field traffic. In *Proceedings of Queensland. Department of Primary Industries Soil Compaction Workshop*, Toowoomba, Australia (pp. 13–25).

Ziyaee, A., & Roshani, M. R. (2012). A survey study on Soil compaction problems for new methods in agriculture. *International Research Journal of Applied and Basic Sciences*, *3*(9), 1787–1801.

Chapter 7
Performance Improvement Study of Linear Photovoltaic Systems With Concentration of Solar Radiation

Petr Alexandrovich Nesterenkov
Al-Farabi Kazakh National University, Kazakhstan

Alexander Gennadievich Nesterenkov
Kazakh Research Institute of Power Engineering, Kazakhstan

ABSTRACT

A new type of linear cooled photodetectors is considered, of which in the focal region of the optical concentrator mirrors is installed an array of solar cells operating with the low-ratio solar concentration. This work is focused on the theoretical and experimental substantiation of the efficiency increase of photodetectors under conditions of an optimal combination between solar radiation concentration and heat transfer intensity of photovoltaic cells with heat carriers. The problem of obtaining a high temperature liquid due to the limitations of solar cells is solved by organizing the flow of fluid within the thermal collector channels in the focal region of an additional optical concentrator. A mathematical model of engineering characteristic calculation of the Λ *- shaped photodetectors and cost calculation of electrical and thermal energy generation is presented. The research results are used in the development of industrial prototypes of photodetectors with a concentration of solar radiation and low production costs.*

INTRODUCTION

Long-term studies on photovoltaic led to the creation of a technology for the production of silicon solar cells with an efficiency close to the theoretical limit ≈ 25%. Commercial photovoltaic modules (PV modules) with an efficiency of ≈ 16% have appeared on the market, which dissipate most of the incoming solar energy as heat into the environment. In order to solve this issue, intensive research is being carried out, which allow receiving simultaneously electric and thermal energy. The technology

DOI: 10.4018/978-1-5225-9420-8.ch007

and production line of photovoltaic-thermal (PVT) systems was developed, a detailed review of which was presented by Chow (Chow, 2010), and methodology for calculating performance was proposed by Duffe and Bechman (Duffie & Beckman, 2006). The first samples of PVT systems on silicon photocells had a technical limitation on the operating temperature, which conflicted with the aim to obtain a high temperature of heat carriers at the outlet for economic use. The problem was partially solved with the transition to high-temperature GaAs photovoltaic cells, capable of efficiently generating electrical and thermal energy. However, they turned out to be scarce and much more expensive than silicon solar cells, so there was a need to use solar concentration ratio in more than 50-fold to ensure payback.

In Israel, is developing a cogeneration plant based on a paraboloid with an area of ≈ 1m^2 and 400-fold concentration ratio (Kribus et al., 2006). Three-junction InGaP/InGaAs/Ge solar cells are used, which make it possible to obtain a specific electric power of ≈ 170 W/m^2 and thermal ≈ 530 W/m^2 with a ratio of 530/170 ≈ 5.3 and an output coolant temperature ≈ 70 °C. Such systems are too small in power to provide the individual consumer with enough electricity and heat energy.

In Zurich, a three-year project of developing a "DSolar Sunflower" pilot system with a parabolic optical hub of 36 mirrors was completed in 2017 (Airlight, 2017). For a very high (>1500-fold) optical concentration of solar radiation on the limited surface of three-junction GaAs solar cells, it is necessary to use a relatively expensive precision sun tracking system and a high-pressure station with liquid purification filters for cooling the solar cells (Escher, Paredes, Zimmermann, Chin, 2012; Zimmermann et al., 2013). The complex process of manufacturing elliptical mirrors, the low reliability of the cooling system of the solar cells, and the large weight ≈ 18.5 tons reduce the potential for commercialization of the technology.

In the work of Prof. Xu Ji et al., the efficiency of a linear photodetector with silicon solar cells at different radiation concentrations was studied (Ji, Li, Lin et al., 2012). It was noted that the thermal efficiency of the photodetector increased linearly in the range of direct radiation 200 - 1200 W/m^2, and the electrical one only up to 500 W/m^2. The decrease in the rate of growth of electrical efficiency is explained by the negative effect of an increase in the internal series resistance on the no-load voltage.

J.S. Coventry et al., investigated the process of converting solar energy at 30-fold solar concentration ratio on silicon solar cells with a width of 40 mm, placed on a cooled photodetector with a width of 80 mm (Coveny, 2005, pp. 211-222). The low series resistance of solar cells ≈ 0.043 Ohm/cm^2 was achieved due to the creation of narrow gaps between the conductive pins, doping with phosphorus in the area of collecting electrons and a large cross section of electrical contacts. In this case, the peak electric power was ≈ 11% and thermal ≈ 57% with the achievement of the ratio 57/11 ≈ 5.2.

Cogenra Solar (SunPower Corporation) has developed a cogeneration system with linear V-shaped photodetectors in the focal region of concentrator flat mirrors (Forbes Solar, 2017). At the latitude of South Carolina state (≈ 35° N) at 9-fold concentration ratio, photodetectors produce specific electric energy ≈ 168 kWh/m^2 and thermal ≈ 657 kWh/m^2 per year, and the coolant output temperature in the autumn period reaches ≈60 °C. For the first time, a five-year payback of capital costs was proved with the ratio of the generated specific heat and electric energy ≈ 3.9. The peak power of photodetectors with optical concentrators from 14 flat mirrors reached ≈ 100 W/m^2 and ≈ 498 W/m^2 heat power. The disadvantages of the system are large heat losses, the inability to obtain a high output temperature of the liquid and a relatively large unit cost.

The most promising method of working with a low concentration of the sun are solar cells made of n-Si, whose advantage over p-Si is due to the low rate of degradation of performance and low sensitivity to impurities and defects (Untila et al., 2012; Milichko et al., 2016). SunPower Corporation achieved the greatest success in the production of commercial solar cells by improving the back contacts and reducing the coherent internal resistance by reducing the thickness of silicon wafers to 180 microns (Bunea et al., 2010). At Amonix corporation, developers reduce the series internal resistance of a solar cell to 3 mOhm/cm^2, which allows increasing the efficiency of solar cells up to 27% under conditions of 100-fold concentration of solar radiation (Slade & Garboushian, 2005). At the Fraunhofer ISE, sequential internal resistance is reduced by laser drilling holes in silicon crystals and mounting wire electrodes in them, which resulted in an increase in the efficiency of photocells up to 21% with a tenfold concentration of the sun (Fellmeth, Ebert, Efinger, 2014).

The key problem of cogeneration systems with a concentration of solar energy is the transportation of generated thermal energy to the consumer. Not solved the problem of increasing the unit cost of equipment and heat loss from the surface of photodetectors.

This work proposes ways to solve these problems and discusses the results of studies of a new type of photodetectors with relatively low optical and thermal losses and a dual-circuit circulation system that transports the generated thermal energy to the heat accumulator in the natural circulation mode of the consumable fluid. The results of designing a new type of photodetectors and calculation of system performance in the southern latitudes of the Republic of Kazakhstan are proposed.

EXPERIMENTS AND CALCULATIONS

Design of Optical System

One of the main working elements of a cogeneration plant with a concentration of the sun is an optical system with flat mirrors, the design of which provides the possibility of end-to-end modernization with the advent of new components and materials on the market (Nesterenkov, Nesterenkov, Nesterenkova, 2016). The assemblies of optical concentrators are designed taking into account convenient transportation to the installation site in a special container, which also acts as a supporting structure and a unit for placement of batteries and control devices.

Using mathematical modeling and on the basis of the geometric optics rules, it was shown that through using flat mirrors, uniform distribution of the reflected radiation of the sun is created along the surface of linear photodetectors (Smirnov, 2010). The optical system is estimated by three technological parameters: the geometric coefficient K_g, equal to the ratio of sum of product of mirrors width and the cosine of the angle between the incident radiation and the perpendicular to the plane of the mirrors and the width of the photodetector; the reflection coefficient K_r, which characterizes the quality of the mirror coatings and the transmittance K_d, which characterizes the optical loss during the passage of light through the glass. To compare the quality of optical systems from different manufacturers, the technical term "optical efficiency" is introduced, defined as the product of the above-mentioned coefficients: $C_o = K_g \bullet K_r \bullet K_d$. Geometric concentration is determined by the ratio of the area of the aperture of the optical concentrator to the area of the photodetector. Approximately geometric concentration is determined by the ratio of the product of the number of mirrors N and the average cosine of all mirrors: $K_g = N \bullet \cos_{cp} \bullet \psi_i$ subject to the approximate equality of the areas of mirrors and photodetectors.

For high power industrial solar systems, optical concentrators for mechanical strength reasons consist of relatively thick mirrors ≈ 4–6 mm, in which, as shown by previous studies, up to ≈ 6% of solar energy is lost (Chea, Hakansson, Karlsson, 2013). The transmittance of solar radiation depends on the thickness of the glass and its iron content; therefore, in this work, special polished glass with a thickness of no more than 2 mm is used in the mirrors. On a lightweight support frame made of galvanized metal profiles, a sun tracking system based on controllers and linear actuators is used. An alternative to glass is thin mirror films such as "Alanod Silver" on an aluminum (or copper) base with a thickness of 0.25 - 0.5 mm, the average reflection coefficient of which for the solar spectrum reaches $K_r \approx 0.94$ (Alanod Corporation, 2019). Protective quartz layers on the surface of the mirror films increase the reliability and availability of their use in optical concentrators.

The transmittance of the radiation reflected from the mirrors and passing through the thermal glasses was increased to $K_d \approx 0.94$ by using composite protective thermal glasses of relatively small area. Reducing the area allows to reduce the glass thickness to ≈ 1.8 mm with preservation of mechanical strength.

The innovation of the optical system consists in the formation of a Λ-shaped focal region from the solar fluxes reflected by the mirrors, in which the frontal walls of the linear photodetectors with silicon photocells are located. Installed along an axial plane at an angle to each other, the frontal walls radiate and simultaneously shield thermal radiation. As a result of mutual shielding of thermal radiation, a significant reduction in heat losses is achieved. In addition, the opposite front walls protect each other from direct contact with hail and other large precipitation.

Let us combine the axis of rotation of the optical concentrators with the coordinate axis Z, and the coordinate axes X and Y will be placed in the plane of the drawing. Direct solar radiation (depicted by dotted lines) hits on flat mirrors and is reflected on to the front surfaces of the photodetectors, whose centers are focuses of parabolic surfaces. Figure 1 shows the optical scheme explaining the process of reflection of the sun rays from the mirrors of an optical concentrator located to the right of the axial plane. The top of the parabolic surface, along which the centers of the mirrors are placed, is located at the point of intersection of the X and Y coordinate axes.

Figure 1. Optical scheme of sun rays reflection by flat mirrors

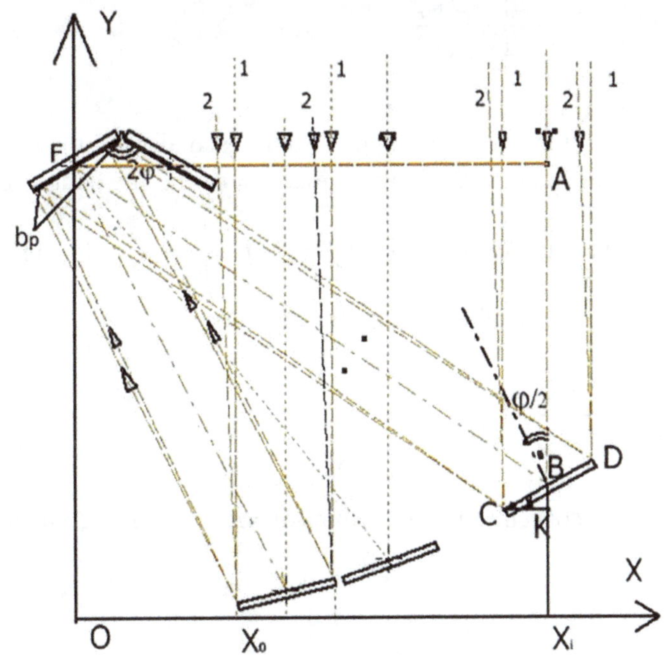

Front walls of width b are set at an angle of 2φ in the focal areas of the respective optical hubs. By definition, the coordinates of the points of a parabolic curve satisfy the mathematical expression: $x^2 = 4 \bullet f \bullet y$, where f is the focus value, i.e. distance from the top of the parabola to the focal point F.

During the tracker's stopping period, the sun's rays are shifted by δ from position 1 to position 2, and the focused light spot is shifted along the width of the front wall by a distance b_p. At this juncture, the photocells always remain in the light zone of the focused solar radiation, and the selective film is in the zone of periodic displacement. As a result, the efficiency of using the area of photodetectors increases, as well as the productivity of heat energy generation.

The geometric concentration of optical concentrators is determined as follows. Denote the width of the flat mirrors by b_Z, and the centers of coordinates by X_i and Y_i, the angle between the rays of the sun hitting on the mirror and the line connecting the focus to the center of the mirrors by φi. According to the laws of geometric optics, the angle of incidence of the rays is equal to the angle of reflection, therefore $\frac{\varphi_i}{2}$ is equal to the angle between the perpendicular to the plane of the mirrors and the incident rays of the sun. From where the magnitude of the projection of the i mirror on the X axis is equal to: $\Delta X_i = b_z \cdot \cos\frac{\varphi_i}{2}$. The projection of the photodetectors on the X axis is: $\frac{3b}{2} \cdot \sin\varphi$. Taking into account the displacement of the sun's rays, we obtain the distance of the edges of the first mirrors from the axial plane: $X_0 = \frac{3b}{2} \cdot \sin\varphi \cdot \delta$. Let δ_1 denote the distance between the edges of successively placed mirrors and the distance from the coordinate axis to the projection of the center of the last mirror is X_N. Since the parabola is the locus of points equidistant from the focus, we have the equation: $FB = f + Y_i$. The second equation is obtained from the triangle ABF: $f - Y_i = FB \cdot \cos\varphi_i$. From the system of two equations with two unknowns, we obtain the coordinates of the i mirror:

$$X_i = 2 \cdot f \cdot \left(\frac{1-\cos\varphi_i}{1+\cos\varphi_i}\right)^{0.5}, \; Y_i = f \cdot \frac{1-\cos\varphi_i}{1+\cos\varphi_i} \tag{1}$$

The coordinates of the last mirror are determined from the condition of parallelism of the reflected sunlight and the surface of the front walls. In this case, $\varphi = \varphi_i$ and the parameters of the mirrors and photodetectors are connected using expressions: $X_N = 2 \cdot f \cdot \left(\frac{1-\cos\varphi_i}{1+\cos\varphi_i}\right)^{0.5}$ and $Y_N = f \cdot \frac{1-\cos\varphi_i}{1+\cos\varphi_i}$. The width of the maximum aperture of the optical concentrator:

$$b_z \cdot \sum\cos\frac{\varphi_i}{2} = X_N + \frac{b_z}{2} \cdot \cos\frac{\varphi_i}{2} - \left(X_0 + \sum\delta_i\right).$$

In this case, the geometric concentration of the optical concentrator is equal to:

$$K_g = \frac{b_z \cdot \Sigma \cos \frac{\varphi_i}{2}}{b} = 2 \cdot \frac{f}{b} \cdot \left(\frac{1-\cos\varphi}{1+\cos\varphi} \right)^{0.5} + \frac{b_z}{2} \cdot \cos\frac{\varphi}{2} - \frac{3}{2} \cdot \sin\varphi \cdot \frac{\delta \cdot (N-1)}{b} \quad (2)$$

Table 1 represents the results of calculating the maximum value of the geometric concentration for different angles between the front walls and the focus of the optical concentrator at b = 0.15m, b_z = 0.155m, δ = 0.006m, $\frac{'(N-1)}{\text{b}} \approx 0{,}4$.

The distance of the first mirror center from the Y axis is:

$$X_1 = X_0 + \frac{b_Z}{2} \cdot \cos\frac{\varphi_1}{2}.$$

To exclude mutual shading of mirrors, the technological distance between their edges δ_1 is introduced, whence we get the distance of the second mirror center from the axial plane:

$$X_2 = X_1 + \frac{b_Z}{2} \cdot \cos\frac{\varphi_1}{2} + \delta_1 + \frac{b_Z}{2} \cdot \cos\frac{\varphi_2}{2} = X_0 + b_z \cdot \cos\frac{\varphi_1}{2} + \delta_1 + \frac{b_Z}{2} \cdot \cos\frac{\varphi_2}{2}.$$

Table 1. The results of the calculation of the geometric concentration and coordinates of the mirrors

φ, °	45	50	55	60	65
sin φ	0,707	0,766	0,819	0,866	0,906
cos φ	0,707	0,643	0,574	0,5	0,423
$\left(\frac{1-\cos\varphi}{1+\cos\varphi}\right)^{0.5}$	0,414	0,466	0,521	0,577	0,636
$\cos\frac{\varphi}{2}$	0,924	0,903	0,887	0,866	0,843
f, m			1,25		
K_g	5.88	6.65	7.47	8.33	9.25
X_N, m	1,038	1,165	1,303	1,443	1,59
Y_N, m	0,215	0,271	0,339	0,416	0,506
f, m			1,75		
K_g	8.64	9.76	10.94	12.18	13.50
X_N, m	1,45	1,63	1,82	2,02	2,23
Y_N, m	0,30	0,38	0,47	0,58	0,71

As a result, we obtain a general expression for determining the coordinates of the projection of the optical concentrator mirrors of on the X axis:

$$X_i = X_0 + (i-1)\cdot\delta_1 + b_z \cdot \sum \cos\frac{\varphi_{i-1}}{2} + \frac{b_Z}{2}\cdot\cos\frac{\varphi_i}{2} \tag{3}$$

In accordance with Figure 1, the triangles ABM and FAC are similar, whence we get the relation $\frac{\Delta Xi}{b_Z} = \frac{AB}{FB}$ and further $\Delta X_i = \frac{f - Y_i}{f + Y_i}$. From the expression (3), we obtain a cubic equation relating the magnitude of the projection of the mirrors with the coordinates of their centers:

$$b_z \cdot \frac{4\cdot f - X_i^2}{4\cdot f + X_i^2} = X_i - \left(X_{i-1} + \frac{\Delta X_{i-1}}{2} + \delta_1\right) \tag{4}$$

Successively solving (4), we obtain the values of the mirrors coordinates and the magnitude of their projections on the X axis.

Figure 2 shows graphs of the geometric concentration on the design parameters of the optical system dependence.

The cosine loss is reduced by limiting the number of mirrors to 12 and increasing the distance to the photodetectors. Due to this, it was possible to increase the geometrical coefficient to the value of $K_g \geq 0.95$ and at the same time reduce the wind load and increase the stability of the supporting structure with mirrors in turbulent winds.

In winter, the surface of the mirrors is covered with frost, which firmly holds fresh snow. In the developed system such issue is solved by orientation of the mirrors at an angle to the horizon and formation of heat flux from the heat exchanger surface, as in Figure 3.

Figure 2. The dependence of the concentration of solar radiation on the parameters of the optical system

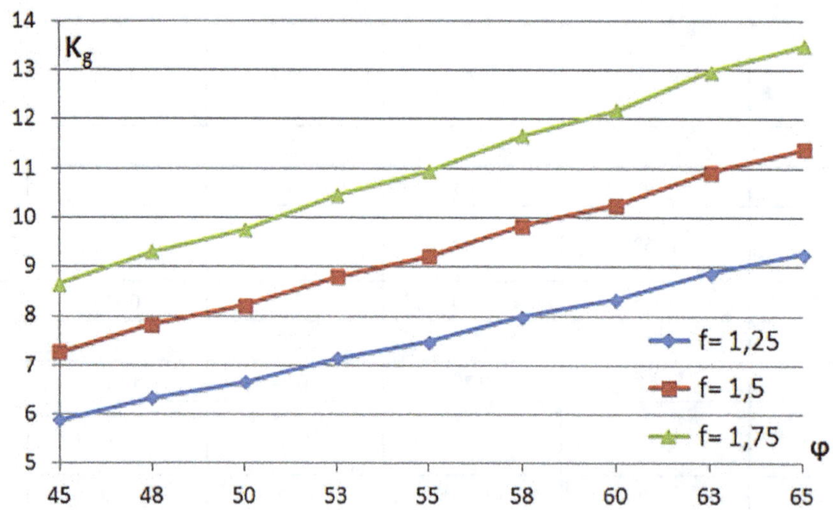

Figure 3. Illustration of snow melting from the surface of laboratory setup mirrors

The issue of increased heat losses at temperatures above ≈ 57 °C is solved by using of a linear thermally insulated collector in the focal area of an additional - adjacent optical concentrator. Figure 4 shows the design of an optical scheme of a solar cogeneration system for the generation of electrical and high-potential thermal energy with a liquid outlet temperature reaching 85 °C.

Linear photodetectors 1 and 2 are placed along the axial plane in the focal region of the mirrors 3, forming the surface of optical parabolic concentrators. Adjacent concentrator with flat mirrors 3, placed along the flow of coolants, reflect the solar radiation on the mirror 4 of a parabolic shape, which in turn reflect solar rays on the linear collector 8. The thermal energy released by the photocells is transferred to the coolant, which is pumped through pumps through the flat channels of photodetectors and coils 6 using pumps 5. The heat energy received by heat carriers is discharged in the countercurrent mode with consumed process water in a linear heat exchanger 7. The consumed water, preheated in the heat exchanger, enters the collector 8, where it increases its temperature in the focal region of the adjacent concentrator and then, through the passage bearings 9, is sent to the heat accumulator 11 placed at a height h relative to the heat exchanger. Through the return pipe 12, the automatic cold water supply valve 13 and the through passage bearing 10, natural circulation of consumed process water is formed.

The mirrors 3 of the main and adjacent concentrators have a transverse dimension b_Z and create the same parabolic cylindrical surface. The secondary parabolic mirrors focus the reflected radiation on the thermal collector surface with a selective coating width b_K. Structurally, secondary mirrors with a width of b_{ZK} are conveniently placed along photodetectors. In this case, $b_{ZK} \approx b$. Using the results of calculating the geometric concentration of the main concentrator $K_g \approx 9$, we find the value of the geometric concentration at the double reflection of radiation $\approx K_g \bullet b/b_K \approx 41$. Optical losses due to double reflection of radiation are approximately $\approx C_{01} \bullet C_{02} \approx 0.83 \bullet 0.83 \approx 0.66$, so the optical concentration can reach $\approx 41 \bullet 0.66$.

Figure 4. Optical scheme of double reflection of solar radiation

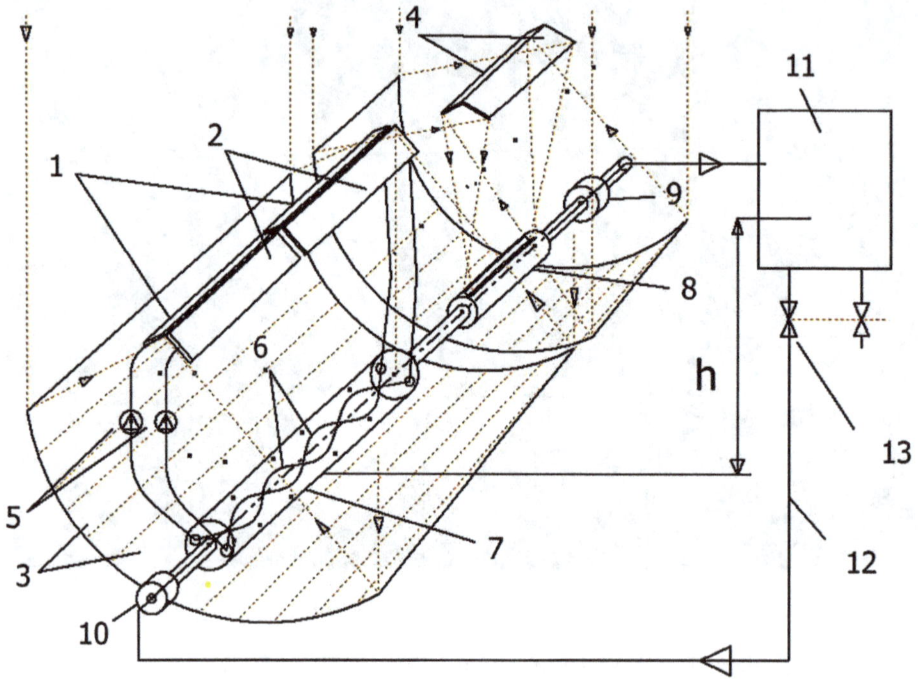

Linear Photodetectors Structure

The sources of electric and thermal energy generation in linear cooled photodetectors are high-quality "Maxeon" silicon solar cells. As shown in Figure 1, during the motion of the sun across the sky with an angular velocity of $\omega \approx 2\pi / 24 \bullet 60$ radian/min during the stop time τ of the supporting platform, there is a shift of the sun rays from position 1 to position 2. In this case, the light spot from the reflected rays shifts across the frontal walls by $b_p = \tau \bullet f \bullet (2\pi / 24 \bullet 60)$. In order for solar cells with a size b to always remain in the light field, the transverse size of the reflected solar radiation $\approx b_x$ must satisfy the condition: $b_x \geq (b + b_p)$. In the zone of periodic displacement of reflected radiation, selective films TiNOX with absorption coefficient β are placed (Almeco Group, 2019). The thickness of the selective films corresponds to the thickness of the solar cell to provide the necessary collinearity of the outer surface during the subsequent joint encapsulation with the use of single thermal glass.

Under standard operating conditions at the point of maximum power, used in research solar cells have a voltage of $\approx$ 0.57 V, a current $\approx$ 5.8 A and a peak power of $\approx$ 3.3 W. Operation with the concentration of the sun is accompanied by an increase in electric current up to $\approx$ 60 A, therefore, it is necessary to increase the cross section of the electrical contacts between the solar cells. This causes a problem with their placement on the cooled wall. To solve it, an innovative design of the flow channel profile of aluminum alloys with longitudinal bridges between the front and rear walls has been developed. At the same time on the front wall opposite the jumpers are made longitudinal grooves, the depth and width of which corresponds to the response section of the electrical contacts. After the metal channel is cast, it is electrochemically treated to obtain a durable and porous electrically insulating coating. It is preferable to conduct microarc oxidation of the walls, which allows to obtain heat-conducting coatings up to 150 microns thick with adjustable porosity up to 50% and breakdown voltage up to 2,000 V.

Figure 5 shows a photodetector (without selective films and thermal glasses), and a scheme of a channel section with encapsulated photocells.

On the front wall of the cooled channel 1 photo cells 2 are installed collinearly with electrical contacts 3, which are completely immersed in the formed longitudinal grooves of the front walls surrounded by a thermoplastic silicone gel. Along the edges of the photocells fixed selective film 4 TiNOX. Due to the collinearity of the rear sides of the photovoltaic cells and selective films, it was possible to ensure their joint encapsulation using a transparent sealant (not shown) and relatively thin thermal glass 5. Reducing the thickness of the silicone sealant reduces thermal resistance, improves cooling and increases the production of electrical energy. Preserving collinearity simplifies the technology of encapsulating photocells under the load of thermal glass without the use of vacuum thermal furnaces.

On the rear wall of the channel through a relatively thick insulation (not shown) installed solar cells 6 which are working from a single solar radiation, and having standard electrical contacts of much smaller section. Photo cells are encapsulated with a transparent silicone sealant and relatively thick thermal glass 8. Thermal resistance on the upper wall can be increased, since the solar cell, operating without concentration of sun radiation, are encapsulated there. In addition, there is no special need to reduce optical losses in thermal glass, so it can be of any thickness and quality, which reduces the overall cost of the system. The flow of solar energy to the front wall is almost ten times higher than the flow of energy to the rear wall of the photodetectors; therefore, the temperature of the front walls T_1 is much higher than the temperature of the rear walls T_2.

It is known that the surface of photovoltaic cells, despite the high-quality anti-reflective coatings, reflect at least 5% of the concentrated solar radiation back. In the developed quasi-closed Λ-shaped focal region, a part of this reflected radiation falls on the oppositely mounted photocell and is absorbed with the exception of the same five percent. In the process of multiple reflections of concentrated solar radiation, the efficiency of solar energy absorption by solar cells increases.

Let us place one of the walls along the X axis, and the second at an angle $2 \cdot \varphi = \frac{2 \cdot \pi}{3}$. The temperature of the counter-mounted walls is the same, so their re-radiation compensates for the heat loss in

Figure 5. Image and scheme of the photodetector in cross section

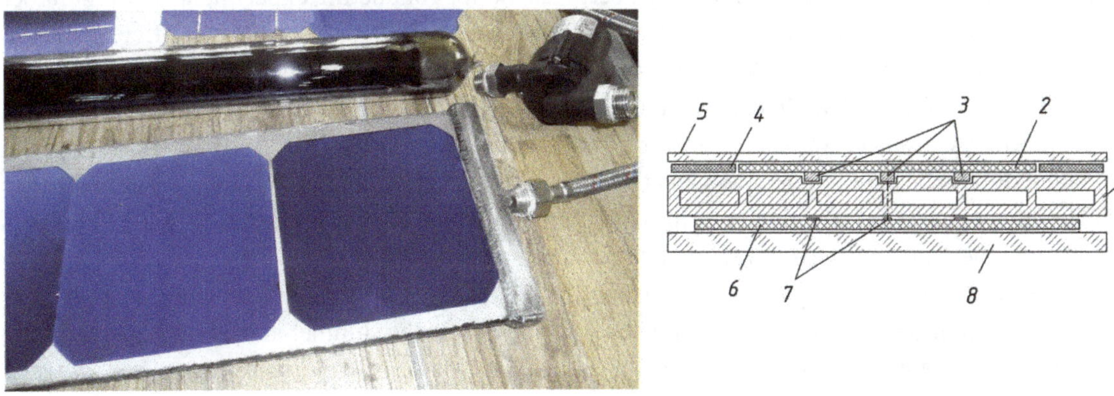

their direction. The area element ΔF of the front walls radiates heat flux into the surrounding space, the value of which is determined by the Lambert law (Mikheev & Mikheeva, 1977):

$$d \cdot Q = \frac{\varepsilon}{\pi} \cdot \sigma_0 \cdot \left(\frac{T}{100}\right)^4 \cdot \Delta F \cdot \cos\varphi \cdot \int_0^{\psi} d\psi \tag{5}$$

where ε and T - the values of blackness and the temperature of the walls; $\sigma_0 = 5.67 \cdot 10^{-8}$ W/(m^2 • K^4) - Stefan-Boltzmann constant; φ is the angle between the perpendicular to the surface and the direction of emitting; ψ - solid angle within which thermal radiation propagates.

From the surface of the Λ - shaped photodetectors, thermal radiation is carried away within the angles of 0-120° due to shielding by the opposite wall, and the photodetectors of the V - shaped form within the angles of 0 - 180°. The ratio of the amount of heat carried by radiation into space for such two cases is:

$$\zeta \approx \frac{\varepsilon_1 \cdot \left(\frac{T_1}{100}\right)^4 \cdot d \cdot F \cdot \cos\varphi \cdot \int_0^{\frac{2\pi}{3}} d\psi}{\varepsilon_2 \cdot \left(\frac{T_2}{100}\right)^4 \cdot d \cdot F \cdot \cos\varphi \cdot \int_0^{\pi} d\psi} = 0.66 \tag{6}$$

An important advantage is noticed: at $\varepsilon_1 \approx \varepsilon_2$ and $T_1 \approx T_2$ and other conditions being the same, the heat losses from the surface of Λ - shaped photodetectors are at 33% less, than in V – shaped case.

Circulating System with Heat Exchanger

To determine the characteristic parameters of the heat exchange of the heat-generating walls of linear photodetectors and thermal collectors with fluid and the external environment, special thermal engineering experiments were carried out on full-size models of channels. The heat flux from photocells was imitated by heating thin metal films glued with heat-resistant glue onto the channel walls. With the passage of electric current through the film from the constant current source along the walls of thermal energy is released. By adjusting the current and the flow rate of the coolant, the conditions for the transfer of heat from the photocells into the liquid and the heat exchange of the walls with the environment were reproduced. Figure 6 shows a mobile heat engineering stand for conducting experiments in the summer and winter, and an experimental linear collector efficiency curve. For comparison, the efficiency curve of a high-quality Vitosol collector (Viessmann Corporation, 2019) is shown next to it.

The efficiency of a two-way linear collector is higher, which is explained by the absence of losses from the thermally insulated walls and, accordingly, heat losses through them. With an increase in the temperature difference between the walls and the environment over ≈ 50 °C, the efficiency curves begin to fall sharply, which is explained by a significant increase in heat loss by radiation. The total heat loss is ≈ 40%, and the ratio between convection and radiation reaches ≈ 5:3.

The processes of solar energy conversion using linear photodetectors and optical concentrators were studied under conditions of different intensity of solar radiation in the summer and winter seasons of the year on laboratory setups. Figure 7 shows one of the variants of laboratory systems with a single-axis

Figure 6. Thermal engineering stand for testing of double-sided thermal collectors

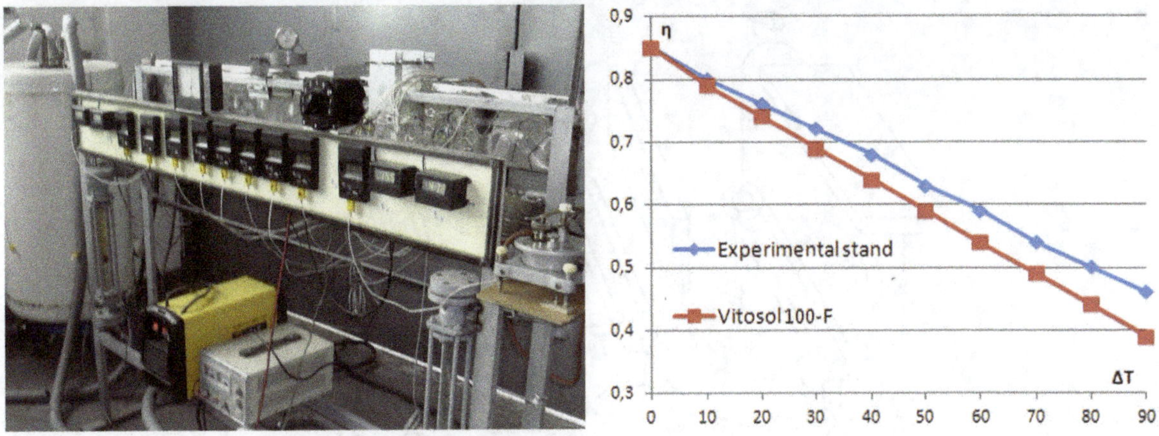

Figure 7. Laboratory setup of cogeneration system

sun tracking system (on the right) that provides rotation of the supporting structure with mirrors and photodetectors using two pass-through bearings.

In this case, the transfer of electricity and heat to the measurement system was carried out using flexible conductors and thermally insulated pipelines. The cooled silicon solar cells with a size of 0.125 • 0.080m and a peak power of 2.8W were placed in serial along the cooled channels symmetrically relative to the central plane, and a flat collector was installed between them. To study the effect of the thermal protective layer, two-sided flat photodetectors were placed below the collector surface. Transportation of hot and cold water with an adjustable flow rate from the stationary tank to the moving photodetectors was carried out using circulation pumps. Figure 8 shows the circulation system of the laboratory setup.

Each photodetector (Φ1, Φ2) has its own circulation circuit with a micro-pump, pumping a mixture of process water and glycol. The heat received from the solar cells was discharged in countercurrent mode of process water in linear heat exchangers (T_1, T_2). Having received a certain amount of thermal energy in them, the process water flowed into the collector (H_K) and increased its temperature. Then it proceeds into the central heat exchanger (T_K), where it discharged the thermal energy of the thermal circulating circuit water to the technical water as in Figure 9.

Figure 8. Laboratory setup circulation system

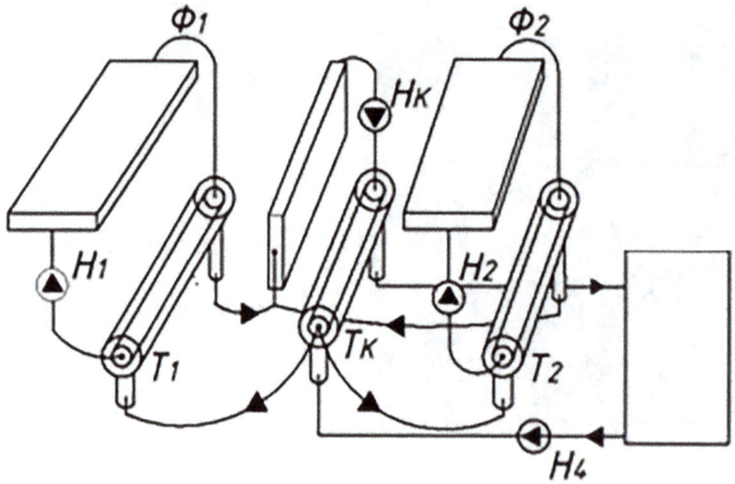

Figure 9. Temperature growth of heat exchanger fluids

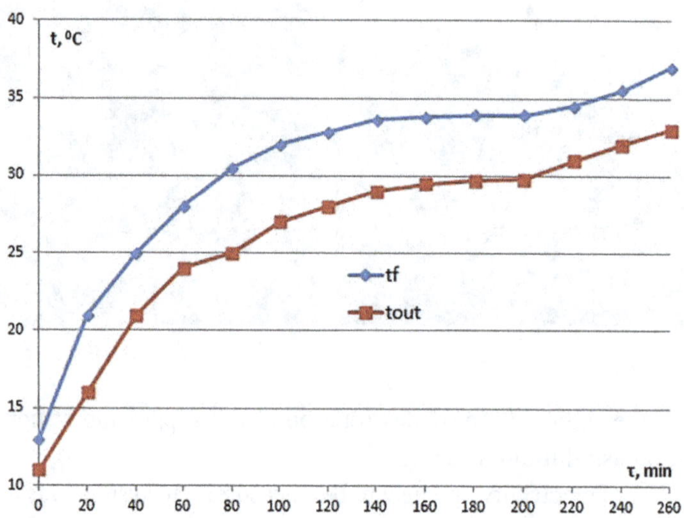

The magnitude of the optical concentration was varied by varying the number of mirrors focused on the surface of the photodetectors. In September, the maximum optical concentration of solar radiation did not exceed 5-fold ratio. Considering that the optical efficiency of the concentrator did not exceed ≈ 0.69, and the direct solar radiation in the autumn period did not exceed 750 W/m^2, a single concentration of direct solar radiation was obtained when the two mirrors were aimed at the photodetector surface.

In a laboratory setup, it was not possible to raise the temperature of process water above 60 °C, due to large heat losses by radiation, therefore a new design concept of a cogeneration system, described above, was implemented. The consumed water is supplied and discharged from movable supporting structure to the fixed source of cold water (heat accumulator) through the thrust bearings. This method of transferring the heated fluid from moving objects to stationary is more reliable, since complex and long flexible hoses with coolants are excluded (Nesterenkov et al., 2015).

The flow of consumed water to the heat accumulator in the mode of natural circulation is carried out due to the temperature difference between the points of heat supply and cold water feed in the return pipe of the circulation loop, those, due to the direct conversion of high-grade thermal energy into mechanical energy through the transport of fluid. When this occurs, the difference in the density of the liquid in the heat exchanger and the heat accumulator leads to the formation of a dynamic pressure value:

$$\Delta P = g \cdot h \cdot \left(p_A - p_T\right) \tag{7}$$

where ΔP - dynamic pressure, Pa; g - gravitational acceleration, 9.81 m/s^2; h - distance between the centers of the heat exchanger and the heat accumulator, m; ρ_T and ρ_A - the average density of water in the heat exchanger and the return pipeline, kg/m^3.

Method of Calculating the Optimal Temperature of Photovoltaic Cells

The efficiency of operation of linear photodetectors is described by the sum of electrical η_e and thermal η_T efficiency:

$$\eta = \eta_e + \eta_T = \left[F \cdot \left(I_{SC} \cdot V_{OC}\right) + Q_T\right] \cdot \frac{n}{\left(C_0 \cdot N \cdot b \cdot L \cdot E\right)} \tag{8}$$

where F is the fill factor; I_{SC} - short circuit current, A; V_{OC} - open circuit voltage, V; Q_T - heat output to the coolant, W; n - number of solar cells connected in series on the channel wall; C_0 - efficiency of the optical system; N - number of reflecting mirrors; E - intensity of direct solar radiation, W/m^2; b and L - respectively, the width and length of the channel of photodetectors, m.

The nominal electric power of solar energy systems operating without solar concentration is determined by the technical characteristics of the solar cells. In terms of the concentration of solar radiation in accordance with equality (8), an internal relationship between the processes of generation of electricity and heat is manifested. An increase in the flow rate of heat carriers contributes to the intensification of heat exchange, a decrease in the temperature of the photovoltaic cells and, accordingly, an increase in electrical efficiency.

The source of energy entering the walls of the channels and then into the liquid is a part of the solar energy converted into thermal energy in accordance with the expression (7). Since the thickness of the photovoltaic cells is much less than the thickness of the remaining layers of the walls, we will adopt a mathematical model in which the temperature of the photovoltaic cells t_f is constant in thickness δ_f. Let us direct the X axis along the flat channel of the photodetector, and the Y axis across the channel. The density of the solar flux is uniform along the plane of the front and rear walls of the channels: q_1 and q_2 ≈ const (x). From the condition of the balance of energy entering the fluid through the walls, we obtain a one-dimensional differential equation (Petukhov, 1967):

$$G \cdot C_p \cdot \frac{dt}{dx} = \left(q_1 + q_2\right) \cdot \left(b + h\right) \tag{9}$$

where G - mass flow rate of the liquid, kg/s; C_P – specific heat capacity, J/kg • K; t - average cross-section temperature of the liquid, °C; b and h - width and height of the channel.

Integration of equation (9) gives a linear increase in the temperature of the coolant along the channel length and, accordingly, a linear temperature distribution of the channel walls, which directly affects the temperature rise of the consistently connected photovoltaic cells. Since they have a negative temperature coefficient of peak power $k_P \approx$ -0.011 W/°C and voltage $k_U \approx$ -1.8 mV/°C, the rise in temperature leads to a decrease in electrical power. As a result, the initial uniformly distributed density of solar energy is transformed into a new internal energy source linearly distributed along the walls, localized in the plane of the photovoltaic cells. Thus, it is necessary to solve the problem of heat transfer of a liquid, along the walls of which a heat flow is formed: $q_1 = q_{01} \cdot \left(1 + \vartheta \cdot X\right)$ and $q_2 = q_{02} \cdot \left(1 + \vartheta \cdot X\right)$, where q_{01} and q_{02} - solar radiation density, q_{01} = N • C_0 • E and q_{02} = (E + E_d); ϑ – coefficient of proportionality, responsible for additional growth of specific heat flux and determined experimentally. The linear distribution of the density of the formed internal heat source and the same temperature of the liquid over the cross section makes it possible to integrate equation (9). For simplicity, the heat exchange between the ends of the channel and the environment is not considered. In this case, with a known inlet temperature of the liquid t_{in}, the magnitude of the temperature difference of the liquid along the length of the channel is:

$$\Delta t = t_{out} - t_{in} = X \cdot b \cdot E \cdot \left(1 + \vartheta \cdot \frac{X}{2}\right) \cdot \frac{N \cdot C_0 + \left(1 + \frac{E_d}{E}\right)}{C_P \cdot G} \tag{10}$$

Figure 10 shows the results of the calculation of the temperature difference and the velocity of the coolant along a flat channel depending on the fluid flow rate, the ϑ parameter, the technical characteristics of the concentrators and solar radiation.

The flow rate of the fluid in the channel is determined from the expression $v = \frac{G}{p \cdot b \cdot h}$, where ρ is the density of the liquid. The system of circulation of coolants at the minimum flow rate ensures efficient heat removal from photocells and automatically maintains their temperature on the walls of the second photodetector at a level of no more than ≈ 55 °C. At different times of the year and time of day with a change in the intensity of solar radiation, an automated system for circulating consumed water in the second circulation loop regulates the flow of water, which changes the intensity of heat exchange of heat transfer fluids with the consumed liquid and the corresponding temperature gradient. Substituting actual values into expression (10), we obtain the minimum flow rate G_{min} = 3.4 • 0.15 • 750 • (1 + 0.02 • 0.5 • 3.4) • [12 • 0.83 + (1 + 0.33)] / 4200 • 20 = 0.043 kg/s. At this flow, and according to graph in Figure 10 specifies the average temperature of the fluid in the channel, the kinematic viscosity and the Reynolds number are specified $\bar{t} = \frac{t_{out} + t_{in}}{2}$. In accordance with the results of calculations in the range of obtained values of the average temperature of the liquid

Figure 10. The graph of the temperature gradient of the fluid along the channel

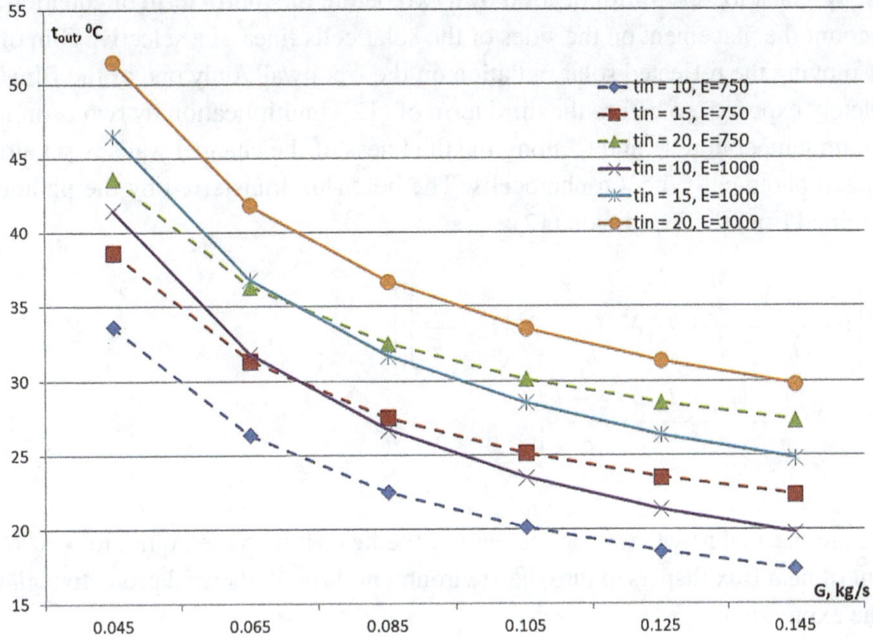

$$\bar{t} \approx 22 - 35°C, \upsilon \approx \left(0.87 - 0.80\right) \cdot 10^{-6} m^2 / s,$$

and Reynolds number $\mathrm{Re} = \dfrac{\upsilon h}{\nu} \approx 340 - 780$. In a given range of flow rates and temperatures of the coolant, the flow regime is laminar.

Note that the wall temperature is not included in the differential equation (9). The accepted assumption that the velocity and temperature of the fluid are constant in the cross section of the channel leads to the fact that the temperature of the inner walls and the heat transfer coefficient of the fluid are the same. To determine the wall temperature average along the channel length, Newton-Richman heat transfer law is used (Isachenko, 1975):

$$G \cdot C_p \cdot \Delta t = 2 \cdot L \cdot b \cdot \bar{\alpha} \cdot \left(t_c - \bar{t}\right) \tag{11}$$

where t_c is the average length of the temperature of the inner walls of the channel, $\bar{\alpha}$ is the average, along the channel length, heat transfer coefficient on the frontal and rear walls, $W/m^2 \bullet K$. In accordance with the energy balance on the walls of the photodetectors, solar radiation is converted by solar cells and selective films into electricity and useful thermal energy and is partially dissipated in the environment as heat:

$$n \cdot E \cdot S_i \cdot \left[N \cdot C_0 + \left(1 + \frac{E_d}{E}\right) + \frac{b_p}{b} \cdot \beta \cdot \left(N \cdot C_0 + 2 \cdot \left(1 + \frac{E_d}{E}\right)\right)\right] = G \cdot C_p \cdot \left(t_{out} - t_{in}\right) + P + W \tag{12}$$

where n is the number of solar cells on the walls; E and E_d - direct and diffuse solar radiation; P - peak electric power; W - heat losses. Multiplication by a two before the fourth term of equation (12) appears taking into account the placement on the sides of the solar cells lines of a selective film of width b_p. In the process of moving the reflected solar radiation on the front wall, only one of the films, or a part of both, is completely exposed, so before the third term of (12), multiplication by two is omitted.

The maximum temperature t_{f1} and t_{f2} along the thickness of the channel walls is set along the plane with encapsulated photo cells, i.e. on photocells. The heat flux transferred by the photocells into the liquid is determined from the expression (12):

$$E\cdot\left\{N\cdot C_0+\left(1+\frac{E_d}{E}\right)+\frac{b_p}{b}\cdot\beta\cdot\left[N\cdot C_0+2\cdot\left(1+\frac{E_d}{E}\right)\right]\right\} \\ -P-W=\left(1+2\cdot\frac{b_p}{b}\right)\cdot\left[\frac{t_{f1}-\bar{t}}{r_1}+\frac{t_{f2}-\bar{t}}{r_2}\right] \tag{13}$$

where r_1 and r_2 are thermal resistances on the path of the heat flow to the liquid, $m^2 \bullet K/W$.

The amount of heat flux dispersed into the environment through thermal protective glasses is determined from the expression:

$$W=2\cdot n\cdot S_i\cdot\left(1+2\cdot\frac{b_p}{b}\right)\cdot\left\{\begin{array}{l}\dfrac{\left(t_{f1}-t_0\right)+\left(t_{f2}-t_0\right)}{r_3}\\ +\varepsilon_1\cdot C_0\cdot\varphi_1\cdot\left[\left(\dfrac{t_{f1}}{100}\right)^4-\left(\dfrac{T_0}{100}\right)^4+\left(\dfrac{t_{f2}}{100}\right)^4-\left(\dfrac{T_0}{100}\right)^4\right]\end{array}\right\} \tag{14}$$

where r_3 is the thermal resistance in the path of the heat flux into the air, $m^2 \bullet K/W$; ε_1 is the reduced value of blackness; C_0 is the blackbody emissivity equal to 5.67 $W/m^2 \bullet K^4$; φ_1 and φ_2 are the angular emission coefficients.

From equations (11) - (14) by the method of iterations determine the temperature of the photo cells t_{f1}. Since $t_{f1} > t_{f2}$, let us calculate for two values of $t_{f2}= t_{f1}/2$ and $t_{f2}= t_{f1}/4$.

The average temperature in the channel section, depending on the flow rate and solar radiation intensity, is determined from the expression:

$$\bar{t}=t_{in}+L\cdot b\cdot E\cdot\frac{N\cdot C_0+\left(1+\frac{E_d}{E}\right)}{C_p\cdot G}=t_{in}+\frac{0.71\cdot E}{G}$$

The temperature level of the photovoltaic cells on the hot front wall can be estimated by adopting a number of assumptions. Firstly, the maximum value of heat loss, taking into account the new Λ - shaped design of photodetectors, does not exceed 30% of the total allocated thermal power; secondly, the magnitude of the peak electrical power is equal to 0.27 of useful thermal power. As a result, from the equations (13) and (14) we can distinguish the value of the temperature of the photovoltaic cells:

$$t_{f1} = \frac{A + \bar{t} \cdot \left(\frac{1}{r_1} + \frac{1}{r_2} \right)}{\frac{1}{r_1} + \frac{1}{2r_2}} \tag{15}$$

Grashof number is relatively low, therefore, natural convection does not have a significant effect on heat transfer. In this case, the formula for the viscous flow of the coolant is used to determine the heat transfer coefficient $Nu = 1.55 \cdot \left(Pe \cdot \frac{h}{L} \right)^{0.33} \cdot \left(\frac{\mu_f}{\mu_c} \right)^{0.14}$ (Blokh, Zhuravlev, Ryzhkov, 1991). The mean temperature of the fluid along the channel is used as the determining temperature $\bar{t} \approx 20.4$ °C and the average wall temperature. Moreover, average wall temperature is chosen by successive approximations and is approximately equal to $\bar{t}_s \approx 31$ °C. At these temperatures, the thermal diffusivity is a = (13.3 and 13.9)•10^{-8} m^2/s, thermal conductivity coefficient equal to $\lambda = 0.559$ W/m • K, dynamic viscosity coefficients μ, respectively (1006 and 650) • 10^{-8} Pa s, Peclet number is $Pe = \frac{v \cdot h}{a} \approx 2210\, and\, 2160$.

From where Nu=5.8 and $\bar{\alpha} \approx \frac{Nu \cdot \lambda}{h} \approx 485$ W/m²•K.

Thermal resistance in the path of heat flow between the surface of a solar cell and the coolant, as well as the environment, is found from the expressions:

$$r_1 = \frac{\delta_{\Pi 1}}{\lambda_{\Pi 1}} + \frac{\delta_O}{\lambda_O} + \frac{\delta_{Al}}{\lambda_{Al}} + \frac{1}{\bar{a}},\ r_2 = \frac{\delta_{\Pi 2}}{\lambda_{\Pi 1}} + \frac{\delta_O}{\lambda_O} + \frac{\delta_{Al}}{\lambda_{Al}} + \frac{1}{\bar{a}},\ r_3 = \frac{\delta_E}{\lambda_E} + \frac{\delta_C}{\lambda_C} + \frac{1}{\bar{a}_2} \tag{16}$$

where $\delta_{\Pi 1}$, $\delta_{\Pi 2}$, δ_O, δ_{Al}, δ_E, δ_C – respectively the thickness of heat-conducting silicone gel layer ("Nomakon KPTD-3") on the front and back walls, oxide coating, aluminum alloy wall, transparent gel (EVA film) and thermal glass, m; $\bar{a}_2$ is the heat transfer coefficient of ambient air, W/m²• K; $\lambda_{\pi 1}$, λ_o, λ_{Al}, λ_E and λ_c - respectively thermal conductivity of silicone gel, oxide coating, aluminum alloy, transparent silicone gel and thermal glass, W/m • K.

The thermal conductivity coefficient of silicone gel is $\lambda_{\pi 1}$, ≈ 2 W/m • K, from where for the layer thickness on the front wall ≈ 0.1 mm and the back wall ≈ 0.5 mm, $\frac{\delta_{\blacklozenge 1}}{\lambda_{\blacklozenge 1}} \approx 0.05 \cdot 10^{-3}$ m² • K/W and $\frac{\delta_{\Pi 2}}{\lambda_{\Pi 1}} \approx 0.25 \cdot 10^{-3}$ m² • K/W. The thermal conductivity of a thermal insulation coating obtained by microarc oxidation is $\lambda_o \approx 2$ W/m · K, for a thickness of ≈ 0.05 mm, gives the value $\frac{\delta_O}{\lambda_O} \approx 0.025 \cdot 10^{-3}$.

The thermal resistance of aluminum alloy walls is equal to $\frac{\delta_{Al}}{\lambda_{Al}} = 0.002/195 = 0.01 \cdot 10^{-3}$. For process water in the range of flow rate ≈ 0,6 - 2 l/min, is equal to $\frac{1}{\bar{\alpha}} \approx 2{,}1 \cdot 10^{-3}$ m²•K/W. Comparing the addends in the expression (17), we can neglect the thermal resistances of the walls of aluminum alloy and the coating of aluminum oxide. As a result, $r_1 = (0.05+0.025+2.1) \cdot 10^{-3}$ m²•K/W and $r_2 = (0.25+0.025+2.1) \cdot 10^{-3}$ m²•K/W.

An EVA film $\delta_E \approx 0.35$ mm thick and thermal conductivity $\lambda_E \approx 0.3$ W / m • K gives thermal resistance $\delta_E/\lambda_E \approx 1.2 \cdot 10^{-3}$. Glass with a thickness of 2 mm and thermal conductivity $\lambda_c = 0.96$ has a thermal resistance $\delta_C/\lambda_C \approx 2.1 \cdot 10^{-3}$. The heat transfer coefficient of air at an average wind speed of 2.5 m/s is $\bar{\alpha}_2 \approx 11$ W/m² • K, whence $\frac{1}{\bar{\alpha}_2} \approx 0.08$. As a result, the thermal resistance in the path of the heat flux from the photocells into the environment is equal to $r_3 \approx (1.2+2.1+80) \cdot 10^{-3}$ m² • K/W.

Figure 11 shows the results of the calculation of the temperature of photovoltaic cells t_{f1} depending on coolant flow rate (G) and the amount of solar radiation (E).

In the most intense summer period of operation, the ambient temperature (on a testing location 43 °N) is ≈ 30 °C, and the surface temperature of the photodetectors is ≈ 55 and 45 °C. The reduced degree of blackness of the photovoltaic cell surfaces is $\varepsilon_1 \approx 0.7$. The angular emissivity of the outer surface of the photodetector varies in the interval $\varphi_1 \approx 0.6 - 0.7$. Substituting the numerical values into equation (14), we obtain the maximum value of the loss of heat emitted by the photoelectric cells: W ≈ 604 W.

The method of calculation allows estimating the interval of technological parameters of photodetectors and the concentration of solar radiation, within which the optimum level of temperature of the photovoltaic cells and fluid flow is ensured.

Figure 11. Average photocell temperature along the channel length

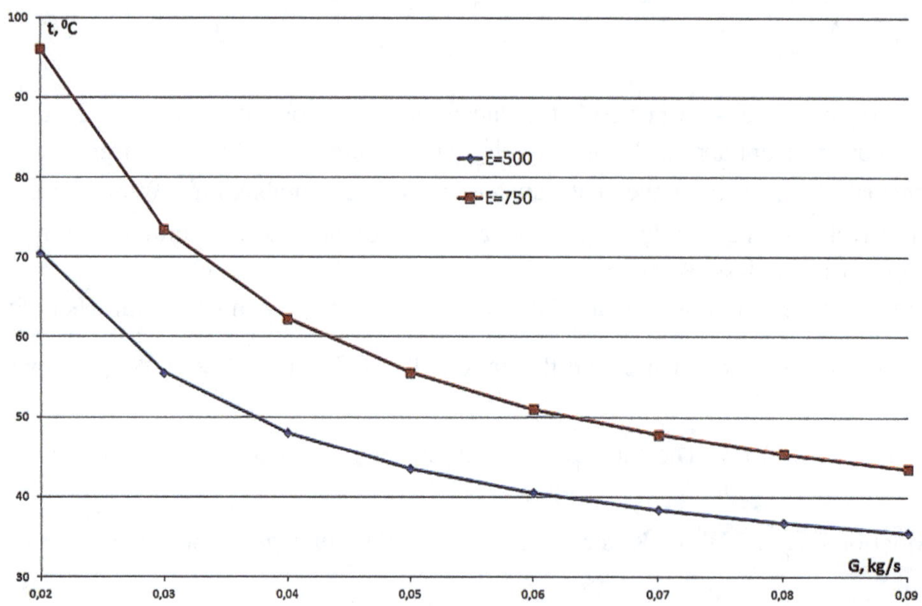

Method of Calculating the Performance of Cogeneration Systems

To determine the productivity of the prototype of the cogeneration energy system, it is necessary, first of all, to correctly estimate the amount and peak thermal power of the solar cells of Λ - shaped photodetectors. At currents ≈ 58 A, the voltage drop across the conductors is $\Delta U \approx 1.2$ V. The optimal process of charging chemical batteries is carried out at a voltage of ≈ 13.5 V, whence the base number of photo cells is determined: $n \approx (13.5 + 1.2) / 0.57 \approx 26$ pieces. This value is adjusted taking into account the temperature loss of the no-load voltage of the photocells. For example, the total value of the voltage loss at the input temperature of the liquid in the first channel ≈ 18 °C and the temperature difference along the channel ≈ 20 °C is $U_T = \sum k_U \bullet \Delta t_j \approx 0.23$ V and ≈ 1.37 V, respectively, for the first and second channel. To obtain the same output voltages in the photodetectors, it is necessary to compensate for the temperature loss of voltage due to the introduction of additional solar cells - by one into each first channels, and by three into each second channels. The length of the channels connected in series with the photocell of size b_i is: $L_1 + L_2 \approx b_i\,[(n+1) + (n+3)]$. The total amount of reduction in peak power due to temperature increases is $P_T \approx \sum k_P \bullet \Delta t_j \approx 4{,}1$ W и ≈ 8,7 W respectively for the first and second photodetectors, or ≈ 4% and ≈ 8% of the total power of the photodetectors.

The efficiency of photocells working with solar radiation concentration will be denoted by η_1, and those working without concentration (sum of direct E and diffuse E_d sun radiation) by η_2. The characteristics of Maxeon photocells with a low concentration of radiation and the temperature set during the measurements are taken from the graphs of experimental power curves obtained using the developed measurement technique (Nesterenkov, Abdullaev, Nesterenkova, 2017). The peak power curve (square points with the indicated measurement error) using the described technique is shown in Figure 12, the round dots and dashed curve correspond to peak power calculated from voltage and current measurements at the maximum power point.

Figure 12. Dependence of peak power on the concentration of solar radiation

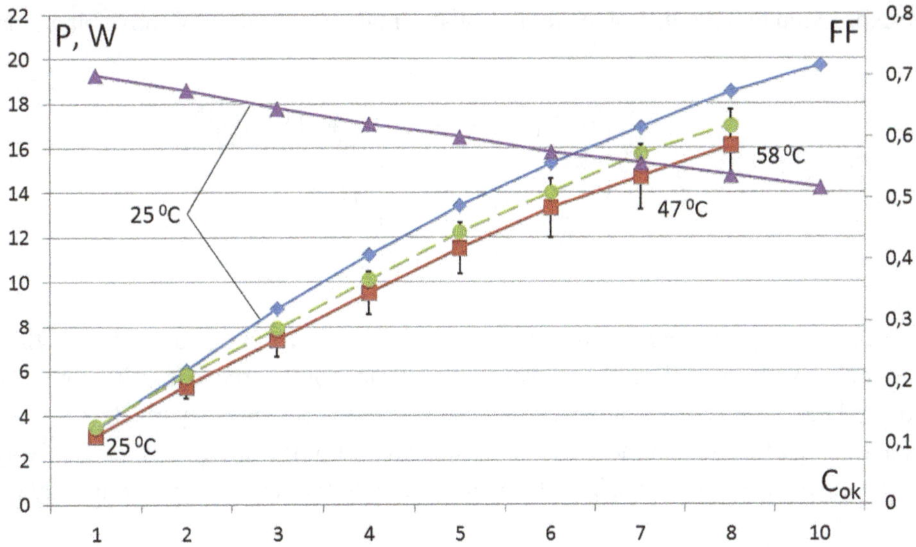

The peak power curve obtained at a fixed temperature is higher in accordance with the lower temperature of the solar cells. Dependence of the fil-factor (almost linear) on concentration ratio results into the corresponding linear dependence of electrical efficiency. The method used to select the peak power of the solar cells for conditions close to actual operating conditions increases the accuracy of the system performance calculations.

Photodetectors with a concentration of the sun produce a nominal peak electrical power $P_1 = E \bullet S \bullet C_{ok} \bullet \eta_1$. Facing the sun, solar cells of the same area with efficiency η_2 produce peak electric power $P_2 = (E + Ed) \bullet S \bullet \eta_2$. As a result, the peak electric power of bilateral photodetectors is:

$$\Sigma P = P_1 + P_2 = E \cdot S \cdot \left[C_{ok} \cdot \eta_1 + \left(1 + \frac{E_d}{E} \right) \cdot \eta_2 \right] \tag{17}$$

As was justified above, the total heat loss to the environment, taking into account the mutual screening of thermal radiation from the surface of the channels, is ≈ 30%. From where received expression for definition of the minimum peak thermal power transferred to heat carrier by solar cells:

$$Q_F = G \cdot C_P \cdot \left(t_{out} - t_{in} \right) = 0.7 \cdot E \cdot S \cdot \left[C_{ok} \cdot \left(1 - \eta_1 \right) + \left(1 + \frac{E_d}{E} \right) \cdot \left(1 - \eta_2 \right) \right] \tag{18}$$

Additionally, heat power is transferred to coolants through a selective film with an area of SΔ: $Q_1 = 0.7 \bullet E \bullet S_\Delta \bullet C_{ok} \bullet \beta$ and from the sun (direct E and diffuse Ed radiation): $Q_2 = 0.7 \bullet E \bullet S_\Delta \bullet (1 + E_d/E) \bullet \beta$. Through the selective surface of the four channels amount of thermal energy diverted to the coolant:

$$Q = Q_1 + Q_2 = 2 \cdot 0.7 \cdot E \cdot \Delta \cdot L_F \cdot \beta \cdot \left[C_{ok} + \left(1 + \frac{E_d}{E} \right) \right] \tag{19}$$

Thus, the expression for calculating the total peak thermal power output of the coolant, has the form:

$$\Sigma Q_T = Q_F + Q \approx 2 \cdot 0.7 \cdot E \cdot L_F \cdot b \cdot \left\{ \begin{matrix} \left[\left[C_{ok} \cdot \left(1 - \eta_1 \right) + \left(1 + \frac{E_d}{E} \right) \cdot \left(1 - \eta_2 \right) \right] \right] \\ + \frac{\Delta}{b} \cdot \beta \cdot \left[C_{ok} + \left(1 + \frac{E_d}{E} \right) \right] \end{matrix} \right\} \tag{20}$$

The product $2 \bullet E \bullet L_F \bullet b_i$ is approximately equal to the area of the FE located on both sides of the channels of four photodetectors. With the help of the coefficient of 0.7, the loss ≈ 30% of thermal energy by convection and radiation is taken into account. The first two terms in curly brackets are related to the fraction of solar energy taken from the heat-generating surface of the solar cells, respectively, from the lower and upper sides of the channels. The third term refers to the heat energy emitted by the selective surface of the lower and upper walls of the channels, respectively. In previous researches (Antoshchenko & Nesterenkov 2010), calculations were made of the flow rate of the coolant in the flat channels of the photodetectors and it was shown that the flow is laminar and the power of the micro-pumps does not exceed 25 W.

The ratio between the generated electric and thermal power in any operating conditions of installations remains constant, and its value is determined from the expression:

$$\frac{\sum Q_T}{\sum P}=\frac{0.7\cdot\left[\left[C_{ok}\cdot\left(1-\eta_1\right)+\left(1+\frac{E_d}{E}\right)\cdot\left(1-\eta_2\right)\right]+\frac{\Delta}{b}\cdot\beta\cdot\left[C_{ok}+\left(1+\frac{E_d}{E}\right)\right]\right]}{C_{ok}\cdot\eta_1+\left(1+\frac{E_d}{E}\right)\cdot\eta_2} \tag{21}$$

The last expression does not depend on the number of photovoltaic cells and depends little on their size. Let us substitute the experimental data into expression (21), take into account the linear drop in the efficiency from temperature, and take the ratio of diffuse radiation to direct radiation for southern latitudes: $E_d/E \leq 0.20$, Table 2 presents the calculation results.

As can be seen from Table 3, the change in the size of linear photodetectors, the parameters of optical concentrators, and the electrical efficiency of photocells has little effect on the ratio between the power and heat productivity. Let us take the average value $\sum Q_T/\sum P \approx 3.7$, whence the peak thermal power is determined: $\sum Q_T \approx 3.7 \bullet \sum P$.

The following is a calculation of performance using the example of a basic design with two pairs of linear photodetectors in the focal region of optical concentrators. The fundamentals of designing such energy systems are described in earlier work (Nesterenkov et al., 2018). The choice of the total area of photocells is justified above and equal to: $S = 2 \bullet S_i \bullet [(n + 1) + (n + 3)] \approx 2 \bullet 0.0156 \bullet 56 = 1.75\ m^2$. Optical concentration ratio amount is: $C_{ok} = C_{ok} \bullet N = 0.81 \bullet 12 \approx 9.7$. The curve extrapolation in Figure 12 gives the peak power value for the sequentially allocated channels, respectively $P_i \approx 18$ W and 17 W. Accordingly, the total peak power generated by them is equal to: $P_1 \approx 2 \bullet [(n+1) \bullet 18 + (n + 3) \bullet 17)] \approx 1960$ W.

With a single solar concentration and an average temperature of ≈ 55 °C, the peak power of one solar cell in accordance with the graphs in Figure 12 is equal to $P_{cp} \approx 3.2$ W, and the total power: $P_2 \approx 2 \bullet [(n + 1) + (n + 3)] \bullet P_{cp} \approx 358$ W. Part of this power goes to the energy needs of the system - the work of circulating pumps and tracker motors. As a result, we obtain the total value of the peak power that goes to the electrical energy storage system: $\sum P \approx P_1 + 0.7 \bullet P_2 \approx 2210$ W. In accordance with expression (19), the value of the total value of the peak thermal power is: $Q_T \approx 3.7 \bullet P \approx 8170$ W.

Table 2. The ratio between the performance of thermal and electric energy

	$b_p/b = 0{,}048$; $C_{ok} \approx 9{,}7$				$b_p/b = 0{,}08$; $C_{ok} \approx 9{,}7$		$b_p/b = 0{,}048$; $C_{ok} \approx 8$	
	η_1	**η_2**	**η_1**	**η_2**	**η_1**	**η_2**	**η_1**	**η_2**
	0,14	**0,18**	**0,17**	**0,22**	**0,17**	**0,22**	**0,16**	**0,19**
$\sum Q_T/\sum P$	3,66		3,69		3,74		3,83	

The total area of the mirrors of optical concentrators is determined from the expression: $S_{2N} \approx 2 \cdot N \cdot b_X \cdot [(n + 1) + (n + 3)] \cdot 0.126 \approx 24.5$ m². Taking into account the geometric coefficient (cosine loss), we obtain the value of the active optical aperture $S_A \approx 0.94 \cdot S_{2N} \approx 23.1$ m².

Specific (hourly) electricity performance is: $W_1 \approx \sum P \cdot \tau / S_A \approx 2210 \cdot 1/23.1 \approx 96$ W•h/m². The specific heat output is $W_1 \approx 354$ W•h/m².

The flow rate of hot water in the heat exchanger channel is: $G = \sum Q_T / C_P \cdot (t_2 - t_1)$. The value of the specific heat of process water is taken at an average temperature in the heat exchanger channel (t2 + t1)/2. Substituting the experimental data with a nine-fold concentration and temperature difference along the length of the heat exchanger ≈ 40 °C, obtain the flow rate of hot water in the heat exchanger channel $G \approx 7990/4200 \cdot 40 \approx 0.048$ kg/s ≈ 177 l/h. At this flow rate, fluid flow is laminar, with low flow resistance. If h ≈ 3 m and the temperature difference between the heat exchanger and the heat accumulator capacity is ≈ 45 °C, then the dynamic head is $\Delta P \approx 470$ Pa, which blocks hydraulic losses and creates natural circulation of the liquid.

Annual productivity depends on the average number of hours of sun J per day. According to the standard of solar radiation density E_c and the value of total annual insolation, one can estimate the standard number of sunshine hours $J_c = \sum E / (\cos \psi \cdot 365\ E_c)$. For example, for southern regions of the Republic of Kazakhstan, the amount of standard hours of sunshine per day is: $J_c \approx 1750$ kW•h/m²/(365•1 kW•h/m²) ≈ 4.8 hours. In this case, the annual generation of electrical and thermal energy is equal to: $W_P = \sum P \cdot 365 \cdot J_c \approx 2.21 \cdot 365 \cdot 4.8 \approx 3\,870$ kW•h и $W_Q = \sum Q_T \cdot 365 \cdot J_c \approx 8.17 \cdot 365 \cdot 4.8 \approx 14\,310$ kW•h. The specific annual generation of electric energy is: $W_{PJ} = W_P / S_A \approx 167$ kW•h/m²/year and heat energy $W_{QJ} = W_Q / S_A \approx 619$ kW•h/m²/year. Figure 13 shows the results of calculating the specific annual performance of electricity (hatched rectangles) and heat at different geographic latitudes. For comparison, the same figure shows data for analogue system C7Tracker (Cogenra Solar & SunPower Corporation).

The experience of manufacturing laboratory setups of solar cogeneration energy system and taking into account the market cost of components makes it possible to predict the cost of an industrial prototype of a commercial installation at the level of ≈ $ 5 300. With the cost of electrical energy at a reduced rate ≈ $ 0.108 /kW•h (Astana Solar, 2019), revenue from the sale of generated electricity is: $D_f \approx 3870$ kW•h • $0.108/kW•h ≈ $ 418/year. With the cost of thermal energy for remote areas at the tariff of $ 0.034 / kWh, the annual income from the sale of thermal energy is: $D_k \approx 14310$ kW•h • $0,034/ kW•h ≈ 486 $. In this case, payback period of production and installation of a cogeneration system is 5300/904 ≈ 5.9

Figure 13. Specific annual capacity of cogeneration plants

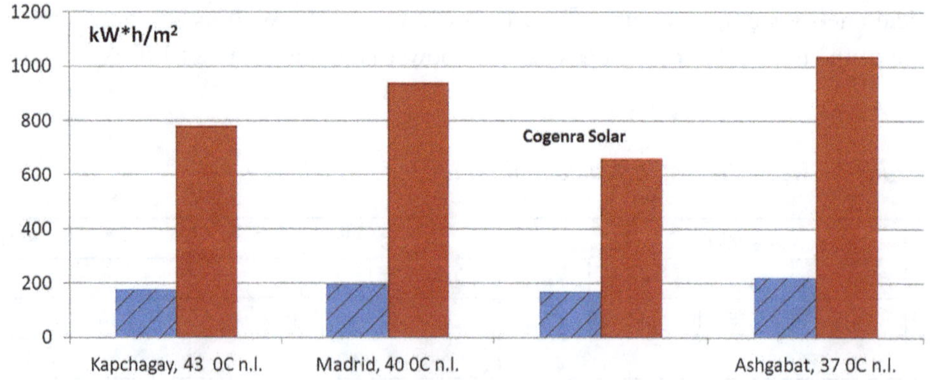

years. Even in case of reducing the cost of heat energy three times, the income from its sales is comparable to the income from sales of electric energy, this conclusion convincingly proves that the thermal energy released by the solar cells should not be emitted into the atmosphere. It is necessary to use an active cooling system and receive additional heat energy in the form of hot water supply for the consumer.

Further studies will be devoted to the search for technological modes of operation of cogeneration systems, at which the cost of generated electricity and heat is minimal.

CONCLUSION

Intensive heat generation of solar cells under conditions of high concentration of radiation increases the temperature level, has a positive effect on the intensification of heat exchange processes and increases the total efficiency of solar cells. In the process of converting solar energy, linear photodetectors have the effect of auto-stabilizing the overall performance, since a decrease in the production of electrical energy automatically leads to an increase in the production of thermal energy.

The use of Λ-shaped photodetectors with a counter-arrangement of the frontal surfaces in the focal region of the mirrors of optical concentrators reduces the heat losses by radiation on ≈ 30% and by convection on ≈ 5% and provides self-defense against large precipitation, including hail. In addition, it became possible to significantly reduce the thickness of protective thermal glasses and, accordingly, optical losses. With this significantly reduces the weight and cost of linear photodetectors.

An innovative photocell encapsulation technology has been developed with a multiple increased cross section of electrical contacts. The collinearity of the surfaces of the photovoltaic cells and a thin layer of thermoplastic sealant reduce the thermal resistance between the cells and the channel walls and improve the cooling of the cells, as well as reduce the resulting mechanical stresses. This contributes to lowering the temperature of photovoltaic cells and increasing useful peak power. Furthermore, collinearity simplifies the technology of polymerization of silicone sealants under load with thermal glass without the use of vacuum thermal furnaces.

The technology of high potential heat transportation from photovoltaic cells to the energy storage system is implemented through an intermediate coolant in countercurrent mode and the formation of the natural flow of hot process water in the heat exchanger circulation loop.

The technique of engineering calculation of linear Λ-shaped photodetectors parameters makes it possible to evaluate the efficiency of cogeneration systems with a low concentration of solar radiation and intensive cooling of the solar cells.

For the first time in research practice of solar cogeneration energy systems, direct conversion of thermal energy into the energy of transporting a consumable fluid from a photodetector to a heat accumulator in the natural flow mode has been realized.

The obtained average annual specific performance of electricity ≈ 167 kW•h/m^2 and thermal energy ≈ 619 kW•h/m^2 ensure recoupment of capital costs for purchase and installation of cogeneration energy system ≈ 5.9 years.

REFERENCES

Airlight Energy. (n.d.). DSolar SunFlower. Retrieved from http://www.airlightenergy.com/high-concentration-photovoltaic-thermal/

Alanod Corporation. (n.d.). Miro-Silver products. Retrieved from https://www.alanod.com/products/

Almeco-Tinox Solar. (n.d.). Tinox Energy. Retrieved from http://www.almecogroup.com/en/pagina/53-tinox-energy-cu

Antoshchenko, V. S., & Nesterenkov, A. G. (2010). Calculation of a combined thermal photovoltaic system with concentrated solar radiation fluxes. *Energy and Fuel Resources of Kazakhstan*, *2*, 27–31.

Astana Solar. (n.d.). Approved tariffs for renewable energy sources. Retrieved from http://astanasolar.kz/ru/news/utverzhdeny-tarify-na-vozobnovlyaemye-istochniki-energii

Blokh, A. G., Zhuravlev, Y. A., & Ryzhkov, L. N. (1991). *Radiation heat transfer: a Handbook*. Moscow, Russia: Energoatomizdat.

Bunea, M., Johnston, K., Bonner, C., Cousins, P., Smith, D., & Rose, D. (2010). Simulation and characterization of high efficiency back contact cells for low-concentration photovoltaics. In *Proceedings of the 35th IEEE Photovoltaic Specialists Conference PVSC* (pp. 823-826). 10.1109/PVSC.2010.5617188

Chea, L. C., Hakansson, H., & Karlsson, B. (2013). Performance evaluation of new two axes tracking pv-thermal concentrator. *Journal of Civil Engineering and Architecture*, *7*(12), 1485–1493.

Chow, T. T. (2010). A review on photovoltaic/thermal hybrid solar technology. *Applied Energy*, *87*(2), 365–379. doi:10.1016/j.apenergy.2009.06.037

Coventry, J. S. (2005). Performance of a concentrating photovoltaic/thermal solar collector. *Solar Energy*, *78*(2), 211–222. doi:10.1016/j.solener.2004.03.014

Duffie, J. A., & Beckman, W. A. (2006). *Solar Engineering of Thermal Processes* (3rd ed.). Hoboken, New Jersey: John Wiley and Son Inc.

Escher, W., Paredes, S., Zimmermann, S., & Chin, L. O. (2012). Thermal management and overall performance of a high concentration pv. In *Proceeding of the 8th International Conference on Concentrator Photovoltaics (CPV-8)* (pp. 239-243). 10.1063/1.4753877

Fellmeth, T., Ebert, M., & Efinger, R. (2014). *Industrially feasible all-purpose metal-wrap-through concentrator solar cells*. In *Proceedings of the 40th IEEE PVSC* (pp. 2106-2110). 10.1109/PVSC.2014.6925340

Forbes Solar. (n.d.). Solar Cogeneration System. Retrieved from https://www.forbesmarshall.com/Download/Solar%20Cogeneration%20System.pdf

Isachenko, V. P. (1975). *Heat transfer processes. Textbook for high schools*. Moscow, Russian: Energy.

Ji, X., Li, M., Lin, W., Wang, W., Wang, L., & Xi, J. (2012). Modeling and characteristic parameters analysis of a trough concentrating photovoltaic/thermal system with gas and super cell arrays. *International Journal of Photoenergy*.

Kribus, A., Kaftori, D., Mittelman, G., Hirshfeld, A., Flitsanov, Y., & Dayan, A. (2006). A Miniature Concentrating Photovoltaic and Thermal System. *Energy Conversion and Management*, *47*(20), 3582–3590. doi:10.1016/j.enconman.2006.01.013

Mikheev, M. A., & Mikheeva, I. M. (1977). *Basics of heat transfer*. Moscow, Russia: Energy.

Milichko, V.A., Shalin, A.S., Mukhin, I.S., Kovrov, A.E., Krasilin, A.A., Vinogradov, A.V., … Simovsky, K.R. (2016). *Solar photovoltaics: current state and development trends. Success in the physical sciences*, *186*(8), 802-852.

Nesterenkov, A. G., Nesterenkov, P. A., & Nesterenkova, L. A. (2015). Patent No. 30003 of 04.13.2015. Republic of Kazakhstan, Astana.

Nesterenkov, P. A., Abdullaev, K. A., & Nesterenkova, L. A. (2017). Cogeneration systems with radiation concentration - a new type of "equipment for solar energy. In *Materials of the World Congress of Engineers and Scientists "Energy of the future: innovative scenarios and methods for their implementation" WSEC-2017* (pp. 264-270).

Nesterenkov, P. A., Nesterenkov, A. G., & Nesterenkova, L. A. (2016). *Fundamentals of designing hybrid concentrator solar systems*. In *Proceeding of the 12th International Conference on Concentrator Photovoltaics (CPV-12)*. 10.1063/1.4962073

Nesterenkov, P. A., Nesterenkov, A. G., & Nesterenkova, L. A. (2018). *Cogeneration solar systems with concentrators of solar radiation. In Handbook of Research on Renewable Energy and Electric Resources for Sustainable Rural Development*. Hershey, PA: IGI Global.

Petukhov, V. S. (1967). *Heat transfer and resistance in laminar flow in pipes*. Moscow, Russia: Energy.

Slade, A., & Garboushian, V. (2005). 27.6% efficient silicon concentrator cell for mass production. In *Proceeding of the 20th European Photovoltaic Solar Energy Conference*, 701.

Smirnov, A. V. (2010). *Improving the efficiency of solar power plants concentrators with high-voltage photovoltaic cells* [Doctoral dissertation].

Untila, G. G., Kost, T. N., Chebotareva, A. B., Zaks, L. B., Sitnikov, A. M., Soldodukh, O. I., & Schwartz, M. Z. (2012). Solar cell of n-type, two-sided, concentrator. *Physics and Technology of Semiconductors*, *46*, 1217–1223.

Viessmann. *Vitosol Solar collector 100-F*. Retrieved 15.01.2019, from https://www.viessmann-us.com/en/residential/solar/flatplate-collectors/vitosol_100-f.html

Zimmermann, S., Helmers, H., Tiwari, M. K., Escher, W., Paredes, S., & Neves, P. (1556). … Michel, B. (2013). *Advanced liquid cooling in hcpvt systems to achieve higher energy efficiencies. AIP Conference Proceedings*, 248–251.

KEY TERMS AND DEFINITIONS

Cogeneration: Use of solar energy system to generate electricity and useful heat at the same time.

Concentrator: System of reflective surfaces used to concentrate a large area of sunlight onto a operation surfaces of solar energy system.

Heat Exchanger: Unit in which heat exchange takes place between the coolant transferring useful thermal energy to the consumers technical water.

Photodetector: Photovoltaic module with two operation surfaces and coolant channel between them.

Selective Film: Thin film with high solar energy absorbance and low emittance properties.

Solar Cells: Semiconductor units that convert light into an electric current using the photovoltaic effect.

Thermal Collector: Module for collecting solar thermal energy carried by visible light and infrared radiation.

Chapter 8
Optimization of Spectral Composition and Energy Economy Effectiveness of Phyto-Irradiators With Use of Digital Technologies

Sergey Mironyuk
Uman National University of Horticulture, Ukraine

Alexander Smirnov
https://orcid.org/0000-0002-9236-2281
Federal Scientific Agroengineering Center VIM, Russia

Alexander V. Sokolov
Federal Scientific Agroengineering Center VIM, Russia

Yuri Proshkin
Federal Scientific Agroengineering Center VIM, Russia

ABSTRACT

It is known that in the case of technology use of the supplementary lighting, an irradiation spectral composition heavily influences the effectiveness of the photosynthesis processes, development and productivity of vegetable crops. Hence, the definition of general points at development and projecting of modern phyto-irradiators is one of high-priority tasks in techniques development for plants growing in conditions of protected ground. The research is aimed at reviewing and assessing the effectiveness of existing sources of illumination used in modern systems of supplementary lighting and at deduction of general points of development and projecting of phyto-irradiators based on results of laboratory investigations with the use of modern digital technologies of monitoring and data analysis. The results of the comparative tests of light emitting diodes-based phyto-irradiators showed that the energy consumption per product kilogram is less than in the case of LED-irradiators. Based on the research results, general points were deducted for use at development of modern LED-phyto-irradiators.

DOI: 10.4018/978-1-5225-9420-8.ch008

INTRODUCTION

First of all, a development of equipment and new techniques for vegetable crops cultivation in conditions of protected ground is oriented to reduction of operation costs and land productivity rise. In modern greenhouse facilities, an energy consumption decrease takes place mainly due to reduction of heat losses. With this purpose, constructions are used with limited indoor-outdoor air exchange; besides, systems are used with precise climate control.

Separately it is worth to consider systems of artificial illumination. It is well known that at use of the supplementary lighting technology, an irradiation spectral composition influences much the photosynthesis processes effectiveness and vegetable crops development and productivity (Tikhomirov et al., 2000; Vasiliev et al., 2017). Hence, the establishment of general points at modern phyto-irradiators development and projecting is one of high-priority problems in technologies development for plants growing in the conditions of protected ground.

Photo-synthetic apparatus of plants is a circumspect catalytic mechanism, action principle of which bases primarily on barriers overcoming of chemical transformations activation and photochemical reactions induction with use of radiant energy (Tyutereva et al., 2014). Now, the irradiation spectral composition influence on plants photo-synthetic apparatus functioning is well studied; that is why a special concern is attracted to researches aimed at study of non-photosynthetic pigments.

Studies conducted in the frame of non-photosynthetic pigments investigations show presence of a sophisticated photo-sensory mechanism at plants. Now, three major groups of photoreceptors are known: phytochromes, cryptochromes (phototropins), superchromes (photoreceptor UV-B) (Golovatskaya, 2005).

The photochromes (PHYA - PHYE) absorb long-wave part of the spectrum, i.e. red and far-red irradiation. Phototropins (PHOT1 - PHOT2) react on blue spectrum area. Cryptochromes (CRY1 - CRY5) show reaction on blue and ultraviolet irradiation. Superchromes (PHY3) absorb red and blue irradiation. As it turned out, photoreceptors allow triggering stress-defense mechanisms of plant photosynthetic apparatus under action of irradiation of high intensity. Also, they control processes of growth and development, blossom and vegetation periods (Golovatskaya, 2016; Chupackina et al., 2011).

BACKGROUND

The plants cultivation effectiveness at artificial illumination and influence of irradiation spectral composition on crops growth and productivity has been studied for quite long time. Being ones of the first in this area as early as in 1865, Russian scientists A.S. Famintsin and I.P. Borodin conducted systematic experiments on artificial illumination action on plants. As light source, petroleum-lamps acted. As object for watching, the water plant spirogyra was taken. In the course of the experiment, formation of starch was observed in chloroplasts.

In 1895 with use of voltaic arc, French botanist Bonnier sprouted seeds and rootstocks of herbs. In the same year by researcher F.W. Rane, first attempts were done of incandescent lamps use as artificial illumination for plants growing. In 1922, American scientist Harvey raised a plant based on completely artificial illumination. In USSR, this experiment was reproduced by Maximov with use of wolfram incandescent lamps.

One of first significant works on influence study of spectral composition of artificial irradiation on photosynthesis effectiveness was made by Hoover in 1937. In his studies, he compared results of wheat cultivation under various light sources: incandescent lamps, mercury lamps and solar light. In the course of the studies, it was established that in the course of photosynthesis process, plant absorbs only a part of solar spectrum, which was called photosynthetic active radiation (PAR). For every plant, PAR parameters are unique and can vary much at different photo-culture.

A large volume of studies for PAR parameters finding out for different vegetable cultures was conducted by McCree and Inada (Inada, 1978). One of results of these works was obtainment of an averaged curve of action spectrum of green leaf photosynthesis.

Usually on the photosynthesis action spectrum curve, it is possible to see two peaks, one in the red area and second in the blue. Sometimes at some seed plants and water plants, only one peak can be found in the red area. It is known that in the case of red color prevailing, the following phenomena run in the plant: an intensive growth of lead square area and a plant stem elongation. It has been established that the red light has a paramount significance for photosynthesis running and for regulation of growth and development of vegetative organs.

Along with comparatively large content of blue color in artificial irradiation, there take place plant stem and leaves slowing-down and leaf specific density increase. It was found out that the blue color is necessary for photosynthesis, plant growth regulation and generative organs formation.

In a case of green color prevailing, the formation takes place of thin leafage with small quantity of chloroplasts. It was established that the green color is necessary for photosynthesis running in lower layers of leafage or in the case of high frequency of sowing. This is due to the high penetration strength.

MOTIVATION

The results of the published papers devoted to investigation of photo-sensory mechanisms of plants (Snowden et al., 2016; Nicole et al. 2016; Smirnov, 2017) allow determining an optimal irradiation spectral composition necessary for prolific growth and development of vegetable crops cultivated in the conditions of protected ground. Besides they provide the possibility of modeling and projecting of light fixtures meeting these requirements. The use of the modern high-effective light emitting diodes as irradiation sources would allow a reduction of energy consumption by artificial illumination systems in greenhouses.

The aim of this research is review and effectiveness evaluation of existing light sources used in modern systems of supplementary lighting as well as general points deduction for development and projecting of phyto-irradiators based on results of the conducted laboratory tests with use of modern digital technologies of monitoring and data analysis.

MODERN SOURCES OF OPTICAL IRRADIATION USED IN ARTIFICIAL ILLUMINATION SYSTEMS FOR PLANTS

Nowadays for supplementary illumination in greenhouse facilities most often, natrium light sources of high pressure (HPS) are used and much more seldom LED- irradiators.

The use of the natrium lamps is conditioned with the following advantages in comparison with luminescent and gas-discharge light sources of low pressure (Izenberg, 1983):

- high energy economy effectiveness;
- quite long operation life;
- high luminous efficiency;
- stable light flux.

Among basic shortages of natrium lamps, one can list as follows:

- presence of mercury;
- explosion danger;
- for making a connection to a network, it is necessary to use PSD (pulse starting device) and SCD (start-controlling device).

If to pay attention to researches revealing influence of irradiation spectral composition on photosynthesis effectiveness and productivity of vegetable crops, among shortages of natrium lamps, it is possible to say also about the non-balanced irradiation, which is not able to promote plants considerable growth and development.

Light emitting diode fixtures are used much more seldom than the natrium lamps. Nevertheless, recently in the entire world, a gradual introduction takes place of the light emitting diodes into plant cultivation. For a short period of time, the light emitting diodes (LEDs) travelled a long way of development from low- effective devices up to the modern, compact and distinguished with high brightness irradiation sources. The light sources relied on super-bright light emitting diodes are considered to be promising for plants growing in conditions of artificial illumination.

The electric ray treatment based on light emitting diode sources of light has the following advantages: the low energy consumption (it reduces energy consumption down to 50% as compared to the traditional solutions); the absence of a special system of utilization; the compact sizes; the full color gamma of irradiation; the high power efficiency; the high mechanical strength; the long operation time; the small response time; the safe voltages in feeding line; the non-sensitivity to low temperatures (Singh et al., 2015).

Depending on their intention virtually, it is possible to divide the light emitting diodes (LEDs) onto two groups:

- Light emitting diodes intended for illumination. For them, white light and large scattering angle are typical. Usually the emitting crystal is made based on gallious nitride and covered by luminophor;
- Special light emitting diodes. They have irradiation spectrum from ultraviolet range 210 nm up to average infrared range 760 nm. Usually the emitting crystal is made of direct semiconductors.
- If to say about the consumed power, all LEDs are divided into three groups:
- Low power light emitting diodes (lower than 1 W);

- Ones of middle power (from 1 to 5 W);
- Light emitting diodes with COB (chip-on-board) (from 5 to 180 and more W).

Let us consider most popular types of SMD light emitting diodes used in constructions of phyto-irradiators.

Now the light emitting diode fixtures find application not only in greenhouse facilities for supplementary lighting or plants cultivation at fully artificial ray treatment but also in researching vegetation-complexes and cosmic orchard-houses. Most widely spread are the light fixtures based on narrowband red-blue LEDs, sometimes with addition of narrowband LEDs in other parts of spectrum. Most recently often for plants cultivation, the white LEDs are used regardless whether with addition of narrowband red or blue LEDs or without them.

The company OSRAM offers the wide selection of light emitting diodes for plants cultivation: 450 nm (dark-blue), 660 nm (red) and 730 nm (far red). To the family OSLON®, the important wave lengths belong with three irradiation angles: 80°, 120° and 150°. It guarantees a perfect illumination for all the types of plants and colors, which allows adapting the light to needs of various vegetable cultures. The ceramic body of the LED is able to sustain very high temperatures, up to Tmax = 135°C. Also, LEDs are produced with the power values 0.3, 1.0, 2.0 W.

The company Edison Opto produces the low power light emitting diodes in body 2835 distinguished with the irradiation in the range PAR 450, 455, 660, 730 nm and also the LEDs of white light with the high luminous efficiency (up to 181 lm/W for LEDs with color-dependent temperature of irradiation 4000K). Those LEDs' power makes 0.5W. It is reasonable to use these LEDs in construction of irradiators for racking cultivation of green cultures. The small single power of the LEDs would allow guaranteeing a plants homogeneous ray treatment from a small distance. The irradiation angle of these LEDs is 120°.

For special attention, analysis is worth of the LEDs produced by the company Cree, one of the today's leading world producers of LEDs. The model line Xlamp includes mono-crystal and multi-crystal products. For this model line, the irradiation distribution on device edges is typical. This innovational solution allowed producing the light fixtures with the big angle of light scattering despite of a minimal quantity of crystals. The LEDs of the model lines XQ-E and "XP-E High Efficiency" possess the light angle 100 – 145 degrees. At comparatively small geometrical parameters 1.6 x 1.6 mm, such light emitting diodes have the power 3 W at the light flux 330 lm (for white LEDs).

In 2016, the company Cree announced the introduction of the light emitting diodes "XQ-E High Efficiency Photo Red" that provide the luminous efficiency by 21% higher than previous generations XQ-E and XP-E, which enabled fruits cultivation light fixtures producers supplying more effective products, reducing light fixture sizes and cutting the system cost. The family XQ-E provides the combination of ultra-compact body, high productivity and wide spectrum of colors optimized for plants cultivation. The XQ-E High Efficiency Photo Red LEDs guarantee the level of photosynthetic photon flux (PPF) up to 5.39 mcmol/s at 85°C of the entire body volume 1.6 x 1.6 mm. The light emitting diodes XQ and XP for orcharding are relied on high-productive technology of Cree the company, which is able to guarantee the operation life of R90 for more than 100,000 hours and even at as extreme temperature as 105°C.

Now let's speak about the light emitting diodes based on the technology TRI-R. In 2017, the specialized-on production of light emitting diodes company Seoul Semiconductor announced the production start of intended for illumination light emitting diodes SunLike. In 2018, this type of light emitting diodes received its international safety certificate. In this type of light emitting diodes, crystals of violet irradiation (380-410nm) covered with luminophor are used. As a result, the irradiation spectrum of the LEDs is close to the spectrum of the solar light, which opens the possibility of use of these types of LED in irradiators for plants cultivation. The spectrum of these LEDs covers the entire range of PAR (from 380 to 780 nm) and has no considerable "gapes" in the irradiation. As for today, the LEDs are produced with the power from 6.8 to 85W and the luminosity up to 4595 lm. These LEDs effectiveness makes 85 lm/W.

The company Valoya (Finland) has developed phyto-irradiators based on luminophor-covered light emitting diodes with their patented spectrum of irradiation. Those phyto-irradiators contain one or several blue light emitting diodes having their own irradiation in the areas 400-500 and 440-500 nm with peaks at 450 and 470 nm respectively. Partly, the irradiation of these light emitting diodes is transformed with use of luminophors in the irradiation within the ranges 500-800 and 600-800 nm. The preferable part of irradiation transformation of blue light emitting diodes makes 35-65%. For this, used are one or more materials (luminophors) providing irradiation transformation of the light emitting diodes and of light emitting diodes placed close to them. As luminophors, some organic and/or non-organic materials are used, technology of production and spreading of which is quite sophisticated and labor-consumed.

Among shortages of the light emitting diodes, it is possible to list the following ones: the low minimum temperature; the use of an external radiator for cooling-down; and the use of low-volt feeding sources of direct current (Alferova et al., 2015). There are also other non-significant shortages and among them the time-depending changes of basic light-technical and operational characteristics (light flux, consumed power and other electric and light parameters). The light emitting diodes operation life is quite long and can reach 100 thousand hours with light characteristics worsening by 40% and more.

The use of point-like compact light emitting diodes in plant cultivation-intended lamps allows creating different combinations of colors in the same light fixture (Yuferev, & Sokolov, 2018). Along with this, it is possible to model their spectral composition with quite high precision.

DEVELOPMENT OF ENERGY-SAVING LIGHT EMITTING DIODES FOR PHYTO-IRRADIATORS

By the studies conducted earlier, it was established that for tomato, the optimal combination of PAR spectrum is the following ratio: blue area of spectrum (400÷500nm) 20%, green (500÷600nm) 20%, red (600÷700nm) 60%. For cucumber, there was obtained the following ratio of the PAR spectrum: Bl 20%, Gr 40%, Rd 40% (Tikhomirov et al., 2000).

Also, the influence was found out of an irradiation spectral composition on internodes elongation and leaf square area increase, which is explained by phytochrom reactions of plants on red and far-red light (Hao et al., 2016). The action of the red and far-red irradiation has a sinergistical nature, which is explained by Emerson's effect (Medvedev, 2013). That is why, the addition of the far-red (FR) component (700÷750nm) to the PAR irradiation can promote the leaves square area growth and enhance the photosynthesis.

Based on the given above data, there were developed the combined light emitting diode-based phyto-irradiators with the increased photosynthetic activeness and the low energy consumption. Depending on PAR spectrum bands combination, the projected irradiators can be divided into three types intended for cultivation of the following cultures: tomato, cucumber and lettuce.

For development of irradiators with desirable ratio of PAR spectrum, a mathematical modeling was conducted. The irradiation spectral density of the developed light fixtures, it is possible to describe with the following formula:

$$\varphi\,() = \sum_{i=1}^{n} q_i \times \varphi_i; \tag{1}$$

where: qi is quantity of light emitting diodes of ith color; φi is irradiation spectral density of light emitting diode of ith color; n is quantity of light emitting diodes in a light fixture.

On order to reveal, whether or not an irradiation spectral composition of an irradiator coincides with PAR of a photo-culture, it is possible to use the following formula (Yuferev, Sokolov, 2014):

$$f_i = \frac{\int_{\lambda_m^{/}}^{\lambda_m^{//}} \left[\sum_{i=1}^{n} K_i \cdot \varphi_i\left(\lambda\right)\right] \cdot d\left(\Delta\lambda_m\right)}{\int_{380}^{700} g\left(\lambda\right) \cdot d\lambda} 100\%, \tag{2}$$

where: f_i is function of irradiator spectral distribution; $\int_{380}^{700} g\left(\lambda\right) \cdot d\lambda$ is cumulative value of the function spectral distribution based on action spectrum of photosynthesis in relative units; $\int_{\lambda_m^{/}}^{\lambda_m^{//}} \left[\sum_{i=1}^{n} K_i \cdot \varphi_i\left(\lambda\right)\right] \cdot d\left(\Delta\lambda_m\right)$ is cumulative value of spectral distribution of energy of a light source in relative units in the spectral range of PAR; $i = 1...n$ is an irradiation source with a certain spectrum; $K_i = g(\lambda_n)$ is coefficient for the curve of photosynthesis action on the working wave length of each light emitting diode; m=1 is the range 380 to 500 nm; $m = 2$ is the range 500 to 600 nm; $m = 3$ is the range 600 to 700 nm.

As spectral densities of irradiation of separate light emitting diodes were known, the approximation of the irradiation curves was conducted (Sokolov, & Yuferev, 2014), after which the expected light-technical indexes were obtained of projected plants-cultivation-intended lamps (Sokolov, & Yuferev, 2015).

The designed phyto-irradiators include the combination of four types of light emitting diodes with maximum peaks of irradiation bands in limits of the following areas of PAR spectrum: blue area of spectrum 434-450 nm, red one 630-632 nm and 660-670 nm, and far-red one 730-735 nm. Notably, the blue light emitting diode is combined with luminophor and re-radiated a part of its light flux in the green part of the spectrum (500-600 nm).

The photos and the irradiation spectral composition of the developed phyto-irradiators are given in the Figures 1, 2, and 3.

It is known that on synthesis of anthocyans (red pigment) of red-leafed green cultures, the essential influence is rendered by irradiation in the UV and blue ranges of wave lengths (320-470nm) (Goto et al., 2016; Nicole et al., 2016). Nevertheless, the large part of the blue irradiation leads to leaf plates growth retardation, which results totally in reduction of land productivity of green cultures. That is why for red-leaf lettuces, the phyto-irradiator was designed with the differentiated mode of ray treatment so that stages are separated of green mass augmentation and biosynthesis of anthocyans (Figure 5).

For green mass augmentation, the ray treatment mode is used based on a combination of white, red and far-red light emitting diodes. For biosynthesis of anthocyans, the ray treatment mode was used based on a combination of blue and red-light emitting diodes in the ratio 50:50.

Figure 1. Averaged curve of action spectrum of green leaf photosynthesis
(according to McCree's data, 1972)

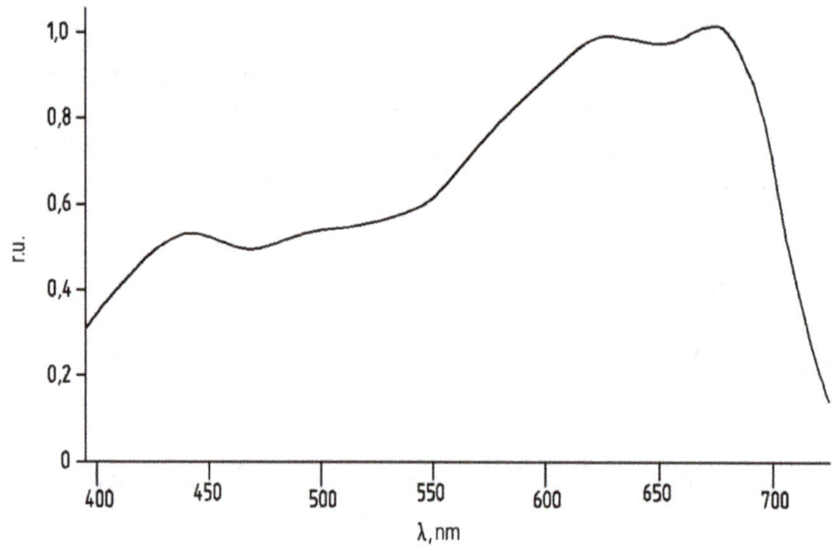

Figure 2. The combined light emitting diode irradiator intended for cucumber cultivation consists as follows: white LEDs in amount of 14 pieces, red 630nm LEDs 10 pieces, red 660nm LEDs 30 pieces, IR 730nm LEDs 16 pieces. a) expected spectral density of a light fixture; b) measured spectral density of a light fixture; c) appearance of light fixtures

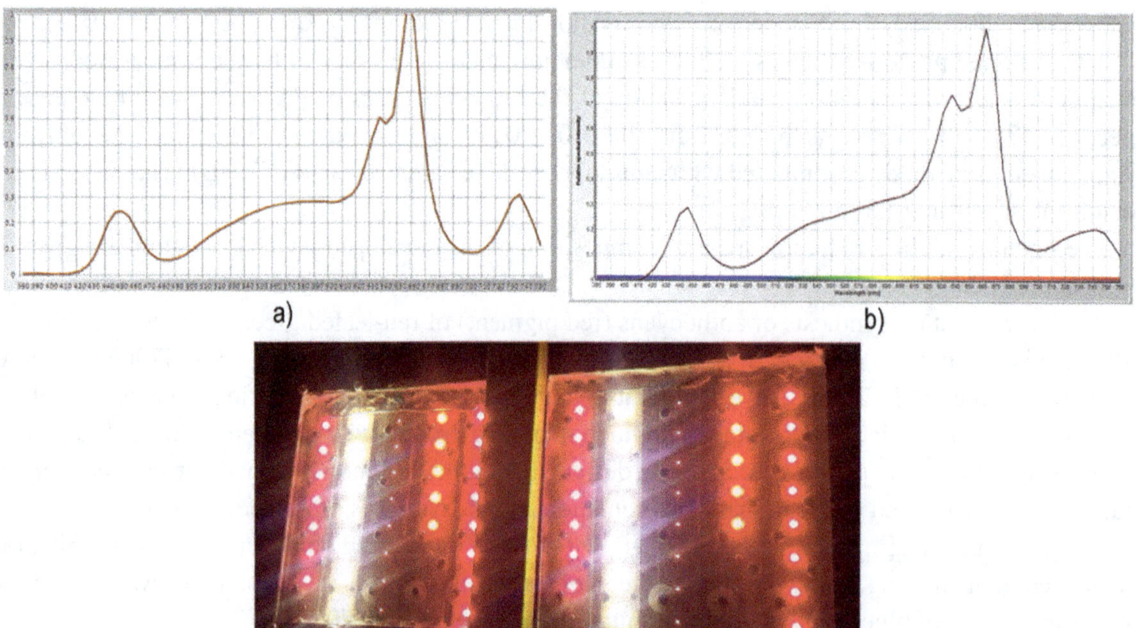

Figure 3. Combined light emitting diode-based irradiator intended for cucumber cultivation consists as follows: white LEDs 40 pieces, red 660nm LEDs 16 pieces. a) expected spectral density of a light fixture; b) measured spectral density of a light fixture; c) appearance of a light fixture

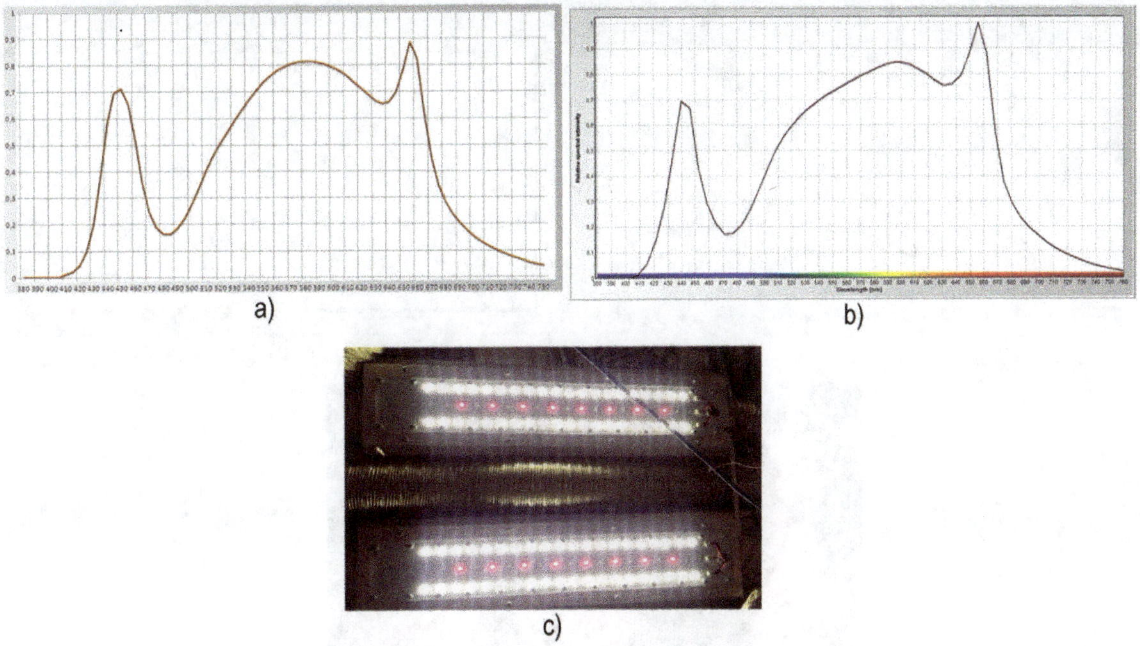

In the course of these investigation conduction and LED phyto-irradiators operation, monitoring was conducted of luminance spreading on areas and heights with use of the spectrophotometer TKA Spectrum. Reference measures were conducted for each type of LED-phyto-irradiator. At the same time, measures were conducted of temperature and carbon dioxide content in the phyto-chambers with aid of digital technologies of monitoring; the digital database was composed. Effectiveness of photosynthesis in plants of cucumber, tomato and lettuce in the course of their vegetation was controlled by way of both chlorophyll content testing in plants leaves with use of the device Apogee MC-100 and direct measures of pigment contents (of chlorophyll a and b, carotenoids and anthocyans) by photometric method on known methodic procedure with use of spectrophotometer Specord 205.

RESULTS OF EXPERIMENTAL INVESTIGATIONS OF DEVELOPED LED-PHYTO-IRRADIATORS

There were conducted comparative tests of light emitting diode-based irradiators of various spectral composition and natrium light source of high pressure (HPS). Also, the influence was studied of combinations of LEDs with irradiation in blue, green, red and far-red areas of the spectrum on pigments biosynthesis intensity, growth, nutritional value and productivity of ordinary tomato (Solanum lycopersicum).

The researches were conducted in the testing laboratory in FSAC VIM. For investigation, there was chosen the indeterminate tomato hybrid T-34 F1, the selection product of the company *"Gavrich"* intended for cultivation in greenhouses in Autumn-Winter period with artificial supplementary lighting

Figure 4. Combined light emitting diode-based -irradiator intended for tomato cultivation consists as follows: white LEDs 12 pieces, red 630nm LEDs 7 pieces, blue LEDs 2 pieces, red 660nm 23 pieces, IR 730nm LEDs 12 pieces. a) expected spectral density of a light fixture; b) measured spectral density of a light fixture; c) appearance of a light fixture

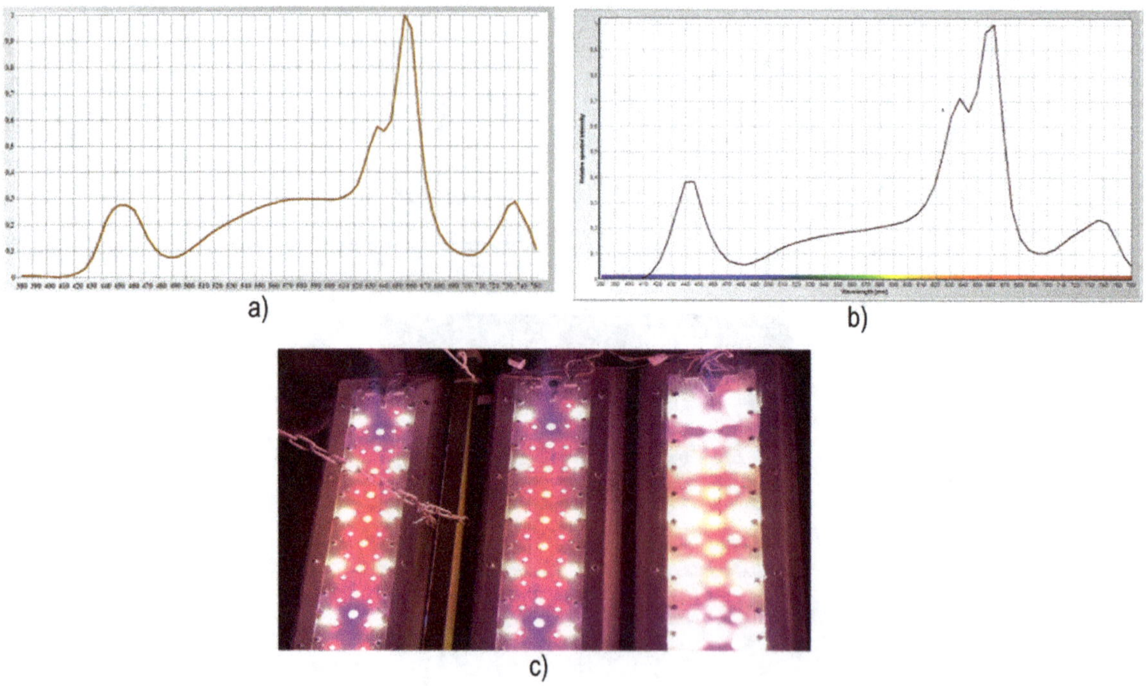

Figure 5. Spectral composition of different modes of ray treatment

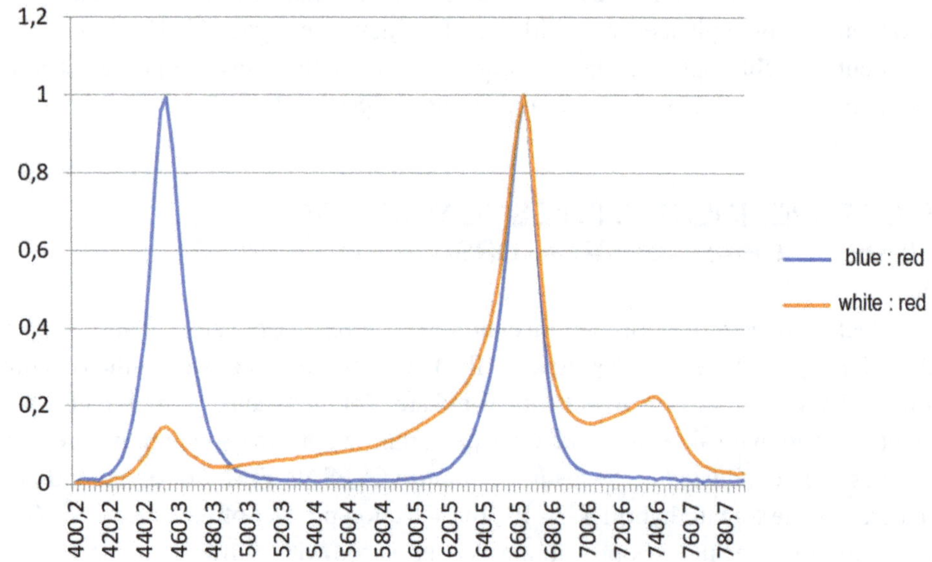

(photo-culture). The tomatoes were grown in three phyto-chambers on the mineral wool GRODAN with drip watering. Sowing of the tomatoes was made into cubicles on 12.01.2018, while landing on mat on 19.02.2018; beginning of mass fruitage was observed on 25.04.2018.

As irradiation sources, the irradiators were used based on LEDs with blue (Bl), white (Wh), red (Rd) and far-red (FR) irradiation in various combinations (see Table 1) and as reference, the light fixtures were used with the natrium lamps of high pressure (HPS). The PAR luminance in all phytotrons was set ~45 W/м^2 during 16 hours-long lighting period and at air temperature 25/18°C (day/night).

The consumed power of the both light fixtures LED and HPS made 220 and 440 W respectively. For measuring of photosynthetic active radiation (PAR) and for analysis of irradiation spectrum, we made use of the spectrophotometer TKA-Spectrum. For control of heating-up, ventilation, enriching with carbon dioxide, humidity, illumination and registration of ambient conditions, the climatic computer was used.

The photosynthesis intensity was evaluated on pigments content in tomato leaves; the content of the following pigments was measured: chlorophyll *a* and *b* (Chl), carotenoids (k) and anthocyans (An). Chl and k were extracted from leaves by milling them and exposing to action of 95% ethanol (Eth) in water bath at 50-70°C. The density of absorption (D) of Chl solution in Eth was measured on the wave lengths 440.5; 664 and 649 nm with use of the spectrophotometer Specord M-40 in a cuvette 1cm thick. The content of Chl*a* (C_a), Chl*b* (C_b) and carotenoids (C_k) in mg/ml in Eth was calculated on the formulas (1), (2) and (3) according to (Lichtenthaler, Buschmann, 2001)

$$C_o = 13.36 A_{664.1} - 5.19 A_{648.6} \quad (1)$$

$$C_b = 27.43 A_{648.6} - 8.12 A_{664.1} \quad (2)$$

$$C_{:} = 4.68 A_{440.5} - 0.268 !_{0+b} \quad (3)$$

The Chl and k content in mg/g of dry weight of this vegetative material was calculated on the formula:

$$A = \frac{C \cdot V}{P \cdot 1000}, \quad (4)$$

Table 1. Spectral characteristics of test irradiators

Irradiator Type (LED Type)	The Ratio of Fluxes Bl: Gr: Rd + FR (% of the Total flow)	Irradiance UV-A, W/m^2
No.1 LED-1 (Bl, Wh, Rd, FR)	20:20:60+14	0.01
No.2 LED-2 (Bl, Wh, Rd, FR)	20:20:60+3	0.01
No.3 HPS	9:67:26+5	0.05

where, *C* is pigment concentration in mg/l; *V* is pigment extract volume in ml; *P* is charge of vegetative material in mg; *A* is pigment content in the vegetative material in mg per 1 g of dry or raw weight.

Anthocyans (An) were extracted from leaves in the following aqueous solution: Eth (1:1) + HCl (1%). The anthocyans content was calculated on the formula:

$$C_{an}(mg/g) = (D_{530} \cdot V\mu)/(P\varepsilon) \quad (5)$$

where D_{530} is *An* solution optical density on the wave length 530 nm; μ is molecular weight *An* (449); ε is molar coefficient of extinction equal to 27000 l/(mole cm). For the calculations, the Excel program was used.

The appearance of the tomato sprouts grown under different irradiators is represented in the Figure 1. The distribution analysis of contents of dry materials of monosaccharides, ascorbic acid and nitrates in the tomato fruits was conducted in the testing laboratory of FGBNU VNIISSOK on standard methodic procedures. These results are given in the Table 2 with specification of scatter in tests line for every sample. Fruit collection was conducted every time when biological ripeness was reached.

The pigments distribution in the leaves samples depended on a plant vegetation period, a leaf location on a bush and a leaf test state for each irradiator. For analysis, we selected tomato leaves under last blooming raceme (Table 3). In the Table 3, results are given of the chemical analysis of tomato leaves in the beginning of fructification stage.

The influence is well-known of FR irradiation on internodes elongation and leaves square area increase, which usually is explained by phytochrome reactions of plants on Rd and FR irradiation (Hao et al., 2016). For example, in the samples No.1, where in the irradiator spectrum, FR irradiation made 14%, the sprouts were distinguished with longer stems and larger leaves square areas than ones in the samples No. 2 and No. 3. Also, the blooming racemes of the sample No. 1 began to be formed earlier, which accelerated the fructification beginning.

Table 2. The results of biochemical analysis fetuses

Sample of Tomato (Illuminator)	Dry Matter, %	Monosaccharides, %	Ascorbic Acid, mg% (or mg / 100 g)	Nitrates NO_3^-, mg / kg
No.1 (LED-1)	5.52±0.05	2.57	24.64	118.8±16
No.2 (LED-2)	5.01±0.07	1.29	26.40	138.0±29
No.3 (HPS)	5.12±0.035	2.14	26.40	117.8±13

Table 3. Leaf analysis results

Sample of Tomato (Illuminator)	Chlorophyll (mg/g)		Carotinoides C_k	Anthocyans
	C_a	C_b	(mg/g)	
LED-1	2,09	0,81	0,62	0,8
LED-2	2,43	1,04	0,71	2,6
No.3 (HPS)	2,78	1,09	0,75	<0,1

Table 4. Yield of one tomato plant

Sample of Tomato (Illuminator)	Yield, g	Average Fruit Weight, g	Number of Fruits, pcs.
LED-1	2300	105	44
LED-2	2400	110	45
No.3 (HPS)	1800	91	40

Figure 6. Tomato plants on the 60th day of growth under different irradiators. a- LED-1, b-LED-2, c- HPS

In the nature usually, the ray treatment enriching by FR light takes place under plants curtain. In lower storey, FR share grows as leaves absorb intensively the red light in the process of photosynthesis. However, a long exposure to the FR irradiation leads to excessive thickness of cenosis. Staying in shadow of wide upper leaves gradually, lower leaves start dying away; a plant photosynthesis is worsened, which results in yield reduction.

One can see the difference in pigment contents (in leaves) and productivity of tomato plants grown under light emitting diode irradiators in comparison with the samples irradiated by the lamp HPS. It is possible to connect these results with the influence of PAR composition (Tables 3 and 4). The sample No. 2 was distinguished with a violet tint of the leaves and stems. This tint is conditioned with emergence in the plant of not only chlorophyll but also of anthocyans, flavonoids derivatives. (Raymond, 2009). It is considered that anthocyans in cells play a role of a filter protective against destroying photodynamic action of the solar light and especially of the UV radiation. The strongest absorption maximum of anthocyans falls within the ranges 265-280 nm and 510-560 nm. In the HPS spectrum, the blue band has its maximum on 460 nm, while the maximums of the irradiation blue bands of LEDs and of chlorophyll absorption band are virtually the same. The HPS irradiation intensity in the blue area is only 1.5-2 times less than the LEDs irradiation flux intensity in this area (Table 1), while the *An* synthesis efficacy of the sample No. 2 is higher by order than that of the samples No. 1 and No. 3. Taking into consideration this fact as well as the correlation of *An* content with the intensity of LED irradiation blue band (Table 3), one can suppose that the anthocyans synthesis initiates a light causing the chlorophyll molecule transition on the new energetic level $S_o \rightarrow S_2$ (Kholmanskiy et al., 2018). Such action of the blue irradiation is qualified as a stress factor. (Giliberto et al., 2005). However, the FR irradiation decreases this stress-causing action. For example, in the sample No.1, the anthocyans content was much less than in the sample No.2, while the contents of chlorophylls and carotenoids was comparable. The absence of UV-A in the spectra of LEDs and HPS (Table 1) excludes UV irradiation as a stress-factor for tomato (Ebisawa et al., 2008; Smirnov, 2017).

The narrowband irradiation of red and blue light emitting diodes leads to inhibition of plant generative development; this was confirmed by the investigations (Smirnov, 2018). As compared with tomato productivity under action of monochromatic light, the ray treatment in the case of the combination of white and red-light emitting diodes results in Chl and carotenoids content growth, while FR irradiation addition neutralizes the action of the light of high intensity as a stress factor.

Our comparative tests of light emitting diode-based phyto-irradiators showed that at a comparable productivity and nutrient qualities of tomato fruits grown under HPS- and LED-irradiators, the energy consumption per a product kilogram in the case of LED-irradiators is twice less (Smirnov, Kholmansky, 2017).

Investigations were conducted for influence studying of ray treatment modes with changed spectral composition on red-leafed lettuce of the Anthony and Lollo Rossa sort. For a green mass augmentation, the ray treatment mode was used based on a combination of white and red-light emitting diodes (Fig. 5). For anthocyans biosynthesis, the ray treatment mode was used based on the combination of the blue and red-light emitting diodes in the following ratio 50:50. In both modes, the PAR luminance made 40 W/m^2.

The conducted researches show that by a reasonable selection of a spectrum separately for vegetation and for anthocyans synthesis before harvesting, it is possible to increase the product quality (Table 5, Fig. 7). The anthocyans biosynthesis in the red-leafed lettuce is launched in presence of the irradiation 445nm of the blue color with its share not less than 20% of the total irradiation in PAR area. The best effect is reached at the ratio Rd/Bl 50/50%. The maximal land productivity of the studied lettuce sort was observed at the irradiation Bl:Gr:Rd with the ratio 22:1:77.

It was established in the investigation that with use of changes (with time) of irradiation spectral composition, it is possible to launch the anthocyans biosynthesis mechanism. On the stage of the green mass augmentation, the percentage of the blue component in an irradiation spectrum must be not more

Table 5. Pigments content in leaves of Lollo Rossa lettuce

	Chl *a*, mg/g	**Chl *b*, mg/g**	**Chl *a+b*, mg/g**	**Chl *a/b*,**	**An, mg/g**
Prior to ray treatment	199.7	123.1	322.8	0.62	0.22
On 8th day of ray treatment	136.9	90.4	227.2	0.66	1.01

Figure 7. Anthocyans biosynthesis dynamics under LED irradiation 445nm and 660nm with ratio 50/50 on 1st, 8th and 14th day respectively

than 25%. Besides, in order to obtain the red color of the lettuce leaves, it is necessary to increase the share of the blue component up to 50% and render the exposure not less than seven days.

Based on the researches results, there were established general rules applicable at development and designing of modern phyto-irradiators based on light emitting diodes. The determination of light-technical and operational characteristics of phyto-irradiators based on LEDs must be done relying on the following requirements:

- The LED-phyto-irradiator spectral composition must correlate with the absorption spectra of pigments and chromephores taking part in work of photosynthetic apparatus of every type of plants;
- The intensity of irradiation bands in a LED-phyto-irradiator spectrum must correlate with percentage and the role in the photosynthesis of absorption of corresponding pigments and chromephores of phyto- and cryptochrome systems;
- Time-connected correlations of LED-phyto-irradiator spectrum must correspond to vegetation periods of growth of the plant;
- Energy consumption of the LED-phyto-irradiator must be much less in comparison with that of standard natrium lamps at the same climatic conditions;
- For plants grown under LED-phyto-irradiators, the nutrient elements contents in fruits, their ripening periods and land productivity must be comparable with the similar parameters of plants cultivated in conditions of ray treatment under a natrium lamp;
- Use of LED-phyto-irradiators must be conducted in conditions of automated control in concord with changes of climatic and nutritional conditions;
- A luminance distribution in space of a greenhouse or a phyto-chamber must be balanced on power and spectral composition depending on height, vegetation period and optimal temperature for each kind of plant.

For confirmation of importance of LED-phyto-irradiator spectral composition and its influence on the plant growth and development process, the test series were conducted and they allowed evaluating the influence of an irradiation spectral composition on the chlorophyll photoreaction in the plant's photosynthetic apparatus as well as influence evaluating of the irradiation spectral composition on a content of microelements and pigments in leaves of a plant.

INFLUENCE ASSESSMENT OF IRRADIATION SPECTRAL COMPOSITION ON CHLOROPHYLL PHOTOREACTION IN PLANTS PHOTOSYNTHETIC APPARATUS

For studying of orbital structure of photochemical active states (PCAS) of biomolecules and finding out their influence on Chl*a* and Chl*b* photochemistry, dependencies were investigated of absorption spectra on the function of universal interaction and their influence on the reaction process of their photo-discoloration by a dissolvent, oxygen, *An* and HCl.

Chl was extracted from leaves of studied cultures (cucumber, tomato, lettuce) and exposed to 95% ethanol. *An* was obtained from leaves of red-leafed lettuce in aqueous solution of ethanol (1:1) with addition of 1% HCl. After *An* drying and extraction, portions of *An* were entered into Chl solutions as proton acceptor supplement. For modeling of charge transfer process in the fragment Chl*b*, spiropyran was used. The absorption spectra were determined by quantity and locality of OH-groups in the *An* structure. In the course of the conducted laboratory studies for *An* solution color, the cation (cyanidin-3-glucoside) was responsible (R5 = R7 = OH; R6 = R`3 = R`5 = H); maximum of the first absorption band was in the range 530-540 nm.

For registration of spectra of Chl and *An* solutions, the spectrophotometer Specord 205 was used. The results are given in the Figure 8. The Chl and carotinoides contents in ethanol were calculated on known formulas (Lichtenthaler, &Buschmann, 2001).

Figure 8 – Chlorophyll absorption spectra in 95% ethanol. (1) Original solution; (2) After 1 hour-long ray treatment by light of red LED; (3) After 0.5 M HCl addition to the solution

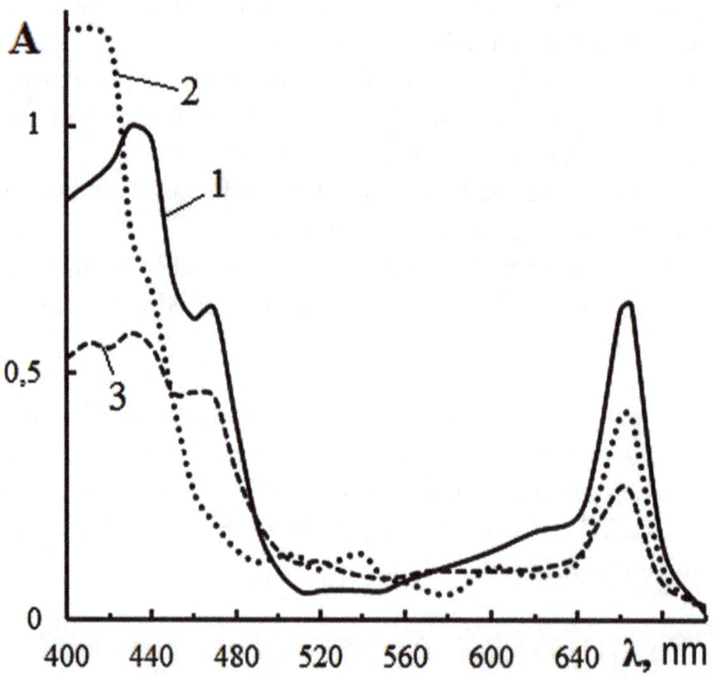

$$C_{Cla} = 13.36 A_{664.1} - 5.19 A_{648.6},\ C_{Clb} = 27.43 A_{648.6} - 8.12 A_{664.1},\ C_{Car} = 4.68 A_{440.5} - 0.268 ! _{Cla+Clb} \quad (6)$$

The solutions were exposed to blue and red-light emitting diode irradiation with wave lengths λ ~ 434 nm, Δλ1/2 ~ 23 nm and λ ~ 670 nm, Δλ1/2 ~ 30 nm respectively. The carotenoides content did not exceed 15% of $C_{Chla+Chlb}$; carotenoids extinction coefficient was much less than that of Chl; hence, their impact on the obtained results was negligible.

The discoloration process of Chl solution under blue and red-light emitting diodes is represented in the Figure 9.

Figure 9. Discoloration process of Chla (solid lines) and Chlb (dashed lines) under light of red diode in 95% ethanol; logarithmic anamorphosis (3) of curve 1 (see a) and of curves 1, 1' (see b). a) Curves 1' and 2': into the solution, anthocyan was added; b) Curves 1', 2': the solution is bubbled by nitrogen.

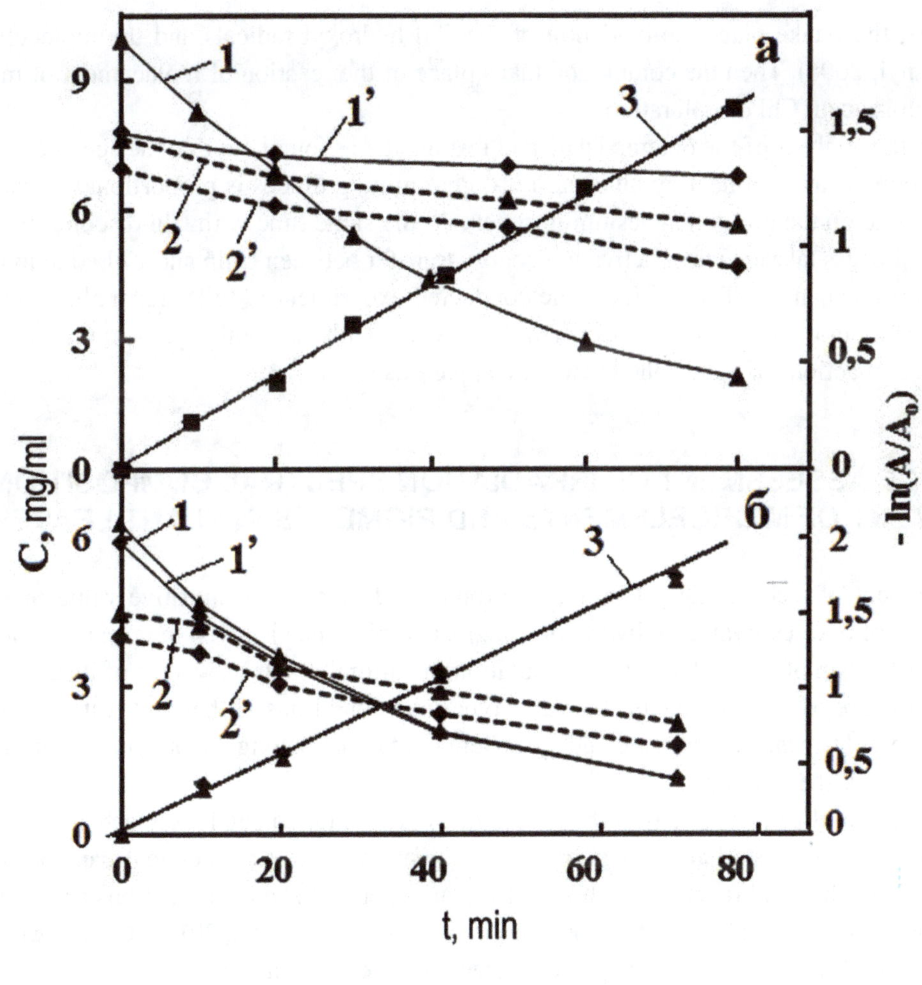

At Chl solution bubbling with helium and nitrogen during about 5 minutes, the discoloration process speed virtually did not change. The highest speed of the discoloration was observed at HCl addition.

The obtained results show that the Chl photoreaction speed, at which magnesium atom replacement on two atoms of hydrogen takes place, depends primarily on Chl structure, ray treatment wave length and dissolvent. In general case, the discoloration speed constant value depends on charge change in Chl transitions and deactivation speed expressed in the form of fluorescence with subsequent release of heat energy. Notably, the activation energy transfer from Chl*b* and Chl*a* is fulfilled on inductive-resonance mechanism due to dipole-dipole interaction; and it is conditioned by the coincidence of the Chl*b* fluorescence spectrum with the Chl*a* absorption spectrum.

In accordance with the first order of the reaction, Chl interacts with a hydrogen donor, which (in the ethanol solution) are H_2O extraligandes and ethanol molecules. The mechanism of hydrogen phototransfer on Chl for H_2O extraligandes has the following form:

$$Cl^{*} + H_2O \rightarrow Cl^{-} + H_2O^{+} \rightarrow ClH + OH^{\cdot} \tag{7}$$

After this, there take place the oxidation of the Chl hydroxyl radicals and the molecular oxygen release (Prisnyi, 2008). Then the completion takes place of the reaction of replacement of magnesium atoms and subsequent Chl discoloration.

Based on the analysis of the obtained data and results, it was found out that the speed of Chl discoloration reaction at ray treatment by blue and red-light emitting diodes is proportional to charge value in activated state on the group magnesium-nitrogen. At the same time with Chl discoloration reaction running, the process takes place of activation energy transfer between Chl*b* and Chl*a* due to the inductive-resonance mechanism. The results of the conducted experiment in fuller form allow representing and comprehending the role and the specialization of Chla, Chlb and anthocyans in the mechanism of photochemical reactions of plants photosynthetic apparatus.

INFLUENCE ASSESSMENT OF IRRADIATION SPECTRAL COMPOSITION ON CONTENT OF MICROELEMENTS AND PIGMENTS IN PLANT LEAVES

For improvement of technologies of protected ground and increase of nutritive value of vegetables and especially of ones cultivated on hydroponics, great significance is given to researches of physical-chemical mechanism of microelements accumulation in cultural plants as well as of its dependence on electronic structure of the microelements, plants types and conditions of their cultivation. To the latter first of all, the light-technical, climatic and agrochemical factors belong influencing plant metabolism on all the stages of their ontogeny.

It is known that the plants photosynthetic apparatus is well tuned catalytic mechanism, work principle of which is based first of all on overcoming of activation barriers of chemical transformations and inducing of photochemical reactions with use of radiant energy. Speeds of reactions running in plants depend primarily on physical-chemical properties of water (Kholmanskiy, 2015, 2016) as well as on the orbital structure of the photosynthetic apparatus elements (Feodorova, 2010).

In the course of the studies conduction, the analysis was done of effectiveness of biosynthesis of pigments and accumulation of micro and macro elements in the cultivated cultures (cucumber, tomato, lettuce). The analysis was aimed at the influence determination of photosynthetic irradiation spectral composition (Smirnov et al., 2018). It was found out that the absolute content of pigments, micro and macro-elements in cucumber and tomato is higher than that in lettuce. The effectiveness of micro and macro-elements consumption from nutritive solution is similar. Notably, for all plants being under observation, the consumption of most part of micro- and macro-elements depended in a linear way on coordinational number of each element and reciprocal numbers of its atomic mass and second potential of ionization. For the watched plants in the similar way, this dependence changed with varying of ray treatment spectrum. The assumption was put forward that an effectiveness of accumulation by plant of micro- and macro-elements depends on not only chemical activeness of elements of photosynthetic apparatus but also on spectral characteristics of pigments; and the effectiveness is conditioned by the mechanism of effects of micro-polarization.

The dependence of extraction effectiveness of mineral elements by plant leaves and fruits on a plant type and spectral composition of used irradiator is possible to characterize in adequate way with use of average coefficients of accumulation of micro- and macro-elements in them. Our comparative analysis of accumulation coefficients for specific elements allows revealing and explaining their functional peculiarities in different plants metabolism.

Connecting the effectiveness of the extraction mechanism with extent of demand in the microelements in plant metabolism, we made the assumption that this extent is proportional to chemical activeness and electric potential of the microelement ion. These factors together with gravitation factor were expressed by the corresponding combination of values of charge, radius, mass and coordinational number of a microelement ion. The linear trends of the dependences of extraction coefficients on this combination of values confirmed that the proposed mechanism of extraction does depend on a spectrum of ray treatment and acts in tomato twice more effectively than in cucumber; however almost in no way it manifests itself in lettuce.

CONCLUSION

The investigations of dependence of pigments biosynthesis effectiveness in plants of cucumber, tomato and lettuce on spectral composition of phyto-irradiators allowed substantiating the optimal spectral composition of phyto-irradiators for vegetable crops and optimizing the ray treatment modes. The comparative researches of productivity dependences of cucumber and tomato on a ratio of LED ray treatment intensity in combination of phyto-irradiators for cucumber and tomato allowed developing an original phyto-irradiator based on light emitting diodes for cultivation of cucumber and tomato of certain sorts, which provides with considerable economy in energy consumption and along with it conservancy of land productivity and improvement of nutritional properties in comparison with natrium lamp use. At development of ray treatment modes for red-leafed green cultures, it is necessary to take into account sort peculiarities of every type of plant.

By way of a combination of spectra of light emitting diodes, it is possible to vary a PAR spectral composition and in so doing to increase effectiveness of photosynthesis and land productivity without a loss of nutritional quality of tomato fruits in comparison with corresponding characteristics of PAR of natrium lamps. The application of the light emitting diodes as light sources in greenhouse irradiators will allow rising quality of cultivated vegetable products with simultaneous costs reduction on electric energy down to twice.

REFERENCES

Alferova L.K., Yuferev L.Y., & Yufereva A.A. (2015). K metodike rascheta svetodiodnogo osveshcheniya [More about methodic procedure of calculation of light emitting diode-based illumination]. *Innovatsii v sel'skom khozyaystve [Innovations in agrarian sector], 2*(12), 31-34. (in Russian)

Chupackina, G. N., Maslennikov, P. V., & Skrypnik, L. N. (2011). Prirodnyye antioksidanty (ekologicheskiy aspect) [Natural anti-oxidants (ecological aspect)]. (in Russian)

Ebisawa, M., Shoji, K., Kato, M., Shimomura, K., Goto, F., & Yoshihara, T. (2008). Supplementary ultraviolet radiation B together with blue light at night increased quercetin content and flavonol synthase gene expression in leaf lettuce (Lactuca sativa L.). *Environment Control in Biology*, *46*(1), 1–11. doi:10.2525/ecb.46.1

Feodorova, O. A. (2010). *Supramolekulyarnaya khimiya* [Supramolecular chemistry]. Moscow: RHTU. (in Russian)

Giliberto, L., Perrotta, G., Pallara, P., Weller, J. L., Fraser, P. D., Bramley, P. M., ... Tavazza, M. (2005). Giuliano Manipulation of the blue light photoreceptor cryptochrome 2 in tomato affects vegetative development, flowering time, and fruit antioxidant content. *Plant Physiol.*, 199–208. doi:10.1104/pp.104.051987

Golovatskaya, I. F. (2005). Rol' kriptokhroma 1 i fitokhromov v regulyatsii fotomorfogeneticheskikh reaktsiy rasteniy na zelenom svetu [Role of cryptochrome 1 and phytochromes in regulation of plants photomorphogenetic reactions on green light]. Fiziologiya rasteniy [Plants physiology], 6, 822-829. (in Russian)

Golovatskaya, I. F. (2016). Morfogenez rasteniy i yego regulyatsiya. Chast' 1: Fotoregulyatsiya morfogeneza rasteniy [Morphogenesis of plants and its regulation. Part 1: Photoregulation of plants morphogenesis: Educational manual]. Tomsk: Publishing house of Tomsk State University. (in Russian)

Goto, E., Hayashi, K., Furuyama, S., Hikosaka, S., & Ishigami, Y. (2016). Effect of UV light on phytochemical accumulation and expression of anthocyanin biosynthesis genes in red leaf lettuce. *Acta Horticulturae*, (1134), 179–186. doi:10.17660/ActaHortic.2016.1134.24

Hao, X., Little, C., Zheng, J. M., & Cao, R. (2016). Far-red LEDs improve fruit production in green house tomato grown under high-pressure sodium lighting. *Acta Horticulturae*, (1134): 95–102. doi:10.17660/ActaHortic.2016.1134.13

Inada, K. (1978). Spectral dependence of photosynthesis in crop plants. *Acta Horticulturae*, *87*(87), 177–184. doi:10.17660/ActaHortic.1978.87.18

Izenberg, Yu. B. (1983). *Reference book on light-techniques*. Moscow: Energopromizdat. (in Russian)

Kholmanskiy, A. (2015). Activation energy of water structural transitions. *Journal of Molecular Structure*, *1089*, 124–126. doi:10.1016/j.molstruc.2015.02.049

Kholmanskiy, A. (2016). Chirality anomalies of water solutions of saccharides. *Journal of Molecular Liquids*, *216*, 683–687. doi:10.1016/j.molliq.2016.02.006

Kholmanskiy, A., Smirnov, A., & Parmon, V. (2018). Bleaching of solutions of chlorophyll a and b with blue and red light. *High Energy Chemistry*, *52*(1), 6–13. doi:10.1134/S001814391801006X

Lichtenthaler H. K., Buschmann C. (2001). Chlorophylls and Carotenoids: Measurement and Characterization by UV-VIS Spectroscopy. *Current Protocols in Food Analytical Chemistry*.

Medvedev, S. S. (2013). Plants physiology. St. Petersburg: BHV Petersburg. (in Russian)

Nicole, C. C. S., Charalambous, F., Martinakos, S., van de Voort, S., Li, Z., Verhoog, M., & Krijn, M. (2016). Lettuce growth and quality optimization in a plant factory. *Acta Horticulturae*, (1134), 231–238. doi:10.17660/ActaHortic.2016.1134.31

Prisnyi, A. A. (2008). *Biophysics: lectures course: manual*. Belgorod. (in Russian)

Keller, R.B. (2009). Flavonoids: Biosynthesis, Biological Effects and Dietary Sources. Nova Science Publishers.

Singh, D., Basu, C., Meinhardt-Wollweber, M., & Roth, B. (2015). LEDs for Energy Efficient Green house Lighting. Hannover Centre for Optical Technologies.

Smirnov A., Kholmanskiy A., Ukhanova V. (2018). Optimization of lighting spectrum of greenhouse vegetables by using light emitting diodes. *International Journal of Research in Pharmacy and Biosciences*, *5*(4), 11-17.

Smirnov A.A. (2017). Vliyaniye UF-A radiatsii na biosintez antotsianov krasnolistnogo salata [Influence of UV-A radiation on anthocyans biosynthesis of red-leafed lettuce]. *Innovatsii v sel'skom khozyaystve [Innovations in agrarian sector]*, *1*, 6-11. (in Russian)

Smirnov A.A. (2018). Zavisimost' biosinteza pigmentov i produktivnosti tomata ot spektral'nogo sostava izlucheniya [Dependence of pigments biosynthesis and productivity of tomato on spectral composition of irradiation]. *Innovatsii v sel'skom khozyaystve [Innovations in agrarian sector]*, *3*(28), 78-86. (in Russian)

Smirnov A.A., Kholmansky A.S. (2017). Zavisimost' fotosinteza pigmentov i produktivnosti tomata ot spektral'nogo sostava obluchatelya [Dependence of pigments photosynthesis and productivity of tomato on spectral composition of irradiator]. *Nauchnaya zhizn' [Scientific life]*, *10*, 14-19. (in Russian)

Snowden, M., Cope, K., & Bugbee, B. (2016). Sensitivity of Seven Diverse Species to Blue and Green Light: Interactions with Photon Flux. *PLoS One*, *11*(10), e0163121. doi:10.1371/journal.pone.0163121 PMID:27706176

Sokolov A.V., Yuferev L.Yu. (2014). Modelirovaniye spektrov svetodiodnykh matrichnykh svetil'nikov [Modeling of spectra of light emitting diode-based matrix light fixtures]. *Innovatsii v sel'skom khozyaystve [Innovations in agrarian sector]*, *2*(7), 65-72. (in Russian)

Sokolov A.V., Yuferev L.Yu. (2015). Modelirovaniye spektra osveshcheniya svetodiodnym obluchatelem [Modeling of spectra of illumination by light emitting diode irradiator]. *Mechanization and electrification of agriculture*, *8*, 22-24. (in Russian)

Tikhomirov, A. A., Sharupich, V. P., & Lisovsky, G. M. (2000). Svetokul'tura rasteniy: biofizicheskiye i biotekhnologicheskiye osnovy [Photoculture of plants: biophysical and biotechnological basics]. Novosibirsk: Siberian department of Russian Academy of Sciences. (in Russian)

Tyutereva E.V., Ivanova A.N., Voitsekhovskaya O. V. (2014). K voprosu o roli khlorofilla b v ontogeneticheskikh adaptatsiyakh rasteniy [More about role of chlorophyll-b in plants ontogenetic adaptations]. *Successes of modern biology, 134*(3), 249-256. (in Russian)

Vasiliev, A. N., Budnikov, D. A., Krausp, V. R., Dubrovin, V. A., Toporkov, V. N., Nurgaliyev, I. S., . . . Kazakova, V. A. (2017). Metody energeticheskogo vozdeystviya na semena prioritetnykh zernovykh i ovoshchnykh kul'tur razlichnykh sortov, rasteniya i sel'skokhozyaystvennyye materialy. kontseptsiya ispol'zovaniya elektrotekhnologiy dlya obrabotki kormov, udobreniy, otkhodov rasteniyevodstva. Nauchno obosnovannyye parametry energosberegayushchikh kombinirovannykh ustanovok dlya obezzarazhivaniya vozdukha i poverkhnostey. [Methods of energetic action on seeds of high- priority grain and vegetable crops of various sorts; plants and agrarian materials. Concept of use of electric technologies for processing of fodder, fertilizers, wastage of plants cultivation. Scientifically substantiated parameters of energy-saving combined installations for detoxication of air and surfaces]. Federal scientific agro-engineering center of VIM. (in Russian)

Yuferev, L., & Sokolov, A. (2018). Energy-Efficient Lighting System for Greenhouse Plants. In V. Kharchenko & P. Vasant (Eds.), Handbook of Research on Renewable Energy and Electric Resources for Sustainable Rural Development (pp. 204–229). Hershey, PA: IGI Global. doi:10.4018/978-1-5225-3867-7.ch009

Yuferev L.Yu, Sokolov A.V. (2014). Energosberegayushchaya sistema osveshcheniya dlya teplichnykh rasteniy [Energy-saving illumination system for greenhouse plants]. *Innovatsii v sel'skom khozyaystve [Innovations in agrarian sector], 4*(9), 78-81. (in Russian)

Chapter 9
Energy-Saving Technologies for Pre-Sowing Seed Treatment in a Magnetic Field

Volodymyr Kozyrskiy
National University of Life and Environmental Sciences of Ukraine, Ukraine

Vitaliy Savchenko
National University of Life and Environmental Sciences of Ukraine, Ukraine

Oleksandr Sinyavsky
National University of Life and Environmental Sciences of Ukraine, Ukraine

Vasyl Bunko
Berezhansk Agrotechnical Institute, Ukraine

ABSTRACT

The purpose of the research was to establish the mechanism of the magnetic field impact on seeds to determine the most effective mode of pre-sowing treatment of seeds in a magnetic field and design parameters of the device for magnetic treatment of seeds. It is established that under the influence of a magnetic field the rate of chemical reactions occurring in plant cells is accelerated, solubility of salts and acids increases, and permeability of cell membranes accelerates the diffusion of molecules and ions through them. This leads to an increase in the concentration of ions in the cell and oxygen molecules and the growth of water absorption of seeds. Pre-sowing seed treatment promotes increased germination by 25-40%, and germination by 30 - 35%. The most effective pre-sowing treatment of seeds in a magnetic field is a magnetic induction of 0.065 Tl with four reversal magnetization, a pole division of 0.23 m and a seed movement speed of 0.4 m/s. With this mode of treatment, crop yields increase by 20–25%.

DOI: 10.4018/978-1-5225-9420-8.ch009

INTRODUCTION

The use of electrotechnological methods of pre-sowing seed treatment makes it possible, without the use of chemical agents, to increase the yield of agricultural crops, reduce the incidence of plants, and improve the quality of products and storage periods.

Pre-sowing seed treatment in a magnetic field compared with other electrophysical methods is characterized by high plant productivity, low energy consumption, and is safe for the environment and maintenance personnel.

Many researchers have established the positive influence of the magnetic field on seeds of crops, which manifests itself in improving the sowing quality of seeds, the growth of biometric indices and the decrease in plant morbidity, increase in crop yields and product quality.

Since the seed material has certain biological properties, it is therefore necessary to use such regimes that correspond to the specific biological properties of the seed material and give maximum effect to its treatment.

Nowadays, the mechanisms and regularities of the the magnetic field impact on aqueous solutions and biological objects are not fully disclosed. The absence of an explanation of the magnetic field effect on the processes occurring in the seeds makes it impossible to establish all the active factors when it is processed in a magnetic field and to determine their optimal values.

The purpose of research is to establish the mechanism of the magnetic field impact on seeds, to determine the most effective mode of presowing seed treatment in a magnetic field and the design parameters of a device for magnetic seed treatment.

BACKGROUND

Synthesis of organic matter in plants comes from water, mineral salts and carbon dioxide. All biochemical processes in plant cells occur in an aqueous medium. The transport of the elements of the power into the cell is provided by two autonomous mechanisms - the passive flow of matter with the electrochemical gradient and their active transport against the electrochemical gradient (Yagodin, Zhukov, & Kobzarenko, 2002).

The ions pass through the membrane as a result of the diffusion process or together with the solvent. The membrane also provides direct passage of water and dissolved substances in it. The constant passage of ions through the membrane results in a continuous flow of new ions into it to equalize the concentration.

Since the ions have an electric charge, the process of their passage through the membrane depends on the difference in concentration and the magnitude of the diffusion potential arising between the two solutions with different concentrations of matter separated by the membrane.

The electromagnetic field affects the transport of ions and molecules of substances in the cell, contributes to increasing the sowing quality of seeds and yields of crops. In recent years, electrophysical methods have been developed for the seed treatment of certain crops in the electric field of corona discharge, magnetic field, electromagnetic radiation, etc. Based on the analysis of the electrophysical methods of seed treatment, one can conclude that magnetic seed treatment is a promising method that has significant advantages over other methods. (Zholobova, 2012).

Magnetic field affects aqueous solutions. It has now been experimentally established that magnetic treatment of water changes its physical and chemical properties: coagulation and absorption are accelerated, solubility of salts and concentration of gases, crystallization and wetting, magnetic susceptibility, viscosity, hydration of ions, kinetics of chemical reactions change. (Malkin, Zhuravskaya, & Kovalenko, 2015; Klassen, 1982).

Permanent magnetic field is used for the seed treatment. The seed treatment by a variable magnetic field in many appeared to be ineffective, so was supplemented by other types of treatmen (ultraviolet, corona discharge field). As a result, the cost of installations with electromagnets far outweighs the cost of installations with permanent magnets, increasing operating costs and energy intensity of the process of seeds processing.

The simplest seed treatment plants consist of two permanent magnets, which create a periodic field (Floreza, Carbonella, & Martinez et al., 2007). In some installations, magnetic modules with variable polarity are installed above the conveyor belt (Sidortsov, 2007). Passing through a magnetic field, the seeds carry a stimulatory effect, manifested in increasing the permeability of the membrane, regrouping the ions in the cells, the appearance of an electric field as a result of the displacement of the liquid and particles in the magnetic field, etc. (Kolin, Sergeyev, & Gorbatsevich, 2008). In other plants, seed treatment is carried out in a magnetic module, which is a cylinder inside which cylindrical magnets are fixed with bolts and dielectric inserts. +

Plants with electromagnets provide magnetic induction control, but they are more expensive than permanent magnets. Therefore, installations with electromagnets did not find application for magnetic treatmen of seeds and aqueous solutions.

Experimental studies on the treatment of seeds and aqueous solutions have been carried out at installations providing a small value of magnetic field induction (up to 30 mTl). These studies have proven a positive effect of the magnetic field on the seed quality. (Martinez, Pacheco, Aguilar, Pardo, & Ortiz, 2014; Grigor'yeva, 2014; Amaya et al., 1996), biometrics and yield (Sunita, Lokesh, Kadur, & Guruprasad, 2017; Massimo E. Maffei, 2014), storage of agricultural crops (Lysakov, & Ivanov, 2014), a decrease in the incidence of plants (Nizharadze, 2013) and Improving the quality of plant products (Purygin, Vasil'yeva, Purygin, Sovetkin, & Tsaplev, 2015; Ramalingam, & Radhakrishnan, 2018).

Long-term field tests have shown that pre-sowing treatment in the magnetic field of barley seeds, wheat, maize, soybeans, tomatoes, peppers, cucumbers increases the yield by 10-20%.

Water treated in a magnetic field is successfully used for soaking seeds, watering plants, soil soils (Malkin, Furtat, Zhuravska, & Usachov, 2014; Amaya et al., 1996).

Seeds soaking and plants watering with magnetically activated water provided for increase of yield by 10-20% of peas, mustard, cabbage, maize, onions, alfalfa, carrots, cucumbers, sunflower, millet, wheat, radish, rye, rice, lettuce, beets, soy, tomato, beans, barley and potatoes (Mahmood, & Usman, 2014; Teixeira-da-Silva, & Dobránszki, 2014).

Plants watering with magnetically activated water helps to convert nitrogen, phosphorus, potassium into the form consumed by plants, therefore the content of these substances in plants increases by 10-15%. Such water has fungicidal properties, suppressing the process of spore formation of phytopathogenic fungi (Klassen, 1982).

However, the lack of explanation of the mechanism of magnetic field impact on the seeds and water solutions does not make it possible to establish all the regime parameters of the treatment and determine their optimal values, which restrains the introduction of this technology in production.

Attempts to give an exhaustive explanation of the process of physical influence of a magnetic field on biological systems have not led to the creation of a generally accepted theory, although some authors have suggested that water plays a significant role in the mechanism of this influence (Kislovskiy, 1982; Kiva, & Khodurskyi, 2010).

MAIN FOCUS OF THE CHAPTER

Influence of a Magnetic Field on Aqueous Solutions

Water plays an important role in biological objects of plant origin, since all biochemical processes in plant cells occur in an aquatic environment. Water is used in all chemical reactions, carries nutrients, maintains the elasticity of the organs.

Under the influence of a magnetic field the rate of chemical reactions increases, which increases the concentration of reaction products in an aqueous solution:

$$dC_i = \omega dt \tag{1}$$

C_i – the concentration of a substance, mol/l; ω – chemical reaction rate, mol/(l·s); t – time, s.

The rate of chemical reactions in the magnetic treatment of solutions is determined by the expression (Kozyrskyi, Savchenko, & Sinyavsky, 2018).

$$\omega_m = \omega \exp m(K^2B^2 + 2KBv)N_a \,/\, 2RT, \tag{2}$$

ω – the rate of the chemical reaction without the influence of the magnetic field, mole/(l·s); m – resulted mass of ions, kg; B – magnetic induction, Tl; v – ion movement speed, m/s; K – coefficient depending on the concentration and type of ions, as well as the number of re-magnetization, m/(s·Tl); N_a – number Avogadro, molecules /mole; R – universal gas constant, J/mol·K; T – temperature, K.

As a result of the action of the magnetic field on the solution, the chemical reaction of ion formation is accelerated, which is characterized by the degree of electrolytic dissociation:

$$\alpha = \frac{n}{N} = \frac{\omega t}{C}, \tag{3}$$

n – the number of molecules that splits into ions; N – total number of molecules; ω – reaction rate of the electrolytic dissociation, mol/(l·s); t – reaction time, c; C – total concentration of the solution, mol/l.

Taking into account (2), the expression for changing the degree of electrolytic dissociation during the magnetic treatment of the solution has the form:

$$\alpha_m = \alpha e^{\frac{m\left(K_i^2B^2 + 2K_iBv\right)}{2RT}} \tag{4}$$

α_m and α – the degree of electrolytic dissociation after and before treatment in a magnetic field.

The change in the coefficient of electrolytic dissociation can be determined experimentally by changing the specific electrical conductivity of the solution. Under the influence of a magnetic field, due to the growth of the solubility of salts and acids, the specific conductivity of the solution increases:

$$\gamma = \sum_{i=1}^{k} f_i \alpha_i \beta_i \lambda_i^0 C_i e^{\frac{m\left(K_i^2 B^2 + 2K_i Bv\right)}{2RT}} \tag{5}$$

f_i – the coefficient of conductivity; β_i – stoichiometric coefficient of reaction; λ_i^0 – mobility of the ion, Sm·m²/mol.

Experimental studies of the change in the specific conductivity in the treatment of the aqueous solution in a magnetic field were carried out with a solution, which by composition and concentration of substances are similar to the cell solution of plants: ammonium nitric acid – 11,36 g / l; Sodium nitric acid – 0,12 g / l; potassium sulfuric acid – 1, 56 g / l; Calcium Diphosphate – 6.6 g/ l; Magnesium sulfate – 1.05 g / l. The solution was passed through a magnetic field, which was created by permanent magnets, at a speed of 0.4 m/s. The temperature of the solution was 20 °C. The specific electrical conductivity was determined by the laboratory conductometer КЛБ-1М.

It was established that when the magnetic induction is changed from 0 to 0,065 Tl, the specific electrical conductivity of aqueous solutions increases, and with further increase of magnetic induction begins to decrease. Since the treatment of an aqueous solution in a magnetic field increases its specific electrical conductivity, respectively, the solubility of salts and acids improves.

The change in the rate of chemical reaction, as well as the solubility of salts, affects the pH and oxidation-reducing potential of the aqueous medium.

The pH value with calculating (1) is determined in the following way:

$$\Delta pH = \lg C_{H_1^+} - \lg C_{H_2^+} = \lg \omega_{H_1^+} - \lg \omega_{H_2^+}, \tag{6}$$

C_H – the concentration of hydrogen ions, mol/l.

Substituting in equation (6) the expression for the rate of the chemical reaction (2), we obtain:

$$\Delta pH = \frac{mN_a K}{2,3RT}\left(\frac{KB^2}{2} + v_n B\right), \tag{7}$$

or

$$\Delta pH = A_1 B^2 + A_2 Bv, \tag{8}$$

A1 and *A2* – coefficients.

The change in the oxidation-reduction potential (ORP) of the solution is determined by the Nernst equation:

$$\Delta ORP = 2,3\frac{RT}{zF}\left(\lg\omega_2 - \lg\omega_1\right) \tag{9}$$

z – the valence of the ion; F – Faraday number, Cl / mole; ω_1 – the rate of the chemical reaction to the magnetic treatment of the solution, mol / (l·s); ω_2 – the rate of chemical reaction after magnetic treatment, mol/(l·s).

Taking into account (2) we can write:

$$\Delta ORP = -\frac{mN_a K}{zF}\left(\frac{KB^2}{2} + vB\right) \tag{10}$$

or

$$\Delta ORP = A_3B^2 + A_4Bv, \tag{11}$$

$A3$, $A4$ – coefficients.

The coefficients entering into equations (8) and (11) were determined experimentally, since it is impossible to determine them analytically.

Experimental studies of changes in the parameters of aqueous solutions of salts during their treatment in a magnetic field were carried out in a laboratory setup. The solutions were passed through a magnetic field, which was created by permanent magnets from the NdFeB intermetallic composite, installed in parallel above and below the conveyor belt with variable polarity. The pole division was 0.23 m. The temperature of the solutions was 20 °C.

The magnetic induction was controlled by a change in the distance between the magnets and the Teslameter 43205/1 was measured. The speed of movement of the conveyor belt with the Petri dish placed on it with the solution was controlled by changing the speed of the drive motor of the conveyor using a frequency converter

ORP and pH of water and solutions were determined in the magnetic processing of the solution and after it with the help of the И-160M ionometer.

The studies were carried out using the experiment planning method (Adler, Markova, & Granovskiy, 1976; Vasant, Weber, & Dieu, 2016). The value of bottom, main and top levels of the factor were 0; 0.065 and 0.13 Tl for magnetic induction, for the speed of movement of aqueous solutions were 0.4, 0.6 and 0.8 m/s.

It was established that when the magnetic induction is changed from 0 to 0,065 Tl, the pH of water and saline solutions increases, and with further increase of magnetic induction begins to decrease (Figure 1, A). Increasing the speed of the solution reduces the effect of magnetic treatment.

As a result of the multivariate experiment, regression equations were obtained, which in physical quantities have the form:

for water

$$\Delta pH = 3.8B - 0.05Bv - 21.17B^2, \tag{12}$$

Figure 1. Changing the pH of aqueous solutions under the action of a magnetic field: A- water; B - a solution of potassium chloride; C - potassium nitrate; D - calcium nitrate; E - ammonium nitrate; F - magnesium sulfate; G – potassium phosphate; E - potassium sulfate.

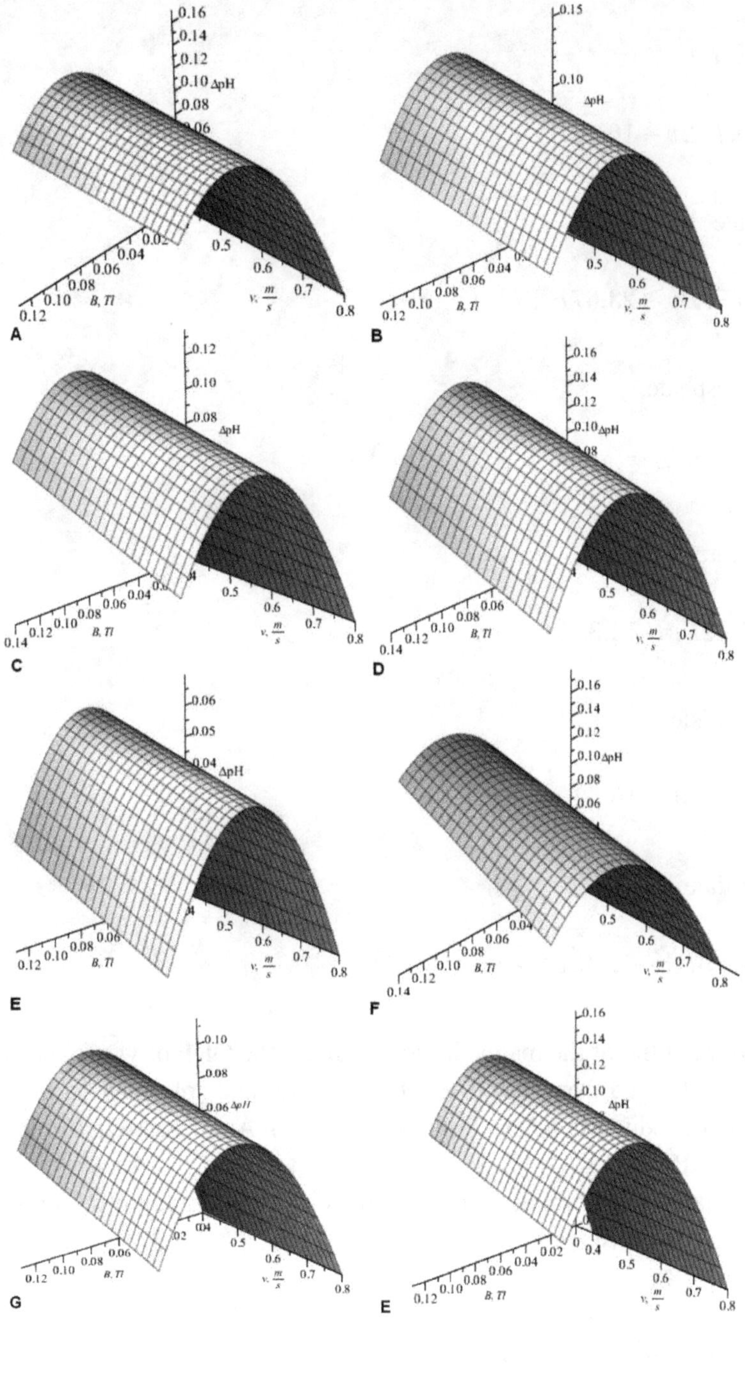

for potassium chloride

$$\Delta pH = 3.92B - 0.39Bv - 23.01B^2; \tag{13}$$

for potassium nitrate

$$\Delta pH = 3.26B - 0.77Bv - 16.17B^2; \tag{14}$$

for calcium nitrate

$$\Delta pH = 4.41B - 0.77Bv - 23.67B^2; \tag{15}$$

for potassium phosphate

$$\Delta pH = 3.1B - 0.64Bv - 17.88B^2; \tag{16}$$

for ammonia nitrate

$$\Delta pH = 3.59B - 1.22Bv - 14.33B^2; \tag{17}$$

for magnesium sulfate

$$\Delta pH = 3.83B - 1.22Bv - 15.52B^2; \tag{18}$$

for potassium sulfate

$$\Delta pH = 4.18B - 0.79Bv - 22.49B^2. \tag{19}$$

Under the influence of the of the magnetic field changes the ORP of water and solutions. When the magnetic induction is changed from 0 to 0.065 Tl, the water and solutions decrease, and with further increase of the magnetic induction begins to grow (Figure 2). An increase in the speed of water causes a smaller change of ORP.

The regression equations describing the change in the ORP of water and solutions in magnetic treatment in physical quantities have the form:

for water

$$\Delta ORP = -276.92B + 38.46Bv + 1538B^2, \tag{20}$$

Figure 2. Changes in ORP of aqueous solutions under the action of a magnetic field: A - water; B - a solution of potassium chloride; C - potassium nitrate; D - magnesium sulfate

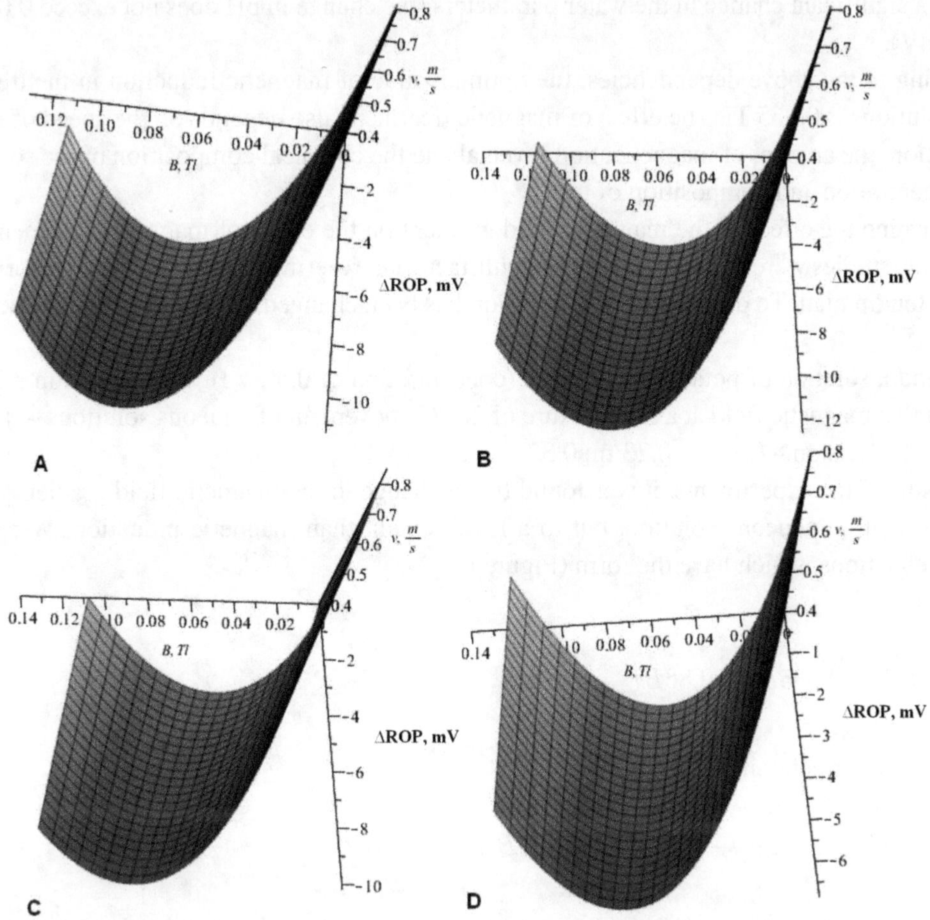

for potassium chloride

$$\Delta ORP = -292.31B + 38.46Bv + 1538B^2, \tag{21}$$

for potassium nitrate

$$\Delta ORP = -236,33B + 70.51Bv + 1065B^2; \tag{22}$$

for magnesium sulfate

$$\Delta ORP = -170.51B + 32.05Bv + 907B^2. \tag{23}$$

An increase in the number of reversal magnetization increases the effect of magnetic water treatment (Figure 3). An optimal one can consider a fourfold reversal magnetization, since its growth does not lead to a significant change in the water parameters (the change in pH does not exceed 0.01, and the ORP is 1 mV).

According to the above dependencies, the optimal value of magnetic induction in the treatment of aqueous solutions is 0.065 Tl. The effect of magnetic treatment also depends on the speed of movement of the solution, the amount of magnetization reversal and the chemical composition of the solution, that is, the concentration and composition of ions.

To determine the effect of the magnetic field gradient on the effect of magnetic treatment of aqueous solutions, studies were carried out using a multifactorial experiment. Studies were performed using the Box – Benkin plan. To do this, the pole division has been changed, that is, the distance between the magnets.

Water and a solution of potassium chloride (concentration of 0.5 g / l) were passed in a laboratory setup through a magnetic field at a temperature of 20 °C, movement of aqueous solutions – 0.4, 0.6 and 0.8 m/s /, pole division – 0.14 m; 0.23 m; 0.32 m

As a result of the experiments, it was found that a change in the magnetic field gradient affects the change in pH of an aqueous solution, but to a lesser extent than magnetic induction. Were obtained regression equations, which have the form (Figure 4):

for water

$$\Delta pH = 3.97B - 0.385Bv - 0.86B\tau - 21.3B^2, \tag{24}$$

for a solution of potassium chloride

$$\Delta pH = 4.044B - 0.385Bv - 0.86B\tau - 22.49B^2. \tag{25}$$

Reducing the pole division contributes to a larger change in the ORP of the aqueous solution (Figure 5). The regression equations for an ORP solution under magnetic treatment have the form:

for water

Figure 3. Dependence of changes in pH (A) and ORP (B) of water during magnetic processing on the number of reversal magnetization.

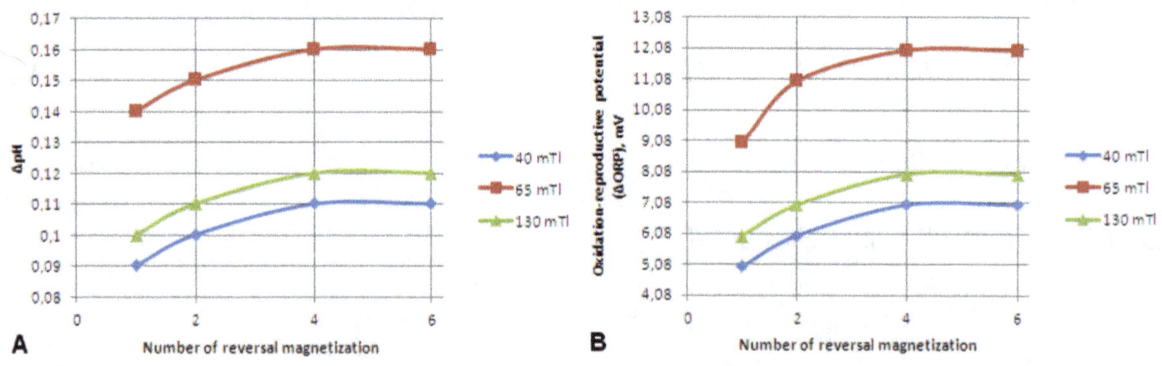

Figure 4. Changing the pH of water with magnetic treatment at pole division A – 0.14 m; B – 0.23 m; C – 0.32 m

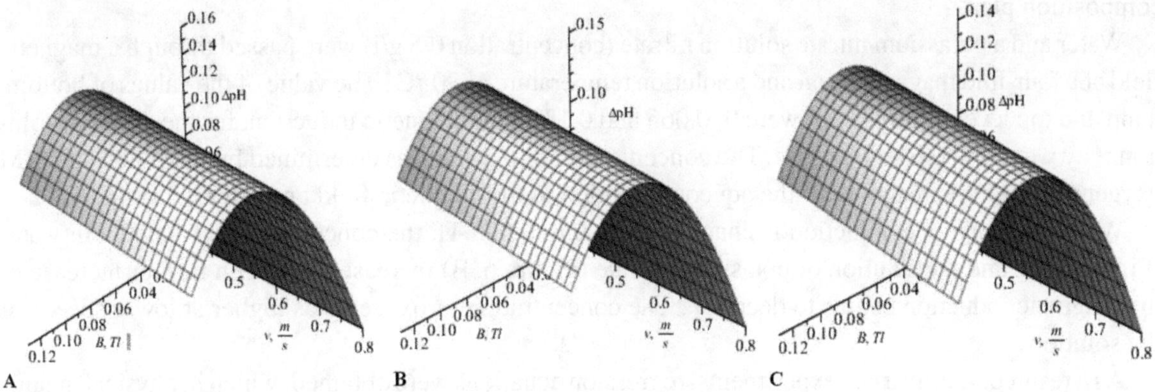

Figure 5. Change in ORP of water with magnetic treatment at pole division A – 0.14 m; B – 0.23 m; C – 0.32 m

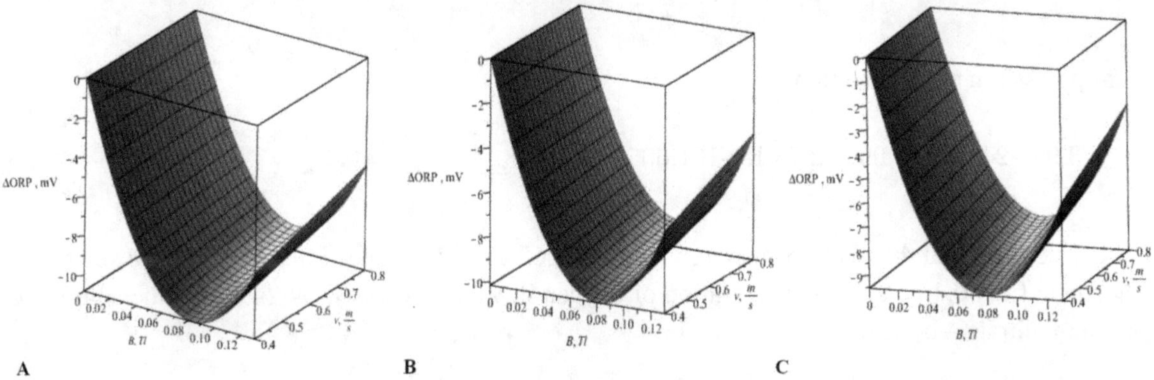

$$\Delta ORP = -285.1B + 38.46Bv + 85.47B\tau + 1538.46B^2, \tag{26}$$

for a solution of potassium chloride

$$\Delta ORP = -296.85B + 38.46Bv + 85.47B\tau + 1538.46B^2; \tag{27}$$

Under the the influence of a magnetic field, the solubility of oxygen in solution increases:

$$C_{CO_2B} = C_{CO_2} e^{K_{1i}^2 B^2 + K_2 Bv}, \tag{28}$$

C_{CO_2B} i C_{CO_2} – the oxygen concentration in the aqueous solution after and before treatment in a magnetic field, mg/l; K_1 and K_2 – coefficients.

Experimental studies of the influence of the magnetic field on the change in the concentration of oxygen in aqueous solutions were carried out by planning the experiment using an orthogonal central composition plan.

Water and a potassium nitrate solution nitrate (concentration 0.5 g/l) were passed through a magnetic field at a four-fold magnetization and a solution temperature of 20 °C . The value of the values of bottom, main and top levels of the factor were 0; 0.065 and 0.13 Tl for magnetic induction, for the speed of solution they were 0.4; 0.6 and 0.8 m/s. The concentration of oxygen was determined by the АЖА-101.1М oxygenmeter to the treatment of the aqueous solution in the magnetic field and after it.

When the magnetic induction is changed from 0 to 0,065 Tl, the concentration of oxygen in water (Figure 6, A) and the solution of potassium nitrate (Figure 6, B) increases, and with further increase of the magnetic induction begins to decrease. The concentration of oxygen was higher at lower speeds of the solution.

As a result of multivariate experiments, regression equations were obtained, which in physical quantities for the concentration of dissolved oxygen have the form:

for water

$$C_{0_2} = 4.1+23.4\text{ B-}0.1\text{ v-}2.37\text{ Bv-}154.37\ B^2; \tag{29}$$

for potassium nitrate solution

$$C_{0_2} = 3.99+24.62\text{ B-}0.09\text{ v-}2.76\text{ Bv-}161.60\ B^2. \tag{30}$$

It was established that with magnetic induction 0.065 Tl and the speed of motion of the solution in a magnetic field 0.4 m /s, the concentration of oxygen in water increased by 24%, and the solution of potassium nitrate – by 25%.

Figure 6. Dependence of dissolved oxygen concentration on magnetic induction and speed of water (A) and potassium nitrate (B) in a magnetic field

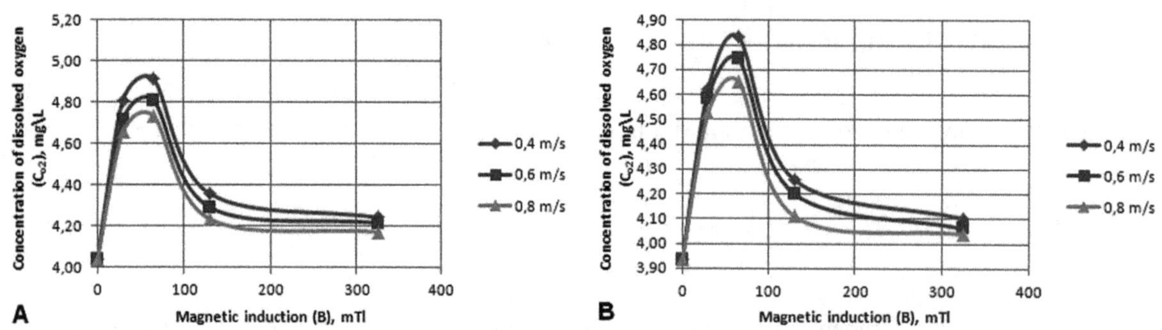

Influence of Magnetic Field on Seeds of Agricultural Crops

Under the influence of a magnetic field, the rate of chemical and biochemical reactions in seeds of crops increases (Kozyrskyi, Savchenko, & Sinyavsky, 2018), which are predominantly redox. As a result, the concentration of the reaction products increases and seed stimulation occurs.

At the same time, the solubility of salts and acids in the plant cell increases.

The change in the rate of chemical and biochemical reactions occurring in the plant cell, as well as the solubility of salts, affects the biopotential of the seeds and the pH of the medium. As for aqueous solutions, the change in pH will be determined by the equation:

$$\Delta pH = A_1 B^2 + A_2 Bv. \tag{31}$$

Under the influence of the magnetic field, the ORP also changes:

$$\Delta ORP = A_3 B^2 + A_4 Bv. \tag{32}$$

For the study of biological objects, A. Saint-Girdier proposed to use the concept of biopotential, which is associated with the ORP by the relation:

$$BP = 820 - ROP. \tag{33}$$

Then the change in the biopotentials in the processing of seeds in a magnetic field

$$\Delta BP = A_5 B^2 + A_6 Bv, \tag{34}$$

A5 and *A6* – coefficients that are determined experimentally.

The magnetic field contributes to accelerating the diffusion of molecules and ions through a cell membrane. The change in the concentration of matter that has passed through diffusion through the membrane is determined by the equation (Kozyrskyi, Savchenko, & Sinyavsky O., 2018):

$$\Delta C = \frac{C_1 - C_2}{2}\left(1 - e^{-\frac{2k_d (a + K_m B/\tau)^2 e^{-\frac{E_a}{kT}}}{L^2} t}\right) \tag{35}$$

k_d – the coefficient c^{-1}, E_a – the activation energy of diffusion, J; k – the Boltzmann constant, J/K.

The growth of the permeability of the membrane and the solubility of oxygen in magnetic processing leads to an increase in its concentration in the cell:

$$\Delta C = \frac{\left(C_{1O_2} - C_{2O_2}\right) e^{K_{1i}^2 B^2 + K_2 Bv}}{2}\left(1 - e^{-\frac{2k_d (a + K_m B/\tau)^2 e^{-\frac{E_a}{kT}}}{L^2} t}\right) \tag{36}$$

The growth of oxygen concentration in cells suppresses the process of spore formation of phytopathogenic fungi and contributes to increasing the productivity of crops.

In the seed magnetic treatment, the transport of water into the cell increases due to the increase in the permeability of biological membranes, which causes the growth of water absorption of seeds (Kozyrskyi, Savchenko, & Sinyavsky, 2018):

$$\Delta m = \frac{C_1 - C_2}{C_1 + C_2} \rho V \left(1 - e^{-\frac{k_d \left(a + K_m B / \tau\right)^2 e^{-\frac{E_a}{kT}}}{L^2} t} \right) \tag{37}$$

ρ – the density of water, kg/m³; V – the volume of water that passed through the membrane, m³.

Under the influence of the magnetic field as a result of the Lorentz force, ion transport through the cell membrane is enhanced, increasing the concentration of mineral substances in the cell involved in chemical reactions (Kozyrskyi, Savchenko, & Sinyavsky, 2018):

$$\Delta C_{i_2} = C_{i_1} f_i l_i \frac{\phi}{L} \left(a + 2K_m B / \tau\right) \left(a + 2K_m B / \tau + K_k K_b B v\right) e^{\frac{m\left(K_i^2 B^2 + 2K_i B v\right)}{2RT}} \tag{38}$$

C_{i1}, C_{i2} – the concentration of substances in plant cells, separated by a membrane, mol/l; l_i – the path of ion motion in a magnetic field, m; φ – diffuse potential, V; L – membrane thickness, m; a – pore size in the membrane, m; K_m, K_b, K_i – coefficients; τ – polar division, m.

From the above analytical expressions (31)–(38), it can be concluded that the main factors in the treatment of seeds in a magnetic field are magnetic induction and its gradient (pole division), as well as the speed of movement of the seeds.

Therefore, pre-sowing seed treatment must be carried out in a non-uniform magnetic field, and the use of a periodic magnetic field enhances the processing effect. The change in the physical and chemical parameters of the seed under magnetic processing depends on the square of the magnetic induction and the speed of its movement in the magnetic field.

Influence of the Magnetic Field on pH, Biopotential and Yield of Bulbous Crops

Experimental studies of the magnetic field influence on the pH and biopotential were carried out with theonion "Luganskiy" and the garlic "Parus". Bulbs have been sent to transport through a magnetic field, which was created by four pairs of permanent magnets, set with variable polarity. The biopotential and pH of the onion and garlic were measured by the И-160 ionometer before treatment in the magnetic field and after it directly in the onion and garlic teal. The biopotential was determined using a platinum measuring electrode, pH - a glass electrode. As an auxiliary chlorine silver electrode was used.

The studies were carried out by the method of planning an experiment using an orthogonal central compositional plan.

Magnetic induction and seed movement speed were taken as factors, and the output value was pH and biopotential. The values of bottom, main and top levels of the factor were 0; 0.065 and 0.13 Tl for magnetic induction, and for the speed they were 0.4; 0.6 and 0.8 m/s.

The experimental dependences of changes in the pH of the onion and garlic on the magnetic induction and the speed of movement of the bulbs in a magnetic field are shown in Figure 7, and for biopotentials – In Figure 8. Due to the abovementioned dependences, with a change in magnetic induction from 0 to 0.065 Tl, the value of pH and biopotentials increases, and with further increase in magnetic induction it begins to decrease. An increase in the speed of movement of the bulb reduces the effect of magnetic treatment, but in the speed range of 0.4–0.8 m/s, it is less significant factor than magnetic induction.

As a result of the conducted studies, the regression equations obtained have the following form:

for a pH onion

$$\Delta pH = 0{,}02 + 8{,}85B - 0{,}04v - 0{,}64Bv - 22{,}22B^2; \tag{39}$$

Figure 7. Dependence of the change of the pH for the onion (A) and garlic (B) on the magnetic induction and the speed of motion in the magnetic field.

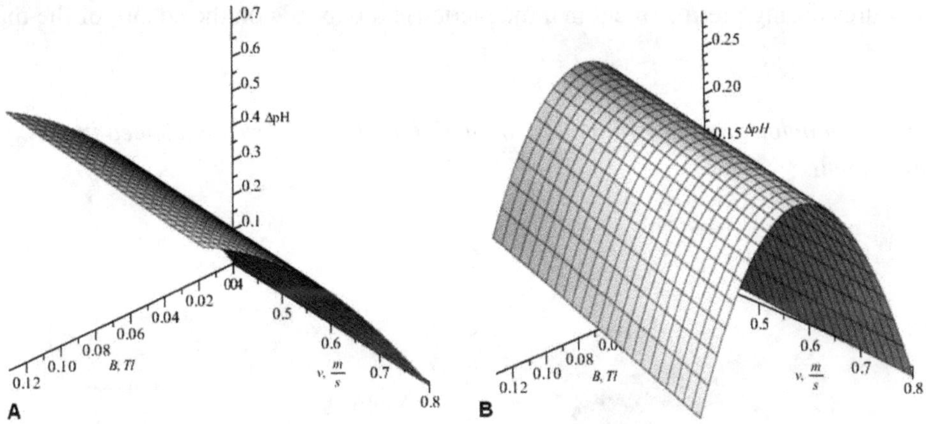

Figure 8. Dependence of the change of the biopotential for the onion (A) and the garlic (B) on the magnetic induction and the speed of motion in the magnetic field.

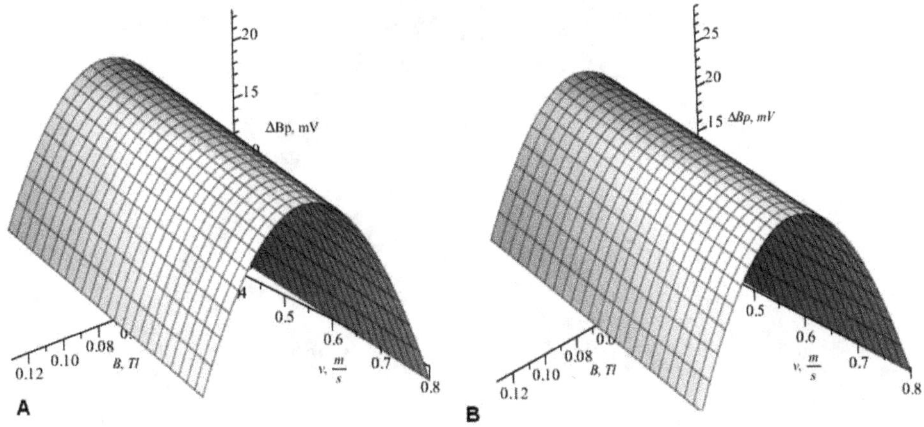

for pH garlic

$$\Delta pH = 0{,}03 + 7{,}93B - 0{,}05v + 1{,}54Bv - 48{,}39B^2; \tag{40}$$

for the biopotential of onion

$$\Delta Bp = 3{,}79 + 598{,}72B - 6{,}25v - 70{,}51Bv - 3800B^2; \tag{41}$$

for the biopotential of garlic

$$\Delta Bp = 1{,}54 + 787{,}19B - 2{,}5v - 128{,}21Bv - 4747B^2. \tag{42}$$

Pre-sowing treatment of onions helps to increase the yield of onions and garlic. Experimental dependence of onion yield on magnetic induction and speed is shown in Figure 9

On the basis of the conducted researches it was established that the change in the pH and biopotential of bulbous cultures during pre-treatment in a magnetic field depends on the square of the magnetic in-

Figure 9. Change of onion yield from magnetic induction and bulb movement speed in a magnetic field during pre-treatment.

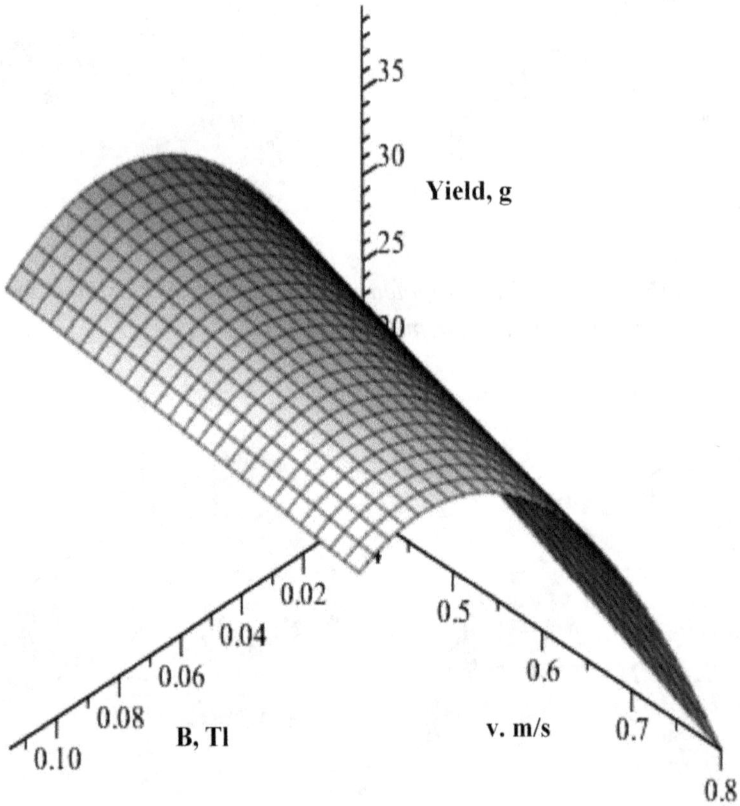

duction and the speed of the bulb movement in the magnetic field. By changing the pH and biopotential of the bulbs it is possible to evaluate the effectiveness of their pre-treatment.

It has been established that the most effective treatment mode takes place with a magnetic induction of 0.065 Tl and the bulbs movement speed of 0.4 m/s. In this pre-sowing treatment mode in a magnetic field the onion yield increased compared with the control by 80%.

The Influence of the Magnetic Field on the Sowing Qualities of Grain Seeds

As a result of the magnetic field, the germination energy and seeds germination increase.

Experimental studies of the magnetic field influence on the germination energy and seeds germination were carried out with wheat seed «Natalka», rye «Kharkiv 98», barley «Solntsedar», maize «Zoria 123», oats «Desnyansky», pea «Adagumsky», bean «Gribovsky», beet «Detroit», zucchini «Biloplodnyi».

The seeds were moved on a conveyor through a magnetic field created by four pairs of permanent magnets installed in parallel above and below the conveyor belt with variable polarity.

Magnetic induction was controlled by changing the distance between the magnets in the range of 0–0.5 Tl and was measured by teslameter 43205/1. The pole division was changed by changing the distance between the magnets. The seed movement speed through the magnetic field was regulated by changing the rotational speed of the conveyor belt driving motor using a frequency converter.

Seeds processed in a magnetic field were germinated and the germination energy and germination were determined (Kozyrskyi, Savchenko, & Sinyavsky, 2018).

The studies were carried out by the experiment planning method. For this, the Boxing-Benkin plan was used. Single-factor experiments show that the values of bottom, main and top levels of the factor were 0; 0.065 and 0.13 Tl for magnetic induction, for the speed of seeds they were 0.4; 0.6 and 0.8 m/s, and for pole division of 0.14; 0.23 and 0.32 m.

With a change in the magnetic induction from 0 to 0.065 Tl, the seed germination energy increases, and with a further increase in the magnetic induction it begins to decrease. With magnetic induction exceeding 0.13 Tl, the germination energy changes insignificantly, but there is more than in the control.

Germination energy is also influenced by the seed speed and pole division (magnetic field gradient), although they are less significant factors than magnetic induction.

According to the results of a multifactorial experiment, the regression equations for the germination energy of oat seeds, which in physical terms has the form:

$$E = 42.6 + 6781B - 2.5v - 9.7\tau - 153.9Bv - 213.7B\tau - 3461.5B^2 \quad (43)$$

When pre-sowing seed treatment in a magnetic field with an ind 0.065 Tl, a pole division of 0.23 m and a seed movement speed of 0.4 m / s, the germination energy of wheat seeds increased by 50%, barley by 42%, rye by 30%, maize by 24%, oat by 26%, pea by 54%, bean by 26%, beet by 40%, zucchini by 26%.

The dependences of the oat seeds germination on magnetic induction and movement speed in a magnetic field are shown in Figure 10. With a change in the magnetic induction from 0 to 0.065 Tl, the seeds germination increases, and with a further increase in the magnetic induction it begins to decrease. With magnetic induction of more than 0.130 Tl, seed germination practically did not change, but was higher than in the control. Seed speed and magnetic field gradient are less significant factors than magnetic induction.

Figure 10. Dependence of oat seeds germination on magnetic induction, pole division and seed movement speed in a magnetic field

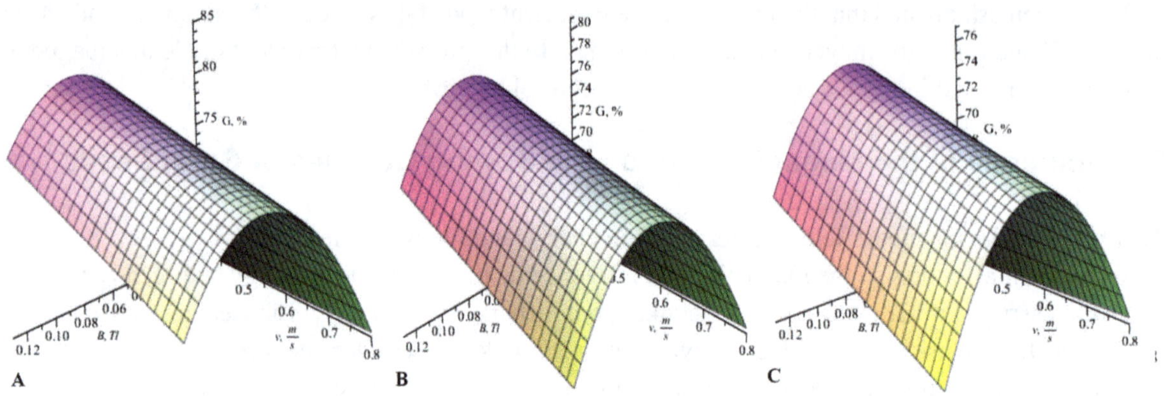

According to the results of a multifactorial experiment, the following equations were obtained for the oat seeds germination in regard to the physical values:

$$G = 64.5 + 636.5B + 1.25v - 16.7\tau - 192\ Bv - 341.9B\tau - 2899.4B^2 \quad (44)$$

When pre-sowing seed treatment in a magnetic field with an ind 0.065 Tl, a pole division of 0.23 m and a seed movement speed of 0.4 m/s, the germination of wheat seeds increased by 22%, barley by 38%, rye by 26%, maize by 18%, oat by 20%, pea by 30%, bean by 26%, beet by 38%, zucchini by 26%.

It was established that the germination energy and seeds germination have a maximum value with a magnetic induction of 0.065 Tl, a polar division of 0.23 m and a seed movement speed of 0.4 m/s.

To determine the duration of the magnetic treatment effect, studies have been carried out on the change in germination energy (Figure 11) and the rye seeds germination (Figure 12) within a month after treatment in a magnetic field. To do this, rye seeds were processed in a magnetic field and the germination energy and seeds germination were determined every seven days.

Figure 11. The change in time of the rye seeds germination energy treated in a magnetic field

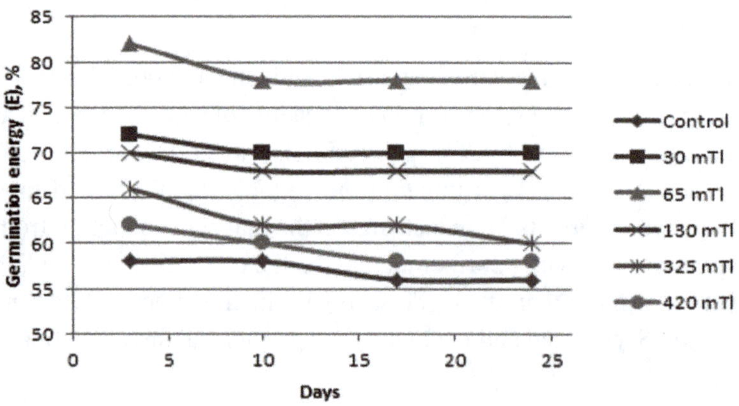

Figure 12. The change in time of the rye seeds germination treated in a magnetic field

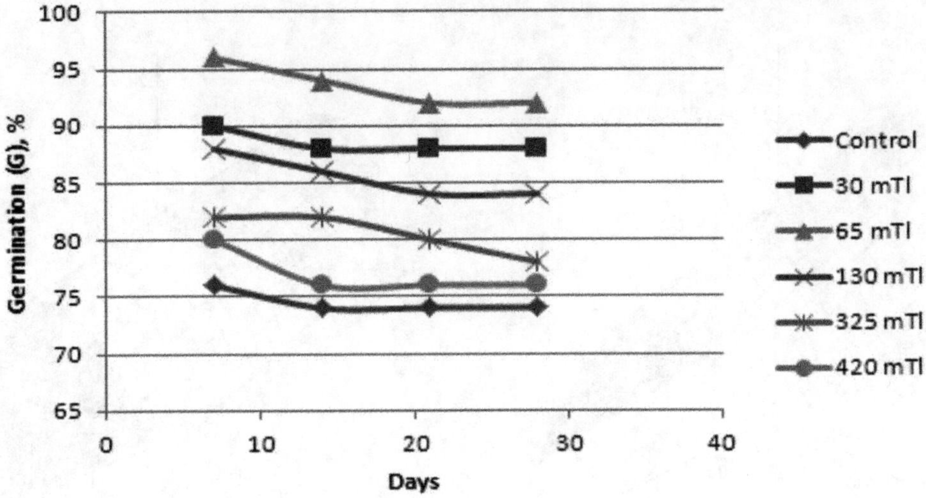

It was established that the effect of seeds magnetic treatment persists for 30 days after treatment.

According to the results of multifactor experiments obtained for the germination energy and rye seeds germination, the regression equation in physical terms is:

$$E = 55,5 + 821B - 3,61v - 153,85Bv - 4536B^2; \qquad (45)$$

$$G = 71,61 + 721,82B - 3,61v - 153,85Bv - 4339B^2; \qquad (46)$$

Under the influence of a magnetic field, water absorption of seeds changes (Figure 13)

The dependences of the specific water absorption of seeds from the processing regime parameters look like:

for wheat seeds

$$Y = 8,9 + 154,53B - 1,03v\text{-}33,33Bv - 844,18B^2; \qquad (47)$$

for barley seeds

$$Y = 9,84 + 298,33B - 0,49v\text{-}75Bv - 1562B^2. \qquad (48)$$

The most specific water absorption seeds happens at a magnetic induction of 0.065 Tl and a seed movement speed of 0.4 m/s. At the same time, the specific water absorption of wheat seeds increased by 6%, barley by 11%.

Figure 13. The change of specific water absorption capacity when wheat seeds (A) and barley seeds (B) are processed in a magnetic field

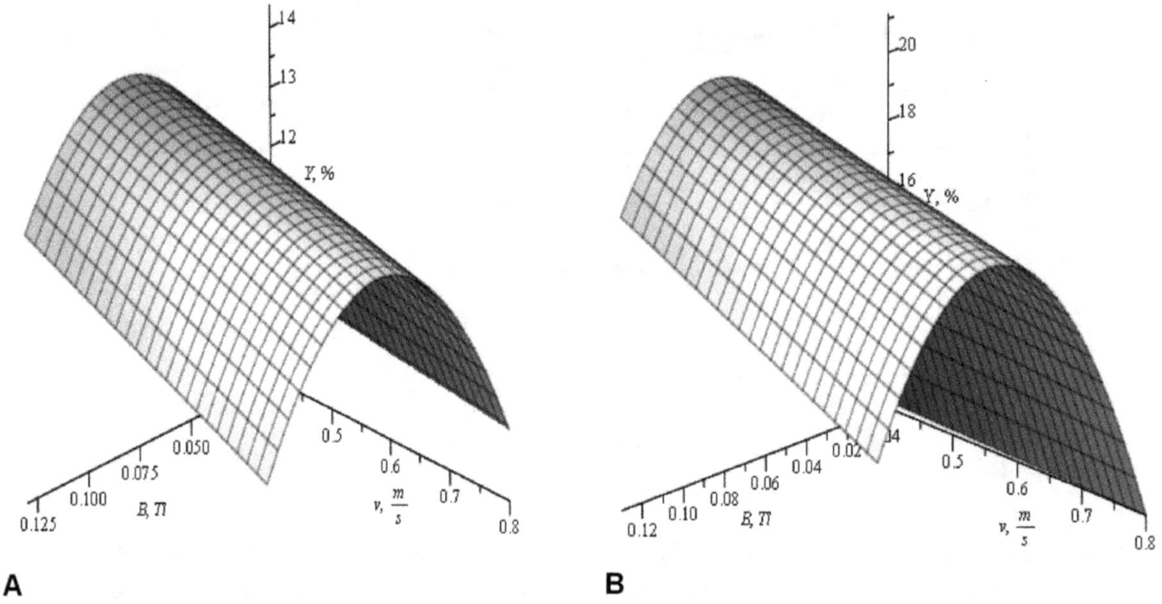

The Influence of a Magnetic Field on Agricultural Crops Biometric Indicators and Productivity

Presowing seeds treatment in a magnetic field has a positive effect on the growth and development of plants.

This improves the biometric parameters of plants. Figure 14 shows zucchini and pea sprouts when treating seeds in a magnetic field with different magnetic induction. The best results were obtained with a magnetic induction of 0.065 Tl and a movement speed of 0.4 m/s.

The pea plants biometric indicators were also the largest in the presowing treatment of pea seeds in a magnetic field with a magnetic induction of 0.065 Tl and a seed speed of 0.4 m/s (Figure 15, A).

Figure 14. Pea sprouts (a) and zucchini (b) seeds treated in a magnetic field with magnetic induction: 1 – control; 2 – 0.03 Tl; 3 – 0.065 Tl; 4 – 0.13 Tl; 5 – 0.325 Tl; 6 – 0.42 Tl

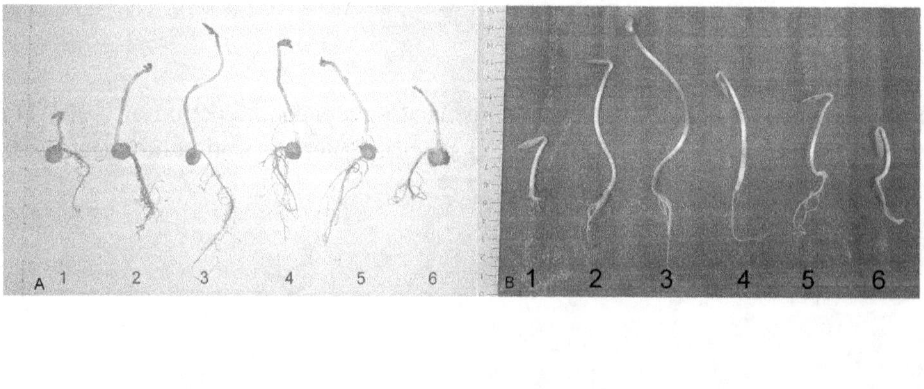

Figure 15. Biometric indicators of plants 7 weeks after planting (A) and the yield of peas (B) after pre-sowing seed treatment in a magnetic field

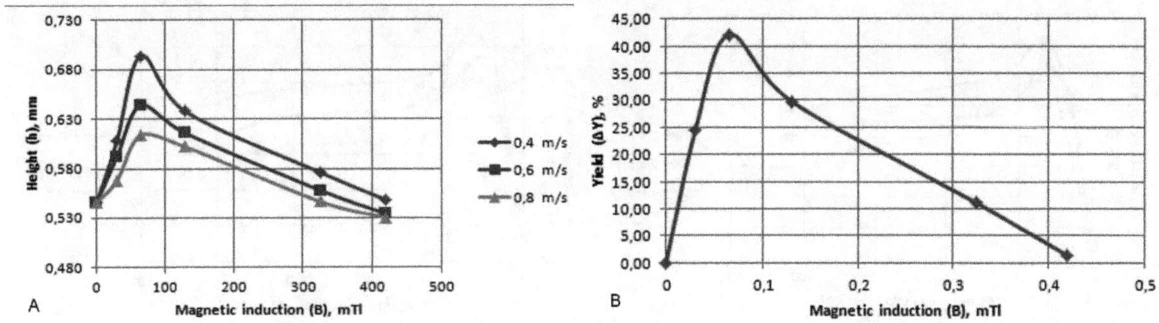

Pre-sowing seeds treatment in a magnetic field helps to increase crop yields. Experimental dependences of pea yield on magnetic induction at a movement speed in a magnetic field of 0.4 m/s are shown in Figure 15, B. The highest yield of agricultural crops was with a magnetic induction of 0.065 Tl.

With the pre-sowing seeds treatment in a magnetic field with a magnetic induction of 0.065 Tl and a seed movement speed of 0.4 m/s, the yield of peas in the experimental plots increased by 42%, grain crops by 20-25%.

Soaking Seed Crops in Magnetized Water

On the basis of the conducted research on the influence of the magnetic field on changes in pH, redox potential, electrical conductivity of an aqueous solution and dissolved oxygen concentration in it, it has been established that the most effective magnetic treatment mode of an aqueous solution takes place with a magnetic induction of 0.065 Tl, a fourfold magnetization reversal and a speed of 0.4 m/s.

Soaking seeds in magnetized water improves its sowing qualities. Studies were conducted on oat seeds, treated in a magnetic field. Seeds were soaked in tap water and in water treated in a magnetic field with a magnetic induction of 0.065 Tl and a movement speed of 0.4 m/s.

It has been established that the germination energy of oat seeds treated in a magnetic field and soaked in tap water increased by 24% compared to untreated seeds (control), germination by 20%. For oat seeds treated in a magnetic field and soaked in magnetically activated water, these indicators were higher: germination energy increased by 34% compared to untreated grain (control), germination by 30% (Figure 16).

The Energy Dose of Presowing Treatment of Seeds in Magnetic Field

To calculate the energy efficiency of seed treatment in magnetic field we need to calculate the energy dose of the treatment.

The energy dose of the treatment can be calculated by the following formula:

$$D = \int \frac{W}{m} dt, \tag{49}$$

Figure 16. Dependence of oat seeds germination energy (a) and germination (b), soaked in magnetically activated water, from magnetic induction and seed movement speed in a magnetic field

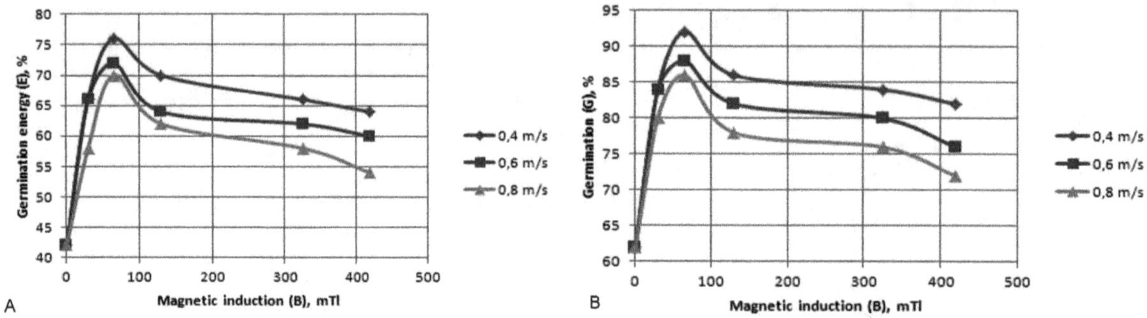

W – energy of magnetic field, J; t – time of the treatment, s; m – the mass of seeds, kg; or

$$D = \int \frac{B^2 dt}{2\mu\mu_0\rho} \tag{50}$$

B – magnetic induction, Tl; μ – relative magnetic penetration; μ_0 – magnetic constant, H/m; ρ – seed density, kg/m^3.

When we substitute dt by dl:

$$dt = \frac{dl}{v}, \tag{51}$$

l – the path which seeds take in magnetic field, m.

Then we receive the energy dose of the treatment:

$$D = \int \frac{B^2 dl}{2\mu\mu_0\rho v} \tag{52}$$

We need to figure out the change of magnetic induction in an air gap to calculate the energy dose of the treatment.

Experimental studies of the device for magnetic treatment of potatoes were conducted by measuring magnetic induction with the help of teslameter 43205 in the centre of an air gap along the axis of the conveyer.

The received experimental dependence of the change of magnetic induction in the centre of an air gap along the axis of the conveyer (Figure 17) enables us to calculate the energy dose of seed treatment in magnetic field.

We used the trapezoidal rule to approximate integral (52). We approximate linear functions to some areas of the dependence which is presented in Figure 17. We did F-test to check the efficiency of this approximation.

Figure 17. The dependence of the change of magnetic induction in the centre of an air gap along the axis of the conveyer

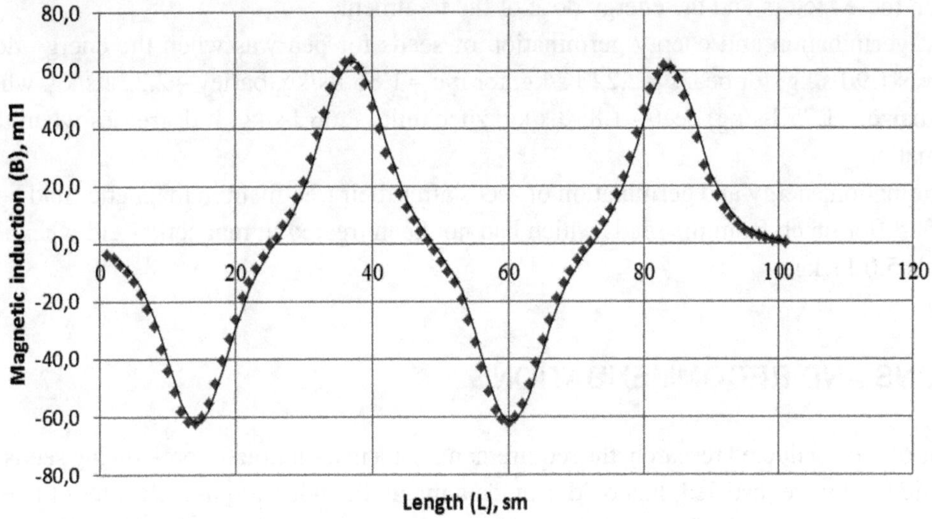

$$\int_0^L B^2 dl = \int_0^{l/8}\left(-\frac{8B_m}{L}l\right)^2 dl + \int_{L/8}^{3L/8}\left(-2B_m + \frac{8B_m}{L}l\right)^2 dl + \int_{3L/8}^{5L/8}\left(4B_m - \frac{8B_m}{L}l\right)^2 dl +$$
$$+\int_{5L/8}^{7L/8}\left(-6B_m + \frac{8B_m}{L}l\right)^2 dl + \int_{L/8}^{L}\left(8B_m - \frac{8B_m}{L}l\right)^2 dl, \qquad (53)$$

B_m – the value of magnetic induction in the plane of installed magnets, Tl; L – the distance of seeds in magnetic field, m.

We receive the following formula, after we approximated the value of integrals which are included in equation (52):

$$\int_0^L B^2 dl = \frac{B_m^2 L}{24} + \frac{B_m^2 L}{12} + \frac{B_m^2 L}{12} + \frac{B_m^2 L}{12} + \frac{B_m^2 L}{24} = \frac{B_m^2 L}{3} \qquad (54)$$

Then the value of the energy dose of the treatment is

$$D = \frac{B_m^2 L}{6\mu\mu_0 \rho v} \qquad (55)$$

We can use formula (54) to calculate the energy dose of seeds treatment in magnetic field when we know magnetic induction and the speed of seeds.

Experimental dependences of the germination energy and germination of seeds of agricultural crops on magnetic induction and the speed of seeds in magnetic field enable us to figure out the interconnection between these factors and the energy dose of the treatment.

The best germination and energy germination of seeds for pea was when the energy dose of the treatment was 1.9 J·s/kg; for beans – 2.22 J·s/kg, for rye – 1.86 J·s/kg, barley – 2.22 J·s/kg, wheat – 1.73 J·s/kg, for maize – 1.75 J·s/kg, beet – 1.85 J·s/kg, zucchini – 2,66 J·s/kg. it decreases when the dose is larger or smaller.

The germination energy and germination of seeds after their treatment in magnetic field before sowing does not differ much from the seeds which had not been treated in magnetic field when the energy dose exceeds 5.0 J·s/kg.

SOLUTIONS AND RECOMMENDATIONS

On the basis of the conducted research, the requirements for a installation for pre-sowing seeds treatment in a magnetic field were justified. It should provide a magnetic induction in the center of the air gap of 0.065 Tl, a periodic magnetic field with fourfold reversal magnetization and polar division of 0.23 m and a seed movement speed of 0.4 m/s.

The installation has been developed (Figure 18) and consists of a floor conveyor with an electric drive, a device for magnetic treatment, where the seed treatment itself happens, a loading conveyor and a control panel take place.

In an installation for magnetic treatment of agricultural crops seeds, four pairs of magnets with an intermetallic composite NdFeB are used, which are installed in parallel above and below the conveyor belt with variable polarity (Figure. 18). Magnets are glued to a steel plate. The gaps between the magnets are filled with textolite. From the front sides of the plate are sheathed in stainless steel.

Figure 18. Installation for magnetic seed treatment: a - general view; b - functional diagram: 1 - loading conveyor; 2 - conveyor installation for pre-sowing seeds treatment in a magnetic field; 3 - textolite inserts; 4 - permanent magnets; 5 - steel plate; 6- processing object; 7 - container

A

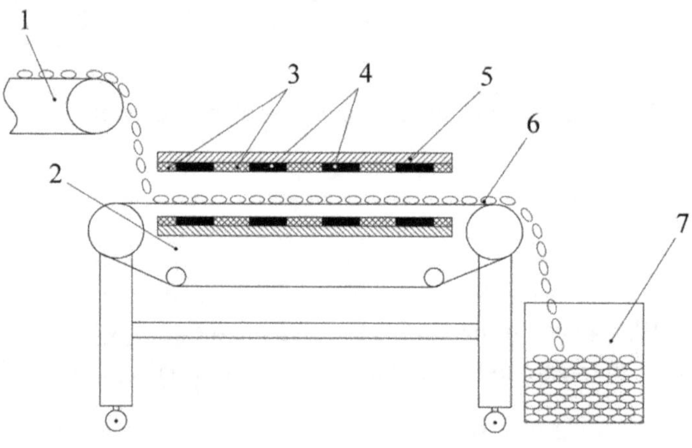

B

The conveyor drive is carried out from a three-phase asynchronous electric motor with a power of 0.25 kW through a reduction gear.

Using the developed simulation ELCUT-model, the analysis of the magnetic seed treatment paramentres has been performed. As a result, it has been found that a magnetic induction of 0.065 Tl in the center of the air gap is provided by using four pairs of permanent magnets from N38SH intermetallic composite NdFeB, which are placed on a plate of Ст3 steel with a thickness of 10 mm. The pole division is 230 mm, the air gap is 90 mm.

Experimental studies of the device for magnetic treatment of seeds were carried out by measuring the magnetic induction teslameter 43205 at different points in the air gap (Figure 19).

On the basis of the conducted studies of changes in the magnetic induction in the air gap, it can be concluded that the deviation of the magnetic induction in the working zone from the specified value does not exceed 5%.

Magnetic induction changes as seeds move through a magnetic field. It reaches its greatest value in the plane of the installation of permanent magnets. When seeds move from one magnet to another, induction comes, approaching zero in the middle between the magnets, and then, changing the sign, begins to grow (Figure 17).

FUTURE RESEARCH DIRECTIONS

Pre-sowing seeds treatment of grains, leguminous plants and vegetable crops in a magnetic field has shown its effectiveness. Conducted research studies with the seeds of forest crops have shown the promising application of this technology. One of the promising areas of research is the treatment of seeds and flower bulbs in a magnetic field.

Research should be aimed at determining the effective factors in treating the seeds of forest and flower crops and determining the most effective treatment regime.

When developing plants for pre-sowing seeds treatment in a magnetic field, it is necessary to conduct research aimed at optimizing the characteristics of the magnetic field created by permanent magnets, which will reduce their cost.

CONCLUSION

1. When treating seeds in a magnetic field, the rate of chemical and biochemical reactions occurring in plant cells changes, resulting in the stimulation of growth and development of plants.
2. Stimulation of seeds is related to the effect of the magnetic field on aqueous solutions that are located in a biological object. When treating aqueous solutions in a magnetic field, the solubility of salts and acids improves and the oxygen concentration increases. An increase in the concentration of oxygen suppresses the process of spore formation of phytopathogenic fungi and reduces the incidence of agricultural plants.
3. The change in the rate of chemical reactions in the magnetic treatment of solutions, as well as the solubility of salts and acids causes a change in the pH and oxidation-reduction potential of the aqueous medium. When the magnetic induction is changed from 0 to 0.065 Tl, pH, specific electrical conductivity and oxygen concentration increases, and then begins to decrease. The processing

Figure 19. Change of magnetic induction in the center of the air gap at the place of installation of the magnets in the vertical plane (A) and across the conveyor belt (B, C) with different polarities of the magnets: B – N-S; C – S-N.

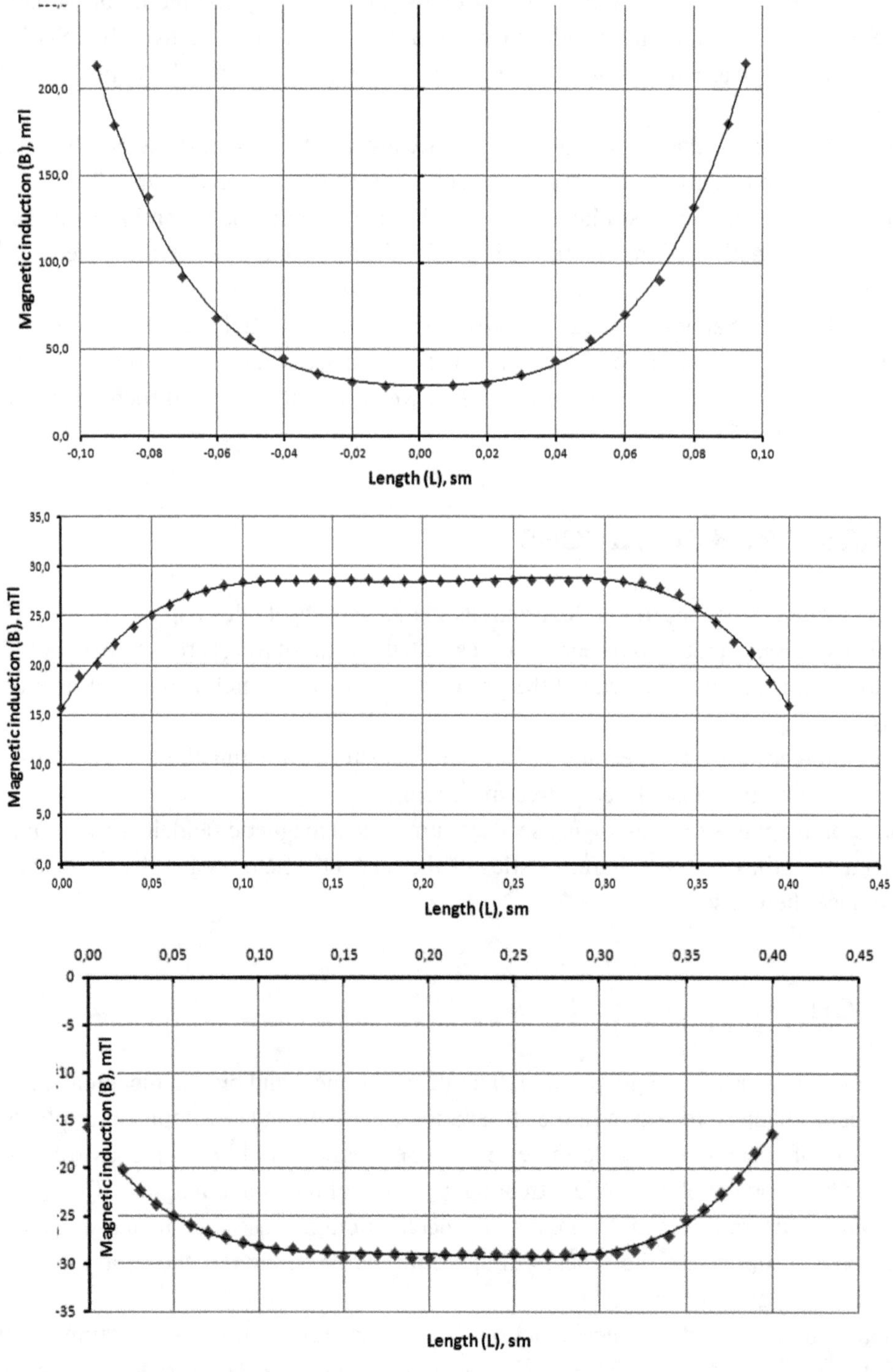

effect also depends on the gradient of the magnetic field (pole division), the speed of the solution and its chemical composition. Soaking seeds in magnetically activated water helps to increase the crop quality of seeds and yields of agricultural crops.

4. Under the influence of the magnetic field on the cell membrane, their permeability increases, which causes the growth of diffusion through the cell membrane of molecules and ions. This increases water absorption of seeds, oxygen solubility and accelerates the diffusion of its molecules through a cell membrane. By the action of Lorentz, the transport of ions accelerates and the concentration of mineral elements entering the cell increases.
5. It was established that the main factors in the seed treatment in a magnetic field are the magnetic induction and its gradient, as well as the speed of the seed in a magnetic field. The effect of seed treatment depends on the square of the magnetic induction and the speed of the seed in a magnetic field.
6. Under the influence of the magnetic field, the quality of the seed is improved. The greatest values of germination energy and germination were obtained at a magnetic induction of 0.065 Tl, a fourfold reversal magnetization, a pole division of 0.23 m, and a seed speed of 0.4 m/s. Under this treatment, the seed germination energy increased compared with the control by 26-50%, and the germination - by 20-30%. With magnetic induction exceeding 0.13 Tl, the germination energy and the germination of the seeds vary insignificantly, but are larger compared to the control.
7. The relationship between the energy dose of seed treatment in a magnetic field and the germination energy and the germination of the seeds are established. The optimal mode of pre-sowing seed treatment in a magnetic field takes place at an energy dose of treatment of 1.7-2.7 J·s/kg.
8. According to the results of studies on changes in the sowing quality of seeds, its water absorption and biopotential during pre-sowing treatment in a magnetic field, it has been established that the most effective treatment is taking place at a magnetic induction of 0.065 Tl, a pole division of 0.23 m, fourfold reversal magnetization and seed speed of 0, 4 m/s. Under such a treatment of seeds in a magnetic field, grain yield increased by 20-25%, peas - by 40%, onions - by 80%.
9. Pre-sowing treatment of grain crops in a magnetic field provides a net profit of $ 70100 per 100 hectares. The installation for magnetic treatment of seeds pays off in the first year of operation, the profitability index is 2.6.

ACKNOWLEDGMENT

The authors are deeply grateful to the editors of the book for the Prof. Valeriy Kharchenko and Pandian Vasant and the reviewers for valuable comments that were taken into account in the preparation and finalization of the chapter.

REFERENCES

Adler, Yu. P., Markova, E. V., & Granovskiy, Yu. V. (1976). Planirovaniye eksperimenta pri poiske optimal'nykh usloviy [Planning an experiment when searching for optimal conditions]. Moskow: Nauka.

Amaya, J. (1996). Effects of stationary magnetic fields on germination and growth of seeds. *Hortic. Abst*, *68*, 1363.

Flórez, M., Carbonell, M. V., & Martínez, E. (2007). Exposure of maize seeds to stationary magnetic fields: Effects on germination and early growth. *Environmental and Experimental Botany*, *59*(1), 68–75.

Grigor'yeva, O. (2014). Sposoby podgotovki semyan k posevu [Methods for preparing of seeds for sowing]. *LesProm*, *6*(104), 176–177.

Hyland, G. J. (2003). Bio-Electromagnetism. In F.A. Popp, & L.V. Beloussov (Eds.), Integrative Biophysics-Biophotonics (pp. 117-148). Dordrecht: Kluwer Academic Publisher.

Kataria, S., Baghel, L., & Guruprasad, K. N. (2017). Pre-treatment of seeds with static magnetic field improves germination and early growth characteristics under salt stress in maize and soybean. *Biocatalysis and Agricultural Biotechnology*, *10*, 83–90. doi:10.1016/j.bcab.2017.02.010

Kislovskiy, L. D. (1982). Rol' vody v labil'nosti poverkhnostnykh struktur [The role of water in the lability of surface structures]. Moskow: VINITI.

Kiva, O. V., Khodurskyi, V. Y. (2010). Doslidzhennia ta rozrobka prystroiu dlia peredposivnoi obrobky nasinnia tsukrovoho buriaku [Research and development of the device for pre-sowing processing of seeds of sugar beet]. *Visnyk Poltavskoi derzhavnoi ahrarnoi akademii*, 4, 176 – 178.

Klassen, V. I. (1982). Omagnichivaniye vodnykh system [Magnetization of water systems]. Moskow: Khimiya.

Kolin, A. R., Sergeyev, V. V., & Gorbatsevich, N. A. (2008). Vozdeystviye gradiyentnym magnitnym polem na posadochnyy material i vegetiruyushchiye kartofel'nyye rasteniya [Effect of a gradient magnetic field on planting material and vegetative potato plants]. Russkiye vysokiye tekhnologii. Retrieved from http://skutis.ucoz.ru/publ/26-1-0-13

Kozyrskyi, V., Savchenko, V., & Sinyavsky, O. (2018). Presowing Processing of Seeds in Magnetic Field. In Handbook of Research on Renewable Energy and Electric Resources for Sustainable Rural Development (pp. 576–620). Hershey, PA: IGI Global. doi:10.4018/978-1-5225-3867-7.ch024

Lobyshev, V. I. (2005). Water is a sensor to weak forces including electromagnetic fields of low intensity. *Electromagnetic Biology and Medicine*, *24*(3), 449–461. doi:10.1080/15368370500382248

Lysakov, A. A., Ivanov, R. V. (2014). Vliyaniye magnitnogo polya na sokhrannost' kartofelya [The influence of the magnetic field on the conservation of potatoes]. *Uspekhi sovremennogo estestvoznaniya*, *8*, 103-106.

Maffei, M. E. (2014). Magnetic field effects on plant growth, development, and evolution. *Frontiers in Plant Science*, *5*, 445. doi:10.3389/fpls.2014.00445 PMID:25237317

Mahmood, S., & Usman, M. (2014). Consequences of Magnetized Water Application on Maize Seed Emergence in Sand Culture. *Journal of Agricultural Science and Technology*, *16*(1), 47–55.

Malkin, Y.S., Furtat, I.Ye., Zhuravska, N.Y., Usachov, V.P. (2014). Perspektyvy stvorennia resursozberihaiuchykh tekhnolohii shliakhom mahnitnoi obrobky vody ta vodnykh rozchyniv [Prospects for resource-saving technologies through magnetic treatment of water and water solutions]. *Ventyliatsiia, osvitlennia ta vodopostachannia*, 17, 120-127.

Malkin, Y.S., Zhuravska, N.Y., Kovalenko, N.O. (2015). Protses obrobky vody v mahnitnykh poliakh [The process of water treatment in magnetic fields]. *Ventyliatsiia, osvitlennia ta vodopostachannia*, *18*, 70-74.

Martinez, E., Florez, M., & Carbonell, M. V. (2017). Stimulatory Effect of the Magnetic Treatment on the Germination of Cereal Seeds. *International Journal of Environment Agricultural Biotechnology*, *2*(1), 375–381.

Martinez, F. R., Pacheco, A. D., Aguilar, C. H., Pardo, G. P., & Ortiz, E. M. (2014). Effects of magnetic field irradiation on broccoli seed with accelerated aging. *Acta Agrophysica*, *21*(1), 63–73.

Nizharadze, T. S. (2013). Vliyaniye ekologicheskikh priyemov predposevnykh obrabotok semyan yachmenya na porazhennost' listosteblevymi boleznyami [The impact of environmental practices of presowing treatment of barley seeds on the incidence of leaf stem diseases]. *Izvestiya Orenburgskogo gosudarstvennogo agrarnogo universiteta*, *6*(44), 56-58.

Podleoeny, J., Pietruszewski, V., & Podleoena, A. (2004). Efficiency of the magnetic treatment of broad bean seeds cultivated under experimental plot conditions. *International Agrophysics*, *18*, 65–71.

Purygin, P. P., Vasil'yeva, T. I., Purygin, V. A., Sovetkin, D. A., Tsaplev, D. A. (2015). Vliyaniye predposevnoy obrabotki semyan l'na na rost i biokhimicheskiye pokazateli prorostkov [Effect of presowing treatment of flax seeds on the growth and biochemical parameters of seedlings]. *Estestvennonauchnaya seriya*, *10*(132), 166–173.

Ramalingam, R. (2018). Seed pretreatment with magnetic field alters the storage proteins and lipid profiles in harvested soybean seeds. *Physiology and Molecular Biology of Plants*, *24*(2), 343–347. doi:10.100712298-018-0505-8 PMID:29515328

Sidortsov, I. G. (2007). Ustanovka dlya predposevnoy obrabotki semyan [Plant for presowing seed treatment]. *Tekhnika v sel'skom khozyaystve*, *3*, 61-62.

Teixeira-da-Silva, J., & Dobránszki, J. (2014). Impact of magnetic water on plant growth. *Environmental and Experimental Biology*, *12*, 137–142.

Vasant, P., Weber, G. V., & Dieu, V. N. (2016). Handbook of research on modern optimization algorithms and applications in engineering and economics. Hershey, PA: IGI Global. doi:10.4018/978-1-4666-9644-0

Yagodin, B. A., Zhukov, Y. P., & Kobzarenko, V. I. (2002). Agrokhimiya [Agrochemistry]. Moskow: Kolos.

KEY TERMS AND DEFINITIONS

Biopotential: The difference of potentials between two points of tissue which reflects its bioelectric activity.

Diffusion: The process of mutual penetration of molecules or atoms of one substance among other atoms or molecules, which usually leads to equilibrium of their concentrations in the occupied volume.

Germination: The ability to form well developed seed sprouts.

Germination Energy: Seed capacity to germinate fast and together.

Magnetic Field: The component of electromagnetic field, due to which the interaction occurs between moving electrically charged particles.

Magnetic Induction: A vector physical value, the main characteristic feature of the magnitude and direction of the magnetic field.

Oxidation-Reduction Potential (redox): is a measure of the chemical activity of elements or their compounds in reversible chemical processes associated with a change in the charge of ions in solutions

pH (Hydrogen): The value which shows the extent of the activity of hydrogen ions ($H+$) in a solution.

Rate of Chemical Reactions: The change of the concentration of reactants or a reaction product per unit time

Chapter 10
Mobile Measuring Complex for Conducting an Electric Network Survey

Alexander Vinogradov
Federal Scientific Agroengineering Centre VIM, Russia

Vadim Bolshev
https://orcid.org/0000-0002-5787-8581
Federal Scientific Agroengineering Centre VIM, Russia

Alina Vinogradova
Federal Scientific Agroengineering Centre VIM, Russia

Maxim Borodin
Orel State Agrarian University, Russia

Alexey Bukreev
Orel State Agrarian University, Russia

Igor Golikov
Orel State Agrarian University, Russia

ABSTRACT

An energy audit of the electrical network is required in the process of constructing new electrical networks as well as in justifying the reconstruction need of existing ones. In this chapter, the structure of a mobile measuring complex has been developed to conduct an electrical network survey without disconnecting consumers. The complex can be used to inspect 0.4 kV electrical networks and microgrids of the same voltage class and allows data collection on voltage losses and electric power losses in network elements such as a power lines (electric transmission line), and power transformers. The energy audit is conducted without disconnecting consumers in order to avoid an undersupply of electricity as well as to determine the real operating modes of power networks. Ultimately, the use of the developed measuring complex will increase the reliability of the power supply to consumers and ensure the required quality of the electricity supplied to them.

DOI: 10.4018/978-1-5225-9420-8.ch010

INTRODUCTION

An energy audit of the electrical network is required in the process of construction of new electrical networks as well as in justifying the reconstruction need of existing ones. At the same time, it is necessary to obtain data on voltage losses and electric energy losses in network elements such as electric transmission lines (ETL) and power transformers. It is desirable to obtain such a survey without disconnecting consumers in order to studying the real network operating modes as well as to avoid undersupply of energy to them. Of particular relevance is the examination of microgrids containing renewable energy sources (RES) and distributed generation (DG) (Kharchenko et al., 2019). In this case, it is important to understand how the parameters of the network operating modes change at different loads and operation modes of energy sources. This will allow to correctly set up the generating equipment and to choose equipment for the storage of electricity, protective devices. Thus, the development of a mobile measuring complex is required, which allows for the examination of microgrids and networks in various operating modes without disconnecting consumers.

The issue of electrical network surveys is also closely related to the issues of ensuring the quality of the supplied electrical energy. Issues of improving power quality (PQ) as well as problems associated with PQ inconsistency time with regulation requirements have been repeatedly considered in the works of Sudnova et al. (2007), ZHelezko (2002), Kartashev (2001), Vinogradov et al. (2018), Vinogradov et al. (2019a) and Vinogradov et al. (2019b). However, they did not propose solutions for the creation of mobile measuring complexes (MIC) for energy audit of electrical networks.

Taking into account the fact that the length of electric networks of voltage class of 0.4 (0.38) kV across Russia is more than 770 thousand kilometers (Standards PJSC Rosseti, 2017), it is not possible to equip all power lines of this voltage with systems of continuous monitoring of operating modes in a short time. This is also the rationale for the development of the MIC since a MIC allows for surveys of a significant number of electrical networks. In addition, a large number of 10/0.4 of kV power transformers are operated in distribution electrical networks, many of which have already worked out the standard service life and have excessive no-load losses due to repairs and a long service life (Vinogradov et al., 2015). Periodic assessment of the level of no-load losses (steel losses) of these transformers will make it possible to make timely decisions on their replacement in case of excessive losses.

An analysis was made of existing devices and technical solutions that allow energy audit in electrical networks. The analysis showed that the technical level of the existing energy audit means does not allow the automatic collection of electrical network parameters at various locations. In addition, the existing solutions of these means are not mobile enough and are not able, in particular, to analyze the estimate of electric power losses in 10/0.4 kV power transformers without disconnecting from the load.

Automation of measuring systems allows to minimize the labor costs related to conducting surveys of electrical networks and significantly simplify the energy auditor work while improving the measurement accuracy. It becomes also possible to obtain a kind of passport of the examined electrical network due to the reading from the studied electrical network in real time. Technical solutions for creating a mobile measuring complex for inspecting electrical networks should be able to process data collected from network control points not only separately, but also as general interconnected information. Existing devices analyzing the power quality do not fully possess the necessary properties since they are used for other purposes.

The aim of the work is to develop a mobile measuring complex (MIC) for conducting an electrical network survey without disconnecting consumers, which can be used to inspect 0.4 kV electrical networks and microgrids of the same voltage class.

MIC PROJECT

The main tasks to be solved with the help of the developed MIC should be:

- Measurement and analysis of load graphs of the network and consumers connected to it
- Measurement and analysis of power losses in the network
- Measurement of the actual voltage value and analysis of voltage loss at different points of the network
- Measurement and analysis of the maximum values of currents in various points of the network, including branch lines consumers connected to
- Assessment of electric power losses in power transformers without disconnecting them from the network and ranking for short-circuit losses, load losses and no-load losses

The solution of these tasks imposes a number of requirements both on the structure of the whole complex and on its constituent elements. Thus, the developed complex should have increased mobility due to its placement on the basis of the vehicle. It should also provide the ability to conduct energy audit of the electrical network without disconnecting consumers. The sensors used in the complex should make it possible to carry out measurements and transfer measured values online to an information processing unit that performs the function of comparing information from sensors, analyzing, visualizing and performing the necessary calculations.

The structure of the complex is shown in Figure 1. The mobile measuring complex for conducting preliminary energy audit of rural electrical networks without disconnecting consumers contains a multifunctional information processing unit (MIPU) based on PC and located on the vehicle chassis, sensors fixing the basic parameters of power quality including voltage and amperage (timer-electric meter) 3. Timer-electric meter uses various technologies for data transmission depending on the specific situation and organizes the channel for the transfer and exchange of information as well as measuring the

Figure 1. Option to build the structure of a mobile measuring complex for energy audit of rural electrical networks without disconnecting consumers

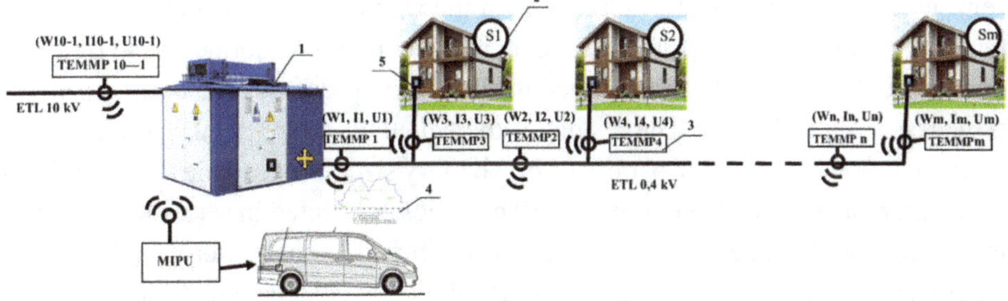

indicators of a power transformer installed at a transformer substation 1 and the consumers of electrical energy 2. Data on voltage, power current and other data, including for constructing load graphs (4), are transmitted via communication channels to the MIPU. The functioning of the complex depends on the version of the sensors installed in the network and transmitting data to the MIPU. There are two main variants of such sensors. The first option allows measuring only the current strength without violating the integrity of the wire insulation and transmit data to the MBOI. It can also estimate the amount of consumed electricity, the time during which the measurements were taken, the time during which the consumption was or not. Due to the specifics of the functions performed by this sensor, it is called TEMMP (timer-electric meter mobile portable). TEMMP can be used both as part of the MIC complex and separately, as it is equipped with an LCD display with basic data output and can transmit measured information via various types of channels, including SMS messages.

Thus, the mobile and portable timer-electric meter is intended for mobile measurement of the operating time of an electrical consumer (or network section) and estimating the amount of consumed electricity without disturbing the insulation of the supply wire. It can be used to solve problems:

- Detecting electricity theft
- Compiling energy balance of objects during an energy audit
- Determination of the coefficients characterizing the time of the equipment use, its operation modes
- Clarification of the correct choice of wires and cables
- Study of daily load schedules

Within the framework of the MIC project, it is planned to develop the following sample of a TEMMP equipped with relevant functions:

- Measurement of the operating time of the electric consumer (network section)
- Measurement and fixation of the maximum current consumption
- Measurement and fixation of the maximum power consumption
- Metering of electricity consumed by an electrical consumer or a network section
- Fixation of the periods of switching on and off of the electric consumer or the network section
- Manual and remote-control device
- Implementation of communication with a personal computer (MIPU) and a smartphone in order to monitor readings in real time

Functional diagram of the device, explaining the principle of operation, is presented in Figure 2.

According to figure 1, 1 is a current sensor, for example, 100A SNS-CURRENT-CT013-100A; 2 is a conductor; 3 is a microcontroller board, 4 is an LCD display.

The current sensor due to its design is set on one of the wires (phase or neutral wire). When an alternating current flow through the wire, an emf appears in the coil, the sensor begins to transmit an analog signal to the microcontroller board. The microcontroller board processes the current sensor signal and performs the necessary calculations. The results of the calculations are displayed on the liquid crystal display. The device is powered by a 9 V battery of PP3 battery size (it is possible to power the device from a 12-volt li-ion battery assembled from three 18650 batteries connected in series for greater autonomy).

The main component of the device is a microcontroller board, in the prototype it is made by Arduino Uno based on ATmega328P. It has 14 digital I/O pins (of which 6 can be pulse width outputs), 6 analog

Figure 2. Functional diagram of the device TEMMP

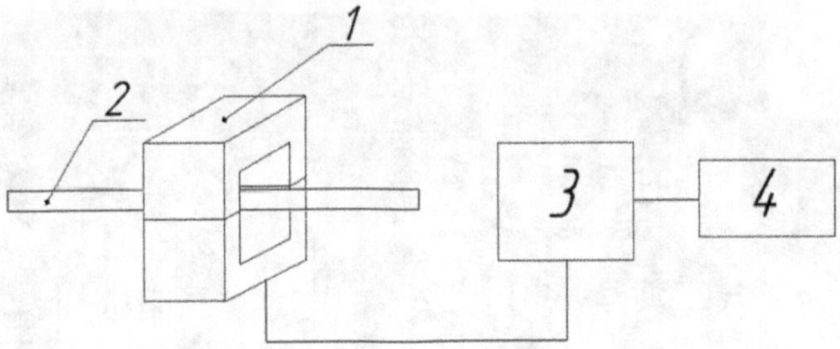

input pins, a 16 MHz quartz crystal, a USB connection, a power connector, an ICSP comb, and a reset button. The electrical circuit of the microcontroller is shown in Figure 3, and its appearance is shown in Figure 4.

LCD display is illuminated and with I2C/SPI converter. Unlike a conventional display, it communicates with an Arduino controller by means of 2-wire communication. It helps to save the controller's digital buses for connecting additional peripherals. There is also a potentiometer for adjusting the backlight brightness on the I2C/SPI converter. The PCF8574 chip port extender is used to connect the display to the I2C bus. Display characteristics are

- Backlight color blue
- Number of characters in line 20
- Number of lines 4

Figure 3. The electrical circuit of microcontroller

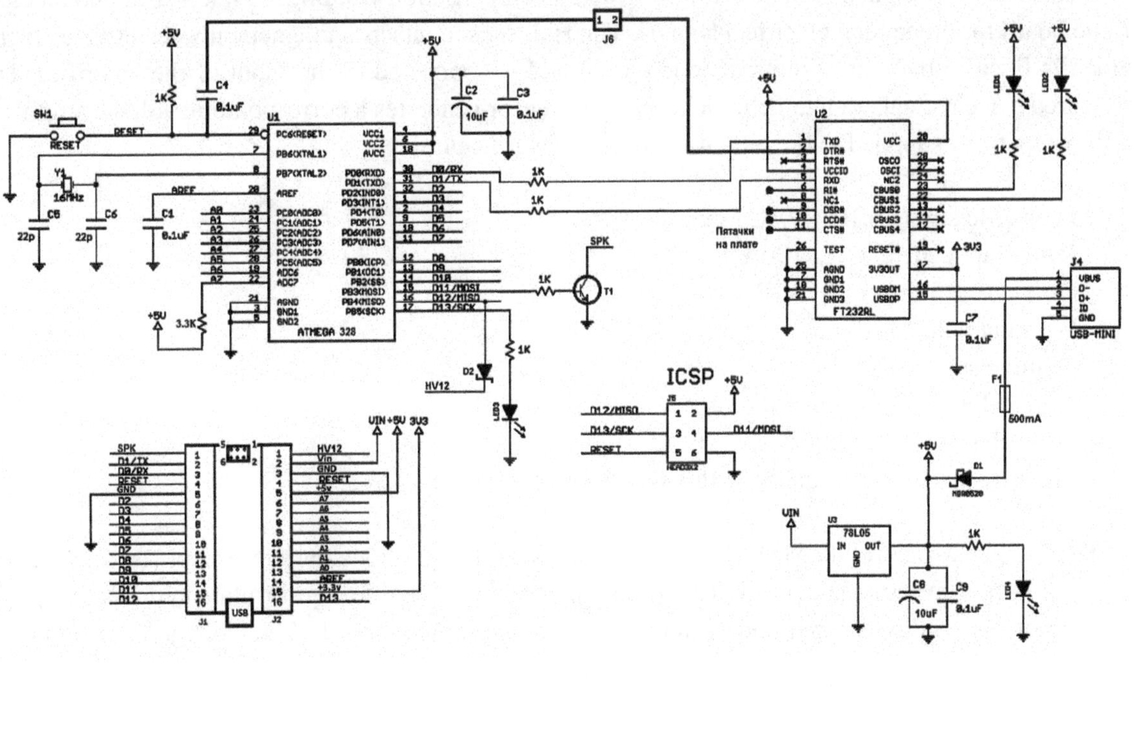

Figure 4. The appearance of microcontroller

- Language: by default, it supports Latin
- Interfaces IIC / I2C / TWI
- Supply voltage: 5V

The electrical circuit of the LCD is shown in Figure 5 and its appearance is shown in Figure 6.

Current sensor is SNS-CURRENT-CT013-100A allowing to measure current up to 100A. The current sensor is designed to determine the strength of an alternating or direct current in electrical circuits. Its appearance is shown in Figure 7.

It consists of a magnetic circuit with a gap and a compensation winding, a Hall sensor and an electronic board for processing electrical signals. The Hall sensor acts as a magnetically sensitive element, which is fixed in the gap of the magnetic circuit and is connected to the input of the amplifier. The measured current creates a magnetic field, the Hall sensor generates a corresponding voltage amplified at the output and supplied to the output winding. Sensor features:

- Dielectric strength 1000VAC / min
- Rated input current 0 ... 100A
- Output 0 ... 50mA or 0 ... 1B
- Working hole 13x13mm
- Nonlinearity ± 3%
- 1.5m cable, with 3.5mm connector
- Operating temperature -25 ... + 70 °C
- Fire resistance in accordance with UL94-VO

The second variant of TEMMP is almost identical to the previous variant except for the fact that it has an output for voltage control. This output can be connected to non-insulated conductive parts, for example, to consumer input nodes. In this case, an additional control of the voltage in the circuit is

Figure 5. The electrical circuit of the LCD

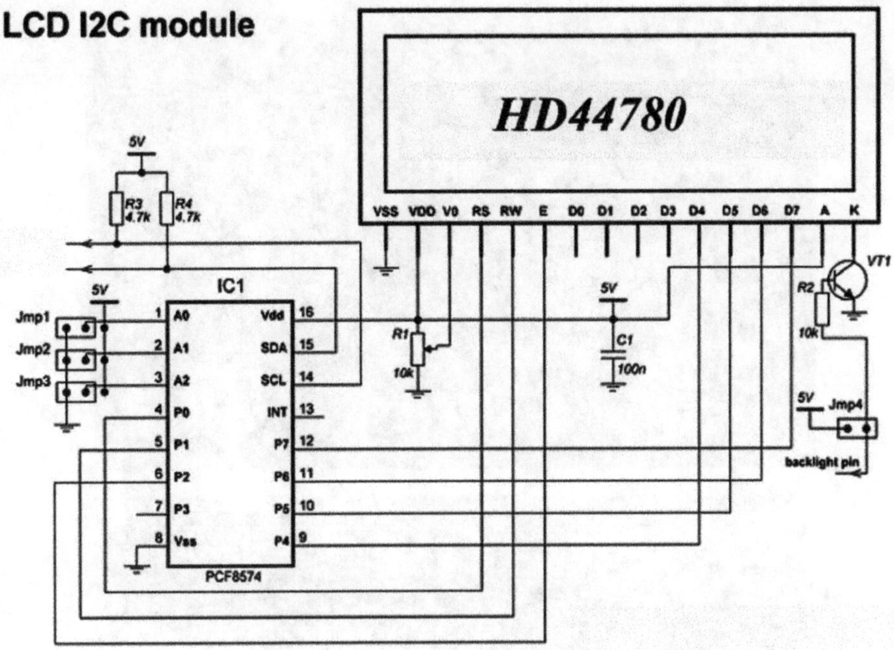

provided in order to increase the accuracy of metering the amount of electricity and determining the power consumption.

Data transmission from a TEMMP to an MIPU can be carried out via a Wi-Fi channel, via a radio channel depending on the distance between the MIPU and the sensors.

Figure 8 shows the diagram of a mobile measuring complex for conducting a preliminary energy audit of rural electrical networks without disconnecting consumers.

According to figure 8, T1 is a power transformer (1), CSt, VS1 … VSn are measuring sensors (2), MMD (3) is a multifunctional measuring device, S1- Sn are consumers of electrical energy (4).

MEASUREMENT AND ANALYSIS OF POWER LOSSES IN THE STUDIED NETWORK

The installation of TEMMP sensors at various points of the studied network allows determining electric power losses in various parts of the network by comparing data from sensors.

For example, the power loss in the area between the sensors TEMMP 1 and TEMMP 2 (Figure 1) can be determined as follows:

$$\Delta W_{1-2} = W_1 - W_2 - W_3 \quad (1)$$

Figure 6. The appearance of the LCD

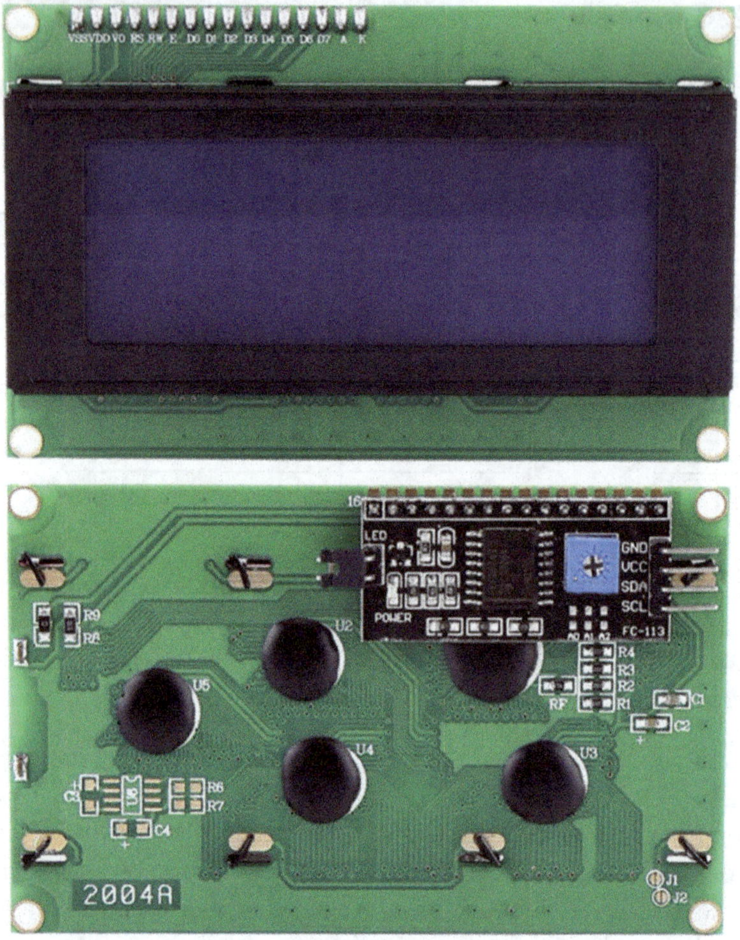

where W_1 is the amount of electricity consumed for a specified period of time according to the readings of the sensor TEMMP 1.

W_2 is the amount of electricity consumed for a specified period of time according to the readings of the sensor TEMMP 2.

W_3 is the amount of electricity consumed for a specified period of time according to the readings of the sensor TEMMP 3.

With the help of TEMMP sensors, an assessment of the correctness of meter readings installed at consumers and the identification of unaccounted consumption can be also carried out by comparing the readings of the meter data with the readings of the corresponding TEMMP sensor.

Figure 7. Current sensor SNS-CURRENT-CT013-100A

Figure 8. The diagram of a mobile measuring complex for conducting a preliminary energy audit of rural electrical networks without disconnecting consumers

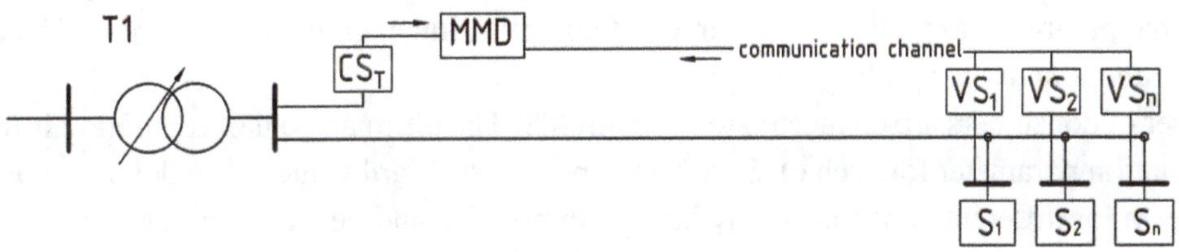

Measurement of the Actual Voltage Value and Analysis of Voltage Loss at Different Points of the Network

The information about the voltage at different points of the network allows to properly configure the no-load tap changer (NLTC) or on-load tape changer (OLTC) of a transformer in order to ensure the required voltage level at the consumer inputs.

In particular, the following method for setting the voltage at a transformer substation can be implemented when using TEMMP to control the voltage at the consumer inputs. A similar method is given in Golikov and Vinogradov (2017), which used for adaptive automatic voltage regulation in the 0.4 kV network in the presence of automatic means of voltage regulation at the transformer substation. This method can be used in a slightly modified form in the event when there are no such means.

In the method given in the work of Golikov and Vinogradov (2017) it is necessary to set the relation between the voltage of the secondary winding of the transformer of the electrical substation and the voltage of the nearest and most distant consumers receiving power from an outgoing line of the electrical substation where the voltage is regulated. The actual voltage of the secondary winding of the transformer must be measured as well as the actual voltage of the nearest and most distant consumers. After that they must be compared with the standard values. In case of difference between the consumer actual voltage and normative values, the actual deviation of the ratio between the desired and actual voltages is defined. The obtained voltage deviation is used as correction signal for the voltage regulation.

The essence of the method is illustrated by figure 9 showing a block diagram of a device that implements the method. According to Figure 9, T1 is a transformer with on-load tape changer (T1), communication and data transmission devices; WO is a working body; ED is an executing device; A is an amplifier; CS is a control system; VS_{bus} is a voltage sensor on the transformer buses; $VS_{rem.con}$. and $VS_{prox.con}$ are voltage sensor of the most remote and nearest consumers; DU is a driver unit.

The work of automatic voltage regulation at an electrical substation is provided as follows. With the help of the driver unit, the charger sets the relation between the voltages of the nearest and most distant consumers and the voltage of the secondary winding of the transformer of this electrical substation is set. Next, the automated collection of dynamic voltage data is carried out from the voltage sensors $VS_{rem.con.}$ and $VS_{prox.con}$ of the nearest and most distant consumers of electricity S_{rem}-S_{prox}, and the voltage sensor VS_{bus} installed on the low side of the transformer substation through the communication channel of modern wired/wireless communications and data transmission. The collected data from consumers, voltage sensor of the transformer and the driver unit are analyzed by the control system device CS for the presence of deviations of actual voltage at consumers from standard values. The resulting voltage deviation is used as the correction signal processed by the amplifier A. Then the signal goes to the execution device ED, which in turn affects the working body WO.

Thus, the normal parameters of electric power quality are maintained by using only linear voltages as a monitored parameter as well as the accuracy of energy voltage regulation is increased by expanding the method functionality for polling voltage data.

A number of consumers are connected to the same TS. The ultimate goal of regulation is to bring the voltage deviation parameter for each of these consumers to standard values. The deviation of this value from the norm for different consumers may have both positive and negative values. Regulation should be made in the event that the level of steady-state voltage deviation for at least one of the consumers is beyond the standard values. When the steady-state voltage deviation exceeds the standard values, it is necessary to choose which value of the regulation coefficient should be changed. At the same time this should be done in such a way as to minimize the current voltage deviations from the standard values for all consumers. Thus, having revealed the presence of deviations it is necessary to determine the total value of the deviation for all consumers. If the sum of deviations takes a positive value, the voltage should be adjusted in the direction of decreasing its value and vice versa. If the sum of deviations of the

Figure 9. The method for automatic voltage regulation at a transformer substation

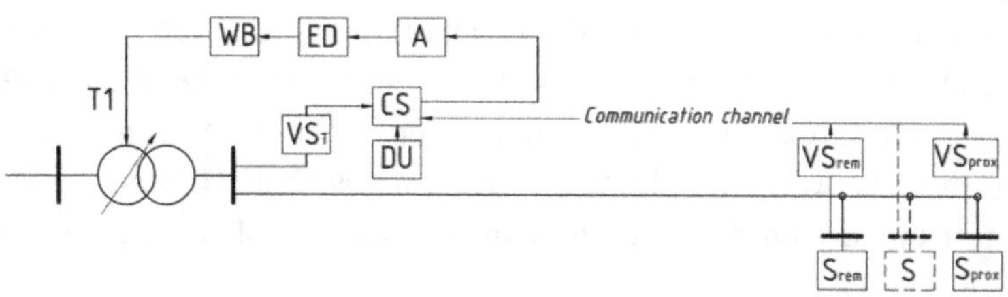

effective voltage is zero, even in the case when the voltage deviation is greater than the standard value, then the regulation should not be carried out since this will lead to even greater deviations. Next, the regulation coefficient should be determined. With a positive total deviation, the following equation is solved to this end:

$$B_1 = \left(\frac{\sum_{i=1}^{n} A_n}{\sum_{i=1}^{n} \left(\frac{\Delta U_d}{100} + 1 \right)} - 1 \right) \cdot 100 \tag{2}$$

where B_1 is coefficient indicating the value necessary to change the deviation level of the steady-state voltage at the transformer substation with its positive deviation from the standard value, %;

ΔU_d is the deviation level of the steady-state voltage value at the border of balance delineation between a consumer and a power supply company, % (Sudnova, 2000).

A_n is N-th point of balance declination of a consumer and a power supply company, pcs.

In this case B1 takes a value from -1 to 0, a negative value indicates that the regulation must be made in the direction of decreasing the voltage value.

With a negative deviation of the steady-state voltage, the coefficient indicating the value necessary to change the deviation level of the steady-state voltage will take the form:

$$B_2 = \left(\frac{\sum_{i=1}^{n} A_n}{\sum_{i=1}^{n} \left(\frac{\Delta U_d}{100} - 1 \right)} + 1 \right) \cdot 100 \tag{3}$$

where B_2 is coefficient indicating the value necessary to change the deviation level of the steady-state voltage at the transformer substation with its negative deviation from the standard value, %;

In this case B2 takes a value from 0 to 1, a positive value indicates that the regulation must be made in the direction of increasing the voltage value.

When using technical means to increase the voltage level, it is necessary to take into account the characteristics of the specific technical device that performs the regulation. The impact on a technical tool does not always lead to a proportional change in voltage. Therefore, it is necessary to use the proportionality factor taking into account the characteristics of each of the used devices. In this case, it is possible to determine the regulation coefficient, which will take the form:

With the positive total deviation of steady-state voltage:

$$K_{1reg.} = K_{TD} \cdot B_1 \tag{4}$$

With the negative total deviation of steady-state voltage:

$$K_{2reg.} = K_{TD} \cdot B_2 \tag{5}$$

where K_{1reg} and $K_{1reg,}$ are regulation coefficient with the positive and negative total deviation of steady-state voltage, %;

K_{TD} – is coefficient of proportionality depending on the type of control used device.

In works of Golikov & Vinogradov (2017) and Sudnova (2000) an algorithm (Figure 10) for the practical implementation of the method of automatic voltage regulation at the TS is developed. The algorithm was compiled using the principles described in Smith & Corripio (2012), Ebel et al. (2008), Groover (2007). The beginning of the algorithm involves the input of the normalized voltage value ΔU_{norm}. and

Figure 10. Algorithm of voltage regulation at the TS using the regulation factor

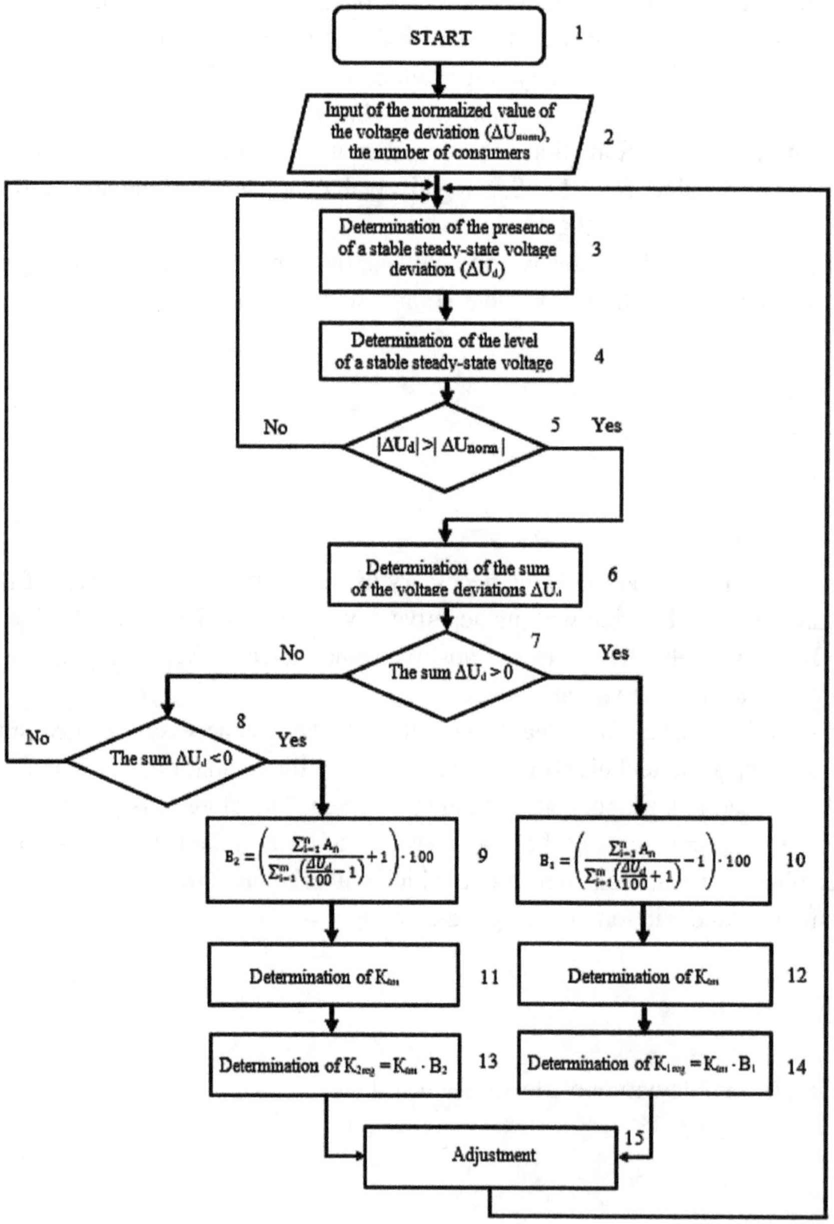

the number of consumers. The specified data is stored in block 2. Block 3 determines the presence of a steady-state voltage deviation ΔUd. Block 4 determines the level of voltage deviation. Block 5 compares the level of steady-state voltage deviation with the normalized value. When the voltage level deviates more than normalized, the signal goes to block 6, which determines the sum of voltage deviations. If the deviation is less than the normalized value, the signal goes to the beginning of block 3. Blocks 7 and 8 perform the function of determining the direction of control. If the total deviation is greater than zero, the signal arrives at block 10, which determines the coefficient indicating the value the steady-state voltage deviation needs to be changed. Blocks 12 and 14 determines the coefficient of proportionality and the coefficient of regulation. Information from block 14 is fed to block 15, which gives the executive signal at the start of regulation. If the condition of block 7 is not met, the signal goes to block 8. In this case, the circuit operates according to a similar algorithm using blocks 9,11 and 13. If the conditions of blocks 7 and 8 are not met, the signal from block 8 goes to the beginning of block 3. Thus, this algorithm allows to automatically determine the control factor and apply appropriate technical devices to regulate the voltage deviation at the TS (Purdum & Levy, 2012; Morton, 2002).

The use of the above method in the case of the implementation of the mobile measuring complex (MIC) for adjusting the voltage at the TS 10/0.4 kV is as follows. The data from the TEMMP sensors about the voltage in the network control points (at the inputs of the consumers) are collected in the MIPU. This data is processed using the algorithm shown in Figure 10. Measurements are made during the day or within seven days (which is preferable for more precise voltage setting) depending an option. Then the measurement results processed by the algorithm are analyzed. In fact, the analysis is reduced to the determination of the weighted average value of the regulation coefficient. This coefficient is determined using the expression:

$$K_{w.aver.} = \frac{K_{reg.t1\%} \cdot t1\% + K_{reg.t2\%} \cdot t2\% + \ldots + K_{reg.tn\%} \cdot tn\%}{t1\% + t2\% + \ldots + tn\% = 100\%} \quad (6)$$

where $K_{reg.t1\%} \ldots K_{reg.tn\%}$ are regulation coefficients obtained and recorded during the corresponding time intervals $t_1 \ldots t_n$. The coefficients can have both positive and negative values;

$t_1 \ldots t_n$ are time intervals during which the same corresponding value of the regulation coefficients were observed, h;

$t_{1\%} \ldots t_{n\%}$ is the proportion of time as a percentage of the measurement time during which the same corresponding value of the regulation coefficients were observed, h.

These time intervals are calculated by expression 7:

$$tn\% = \frac{tn}{\sum_{1}^{n} ti} \cdot 100\% \quad (7)$$

A simple example is given. Suppose, within the measurement process (measurement time 24 hours) and processing the obtained data, the following results were recorded for the recommended regulation coefficients and the corresponding time intervals: t_1 = 15h, K_{regt1} = + 2%; t_2 = 5h, K_{regt2} = + 3%; t_3 = 4h, K_{regt2} = 0%. Then $t_{1\%}$ = (15/24) · 100 = 62.5%, $t_{2\%}$ = 20.8%, $t_{3\%}$ = 16.7%. In this case, the weighted average coefficient will be:

$$K_{w.aver.} = \frac{+2\%\cdot 62,5\% + \left(+3\%\right)\cdot 20,8\% + 0\%\cdot 16,7\%}{100\%} = +1,87\%$$

Thus, it is concluded that the no-load tap changer (NLTC) at the TS should be in the position closest to +1.87 (probably it is necessary to set to + 2% with the steps of NLTC +05, +1, +1.5, +2, + 2.5%) to ensure the most rational voltage level at the inputs of all consumers.

Measurement and Analysis of Load Graphs of the Studied Network and Consumers Connected to It

Measurement and analysis of load graphs of the studied network and consumers connected to it are as follows. First, the topography of the studied network is applied to the corresponding developed software product in MIPU. The locations of the intended installation of the TEMMP sensors are indicated there as well. The next step is the actual installation of the sensors and their configuration including setting up and checking the data transfer channels. After this, the of measuring and processing of the data received from the TEMMP sensors begins. The software product collects information about the currents (and voltages) at the sensor installation points and records them in the form of tables and graphs at these points.

The total load chart is based on the data obtained from the TEMMP installed at the transformer substation (TS) (Dm in Figure 8). This sensor monitors the current and voltage either on a specific outgoing line, if one line is audited, or on the TP input, if an audit is carried out for all lines departing from the TS. According to data obtained from other sensors, load graphs for the points of the network where they are installed are accordingly constructed.

The construction of these graphs makes it possible to estimate the load of all sections of power lines extending from the transformer substation, to draw conclusions about the conformity of the wire sections, the need to replace the power transformer, etc.

The analysis of the power line load gives an idea of the transmission capacity of the studied electrical network. This item is also an important step in achieving energy efficiency of electrical networks. The free transmission capacity of power lines is calculated as the difference between the maximum and current transmission capacity. When calculating the transmission capacity, the requirements for limiting the current density in the wires, limiting voltage fluctuations, and ensuring the stability of the regime must be also checked (Aleksandrov, 2018).

The transmission capacity of ETL is a quantity that depends on the voltage, current and reactive resistance of the line. The largest transmitted power is determined by the expression:

$$P_{larg} = \frac{U_1 \cdot U_2}{Z_c \cdot \sin\alpha_o \cdot L} \tag{8}$$

where U_1 is the voltage modulus at the beginning of the line; U_2 is the voltage modulus at the beginning of the line; Z_c is line wave impedance; α_0 is line wave distance.

Wave impedance is possible to determine as follows:

$$Z_c = \sqrt{\frac{x_0}{b_0}} \tag{9}$$

where x_0 is line specific reactance; b_0 is line capacitive susceptance (Fayzullin & Starikov, 2016).

The formula for calculating the current capacity at the power line section:

$$P = \frac{\sqrt{3} \cdot U \cdot I \cdot \cos\varphi}{1000} \tag{10}$$

Before the beginning of the capacity assessment, the electrical network parameters for each electrical network section are entered into the MIPU using a driver unit. ETL length, the type of the wire and its installation method (taking into account the limitations of the current density in the wires) are indicated there.

Reference data on permissible load currents and other factors are in the MIPU to ensure the functionality of the system.

During operation, the MIPU receives data on the operation mode of the electrical network and parameters such as voltage and current in each section of the electrical network where the TEMMP sensors are located. The maximum transmitted power of the network is calculated during the measurement period and then the current transmission capacity of the transmission lines is defined. In this case, the transmission capacity is calculated as the difference between the largest transmitted power and the current transmission capacity of the transmission lines. The obtained results on the assessment of the power line workload can be displayed on the display of the mobile measuring complex, stored in the device's memory and transmitted via xml format.

Figure 11 shows the algorithm for determining the estimate of the ETL workload. The input of primary data for the operation of the system starts in block 2 where there is the input of the electrical network sections, the lengths of these sections, the current time t and the tables of capacitive susceptance b_0 and inductive reactance x0. Then, in block 3 the voltage is determined at the beginning of U1 and at the end of U2 of the investigated line section. The wave impedance of the line is calculated in block 4. The line wave distance is found in block 5. The largest transmitted power for the considered electrical network is found in block 6. The current of the network section is defined in block 7 and then the power transmitted at the moment is getting in block 8. As a result, the difference between the greatest transmission capacity of the considered electrical network and the power transmitted at the moment is calculated in block 9. The resulting value is the indicator of the electrical network workload expressed as a percentage relative to the maximum transmitted power. The obtained value is transmitted to the MIC monitor 10.

Measurement and Analysis of the Maximum Values of Currents in Various Points of the Network and the Branch Line of the Consumers Connected to it

TEMMP sensors allow to select and record from the measured values the maximum currents at the point of their installation. These data allow to solve several questions. These are the tests of the wire cross sections and line transmission capacity, the control of values of inrush current in the network, the identification of consumers with maximum load currents etc.

Figure 11. Algorithm for determining the assessment of ETL workload

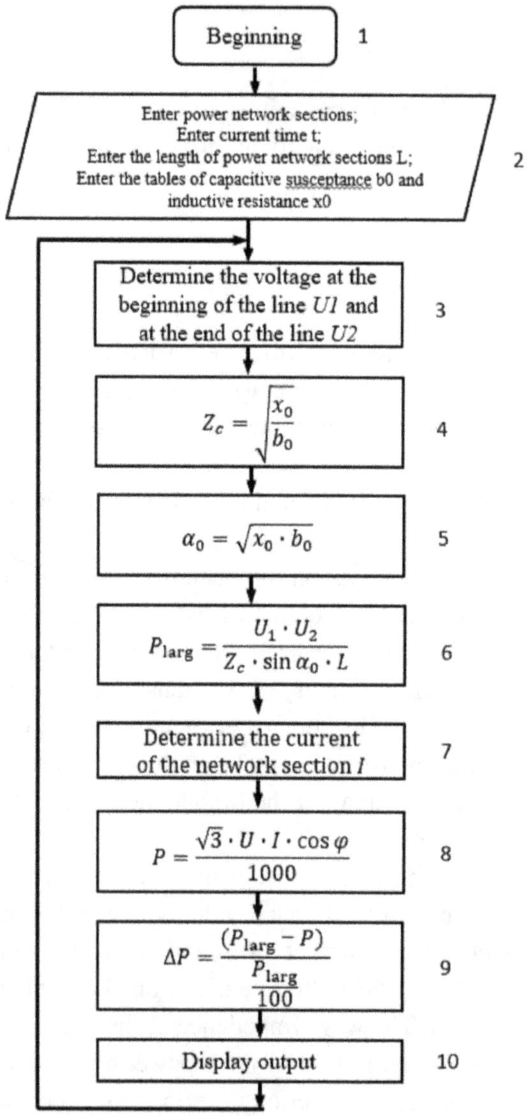

Estimation of Electricity Losses in Power Transformers Without Disconnecting Them From the Network and Ranking for Short Circuit Losses, Load Losses and No-Load Losses

The structure of the complex for assessing electricity losses in a power transformer has the form shown in Figure 12.

The mobile measuring complex for measuring electric power losses in power transformers without disconnecting consumers contains the multifunctional information processing unit MIPU 1 placed on the chassis of the vehicle 6, voltage and current sensors installed on the power transformer sides of high (HV) 4 and low (LV) voltages 3, temperature sensor 2 installed in the oil tank, variable active load unit

Figure 12. The structure of the mobile measuring complex for measuring electrical energy losses in power transformers of rural electrical networks without disconnecting consumers

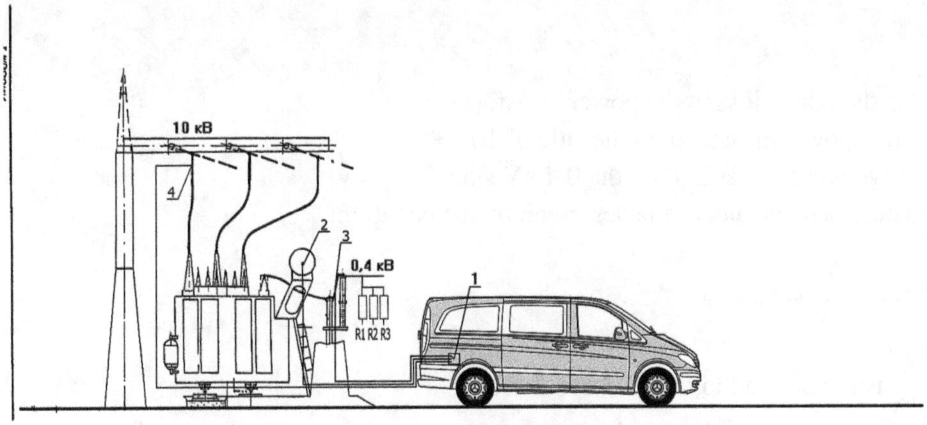

5 (set of resistors R1, R2, R3). Data on voltage, current and oil temperature from HV (4) and HH (3) are transmitted to MIPU (1) via communication channels.

Figure 13 shows a diagram of a mobile measuring complex for measuring electric power losses in power transformers without disconnecting consumers. According to Figure 13, MIPU is the multifunctional information processing unit, VS1 ... VS2 are voltage sensors, CS1 ... CS2 are current sensors, TS1 is a temperature sensor (thermocouple), R_{11} ... R_{13} - variable active load unit, T1 is a power transformer, S_1-S_n are electrical energy consumers.

The procedure for assessing the electrical energy losses in power transformers 10/0.4 kV without disconnecting them from the load is as follows.

The mobile measuring complex located on the vehicle chassis is connected by TEMMP sensors (in Figure 13 it is by VS voltage sensors and CS current sensors) from the high and low sides of the power transformer and to the transformer oil temperature sensor TS. A variable resistive load set is installed on the low side of the transformer. During the measurement of transformer losses, the change in current and voltage from both the high and low sides of the power transformer and its temperature as the resistance of the variable load changes is recorded. In the process of measuring the load increases in steps. According to the obtained values of the active components of current and voltage (in fact, active power), further calculation of transformer losses is performed according to the algorithm below.

Figure 13. Structure diagram of the mobile measuring complex for measuring electrical energy losses in power transformers of rural electrical networks without disconnecting consumers

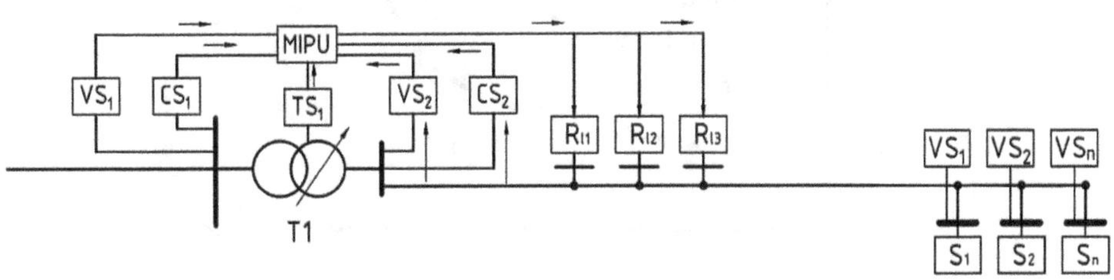

For each measurement, the power loss in the transformer is determined by the formula:

$$\Delta P_{PT} = P_{10kV} - P_{0,4kV} \tag{11}$$

Where ΔP_{PT} is the power loss in the power transformer;

P_{10kV} is active power measured on the side of 10 kV;

$P_{0,4kV}$ is active power measured on the 0.4 kV side.

Thus, the obtained dependence is described by the equation:

$$\Delta P_{PT} = \Delta P_{NLL} + \beta^2 \Delta P_{SCL} \tag{12}$$

where ΔP_{NLL} is the no-load losses of a transformer;

ΔP_{SCL} is the short-circuit losses of a transformer depending on the degree of PT loading β

$$\beta = \frac{P_{PTloading}}{P_{PTnom}} \tag{13}$$

where P_{PTnom} is the nominal active power of the power transformer.

The actual dependence $\Delta P_{PT} = f(\beta)$ is constructed in the form of a graph shown in Figure 14. Having obtained the character of dependence $\Delta P_{PT} = f(\beta)$ for the transformer under study, the graph of this dependence is mathematically matched to the value $P_{PTload} = 0$.

In this case, $\Delta P_{PT} = \Delta P_{NLL}$, this value is the value of the no-load losses of the investigated power transformer at the current transformer temperature

Reduction of the obtained value of no-load losses to normal conditions is made taking into account the measured value of the oil temperature of the transformer or transformer tank. If a winding or oil

Figure 14. Graph of power losses in a power transformer on its load factor $\Delta P_{PT} = f(\beta)$

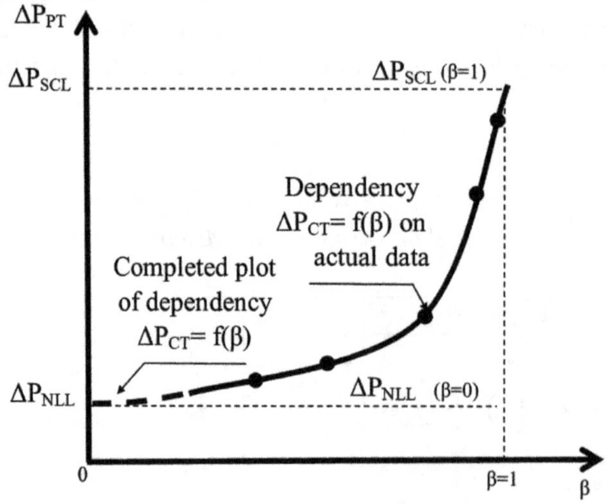

temperature sensor is pre-installed in a transformer, data from these sensors is used. If such sensors are not available and the measurement is carried out without disconnecting consumers, measurements can be performed using a digital pyrometer, the data from which should also be received to the MIPU.

The greatest accuracy can be achieved using winding temperature sensors, the lowest one is gotten using a pyrometer since a pyrometer can determine the transformer tank temperature instead of the winding temperature.

For this purpose, a correction factor $k_{correct}$ taking into account the dependence of losses on temperature is introduced. The value of this factor depends on the method of temperature measurement (the temperature of the oil, the winding or the tank is measured) and the characteristics of the transformer. To obtain the values of this factor it is necessary to conduct additional studies.

In this case, the value of no-load losses is reduced to normal conditions as follows:

$$\Delta P_{NLLnorm} = \Delta P_{NLL} \cdot k_{correct} \tag{14}$$

Next, the short-circuit losses of the power transformer are calculated in accordance with the following expression:

$$\Delta P_{SCL} = \frac{\Delta P_{PT} - \Delta P_{NLL}}{\left(\frac{P_{PTloading}}{P_{PTnom}}\right)^2} \tag{15}$$

The MIPU stores and can displays (if necessary) the following information on the monitor:

1. Total power loss in PT (ΔP_{PT}), kW, as well as losses of electrical energy in PT (ΔPW_{PT}) kWh
2. Power loss of the transformer idling (ΔP_{NLL}), kWh, and losses of electrical energy of the transformer idling (ΔW_{NLL}), kWh.
3. Loss of transformer short-circuit power (ΔP_{SCL}) kW, and loss of transformer short-circuit electrical energy (ΔW_{SCL}), kWh.

The transformer losses obtained during the study is compared with the passport values. A conclusion about the state of the transformer is made. It should be noted that the above method allows only estimating the power transformer losses with a certain accuracy.

Currently, a large number of power transformers with a standard service life are operated in rural distribution networks (Cicorin, 2012). During operation of power transformers, no-load losses increase due to a result magnetic system aging, metal structure changes, interlayer insulation deterioration, the weakening of the transformer core pressing (Cicorin, 2012; Janardhan & Galloway, 2001; Beckley, 1999). Previously it was assumed that no-load losses for the entire service life (35-40 years) in serviceable transformers will increase by no more than 5% compared with the results obtained during factory tests. In the works of Cicorin (2012) and Konstantinov et al. (2007) it was shown that in reality the no-load transformer losses increase to a greater extent, what is the source of the increased costs for the electric grid organizations to pay for the losses of electricity. At the same time, the difference between actual and estimated no-load losses refers to commercial losses, which makes it impossible to justify a higher electricity tariff and directs the grid organization in the wrong direction to search for lost kilowatt hours

Existing methods for estimating the no-load losses of power transformers do not allow it to be in operation without disconnecting the load (Cicorin, 2012; Konstantinov et al., 2007). Therefore, it is necessary to develop new methods for estimating no-load losses making it possible to carry out a rapid assessment without removing the load from the transformer, therefore, avoiding unreasonable interruptions in the power supply of electricity consumers.

The statistical data on the transformers of the Kromsky branch of Oreloblenergo JSC were evaluated using traditional methods in accordance with the requirements of Standards Russia (1988) and the no-load transformer losses were measured. The obtained data are shown in Table 1. Also, Table 1 shows the passport values of the no-load transformer losses.

From the analysis of the presented statistical data it can be concluded that the no-load losses of the operated power transformers on average significantly exceed the passport ones. Their difference (sometimes by more than 50%) must be taken into account when calculating the standard level of electric power losses in electric networks. Increasing the no-load losses by 10% will increase the total transformer losses by 1-2%. In addition, most rural power transformers operate at low loads so the share of no-load losses in the total losses of transformers is even more (Cicorin, 2012).

Various authors, scientific and production teams are working to improve the design of power transformers in order to reduce no-load losses. So, the quality of electrical steel is continuously improving. The level of specific losses for widely used grades of cold-rolled, oriented, high magnetic permeability of steel with high silicon content is about 1.05-1.10 W/kg at 50 Hz and 1.7 T. The best grades of steel have specific losses of about 0.85 W/kg. The use of sheets of smaller thickness also reduces the losses. Thus, steel with a thickness of 0.23 mm has a specific loss of 20% less than steel with a thickness of 0.3 mm (Beckley, 1999).

An effective technology for processing steel is laser scribing with a decrease of oriented crystal length. In this way, the level of specific losses of 0.5 W/kg was obtained in combination with the use of plates of reduced thickness. A decrease in specific losses is predicted to be 0.3 W/kg with a decrease in the thickness of the sheets to 0.18 mm (Baehr, 2001).

Recent developments in the field of transformer engineering are based on the use of steel with a silicon content of 3% what has a low magnetostriction and allowable losses. The use of reduced induction in the core allows not only to decrease losses, but also significantly reduce the noise level of the transformer. The decision is made on the basis of technical and economic considerations.

Fundamentally new way to reduce losses in the transformer is the use of superconducting materials for winding. Low-temperature superconductors, on which the first prototypes of transformers were made, are uncompetitive with high-temperature superconductors (HTSC). Progress in the creation of high-temperature superconducting materials makes it possible to consider the economic prospects of such developments unquestionable. The advantages of HTSC transformers are the following: reduction of load losses by 90%, reduction of mass to 40%, limitation of short-circuit currents, reduction of reactance, good overload capacity (up to 100% long), low noise level. With the development of production such a transformer is 20% cheaper than usual one with the same power. All the above solutions can reduce no-load losses in newly created transformers. The problem of excessive losses of the transformer electrical networks installed at the enterprise remains unresolved.

Table 1. No-load losses for various types of transformers operated in the Kromsky branch of Oreloblenergo OJSC

Type of Transformer	Year of Manufacture	Lifetime, Years	Overhaul	No-Load Losses, kW		The Increase of No-Load Losses for the Lifetime,%
				Passport	Measured	
TM-315/10/0,4	1967	47		1,9	3,2	68,42
TM-315/10/0,4	1967	47		1,9	3,2	68,42
TM-400/10/0,4	1989	25		1,08	1,59	47,22
TM-320/10/0,4	1988	26	1999	1,9	2,2	15,79
TM-250/10/0,4	1999	15	2008	1,05	1,3	23,81
TM-160/10/0,4	1970	44		0,54	1,8	233,33
TM-320/10/0,4	1963	51	1996	1,9	2,5	31,58
TM-320/10/0,4	1963	51	2005	1,9	2,3	21,05
TM-160/10/0,4	2008	6		0,54	0,98	81,48
TM-250/10/0,4	1983	31	2008	1,05	1,4	33,33
TM-250/10/0,4	1978	36		1,05	2,1	100
TM-100/10/0,4	1978	36	2004	0,365	0,79	116,44
TM-400/10/0,4	1980	34	1996	1,08	1,9	75,93
TM-160/10/0,4	1976	38	1999	0,54	1,1	103,7
TM-400/10/0,4	1981	33		1,08	1,9	75,93
TM-400/10/0,4	1981	33		1,08	2,01	86,11
TM-160/10/0,4	1981	33		0,54	0,83	53,7
TM-400/10/0,4	2011	3		1,08	1,7	57,41
TM-250/10/0,4	1985	29		1,05	2,6	147,62
TM-160/10/0,4	1984	30	2003	0,54	0,73	35,19
TM-400/10/0,4	1990	24		1,08	1,91	76,85
TM-400/10/0,4	1990	24		1,08	2,035	88,43
TM-250/10/0,4	1988	26		1,05	1,87	78,1
TM-160/10/0,4	1984	30		0,54	1,1	103,7
TM-160/10/0,4	1985	29		0,54	1,16	114,81
TM-160/10/0,4	1987	27	2006	0,54	0,85	57,41
TM-160/10/0,4	1963	51	2008	0,54	0,83	53,7
TM-160/10/0,4	2012	2		0,54	0,65	20,37
TM-160/10/0,4	2012	2		0,54	0,6	11,11

At the same time, today there are no measuring complexes or devices on the market that allow estimating the no-load losses in power transformers during operation at their installation sites without disconnecting the load. Method of measurement and electrical laboratory developed by A.N. Tsitorin (Cicorin, 2012), allows measuring no-load losses during operation, but requires load disconnection during the measurements. This leads to undersupply of electricity to consumers.

The developed mobile measuring complex assumes the possibility of estimating no-load losses, short-circuit losses in power transformers without disconnecting the load. MIC allows to achieve the following results:

- Reducing energy losses in power transformers and reducing the cost for this by identifying and replacing transformers with excessive idle losses
- Improving the organization of work on the replacement of power transformers through the effective planning and distribution of investments
- Improving the accuracy of calculating the loss of electricity in power transformers and the correctness of the classification of losses as technical and commercial ones
- Providing an opportunity to rely on real measurements and assessments of electric power losses when calculating electricity tariffs
- Prevention of unjustified interruptions in the power supply to consumers and reduction of damage from undersupply of electricity

The economic effect of the MIC use consists of many of the above components. One of the main components is the reduction of electric power losses in power transformers due to the replacement of transformers with increased no-load losses. There is an example. During the year, one hundred transformers with a capacity of 25 kVA having no-load losses higher than the passport ones on average 50% result electric power losses in the following increase:

$$\Delta P = 100 \times 8760 \times \left(\Delta P_{NLL} - \Delta P_{NLL(passport)}\right) \tag{16}$$

where 100 is the number of transformers, pcs; 8760 - the number of hours per year, h; ΔP_{NLL} is the no-load losses of an old transformer, 0.195 kWh; $\Delta P_{NLL(passport)}$ is the no-load power losses of a new transformer, 0.13 kWh.

$$\Delta P = 100 \times 8760 \times \left(0.195 - 0.13\right) = 56940 \text{ kWh}$$

This value of losses requires an additional payment in the amount of 228 thousand rubles if the cost of electricity is 4 rubles per kWh. It is equivalent to the cost of purchasing three transformers each of which cost more than 60 thousand rubles. This “effect on volume” can be used when planning replacements for transformers whose service life is coming to an end. Currently, the replacement for transformers by the criterion of the level of no-load losses is not carried out.

There is an effect when replacing individual transformers, the service life of which has not yet been exhausted, but the no-load losses exceed passport values. Calculations show that the profitability of transformer replacement increases with increasing power. Thus, it is cost-effective to replace transformers with a power of 25 kVA when no-load losses are exceeded the passport ones by 70% and for transformers with a capacity of 100 kVA when they are by 40%.

FUTURE RESEARCH DIRECTIONS

Further studies on the creation of a mobile measuring complex for conducting an electrical network survey should be aimed at creating the necessary component parts of the complex. Particular attention should be paid to the design of TEMMP sensors taking into account the possibility of their installation on the power line with the fulfillment of safety requirements. First of all, it concerns sets of TEMMP for voltage of 10 kV. Practical studies are required to verify the accuracy of the results obtained using the method described above for estimating the no-load losses of power transformers. It is required to conduct experiments to determine the values of the coefficient, which allows to bring the values of no-load losses of the transformer to standard conditions. It is necessary to create specialized software to equip them with a multifunctional information processing unit. Testing of the complex is required in order to select the most rational communication channels between the TEMP sensors and the MIPU.

CONCLUSION

1. A concept of a mobile measuring complex was developed for conducting a survey of electrical networks and estimating losses in power transformers during the study;
2. The developed methods, algorithms, technical solutions for creating a mobile measuring complex allow measuring the parameters of the electrical network operation modes without disconnecting consumers, getting a passport of the electrical network modes and using it to determine losses in the network and its elements, configuring network protection tools and decide other practical tasks.
3. The developed structural and functional schemes make it possible to carry out the selection of the necessary sensors, information processing units of the mobile measuring complex for conducting electrical network survey without disconnecting consumers, to formulate requirements for the accuracy of measurements and the design of all complex elements.

REFERENCES

Aleksandrov, G. (2018). *Rezhimy raboty vozdushnyh linij ehlektroperedachi* [Modes of operation of overhead power lines]. Retrieved from http://www.cpk-energo.ru/metod/AlexandrovLEP.pdf

Baehr, R. (2001). Transformer technology state-of-the-art and trends of future development. *ELECTRA-CIGRE*, 13-19.

Beckley, P. (1999). Modern steels for transformers and machines. *Power Engineering Journal, 13*(4), 190–200. doi:10.1049/pe:19990403

Cicorin, A. (2012). *Ocenka poter' holostogo hoda v sel'skih silovyh transformatorah pri ih ehkspluatacii* [Estimation of no-load losses in rural power transformers during their operation]. Unpublished doctoral dissertation.

Ebel, F., Idler, S., Prede, G., & Scholz, D. (2008). *Fundamentals of Automation Technology, technical book*. Germany: Denkendorf.

Fayzullin, R., & Starikov, V. (2016). *Propusknaya sposobnost' lehp i meropriyatiya po eyo povysheniyu* [Capacity of power lines and measures for its improvement]. Ural State Mining University. Retrieved from http://science.ursmu.ru/upload/doc/2016/06/15/11_elektrotehnicheskie_kopleksy_i_sistemy.pdf

Golikov, I., & Vinogradov, A. (2017). *Adaptivnoe avtomaticheskoe regulirovanie napryazheniya v sel'skih ehlektricheskih setyah 0,38 kV* [Adaptive automatic voltage regulation in rural power networks of 0.38 kV]. Orel, Russia: FGBOU VO Orel GAU.

Groover, M. (2007). *Automation, production systems, and computer-integrated manufacturing*. Prentice Hall Press.

Janardhan, V. & Galloway, D. (2001). Thinking about the box: Ten ways to mitigate distribution transformer losses. *ElectricLight& Power, 79*(10).

Kartashev, I. (2001). *Upravlenie kachestvom ehlektroehnergii* [Power Quality in Power Supply Systems]. Moscow, Russia: Izdatel'skij dom MEHI.

Kharchenko, V., Gusarov, V., & Bolshev, V. (2019). Reliable Electricity Generation in RES-Based Microgrids. In Handbook of Research on Smart Power System Operation and Control (pp. 162–187). Hershey, PA: IGI Global. doi:10.4018/978-1-5225-8030-0.ch006

Konstantinov, I., Zakonov, I., Vakatov, A., Korovin, S., & Abdrahmanov, A. (2007). Vnedrenie metodiki rascheta tekhnicheskih poter' ehlektroehnergii vo vnutristancionnyh ehlektricheskih setyah generiruyushchih predpriyatij OAO «Tatehnergo» [Implementation of the methodology for calculating technical losses of electricity in the internal electrical networks of generating enterprises of JSC "Tatenergo"]. In Sbornik dokladov 5-go nauchno-tekhnicheskogo seminara-vystavki «Normirovanie i snizhenie poter' ehlektricheskoj ehnergii v ehlektricheskih setyah». Moscow, Russia: DialogEHlektro.

Morton, J. (2002). *AVR: an introductory course*. Elsevier.

Purdum, J. J., & Levy, B. (2012). *Beginning C for Arduino*. Apress. doi:10.1007/978-1-4302-4777-7

Smith, C. A., & Corripio, A. B. (2012). *Principles and Practice of Automatic Process Control*. Editorial Félix Varela.

Standards PJSC Rosseti. (2017). STO 34.01-21.1-001-2017. Raspredelitel'nye ehlektricheskie seti napryazheniem 0,4-110 kV. Trebovaniya k tekhnologicheskomu proektirovaniyu [STO 34.01-21.1-001-2017. Distributive electric networks of 0.4-110 kV. Requirements for technological design]. Russia: PJSC Rosseti

Standards Russia. (1988). *GOST 3484.1-88. Transformatory silovye. Metody ehlektromagnitnyh ispytanij [GOST 3484.1-88. Power transformers. Electromagnetic test methods].*

Sudnova, V. (2000). *Kachestvo ehlektricheskoj ehnergii [Power quality]*. Moscow, Russia: EHnergoservis.

Sudnova, V., Prigoda, V., & Hakimov R. (2007). Principy postroeniya AIIS monitoringa PKEH i upravlenie kachestvom ehlektroehnergii [Principles of building monitoring PQ indexes and quality management of electric energy]. *Promyshlennaya ehnergetika, 3*, 37-42.

Vinogradov, A., Bolshev, V., Vinogradova, A., Kudinova, T., Borodin, M., Selesneva, A., & Sorokin, N. (2019a). A System for Monitoring the Number and Duration of Power Outages and Power Quality in 0.38 kV Electrical Networks. In P. Vasant, I. Zelinka, & G. W. Weber (Eds.), *Intelligent Computing & Optimization. ICO 2018*. Cham: Springer.

Vinogradov, A., Borodin, M., Bolshev, V., Makhiyanova, N., & Hruntovich, N. (2019b). Improving the Power Quality of Rural Consumers by Means of Electricity Cost Adjustment. In *Advanced Agro-Engineering Technologies for Rural Business Development*. Hershey, PA: IGI Global.

Vinogradov A., Kopylov, R., & Matveev, A. (2015). Obosnovanie sozdaniya mobil'nogo izmeritel'nogo kompleksa po ocenke poter' ehlektroehnergii v silovyh transformatorah [The rationale for the creation of a mobile measuring complex for the assessment of electricity losses in power transformers]. *Agrotekhnika i ehnergoobespechenie, 2*(6), 36-43.

Vinogradov, A., Vasiliev, A., Bolshev, V., Semenov, A., & Borodin, M. (2018). Time Factor for Determination of Power Supply System Efficiency of Rural Consumers. In Handbook of Research on Renewable Energy and Electric Resources for Sustainable Rural Development (pp. 394–420). Hershey, PA: IGI Global. doi:10.4018/978-1-5225-3867-7.ch017

ZHelezko, YU. (2002). O normativnyh dokumentah v oblasti kachestva ehlektroehnergii i uslovij potrebleniya reaktivnoj moshchnosti [About normative documents in the field of power quality and consumption conditions of jet capacity]. *EHlektricheskie stancii, 6*, 18–24.

KEY TERMS AND DEFINITIONS

Consumer: A legal entity or a private person exercising the use of electric energy (capacity) on the basis of a concluded contract.

Electrical Network: A set of electrical installations designed to transmit and distribute electricity from the power plant to the consumer.

Microgrid.: A localized group of electricity sources and loads that can operates both connected to the centralized power network and function autonomously as physical or economic conditions dictate.

Mobile Measuring Complex for Conducting A Survey Of Electric Networks: A set of equipment including sensors for operating modes of the electrical network and the multifunctional information processing unit, means for transmitting data from sensors to the complex and other necessary equipment for analyzing the operating modes of the electrical network located on car or another mobile vehicle.

Mode of the Electrical Network: The electrical state of the electrical network, characterized by basic parameters such as frequency and voltage as well as losses of electrical energy, voltage losses etc.

Power Supply Reliability: The ability of the power supply system to transmit and distribute the required amount of electricity from sources to consumers at standard voltage levels and in accordance with a specified load schedule

Power Quality: The conformity degree of electrical energy characteristics at a point in the electrical system with a set of standardized indicators

Chapter 11
Technology of Managing Reactions of Biological Objects at Anthropogenically Transformed Territories

Maria Belitskaya
FSC Agroecology RAS, Russia

Irina Gribust
FSC Agroecology RAS, Russia

Elena Nefed'eva
Volgograd State Technical University, Russia

Valeriy Drevin
Volgograd State Agrarian University, Russia

Soumana Datta
University of Rajasthan Jaipur, India

Igor Yudaev
https://orcid.org/0000-0001-9120-7637
Don State Agrarian University, Russia

ABSTRACT

Solving the problem of increasing plant resistance, the development of environmentally-friendly technologies is particularly important, which also contribute to the reduction of resource costs for production and load on the environment. The research results indicate a positive effect from the treatment of plants and seeds with electrochemically activated (ECA) water, electric fields, and impulse pressure (IP). Pre-sowing treatment of seeds with ECA water increases the germination rate and seed germination energy, improves the development of plants, improves morphological parameters, etc. The reactions of economically dangerous pests and causative agents of infectious plant diseases to the use of ECA water are identified. The combination of pre-sowing seed treatment with the treatment of vegetative plants provides the highest possible result.

DOI: 10.4018/978-1-5225-9420-8.ch011

INTRODUCTION

Crop production is currently accompanied by a high level of energy and economic costs, as well as the adverse effects of technological processes and certain operations on the environment. Currently, the trend of greening food production on the basis of alternative farming systems with the minimization of environmental pollution is of particular relevance.

In most cases, the yield and resistance of the crop are inversely related, so the optimal balance between the resistance of plants to adverse environmental conditions and their resistance to pests and diseases is important. The creation and development of efficient environmentally friendly technologies and technical means for the management of agro-ecosystems are of particular importance. A simultaneous increase in the productivity and stability of plants with the help of stimulating physical or chemical effects is accompanied by an increase in product quality and a decrease in cost, which also ensures the growth of its competitiveness. Physical stimulants reduce the risk of overdose, leading to ecological imbalances, mutations, etc., and besides that they are harmless to the persons performing the treatment, technologically effective, productive and economically viable.

The concept of environmentally friendly farming involves the regulation of components of biocenoses, which reduces the number of pest populations to an ecological and economic imperceptible level that does not exceed the economic threshold of harmfulness.

The implementation of this concept will allow optimizing plant growth, changing the biocenotic situation in the direction of worsening conditions for the development of pests, reducing their numbers and reducing damage with minimal environmental pollution. To achieve this, the task is to improve the technology, taking into account the adaptive potential of plants, increasing the efficiency of the natural regulation of the biological component of agrocenoses, as well as direct impact on pests with environmentally safe methods and means.

The goal of the research is the scientific substantiation of the application of the electric field and electrochemical means for optimizing the processes of controlling the reactions of biological objects in agro-ecosystems.

BACKGROUND

Plant organisms are highly sensitive to biotic, physical, and chemical factors (Vartapetian, 1985). Their responses include a cascade of events at the molecular, cellular, tissue, organ, organism, and population levels (Polevoy, 2001). Functional and structural changes, at the molecular level, manifest themselves in changes in plant growth, development, stability, viability and productivity (Tarchevsky, 2002).

The disclosure of plant response mechanisms to such effects is of research and practical interest. Electromagnetic fields (Pietruszewski, Muszynski & Dziwulska, 2007; Kasakova et al., 2018), structured, ionized and magnetically activated water (Booth, 1992) are effective plant growth stimulators that affect the physiological spectrum processes occurring in them (Opritov, Pyatygin & Retivin, 1991).

At present, ideas about the effect of doses of a stressor are justified (Pakhomova, 1995). If we consider resistance as the proportion of surviving individuals in pro cents of the control, then it follows that small exposures increase resistance, a greater number of individuals survive than in the control (the hormesis phenomenon). Impacts that lead to the death of less than 50% of organisms contribute to the selection of the most viable individuals in the population (or sample), and the effects leading to the death of more than 50% of individuals contribute to the emergence of genetic changes in survivors (Pakhomova, 1995).

The site in the range of small doses is a double-vertex (Dmitriev & Stratskevich, 1986). The first maximum is associated with a general increase in vital activity, and the second is characterized by a change in the ratio of the growth of individual parts and organs of the plant. The area of large doses is determined by the suppression of various processes, the increase in mortality, up to one hundred percent, and the genetic change in the offspring (Dmitriev & Stratskevich, 1986). With extreme exposure, selection of resistant forms, fixation of changes in the genome, and formation of a subspecies or variety resistant to this factor are possible (Kefeli & Tom, 1985).

When using different exposure doses, research results vary. For example, exposure to electromagnetic, electric, and magnetic fields in optimal doses helps improve seed sowing qualities (Mikitinets & Kanibolotsky, 1984; Goncharov, Berezhnaya & Gursky, 1994; Pietruszewski, Muszynski & Dziwulska, 2007; Kasakova et al., 2018 Vasilev et al., 2018). The increase in germination is associated with an increase in water absorption of seeds, both in ionized water and under the action of an electric field (Larionova, 1993). When processing the corona discharge with the electric field, the density of the seed coat changes - it looked “swollen”. In the seed, long, irregularly shaped, thin light lines - microcracks (Larionova, 1993) were visible. The increase in water absorption in the first hours of swelling is associated with an increase in the microcapillary structure of the seed, and later with the activation of hydrolases, the accumulation of osmotically active compounds and an increase in the sucking power.

Germination is influenced by post-harvest time between processing and sowing. When the electric field acted on the seeds, the sowing qualities changed slightly on the first day of storage, increased by 35% on the 35-50th day, then did not change (Larionova, 1993). Seeds with poor sowing qualities (50% or less) and with high (first class) did not provide a stable effect after stimulation. Good results yielded seeds with a slight decrease in crop quality (Dmitriev & Strats-kevich, 1986). Thus, the effect of various factors on the germination of seeds depends not only on the dose of the disturbing factor, but also on the condition of the seeds, that is, sensitivity to the effects.

Seed treatment with electromagnetic fields (Goncharov, Berezhnaya & Gursky, 1994; Pietruszewski, Muszynski & Dziwulska, 2007; Kasakova et al., 2018), electrochemically and magnetically activated water (Labutina, 1988) enhances growth and organ-forming processes in including at later stages. The magnetic field affects cell division (Novitsky, Strekova & Tarakanova, 1971), their stretching differentiation (Novitsky, 1973; Novitsky, 1987), accelerates seed germination (Vasilev et al., 2018).

When seeds were treated with an electromagnetic field, an increase in yield and an increase in the protein content in the grain, starch in potatoes, and sugar in beets were observed. Processing with high voltage pulses allowed increasing onion yield by 42%, potatoes by 15 ... 17%. When watering plants with magnetised water, the yield of radish increased several times, and of tomatoes - by 10 ... 24%. The method of presowing seed treatment with a homogeneous magnetic field allowed to get 10-30% increase in the production of millet, wheat, buckwheat, tomatoes, lettuce (Mikitinets & Kanibolotsky, 1984; Labutina, 1988; Kasakova et al., 2018).

The results show that physical factors, depending on the dosage, can be used to control plant growth and development. By acting on plants, we “force” the body to include its own defenses, to launch and independently control regulatory changes. Since, depending on the dosage, these factors may inhibit plants or, conversely, produce a stimulating effect, the disclosure of the mechanism.

USE OF ELECTROACTIVATED LIQUID SYSTEMS FOR MANAGING THE RESPONSES OF PLANTS AND HARMFUL ORGANISMS

The priority direction of phytosanitary optimization of agro-ecosystems based on an agro-technical method of plant protection is the use of various alternative chemical pesticides (microbiological preparations, entomophages, biologically active substances, etc.) and their combinations. The most important component of modern plant protection is a complex of resource-saving and environmental measures, including through the use of electrical technologies.

Extremely promising in this respect is the use of biologically active substances of a non-biocidal nature, which stimulate growth and plants and their protective functions in relation to economically dangerous types of pests and infectious diseases pathogens — electrochemically activated (ECMA) water and electrolytic nano-structured solution based on bischofite (ENSSB).

According to literary data, the treatment of seeds with electrochemically activated liquid systems improves their sowing qualities and improves the growth and development of plants (Dzhurabaev, 1986; Pyndak, Lagutin & Yushkin, 2001; Osadchenko, Kharchenko & Churzin, 2009; Belitskaya & Gribust, 2009; Belitskaya & Gribust, 2011; Goryagina, 2011). These funds contribute to the formation of plant resistance, regulate the activity of various stages of plant growth and development without significant changes in morphological features (Dzhurabaev, 1986; Pyndak, Lagutin & Yushkin, 2001; Goryagina, 2011). These properties make it possible not only to overcome a deep organic dormancy with aging seeds, the shelf life of which expired or the germination period does not correspond to the vegetative period of a given year (Belitskaya, Gribust & Nefedieva, 2012).

Electrochemically activated liquid systems not only stimulate biological life processes of biota, but also affect harmful organisms, which makes it possible to use these means to inhibit pests of disease (Osadchenko, Kharchenko & Churzin, 2009). However, these biologically active systems are among environmentally friendly and affordable tools. In this regard, it is advisable to assess the possibility of optimizing the phytosanitary condition of seeds and crops, the yield of grain crops and its quality through the use of biologically active electro-activated liquid systems for sowing seed treatment and foliar treatment of crops.

Prior to the start of our research, no work was done on the effectiveness of the use of ENSSB in crop production.

For research purposes, an improved prototype of the irrigation water activation module has been developed and manufactured, based on the results of tests that substantiate the main technological parameters for obtaining components of activated water - “catholyte” and “anolyte”.

Figures 1 and 2 show the construction drawing and a general view of the improved irrigation water activation module, as well as the electrical circuit diagram of the installation.

The improved irrigation water activation module (Fig. 1a) includes the following basic assembly units: building 1 with compartment for electrical equipment 2 and section for equipment for activating water 3, in which reactor 4 is located; power cord with plug 5; switch 6; current regulator 7; ammeter 8; anoly-consumption valve 9; catholyte flow valve 10 and water supply valve 11.

The improved irrigation water activation module consists of a reactor - a modular type electrolytic cell placed in a special frame and made in the form of a diaphragm cylindrical electrolyzer with inlet and outlet fittings and a water supply system equipped with valves for controlling the flow of water from a current source connected to the anode and cathode through the switching node, which includes an isolation transformer (T), a diode bridge (D), a power regulator (U), a resistor (R1), a shunt (R2), an

Figure 1. An improved prototype of the irrigation water activation module: a - construction drawing: 1 - housing; 2 - compartment for electrical equipment; 3 - water treatment compartment; 4 - reactor; 5 power cord with plug; 6 - switch; 7 - current regulator knob; 8 - ammeter; 9 - anolyte valve; 10 - catholyte valve; 11 - water supply valve; b - general view

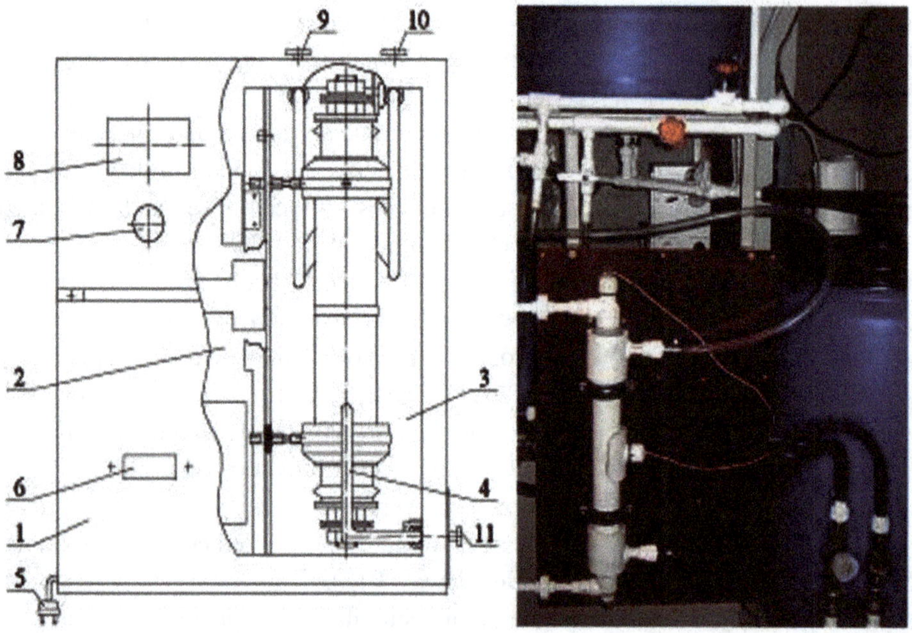

a *b*

ammeter (P) (Figure 2). To ensure electrical safety, the electrical circuit of the irrigation water activation module provides two differential automatic machines (S1, S2) with built-in safety devices, and the activation module frame is grounded and sheathed with textolite with high dielectric properties.

The enhanced irrigation water activation module (Figure 1) works as follows. The treated water from a pressure source under a working pressure from 0.2 MPa to 0.8 MPa is supplied through fittings into the cathode and anode chambers. When electrical potential is applied in the chambers, the process of electrolysis of water begins. Under the action of electric current at the anode, water molecules are separated to form H + and gaseous oxygen. At the cathode, the separation of water molecules with the formation of OH-ions and gaseous hydrogen. As a result, anolyte with pH 2.7 ... 4.5 and oxidation-reduction potential (ORP) equal to + 150 ... 1000 mV is obtained in the anode chamber, and catholyte with pH 7.5 ... 12 and ORP within 0 ... - 1000 mV.

Research results indicate a pronounced positive effect of electrochemically activated media on plants. In this case, a comparative assessment of research results shows an ambiguous picture. This is manifested in the early stages of development, beginning with the formation of a cone of growth, which stimulates the growth of vegetative organs and the formation of a larger number of primary roots, creating a good basis for subsequent tillering and rooting of plants.

To confirm the validity of the revealed patterns, since 2005, studies have been conducted to assess the complex protective-stimulating effect of electro-chemical agents on grain agrocenoses. A wide

Figure 2. The electric circuit diagram of the improved test activation of irrigation water: X - plug connector; S1, S2 - automatic differential protection with built-in protective cutout; T –transformer of galvanic isolation; U - power regulator; R1 - load resistor; R2 - current shunt; RA - ampermatr; D - diode rectifier bridge; E – reactor

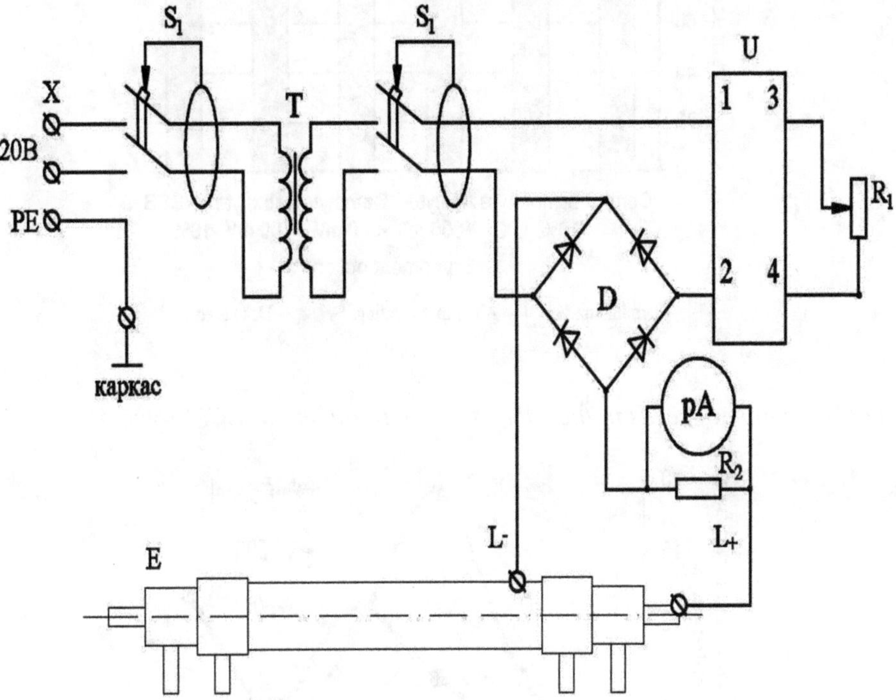

production test of the effectiveness of ECMA water and ENSSB confirmed their positive effect on the immune status and the initial growth of plants, especially of the secondary root system.

Electro-activated solutions affect the germination of seeds of winter wheat, in particular, ECA water with a redox potential of –500 mV contributed to a decrease in germination rate by 25.7%, and with ORP 700 mV and ENSSB (10%) provide an increase in germination rate of 5.9% and 4.7%, respectively (Figure 3).

The reasons for the change in germination can be explained using the phosphorescence index at room temperature (PRT) (Figure 4), determined by the method developed by T.V. Veselova and V.A. Veselovsky (MSU). The correlation coefficient between germination and PRT is -0.94÷-0.98. In batches, there are three fractions with increasing PRT: 1 - seeds from which normal sprouts grow, 2 - seeds from which sprouts grow with morphological defects, 3 - mostly dead seeds. The seeds of the second fraction are labile and, under certain conditions, can behave either as seeds of the first fraction, or as the second.

The distribution in the control had the form of a curve with a maximum (40 ... 59 rel. units), which corresponded to the seeds of the first fraction. The weakened seeds of the second fraction belonged to the class 80 ... 99, and the dead seeds of the third fraction belonged to the class 100 ... 119. PRT distribution of seeds treated with ECMA water (+500 mV), according to criterion 2, corresponded to the control. The FRT distribution of seeds treated with ECMA water (–500 mV), according to criterion 2, differed from the control (P_{05}). The reason was the transfer of normal seeds from the first fraction to the second

Figure 3. The effect of electro-activated funds on the germination of wheat seeds

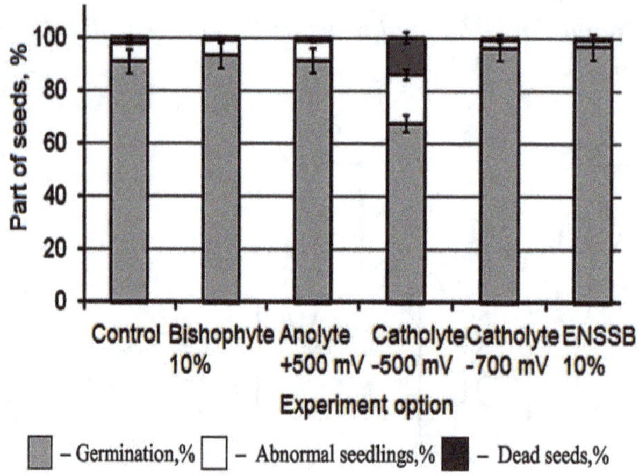

Figure 4. The distribution of seeds in the PRT after the treatment of ECMA water

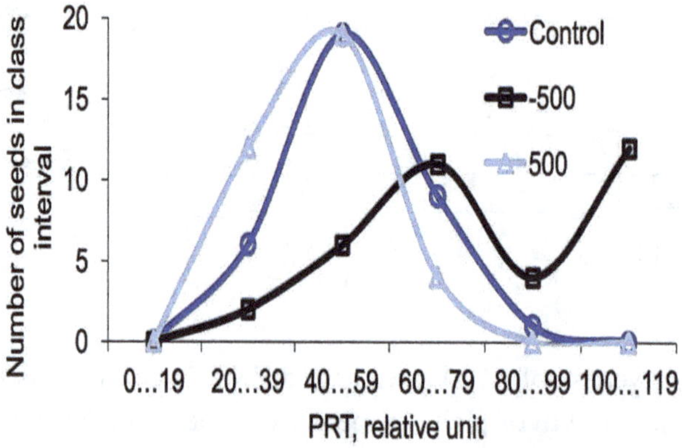

or third fraction. Increase germination can only be due to live seeds. Consequently, under the action of water with ORP - 700 mV (PRT was not determined), increasing germination, some of the seeds of the second fraction passed into the first fraction. These seeds are called "improved."

The treatment of wheat seeds with ECMA water and ENSSB had an after-effect on the morphological features of the seedlings. The ECA water with an ORP of -500 mV and -700 mV, as well as ENSSB, contributed to an increase in the length of the shoot of the seedlings and the predominant accumulation of dry substance in the shoots (Figure 5, Figure 6). The length and dry weight of roots in all variants of the experiment practically did not differ, as did the dry weight of endosperm.

The morphological features of seedlings after the action of electrochemical agents on seeds were the basis for increasing the yield of plants, because on the eighth day the seedlings of vegetative organs formed.

The results of laboratory studies served as the basis for carrying out production tests. The pre-sowing treatment of seeds with ECA water contributed to a uniform emergence of shoots on the field 2 ... 4 days earlier than on the variant using a chemical treater (Dividend Star) and in control (tap water). In the future, the difference in the length of interphase periods was 1 ... 2 days.

Figure 5. The post-effect of treatment of wheat seeds with ECMA water on the length of the root and shoot of 8-day-old seedlings

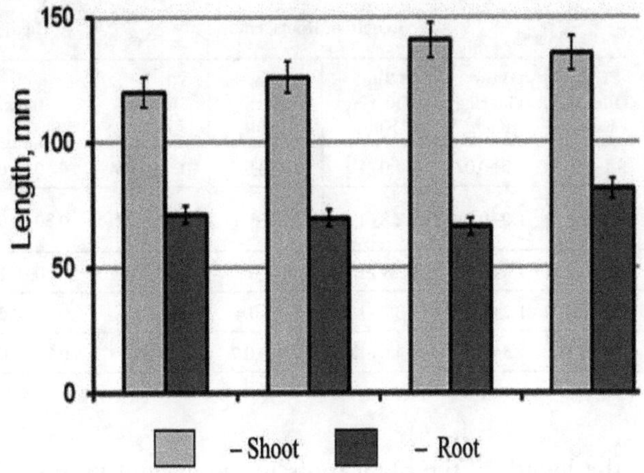

Figure 6. The post-effect of treatment of wheat seeds with ECMA water on the distribution of dry weight in parts of 8-day-old seedlings

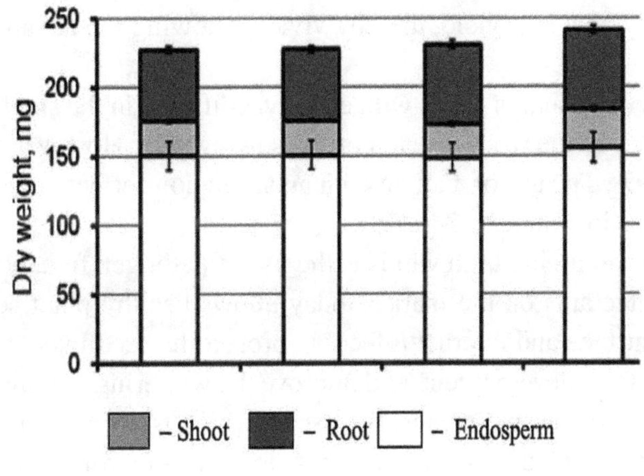

Vital organ in winter wheat plants is a tillering knot, which is a kind of storage of energy resources and thus gives the opportunity for full growth and development of cereals. Observations during the pre-winter period showed that the formation of tillering shoots and nodal roots in winter wheat plants is uneven in variants (Table 1).

Plants whose seeds were treated with catholyte before sowing turned out to be at the control level or even slightly lower according to some indicators. The effect of the anolyte on the development of the roots and the aerial part of winter wheat is somewhat more positive, which is observed already in the early stages of plant development, the average height of which is 15% higher than that in the control plot. An increase in the dry mass of the main wheat vegetative organs was also found on average by 18%.

The productivity of tillering nodes in the autumn-winter period of the sowing year in this variant was higher - the number of stalks of the upper tillering node increased by 23.5%, and the lower one - by

Table 1. Effect of electrochemically activated agents in seed treatment on the morphometric parameters of Triticale winter wheat

Option	Plant Height, cm	Number of Stems, Pieces		Length of Roots, cm		Mass, gr			Tillering Coefficient
		Top Tillering Knot	Low Tillering Knot	Top Tillering Knot	Low Tillering Knot	Leaves	Roots	Tillering Knots	
Control	14.7±0.5	5.1±0.9	1.0±0.02	14.0±1.1	10.0±0.9	1.7±0.06	0.70±0.01	0.70±0.01	2.92±0.2
Dividend Star (standard)	17.7±0.3	5.5±0.4	1.0±0.01	12.8±1.3	7.3±0.6	1.0±0.02	0.30±0.01	0.70±0.1	2.7±0.1
Anolyte, +500 mV	16.9±0.4	6.3±0.2	1.5±0.02	13.9±1.2	9.5±0.8	2.0±0.01	1.0±0.1	1.0±0.09	3.92±0.2
Catholyte, -300 mV	14.1±0.3	5.8±0.3	1.2±0.01	13.2±1.2	9.6±0.4	1.7±0.2	0.67±0.03	0.67±0.02	3.56±0.2
ENSSB	16.4±0.1	6.0±0.3	1.5±0.03	14.6±1.4	10.1±0.7	2.3±0.1	0.70±0.02	1.10±0.2	4.10±0.3

50.05%. At the same time, the length of the plant roots in the variant with seed treatment with anolyte was 0.4 ... 0.5 cm below the control indicator.

Morphometric parameters of young wheat plants on the standard were slightly better than the controls, and the tillering factor was even lower, indicating a decrease in tillering intensity. This circumstance directly influences the formation of spikelet tubercles, the further development of the spike, its aquiferousness and, accordingly, the crop yield, thereby vividly showing the advantages experienced in the agrocenosis of ECA.

The plants with the pretreatment of seeds with catholyte differed in the smallest activation of growth and physiological processes in the winter period of the sowing year. However, there was an increase in the tillering factor by 21.9%. The use of anolyte with an ambiguous effect on young plants has led to an increase in the value of this indicator by 34.3%.

The main condition for obtaining high yields is the use of pathogen free, healthy seeds. Pre-sowing seed treatment with disinfectants on the market today allows healthy plant seedlings to be obtained. These funds destroy the surface and internal infection, protect the seedlings from mold and soil pathogens in the early period of their development, and improve the wintering of winter grain crops. However, disinfectants are among the expensive means, the use of which is accompanied by a decrease in seed germination by 2 ... 4% and they adversely affect the state of the environment.

In the course of research, a search experiment was conducted to find out the possibility of suppressing pathogens of infectious diseases through the use of environmentally safe biologically active etroactivated liquid systems. Anolyte (ORP + 400 ... + 700 mV), catholyte (ORP-300 ... -700 mV) and ENSSB (10%) were used. The standard was an aqueous solution of bischofite, 10%.

The results of the microbiological analysis showed that the treatment of seed with biologically active agents contributed not only to the intensification of plant growth, but also caused a decrease in the development of seed diseases by 76.2 ... 92.0% (Figure 7).

Higher results were recorded on the variants with the use of anolyte (+ 500 mV), ENSSB (10%) and catholyte (-300 mV). Based on the results of this experiment, it can be concluded that the treatment of seeds with electroactivated solutions has an inhibitory effect on the seed infection, which ultimately leads to an increase in the number of grown seeds and to accelerate the growth of seedlings.

The high impact of electroactivated liquid systems is due to the activation of the course of biological processes in the semen due to the production of additional energy and nutrients during processing.

Figure 7. The effect of electrochemically activated water solutions on Triticale seed infection: a - control; b - anolyte, +500 mV; c - anolyte +700 mV; d - catholyte -300 mV; e - catholyte -500 mV; f - catholyte -700 mV

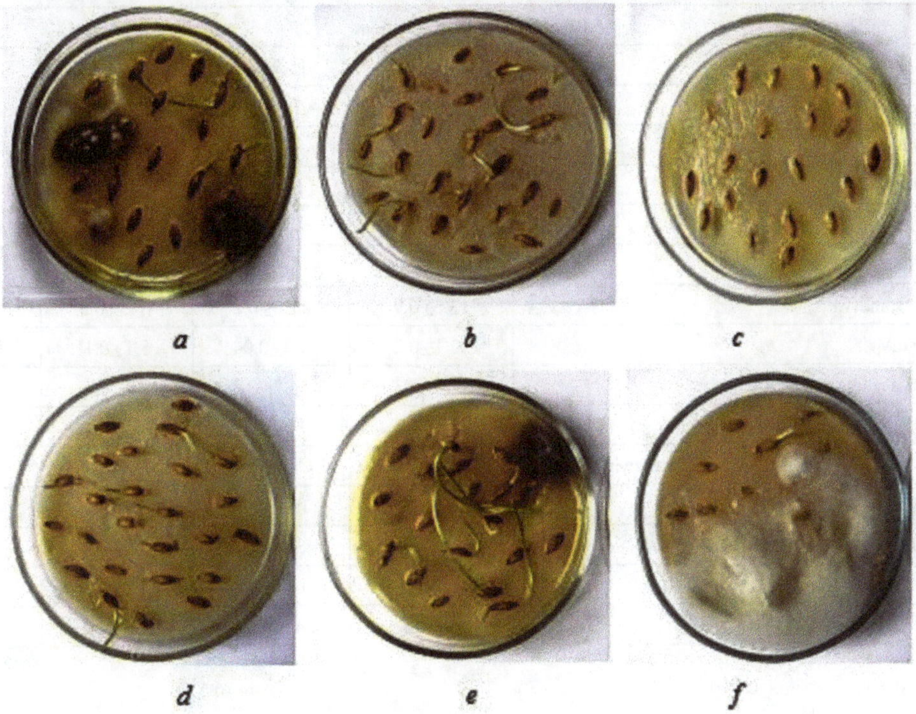

Observations showed that the treatment of seeds before sowing with electrochemically structured means contributed to reducing the development of the most important infectious plant diseases (Table 2). The greatest effect was shown by ENSSB and anolyte: the development of powdery mildew, rust and bacteriosis of the ear was restrained by 2.3 ... 3.6 times, root rot - by 3.0 ... 4.0 times. The use of catholyte provided a significantly lower result - 1.6 ... 2.5 times and 2.4 times, respectively.

The use of electroactivated solutions especially for seed treatment inhibits phytopathogens, starting from the earing phase. To a greater extent, the anolyte and ENSSB increase the immunity and resistance of plants to powdery mildew. The similar result was shown by these agents against bacteriosis, and the anolyte turned out to be more effective for this disease.

Spraying of crops with activated solutions had a positive impact on the phytosanitary condition of plants. At the same time, by their fungicidal properties with respect to the pathogen complex, electrochemical agents were almost at the same level.

When a combination of pre-sowing seed treatment with spraying of vegetating plants, a pronounced protective effect was noted on the variants with the use of anolyte and ENSSB. If in the control the development of diseases of the assimilation apparatus fluctuated at the level of 11.2 ... 19.9%, then on the variants with the use of these funds 3.9 ... 5.2% and 4.0 ... 7.3% respectively. The effectiveness of catholyte was significantly lower than 5.6 ... 11.5%.

Of all the pests that inhabit crops, the biocenotic links of plants with a pest such as wheat thrips are most adversely affected by the use of electro-activated liquid systems. This is expressed in a sharp de-

Table 2. Effect of electrochemical agents for different methods of application for the development of diseases of winter wheat Triticale, %.

Options	Mealy Dew	Rust	Bacteriosis	Root Rot
Presowing Seed Treatment				
Control	19.9±0.4	15.8±0.9	11.2±0.8	1.2±0.2
Bishofyte, 10% – standard	16.5±0.6	7.7±0.3	6.0±0.5	0.3±0.001
Anolyte, + 500 mV	7.0±0.8	4.6±0.2	4.3±0.3	0.3±0.002
Catholyte, - 300 mV	12.4±0.4	6.4±0.3	5.9±0.6	0.5±0. 003
ENSSB, 10%	8.1±0.3	4.4±0.1	4.8±0.4	0.3±0.001
Spraying of Crops				
Bishofyte, 2% – standard	13.6±0.9	11.0±0.2	10.3±0.3	1.3±0.1
Anolyte, + 500 mV	12.1±0.2	9.3±0.04	8.1±0.1	0.8±0.03
Catholyte, - 300 mV	16.4±1.3	14.7±0.2	7.4±0.1	0.7±0.01
ENSSB, 2%	12.7±0.7	9.9±0.07	9.2±0.4	1.0±0.06
Presowing Seed Treatment + Spraying of Crops				
Bishofyte, 10% + Bishofyte, 2%	12.9±02	10.5±0.4	5.0±0.1	0.4±0.01
Anolyte (+ 500 mV) + Anolyte (+ 500 mV)	5.2±0.1	3.9±0.1	4.1±0.03	0.3±0.1
Catholyte (- 300 mV) + Catholyte (- 300 mV)	11.5±0.3	6.2±0.4	5.6±0.2	0.4±0.01
ENSSB, 10% + ENSSB, 2%	7.5±0.2	4.0±0.6	4.2±0.01	0.3±0.02

crease in the pest density on the experimental variants - 77.8 ... 86.7%. The number of bread-flea in the early spring period decreased by 29.3 ... 62.5%. In the summer, a decrease in the density of a harmful bug was noted by 21.8 ... 54.0%, meromiza by 69.9 ... 73.0%.

The pre-sowing treatment of Triticale seeds with test solutions helps to optimize the entomological situation on the crops during the entire growing season. The anolyte (+ 500 mV) increases plant resistance to water inhibitors to a greater degree (Table 3). Their number on this option is reduced by an average of 62.4%. The effectiveness of the use of catholyte (- 300 mV) was almost 17% lower.

Table 3. Effect of presowing seed treatment of winter wheat seeds with Triticale ECA with water on the number of pests

Pests	Control, Pieces per Count Unit	The Decrease in the Number, %			
		Bishofyte, 10%	Anolyte, + 500 мВ	Catholyte, - 300 мВ	ENSSB, 10%
Eurygaster integiceps	16.4	39.5±0.9	54.0±1.1	21.8±0.3	36.3±1.5
Anisoplia austriaca	4.0	33.7±0.5	40.0±3.1	25.0±0.5	37.9±1.2
Chrysomelidae: Chaetocnema, Phyllotreta	78.3	23.1±0.3	62.5±2.3	29.3±0.2	48.9±1.8
Haplothrips tritici	192.0	78.1±24	86.7±3.1	81.3±2.1	77.8±2.4
Meromyza nigriventris	11.5	63.3±1.7	72.8±2.9	69.9±1.3	73.0±2.1

Anolyte (+500 mV) has a significant inhibitory effect on pests when spraying vegetative plants. Putting it into the agrocenosis during the tillering-booting period reduces the number of harmful turtles by 9.6 ... 33.4%. There was also a decrease in the abundance of tsikadok by 55.1%. At the same time, the grain beetles practically do not react to the introduction of water into the stemmed echo.

Analytic solutions with an anolyte with a redox potential from +800 mV to +900 mV have an entomocidal effect on wheat thrips when spraying sowing. Processing of crops by them reduces the abundance of larvae of the pest by almost three times relative to control. The number of adult insects is practically unchanged.

The effect of ECA water on the resistance of crops to harmful insects is connected, in our opinion, with the following factors: the presence of structural-morphological, anatomical and biochemical (mechanical) barriers caused by the structure of the covering and internal tissues; a mismatch between the phenology of plants and harmful insects. The reliability of the barriers is determined by the acceleration of the process of differentiation of the cone of growth, the tightness of the leaf sheaths and spikelet scales, the increased fiber content - a mechanical barrier to the colonization of plants by harmful insects, their nutrition, egg laying, etc.

During the observations, the positive effect of activated agents on the abundance of useful biota was noted. The accumulation and preservation of entomophages in the agrocenosis is more pronounced in the first stages of plant vegetation, when the number of pests in the phytocenosis is maximum.

The effectiveness of the use of biologically active liquid systems can be judged by the structure of the crop. A biometric analysis of the winter wheat harvest structure revealed a positive effect of pre-sowing treatment of ECA with water on the main elements forming the crop.

On the experimental variants, an increase in the number of plants, including of productive stems, was recorded. The best results showed ENSSB (10%) and anolyte (+500 mV). This phenomenon may be associated with an increase in plant resistance to adverse environmental factors, since the initial germination on these variants did not exceed the control.

Under the influence of biologically active substances, the height of plants and the mass of straw increased, which could contribute to the formation of more stocky plants and prevent their lodging (Table 4). The yield of plants (grain mass) increased not due to the number of grains in the ear and the mass of 1000 grains, but due to an increase in the number of productive stems. The increase in yield was 10.1% and 21.8% for ECA water with ORP +500 mV and -500 mV, respectively, with the base crop in the control being 9.04 centners / ha.

The combination of sowing material treatment with spraying of crops turned out to be even more effective, while the yield increased by 0.5 ... 1.6 c / ha or by 10.8 ... 15.3%. For all other elements of the crop structure, the improvement in performance was somewhat less pronounced.

The obtained data are in good agreement with the results when studying the effect of ECA water on other cereals. The increase in grain yield during the processing of this seed material was at the level of 22.6 ... 52.1%, when combined with the non-root treatment of crops - 25.7 ... 50.2%.

The positive effect of ECA water on product quality is revealed. The nature of the grain increased by 10 and 15 g / l. Anolyte and catholyte contribute to the change and other qualitative indicators of grain.

Thus, the use of electro-activated water for pre-sowing seed treatment and the combination of this technique with spraying of crops is an effective environmentally friendly and resource-saving way, successfully competing with the use of chemical plant protection products and growth stimulants.

Table 4. Economic efficiency of electroactivated means in the field experiment on winter Triticale

Options		Number of Productive Stems	Height of Plants, cm	Spike Length, cm	Number of Grains in a Spike, Pieces	Weight of 1000 Grains in a Spike, Pieces	Productivity Centers per Hectare
Presowing Seed Treatment	**Spraying of Vegetative Plants**						
Control without treatment		232±8.5	51.0±1.2	5.5±0.1	28.6±2.1	27.4±0.2	9.04±0.3
Bishofyte, 10%		248±8.2	56.0±1.0	6.7±0.2	30.3±1.9	29.6±1.1	9.70±0.4
Anolyte, + 500 mV		296±9.1	61.2±1.3	8.3±1.1	33.8±2.4	29.3±0.6	10.9±1.0
Anolyte, + 500 mV	Anolyte, + 500 mV	300±9.3	65.9±1.7	8.6±1.4	33.4±2.3	29.7±1.3	12.4±0.6
Catholyte, +300 mV		264±8.5	57.3±2.1	8.2±0.9	32.8±2.5	28.0±1.1	11.3±0.4
Catholyte, +300 mV	Catholyte, +300 mV	272±8.1	58.2±1.9	8.2±0.6	33.9±1.8	29.8±1.6	12.0±0.9
ENSSB, 10%		268±8.6	58.0±1.5	7.3±0.2	30.4±1.2	30.1±1.7	10.2±1.1
ENSSB, 10%	ENSSB, 2%	266±7.9	58.0±1.7	7.8±0.3	31.0±1.5	30.5±1.8	11.8±0.6

Modern technologies of cultivation of agricultural crops provide for the stable production of high-quality products. This is largely dependent on the means used. Our tests showed that on average over the years of research on the variant without the use of electrochemical agents, the protein content in the grain averaged 13.25% (Table 5). The maximum increase in this indicator contributes to the processing of seed material ENSSB - 17.96%.

According to our data, these funds have a significant impact on the amount of raw gluten. So, when using catholyte and ENSSB, the content of raw gluten increases by 5.2% and 5.0%, respectively, the anolyte was less effective - only 4.4%.

APPLICATION OF ELECTRIC IMPACTS ON AGRICULTURAL CULTURAL SEEDS

The recent trend towards finding alternative ways of ecologizing agricultural production calls for finding new ways of pre-sowing seed treatment in order to increase plant resistance to harmful factors and to consistently produce high-quality products. Currently, there are many methods of pre-sowing treatment. These include: maceration, freezing, draining, stratification, scarification, etching, etc. Electrophysical methods are also used, such as grain processing by magnetic and electromagnetic radiation (Goncharov, Berezhnaya & Gursky, 1994; Pietruszewski, Muszynski & Dziwulska, 2007; Kasakova et al., 2018; Vasilev et al., 2018). The use of, for example, electropulse stimulation in presowing treatment increases heat resistance, seed resistance to harmful factors, due to a decrease in electrical conductivity (Bobryshev, Starodubtseva & Popov, 2000). We studied the reducibility of the action of a pulsed electric field, an alternating current and a direct current electric field.

Electrical treatment allows the physiological state of the embryo to be transferred, up to a change in the chemical composition, from a state of "rest" to a state of "awakening" and active growth (Kasakova et al., 2018). At the time of exposure to an electric field inside the seed, a redistribution of electric charges occurs, which, in turn, somewhat changes the course of physicochemical processes that affect the subsequent growth and development of plants.

Table 5. Effect of electrochemical agents on product quality

Options	Protein, %	Gluten, %	Gluten Strain Index, Units.	Nature, g / l
Control	17.17	27.6	77.0	765
Presowing Seed Treatment				
Bishofyte, 10%	13.25	22.8	80	780
ENSSB, 10%	17.96	27.6	78	760
Catholyte	16.99	28.0	70	765
Anolyte	17.55	27.2	81	775
Anolyte + Bishofyte	16.40	28.0	65	780
Presowing Seed Treatment + Spraying of Crops				
Bishofyte, 10%	13.87	24.0	85	780
ENSSB, 10%	14.39	25.6	80	760
Anolyte	15.85	27.2	70	780
Anolyte + Bishofyte	17.61	27.6	77	760

The task of conducting research on the effects of electrophysical methods of treatment on seed material included identifying doses, the type and nature of the effects, causing resistance to diseases and pests, as well as endurance to adverse factors, strengthening plants, both in the early stages of development and in subsequent stages. periods of growth. During the treatment, the electrophysical parameters of the same portions (weights) of the grain were monitored, the voltage, current and exposure time were monitored, which ultimately revealed the causes of changes in the embryo tissues and selected the optimal electrostimulation modes of the seeds. Private experiments were developed for experimental studies, high-voltage power supply units were manufactured and tested, the structure was justified and circuit solutions were assembled to create high-voltage effects in a dc and ac electric field, as well as in a pulsed electric field (Figure 8).

Electrotechnological installations for each of the variants of the studied electrophysiological impact were collected on the basis of a high voltage source — an industrially manufactured apparatus SKAT-70, at the output of which it was possible to obtain 70 kV dc voltage and 50 kV ac voltage with a frequency of 50 Hz. The microprocessor control system for the device's operational performance and safety, made it possible to set the technologically necessary processing voltage with an accuracy of a few volts.

The measuring cells in which the electropulse treatment of the inoculum was carried out are represented by two structures: the first one is a closed container made of an electrically insulating material - fiberglass, with stainless steel electrodes located in the planes of the bottom and a removable cover, i.e. condenser type container; the second is an open container made of an electrically insulating material, Getinax, with brass electrodes installed at the top, along the long walls, between which high-voltage discharges were formed.

Electrotechnological installations allow changing processing voltage, selecting processing modes: "direct current", "alternating current", "pulse discharges", changing the pulse repetition frequency and, as a result, adjust the amount of energy supplied to the object, i.e. exposure dose, which gives an additional opportunity to choose the optimal modes of electrical treatment of seed.

Figure 8. Structural diagrams of experimental installations for treating seeds in an electric field of alternating and direct current (a): 1 - high voltage source - Skat-70; 2 - processing cell; 3 - kilovoltmeter spectral digital CEC-120; 4 - oscilloscope; electric pulse field (b): 1 - high voltage source - Skat-70; 2 - high voltage capacitor; 3 - controlled discharger - trigatron; 4 - processing cell; 5 - kilovoltmeter spectral digital CEC-120; 6 – oscilloscope

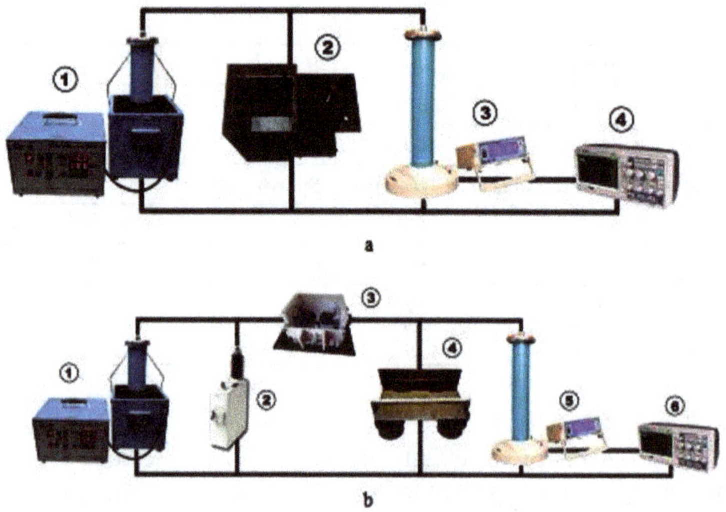

Under laboratory conditions, exploratory studies were conducted to determine the effect of ac stimulation by an electric field of an alternating current and a pulsed field on the germination energy and germination of Triticale seeds. When processing seeds in an electric field of direct and alternating current, the voltage on the electrodes was maintained the same and was 15 kV, and for the variant with a pulsed electric field - 25 kV.

As a result of the research conducted, it was noticed that the speed and friendly nature of germination depends on the impact parameters. A very interesting picture of the dynamics of the germination energy at an early stage of the experiment was revealed. The smallest number of abundantly germinated seeds on the second day was recorded in the variant using the electric field of alternating current with a distance from the top electrode to the seed layer being treated 7.0 cm (variant 1) - 4.3%. In variants with treatment in an electric field of alternating current with a distance to the seed layer of 1.0 cm (option 2) and in a pulsed electric field at the same distance (option 3), this indicator turned out to be significantly higher compared to the control, and amounted to - 23.0 and 28.5%, respectively; at the control - 7.0%.

On the third day of observation, the germination rate of the treated seeds increased on average by 2.1 ... 2.6 times relative to the seeds that were not subjected to electrical impact.

On the fourth day of the experiment, the speed and friendliness of germination of the inoculum, which was affected by the electrical influence, leveled out somewhat. The optimal mode in this regard turned out to be option 2, where the amicability of seed development was 96.0%. In variants 1 and 3, the germination energy was 92.5 and 92.0%, respectively, while on the control variant this indicator was twice as low.

Analysis of data on changes in seed germination revealed the following results. The most responsive were the seeds irradiated by an alternating electric field with a distance of 1.0 cm from the layer (Figure 9). Here, the number of normally germinated seeds in the sample was 98.7%, which is 4% more compared to the control variant. The germination of the processed grain in the other two variants was only slightly higher than that in the control.

The results of laboratory studies allowed to evaluate the modes of action of the electric field on the seed. The high result of the stimulating effect is explained by the acceleration of the course of biological processes in the semen due to the production of additional energy during processing.

Activation of the initial stages of development of seeds exposed to an electric field leads to a change in the morphological characteristics of seedlings (Table 6, Figure 7). In vegetative plants, the first internodes lengthen, a more powerful root system is formed compared to the control. Very important for the conditions of the arid zone is the fact that the growth of roots contributes to a more rapid rooting of plants, their better use of spring moisture and nutrients.

The degree of influence of electrostimulation depends on the processing mode. The best results were recorded when processing in a DC electric field and pulsed field. When comparing the processing time for all options, the best performance has an impact of 120 seconds. The increase in morphometric parameters of seedlings observed in this case undoubtedly contributes to an increase in yield.

This technique allows you to transfer the physiological state of the embryo, up to a change in the chemical composition, from the state of "rest" to the state of "awakening" and active growth. At the time of exposure to an electric field inside the seed, a redistribution of electric charges occurs, which, in turn, somewhat changes the course of physicochemical processes that affect the subsequent growth and development of plants.

Figure 9. The effect of irradiation of Triticale seeds by an electric field on germination: 1 - control; 2 –electric field of alternating current with a distance to the treated layer of 7.0 cm; 3 - alternating current electric field with a distance of 1.0 cm to the processed layer; 4 - pulsed electric field with a distance of 1.0 cm to the treated layer

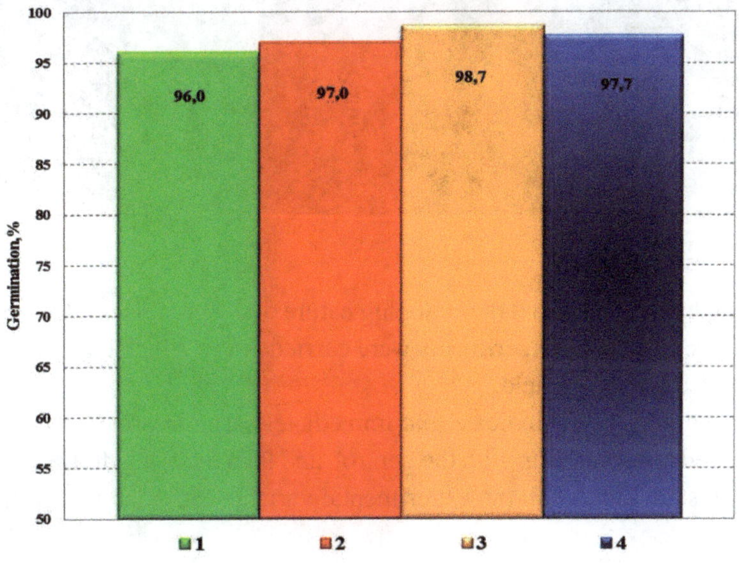

Table 6. Features of changes in morphometric parameters of barley seedlings under the influence of an electric field

Options		Number of Roots in One Plant	Length of Roots, cm	The Height of Seedlings, cm
Type of Electrical Impact	Time of Processing, s			
Control without treatment		5.1±0.4	9.4±0.4	6.3±0.3
Electric pulse field	60	5.6±0.9	11.3±0.2	8.2±0.4
	120	5.6±0.2	11.9±0.3	8.9±0.4
DC electric field	60	5.9±0.2	11.2±0.2	7.4±0.6
	120	6.0±0.7	12.1±0.5	7.2±0.8
AC electric field	60	5.1±0.6	8.9±0.2	5.0±0.3
	120	5,1±0.4	10.4±0.4	8.5±0.5

Figure 10. Results of pre-sowing treatment of barley seeds with an electric field - the phase for tillering

To test the existing hypotheses about the changing nature of the processes taking place in the internal structures of plant cells, a series of experiments were carried out in which the grain of various cultures was processed in an electric pulsed field.

The results obtained (Figure 11) indicate that immediately after treatment, the moisture content of the grain mass increased in comparison with the control, and then relatively fell down. At the same time, the value of grain moisture indices on the experimental variants exceeded that of the untreated grain.

Figure 11. The humidity dependence of the processed grain of agricultural crops from the dose of the acting energy

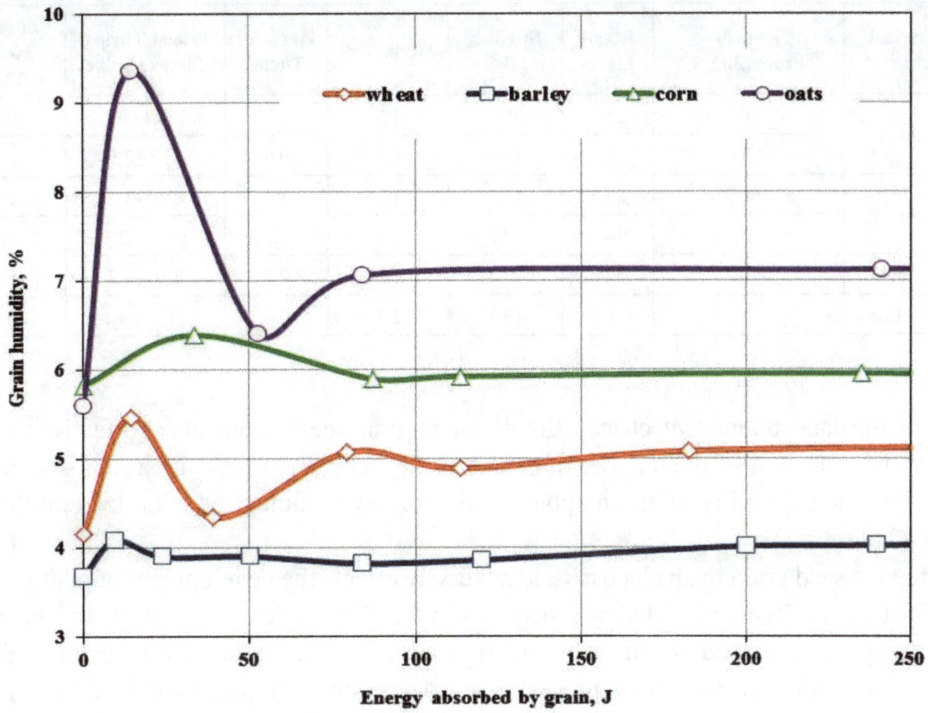

These observations indicate that under the influence of electrical effects changes occur in the cells of the seeds, as a result of which the bound moisture is mixed with the free.

In the modern literature there is a material testifying to the possibility of managing the phytosanitary condition of agrocenoses and the formation of elements of crop productivity through the use of electrophysical methods. These include pre-sowing seed treatment in an electric field. The use of this technique achieves an increase in seed resistance to stress factors, due to a decrease in electrical conductivity (Bobryshev, Starodubtseva & Popov, 2000).

The stimulating effect of the electric field on seeds includes regulatory mechanisms at all levels of the organization of the plant organism. The consequence of such changes is the formation of plant resistance to a complex of harmful factors, including to economically dangerous pests and infectious diseases.

During the field observations, it was established that the stimulation of inoculum contributes to a decrease in the number and harmfulness of phytophages by 46.5 ... 68.3% and 13.1 ... 88.5%, respectively (Table 7). Moreover, the effectiveness of this technique in protecting plants from various pests is directly dependent on the treatment mode. The highest results showed:

- against bread flea and mermoise - AC electric field with an exposure of 120 seconds;
- against the swedish fly - an alternating current electric field with an exposure of 60 seconds.
- against wheat thrips and harmful turtles - electric pulse field with an exposure of 60 seconds.

Table 7. The effectiveness of pre-sowing treatment of seeds of winter Triticale by an electric field

Processing Mode		Damaged Plants, %				Number, Pieces / Unit Accounting	
Type of Electrical Impact	Time of Processing, s	Bread Flakes	Swedish fly	Meromeisa	Harmful Turtle	Wheat Thrips, Pieces / Spike	Entomophagous, Pieces / Unit Accounting
AC electric field	60	25.0	9.4	1.5	1.1	103.1±3.8	6.1±0.05
	120	21.1	13.3	0.3	0.9	78.3±2.9	6.4±0.03
DC electric field	60	28.6	15.5	1.2	1.4	88.0±1.4	6.0±0.02
	120	34.8	17.2	0.7	1.2	118.7±2.1	6.1±0.09
Electric pulse field	60	52.6	12.4	1.0	0.7	61.0±1.3	6.8±0.07
Control without treatment		78.3	19.8	2.6	6.4	192.4±4.1	6.0±0.04

Judging by the data obtained, electrical stimulation of grain seeds is one of the effective methods that contribute to the reduction of the number of economically dangerous pests in agrobiosis, maintaining biodiversity and the possibility of entomophage activation by reducing the toxic load on the agroecosystem as a whole.

Processing the seed grain in an electric field adversely affects the development of pathogens of seed infection. The best results are provided by processing in the electric field of an alternating current for 60 seconds - the suppression of pathogens is observed at 5.5 ... 27.2%. To the greatest extent, this mode of pre-sowing seed treatment affects the representatives of the genus Helmintosporium, Fusarium. Seven days after irradiation, the mycelium of these fungi was sterile and no further sporulation was observed.

Seed treatment in an electric field contributes to an increase in the immunity and resistance of crops sown to economically dangerous infectious diseases (Table 8-9). The sensitivity of plants to leaf-stem pathogens is most increased. So, when the leaves are damaged by helminthosporia and Septoria, a small number of necrotic spots formed on them, and, in the further development of these diseases, it was not observed. The action of the electric field of alternating current with an exposure of 60 seconds turned out to be more effective - the intensity of development of leaf-stem infections decreased in barley by 5%, in triticale - by 5.4 ... 10.8%. The electric field of direct current to a lesser extent affected the immunity of plants. However, it should be noted that an increase in exposure to 120 seconds leads to a decrease in the intensity of development of leaf-stem pathologies by 4.2 ... 11.5%.

Root pathogens are the least responsive to electrical seed treatment. Impact on seeds by electric fields of different time exposures provides a reduction in the development of the incidence of vegetative plants by 3.1 ... 4.2%.

Along with the optimization of the phytosanitary condition of crops in the pre-sowing treatment of seeds by an electric field, its stimulating effect on the structure of the crop of grain crops was found (Table 10)

Analysis of the data obtained shows that electrical treatment of seeds contributes to an increase in productivity, tillering and the formation of a more dense productive stem-up to 500 ears / m^2.

Electrical stimulation has a positive effect on such a hard-to-adjust parameter as the earliness of ears - the number of grains in an ear increases by 13.8 ... 31%, the weight of 1000 grains slightly increases (3 ... 6.2%). The yield of grain to the total mass increased by 40.3 ... 57.8%.

Table 8. The effect of presowing treatment of seeds of grain crops by an electric field on the development of diseases, %

Barley							
Type of Electrical Impact	Time of Processing, s	Mealy Dew	Septoria	Dark Brown Spot	Bacteriosis	Helminthic Spore Fusarium Rot
Control without treatment		30.0	15.0	24.5	35.4	9.0
AC electric field	60	18.2	10.0	16.2	20.6	5.8
	120	20.9	12.1	18.6	23.1	7.5
DC electric field	60	20.6	13.5	20.7	23.9	7.5
	120	19.4	10.8	21.1	27.2	7.3
Electric pulse field	60	-	-	17.5	23.6	6.2
	120	-	-	18.3	23.5	6.3

Table 9. The effect of presowing treatment of seeds of grain crops by an electric field on the development of diseases %

Triticale					
Type of Electrical Impact	Time of Processing, s	Mealy Dew	Rust	Bacteriosis	Helminthic Spore Fusarium Rot
Control without treatment		19.9	15.8	11.2	1.2
AC electric field	60	12.5	5.0	5.8	0.4
	120	17.2	5.8	5.0	0.5
DC electric field	60	16.8	6.1	7.0	0.6
	120	17.0	7.6	8.2	0.7
Electric pulse field	60	16.5	5.4	8.3	0.4

Table 10. Influence of the type of electrical treatment of triticale seeds on the crop structure

Options		The Number of Productive Stems, Pieces / m²	Mass of Straw, g / m²	Spit Length, cm	The Number of Grains per Ear, pcs.	1000 Grain Weight	Yield Increase,%
Type of Electrical Impact	Time of Processing, s						
Control without treatment		64.0±2.1	61.1±1.9	7.4±0.3	46.5±2.3	37±3.1	-
AC electric field	60	96.0±5.0	102.6±3.1	8.5±0.6	42.2±1.5	36.7±3.0	+43.6
	120	73.0±1.1	80.9±2.3	8.4±0.4	41.8±1.6	37.7±2.4	+24.0
DC electric field	60	91.3±5.2	83.4±2.1	7.8±0.3	38.2±1.2	39.5±2.1	+21.6
	120	111.3±6.8	101.8±2.9	7.8±0.3	35.4±1.1	41.1±1.3	+29.8
Electric pulse field	60	104.3±6.2	91.0±2.0	8.1±0.4	41.1±1.3	37.9±1.7	+21.98

Thus, the processing of seed in an electric field has a positive effect on all elements of the crop structure. The greatest effect, combined in all respects, is achieved using a direct current electric field.

Considering the unfavorable weather conditions during the grain loading period, higher mass indices of 1000 grains on the variants with the use of a direct current electric field indicate an increase in the general level of plant drought resistance.

SOLUTIONS AND RECOMMENDATIONS

The research results indicate a positive effect of the treatment of plants and seeds with electrochemically activated water, an electric field, etc.

Presowing treatment of seeds with electrochemically activated water increases the germination and energy of seed germination, improves the development of plants, their morphological indicators, and others. The most pronounced positive effect is achieved through a combination of pre-sowing seed treatment with the treatment of vegetative plants provides the highest possible result.

FUTURE RESEARCH DIRECTIONS

The data obtained allow us to expand the boundaries of the use of resource-saving technology for managing reactions of biological objects, for example, at the expense of woody plants. This will reduce the level of anthropogenic impact, limit the use of chemical pesticides and significantly limit the population of trees and shrubs by pests and, as a result, the preservation of plant objects, increasing their sustainability and optimizing the environment.

CONCLUSION

Biologization of plant growing is the basis of energy-efficient production of agricultural products and allows you to solve many environmental problems. An important role in the formation of optimal growing conditions for plants is favored by the use of electrical effects and biologically active electroactivated agents.

Under the influence of these funds there is an improvement in the growth and development of plants, the suppression of pathogens and harmful insects. The analysis of the results indicates that their biota has an unequivocal effect, which manifests itself from the first stages of development, starting with the formation of a growing cone, which stimulates the growth of vegetative organs and the formation of a larger number of primary roots, creating a good basis for subsequent tillering and rooting of plants, stimulating and antistress effect on plants and suppress harmful organisms.

A characteristic feature of the use of tested tools is their ability to positively affect physiologically old and substandard seeds, determining not only the most acceptable parameters of crops, but also obtaining optimal yields.

The combination of presowing treatment of seeds with the treatment of vegetative plants provides the highest possible result.

REFERENCES

Batygin, N.F. (1980). Biologicheskiye osnovy predposevnoy obrabotki semyan i zony yeye effektivnosti [Biological basis of presowing treatment of seeds and the zone of its effectiveness]. *Sel'skokhozyaystvennaya biologiya [Agricultural Biology], 15*(4), 504-509.

Belitskaya, M. N., & Gribust, I. R. (2009). Stabilizatsiya i produktivnost' zernovykh agrotsenozov pri ispol'zovanii nanostrukturirovannogo rastvora bishofita [Stabilization and productivity of grain agrocenoses using nanostructured bischofite solution]. In Nanotekhnologii i nanomaterialy: sovremennoye sostoyaniye i perspektivy razvitiya v usloviyakh Volgogradskoy oblasti [Nanotechnologies and nanomaterials: current state and development prospects in the conditions of the Volgograd region] (pp. 62-65). Volgograd: Izdatel'stvo VolgGU [VolgGU Publishing].

Belitskaya, M. N., & Gribust, I. R. (2011). Izucheniye vozmozhnostey ispol'zovaniya nanostrukturirovannogo rastvora bishofita v rasteniyevodstve [Study of the possibilities of using nanostructured bischofite solution in plant growing]. In Nanotekhnologii i nanomaterialy: sovremennoye sostoyaniye i perspektivy razvitiya v usloviyakh Volgogradskoy oblasti [Nanotechnologies and nanomaterials: current state and development prospects in the conditions of the Volgograd region] (pp. 42-46). Volgograd: Izdatel'stvo VolgGU [VolgGU Publishing].

Belitskaya, M. N., Gribust, I. R., & Nefedieva, E. E. (2012). Ustanovleniye reglamentov primeneniya EKHA vody v zernovykh agrotsenozakh [Establishment of regulations for the use of ECA water in grain agrocenoses]. *Izvestiya Nizhnevolzhskogo agrouniversitetskogo kompleksa: Nauka i vyssheye professional'noye obrazovaniye* [Proceedings of the Nizhnevolzhsky agrouniversity complex: Science and higher professional education], 2 (26), 3-8.

Bobryshev, F. I., Starodubtseva, G. P., & Popov, V. F. (2000). Effektivnyye sposoby predposevnoy obrabotki semyan [Effective methods of pre-sowing seed treatment]. *Zemledeliye, 3*, 45.

Booth, A. I. (1992). *Elektronno-ionnyye protsessy vodnykh struktur zhivykh organizmov i produktov ikh pererabotki* [Electron-ion processes of aquatic structures of living organisms and their products]. Moscow.

Cruz, J. A., Salbilla, B. A., Kanazava, A., & Kramer, D. M. (2001). Inhibition of plastocyanin to P_{700} electron transfer in Chlamydomonas reinhardtii by hyperosmotic stress. *Plant Physiology, 127*(3), 1167–1179. doi:10.1104/pp.010328 PMID:11706196

Dmitriev, A. M., & Stratskevich, L. K. (1986). *Stimulyatsiya rosta rasteniy* [Stimulation of plant growth]. Minsk: Uradzhay.

Dzhurabaev, M. (1986). Primeneniye elektroaktivirovannoy vody v sel'skom khozyaystve [Use of electro-activated water in agriculture]. *Mekhanizatsiya i elektrifikatsiya sel'skogo khozyaystva [Agricultural mechanization and electrification], 11*, 51-53.

Goncharov, A. A., Berezhnaya, I. E., & Gursky, I. G. (1994). *Vliyaniye fizicheskikh metodov predposevnoy obrabotki semyan sel'skokhozyaystvennykh kul'tur na ikh semennyye kachestva* [The influence of physical methods of pre-sowing treatment of seeds of agricultural crops on their seed qualities]. Zelenograd.

Goryagina, E. B. (2011). Predobrabotka nezrelykh zarodyshey gibridov rapsa yarovogo i gorchitsy beloy na etape vvedeniya v kul'turu in vitro [Pre-treatment of immature embryos of spring rape and white mustard hybrids at the stage of introduction into culture in vitro]. In *Materialy VI mezhdunarodnoy konferentsii molodykh uchenykh i spetsialistov [Proceedings of the VI International Conference of Young Scientists and Specialists]* (pp. 66-69). VNIIMK.

Kasakova, A. S., Yudaev, I. V., Fedorishchenko, M. G., Mayboroda, S. Y., Ksenz, N. V., & Voronin, S. M. (2018). New approach to study stimulating effect of the pre-sowing barley seeds treatment in the electromagnetic field. *Online Journal of Biological Sciences*, *18*(2), 197–207. doi:10.3844/ojbsci.2018.197.207

Kefeli, V. I., & Tom, S. I. (1985). *Rost rasteniy i yego regulyatsiya (geneticheskiye i fiziologicheskiye aspekty)* [Plant growth and its regulation (genetic and physiological aspects)]. Chisinau: Shtiintsa. [Shtiints]

Labutina, E. V. (1988). *Vozdeystviye sochetaniya magnitoaktivirovannoy vody, vodno-go i pitatel'nogo rezhimov pochvy na formirovaniye urozhaya tomatov pri dozhdeva-nii v usloviyakh Volgo-Akhtubinskoy poymy* [The impact of a combination of magnetically activated water, water and nutrient regimes of the soil on the formation of a tomato crop during rainfall under the conditions of the Volga-Akhtuba floodplain] [dissertation]. Volgogradskiy gosudar-stvennyy universitet [Volgograd State University], Volgograd.

Larionova, A. P. (1993). Biologicheskoye obosnovaniye obrabotki semyan zlakovykh kul'tur i donnika elektricheskim polem koronnogo razryada [Biological substantiation of the treatment of seeds of cereals and clover electric field of the corona discharge]. In Sovershenstvovaniye nauchno-teoreticheskogo i metodicheskogo urovnya prepodavaniya fiziologii rasteniy [Improving the scientific-theoretical and methodological level of teaching plant physiology] (pp. 63). Smolensk.

Mikitinets, Z. G., & Kanibolotsky, V. G. (1984) Vliyaniye omagnichennoy vody na fi-ziologicheskiye protsessy rasteniy [The effect of magnetized water on the physiological processes of plants]. In Vsesoyuznoye nauchno-proizvodstvennoye sove-shchaniye po primeneniyu opticheskogo izlucheniya v sel'skokhozyaystvennom proizvodstve pri vypolnenii Prodovol'stvennoy programmy [All-Union Scientific and Production Meeting on the use of optical radiation in agricultural production during the implementation of the Food Program] (pp. 53). L'vov [Lviv].

Novitsky, Yu. I. (1973). Magnitnoye pole v zhizni rasteniy [Magnetic field in plant life]. In. V.N. Chernihiv (Ed.), Problemy kosmicheskoy biologii [Problems of space biology] (pp. 164-188). Moscow: Nauka.

Novitsky, Yu. I. (1987). Reaktsiya rasteniy na magnitnyye polya. [The reaction of plants to magnetic fields] In Yu. A. Kholodov (Ed.), *Reaktsiya biologicheskikh sistem na magnitnyye polya* [The Reaction of Biological Systems to Magnetic Fields] (pp. 117–130). Moscow: Nauka.

Novitsky, Yu. I., Strekova, V. Yu., & Tarakanova, G. A. (1971). Deystviye postoyannogo magnitnogo polya na rost rasteniy. [The effect of a constant magnetic field on plant growth] In Yu. A. Kholodov (Ed.), *Vliyaniye magnitnykh poley na biologicheskiye ob"yekty* [Effect of magnetic fields on biological objects] (pp. 69–88). Moscow: Nauka.

Osadchenko, I. M., Kharchenko, O. V., & Churzin, V. N. (2009). Povysheniye posevnykh kachestv semyan arbuza, dyni i kabachka s primeneniyem biologicheski aktivnykh veshchestv [Increase of sowing qualities of seeds of watermelon, melon and zucchini with the use of biologically active substances]. In *Izvestiya Nizhnevolzhskogo agrouniversitetskogo kompleksa: Nauka i vyssheye professional'noye obrazovaniye* [Proceedings of the Nizhnevolzhsky agrouniversity complex: Science and higher professional education] (pp. 49-53).

Pakhomova, V. M. (1995). Osnovnyye polozheniya sovremennoy teorii stressa i nespe-tsificheskiy adaptatsionnyy sindrom u rasteniy [The main provisions of the modern theory of stress and non-specific adaptation syndrome in plants]. *Tsitologiya, 37*(1), 66–91.

Pietruszewski, S., Muszynski, S., & Dziwulska, A. (2007). Electromagnetic field and electromagnetic radiation as non-invasive external stimulants for seeds (selected methods and responses). *International Agrophysics, 21*, 95–100.

Polevoy, V. V. (2001). Fiziologiya tselostnosti rastitel'nogo organizma [Physiology of the integrity of the plant organism]. *Fiziologiya rasteniy, 48*(4), 631–643.

Putintsev, A. F. (1997). Obrabotka semyan elektromagnitnym polem [Seed treatment with electromagnetic field]. *Zemledeliye, 4*, 45.

Pyndak, V. I., Lagutin, V. V., & Yushkin, A. V. (2001). Perspektivy primeneniya ekologicheski chistykh aktivirovannykh vodnykh rastvorov v rasteniyevodstve [Prospects for the use of environmentally friendly activated aqueous solutions in crop production]. Povolzhskiy ekologicheskiy vestnik [Volga ecological journal], 8, 119-122). Volgograd: Izdatel'stvo VolgGU [VolgGU Publishing].

Savin, V. N. (1981). *Deystviye ioniziruyushchego izlucheniya na tselostnyy rastitel'nyy organizm* [The effect of ionizing radiation on an integral plant organism]. Moscow: Energoizdat.

Tarchevsky, I. A. (2002). *Signal'nyye sistemy kletok* [Cell signaling systems]. Moscow: Nauka.

Urmantsev, Yu. A., & Gudskov, N. L. (1986). Problema spetsifichnosti i nespetsifichno-sti otvetnykh reaktsiy rasteniy na povrezhdayushcheye vozdeystviye [The problem of specificity and nonspecificity of plant responses to damaging effects]. *Zhurnal Obshchei Biologii [Journal of General Biology, 48*(3), 337–349.

Vartapetian, B. B. (1985). *Anaerobioz i strukturno-funktsional'nyye perestroyki rastitel'noy kletki* [Anaerobiosis and structural and functional rearrangements of the plant cell]. (pp. 175–198). Moscow: Nauka.

Vasilev, S. I., Mashkov, S. V., Syrkin, V. A., Gridneva, T. S., & Yudaev, I. V. (2018). Results of studies of plant stimulation in a magnetic field. *Research Journal of Pharmaceutical, Biological and Chemical Sciences, 9*(4), 706–710.

Velicheva, Z. M., & Velichev, P. G. (1993). Vliyaniye oblucheniya semyan lyupina svetom He-Ne lazera na nekotoryye fiziologo-biokhimicheskiye osobennosti semyan i rasteniy, ikh produktivnost' i kachestvo urozhaya [The effect of irradiation of lupine seeds with He-Ne laser light on some physiological and biochemical characteristics of seeds and plants, their productivity and crop quality]. In Sovershenstvovaniye nauchno-teoreticheskogo i metodicheskogo urovnya prepodavaniya fiziologii rasteniy [Improving the scientific-theoretical and methodological level of teaching plant physiology] (pp. 65).

Veselovsky, V. A., Veselova, T. V., & Chernavsky, D. S. (1993). Stress rasteniy. Biofizicheskiy podkhod [Stress plants. Biophysical approach]. *Fiziologiya rasteniy*, *40*(3), 553–557.

ADDITIONAL READING

Wilson, J. M. (1978). Leaf respiration and ATP levels at chilling temperatures. *The New Phytologist*, *80*(2), 325–334. doi:10.1111/j.1469-8137.1978.tb01565.x

Chapter 12
Potential Evaluation and Best Practices of Solar Power Plant Application in Rural Areas

Yuliia Daus
https://orcid.org/0000-0001-9120-7637
Don State Agrarian University, Russia

Valeriy Kharchenko
https://orcid.org/0000-0003-3725-2976
FNAC VIM, Russia

Igor Yudaev
Don State Agrarian University, Russia

Vera Dyachenko
Zaporizhzhya National Technical University, Ukraine

Shavkat Klychev
Academy of Science, Uzbekistan

Sergey Rakitov
Suburban MPG Branch of PJSC "Volgogradoblelectro," Russia

ABSTRACT

Recreational rest and recovery zones require a daily hot water supply, and constant availability of electricity. Therefore, the need for renewable energy sources usage in the Lower Volga region is obvious and power plants with an environmentally friendly component significant in the region. An analysis of the theoretically calculated potential renewable energy makes it possible to optimistically assert that the region is promising for autonomous renewable energy source implementation. It may be noted that potential wind and solar energy, in spite of the fact that it is distributed unevenly, is sufficient to provide energy for remote rural communities and tourist facilities. An analysis was conducted on the availability of actinometric data required for heliotechnical calculations and identified areas of applicability and accuracy of the information received from various meteorological information sources.

DOI: 10.4018/978-1-5225-9420-8.ch012

INTRODUCTION

Current trends in the development of the global energy industry and the ever-increasing energy consumption dictate the need to search for non-traditional energy sources.

Today, traditional energy resources (hydrocarbon fuels, water resources) and technologies that transform them into available forms of energy are not always able to provide the required level of energy consumption to the world economy and the population's need for it. In addition to the energy requirements, the economic sector also faces the exhaustion of raw materials and numerous environmental problems arising because of the extraction, production and transportation of energy.

The most promising energy supplier today and in the near future is renewable energy, which includes: solar, wind, geothermal energy, energy of sea waves, tides and ocean, biomass, wood, charcoal, peat, cattle, shale, tar sands and the hydropower of large and small watercourses. The economic potential of renewable energy sources is currently estimated at 200 billion tons of fuel equivalent per year, which equals twice the annual global production of all types of fossil fuels.

The main advantage of renewable energy sources is their inexhaustibility and environmental friendliness. The use of power plants that are operated on the fundamental principles of renewability does not change the energy balance of the planet and does not violate the whole picture of the earth's ecology. These qualities caused the rapid development of renewable energy industry in the world and very optimistic forecasts of its development in the coming decade. Renewable energy sources play a significant role in solving the three global problems that humanity faces: energy, ecology, food.

The development of renewable energy in Russia is often declarative, in spite of the fact that at the governmental level the development rates have been approved, and quantitative indicators have been agreed by year. At the same time, the concept of renewable energy is replaced by the concept of using secondary resources of industry and agriculture, which makes it possible to account for the fulfillment of the goals set – the development of hitherto unnecessary energy emitted into the atmosphere and the use of secondary components plunged into the production process at a new level for solving the corresponding tasks. But the development of this innovative industry of renewable energy application is, first of all, not the substitution of concepts, but a real application of modern facilities, modernization of existing ones, estimation of the potential for the use of renewable energy, the attempt to obtain economically beneficial energy that many remote settlements and territories need with quality indicators that can be predicted in advance.

Therefore, at present, experience in renewable energy application in rural areas, in the region of arid farming, in areas where agricultural production is widely developed, especially livestock farming, at remote territories where agricultural producers breed sheep farms, cattle farms, etc., is of particular importance in the country. These settlements are often located in the steppe zone and the ability to be connected to the centralized power supply is absent or requires such a great infusion of funds and moneys, which makes farmers seriously think about stopping the continuation of the normal functioning of these industries. In these conditions renewable energy sources can really help them to solve the most part of described problems.

In order to characterize the region in terms of attractiveness for the renewable energy intrusion, it is necessary to assess the potential of renewable energy in these areas and, above all, assess the solar energy potential, identify regions that are promising for the application of autonomous sources of renewable energy. It can be noted that the energy potential of the Sun, despite the fact that it is distributed unevenly, is sufficient to provide energy to remote rural settlements.

At present, the program for providing renewable energy from remote settlements of agricultural producers is being implemented in the region of southern Russia, and the attractiveness of the Lower Volga region as an extensive remote recreation zone suggests the prospect of introducing such sources in recreation centers, tourist sites and in boarding houses.

BACKGROUND

The application of solar power installations is caused by the inexhaustibility of the primary energy source, the environmental cleanliness of receiving electrical energy, the ease of exploitation and operation of the equipment, and the possibility of ubiquitous distribution throughout the globe. Photovoltaic energy is now characterized by intensive and accelerated development, which is particularly evident in the last ten years, despite the relatively low cost of hydrocarbons and fuel based on it. For example, in ("Annual Energy Outlook 2017"), using the example of the United States, there was substantiated forecast for the period from 2020 to 2040 according to which there is going to be observed increase in the conversion of various types of energy into electricity, which should be characterized by the substantial parallel growth of consumed natural gas and solar energy, with the observed trend of a decline in the processing of coal and fuel oil, and reducing the share of nuclear energy. Also (Ospina, & Quijano, 2016) indicates the record growth in 2016 of the number of input capacities of photovoltaic facilities that is more than 25% in 2015, and the leaders here are countries such as Germany, Japan, China, the United States and others.

Such dynamics, above all, are associated with the decrease in the cost of implementing projects and the price of equipment for solar power plants. All this is achieved mainly by reducing the specific capital costs and improving the efficiency of photovoltaic modules ("Report on the development of renewable energy sources and proposals for the energy strategy of Russia"). The cost of 1 W of installed power of the photovoltaic module over the past 13 years has decreased more than by 2 times, and the efficiency over 22 years has increased for polycrystalline modules in 1,5 times, thin-film – by 20%, monocrystalline – by 25%. (Winter, 1991)

It should also be noted that there is a significant potential not only in further reduction of the cost of photovoltaic modules, but also in reduction of the cost of individual elements of power plants based on them (conversion and storage systems, controllers, installation equipment, etc.), since their cost contribution to the increase in the total cost is significant - the solar modules themselves make up only 45-50% of the total capital investment in the project that is designed for implementation.

Despite the fact that the Russian Federation is not in the list of global solar energy flagships, it is not inferior to them in the relative rates of construction and commissioning of new solar generation facilities. By 2024, it is planned to put into operation solar power plants with a total capacity of 1600 MW with a high degree of localization from 55% to 100% (Amerhanov et al., 2008).

Today, according to various estimates, in Russia a large part of the commissioned solar generation capacity falls on private households and small production facilities, but at the same time more and more often there are described large solar power plants such as the Ltd AltEnergo with a capacity of 100 kW in the Belgorod Region, Bugulchanskaya and Baribayevskaya power plants in the Republic of Bashkortostan with the capacity of 15 and 10 MW, respectively, the Kosh-Agachskaya station of 10 MW in the Republic of Altai, etc.

Until 2020, it is planned to build several photovoltaic power plants with the total installed capacity of 1.52 GW (Popel et al., 2010), here they are: 5 solar power plants with the capacity of 15 MW in the Astrakhan region, 6 power plants with the capacity of 15 MW in the Volgograd region, 4 power plants with the capacity of 15-25 MW in the Republic Kalmykia, 4 power plants with the capacity of 15-50 MW in the Stavropol territory (Popel et al., 2010; Bezrukih et al. (2008). In this case, the predicted cost of 1 kW of installed power is 122-130 thousand rubles.

But, despite major investments and the growing interest in generating facilities of this type, the share of renewable energy sources, including solar power plants, is only 0.2% with the number of hours of use of installed capacity 8.43% of calendar time ("Report on the functioning of the UES of Russia in 2015"). Electricity production by solar power plants in the whole country amounted to 9.91 million kWh / year ("United Energy System of Russia: Interim Results," 2016)

MAIN FOCUS OF THE CHAPTER

All previously submitted data are contained in summarized reports based on integrated performance indicators of the energy sector of the economy. This, above all, includes the so-called "utility-scale" solar power plants connected to the Unified Energy System of the country and generating electrical energy on an industrial scale. Many researchers, calling the main obstacle to the widespread introduction of photovoltaic plants, the high cost of 1 kWh of electrical energy, argue that one of the ways to reduce it is to use low-power plants, both networked and autonomous, used for households, small industrial enterprises, social and health facilities, recreational areas, remote sites and agricultural production. Distributed generation facilities also fall into above categories, the projected growth for the next nine years in the world is estimated at 200%. In ("Next Generation Wind and SolarPower," 2016) it is noted that the era of the race for low cost and high efficiency ends, and there begins the time to search for new approaches to the selection of the composition, parameters of the elements of low-power solar plants,

The purpose is to assess the potential of solar energy in rural areas and analyze the possibility of its implementation on the example of photovoltaic power plants of various capacities and configurations for power supply of consumers in rural areas.

CALCULATION MODEL OF HOURLY SUMS OF THE SOLAR RADIATION COMPONENTS ON THE RECEIVING SURFACE OF VARIOUS SPATIAL ORIENTATION

To estimate the potential of solar energy at the certain geographic point according to (Atwater, & Ball, 1978), the calculation of hourly sums of total solar radiation Rsum for day n at time t under real cloud conditions is performed relatively arbitrarily tilted at the angle β receiving surface with azimuth γ as follows:

$$R_{sum} = R_{dr\beta\gamma} + R_{df\beta\gamma} + R_{rf\beta\gamma} = R_{drH} \cdot \frac{\cos u}{\cos\theta} + R_{dfH} \cdot \left(\frac{1+\cos\beta}{2}\right) + \left(R_{drH} + R_{dfH}\right) \cdot r_7(n) \cdot \left(\frac{1-\cos\beta}{2}\right) \tag{1}$$

where $R_{dr\beta\gamma}, R_{df\beta\gamma}, R_{rf\beta\gamma}$ – hourly sums of direct, diffused and reflected intensity of solar radiation for day n at time t on the receiving surface arbitrarily inclined at the angle β with azimuth γ, kWh/m²; $r_7(n)$ – earth reflection coefficient (albedo), u – the angle of coming flux of direct solar radiation on the receiving surface with azimuth γ arbitrarily inclined at the angle β > 0 is determined according to (Liu, & Jordan, 1961) using the formula:

$$\cos u = \left(\begin{array}{l} \sin\phi\left(\sin\delta\cos\beta + \cos\delta\cos\gamma\cos\omega\sin\beta\right) \\ +\cos\phi\left(\cos\delta\cos\omega\cos\beta - \sin\delta\cos\gamma\sin\beta\right) + \cos\delta\sin\gamma\sin\beta\sin\omega \end{array}\right) \tag{2}$$

According to the Byrd's method (Aronova, & Grilihes, 2006), the calculation of the intensity of direct and diffused solar radiation under clear sky is based on the following system of equations:

$$\begin{cases} R^{cs}_{drH} = E_c \cdot \cos\theta \cdot \tau_{dr}, ec;\ 8\cos\theta > 0 \\ R^{cs}_{dfH} = E_c \cdot \cos\theta \cdot \tau_{df} \\ R^{cs}_{drH} = 0, ec;\ 8\cos\theta < 0 \end{cases} \tag{3}$$

where E_c=1,367 kWh/m² – normal density of solar radiation in space; θ – the angle of coming of sunlight on the receiving surface that is horizontal to the surface of the Earth (hereinafter referred to as the horizontal receiving surface or platform), which depends on the latitude of the terrain φ, time t and the day number n; τ_{dr}, τ_{df} – the coefficient that takes into account the refraction and absorption, respectively, of direct and diffused solar radiation in the atmosphere by ozone, gas mixture, water vapor, and Rayleigh and aerosol scattering of solar radiation.

To take into account the effect on the insolation characteristics of the processes occurring in the atmosphere, coefficients that take into account the actual atmospheric conditions for the direct and diffused radiation components were introduced into the system of equations (3), the values of which will be determined basing on the average monthly values of the direct and diffused solar radiation per day per 1 m² of the horizontal receiving area, obtained on the basis of multiyear actinometric measurements, as well as a result of calculations using the method of applying the daily profile solar radiation captivity under perfectly clear skies:

$$\begin{cases} R_{drH} = R^{cs}_{drH} \cdot K^{r}_{\mathrm{dr\ i}} \\ R_{dfH} = R^{cs}_{dfH} \cdot K^{r}_{\mathrm{df\ i}} \end{cases} \tag{4}$$

where $R_{drH}^{cs}, R_{dfH}^{cs}$ – the intensity of direct and diffuse solar radiation under clear sky, calculated by the formula (3).

Coefficients of actual atmospheric conditions for direct and diffused solar radiation K_{dri}^{r} и K_{dfi}^{r} for i- th month are calculated according to the condition of minimum relative error ($\varepsilon_{dri}, \varepsilon_{dfi}$):

$$\min\left(\varepsilon_{dri}\right) = \min\left\{\left|\frac{DRH_i \cdot \left(n_{2i} - n_{1i}\right) - \sum_{n=n_{1i}}^{n_{2i}} \sum_{t=1}^{24} R_{drH}}{DRH_i \cdot \left(n_{2i} - n_{1i}\right)}\right|\right\} \tag{5}$$

$$\min\left(\varepsilon_{dfi}\right) = \min\left\{\left|\frac{DFH_i \cdot \left(n_{2i} - n_{1i}\right) - \sum_{n=n_{1i}}^{n_{2i}} \sum_{t=1}^{24} R_{dfH}}{DFH_i \cdot \left(n_{2i} - n_{1i}\right)}\right|\right\} \tag{6}$$

where DRHi, DFHi – average monthly values of daily sums of direct and diffused solar radiation, respectively, per 1 m^2 of the horizontal receiving surface, obtained on the basis of actinometric measurements for the i-th month, the source of which is the electronic base of the National Aeronautics and Space Administration of the USA (NASA) ("NASA Atmospheric Science Data Center") kWh/(m^2·day); n_{1i}, n_{2i} – the ordinal number of the day of the beginning and end of the month respectively.

By the method of ordered search, the area of projections of the matrix of the coefficient of actual atmospheric conditions is established: $0 \leq K_{dri}^{r} \leq 1$, $1 \leq K_{dfi}^{r} \leq 2$ to adjust the intensity of total solar radiation (4), corresponding to the grid of geographical coordinates of the NASA base.

The range of data variation was expanded by approximating the obtained values of the coefficients of actual atmospheric conditions. To extend the range of identification of accounting coefficients (5), (6) from geographic coordinates (latitude and longitude), cubic spline interpolation and approximation methods were used for each month (Daus, Kharchenko, & Yudaev, 2016):

$$\begin{cases} K_{dr\,i}^{r} = K_{dr\,i}^{r}(\phi,\lambda) = \sum_{\xi}\left(\sum_{\vartheta} C_{Kdr\vartheta,\xi} \cdot \lambda_i^{\vartheta}\right)\phi_i^{\xi}, \\ K_{df\,i}^{r} = K_{df\,i}^{r}(\phi,\lambda) = \sum_{\xi}\left(\sum_{\vartheta} C_{Kdf\vartheta,\xi} \cdot \lambda_i^{\vartheta}\right)\phi_i^{\xi} \end{cases} \forall i \in \{1,2...12\} \tag{7}$$

where λ – longitude of the terrain,°E, $C_{Kdr\vartheta,\xi}$ and $C_{Kdf\vartheta,\xi}$ – polynomial coefficients for equations of direct and diffused radiation, respectively.

CALCULATION OF SOLAR RADIATION INTENSITY ON VARIOUSLY-ORIENTED SURFACES FOR RURAL TERRITORIES

Figure 1 shows the values of the annual total solar radiation on the horizontal surface (kWh/m^2 per year) for the areas of the Volgograd region according to ("NASA Atmospheric Science Data Center").

Most often, solar power plants are connected to the existing internal networks and communications on the upgraded or designed object, and are placed on the roofs, the shape and design of which can vary from flat to sloping with different angles of inclination (Daus, & Yudaev, 2016; Daus, Kharchenko, & Yudaev, 2018).

Flat roofs are the simplest for design, since their shape repeats the surface of horizontally located heliopoles and for which design requirements are now quite extensively developed (Daus, & Yudaev, 2016; Daus, Kharchenko, & Yudaev, 2018). Therefore, there are considered the configuration of the roofs with pitched constructive solutions.

There were calculated the intensity of solar radiation for the receiving surface that is planned to be placed at the optimum angle of inclination of gable roof (Daus, Yudaev, & Stepanchuk, 2018) 20–50° with orientation to the east, southeast, south, southwest and west, according to certain geographical coordinates on the example of the city of Volzhsky. The calculated angle of inclination of the pitched roof was taken of 30°.

The results of calculating the annual values of the total insolation on differently oriented areas (slopes of roofs) both in relation to the horizon and to the cardinal points are presented at Figure 2.

Figure 1. Annual total solar radiation on the horizontal surface for the rural territories of the Volgograd region

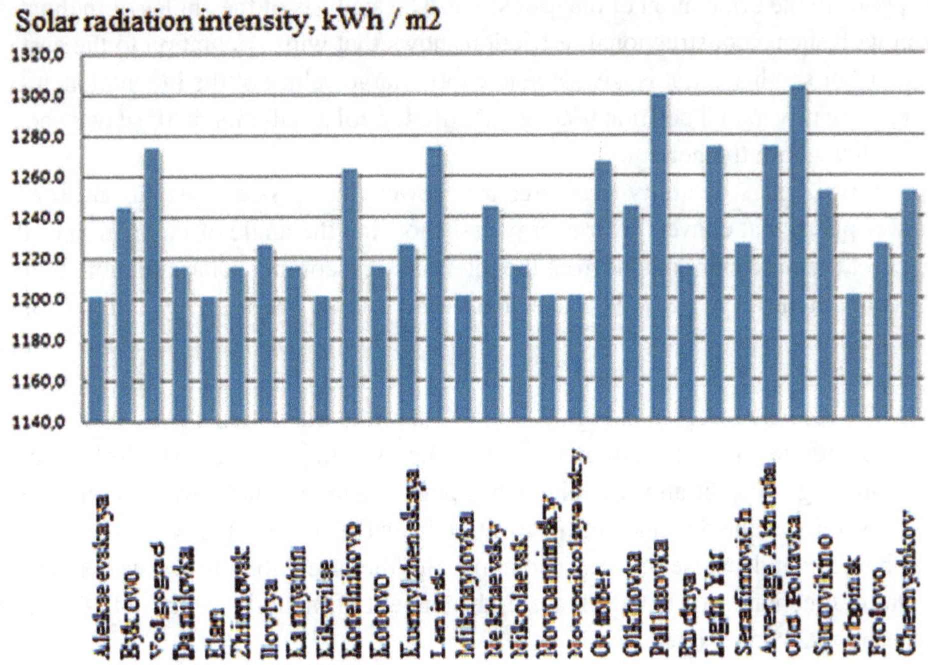

Figure 2. Annual values of solar radiation on variously oriented surfaces

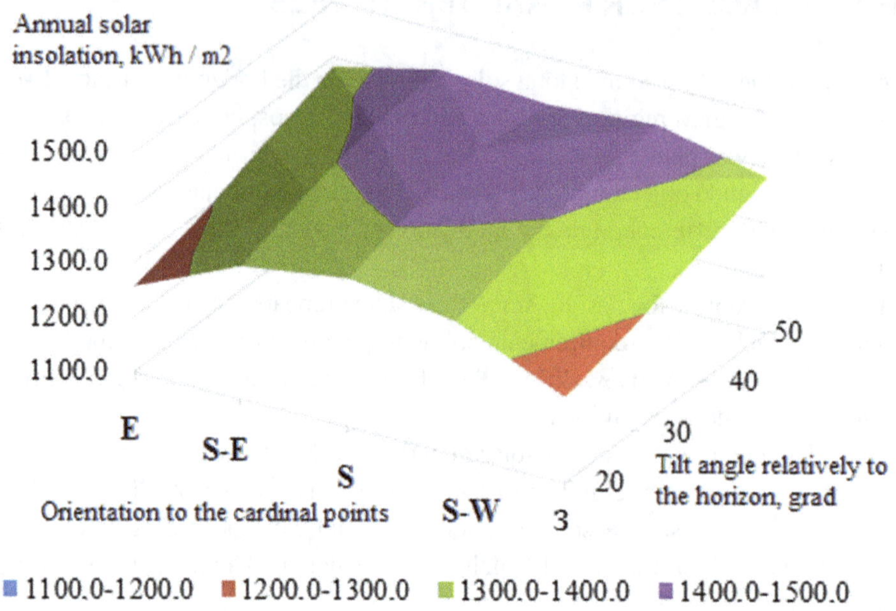

Under the climatic conditions of the city of Volzhsky, the horizontal surface for the year receives 1231.98 kWh/m² of solar radiation. It can be seen from Figure 1 that the potential of the utilized solar energy can be increased by varying the orientation of photovoltaic panels in space. So, the energy flow reaches the maximum value at the south-east or south-west orientation with the tilt angle relative to the horizon of 50°. If the tilt angle is below 50°, the southward position is the most advantageous one. However, when installing solar power plant on the roof of the building, the orientation to the cardinal points corresponds to the orientation of the roof slope. The analysis of the angles of inclination relative to the horizon with such a constructional restriction shows that with orientation to the east or west, as well as south-east or south-west, it is advisable to choose large values of the inclination angles relative to the horizon, since this spatial position is aimed at utilizing solar radiation in the dawn and sunset time when the sun is low above the horizon.

Since preventive repairs of utility structures and power transmission systems are seasonal, when designing power plants that convert solar energy and choosing the angle of inclination of their receiving surface, to get the maximum output from the intensity of incoming solar radiation, there should be analyzed the data obtained on the basis of calculating the sum of the intensity of total solar radiation depending on the time of year. Figure 3 presents the calculated seasonal sums of total solar radiation that is incident on receiving surfaces of different orientation on the cardinal points with a tilt angle of 30°.

Analysis of the results presented at Figure 3 shows that the orientation should be considered as the south orientation with the same angle of inclination of the roof surface relative to the horizon. However, using roofs of buildings to accommodate solar power plants on them is not always possible. The calculation results also show that the deviation of the location from the south to the south-east or south-west is characterized by a slight decrease in solar insolation. But the orientation to the west or east compared with the southern direction leads to a decrease in the amount of incoming solar radiation, respectively, from 7% in spring to 22% in winter.

Figure 3. Values of total insolation by seasons for the receiving surface at the angle of 30° to the horizon with different orientations to the cardinal points

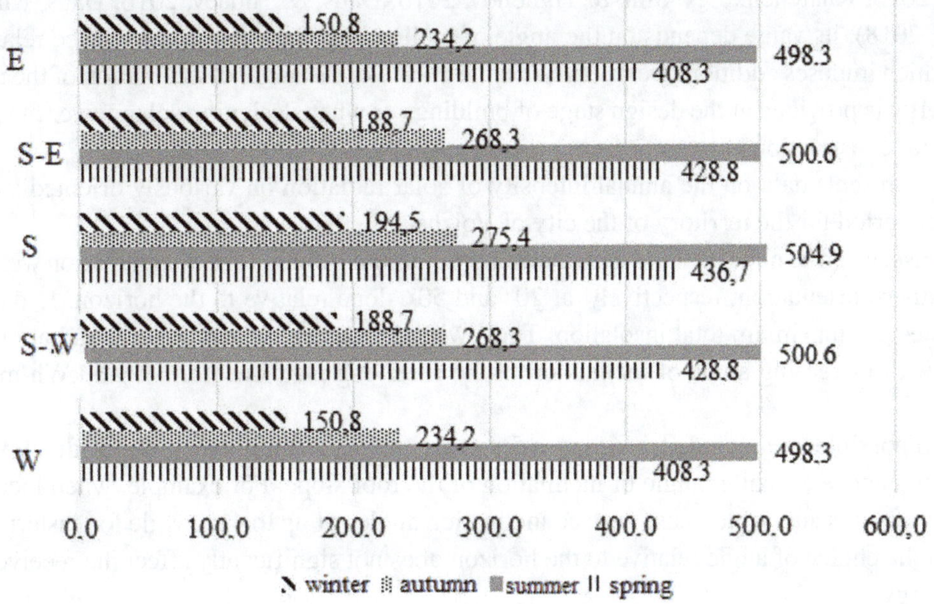

When choosing and completing a solar energy conversion system, it is necessary to know the proportion of incoming direct, diffused and reflected radiation. Such a ratio of the components of solar radiation, which reaches the surface of the receiving site with an inclination angle of 30° relative to the horizon and orientation to the south-east (south-west), taking into account the specific season for the city of Volzhsky, is shown at Figure 4.

Figure 4. Components of total solar radiation falling on the inclined surfaces

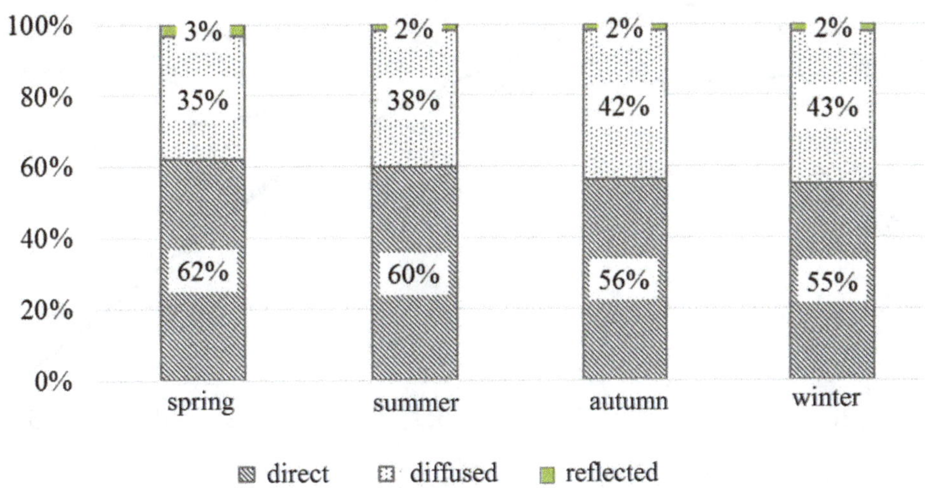

As can be seen from Figure 4, the diffused component ranges from 35% to 43% in different seasons of the year; therefore, it cannot be neglected. According to the calculation formulas (Ekadewi, & Djatmiko, 2013; Kharchenko, Nikitin, & Tikhonov, 2010; Daus, & Yudaev, 2016; Daus, Kharchenko, & Yudaev, 2018), its value depends on the angle of inclination of the receiving surface relative to the horizon, which imposes additional restrictions on the choice of the angle of inclination of the roof slope, if solar energy is provided at the design stage of buildings or when designing solar power plants located on horizontal constructions.

Figure 5 presents data on the annual intensity of solar radiation on variously oriented surfaces for the summer period for the territory of the city of Volzhsky.

In this case, it can be noted that the maximum and minimum values of the total radiation were detected for the southern orientation, respectively, at 20° and 50° slope relative to the horizon. Radiation does not guarantee the maximum total insolation. Thus, with the eastern orientation, the radiation intensity decreases with increasing angle of inclination of the receiving platform from 499.6 kWh/m^2 to 489.1 kWh/m^2.

It is also possible to conclude from Figure 5 that for this orientation along the cardinal points, it is preferable to choose a smaller angle of inclination of the roof slope. For example, when located in the south direction, it is advisable to use surface inclination angles of up to 41°, while for eastern (western) orientation the choice of angle relative to the horizon does not significantly affect the received amount of solar energy.

In addition, when the angle of inclination of the receiving surface is from 20° to 35°, the southern direction can be recommended, and at tilt angle of above 35°, the eastern orientation (western) should be considered as the best orientation. If it is impossible to provide such a solution constructively, then

Figure 5. The dependence of the annual amount of solar radiation on the inclination angle of the receiving surface and its orientation to the cardinal points for the summer period

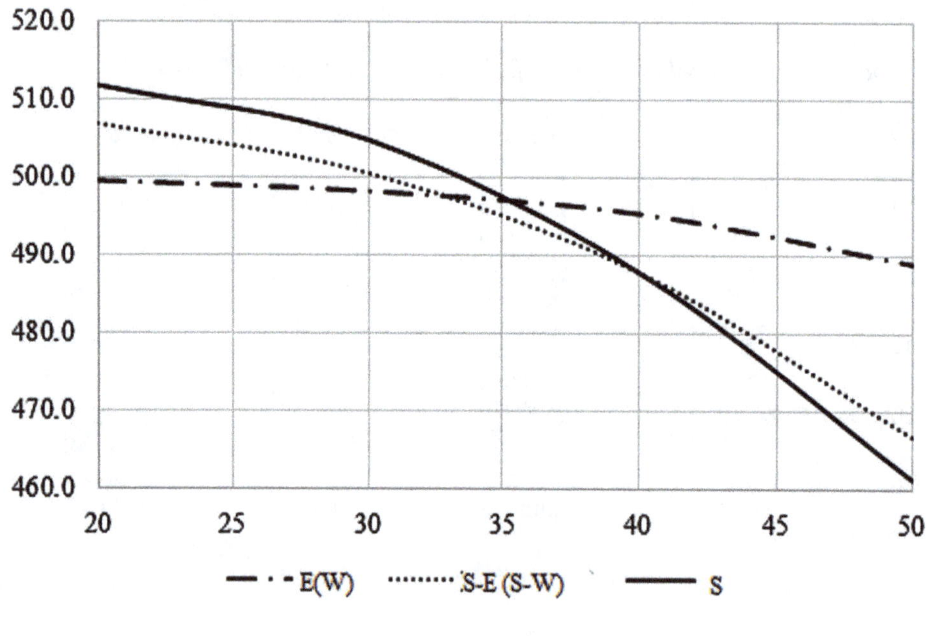

at the inclination angle relative to the horizon of above 41° it is rational to plan and locate the receiving surfaces of power plants oriented to the south-east (south-west).

The result of the calculation according to (Daus, & Yudaev, 2016) is the hourly amounts of direct, diffused and reflected solar radiation at the receiving surface, which is at a given angle of inclination, according to certain geographical coordinates. Figure 6 shows the hourly values of direct and diffused insolation on the horizontal surface, the reflected component of radiation in this case is zero.

It is known that solar power plants are most often connected to already existing internal networks and communications at the facility and mounted on roofs, the design of which can vary from flat to pitched with different angles of inclination. Therefore, data on the intensity of solar radiation, coming only on a horizontal surface, are not exhaustive.

Figure 7 shows the daily sums of total solar radiation incident on receiving surfaces with the tilt angle from 0° to 45°.

At Figure 7 it can be seen that the maximum of the insolation intensity can be utilized in summer — in July at the inclination angle of the receiving surface of the solar power plant of 15°. However, when analyzing the character of the curve, it can be seen that already in August this dependence falls below the curve of the angle of 30°, and in September – 45°. The dependence of the intensity of radiation by day for the horizontal position of the receiving platform lies below all the curves - that is, when positioning power plants on flat roofs, it is advisable to use additional fittings to create the angle of inclination of photovoltaic panels or solar collectors.

Since preventive repairs are seasonal measures, when planning the outage schedule, to obtain the maximum intensity of solar radiation the choice of the tilt angle should be based on the calculation of the sums of the intensity of the total solar radiation over the seasons. The results of the calculations are presented at Figure 8.

Figure 6. Hourly sums of direct (a) and diffused (b) solar radiation on the horizontal surface – Volgograd

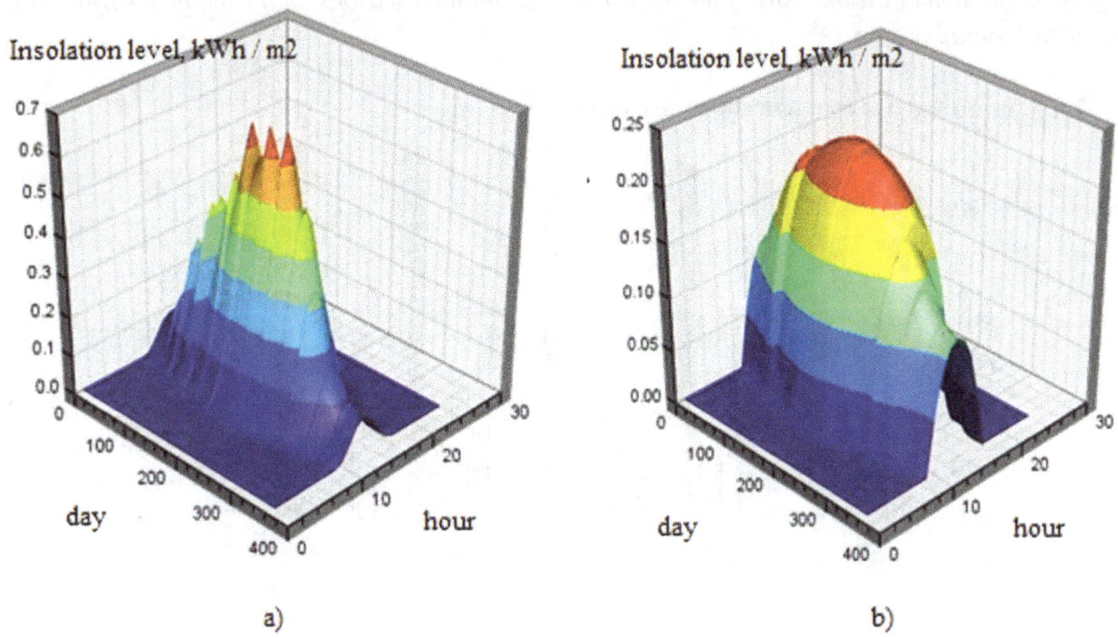

Figure 7. Daily amounts of total solar radiation on variously oriented surfaces for the city of Volgograd

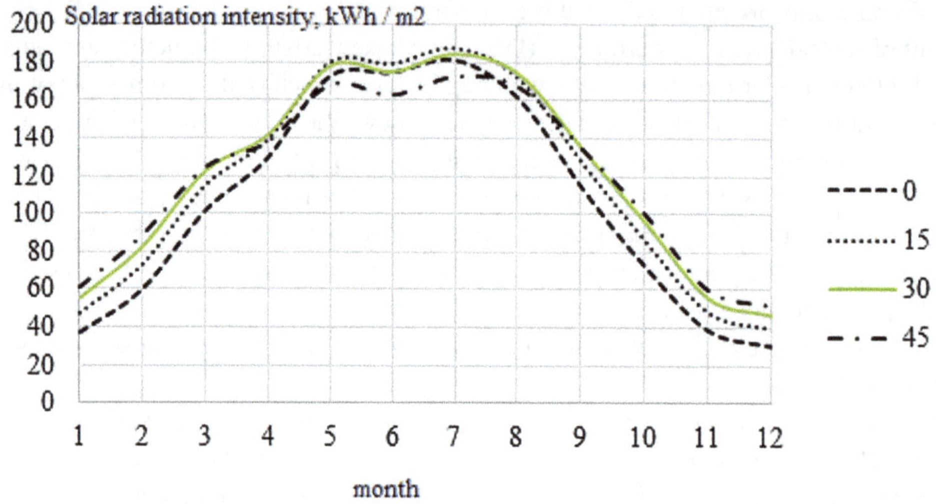

Analyzing the data presented at the diagrams of Figure 8, it can be concluded that the maximum flux of total radiation in autumn and winter, when the Sun is lower over the horizon, is reached at large angles (45° and 60°, respectively). At the same time, in summer, the largest amount of radiation falls on the receiving surface at the angle of 15°. The maximum value of insolation coming to the receiving surface of the solar power plant for the year corresponds to its inclination angle of 30°.

Figure 8. The sum of the intensity of the total solar radiation on seasons on the inclined surface for the city of Volgograd

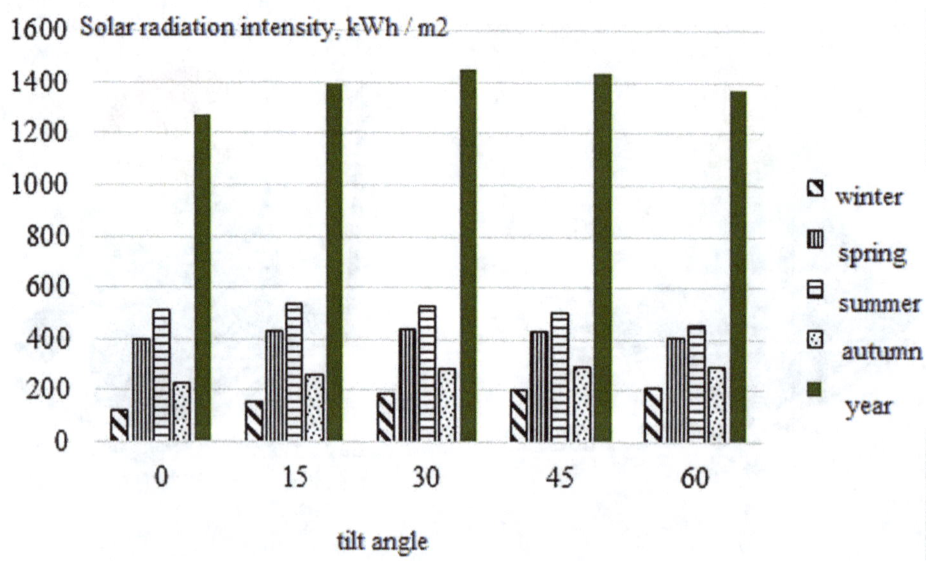

When choosing a system for converting solar energy to any other type, it is important to know the composition of radiation, and the nominal fraction of direct, diffused and reflected radiation. Such a ratio for the annual sums of insolation on variously inclined surfaces is presented at Figure 9.

The analysis of the diagrams presented at Figure 9 makes it possible to note that with the increase in the angle of inclination of the receiving surface, the diffused component decreases, while the reflected one increases. At the same time, it can be noted that the maximum value of direct radiation does not guarantee the maximum total insolation. So, at the tilt angle of the receiving surface of 45°, the intensity of direct radiation is 16.2 kWh/m^2 more than at 30°. However, the maximum of the total radiation is reached as in the data of Figure 3 at the tilt angle of 30°, and its value is greater than the corresponding value at the angle of 45° at 17.1 kWh/m^2. This is due to the fact that with the increase in the angle of inclination from 30° to 45°, the diffused component decreased by 48.4, and the reflected one increased by 15 kWh/m^2.

Thus, the geographical point with the coordinates of the city of Volgograd has a high potential of solar energy: 1276.7 kWh/m^2 for the horizontal surface. However, the analysis of the change in insolation magnitude with the increase in the tilt of the receiving surface from 0° to 45° showed that the specific reserve of using solar energy can be increased by different orientation of the receiving surface of power plants and when positioning power plants on flat roofs, it is advisable to use additional fittings to create a tilt of photovoltaic panels or solar collectors. So the use of the tilt angle of 30° – 45° at the constant value throughout the year allows to increase in the converted energy flux of solar radiation by 12 – 13.5%. When adjusting the angle of the photovoltaic panel relative to the horizon seasonally in summer, it is advisable to use the angle of inclination from 15° to 30°, since already at 45° the insolation value

Figure 9. Components of total solar radiation coming on inclined surfaces

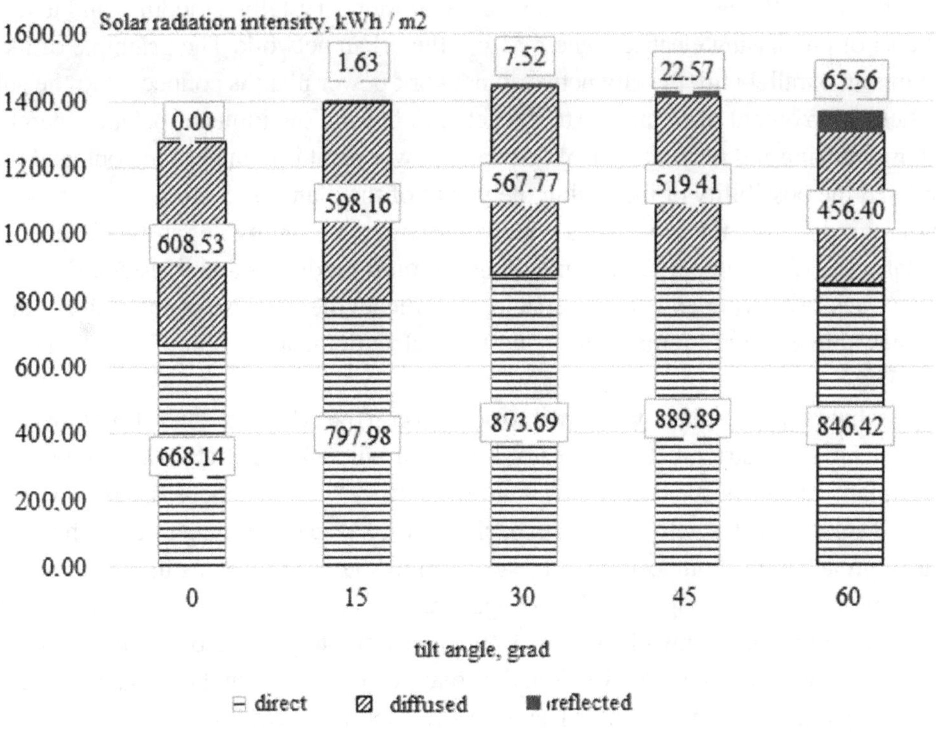

decreases comparing to 15° during this period from 539.8 to 503.4 kWh/m^2. In winter, the maximum insolation intensity is reached at the largest angles of inclination of the receiving surface of the power plant, since the Sun is low relatively to the horizon. Seasonal regulation of the angle of the photovoltaic panel allows to increase the utilized potential of solar energy by 16.8%.

BEST PRACTICES

Photovoltaic energy is being actively developed both in the world and in Russia, as evidenced by the developed and promoted projects of large solar power plants. This is caused by the improvement in the situation when the country receives everything necessary, which makes it economically and technically viable and also affects the continuous growth of electricity tariffs, cost of services and goods for consumers.

The territory of the Russian Federation is characterized by a significant length from east to west and is determined by the distance from the equator, capturing areas up to the Arctic Circle. Therefore, it can be marked that the country's territory is located in different areas of insolation. At the same time, such regions as Kalmykia, Stavropol Territory, Rostov Oblast, Krasnodar Territory, Volgograd Oblast, Astrakhan Oblast and other southwestern regions, as well as Altai, Primorye, Chita Oblast, Buryatia and others in the south-east, are the most promising regions as for utilization of incoming solar energy to Earth.

It is in these territories that the photoelectric conversion of solar energy into electrical energy has been intensively introduced recently.

Objects of solar energy of various capacities are being actively implemented in the south of the Russian Federation. For example, in the Rostov region there is a number of examples of the successful application of low-power solar power plants to supply individual consumers.

Since 2013, solar power plant has been operating in Rostov-on-Don to provide power supply for a business center of 6.5 kW. The power plant includes solar monocrystalline modules and inverter, which reduces the cost of purchasing electrical energy from the urban network. The principle of its operation is the following: in parallel with the city network, network power plant is connected to the subscriber's side, while there is no output of electricity to the network beyond the limits of balance membership. At the moment, monitoring and maintenance of this solar power plant is being carried out, and there is also being considered the possibility of increasing the power of the plant to 15–30 kW.

In the Volgograd region, the program is widely implemented to provide renewable energy sources, including solar energy, for remote settlements of agricultural producers and livestock farms.

Programs on energy saving and energy efficiency indicate the promise of introducing autonomous sources of renewable energy in remote areas and in rural settlements, hospitals and social institutions (Yudaev, 2015).

Autonomous hybrid power plants were put into operation at the "Osipovo" and "Fomin Well" shepherd sites in the Kalachevsky district, in the Nikolaevsk and Surovikinsky districts. The composition of such hybrid power plants includes standardly two wind generators "Electrosphere Breeze" with a capacity of 5 kW each, 8 solar modules with the total capacity of 1.2 kW, a diesel generator with a capacity of 8 kW. Such an approach to the composition of the generation source allowed using it autonomously from the centralized network and saving 75% of diesel-generator fuel. The power plant is designed to operate in the basic mode in the local network and allows the consumer to provide power to 8 kW.

In the city of Kamyshin, Volgograd Region, the solar power plant with the capacity of 72 kW on the roof of a two-story entertainment complex has been successfully operating for 6 years.

One of the most favorable regions for the widespread introduction of solar power is the Astrakhan region. Since 2007, the project "Increasing Energy Efficiency at the Regional Level in the Arkhangelsk, Astrakhan and Kaliningrad Oblasts" has been acting in Russia, according to which the Astrakhan region has been identified as the energy deficient region that buys 30% of electric energy on the wholesale market for electric energy and power.

For the Astrakhan region, the use of energy equipment based on renewable energy sources is also relevant in areas such as agriculture objects, fisheries, tourism and the housing and utilities sector. Many tourist sites in the region are located in the Volga-Akhtuba floodplain or in the Volga delta that are far from populated areas. Laying gas pipelines or electric or cable lines to them is too expensive, economically unprofitable. Therefore, travel companies often use diesel generators for electricity supply, thereby causing harm to the environment.

In addition, most of the recreation centers, fishing and hunting sites in the Volga-Akhtuba floodplain or in the Volga delta are already equipped with wind generators, solar water collectors and photovoltaic modules. Also, at shepherd points, for example, Stepnovsky Village Council, electricity is supplied from solar photovoltaic batteries, etc.

Thus, under conditions of the construction of high-power grid solar power plants, small distributed photovoltaic plants of distributed generation confidently fill their niche, providing electricity only to their consumers, and this requires new approaches to their layout and selection of element parameters.

A single-circuit solar water-heating unit with a volume of 15 m^3 per day was put into operation and successfully is exploited for hot water supply of the Municipal Healthcare Institution "Central Regional Hospital of Leninsky District" during the warm period of the year (from April to October). The solar photovoltaic power plant of 6.0 kW was installed for back-up power supply to the operational and intensive care unit of the municipal institution "TsRB, Leninsky District".

If to talk about big energy, then a wind farm with thean installed capacity of 1 GW, the product of the joint project of the state corporation Russian Technologies and JSC RusHydro, could appear in the Volgograd region. But the decision on its construction has not yet been made, and the implementation period seems to be just a cherished dream. And one has to hope that the journalistic phrase "the project will begin after the government adopts a program to stimulate the production of electricity based on renewable energy sources" may turn out to be prophetic (Yudaev, 2015).

The combination of natural and economic conditions creates favorable prospects for the development of renewable energy in the Republic of Kalmykia, the potential of which, according to experts of the Laboratory of Renewable Energy Sources of Moscow State University, is quite high among Russian regions. First of all, these are rather large wind speeds: at the altitude of 50 meters – more than 6 m/s, and at an altitude of 100 meters - 8–9 m/s, which is determined by the position of the territory in the foothill zone of the Caucasus between the Black and Caspian Sea and creates in Kalmykia natural wind tunnel.

Solar insolation also brings the region to one of the sunniest regions of Russia with an average daily amount of solar radiation of 3.5 kWh/m^2 and a sunshine duration of about 2,200 hours per year, which is more than 180 days (Yudaev, 2015).

Low population density, large unoccupied areas provide ample opportunities for the construction of renewable energy facilities, and a small number of people and small amounts of domestic energy consumption create a theoretical opportunity to fully covering of all domestic energy needs through renewable energy.

In the Republic, the Government adopted Resolution No. 358 of October 21, 2008 "Development of Renewable Energy Sources in the Republic of Kalmykia for 2008–2012" and the corresponding target program. The implementation of the Program was supposed to be implemented within five years (2008–2012).

In the summer of 2012 in Kalmykia, near the village of Peschany, located in 10 km from the capital of the Republic Elista, two wind turbines (produced by the German company Vensys) with a capacity of 1.2 MW each were put into operation. The launch of these units determined the beginning of a large-scale project to create a wind farm with a total capacity of 300 MW. Subsequently, the installation of Vestas wind turbines with a capacity of 3 MW is expected. Wind-driven units are constructions with a height of more than 60 meters with a blade radius of 31 meters, and the following installations with a capacity of 3 MW are even higher – more than 100 meters (Yudaev, 2015).

In addition, in accordance with the agreement between the Republic of Kalmykia, the Czech-Swiss company JSC Falcon Capital, JSC Czech Export Bank and JSC CKD Novaya Energo of June 9, 2007, it was planned to build the 150 MW wind power plant in the Priyutnensky district of the Republic of Kalmykia.

But, despite the developed programs, the large renewable energy in the republic has not yet started working at full capacity. At the same time, there are examples of the implementation of small projects in rural areas.

As the example of an autonomous power supply system, there can be cited animal livestock farm built in the Tselinny district of Kalmykia, for the construction of which local environmentally friendly building materials were used, and wind and sun energy are utilized for everyday energy supply of nomadic herdsmen for the uninterrupted functioning of the circulating water system.

According to expert estimations, over the past 5 years about 50 private houses, personal farms and livestock outlets are equipped with wind generators, about 300 – with solar panels. In particular, about 25 points are equipped with solar panels in the village of Achinera (Chernozemelsky district), several points in the villages of Adyk (Chernozemelsky district) and the farm "Ulan-Heyech" (Yashkulsky district). In most cases, it means supplying by 1–3 solar batteries with a capacity of 80–160 W of a private farm.

Renewable energy sources are widely used in the Republic of Kalmykia and for the organization of telephone communications in remote areas - providing base stations of cellular operators with electrical energy. Thus, in the area of Iki-Chonos village,there is being tested hybrid complex manufactured by Nokia Siemens Networks, including a wind generator, 24 solar panels and specialized batteries. This technical solution was chosen in order to maximize the potential of alternative sources, as well as taking into account the climatic and environmental conditions of the region. Kalmykia is the territory with strong winds and high solar activity (186 sunny days per year). In cloudy, calm weather, the base station is powered by the external network or batteries. In addition to autonomous power supply, there is carried out remote monitoring and control system of energy consumption, thereby ensuring stable power supply to the site. In addition, the autonomous complex is equipped with safety means to prevent theft and damage equipment (Yudaev, 2015).

For several years, the TACIS project "Renewable Energy Sources and the Reconstruction of Low-Capacity Hydroelectric Power Plants" was carried out in Astrakhan Oblast. Based on the analysis of the region's potential, there was designed qualified conclusion on the possibility of developing and utilization of renewable energy sources in the region.

It has been established that solar energy is the most promising direction for the development of renewable energy in the region. The fact is that the average duration of sunshine in the region is 2441 h/year, and the average total amount of solar energy entering the horizontal surface reaches almost 5,000 MJ per year. This means that the Astrakhan region is one of the most promising regions of southern Russia in terms of the potential of using solar energy (Yudaev, 2015).

Another promising direction is wind power. Average wind speeds tend to increase from south to north of the Astrakhan region and vary from 3 to 4 m/s. These indicators are sufficient for the implementation of cost-effective projects of autonomous wind power plants with the capacity of 4 kW and above, as well as small-scale wind power plants.

Experts of the TACIS project estimated the total technological potential of renewable energy in 980 thousand tons of reference fuel, or 47% of the consumed electricity in 2007. This allows planning economically feasible production of electricity on the basis of renewable energy sources to 6% by 2020, or 1.27 billion kWh (Yudaev, 2015).

In the Astrakhan region there has been implemented a number of projects at the expense of private investment. This is primarily the use of heat pumps for heating and cooling, solar energy for heating water and electricity generation. For example, at the "Zelenga" fishing base, the hot water supply system and outdoor lighting are supplied by solar panels. At the recreation center "Larisa", the heating systems and the hot water supply system operate with the help of heat pumps, and the power supply to the floating landing stage is provided by solar collectors. On shepherd points, such as in the shepherd's house of the Stepnovsky Village Council, electricity is supplied from solar photovoltaic batteries, etc. (Yudaev, 2015).

In the Narimanov district of the Astrakhan region there was launched the project "Solar City". On four fields with an area of 2400 m^2, there are 2,200 collectors that convert solar energy to provide hot water. Today it is Russia's largest renewable energy project. In addition to the economy of natural gas, the operation of solar water heaters relieves part of the load from the local boiler house, so that in Narimanov there is no longer any interruption in the supply of hot water. The cost of the solar boiler is 96 million rubles, the annual economy is 18.6 million rubles (Yudaev, 2015).

SOLUTIONS AND RECOMMENDATIONS

At present, the experience of using renewable energy in the region of arid farming, where agricultural production is highly developed, especially livestock farming, which is being bred at remote points of agricultural producers: sheep farms, cattle farms, etc., is acquiring particular significance. These settlements are often located in the steppe zone where there is absent the possibility to connect to the centralized electricity supply system or great infusion of funds is required, which makes farmers seriously think about the continuation of the normal functioning of these industries. Under these conditions renewable energy sources can provide real help to organize power supply of the consumer.

The use of small-scale and, therefore, financially attractive projects based on renewable energy can be technically implemented with a relatively short payback period for energy supply of rural remote areas and recreational zones. This statement is supported by real examples of functioning sources and units of renewable energy.

FUTURE RESEARCH DIRECTIONS

Further increase in the efficiency of the solar power plant is possible through the harmonization of consumption and power generation schedules, as well as in the framework of the comprehensive modernization of the consumer's power supply system, and the choice of the location of the solar power plant in terms of reducing losses in the grid elements.

CONCLUSION

Currently, there are widely and intensively produced agricultural products, especially livestock products, which are grown at remote locations by agricultural producers: sheep farms, cattle farms, etc. That is why there are not required ways for ensuring electricity supply so farmers can provide the normal operation of their industries. Under these conditions renewable energy sources can provide real help to organize power supply of the consumer. The introduction of small power plants helps to reduce the load on the power system, which can significantly reduce the loss of time.

Such power plants can operate in the following modes: autonomous, redundant from the network, connected to the network and transmitting, and also working only in the network. The last two options imply increased requirements for equipment, due to the requirements of energy systems to the voltage level, high-quality electrical energy, etc.

Autonomous photovoltaic plants, power backup, energy saving in power systems, have less cost and implementation time.

To assess the possibilities of using renewable energy, it is necessary to assess renewable energy potential in these areas and, above all, to assess the use of solar and wind energy. Thus, the geographical point with the coordinates of the city of Volgograd (the territories of agriculture production) has high potential of solar energy: 1276.7 kWh/m^2 for the horizontal surface. However, the analysis of changes in the utilized insolation flux with the increase in the angle of inclination of the receiving surface from 0° to 45° showed that it is possible to realize reserve of solar energy application by photovoltaic panels or solar collectors. Such application of the tilt angle of 30° - 45° at constant value over a year makes it possible to increase the converted energy flux of solar radiation by 12 – 13.5%. When adjusting the angle of the photovoltaic panel relatively to the horizon seasonally in summer, it is advisable to use the angle of inclination from 15° to 30°, that is, as already at 45°, the value increases comparing to 15° during this period from 539.8 to 503.4 kWh/m^2. So photovoltaic panels can increase the potential of solar energy by 16.8%.

Analysis of the best practice examoles of using real energy sources and units for the renewal energy resources in the rural territories of southern Russia revealed that these projects can be widely technically implemented at photovoltaic power plant with a relatively short lifespan to ensure the efficiency of remote areas and recreational zones.

REFERENCES

Amerhanov, R. A. (2008). Jenergeticheskij potencial solnechnoj radiacii i jekonomicheskaja celesoobraznost' primenenija gelioustanovok v Krasnodarskom krae i Jakutii [Energy potential of solar radiation and economic feasibility of solar installations in the Krasnodar Territory and Yakutia]. *Works of the Kuban State Agrarian University*, *1*, 26–32.

Annual Energy Outlook. (2017). Retrieved from http://www.eia.gov/outlooks/aeo/

Aronova, E. S., & Grilihes, V. A. (2006). Metodika rascheta real'noj plotnosti solnechnogo izluchenija pri proektirovanii fotojelektricheskih ustanovok [Method for calculating the real density of solar radiation in the design of photovoltaic installations]. *SPbSU Scientific and Technical Statements*, *6*(1), 62–66.

Atwater, M., & Ball, J. T. (1978). A numerical solar radiation model based on standard meteorogical observation. *Solar Energy*, *21*(3), 163–170. doi:10.1016/0038-092X(78)90018-X

Belenov, A. T. Kharchenko, V.V., Rakitov, S.A., Daus, Y.V., Yudaev, I.V. (2016). The Experience of Operation of the Solar Power Plant on the Roof of the Administrative Building in the Town of Kamyshin, Volgograd Oblast. *Applied Solar Energy*, *52*(2), 105-108.

Bezrukih, P. P. (2008). *Resursy i jeffektivnost' ispol'zovanija vozobnovljaemyh istochnikov jenergii v Rossii: uchebnoe posobie* [Resources and Efficiency of Renewable Energy Sources in Russia: Tutorial]. Moscow: Kniga–Renta.

Daus, Y., Kharchenko, V., & Yudaev, I. V. (2018). Solar Radiation Intensity Data as Basis for Predicting Functioning Modes of Solar Power Plants. In *Handbook of Research on Renewable Energy and Electric Resources for Sustainable Rural Development* (pp. 283–309). Hershey, PA: IGI Global.

Daus, Yu. V., & Kharchenko, V. V. (2018). Evaluating the applicability of data on total solar-radiation intensity derived from various sources of actinometric information. *Applied Solar Energy*, *54*(1), 71–76. doi:10.3103/S0003701X1801005X

Daus, Yu. V., Kharchenko, V. V., & Yudaev, I. V. (2016a). Evaluation of Solar Radiation Intensity for the Territory of the Southern Federal District of Russia when Designing Microgrids Based on Renewable Energy Sources. *Applied Solar*, *52*(2), 124–129.

Daus, Y. V., & Yudaev, I. V. (2016). Designing of Software for Determining Optimal Tilt Angle of Photovoltaic Panels. In Proceedings of International Conference on Education, Management and Applied Social Science (EMASS2016). Beijing: DEStech Publications, Inc.

Daus, Yu. V., & Yudaev, I. V. (2016). Designing of Software for Determining Optimal Tilt Angle of Photovoltaic Panels. In *Proceedings of International Conference on Education, Management and Applied Social Science (EMASS2016)*. Beijing: DEStech Publications, Inc.

Daus, Yu. V., Yudaev, I. V., & Stepanchuk, G. V. (2018). Reducing the costs of paying for consumed electric energy by utilizing solar energy. *Applied Solar Energy*, *54*(2), 137–141. doi:10.3103/S0003701X18020056

Ekadewi, A., & Djatmiko, I. (2013). The optimal tilt angle of a solar collector. *Energy Procedia*, *32*, 166–175. doi:10.1016/j.egypro.2013.05.022

Distributed Generation, D. (2013). An Overview of Recent Policy and Market Developments. Retrieved from https://www.publicpower.org/files/PDFs/Distributed%20Generation–Nov2013.pdf

Informatsionnyy obzor. (2016). Operativnyye dannyye [Informational review "United Energy System of Russia: Interim Results]. Retrieved from http://so–ups.ru/fileadmin/files/company/reports/ups–review/2016/ups_review_june16_1.pdf

Kharchenko, V. V., Nikitin, B. A., & Tikhonov, P. V. (2010). Estimation and forecasting of PV cells and modules parameters on the basis of the analysis of interaction of a sunlight with a solar cell material. In *Proceedings of 4th International Conference TAE 2010, Trends in Agricultural Engineering* (pp. 307-310).

Liu, B. Y. H., & Jordan, R. C. (1961). Daily insolation on surfaces tilted towards the equator. *ASHRAE Journal*, *3*(10), 53–63.

NASA Atmospheric Science Data Center. (n.d.). Retrieved from https://eosweb.larc.nasa.gov/sse

Next Generation Wind and SolarPower. (2016). From cost to value. Retrieved from https://www.iea.org/publications/freepublications/publication/NextGenerationWindandSolarPower.pdf

Ospina, A., & Quijano, N. (2016). Distributed control of small–scale power systems using noncooperative games. *International Journal of Electrical Power & Energy Systems*, *82*, 535–544. doi:10.1016/j.ijepes.2016.03.065

Otchet o funktsionirovanii YeES Rossii v 2015 godu [Report on the functioning of the UES of Russia in 2015]. (n.d.). Retrieved from http://www.rcit.su/inform–rf–ees2015.html#inf–ees2015–21

Otchet o razvitii VIE i predlozheniya v energeticheskuyu strategiyu Rossii [Report on the development of renewable energy sources and proposals for the energy strategy of Russia]. (2014). Retrieved from http://ac.gov.ru/files/content/1578/11–02–14–energostrategy–2035–pdf.pdf

Popel, O. S. (2010). *Atlas resursov solnechnoj jenergii na territorii Rossii* [Atlas of solar energy resources in Russia]. Moscow: OIVT RAN.

Sen, Z. (2008). *Solar Energy Fundamentals and Modeling Techniques, Atmosphere, Environment, Climate Change and Renewable Energy* (1st ed.). Switzerland: Springer.

Shyam, S., & Rajeev, K. (2011). Estimation of Hourly Solar Radiation on Horizontal and Inclined Surfaces in Western Himalayas. *Smart Grid and Renewable Energy*, *2*(1), 45–52. doi:10.4236gre.2011.21006

Vissarionov, V.I. Burmistrov, A.A., Deryugina, G.V., Kuznetsova, V.A., Kunakin, D.N., Malinin, N.K., & Pugachev, R.V. (2008). *Solnechnaja jenergetika: Uchebnoe posobie dlja vuzov [Solar energy: A textbook for high schools]*. Moscow: Izdatel'skij dom MJeI.

Winter, C.-J. (1991) Solar Power Plants. Springer. doi:10.1007/978-3-642-61245-9

Yudaev, I. V. (2015). Opyt ispol'zovaniya VIE na sel'skikh territoriyakh i v rekreatsionnykh zonakh v regionakh YuFO [Experience of using renewable energy sources in rural areas and in recreational areas in the regions of the Southern Federal District]. *Don Agrarian Science Bulletin*, *1*, 82–92.

ADDITIONAL READING

Darhmaoui, H., & Lahjouji, D. (2013). Latitude Based Model for Tilt Angle Optimization for Solar Collectors in the Mediterranean Region. *Energy Procedia, 42*, 426–435. doi:10.1016/j.egypro.2013.11.043

Głuchy, D., Kurz, D., & Trzmiel, G. (2013). Studying the Impact of Orientation and Roof Pitch on the Operation of Photovoltaic Roof Tiles. *Przegląd elektrotechniczny, 6*, 281-283.

Jo, J. H., Loomis, D. G., & Aldeman, M. R. (2013). Optimum penetration of utility–scale grid–connected solar photovoltaic systems in Illinois. *Renewable Energy, 60*, 20-26. doi:10.1016/j.renene.2013.04.008

Markvart, T., & Castaner, L. (2003). *Practical handbook of photovoltaics: fundamentals and applications. New York*: Elsevier.

Petrakopoulou, F., Robinson, A., & Loizidou, M. (2016) Simulation and evaluation of a hybrid concentrating–solar and wind power plant for energy autonomy on islands. *Renewable Energy, 96*(A), 863-871.

KEY TERMS AND DEFINITIONS

Actinometrical Data: Results of long-term meteorological observations at weather stations, processed and systematized by specialized organizations in the form of climate reference books and databases.

Optimum Inclination Angle of the Receiving Surface: Inclination angle of the receiving surface relative to the horizon, which allows obtaining the maximum solar radiation flux on its surface for a given period of time.

Photovoltaic Power Plant: Engineering structure that converts solar radiation into electrical energy by photovoltaic modules.

Rural Territories: Place where people work in agriculture.

Renewable Energy Sources: Sources of continuously renewable energy types in the Earth's biosphere. **Agriculture:** Sector of the economy aimed at providing the population with food and obtaining raw materials for a number of industries.

Solar Radiation Intensity: The density of solar radiation (energy illumination), coming per unit area of the photoelectric module.

Chapter 13
Development and Research of PVT Modules in Computer-Aided Design and Finite Element Analysis Systems

Vladimir Panchenko
Russian University of Transport, Russia

Valeriy Kharchenko
https://orcid.org/0000-0003-3725-2976
Federal Scientific Agroengineering Center VIM, Russia

ABSTRACT

This chapter discusses the simulation of solar photovoltaic thermal modules of planar and concentrator structures in computer-aided design systems KOMPAS 3D and finite element analysis ANSYS. To create photovoltaic thermal modules, a method for designing their three-dimensional models in the computer-aided design system has been developed. To study the thermal regimes of the created three-dimensional models of modules, a method has been developed for visualizing thermal processes, coolant velocity, and flow lines of a cooling agent in a finite element analysis system. As a result of calculations in the finite element analysis system using the developed method, conclusions can be drawn about the feasibility of the design created with its further editing, visualization of thermal fields, and current lines of the radiator cooling agent. As an illustration of the simulation results, a three-dimensional model of a photovoltaic thermal planar roofing panel and an optimized three-dimensional model of a photodetector of a solar concentrator photovoltaic thermal module are presented.

DOI: 10.4018/978-1-5225-9420-8.ch013

INTRODUCTION

The principle of creating a single device that provides simultaneous generation of electrical and thermal energy is the creation of a photovoltaic thermal module (PVTM) (Kharchenko, Panchenko, Tikhonov & Vasant, 2018), where solar cells are placed on a heated absorbing surface of a flat solar collector. The absorber in this design performs a double function – firstly, it cools the photovoltaic panel, removing excess energy that is not involved in the generation of electricity, thereby increasing its efficiency, and secondly, it produces thermal energy.

The history of the development of this class of devices goes back several decades, where it is shown that such a combined system is a promising design for further development. In (Zharkov, 2014; Kamilov, Muminov & Tursunov, 2008), the authors studied the modules that heat water and air, identified key concepts and main priorities for the development of cogeneration systems of this type. Improving the efficiency of thermal conversion of solar radiation implies a high operating temperature, which at the same time reduces the efficiency of photoelectric conversion. It is not surprising that in the reviewed papers attention is paid to the development of the optimal design of the heat generating part of the PVTM, since the optimal design of such a module will ensure high efficiency of the photovoltaic plant and high thermal output.

Experimental studies (Hosseini, Hosseini & Khorasanizadeh, 2011; Rawat, Debbarma, Mehrotra et al., 2014; Buonomano, Calise & Vicidimini, 2016) showed that the overall efficiency of the combined module is greater than the efficiency of a conventional solar panel, where the daily thermal efficiency was 50.1%, and the total efficiency of the developed module exceeded 73%. Improving the design of the photovoltaic thermal module (Ibrahim, Jin, Daghigh, Salleh, Othman, Ruslan et al., 2009), where the absorber is made in the form of a rectangle in cross section, it is possible to manufacture an absorber in the form of a V-shaped triangle, which is presented in (Othman, Ruslan, Sopian & Jin, 2009), thereby reducing heat loss and improving heat removal. In (Sevela & Olesen, 2013), a photovoltaic thermal module with a tubular heat exchanger is presented, where the maximum efficiency of a liquid solar collector in the installation was 48% with photoelectric converters turned off, and with simultaneous production of electrical and thermal energy, its value decreased to 42%. A flat photoelectric thermal collector was considered in (Ibrahim, Othman, Ruslan, Mat & Sopian, 2011), in which the radiation receiver is a light-absorbing plate with photoelectric elements and under this plate are tubes with a circulating coolant. Studies (Dubey & Tay, 2012) have shown greater efficiency of a photovoltaic thermal module with rectangular coolant channels compared with a module with a tubular sheet heat exchanger. The disadvantages of photovoltaic thermal modules with tubular heat exchangers include the low efficiency of heat energy transfer due to insufficient thermal contact between the absorber and the substrate of the photovoltaic cells. To solve this problem, a photovoltaic thermal module is presented, in which the photodetector is mechanically pressed against the thermal collector without the aid of any mounting glue. Compared with a photoelectric thermal collector with a tubular radiator, this solution provides better thermal contact between the photoreceiver and the heat exchanger, which increases the efficiency of solar energy conversion, but this solution is problematic due to the fragility of the photoelectric converters and the need for a sealant to maintain high electrical characteristics.

Today, one of the most advanced designs in terms of optimization is the modules of Solimpeks (http://www.solimpeks.com) and Sunsystem (http://www.sunsystem.bg/en/fotovoltaika/PV-T/), which produce photovoltaic thermal modules with a tubular heat exchanger. However, even commercially manufactured modules are distinguished by high material consumption, mass, and, accordingly, cost. In addition, the modules are carried out in two versions for various purposes – maximum electrical efficiency and maximum thermal efficiency, which makes one doubt the universality of this development, since in any case one component of the power supply is reduced. It is also a question of the service life of the electrical part of the module at the level of nominal power, the level of which falls over the years.

BACKGROUND

Analysis of the Structures of Solar Photovoltaic Thermal Modules

PVTM can be classified according to various criteria, such as: type of solar cells used (monocrystalline, polycrystalline or amorphous silicon, thin films, etc.), glazed or unglazed module, type of cooling fluid (water, glycol or air), presence of a concentrator etc. In practice, the most common are air and water heat-transfer PVTM, which can be used to supply hot water and which are created on the basis of a solar collector, which is complemented by solar cells placed on absorber surface. PVTMs with concentrators are also of interest because they help reduce the number of expensive solar cells and increase their efficiency. The best results demonstrate PVTM with single glazing, where the density of the output radiation is lower than that of an open structure.

From the above review and studies (Kharchenko, Panchenko, Tikhonov & Vasant, 2018; Kharchenko, 2014; Kharchenko, Nikitin, Tikhonov & Gusarov, 2013) can distinguish generalizing designs of PVTM: PVTM with an absorber of the type "metal sheet-tube"; PVTM with absorber in the channel for the liquid above the photovoltaic panel; PVTM with absorber in the form of a channel for the liquid under the photovoltaic panel; PVTM absorber with rectangular channels; PVTM with absorber in the channel, partially filled with liquid and located above the photovoltaic panel; PVTM with double absorber and air gap.

PVTM with a metal sheet-tube absorber is a traditional solar collector, on the working surface of which there is a photovoltaic panel (one of the most common types of PVTM, this type of module can be called a photoelectric thermal collector). With additional transparent heat insulating coating increases the reflection of solar radiation, reduced heat loss from the working side of the solar cell, which leads to deterioration of electricity generation.

A PVTM with a channel absorber is a channel with a liquid (water) adjacent to a photovoltaic panel in front or behind. If the fluid channel is in front of the working surface of the solar cell (front side), then the liquid absorbs the part of the solar radiation that passes through it, and thus is part of the absorber. The advantage of this design compared with the metal sheet-tube design is uniform dispersion of solar energy. When using this type of PVTM, it is desirable that the liquid absorbs as much sunlight as possible with the wavelength of that part of the spectrum that solar cells do not convert into electricity, thanks to which the solar cell will work under conditions close to optimal due to the removal of unused energy by the liquid layer (which imposes restrictions on the type of fluid). The choice of a suitable liquid for such a channel is a difficult task, since the liquid absorbs a part of the working spectrum of solar cells and their electrical efficiency decreases when using water in such PVTMs.

The PVTM design may have another configuration with a heat removal channel located behind the photovoltaic panel, i.e., adjacent to the back side of the solar cell. Also, the absorber may consist of a set of vertical rectangular channels that are located behind the photovoltaic panel. This structure forms ribs-absorbers, which increase the surface area to remove heat by the liquid, thereby increasing the heat sink.

STATEMENT OF THE PROBLEM AND RESEARCH METHODS

As a result of the analysis of the literature and the conducted research, it can be concluded that the receivers of solar photovoltaic thermal modules of planar and concentrator installations may have different designs, but it should be noted that the design components used in such modules have a fundamentally similar structure. The design of such modules is complex, but in most cases the complexity of the design allows to achieve maximum module efficiency, which is summed from the electrical and thermal efficiencies, whose contribution to the overall efficiency can vary depending on the need, but often with increasing electrical efficiency decreases thermal and vice versa. The main task of modern research is to find the optimal design, at which the overall efficiency of the module will be maximum.

To carry out such search tasks, it is necessary to create a method that would allow at the initial stage to create such solar photovoltaic thermal modules of various designs in the computer-aided design system in the form of three-dimensional models, after which they were created, such models would be subject to detailed study of the thermal processes occurring in them. This task also requires a calculation method, which describes the sequence of such thermal calculations and visualizations of thermal processes in the system of finite element analysis occurring in the module under various conditions. The result of the simulation is the creation of an optimized model of a solar photovoltaic thermal module, which is recommended for prototyping.

Currently, there are a number of software systems that, as a tool, allow to create both three-dimensional object models, conduct various modeling of the thermal state of the modules while visualizing the results obtained, and create prototypes of such modules using additive technologies, spending relatively little resources, which is very important at the initial design stage.

THREE-DIMENSIONAL SIMULATION OF SOLAR PHOTOVOLTAIC THERMAL MODULES IN THE COMPUTER-AIDED DESIGN SYSTEM

As a tool for creating two-dimensional and three-dimensional models of solar photovoltaic thermal modules, it should be noted computer-aided design systems, for example, the Ascon software complex - Kompas 3D (http://kompas.ru/), which for several decades has established itself as a worthy domestic 3D design tool with full compliance with a single standard design documentation of the Russian Federation.

In the considered method of creating models of solar photovoltaic thermal modules, designs of modules of stationary and mobile power generation are developed, the main differences of which are the size of solar cells, the number of illuminated sides of solar cells (one- and two-sided) and the size of radiator cavities due to different solar flux on the ray-receiving surface (these sizes are optimized in the Ansys finite-element analysis software package (http://www.ansys.com/).

The main task of the developed method is to create models of solar photovoltaic thermal modules of stationary and mobile power generation, the sequence of which will be universal for creating a wide range of such modules for various purposes and different requirements. The developed method is created for two types of solar photovoltaic thermal modules, which are used in planar and concentrator systems.

The first type of solar photovoltaic thermal module for use in a concentrator system is a solar module with a two-sided ray-receiving surface. The number of components used in this type of module is limited by the need to ensure the transparency of both ray-sensing sides of the module in the spectrum of solar radiation, in which the solar cell generates electricity.

The second type of solar photovoltaic thermal module for use in a planar system is a solar module with a single-sided ray-sensing side. In this type of module, the number of components used can be expanded and the designs are complicated (Figure 1).

Both types of receivers of solar photovoltaic thermal modules are also subdivided according to the type of cooling of the radiation-receiving side of the receiver (heat removal) – front, rear, bilateral. Depending on the radiation-receiving sides and the type of heat removal, the module is created according to one of the four models (Figure 1).

The developed method of creating models of solar photovoltaic thermal modules of stationary and mobile power generation allows creating models of photodetectors with:

- One-sided solar cells (http://jasolar.com/site/solar_Mono/564#level-2; https://us.sunpower.com/buy-solar-cells/) and front heat removal (model 1, figure 2 and 4);
- One-sided solar elements and rear heat removal (model 2, figure 6);
- One-sided solar cells and bilateral heat removal (model 3, figure 8);
- Bilateral solar cells (Strebkov, Polyakov & Panchenko, 2013; Strebkov, Mayorov, Panchenko, Osmakov. & Plokhikh, 2013; Panchenko, Strebkov, Polyakov & Arbuzov, 2015) and bilateral heat removal (model 4, figure 9 and 11).

Figure 1. Method of creating three-dimensional models of solar photovoltaic thermal modules with different ray-sensing sides and types of heat removal

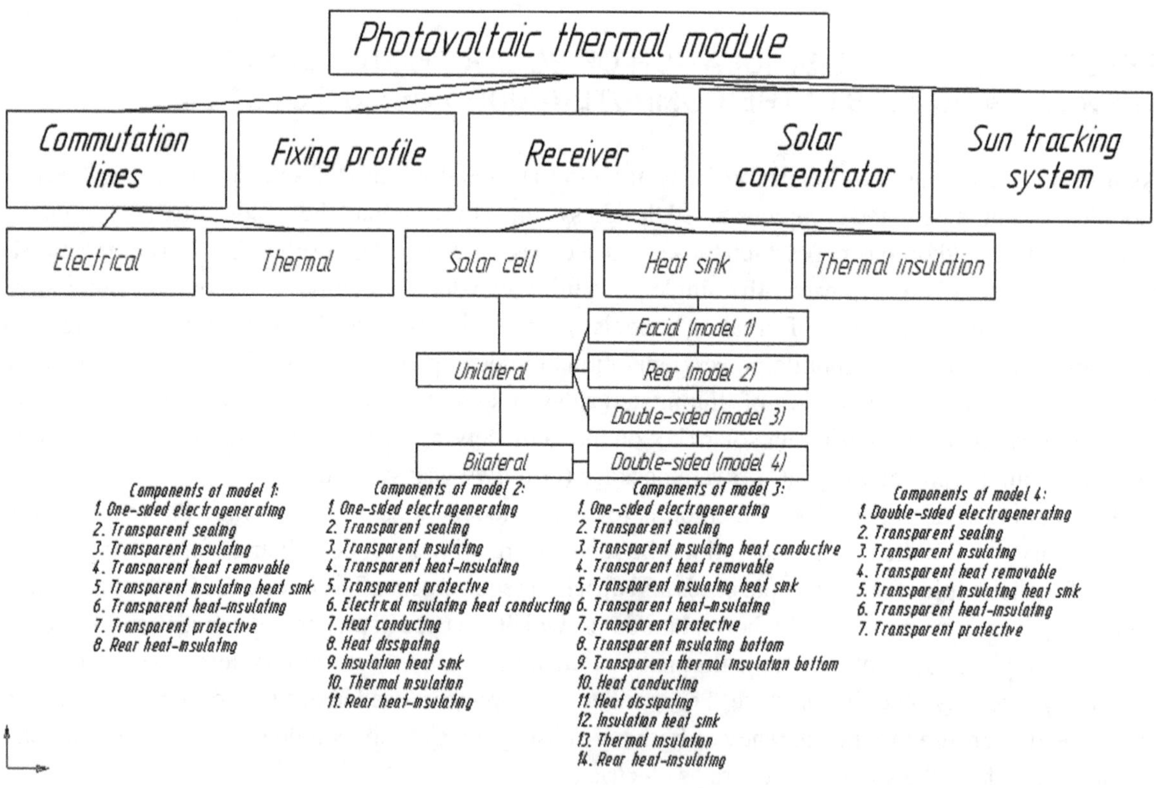

Silicon solar one-sided elements with one- and two-sided contact grids are most widely used as an electrogenerating element. The dimensions of such solar cells are 125 mm and 156 mm, they are made in the form of a square and a pseudo-square (Figure 2). Such elements are taken as the basis for solar photovoltaic thermal modules with one-sided exposure (model 1, 2 and 3, figure 1).

In the process of creating all three-dimensional components included in model 1 with a one-sided solar cell and front heat removal (Figure 3), an assembly unit is formed in the form of a solar photovoltaic thermal module (Figure 4).

It is worth noting that a two-component polysiloxane compound can be used as a sealing component in the solar photovoltaic thermal module assembly, which increases the lifetime of the nominal power of solar cells, is optically transparent, which increases the efficiency of solar cells compared to ethylene-vinyl acetate film and can be used in systems with concentrators, and the efficiency of solar cells does

Figure 2. One-sided solar cells with a two-sided contact grid and dimensions of 156 × 156 mm (left) (http://jasolar.com/site/solar_Mono/564#level-2) and a one-sided contact grid and dimensions of 125 × 125 mm (right) (https://us.sunpower.com/buy-solar-cells/)

Figure 3. Three-dimensional components that are part of model 1, with a one-sided solar cell and front heat sink

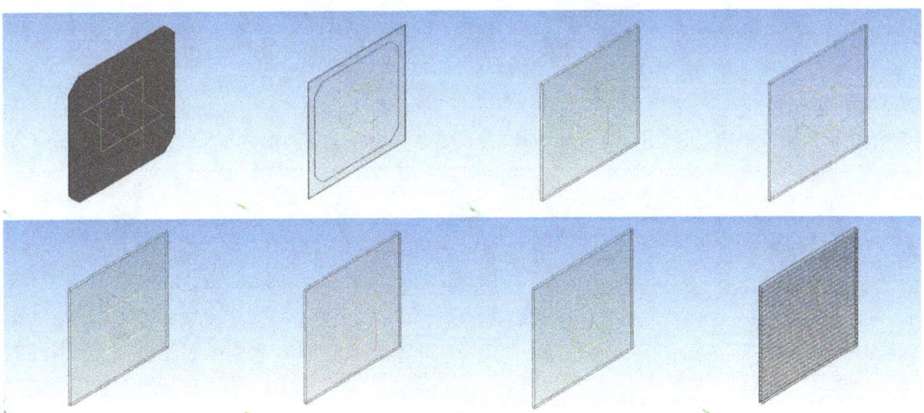

Figure 4. Three-dimensional model of a receiver with a one-sided solar cell and front heat removal (model 1)

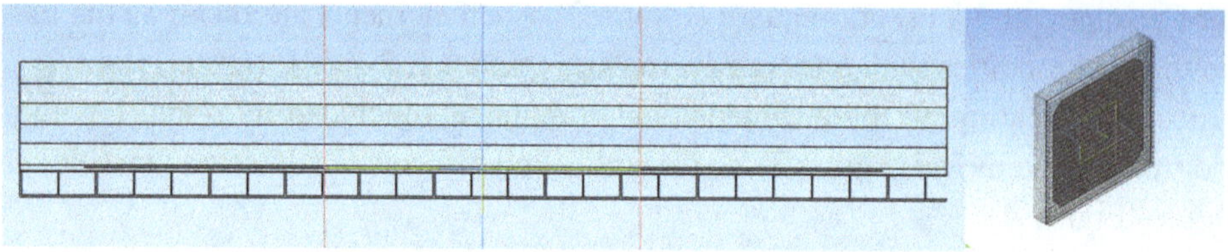

not decrease as with large positive and large negative temperature (Poulek, Strebkov, Persic & Libra, 2012; Strebkov, Persits & Panchenko, 2014).

In the process of creating all three-dimensional components included in model 2 with a one-sided solar cell and a rear heat sink (Figure 5), an assembly unit is formed in the form of a solar photovoltaic thermal module (Figure 6). Due to rear heat removal, the quality of thermal insulation can be improved by using a potentially larger number of components regardless of their transparency.

In the process of creating all three-dimensional components included in model 3 with a one-sided solar cell and two-sided heat sink (Figure 7), an assembly unit is formed in the form of a solar photovoltaic thermal module (Figure 8). This model combines the components used in the creation of model 1 and model 2, thus, the design becomes more complicated, but at the same time there is a possibility of more subtle optimization of the cooling of the two sides of the solar cell.

Figure 5. Three-dimensional components that are part of model 2 with a one-sided solar cell and a rear heat sink

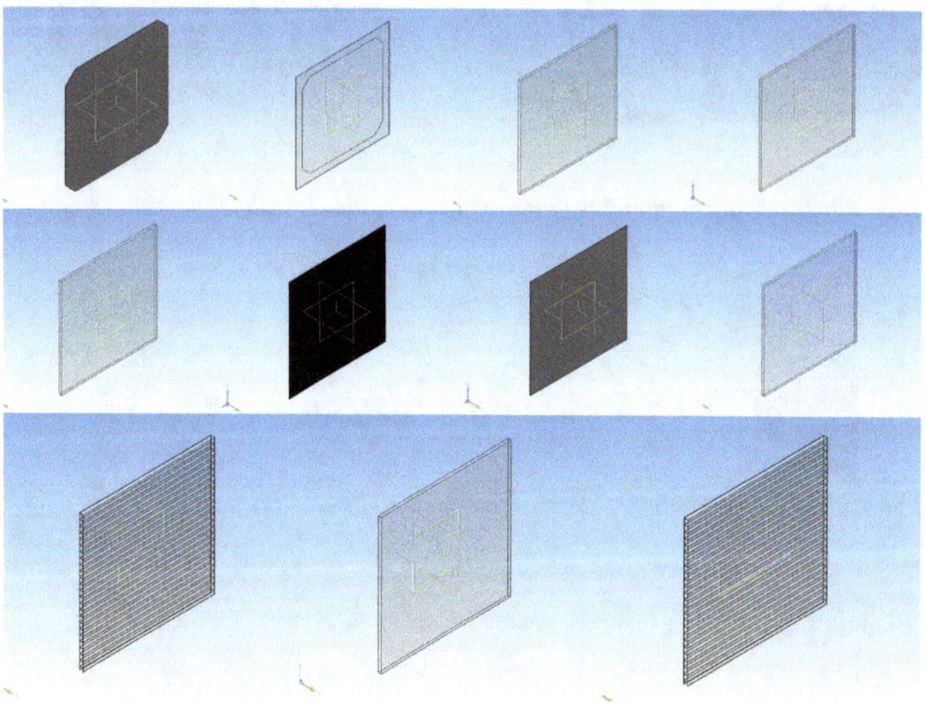

Figure 6. Three-dimensional model of a receiver with a one-sided solar cell and a rear heat sink (model 2).

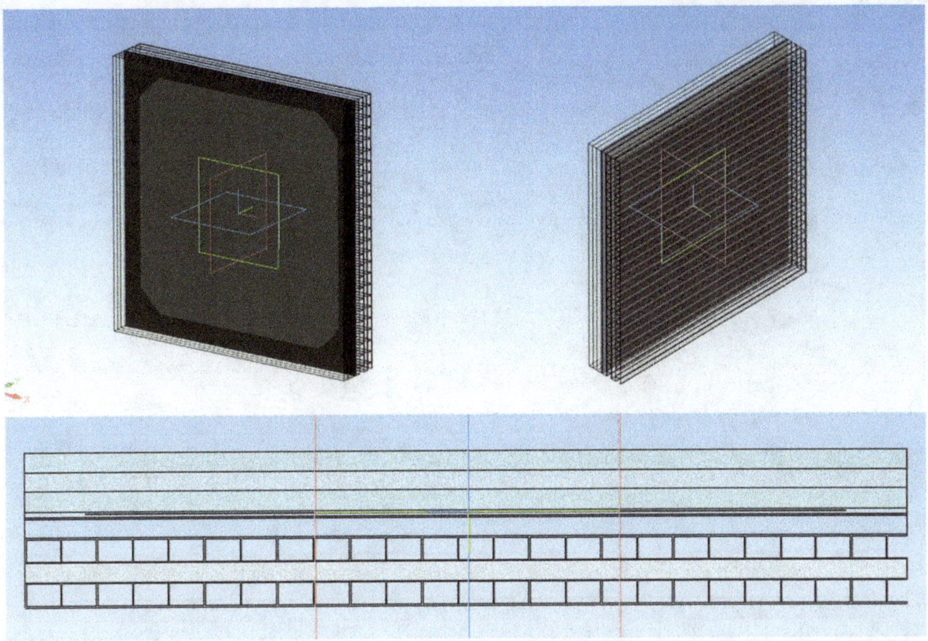

Figure 7. Three-dimensional components that are part of model 3 with a one-sided solar cell and two-sided heat sink

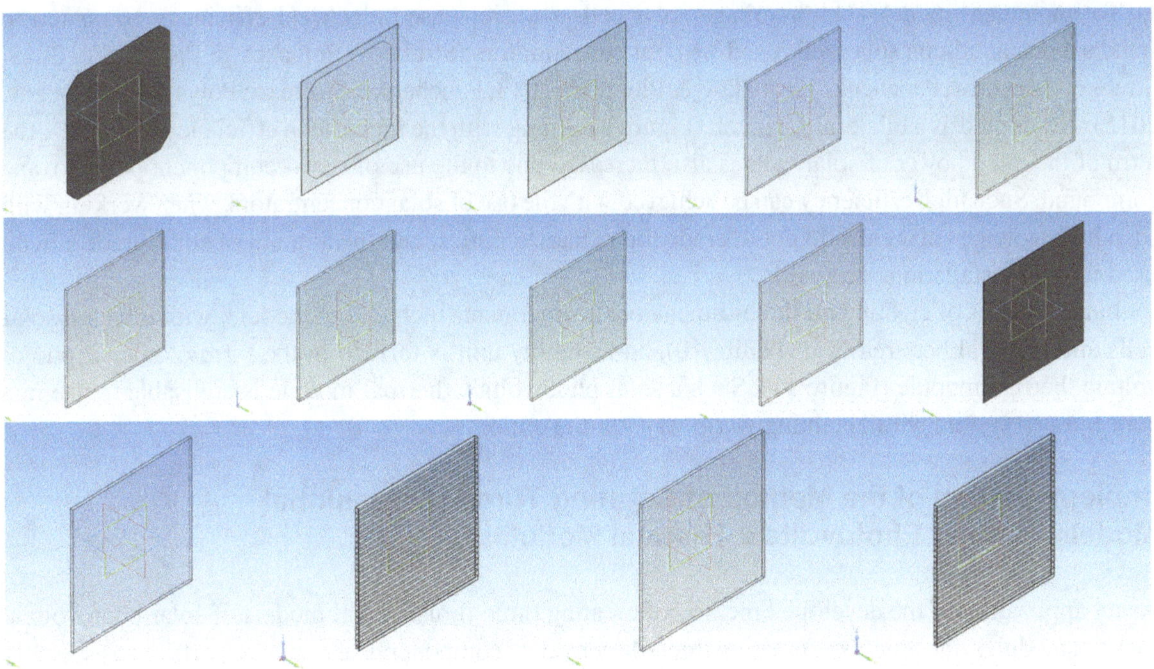

Figure 8. Three-dimensional model of a receiver with a a one-sided solar cell and two-sided heat removal (model 3).

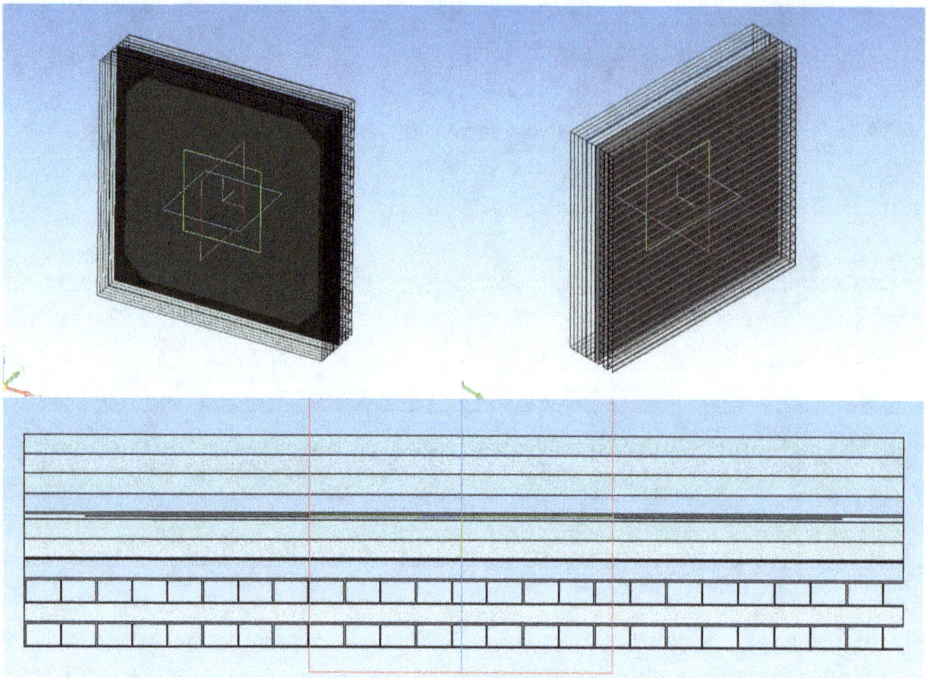

To implement model 4, high-voltage solar cells with increased electrical efficiency compared to standard planar silicon solar cells used without concentrators (Strebkov, Polyakov & Panchenko, 2013; Strebkov, Mayorov, Panchenko, Osmakov. & Plokhikh, 2013; Panchenko, Strebkov, Polyakov & Arbuzov, 2015) are adopted as a bilateral solar cell (Figure 9). Along with the increase in efficiency up to 28%, the term of the rated power of solar cells is also increased due to the use of a two-component polysiloxane compound. Such high efficiency can be achieved with the use of solar concentrators, when working with such high-voltage solar cells do not degrade their characteristics, and the amount of solar-grade silicon used in such installations decreases.

In the process of creating all three-dimensional components included in model 4 with bilateral solar cells and bilateral heat removal (Figure 10), an assembly unit is formed in the form of a solar photovoltaic thermal module (Figure 11). Such a solar photovoltaic thermal module is advisable to use in a concentrator system with obtaining warm water at the outlet.

Implementation of the Method of Creating Three-Dimensional Models of Solar Photovoltaic Thermal Modules

As an application of the developed method of creating three-dimensional models of solar photovoltaic thermal modules, the process of creating three-dimensional components of a solar as parts of an assembly unit (Figure 12) and the assembly unit of a roofing photovoltaic thermal tile (Figure 13) are presented.

To create an assembly in the computer-aided design system Kompas 3D, various components were created in the form of individual parts that make up the module being developed. The assembly unit of

Figure 9. Bilateral solar high-voltage cells in a polysiloxane compound and with a voltage of 1000 V

Figure 10. Three-dimensional components that are part of model 4 with a bilateral solar cell and two-sided heat removal

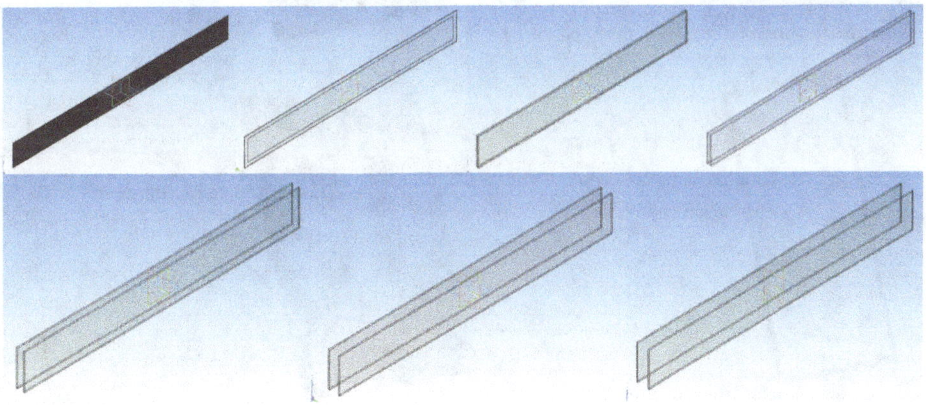

the planar solar photovoltaic thermal module in the form of a roofing panel includes 8 components that perform various functions. The main structural element is the housing to which the remaining components are attached. The cooling agent adopted is water, which washes the aluminum radiator in black. Solar cells are sealed using a two-component polysiloxane compound, a thin front transparent film and black tape around the perimeter. The gas heat insulating region is an air gap that is bounded from the front surface of the module by optically transparent glass.

Figure 11. Three-dimensional model of a receiver with a two-sided solar cell and two-sided heat removal (model 4)

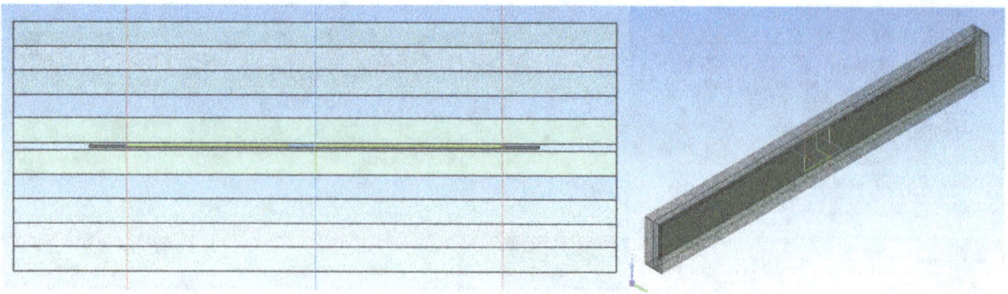

Figure 12. Three-dimensional models of components that make up the three-dimensional assembly unit of a planar solar photovoltaic thermal module in the form of a roofing panel

As a result of the sequential assembly of the components presented, a three-dimensional assembly unit of the planar solar photovoltaic thermal module is formed in the form of a roofing panel (Figure 13).

The created three-dimensional model of a planar solar photovoltaic thermal module in the form of a roofing solar panel is associated with a two-dimensional drawing for obtaining design documentation and the subsequent development of the manufacturing technology of the experimental module (Strebkov, Panchenko, Irodionov & Kirsanov, 2015; Strebkov, Bobovnikov, Irodionov, Kirsanov, Panchenko & Filippchenkova, 2016; Patent of the Russian Federation for invention N° 2557272) (Figure 14).

Figure 13. Three-dimensional assembly unit of a planar solar photovoltaic thermal module in the form of a roofing panel

Figure 14. The drawing of the assembly unit of the planar solar photovoltaic thermal module in the form of a roofing panel, obtained from the created three-dimensional assembly model

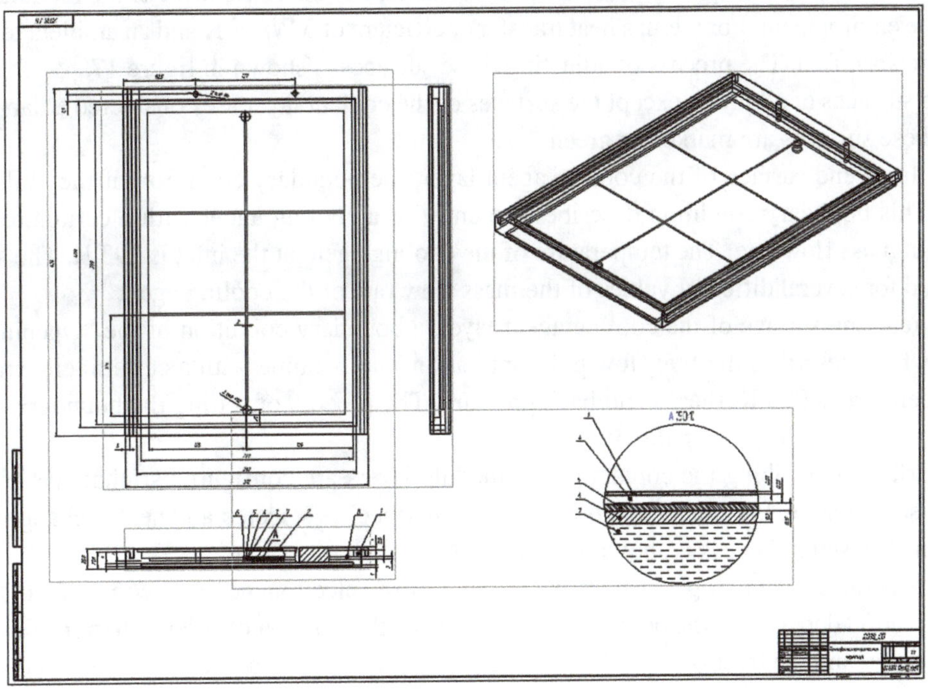

OPTIMIZATION OF THE CONSTRUCTION OF A PLANAR SOLAR PHOTOVOLTAIC THERMAL MODULE IN A FINITE ELEMENT ANALYSIS SYSTEM

Three-dimensional models of solar photovoltaic thermal modules with different parameters of planar and concentrator types created using the developed method should be tested in the Ansys finite element analysis software (http://www.ansys.com/) to optimize the design of receivers, in view of which the method of thermal calculation of solar photovoltaic thermal modules was developed (Figure 15).

As an implementation of the method, the process of modeling the thermal state of a solar photovoltaic thermal module in the form of a roofing panel (Figures 13 and 14) is considered. The dimensions of the investigated area of the module are shown in figure 16. The names of the layers, their materials and properties are listed in table 1.

The size of finite elements is given by the number of partitions of different sides of the layers. The layers of solids are divided into 5 parts, the layers of gas or liquid – into 10 parts. The width of the module is divided into 10 parts, the length – into 50 parts.

Surfaces that separate domains are called interfaces. Through interfaces, heat and matter can be exchanged between domains. Surfaces can describe isothermal or adiabatic boundaries. In addition, interfaces can describe symmetry in the computational domain.

The upper surface of the upper layer and the lower surface of the lower layer of the module exchange heat with the environment. For them, a heat transfer coefficient of 5 $W/m^2 \cdot K$ and an ambient temperature of 293 K are specified. The process of adjusting these surfaces is shown in figure 17.

The end surfaces of all layers except the surfaces of the coolant layer are considered adiabatic. In the figure 18 these surfaces are marked in green.

For the front end surface of the cooling agent layer, the boundary condition of the "Inlet" type is specified. This boundary condition describes the entry of a cooling agent with a certain temperature and a certain mass flow rate. The temperature of the cooling agent at the inlet is 293 K. The simulation is performed for several different values of the mass flow rate of the cooling agent.

For the rear end surface of the cooling agent layer, a boundary condition of the "Opening" type is specified, which describes the free flow point into an infinite volume with a certain temperature. The outlet temperature is 293 K, there is no back pressure. The process of setting the boundary conditions for these surfaces is shown in figure 19.

The interfaces describing the contact of the module layers are configured so that heat transfer between layers is possible. For this, the interface type "Interface" is selected and heat exchange through it is turned on. The setup process is shown in figure 20.

Heat is generated on the upper surface of the layer of the electrogenerating component. The power of heat generation depends on the power of solar radiation, the fraction of solar radiation converted into electrical energy and the fraction of reflected solar radiation. For this model, the surface power of heat dissipation is assumed to be 800 W/m^2. When setting up the model for the upper surface of the electrogenerating component, it is necessary to specify the heat source. The process for setting this boundary condition is shown in figure 21.

For this computational domain, the effect of some system characteristics on the maximum temperature of the electrogenerating component is investigated. Variable parameters: gas gap thickness (component No. 2 (Transparent heat-insulation *air*) in table 1) and mass flow rate of the cooling agent. Temperature

Figure 15. The method of calculating the thermal state of the solar photovoltaic thermal module in the Ansys software package

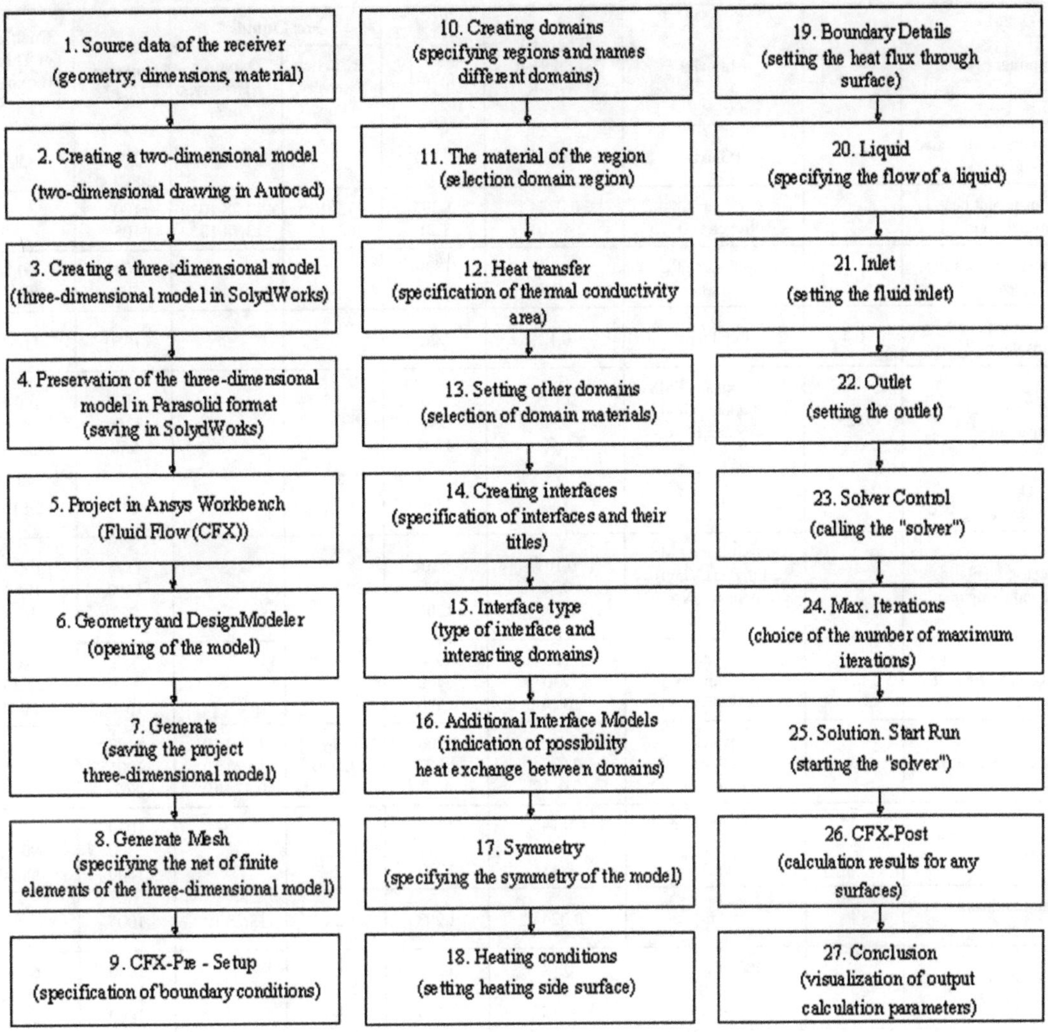

Figure 16. Dimensions of the investigated area of the solar planar photovoltaic thermal module in the form of a roofing panel

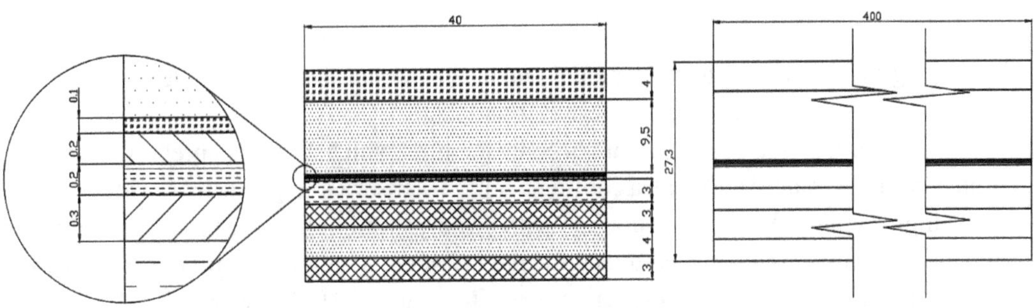

Table 1. Layer thicknesses, materials and their properties (from top to bottom, the transparent side of the module is located on top)

Component Name	Thickness mm	Material	Thermal Conductivity, W/m·K	Density, kg/m^3	For Liquids		Heat Capacity, J/kg·K	Coefficient of Thermal Expansion, 1/K
					Kinetic Viscosity, Pa·s	Dynamic Viscosity, m^2/s		
1. Transparent protective *glass*	4	Glass	0,937	2530			750	$8{,}9\cdot10^{-6}$
2. Transparent heat-insulation *air*	9,5	Air Nitrogen at 0 °C	0,0244 $2{,}43\cdot10^{-2}$	1,293 1,21	$17{,}2\cdot10^{-6}$ $16{,}7\cdot10^{-6}$	$13{,}28\cdot10^{-6}$ $13{,}78\cdot10^{-6}$	1005 1051	
3. Transparent sealing *germ 3*	0,1	Polyethylene PET Glass	0,14 0,937	1330 2530			1030 750	$60\cdot10^{-6}$ $8{,}6\cdot10^{-6}$
4.1. Transparent insulating[*1]	0,4	Butyl tape	0,12	920			1950	$110\cdot10^{-6}$
4.2. Transparent insulating[*2]	0,23	Polysiloxane PMS Ethylene Vinyl Acetate EVA Epoxy	0,167 0,35 0,59	950 931 1200			1175 1400 950	$100\cdot10^{-6}$ $180\cdot10^{-6}$ $55\cdot10^{-6}$
5. One-sided electrogenerating *PVe*	0,2	Silicon	148	2330			714	$2{,}54\cdot10^{-6}$
6. Electric insulating heat-conducting *insul*	0,2	Polysiloxane PMS Ethylene Vinyl Acetate EVA Epoxy	0,167 0,35 0,59	950 931 1200			1175 1400 950	$100\cdot10^{-6}$ $180\cdot10^{-6}$ $55\cdot10^{-6}$
7. Heat conducting *HEx*	0,3	Copper Aluminum Polyethylene	385 230 0,14	8900 2700 1330			383 897 1030	$16{,}6\cdot10^{-6}$ $22{,}2\cdot10^{-6}$ $60\cdot10^{-6}$
8. Heat dissipating *cool*	3	Water Air Freon (gas)	0,569 0,0244 0,0117	1000 1,293 4,39	$1788\cdot10^{-6}$ $17{,}2\cdot10^{-6}$ $11{,}2\cdot10^{-6}$	$1{,}789\cdot10^{-6}$ $13{,}28\cdot10^{-6}$ $0{,}778\cdot10^{-6}$	4182 1005 867	
9. Insulating heat sink *frame 1*	3	Polyethylene ABS PLA	0,14 0,2 0,13	1330 1040 1220			1030 1800 1800	$60\cdot10^{-6}$ $90\cdot10^{-6}$ $68\cdot10^{-6}$
10. Thermal insulation *gap*	4	Air	0,0244	1,293	$17{,}2\cdot10^{-6}$	$13{,}28\cdot10^{-6}$	1005	
11. Rear heat-insulating *frame 2*	3	Polyethylene ABS PLA	0,14 0,2 0,13	1330 1040 1220			1030 1800 1800	$60\cdot10^{-6}$ $90\cdot10^{-6}$ $68\cdot10^{-6}$

Notes: *1. The transparent insulating component 4.1 is located along the external contour of the module, therefore it is not taken into account in the analytical calculation.

*2. The thickness of the transparent insulating component 4.2 is different: in the gap between the electrogenerating elements, but since the conditionally electrogenerating element is considered solid, the thickness of the insulating component 4.2 is assumed to be equal to the gap between the transparent sealing component 3 and the electrogenerating component 5.

distributions along the entire length of the model and at the output of the module for various combinations of variable parameters are shown in figures 22 and 23.

Along with water cooling, a design variant is considered, in which air with a pressure approximately equal to atmospheric pressure is used as a cooling agent. To ensure a large air mass flow rate, its flow rate should be high enough. When air is supplied with density ρ through a hole of radius R with velocity v, the mass flow rate G can be estimated by the formula $G = v \cdot \pi \cdot R^2 \cdot \rho$. For example, when air flows at a speed of 100 m/s with a density of 1,29 kg/m^3, a mass flow rate of 0.015 kg/s will be obtained through an orifice with a radius of 0,006 m. However, the provision of air at such a speed is associated with a

Figure 17. Setting the outer upper and lower surfaces of the module

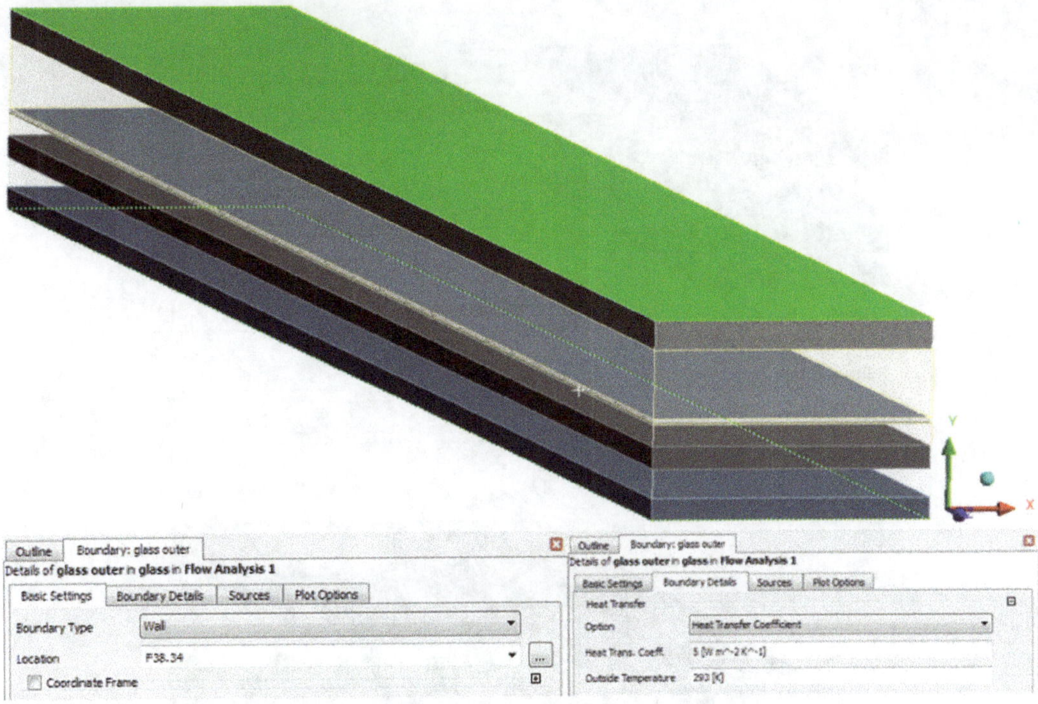

Figure 18. Setting the end adiabatic surfaces of the module

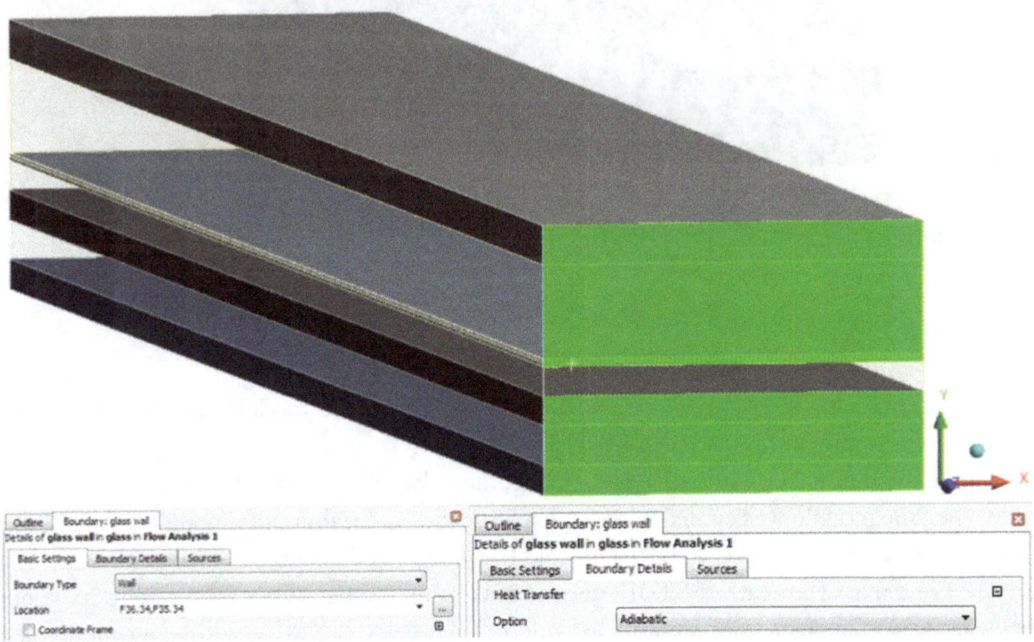

Figure 19. Configuring the end adiabatic surfaces of the module: on the top "Inlet", on the bottom "Opening"

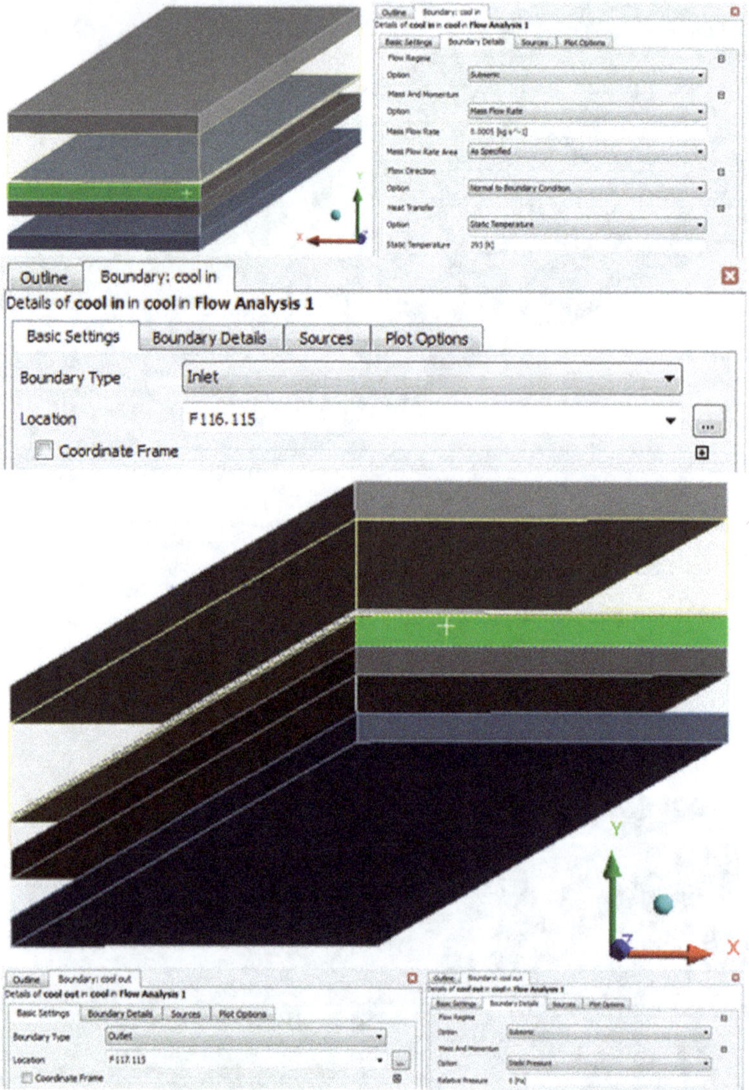

number of problems: the air supply device will be cumbersome, the module will be subject to heavy loads. In addition, significant gas-dynamic losses may occur.

To illustrate the thermal conditions in air cooling, the calculation of the thermal state of the module was made when the module was cooled with atmospheric air. A variant of the module design with an air gap 5 mm and air mass flow rates of 0,01 kg/s and 0,001 kg/s is considered. The results are presented in figure 24. For a mass flow rate of 0,001 kg/s, the maximum temperature of the electricity generating component was 56 °C, which is acceptable for silicon solar cells, but there will be a drop in electrical power relative to electrical power at 20 °C or less.

The obtained distributions of thermal fields in the layers of the module at various expenses were analyzed, and the optimized design of the photovoltaic thermal planar roofing panel was made (Figure 25).

Figure 20. Configuring interfaces between layers

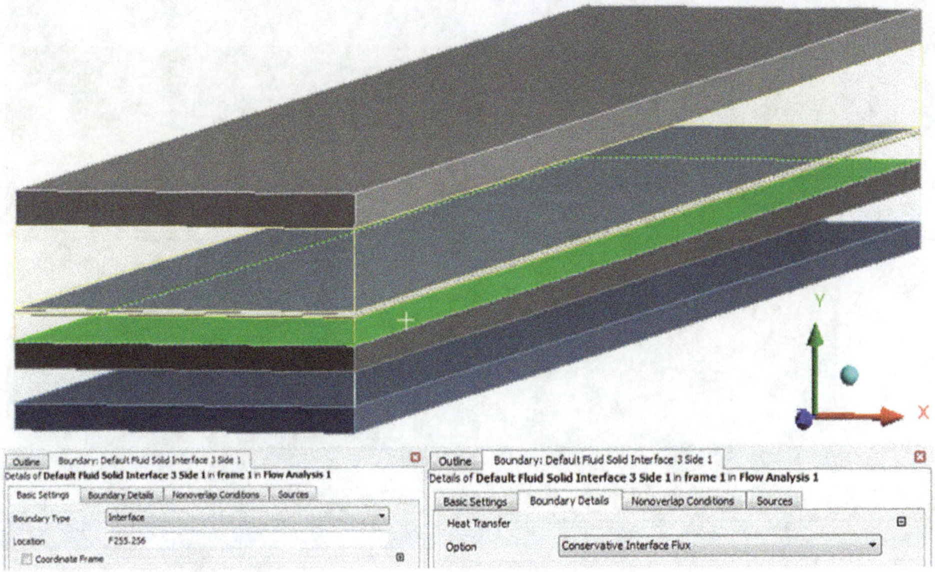

Figure 21. Setting the heat source

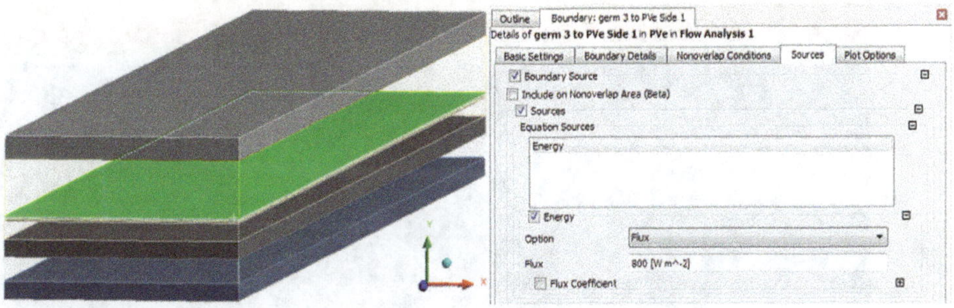

OPTIMIZATION OF THE CONSTRUCTION OF THE CONCENTRATOR SOLAR PHOTOVOLTAIC THERMAL MODULE IN THE SYSTEM OF FINITE ELEMENT ANALYSIS

An example of calculation in the Ansys software package using the developed three-dimensional model of a receiver of a concentrator solar photovoltaic thermal module (Figure 26) (Strebkov, Mayorov & Panchenko, 2013; Patent of the Russian Federation for invention N° 2505755) is presented in figures 27 – 29.

On one part of the thermal photoelectric receiver — on the surface of the cylindrical photoelectric converter, uniform illumination of the concentrated radiation is formed, and on the top of the thermal receiver, the illumination of the concentrated solar radiation is formed to reheat the running water. At the same time, water takes heat away from photovoltaic cells, due to which their efficiency does not decrease and is heated to a certain temperature, which is controlled by the flow rate using a water flow device.

Figure 22. Temperature distribution over the layers of components with a front air insulating gap of 3 mm and a water consumption of 0,0005 kg/s (top) and 0,05 kg/s (bottom).

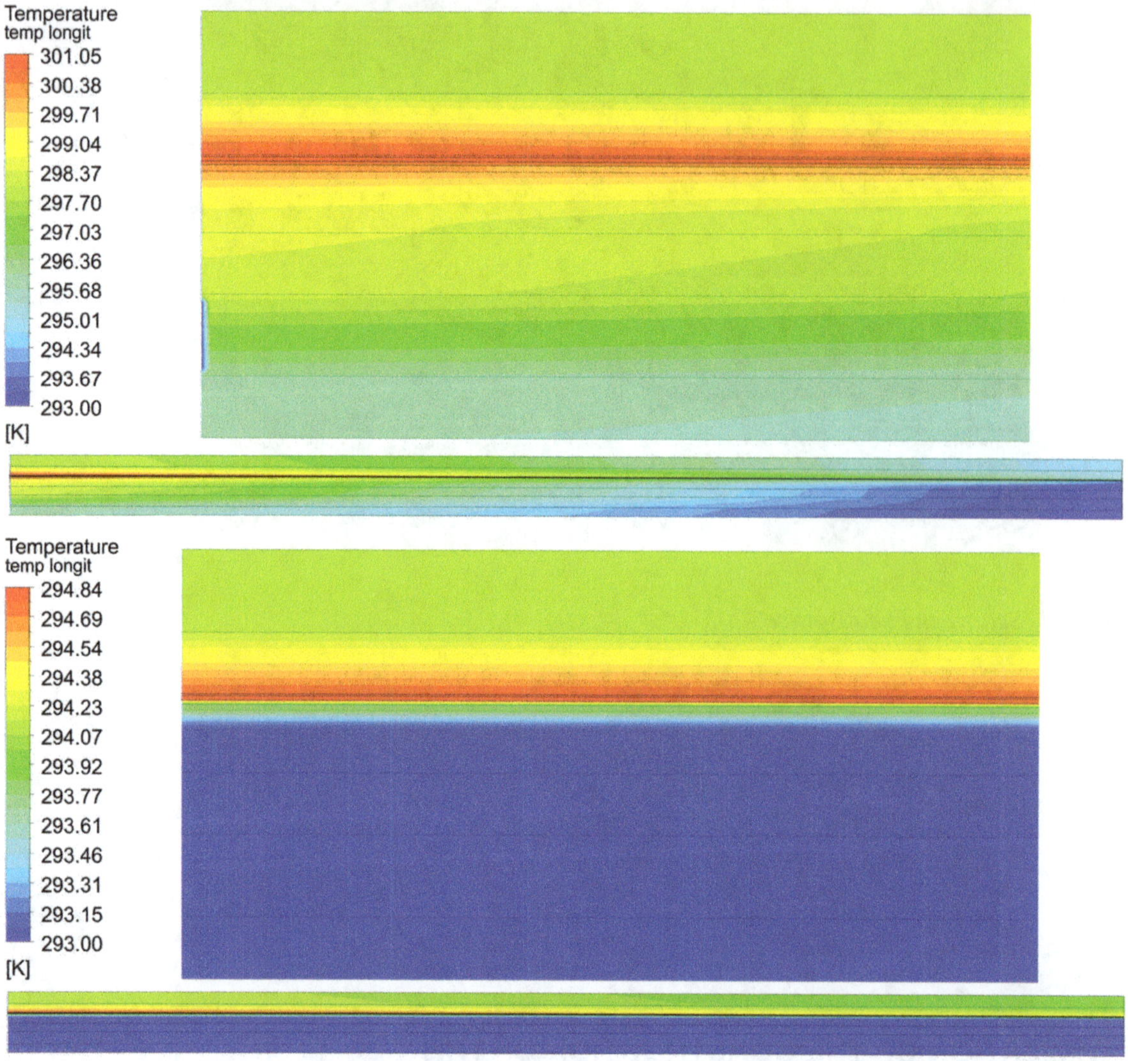

The partition of the three-dimensional radiator model of the receiver of the concentrator solar photovoltaic thermal module into single finite elements for the subsequent thermal calculation is presented in the figure 27.

The definition of the boundary areas with the specification of the conditions of their interaction of the radiator of the receiver of the concentrator solar photovoltaic thermal module is presented in the figure 28.

The derivation of thermal regimes of models and flow lines with visualization of thermal fields, coolant velocities and flow lines according to the developed method is presented in the figure 29.

In the process of optimizing the radiator, its various three-dimensional structures were calculated, after which it was possible to judge the feasibility of using each structure. The radiator optimization criterion was the maximum water temperature at the radiator outlet and the temperature of the radiator

Figure 23. Temperature distribution by component layers with a 9 mm face air insulation gap and a water consumption of 0,0005 kg/s (top) and 0,05 kg/s (bottom)

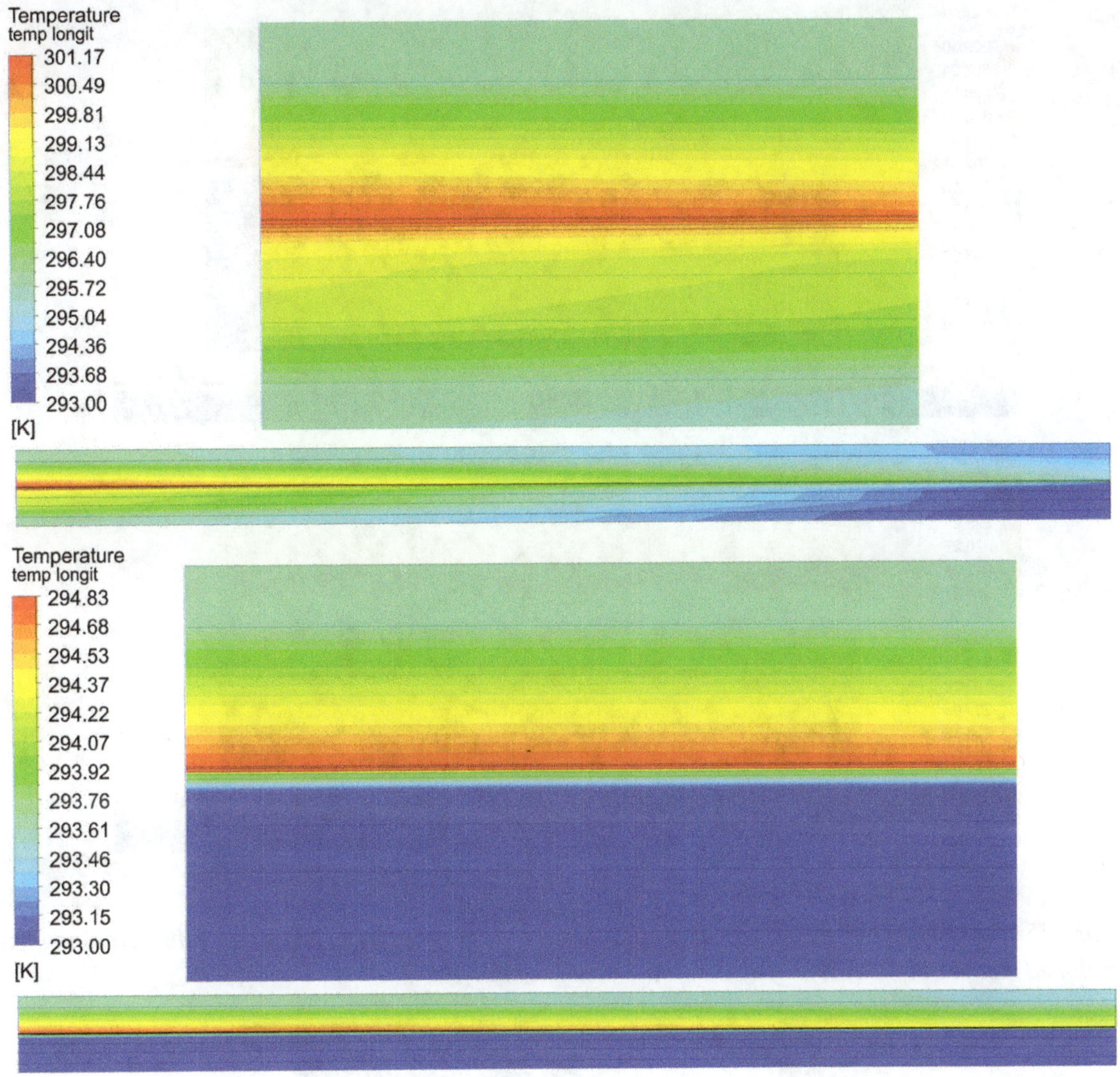

side surface did not exceed the maximum values at which the current-voltage characteristic of solar cells has a rectangular shape.

With the help of the developed method, it is possible to obtain the thermal fields of the developed model, the velocity of the coolant and the flow lines. With the help of visualized models of the thermal state of the radiator, it is possible to make decisions about the need to optimize its design to obtain the required parameters of the thermal state of the radiator itself and the coolant of the solar photovoltaic thermal module.

In the systems of computer-aided design and finite element analysis, various designs of radiators are considered, and for each design a thermal mode of operation is obtained using the developed method. The cylindrical shape of the radiator due to the area of illumination by solar radiation concentrator pa-

Figure 24. Temperature distribution over the layers of the components with a front air insulating gap of 5 mm and air flow of 0,01 kg/s (top) and 0,001 kg/s (bottom)

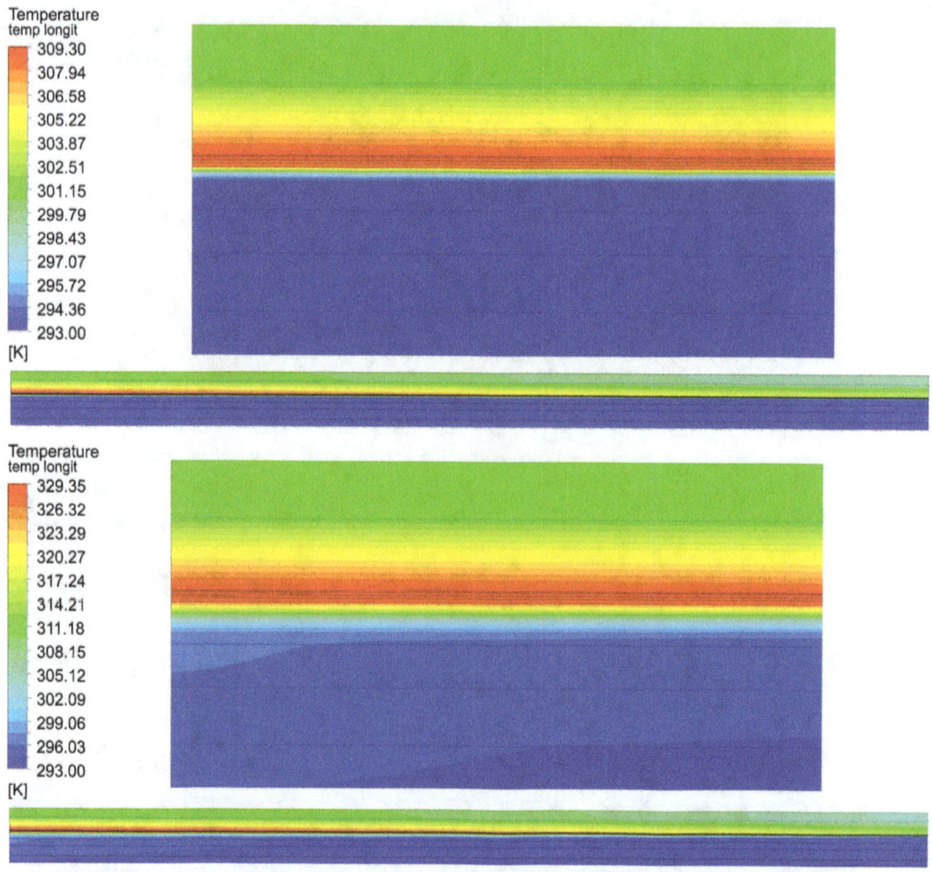

Figure 25. Photovoltaic thermal planar roofing panel, the design of which was optimized using the developed method in the Ansys system of finite element analysis

Figure 26. Concentrator solar photovoltaic thermal module, the radiator of which was optimized in the Ansys system of finite element analysis using the developed method

Figure 27. Partitioning of the three-dimensional model of the radiator of the receiver of the concentrator solar photovoltaic thermal module into single finite elements for subsequent thermal calculation

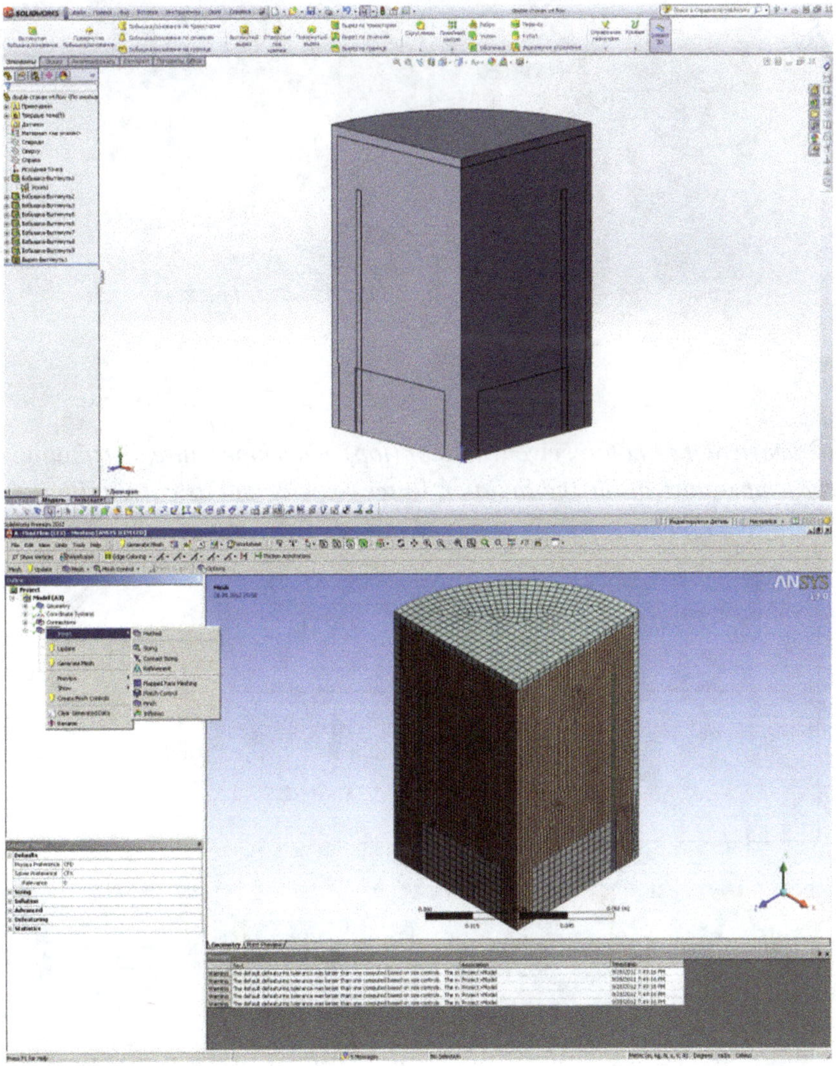

Figure 28. Determination of boundary areas with the specification of the conditions of their interaction radiator receiver concentrator solar photovoltaic thermal module

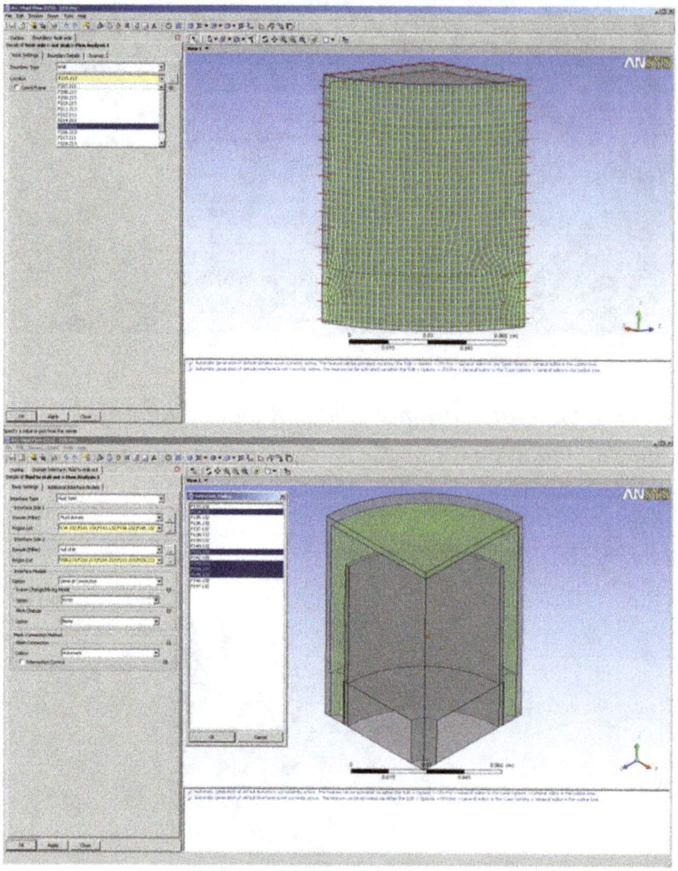

Figure 29. Initial design of the photoreceiver radiator (top) and temperature distribution over the radiator surface, water temperature inside the radiator, water velocity and its current lines

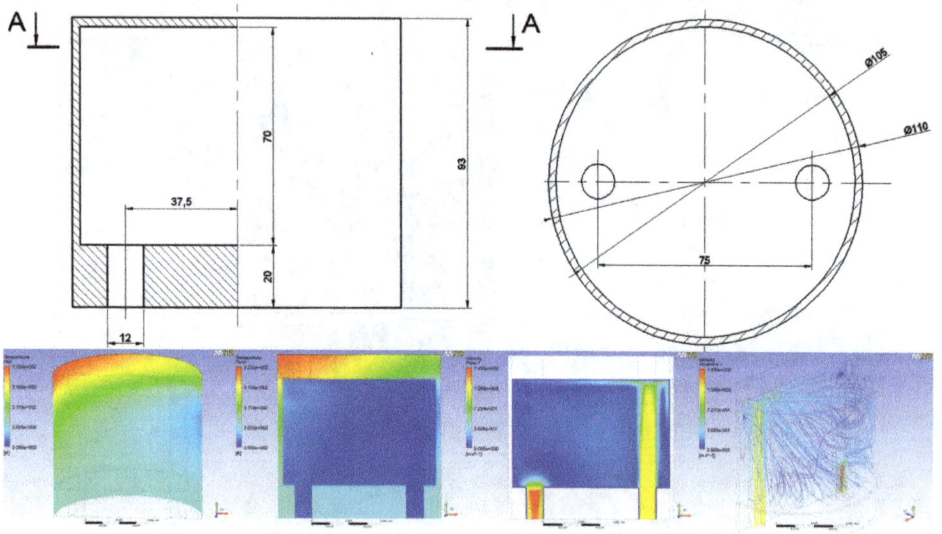

Figure 30. Temperature distribution of photovoltaic cells and outlet water for twelve radiator designs

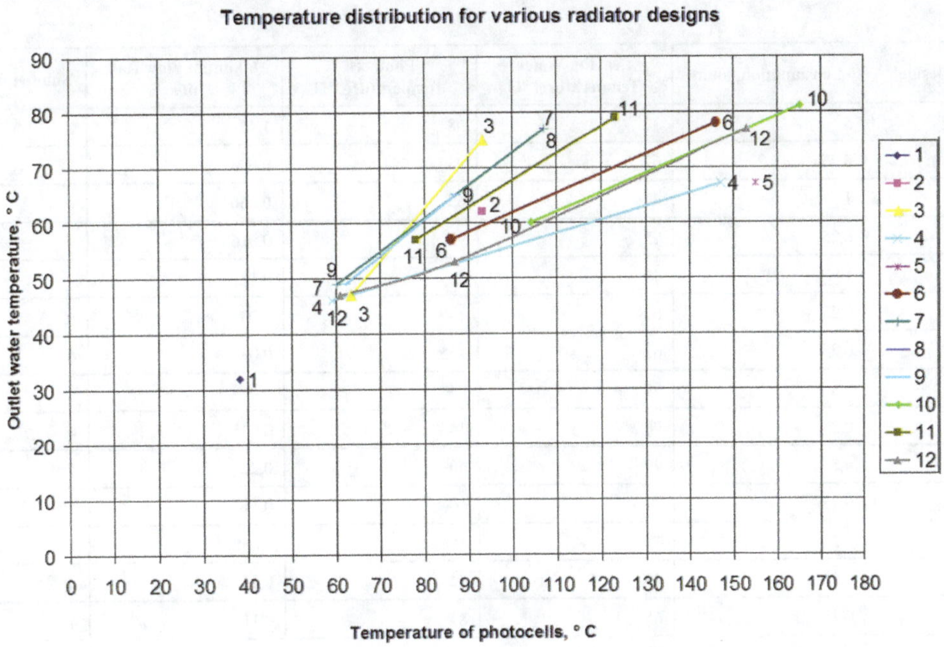

raboloid type. Photovoltaic cells are located on the side of the cylindrical surface and the high electrical efficiency of their operation due to the minimum operating temperature. However, while ensuring the low temperature of the photovoltaic cells, the temperature of the water leaving the radiator also remains low, due to which it is necessary to optimize the thermal modes of operation and the design of the radiator so that the temperature of the side surface of the radiator does not exceed 60 °C and the difference between the temperature the side surface of the radiator and the outlet water temperature is minimal (effective heat sink). A further increase in the temperature of photovoltaic cells leads to a significant decrease in electrical efficiency. In order to increase the electrical and thermal efficiency of the solar module, water that absorbs heat from solar cells is heated by means of the upper surface of the radiator, which is illuminated by concentrated solar radiation. This design of the radiator made it possible not to overheat the photovoltaic cells, but to reheat and increase the output temperature of the coolant using another part of the radiator (top, without photovoltaic cells). Visualization of thermal fields and coolant currents in such a combined model allows analyzing areas with overheating/underheating, the quality of washing the cooled surface and areas with stagnant coolant, resulting in a decision on the need for further changes to optimize the design of the radiator or its thermal mode to ensure high overall solar module efficiency.

The initial type of radiator is not rational from the point of view of heat extraction by water from photovoltaic cells and its further heating (Figure 29). The difference between the inlet and outlet water temperatures is about 5 °C at a flow rate of 5 l/min, and the temperature of the photovoltaic cells is 44 °C. There are areas of overheating and uneven temperature distribution, as well as uneven flow lines of the fluid. In the process of optimizing the design of the radiator, the thermal conditions of their three-dimensional models were considered, after which a decision was made on the expediency of using each design. After the analysis to determine the optimal design using the developed method, twelve different

Table 2. The final table of characteristics of thermal conditions of the twelve options for the design of radiators

Radiator Design	Consumption, l/min	Leaving Water Temperature, °C	Photocell Temperature, °C	Maximum Flow Rate, m/s	Number of Inlets, pcs.
1	5	32	38	1,45	2
2	0,25	62	93	0,028	4
3	**1**	**47**	**63**	**0,066**	**4**
	0,25	75	93	0,0168	4
4	1	46	59	0,12	4
	0,25	67	147	0,03	4
5	0,25	67	155	0,03	4
6	0,5	57	86	0,069	4
	0,25	78	146	0,035	4
7	**0,5**	**49**	**60**	**0,22**	**4**
	0,25	77	107	0,11	4
8	0,25	77	106	0,035	4
9	**0,5**	**49**	**62**	**0,22**	**4**
	0,25	65	87	0,11	4
10	0,5	60	104	0,036	8
	0,25	81	165	0,018	8
11	0,5	57	78	0,1	8
	0,25	79	123	0,05	8
12	**1**	**47**	**61**	**0,13**	**4**
	0,5	53	87	0,065	4
	0,25	77	153	0,036	4

designs of radiators were considered with the aim of improving water heat removal and increasing the overall efficiency of the solar module (Figures 30 and 31, Table 2).

Figure 31 shows four radiator design options (No. 3, 7, 9, 12 in Table 2), which provide such thermal modes of operation when the temperature difference between photocells and outlet water is minimal, and the temperature of the side surface of the radiator (photovoltaic cells) does not exceed 60 °C (in the figure 30, the intersection of the 60 °C line of the temperature of photocells with graphs of thermal modes of operation of radiators). The temperature of the water at the outlet of 49 °C and the temperature of the photovoltaic cells of 60 °C was achieved with water consumption of 0.5 – 1 l/min (Table 2).

CONCLUSION

Three-dimensional modeling of solar photovoltaic thermal modules according to the developed method, implemented in the Kompas 3D computer-aided design system, allows creating modules of various designs for the subsequent determination of their thermal modes of operation. The prepared three-dimensional

Figure 31. Designs of radiators No. 3, 7, 9 and 12, temperature distribution inside the radiators, water velocity inside the radiators and its current line

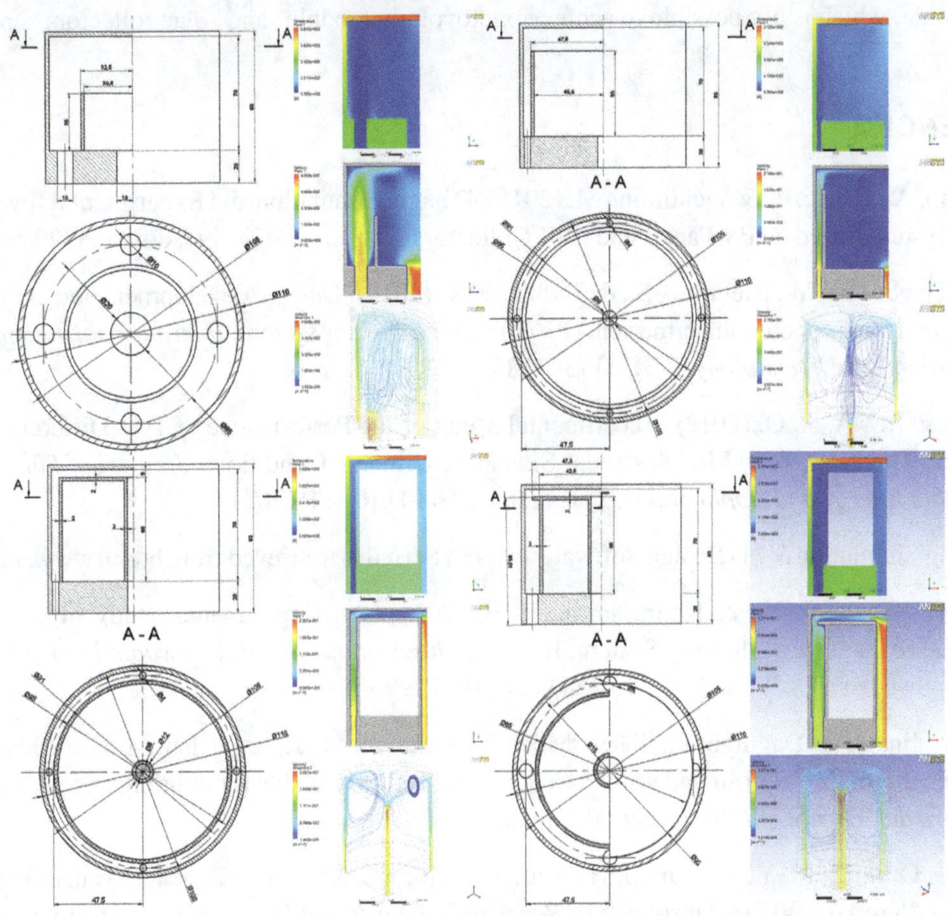

models of solar modules make it possible to prepare design documentation for experimental samples for subsequent production and testing of manufacturing technology.

The developed method of thermal calculation of radiators of photovoltaic thermal solar modules using visualization of processes in the Ansys system of the finite element analysis allows judging the feasibility of the created radiator design and optimizing its design and thermal operating modes. The developed method allows analyzing the thermal regime of the created three-dimensional model, the velocity of the coolant and the flow lines of the coolant of the radiator of the solar photovoltaic thermal module.

Created three-dimensional models of a planar photovoltaic thermal solar module in the form of a roofing panel and a concentrator photovoltaic thermal module allow preparing design documentation for experimental samples of modules. The prepared design documentation of solar photovoltaic thermal modules allows working out the manufacturing technology and making experimental modules for stationary power generation. The developed three-dimensional models allow varying the various parameters of the module and its structural elements, which allows for various experiments to work out the optimized design of the solar photovoltaic thermal module.

Research and optimization of the structures of solar photovoltaic thermal modules make it possible to achieve maximum efficiency of solar modules due to the summation of electrical and thermal energy in one module, which is not possible to achieve photovoltaic modules and solar collectors separately.

REFERENCES

Buonomano, A., Calise, F., & Vicidimini, M. (2016). Design, Simulation and Experimental Investigation of a Solar System Based on PV Panels and PVT Collectors. *Energies*, *9*(7), 497. doi:10.3390/en9070497

Rawat, P., Debbarma, M., Mehrotra, S., & Sudhakar, K. (2014). Design, development and experimental investigation of solar photovoltaic/thermal (PV/T) water collector system. *International Journal of Science. Environmental Technology*, *3*(3), 1173–1183.

Dubey, S., & Tay, A. A. O. (2012). Experimental Study of the Performance of Two Different Types of Photovoltaic Thermal (PVT) Modules under Singapore Climatic Conditions. *Journal of Fundamentals of Renewable Energy and Applications*, *2*, 1–6. doi:10.4303/jfrea/R120313

Engineering Simulation & 3D Design Software – ANSYS. (n.d.). Retrieved from http://www.ansys.com/

Hosseini, R., Hosseini, N., & Khorasanizadeh, H. (2011). An Experimental study of combining a Photovoltaic System with a heating System. In *Proceedings of the World Renewable Energy Congress*, Linkoping, Sweden (pp. 2993-3000). 10.3384/ecp110572993

Ibrahim, A., Jin, G. L., Daghigh, R., Salleh, M. H. M., Othman, M. Y., & Ruslan, M. H.. (2009). Hybrid photovoltaic thermal (PV/T) air and water based solar collectors suitable for building integrated applications. *American Journal of Environmental Sciences*, 618–624.

Ibrahim, A., Othman, M. Y., Ruslan, M. H., Mat, S., & Sopian, K. (2011). Recent advances in flat plate photovoltaic/thermal (PV/T) solar collectors. *Renewable & Sustainable Energy Reviews*, *15*(1), 352–365. doi:10.1016/j.rser.2010.09.024

JA SOLAR. (n.d.). Retrieved from http://jasolar.com/site/solar_Mono/564#level-2

Kamilov, A. G., Muminov, R. A., & Tursunov, M. N. (2008). Ocenka jeffektivnosti sistemy solnechnogo jelementa i kollektora v uslovih zharkogo klimata (Evaluation of system effectiveness solar cell and collector in a hot climate). *Geliotehnika*, *2*, 32–35. (in Russian)

Kharchenko, V., Nikitin, B., Tikhonov, P., & Gusarov, V. (2013). Investigation of experimental flat PV thermal module parametres in natural conditions. In *Conference Proceeding - 5th International Conference, TAE 2013: Trends in Agricultural Engineering*, Prague, Czech Republic, September 2-3 (pp. 51-55).

Kharchenko, V. V. (2014). PVThermal modules. In *Proceedings of XI International Annual Conference "Renewable and small energy"* (pp. 248-257).

Kharchenko, V., Panchenko, V., Tikhonov, P. V., & Vasant, P. (2018). Cogenerative PV thermal modules of different design for autonomous heat and electricity supply. In Handbook of Research on Renewable Energy and Electric Resources for Sustainable Rural Development (pp. 86-119). Hershey, PA: IGI Global. doi:10.4018/978-1-5225-3867-7.ch004

KOMPAS-3D v17. (n.d.). Retrieved from http://kompas.ru/

Othman, M. Y. H., Ruslan, H., Sopian, K., & Jin, G. L. (2009). Performance study of photovoltaicthermal (PV/T) solar collector with V-grooved absorber plate. *Sains Malaysiana*, 537–541.

Panchenko, V.A., Strebkov, D.S., Polyakov, V.I. & Arbuzov, Yu.D. (2015). Vysokovol'tnye solnechnye moduli s naprjazheniem 1000 V (High-voltage solar modules with a voltage of 1000 V). *Alternative energy and ecology*, *19*(183), 76-81. (in Russian)

Patent of the Russian Federation for invention N° 2505755. Mayorov, V.A., Panchenko, V.A. & Strebkov, D.S. Solnechnyj fotojelektricheskij modul' s parabolotoricheskim koncentratorom (Solar photovoltaic thermal module with a parabolic concentrator). Application: 2011153585/28, 28.12.2011. Published: 01.27.2014. Bul. No. 3. [in Russian].

Poulek, V., Strebkov, D. S., Persic, I. S., & Libra, M. (2012). Towards 50 years lifetime of PV panels laminated with silicone gel technology. *Solar Energy*, *86*(10), 3103–3108. doi:10.1016/j.solener.2012.07.013

Sevela P. & Olesen, B.W. (2013). Development and Benefits of Using PVT Compared to PV. *Sustainable Building Technologies,* 90-97.

Solimpeks. Volther Hybrid PV-T Panels. (n.d.). Retrieved from http://www.solimpeks.com

Strebkov, D. S., Bobovnikov, N. Yu., Irodionov, A. E., Kirsanov, A. I., Panchenko, V. A., & Filippchenkova, N. S. (2016). Programma "Odin million solnechnyh krysh" v Rossii [Program "One Million Sun Roofs" in Russia]. *Vestnik VIESH*, *3*(24), 80–83. (in Russian)

Strebkov, D. S., Kirsanov, A. I., Irodionov, A. E., Panchenko, V. A., & Mayorov, V. A. (2015). Patent 2557272. Russian Federation.

Strebkov, D.S., Mayorov, V.A. & Panchenko, V.A. (2013). Solnechnyj teplofotojelektricheskij modul' s parabolotoricheskim koncentratorom [Solar photovoltaic thermal module with a parabolic concentrator]. *Alternative energy and ecology*, 1/2, 35-39.

Strebkov, D. S., Mayorov, V. A., Panchenko, V. A., Osmakov, M. I., & Plokhikh, S. A. (2013). Solnechnaja ustanovka s matrichnymi fotojelementami i koncentratorom (A solar installation with matrix photocells and a concentrator). *Elektro*, *2*, 50–52. (in Russian)

Strebkov, D. S., Panchenko, V. A., Irodionov, A. E., & Kirsanov, A. I. (2015). Razrabotka krovel'noj solnechnoj paneli (Development of a roofing solar panel). *Vestnik VIESH*, *4*(21), 107–111. (in Russian)

Strebkov, D.S., Persits, I.S. & Panchenko, V.A. (2014). Solnechnye moduli s uvelichennym srokom sluzhby [Solar modules with extended service life]. *Innovations in agriculture. Theoretical and scientific and practical journal. Innovations in renewable energy, 3*(8), 154-158. (in Russian)

Strebkov, D. S., Polyakov, V. I., & Panchenko, V. A. (2013). Issledovanie vysokovol'tnyh solnechnyh kremnievyh modulej (Investigation of high-voltage solar silicon modules). *Alternative Energy and Ecology*, *6*(2), 36–42. (in Russian)

SunPower. (n.d.). Retrieved from https://us.sunpower.com/buy-solar-cells/

Sunsystems. (n.d.). Retrieved from http://www.sunsystem.bg/en/fotovoltaika/PV-T/

Zharkov, S. V. (2014). Assessment and enhancement of the energy supply system efficiency with emphasis on the cogeneration and renewable as main direction for fuel saving. *International Journal of Energy Optimization and Engineering*, *3*(4), 1–20. doi:10.4018/ijeoe.2014100101

KEY TERMS AND DEFINITIONS

Computer-Aided Design System: An automated system that implements an information technology for performing design functions, is an organizational and technical system designed to automate the design process, consisting of personnel and a set of technical, software and other automation tools for its activities.

Cooling Radiator: A technical device used for removing heat for a liquid or gaseous coolant from a cooled object in order to cool the object and heat the coolant.

Finite Element Analysis System: A software package based on a numerical method for solving partial differential equations, as well as integral equations arising in solving problems of applied physics, which is widely used to solve problems of deformable solid mechanics, heat exchange, hydrodynamics and electrodynamics.

Modeling: The study of objects on their models; building and studying models of real-life objects, processes or phenomena in order to obtain explanations of these phenomena, as well as to predict phenomena that interest the researcher.

Photovoltaic Thermal Module: A solar module, the design of which consists of photovoltaic solar cells for electrical conversion of solar radiation and heat absorber for their cooling and heat transfer to the coolant.

Receiving Surface: The surface of the photoelectric device/photovoltaic part of the device, which receives solar radiation.

Solar Battery: A combination of photoelectric converters (photocells) – semiconductor devices that directly convert solar energy into direct electric current, in contrast to solar collectors that which heat the coolant.

Solar Concentrator: A technical device designed to focus solar radiation into a focal spot with an increase in the concentration of solar radiation.

Solar radiation Intensity: The density of solar radiation (energy illumination), coming per unit area of the photoelectric module.

Chapter 14
Seed pre-Activation Study by Means of LED Radiation

Alexey Bashilov
Moscow Aviation Institute, Russia

Mikhail Belyakov
https://orcid.org/0000-0002-4371-8042
National Research University "MPEI" in Smolensk, Russia

ABSTRACT

To study the possibilities of pre-sowing seed activation, irradiation with LEDs emitting in the visible, violet, and near-ultraviolet ranges with a maximum of 405 nm was carried out. As a result of the growing experience, it was found that the height of wheat plants grown from the treated seeds significantly exceeds the control indicators except for the period of 45-55 days. To implement the flow, technology of seed activation with LEDs optoelectronic irradiation unit was developed. The advantages of the given installation are the energy efficiency and of seed treatment efficiency, due to of the optimal radiation spectrum selection and treatment doses.

INTRODUCTION

In the modern world there is a constant growth of the population and, as a consequence, the growing demand for quality food. One of the reserves of productivity growth is the use of presowing activation of seeds, including with the help of optical radiation (Kondrat'eva, Krasnolutskaya, Dukhtanova, & Obolensky, 2019, Blaszczak, Aziz, & Gryko, 2017, Hu, Li, & Jiang, 2007, Li, Ji, & Xu, 2013, Wu et al., 2013, Tsai, Huang, Chen, & Yue, 2017, Kakinoki, Kato, Ogawa, Nakao, Okai, & Katsuyama, 2013). Activation of plant seeds by led radiation is energy-efficient, environmentally safe, technological and economically justified. This study is devoted to solving the problem of choosing the optimal modes of led processing of plant seeds, including the optimal processing time, mode (continuous or pulse) and others.

Almost all known sources of radiation from discharge lamps to lasers were used for pre-sowing treatment of plant seeds (Borodin, 1996, Kondrat'eva, 2001, Loginov, 1986, Filippov, Bityuckij, &

DOI: 10.4018/978-1-5225-9420-8.ch014

Fedorov, 1997). However, an increasingly important role in lighting and irradiation techniques begin to play the light-emitting diodes (LEDs) – light sources, the generation of which occurs at the energy expense released by re-combination of carriers – electrons and holes – on the border of semiconductor materials with different character conductivity (Shubert, 2008). Particular interest as radiation sources for pre-sowing plant seeds treatment are LEDs of violet and near ultraviolet ranges (about 250-420nm).

In resistance terms to mechanical loads SD significantly exceed all other radiation sources. The service life of most modern LEDs in nominal mode exceeds 50,000 hours. this parameter is superior to all other LEDs types. Circuit SD is very simple. The advantages of LEDs are also: extremely high reliability, small size, environmental friendliness associated with the absence of mercury and other harmful substances, electrical safety (Ajzenberg, 2006).

Materials and Methods of Research

To study the possibilities of pre-sowing seeds activation, irradiation with LEDs emitting in the visible purple and near ultraviolet ranges with a maximum of 405 nm was carried out, which together with the power supply led module (Gaska, & Zhang, 2007, Bashilov, & Belyakov, 2011) (Fig 1). The optoelectronic module with sixteen LEDs creates irradiation of the working surface of 34 mW/m^2 at a distance of 55 cm.

During irradiation, the time of seed illumination was established and the exposure treatment dose was determined from the expression:

$$H = \int_{0}^{\tau} E(t)dt \qquad (1)$$

where $E(t)$ – is the time dependence of irradiation in the seed treatment zone, τ-processing time.

In the simplest case, when the irradiation is constant during the exposure time, formula (1) takes the form:

Figure 1. Led matrix

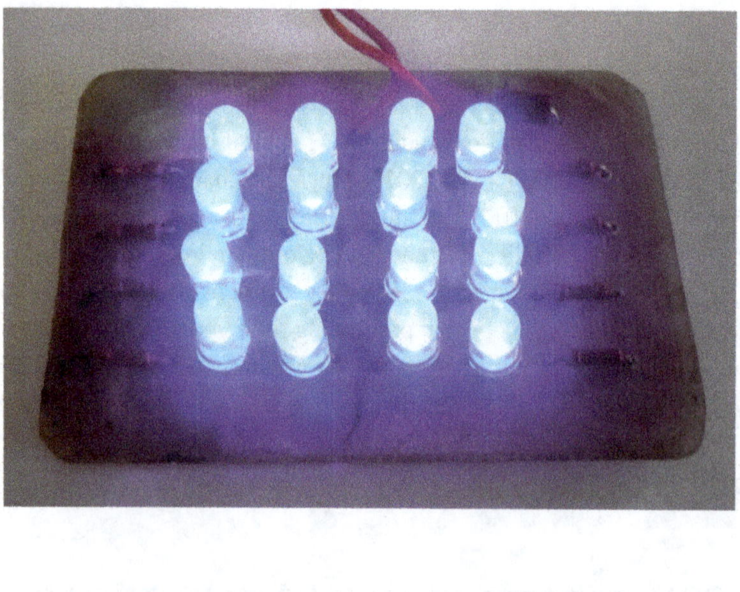

$$H=E\cdot\tau \quad (2)$$

Irradiance was determined from the law of the " square of distance":

$$E = \frac{I_e}{h^2} \cdot \cos\theta \quad (3)$$

where I_e is the radiation force, h – distance from the transducer to the irradiated object, θ - is the angle between the radiation direction incidence and the normal to the seed surface.

Since the seeds were located mainly in the center of the spot, θ=0 and cos θ=1.

As the test seed used seeds of the common grains (wheat), legumes (soybeans), vegetables (radish, cucumber), fodder (Trifolium pratense) and wood (spruce European larch Siberian) plants.

RESULTS AND DISCUSSION

To study the plant dynamics growth was carried out vegetation experiment with seeds of spring wheat Enita. Seeds were treated with continuous or pulsed short-wave LEDs radiation (maximum of 405 nm radiation spectrum) before laying. The lens concentrator was used in some variants of treatment. The treatment time of continuous irradiation was estimated to be about 1 s, the pulse is 0.01 s. Repeated twice with 10 seeds for each growing vessel. The seeds were germinated in a climatic chamber at a temperature of 22-23°C with a lighting period of 13 hours per day. A substrate based on the arable layer of sod-weakly podzolic soil was used. The dynamics of growth and development is presented in the table 1.

When processing seeds of spring wheat (table 1) the option of pulse irradiation with LEDs with a concentrator gives a stable increase in the number of ascended seeds by 21...27% while maintaining the height of the seedlings at the level of control indicators. Options II and III showed no results beyond control.

The results of the final measurements of plant growth are given in table 2.

It can be concluded that pulsed radiation is significantly (20 to 33%) increases the germination rate, while 4-9% reduced the length of shoots. This reduces the likelihood of lodging.

Thus, a tangible effect of radiation of short-wave LEDs on the germination and growth of wheat plants Enita. Pulse radiation has advantages over continuous.

In the course of preliminary laboratory experiments with seeds of spring wheat MIS, the best processing parameters were determined: continuous radiation during $t = 6$ s, or pulsed with a duty cycle Q=4 and a frequency $f = 10$ Hz. The most optimal time for sowing-a day after treatment.

More complete data on the growth dynamics were obtained during germination in the vegetation chamber under fluorescent lighting with maintaining the standard temperature and humidity conditions.

The following options were used in the experiment:

I-control (without treatment);

II-continuous irradiation, $t = 6$ s;

Table 1. Results of wheat Enita germination

Parameter	Variant			
	I	II	III	IV
	Control	LED Continuous Concentrated 1 s	LED Pulse 0.01 s	Pulse Concentrated LED (0.01 s)
Measurement After 13 Days				
Number of seeds grown, %	70	70	75	85
% over control	100	100	107	121
Plant height, cm	20.2	14.3	17.5	20.2
% over control	100	71	87	100
Measurement After 20 Days				
Number of seeds grown, %	75	70	80	95
% over control	100	93	107	127
Plant height, cm	33.4	26.6	27.9	30.7
% over control	100	80	84	92
Measurement After 27 Days				
Number of seeds grown, %	75	70	80	95
% over control	100	93	107	127
Plant height, cm	42.9	36.4	38.2	40.4
% over control	100	85	89	94

Table 2. Results of final measurements of Enita wheat plants 57 days after sowing

Parameter	Variant			
	I	II	III	IV
	Control	LED Continuous Concentrated 1 s	LED Pulse 0.01 s	Pulse Concentrated LED (0.01 s)
Germination, %	75	70	90	95
% over control	100	93	120	127
Length of shoots, cm	57.0	46.5	52.0	51.9
% over control	100	82	91	91
The mass of shoots, g	22.8	11.5	22.3	19.3
% over control	100	50	98	85

III-pulse irradiation, $t = 6$ c, $Q = 4$, $f = 10$ Hz;

IV-pulse irradiation, $t = 24$ c, $Q = 4$, $f = 10$ Hz.

Sowing in 1 day after treatment.

Figure 2. Dynamics of wheat plant growth in the vegetation experiment

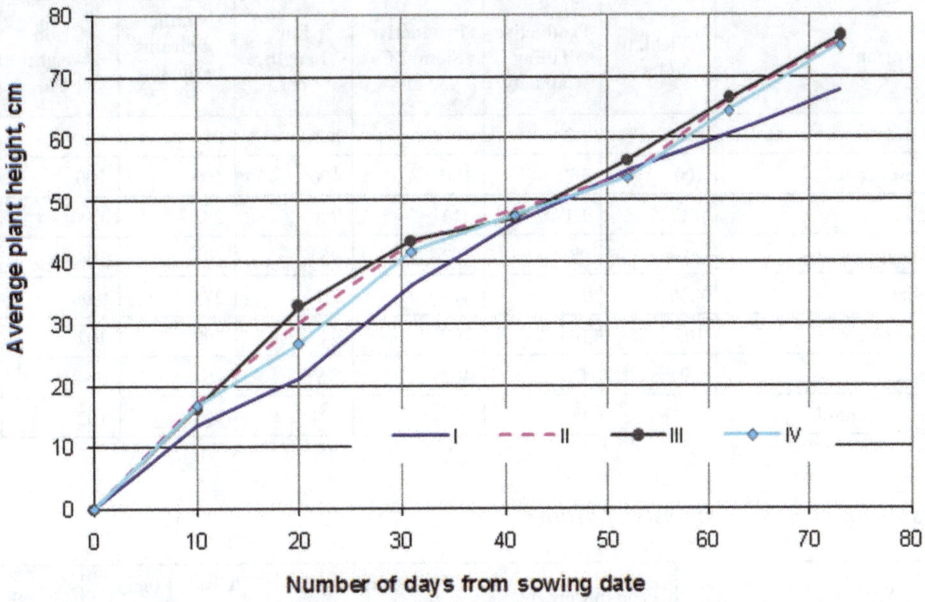

Options II and III compare continuous and pulse processing at the same time, and options II and IV – the same dose for pulse and continuous processing.

As a result of the vegetation experience, it was found that the height of wheat plants grown from treated seeds significantly exceeds the control parameters except for the period of 45...55 days, when due to moisture lack the control and experimental plants were somewhat leveled.

The best indicators of growth dynamics are plants whose seeds were treated continuously. The excess over the control indicators was for these variants at different times up to 42 ... 56% and was noticeable visually.

The dynamics of growth is conveniently illustrated by the graph.

In the future, according to the generally accepted method, a field experiment was conducted with the same options for processing the seeds of spring wheat MIS. The results of intermediate and final measurements are as follows.

In all the variants with presowing seeds treatment germination increased by 7...8%, although the increase is not statistically significant. Leaf area index in all the treatment options on almost all the time measurement exceeds benchmarks for 10...22%, but statistically this difference is only for daily 50...60 plants. In these terms, the highest index in plants of option III. Biomass increases are 8 ... 20% for different treatment options.

All variants of experience with seed treatment (II, III, IV) have a photosynthetic potential higher than the control indicators for all periods of measurement. It should be noted that for option II, the increase over the control decreases from 27% to 12...13% over time, while for pulse processing (options III and IV) the excess over the control increases slightly: from 9% to 12...15%.

The final results of the field experience are presented in the table 3 and 4.

Table 3. Productivity the elements of productivity and yield structure of spring wheat MIS

Variant		Yield, t / ha	Productive Tilling Capacity	Productive Stem, PCs / m^2	Ear Length, cm	Number of Grains in the Ear, PCs.	Grain Weight per Ear, g	Weight of 1000 Seeds, g
I	value	3.87	1.2	405	8.2	27	0.96	35.2
	% over control	100	100	100	100	100	100	100
II	value	4.17	1.1	413	7.8	28	1.00	36.4
	% over control	108	98	102	95	104	104	103
III	value	4.06	1.2	413	7.7	27	0.98	37.0
	% over control	105	104	102	94	100	102	105
IV	value	3.94	1.2	403	8.5	26	0.98	38.3
	% over control	102	116	99	104	96	102	109

Table 4. Results of chemical analysis of grain

Variant		Protein Content, %	The Content of Crude Ash, %	Crude Fiber Content, %	Starch Content, %
I	value	11,1	3,52	14,6	32,7
	% over control	100	100	100	100
II	value	11,4	3,57	15,4	33,0
	% over control	103	101	105	101
III	value	11,6	3,74	16,7	33,3
	% over control	105	106	114	102
IV	value	11,3	3,69	14,8	33,5
	% over control	102	105	101	102

Experienced options slightly higher than the control on biological productivity. At the same time, the best results (an 8% increase) in the option with continuous seed treatment. On the other hand, the highest increases in protein and crude fiber content were observed in the pulse-treated variant III.

For soybean seeds, the possibilities of processing with pulsed and continuous emission of purple LEDs (variants II-IV) were investigated. In variants II and IV, a concentrator based on a collecting quartz lens was additionally used for processing. The growth dynamics is presented in the table 5.

Soybean treatment seeds with continuous concentrated led radiation leads to an increase in germination, but the effect decreases over time from 27% to 17%. Pulse processing of light emitting diodes both with a concentrating lens and without it does not give positive results. The seedlings height increases by 26 ... 29%, but then the effect is reduced.

The final results of the experiment (measurement in 57 days) are presented in the table 6. Field germination, mass of shoots and pods were determined.

Table 5. Dynamics of soybean Rent growth

Parameter	Variant			
	I	II	III	IV
	Control	LED Continuous Concentrated 1 s	LED Pulse 0.01 s	Pulse Concentrated LED (0.01 s)
Measurement After 13 Days				
Number of ascended seeds, %	55	65	60	45
% over control	100	118	109	82
Height of sprouts, cm	7.9	8.0	8.1	11.0
% over control	100	101	103	139
Measurement After 20 Days				
Number of ascended seeds, %	55	70	60	45
% over control	100	127	109	82
Height of sprouts, cm	21.0	19.4	19.4	18.7
% over control	100	93	93	89
Measurement After 27 Days				
Number of ascended seeds, %	55	70	60	45
% over control	100	127	109	82
Height of sprouts, cm	42.4	36.6	36.4	29.1
% over control	100	86	86	69
Measurement After 34 Days				
Number of ascended seeds, %	60	70	60	50
% over control	100	117	100	83
Height of sprouts, cm	77.1	65.6	61.1	55.2
% over control	100	85	79	72

Table 6. Results of measurements of soybean sprouts Rent

Parameter	Variant			
	I	II	III	IV
	Control	LED Continuous Concentrated 1 s	LED Pulse 0.01 s	Pulse Concentrated LED (0.01 s)
Germination, %	60	70	60	50
% over control	100	117	100	83
The mass of shoots, g	24.8	26.3	23.5	24.8
% over control	100	106	95	100
The mass of pods, g	7.1	10.9	5.0	5.1
% over control	100	154	70	72

Analyzing the results, we can conclude that the impulse seed treatment does not give a positive effect – the germination and mass of shoots remain at the level of control and the pods mass even decreases by 28-30%. This can be explained by a relatively small dose of treatment, reduced also by the use of a disk obturator, reflecting and absorbing 99% of the radiation (Belyakov, 2009).

A much better result is the use of continuous concentrated LEDs radiation: an increase in germination by 17%, the pods mass -1.5 times while maintaining the mass of shoots at the control level.

Conducted experience in the processing of radish seeds. Seeds were treated with continuous (variant II) and pulsed (variant III) emission of violet LEDs. In option II, a concentrator based on a collecting quartz lens was additionally used for processing. Repeated twice with 10 seeds for each growing vessel. The results of observations during cultivation are presented in table 7.

The best indicators of germination and height of radish stems for all periods of measurement is the option with irradiation of seeds with continuous LEDs radiation for 1 second using a concentrating lens (table 7). Germination here 113 ... 125% to the control indicators, the height of the seedlings is not worse than the control. The option with the treatment of pulse radiation of LEDs shows an increase in germination by a third only in the later stages of measurement, with the height of the seedlings is

Table 7. Results of radish seed treatment

Parameter	Variant		
	I	II	III
	Control	LED Continuous Concentrated 1 s	LED Pulse 0.01 s
Measurement After 11 Days			
Number of seeds grown, %	40	50	40
% over control	100	125	100
Height of sprouts, cm	6.49	6.18	5.43
% over control	100	95	84
Measurement After 18 Days			
Number of seeds grown, %	40	45	45
% over control	100	113	113
Height of sprouts, cm	9.45	9.40	8.38
% over control	100	99	89
Measurement After 24 Days			
Number of seeds grown, %	40	50	55
% over control	100	125	138
Height of sprouts, cm	10.96	10.50	8.34
% over control	100	96	76
Measurement After 31 Days			
Number of seeds grown, %	40	45	55
% over control	100	113	138
Height of sprouts, cm	11.18	12.55	9.65
% over control	100	112	86

slightly worse than control. Seed treatment of radish radiation water plasma leads to a deterioration of the primary indicators of germination due to suboptimal selection of exposure parameters, including the spectrum of the plasma torch. The final results are presented in the table 8.

According to the results of the experiment, it is obvious that pre-sowing treatment with led radiation significantly increases the field germination of radish seeds (up to 38%). At the same time, the above-ground part of the plants (stems) is less in length than the control indicators, and the length and weight of the underground part, which is productive for the radish, far exceeds the control (up to 124%).

To study the effect of led radiation, a vegetation experiment with cucumber seeds was carried out. Seeds were treated with continuous (variant II) and pulsed (variant III) emission of violet LEDs. In option II, a concentrator based on a collecting quartz lens was additionally used for processing. The intermediate measurements results are presented in table 9.

All variants of irradiation of cucumber seeds have indicators not worse than the control ones. In this case, irradiation with continuous radiation of LEDs gives results rather close to the control in germination and slightly exceeding the control in the length of the seedlings. A significant increase in germination (1.6...1.8 times) gives pulse led processing. The height of the seedlings in this case exceeds the control by no more than 17%, and, over time, the indicators are compared with the control ones.

The final results of the experiment (measurement after 52 days) are presented in the table 10.

As a result, pre-sowing treatment of seeds by continuous radiation of LEDs does not affect germination, increases the length of stems and roots by 12%, increases the weight of roots by 33%. Pulsed led treatment increases germination by 1.8 times, leaving at the level of control of the length of the stems and roots and increasing their weight by 10 and 58%, respectively.

Thus, with the correct selection of processing parameters, LEDs can be used for pre-sowing treatment of vegetable seeds, including cucumber.

The possibility of activation of clover seeds by Smolensky 29 short-wave light-emitting diodes was investigated. To find the optimal treatment time, the seeds were irradiated with continuous radiation of LEDs during 1, 5, 10, 15, 20, 30, 40 and 60s. Irradiation in the treatment area was 34 mW/m^2 .

Table 8. Results of measurements of radish seedlings (60 days after sowing)

Parameter	Variant		
	I	II	III
	Control	LED Continuous Concentrated 1 s	LED Pulse 0.01 s
Germination, %	40	45	55
% over control	100	113	138
Length of stems, cm	12.3	10.5	8.6
% over control	100	85	70
Root length, cm	9.3	12.4	8.3
% over control	100	133	90
Weight of stems, g	6.0	6.7	7.1
% over control	100	112	118
Root weight, g	5.1	11.4	5.6
% over control	100	224	110

Table 9. Cucumber seed germination results

Parameter	Variant		
	I	II	III
	Control	LED Continuous Concentrated 1 s	LED Pulse 0.01 s
Measurement After 11 Days			
Number of seeds grown, %	50	50	80
% over control	100	100	160
Height of sprouts, cm	3.74	3.73	4.38
% over control	100	100	117
Measurement After 18 Days			
Number of seeds grown, %	50	50	80
% over control	100	100	160
Height of sprouts, cm	5.82	6.57	6.50
% over control	100	113	112
Measurement After 24 Days			
Number of seeds grown, %	50	50	90
% over control	100	100	180
Height of sprouts, cm	7.32	8.63	7.85
% over control	100	118	107
Measurement After 31 Days			
Number of seeds grown, %	50	50	90
% over control	100	100	180
Height of sprouts, cm	9.00	9.45	8.85
% over control	100	105	98

Quite good results were obtained at a treatment dose of about 170 mJ/m^2. This is the only statistically reliable result. At the same time, treatment with a higher dose (340 mJ/m^2) gives worse results, and less (68 mJ/m2) - worse for germination and raw weight of plants, but slightly less for the length of sprouts. Taking into account the manufacturability and energy efficiency, it is advisable to use an intermediate dose close to 170 mJ/m^2 in the future, for example, a dose of 136 mJ/m^2 corresponding to the treatment time of 4 seconds.

In the study of pulsed led processing, the best results were obtained when processing with a duty cycle of 2. The same doses in pulse processing give the same results with continuous radiation, except for the length of the seedlings. The best results are obtained at a pulse frequency f=24 Hz.

Researched as well as the timing of incubation of laying the seeds on the germination after treatment. Significant differences between the options for the timing of bookmarks is not observed, but sowing in a day and two days is still better than the rest of the sowing dates for germination.

Taking into account the optimal processing parameters obtained in the course of laboratory studies, the field experience is laid on the experimental field. Its main results are presented in the table 11-14.

Table 10. Results of measurements of cucumber seedlings

Parameter	Variant		
	I	II	III
	Control	LED Continuous Concentrated 1 s	LED Pulse 0.01 s
Germination, %	50	50	90
% over control	100	100	180
Length of stems, cm	19.3	21.6	13.0
% over control	100	112	67
Root length, cm	9.3	10.4	10.0
% over control	100	112	108
Weight of stems, g	21.8	22.2	24.0
% over control	100	102	110
Root weight, g	1.2	1.6	1.9
% over control	100	133	158

Table 11. Iependence of field germination on the treatment dose of clover seeds Smolensky 29 pulsed and continuous emission of LEDs

Parameter		Variant			
Time, s		control	4	4	8
Duty cycle			continuously	2	2
Dose, mJ /m²			136	68	136
Germination, %	average	72	92	82	76
	% over control	100	128	114	106

The germination of seeds and the development of grass clover shoots took place in conditions of excessively wet, but cool June, which was replaced by a somewhat dry July. August was cold, with precipitation exceeding the climatic norm. This contributed to the lodging of shoots and the intensive development of diseases on them. September was very warm and dry, so the cleaning of feed took place in favorable conditions.

From data table 11 it is obvious that all treatment options increase field germination. The greatest increase gives continuous radiation (28%). The increase from pulse processing is less-6 ... 14%.

Seed treatment with continuous radiation gives a small (4...10%) statistically significant increase in plant height, although at the initial stage of measurements it is not noticeable. A similar dose of pulsed radiation gives almost the same results. Pulse radiation with a dose of 68 mJ/m² gives smaller gains.

According to the leaf surface index, the plants from the treated seeds first surpass the control ones, but by the time of harvesting there is a significant decrease in the index compared to the control. This is due to the lodging of a higher herbage of experimental variants and withering leaves. In almost all cases, the differences are statistically unreliable (table 13).

Table 12. Temporal dynamics of plant growth of Trifolium pratense Smolenskiy 29 depending on the dose of seed treatment radiation led LED_{405}

Parameter		Variant			
Time, s		control	4	4	8
Duty cycle			continuously	2	2
Dose, mJ /m^2			136	68	136
period	Value /LSD_{05}				
39 days	height, cm	18.4	17.8 / 0.76	18.4 / 0.71	17.7 / 0.78
	% over control	100	97 / 4	100 / 4	96 / 4
68 days	height, cm	39.2	43.3 / 2.16	42.4 / 2.20	43.5 / 2.02
	% over control	100	110 / 5	108 / 5	111 / 5
80 days	height, cm	43.5	45.2 / 1.51	44.8 / 1.41	44.9 / 1.43
	% over control	100	104 / 3	103 / 3	103 / 3
89 days	height, cm	41.6	44.0 / 1.74	42.6 / 1.59	42.7 / 1.55
	% over control	100	106 / 4	102 / 4	103 / 4

The dry matter content exceeds the control indicators (table 14). Also noted the presence of inflorescences in experimental versions, in contrast to the control, where the inflorescences were not.

Analyzing the results of the field experiment, we can conclude that the led treatment increases the germination of seeds. Doses of continuous radiation 136 mJ/m^2 and pulse radiation 68 mJ/m^2 have no significant impact on yield. The dose of pulse radiation 136 mJ/m^2 causes a significant decrease in yield by 18%. Thus, the effect of led technology is to save up to 28% of the seed material by sowing more germinated activated seeds. At the same time, it is necessary to further improve the technology through the use of shorter-wave UV LEDs, as well as to test the technique on seeds of plants of other cultures and varieties.

Treatment with led seeds radiation of agricultural plants gives a positive result, but the effect on forest tree species is practically not studied. Therefore, it was very promising to study the effect of presowing treatment of spruce seeds and Siberian larch radiation LEDs. This will reduce the consumption of seeds, get stimulated planting material, and in the future and highly productive stands. Currently, the use of stimulated plants in afforestation of unused land is promising (Belyakov, & Rybkina, 2011).

The results of determining the optimal doses for European spruce seeds are presented in table15.

There are two maximum germination and length of seedlings: for relatively small doses (about 170 mJ/m^2) increases are 20 and 21%, respectively, and for relatively large doses (about 1020 mJ/m^2) – 40 and 30%, respectively. Further increase in the dose leads to a decrease in the effect and even inhibition of seedlings.

The use of pulsed light LEDs does not have a stimulating effect on the germination of seeds of European spruce. In all cases of its use germination rates decreased by an average of 24% compared to the control.

When studying the terms of sowing after treatment, it was shown that the most effective sowing is 2-3 days, or 9 days after treatment.

For Siberian larch seeds, two maxima are also observed for small and large doses (treatment times). The first maximum time for 5s, but there is no increase in the length of the seedlings (Table 16). The

Table 13. Time dynamics of changes in the area of the leaf surface of plants clover Smolensky 29 depending on the dose of seed treatment pulse and continuous emission of LEDs LED_{405}

Parameter		Variant				
Time, s		control	4	4	8	LSD_{05}
Duty cycle			continuously	2	2	
Dose, mJ /m²			136	68	136	
period	Value					
68 days	LAI	1.48	2.22	2.84	2.22	1.90
	% over control	100	150	192	150	87
80 days	LAI	3.69	3.55	4.20	3.66	2.51
	% over control	100	96	114	99	67
89 days	LAI	4.64	6.04	5.25	6.10	3.32
	% over control	100	130	113	131	60
111 days	LAI	5.22	2.78	2.78	3.77	4.34
	% over control	100	53	53	72	55

Table 14. The yield of dry matter of red clover Smolenskiy 29 depending on the processing time of the seed pulse and continuous the LEDs radiation LED_{405}

Parameter	Variant			
Time, s	control	4	4	8
Duty cycle		continuously	2	2
Dose, mJ /m²		136	68	136
Dry matter yield, t / ha	2.78	2.73	2.66	2.28
% over control	100	98	96	82
LSD_{05}/ $LSD_{05\%}$	0.23/9			
Content of the main culture, %	87.1	74.4	81.9	81.0
Dry matter content, %	18.2	20.6	20.1	18.6

Table 15. Dependence of the primary parameters of germination of seeds of Norway spruce on the time of treatment.

Parameter		Variant							
Processing time, s		control	1	5	15	20	30	40	60
Dose, mJ /m²			34	170	510	680	1020	1360	2040
Germination, %	average	57	60	69	54	63	80	64	57
	over control	100	105	120	95	110	140	113	100
Length of roots of sprouts	on average, mm	9.1	10.3	11.0	10.6	7.7	11.8	8.0	7.4
	% over control	100	114	121	117	85	130	88	81

Table 16. Dependence of primary parameters of germination of Siberian larch seeds on processing time.

Parameter		Variant							
Processing time, s		control	1	5	15	20	30	40	60
Dose, mJ /m²			34	170	510	680	1020	1360	2040
Germination, %	average	72	73	91	72	90	89	89	85
	over control	100	102	127	100	125	124	124	118
Length of roots of sprouts	on average, mm	10.4	8	10	11	10	10	12	9
	% over control	100	80	100	108	98	99	113	91

second maximum at 38 ... 40s showed an increase in germination by 24% and the length of the seedlings by 13%.

Pulse radiation treatment, as in the case of European spruce, showed worse results than continuous.

Sowing of treated seeds is best done within the first two days after treatment or, in extreme cases, on the ninth day (table 17).

It can be concluded that the electromagnetic radiation of the optical range has a stimulating effect on the primary parameters of germination of seeds of spruce and Siberian larch. In General, the level of seed germination increases and the length of roots of seedlings increases.

We determined the doses that have a stimulating effect on the sowing quality indicators of Siberian larch seeds when using pre-soaking.

Table 17. The Dependence of the primary parameters for seed germination of Siberian larch from the sowing after the treatment.

Variant		Parameter			
		Germination, %		Length of Roots of Sprouts	
		On Average	% Over Control	On Average,mm	% Over Control
sowing on the day of treatment	control	34	100	25.3	100
	experiment	44	129	25.4	100
sowing on the second day	control	40	100	18.8	100
	experiment	47	118	21.2	112
sowing on the third day	control	40	100	18.9	100
	experiment	39	98	18.2	96
sowing on the fourth day	control	36	100	14.2	100
	experiment	26	72	18.2	128
sowing on the eighth day	control	50	100	20.8	100
	experiment	36	72	17.6	85
sowing on the ninth day	control	39	100	19.1	100
	experiment	46	118	20.9	109

At the same intensity of light-emitting diodes and the distance from the emitter to the seeds (15 cm), the exposure dose changed due to changes in the time of exposure to objects. Control samples of seed irradiation was not carried out. Thus, we found the optimal irradiation time and the optimal dose at which the greatest stimulation occurs. The results of the experience are presented in tables.

As can be seen from table 18, there is a decrease in seed germination in all variants relative to control, which averages 25%. The maximum reduction is 33.4% relative to the control. Germination is maintained at the level of control at a dose of 1360 mJ/m² (40 seconds).

Reduction (inhibition) of larch sprouts roots length is also noted in relation to control in all variants. The length of the roots of the seedlings is maintained at the level of control only at a dose of 340 mJ/m² (10 seconds).

Note that the preliminary soaking of Siberian larch seeds increases their germination (table 19). The maximum increase in germination by 50% occurs at a dose of 1360 mJ/m² (40 seconds). The maximum length of roots of seedlings is observed at a dose of 680 mJ/m² (20 seconds). Thus, as a result of the experiments, the optimal irradiation time is 40 seconds, the optimal radiation dose is 1360 mJ/m².

Thus, it can be concluded that soaking the seeds gives a significant increase both in germination and in the length of the roots of the seedlings. Therefore, it is necessary to combine the soaking of seeds and irradiation with LEDs.

In the course of experimental studies, we determined doses that have a stimulating effect on the sowing quality indicators of spruce seeds.

At the same intensity of light-emitting diodes and the distance from the emitter to the seeds (15 cm), the exposure dose changed due to changes in the time of exposure to objects. Control samples of seed irradiation was not carried out. Thus, the optimal irradiation time and the optimal dose at which the greatest stimulation occurs were found.

The results of the experience are presented in tables 20 and 21.

From table 20 it can be seen that the most optimal dose is 1360 mJ/m² (40 seconds) – germination increases to 183.3% of the control, and the length of the roots of seedlings at the same time slightly exceeds the control (by 8.2%).

Analyzing the data of table 21, it can be noted that soaking gives an increase in the length of the roots of seedlings to 173% compared to the control (at a dose of 2040 mJ/m²). Without soaking the length of the roots is significantly less. Germination is kept at the level of control at a dose of 1360 mJ/

Table 18. The dependence of germination of Siberian larch seeds (without pre-soaking) and the length of roots of seedlings on the time and dose of irradiation (measurement after 7 days)

Parameter		Variant						
Processing time, s		control	1	10	20	30	40	60
Dose, mJ /m²			34	340	680	1020	1360	2040
Germination, %	average	12	8	9	8	9	12	8
	over control	100	66.6	75	66.6	75	100	66.6
Length of roots of sprouts	on average, mm	16	13.5	15.1	10.5	12.4	11.7	11.5
	% over control	100	84.4	94.4	65.6	77.5	73.1	71.9

Table 19. The dependence of germination of Siberian larch seeds (with pre-soaking) and the length of roots of seedlings on the time and dose of irradiation (measurement after 7 days)

Parameter		Variant						
Processing time, s		control	1	10	20	30	40	60
Dose, mJ /m²			34	340	680	1020	1360	2040
Germination, %	average	10	14	10	14	10	15	14
	over control	100	140	100	140	100	150	140
Length of roots of sprouts	on average, mm	11.2	14.9	13.9	17.7	12.5	13.1	14.9
	% over control	100	133	124.1	158	111.6	117	133

Table 20. The dependence of germination of European spruce seeds (without pre-soaking) and the length of roots of seedlings on the time and dose of irradiation (measurement after 7 days)

Parameter		Variant						
Processing time, s		control	1	10	20	30	40	60
Dose, mJ /m²			34	340	680	1020	1360	2040
Germination, %	average	6	6	5	8	4	11	4
	over control	100	100	83.3	133.3	66.6	183.3	66.6
Length of roots of sprouts	on average, mm	9.7	7.3	10	10.9	9.0	10.5	6.5
	% over control	100	75.3	103.1	112.4	92.8	108.2	67

Table 21. The dependence of the germination of European spruce seeds (with pre-soaking) and the length of roots of seedlings on the time and dose of irradiation (measurement after 7 days)

Parameter		Variant						
Processing time, s		control	1	10	20	30	40	60
Dose, mJ /m²			34	340	680	1020	1360	2040
Germination, %	average	12	10	11	10	10	11	7
	over control	100	83.3	91.6	83.3	83.3	91.6	58.3
Length of roots of sprouts	on average, mm	6.8	10.1	8.9	8.9	7.6	10.5	11.8
	% over control	100	148.5	130.8	130.8	111.8	154.4	173.5

m² (40 seconds), also at this dose there is an increase in the length of the roots of seedlings to 154.4% in relation to the control.

In this experiment, soaking had a better effect on increasing the roots length of seedlings.

In the study of the influence of electromagnetic radiation of the optical range produced by LEDs with a maximum of 405 nm on the primary parameters of germination of seeds of spruce European and Siberian larch revealed that the optimal for the treatment of seeds of Siberian larch and European spruce are doses of 1360 mJ/m^2, it is observed the maximum germination of seeds. At the same time, small doses do not have a stimulating effect on seed germination, with an increase in the dose, a stimulating effect is manifested, and at doses higher than optimal, the stimulating effect is reduced.

It should also be noted that the soaking of seeds gives a significant increase both in germination and in the length of the roots of seedlings, therefore, it is necessary to combine the soaking of seeds and their processing with LEDs.

To explain the stimulating effect of electromagnetic radiation of the optical range on the primary parameters of germination of spruce and larch seeds, a photoresonance hypothesis is proposed.

Most likely, the main photo-stimulating effect of electromagnetic radiation of the optical range is its effect on the activity of enzymes and the rate of enzymatic reactions.

According to studies, the maximum spectral sensitivity of spruce and larch seeds falls on wavelengths of 360-370 nm, and we used LEDs with a maximum of 405 nm. For further research, it is necessary to process seeds with LEDs of the optimal spectrum, where the maximum is 360-370 nm.

The use of the influence of optical radiation obtained by irradiation with LEDs on sowing qualities and primary parameters of germination of seeds of woody plants is a very promising direction in the study of ways to stimulate the processes of seed germination and growth of seedlings of woody plants.

Thus, while maintaining the current pace of development, in the coming decades, LEDs and led matrix may well displace from agricultural irradiation equipment designed for pre-treatment of seeds, discharge lamps, because of their significantly better energy efficiency, caused by a relatively narrow spectrum, longer service life and excellent performance.

Structural diagram LED opto-electronic system of pre-activation of the seed is shown in Fig. 3.

Before treatment, preliminary monitoring of the initial state and properties of seeds is carried out according to the method (Bashilov, & Belyakov, 2015, Bashilov, & Belyakov, 2008, Bashilov, Gordeev, Belyakov, & Shirokih, 2011): the spectral characteristics of the sensitivity of seeds are measured and the optimal doses of treatment are determined. Further, on the basis of the obtained characteristics and parameters, the required characteristics and parameters of the activation system (the radiation spectrum of the led panel and the exposure dose of treatment) are established. In the course of preliminary monitoring, other seeds parameters can also be controlled: humidity, degree of contamination, etc. The information received is sent to the pre-training device.

After preliminary monitoring, the main batch of seeds enters the pre-sorting device and preparation for processing. It, if necessary, may be drying seeds, removal of litter, as well as layout in one layer before entering the processing.

Opto-mechanical device activation of seeds, in general, may include:

- Block mechanical movement of seeds (conveyor belt or others.);
- Emitters block and control equipment;

Figure 3. Structural scheme of the optoelectronic system pre-activation of the seed

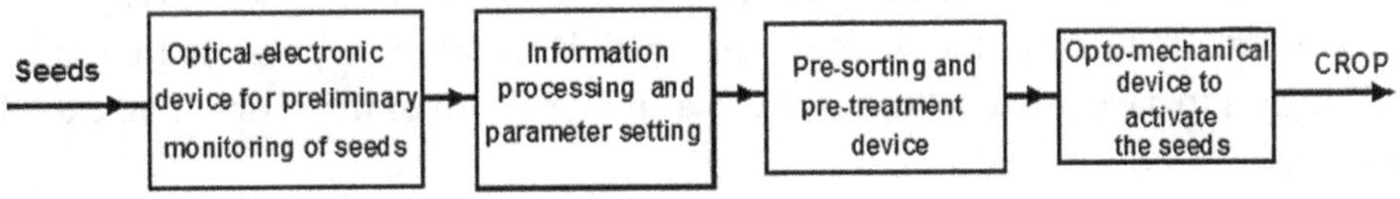

- Control unit of the movement device, emitters, etc.;
- Devices for protection of emitters and maintenance personnel from dust, as well as personnel from radiation.

After processing, a small part of the total batch is allocated, if necessary, to verify the effectiveness of the processing.

The optical-mechanical device of seed activation should provide: the movement of seeds under the emitters for a given processing time with a preliminary uniform placement, the collection of treated seeds, but the main task of the system is irradiation of seeds with radiation of the optimal spectrum with the required irradiation of the surface of the seed layer. The composition of the opto-mechanical system consists of: input hopper and the distributor conveyor tape and tape drive mechanism, the output device Assembly and an output hopper, suspension AI, the system power supply AI, ballasts, lamps, including AI and svetoprelomlyayuschimi devices protection against dust and radiation.

The location and height of the suspension II determine the distribution of irradiance in the working plane of the installation.

To move the seeds under the emitters, it is advisable to use a conveyor belt moved by the rotation of the shafts.

To change the exposure by changing the processing time it is necessary to provide a change in thc speed of the conveyor bclt. It is also necessary to control irradiance in the working plane of the installation. This requires the installation of radiation receivers (two to four) on different sections of the conveyor. The receivers can be small photodiodes with corrective light filters. Technologically, it is advisable to fix them at the edges of the conveyor belt, although only the irradiation of the edge points will be controlled. The value of irradiance from several receivers must be averaged, which should be carried out by a special microprocessor. The average value of irradiance, as well as the speed of the tape (or processing time) and the exposure dose should be displayed on the control panel of the optical-mechanical system (Gaska, & Zhang, 2007).

At the same time, the exposure will be calculated according to the formula (1), and the speed of the tape can be calculated for the performance of the installation for seeds of different cultures.

$$H = E_{aver} \cdot \tau = E_{aver} \frac{l}{v} \tag{4}$$

where E_{aver} – the average irradiance in the working area of the installation, τ – time of seed treatment (time of seeds in the working area), l – the working length of the conveyor belt (length of the working area), v – the speed of the conveyor belt.

The General view of the scheme of optical-electronic system of monitoring and control of biological activity of seeds is presented in Fig. 4.

The advantages of the presented installation is energy efficiency, due to the use of LEDs, and the efficiency of seed treatment, due to the selection of the optimal spectrum of radiation and treatment doses.

The radiation spectrum of the source, if possible, should coincide with the sensitivity spectrum of the seeds. The maximum efficiency of the source is obtained. However, it is possible for seed sensitivities only in the UV region to have a source that captures a small portion of the visible range to visualize the treatment process.

Figure 4. Functional diagram of the seed activation optoelectronic system (continuous lines show the movement of seeds, intermittent-information)

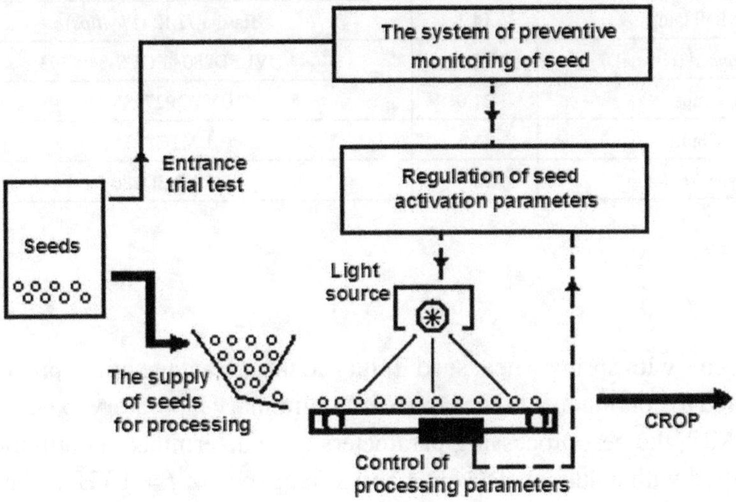

Taking the averaged curve (Belyakov, 2008) as the spectral sensitivity of wheat seeds, we calculate the effective radiation efficiency for different types of lamps and LEDs. Considering that the radiation spectrum is the same in all directions and the irradiance is proportional to the radiation flux, the effective radiation efficiency is determined by the formula:

$$\eta_{eff} = \frac{\Phi_{eff}}{\Phi_e} = \frac{E_{eff}}{E_e} = \frac{S_{\max}\left(\int_0^{\infty} e(\lambda)s(\lambda)d\lambda + \sum_{i=1}^{n} E_{\lambda,i}s_{\lambda}\right)}{\int_0^{\infty} e(\lambda)d\lambda + \sum_{i=1}^{n} E_{\lambda,i}} \tag{5}$$

where Φ_e and Φ_{eff} are, respectively, the total and effective radiation fluxes of the sources, E_e and E_{eff} are the total and effective irradiance generated by the radiation source on the working surface, S_{max} is the maximum value of the seed sensitivity, $s(\lambda)$ is the spectral sensitivity of the seeds, $e(\lambda)$ is the spectral irradiation density of the working surface (Shubert, 2008, Ajzenberg, 2006).

Similar to the systems of bactericidal and erythemic action of radiation, we take $S_{max}=1$ and the effective radiation flux will be determined in effective watts. Let us present the results of calculations in table 22.

Thus, the highest effective output radiation are the UV LEDs UVLED280TO46FW, UVTOP275, UVTOP270SMD range in which a significant part is located in the region of sensitivity of the seeds. The use of a combination of uvled280 LEDs (maximum emission spectrum 280 nm) EOLS-340-393 and (maximum emission spectrum 338 nm), covering both the maximum sensitivity spectrum of seeds has not led to an increase in effective returns.

Net profit in the application of pre-sowing activation is 125 USD per 1 ha for spring wheat seed.

Table 22. Effective output of some sources

Radiation Source	η_{eff}, %	Radiation Source	η_{eff}, %
Mercury Arc Ball lamp	14.1	Лампа *TL12* (*Phyllips*)	33.1
Halide lamp *Ultramed* (*Osram*)	17.0	UVLED280+EOLS-340-393	77.8
Erythemal lamps	21.0	UVTOP270SMD	82.9
LED (λ_{max}=338nm)	21.6	UVTOP275	86.2
Mercury tube type Arc lamp	23.3	UVLED280	90.3

SUMMARY

1. In pot experiments with spring wheat seed Enita had the advantage of the processing pulsed radiation as compared to continuous. In the course of preliminary laboratory experiments with seeds of spring wheat MIS, the best processing parameters were determined: continuous radiation during $T = 6$ s, or pulsed with a duty cycle $M = 4$ and a frequency of $f = 10$ Hz. The most optimal time for sowing-a day after treatment. The best indicators of growth dynamics are plants whose seeds were treated continuously. The exceedance over the benchmarks was up to 42 for these options at different times...56%.
2. When processing soybean seeds, the germination Rate increases by 17%, the mass of pods – 1.5 times.
3. Pre-sowing treatment with led radiation significantly increases the field germination of radish seeds (up to 38%). At the same time, the above-ground part of the plants (stems) is less in length than the control indicators, and the length of the mass and the underground part, which is productive for the radish, far exceeds the control (up to 124%). Pre-sowing treatment of cucumber seeds by continuous radiation of LEDs does not affect germination, increases the length of stems and roots by 12%, increases the weight of roots by 33%. Led Pulse processing increases germination by 1.8 times, leaving at the level of control of the length of the stems and roots and increasing their weight by 10 and 58%, respectively.
4. Led treatment increases the germination of clover seeds Smolensky 29. Doses of continuous radiation 136 mJ/m^2 and pulse radiation 68 mJ/m^2 have no significant impact on yield. The dose of pulsed radiation 136 mJ/m^2 causes a significant decrease in yield by 18%. Thus, the effect of led technology is to save up to 28% of the seed material by sowing more activated germinating seeds.
5. In the processing of seeds of Norway spruce allowances for germination is 30...40%. Sowing Is most effective for 2-3 or 9 days after treatment. For an increase in germination of Siberian larch seeds reach 24%. Sowing of treated seeds is also best done within two days.
6. It was found that pre-sowing treatment of woody plants seeds in combination with pre-soaking gives significantly better results. For Siberian larch seeds germination increases to 150% of the control, sprouts roots length – up to 158%. For spruce seeds, the European increase in the roots length of seedlings - up to 173%, but germination – at the level of control or worse.
7. Optical-electronic system of presowing treatment of seeds includes opto-electronic device of preliminary monitoring of the seeds, after which there is the establishment of optimal processing parameters. This is followed by pre-sorting of seeds and preparation of the installation for operation, then the actual pre-sowing treatment of seeds with led radiation.

8. LEDs and led matrix may well displace from agricultural irradiation equipment discharge lamps, because of their significantly better energy efficiency, caused by a relatively narrow spectrum, longer service life and excellent performance. The most optimal for effective returns are LEDs UVLED280TO46FW (η_{eff}=90.3%), UVTOP275 (η_{eff}=86.2%), UVTOP270SMD (η_{eff}=82.9%). Net profit in the application of pre-sowing activation is 125 USD per 1 ha for spring wheat seed.

REFERENCES

Ajzenberg, Y. U. B. (Ed.). (2006). *Spravochnaya kniga po svetotekhnike [The reference book on light engineering].* Moscow, Russia: Znak.

Bashilov, A. M., & Belyakov, M. V. (2008). Metodika monitoringa chuvstvitel'nosti semyan k vozdejstviyu opticheskogo izlucheniya [Method of monitoring the sensitivity of seeds to the effects of optical radiation]. *Vestnik FGOU VPO MGAU Seriya. Agroinzheneriya, 2*(27), 22–24.

Bashilov, A. M., & Belyakov, M. V. (2011). Optiko-ehlektronnaya aktivizaciya semyan izlucheniem svetodiodov [Opto-electronic activation of the seed radiation of the LEDs]. *Vestnik FGOU VPO MGAU Seriya. Agroinzheneriya, 1*(46), 7–9.

Bashilov, A. M., & Belyakov, M. V. (2015). *Spektral'nye harakteristiki lyuminescencii i otrazheniya semyan agrokul'tur [Spectral characteristics of luminescence and reflection of agricultural seeds].* Moscow, Russia: FBGNU VIEHSKH.

Bashilov, A. M., Gordeev, Y. A., Belyakov, M. V., & Shirokih, T. V. (2011). Svetodiod uvelichivaet vskhozhest [The led light increases the germination rate]. *Sel'skij mekhanizator, 10*, 32-33.

Belyakov, M. V. (2008). *Optiko-ehlektronnaya tekhnologiya i sredstva upravleniya biologicheskoj aktivnost'yu semyan* [Optoelectronic technology and means of controlling biological activity of seeds] (Unpublished doctoral dissertation). FGOU VPO MGAU, Moscow, Russia.

Belyakov, M. V. (2009). Primenenie opticheskoj stimulyacii dlya semyan soi [Application of optical stimulation for soybean seeds]. *Proceedings of Conference Activation of the role of young scientists-the way to the formation of innovative potential of agriculture. Collection of materials of the international scientific-practical conference devoted to the 70th anniversary of Professor M. Gordeev.*

Belyakov, M. V., & Rybkina, S. V. (2011). *Vozdejstvie opticheskogo izlucheniya na semena drevesnyh rastenij dlya stimulyacii rostovyh processov [The effect of optical radiation on the seeds of woody plants to stimulate growth processes].* Smolensk, Russia: Universum.

Blaszczak, U., Aziz, D., & Gryko, L. (2017). Influence of the spectral composition of LED lighting system on plants cultivation in a darkroom. *Proc. SPIE 10445, Photonics Applications in Astronomy, Communications, Industry, and High Energy Physics Experiments.*

Borodin, I. F. (Ed.). (1996). Ispol'zovanie kogerentnogo ehlektromagnitnogo izlucheniya v proizvodstve produkcii rastenievodstva [The use of coherent electromagnetic radiation in crop production]. *Doklady RASKHN, 6*, 41–44.

Filippov, A. K., Bityuckij, N. P., & Fedorov M. A. (1997). *Ustrojstvo dlya plazmennoj obrabotki semyan rastenij* [Device for plasma treatment of plant seeds]. Patent RF. no 2076555.

Gaska, R., & Hang Dzh, Z. (2007). Ul'trafioletovo izuchayushchie diody (UV-emitting diodes). *Svetotekhnika*, *6*, 38–39.

Hu, Y., Li, P., & Jiang, J. (2007). Developing a new supplemental lighting device with ultra-bright white LED for vegetables. *Proc. SPIE 6486, Light-Emitting Diodes, Research, Manufacturing, and Applications*, *11*, 64861A.

Kakinoki, Y., Kato, Y., Ogawa, K., Nakao, A., Okai, Z., & Katsuyama, T. (2013). Efficient plant growth using automatic position-feedback laser light irradiation. *Proc. SPIE 8881. Sensing Technologies for Biomaterial, Food, and Agriculture, 2013*. doi:10.1117/12.2030575

Kondrat'eva, N. P. (2001). Vliyanie predposevnoj obrabotki semyan yarovoj pshenicy na urozhajnost' [Influence of presowing treatment of spring wheat seeds on yield]. *Mekhanizaciya i ehlektrifikaciya sel'skogo hozyajstva, 12*, 17.

Kondrat'eva, N. P., Krasnolutskaya, M. G., Dukhtanova, N. V., & Obolensky, N. V. (2019). Effect of ultraviolet radiation the germination rate of tree seeds. *IOP Conf. Series: Earth and Environmental Science, 226*. 10.1088/1755-1315/226/1/012049

Li, Z., Ji, J., & Xu, M. (2013) Designation of rapid detection system for chlorophyll fluorescence parameters based on LED irradiation. *Proc. SPIE 8768, International Conference on Graphic and Image Processing (ICGIP 2012)*. 10.1117/12.2010885

Loginov, M. I. (1986). Predposevnaya obrabotka semyan ul'trafioletovymi luchami [Pre-sowing seed treatment with ultraviolet rays]. *Lyon i konoplya*, *2*, 28-29.

Shubert, F. (2008). *Svetodiody (LEDs)*. Moscow, Russia: Fizmatlit.

Tsai, C., Huang, C., Chen, C., & Yue, C. (2017). The study of LED light source illumination conditions for ideal algae cultivation. *Proc. SPIE 10107. Smart Photonic and Optoelectronic Integrated Circuits, 19*.

Wu, S., Zhu, L., Zhao, F., Yang, B., Chen, Z., Cai, R., & Chen, J. (2013). Effect of LED lamping on the chlorophylls of leaf mustard. *Proc. SPIE 8761, PIAGENG 2013: Image Processing and Photonics for Agricultural Engineering*.

Chapter 15
Improvement of Technology of Electrical and Magnetic Stimulation of Seeds and Crop Plants

Igor Yudaev
Don State Agrarian University, Russia

Sergey Mashkov
Samara State Agricultural Academy, Russia

Sergey Vasilyev
https://orcid.org/0000-0003-4368-3123
Samara State Agricultural Academy, Russia

Vladimir Syrkin
Samara State Agricultural Academy, Russia

Sergey Shevchenko
Samara State Agricultural Academy, Russia

Kiril Sirakov
University of Ruse, Bulgaria

ABSTRACT

Use of a variety of electrotechnics is a technologically-efficient and environmentally-friendly technique that increases the productivity of cultivated plants. Stimulation of green plants and vegetable crops in electric field with the intensity of 5-50 kV/m made it clear that the maximum efficiency is observed in the growth period – an increase of up to 30%, compared to the control. Plants have been subjected to stimulation for 3 hours twice a day (in the morning and in the evening). Analysis of studies on the pre-seeding seed stimulation showed that the exposure to pulsed magnetic field improves the dynamics of germination and plant growth at the early stages of development by an average of 10-20%, and more uniform germination helps to ensure high yields.

DOI: 10.4018/978-1-5225-9420-8.ch015

INTRODUCTION

There is a need to find and use environmentally friendly (organic) technologies to increase the crop yield and improve the product safety. The most promising technology is electrical electrotechnics, designed to stimulate seeds and plants using the electrical and magnetic fields.

The research focuses on improvement of the technology of electric and magnetic stimulation of seeds and plants to obtain high-quality crop products in the required amount.

To achieve the purpose, the following objectives were solved:

- theoretical research of interaction of electric and magnetic fields with plants;
- experimental research of justification of parameters of electric and magnetic fields that give the greatest effect in stimulation and exposure of plant bodies.

BACKGROUND

Currently, there are many studies (mostly carried out with positive result) of application of heat, light, electromagnetic and other physical impacts on seeds and plants for the purposes of improving germination, increasing germinating energy, increasing harvesting and quality of harvested products. It should be noted that such use of given factors for treatment and stimulation of plants allows them to be classified as environmentally friendly methods, thereby giving them additional benefits for use in agriculture against the background of increasing pollution of the environment with pesticides and other chemicals (Marinković, B. et al., 2008; Aladjadjiyan, 2012; Bilalis et al., 2013).

Today there are various ways for the stimulating treatment of seed grains and improvement of growing functions of plants using laser and ultrasonic radiation, cold plasma, electric, magnetic and electromagnetic fields and other impacts of diverse physical nature (Dardeniz, Tayyar & Yalcin, 2006; Spirov et al., 2008; Yan et al., 2009; Cwintal, Dziwulska-Hunek, & Wilczek, 2010; Goussous et al., 2010; Hernandez et al., 2010; Yang & Shen, 2011; Maffei, 2014; Mihai et al., 2014; Jiang et al., 2014; Jedlička, Paulen & Ailer, 2014; Kasakova et al., 2018).

The positive effect of preseeding treatment of seeds was studied and shown in vegetable, grain, industrial and decorative crops in electrostatic field (Yang & Shen, 2011), constant magnetic field (De Souza et al., 2006; Flyrez, Carbonell & Martínez, 2007; Martínez et al., 2009), alternating magnetic field (Racuciu, 2011; Radhakrishnan & Kumari, 2012) and electromagnetic field of industrial frequency (Pietruszewski, Muszynski & Dziwulska, 2007; Mahmood et al., 2011; Molamofrad et al., 2013; Jedlička, Paulen & Ailer, 2014; Kasakova et al., 2018).

Seed treatment using the electromagnetic fields (Goncharov, Berezhnaya & Gursky, 1994; Pietruszewski, Muszynski & Dziwulska, 2007; Kasakova et al., 2018) enhances the growth and organ-forming processes, particularly at later stages of plant development. Magnetic field significantly affects cell division (Novitsky, Strekova & Tarakanova, 1971; Strekova, 1973), their extension (Strekova, 1973), differentiation (Novitsky, 1973; Novitsky, 1987), as well as accelerates seed germination (Vasilev et al., 2018).

Response of the object depends not only on the dose, but also on the environment, state of the object, etc., but it is difficult to record it, because the plants are impacted for a limited time, following which they are placed into optimal conditions. This ensures repair system deployment, and the situation becomes unclear (Batygin, 1986). In case of seed treatment using the electromagnetic field, the increased yields and increased content of protein in grain, starch in potatoes and sugar in beets have been observed. Treatment using high voltage pulses improved the yield of onion by 42% and yield of potatoes by 15-17%. Preseeding treatment of seeds using uniform magnetic field allowed to get 10-30% increase of yield of millet, buckwheat, wheat, tomatoes and lettuce (Sokolov, 1983; Kasakova et al., 2018; Vasilev et al., 2018).

Research also focused on technologies of treatment of plant matters using, for example, low-frequency electric field for growing grapes, which have been developing better and have been high-yielding (Dardeniz, Tayyar & Yalcin, 2006). Impact by pulse low-frequency electric field on potato tubers causes an increase in the number and weight of tubers of one plant, but at the same time it showed a lack of credible and expressed impact of treatment on such morphometric indicators as the height of potato plants, the number of leaves on the stem and crude weight of tops (Statsyuk et al., 2016). Besides, potato tubers have been treated with alternating electromagnetic field of industrial frequency (Gut, 2007) and alternating magnetic field (Marks & Szecuwka, 2010), that led to an increase in the number of stems and higher yields.

ELECTRICAL AND ELECTROMAGNETIC STIMULATION OF VEGETABLE CROP PLANTS

Factors, influencing plants, include direct and indirect effect of electricity. It is known that weak electric current, transmitted through the soil, affects the vital activity of plants and beneficial soil microflora (Ramnek, 1911; Voytova, Yukin, & Ubiraylova, 2016). It has been also found that the effects of the electrical nature change the movement of different types of soil moisture, contribute to decomposition of a number of substances, not easily assimilable for plants, provoke a variety of chemical reactions that change the reaction of the soil solution. The study also identified the parameters of electric current density, optimal for a variety of soils: DC - 0.02-0.6 mA/cm^2, AC - 0.25-0.50 mA/cm^2 (Gordeev & Shershnev, 1991).

Periodic stimulation of seeds and crop plants in electric or electromagnetic field is, as mentioned earlier, an environmentally friendly, low-cost and relatively simple way to increase the yield of green and vegetable crops, grown both in closed and open soil. But, at the same time, there are a number of outstanding problems, which limit the use of this method:

- Lack of clear and expressly identified parameters of the affecting electric or magnetic fields (tension, frequency, supply voltage curve shape, etc.);
- Lack of justification of duration of the stimulation - duration of each cycle of stimulation, number of cycles and their distribution during the day, etc.;
- Lack of justification of alternation of cycles of stimulation and relaxation;
- Lack of reasoned justification of the direction of electric and magnetic fields relative to stimulated plants and seeds.

Failure to address the specified problems significantly reduces the effectiveness of stimulation, and, in some cases, leads to the opposite effect - inhibition of plants.

Scientists at the Timiryazev Institute of Plant Physiology established that photosynthesis goes quicker the more the potential difference between plants and atmosphere is. So if, for example, a negative electrode is kept near the plant and the voltage is being gradually increased (500 V, 1,000 V, 1,500 V, 2,500

V), the intensity of photosynthesis will increase (up to certain limits). If electric potentials of the plant and the atmosphere are close by the value, the plant ceases to absorb carbon dioxide.

When the electric current passes through the plant, not only the photosynthesis, but the root absorption can be regulated since all elements, required by the plant, are typically transferred in the form of ions (Polunin et al., 2009).

Electric field affects not only adult plants, but also seeds. If they are placed into an artificially created electric field for some time, they will come up evenly faster. This happens due to the fact that some chemical bonds inside the seeds are broken as a result of exposure to electric field, causing the generation of free radicals. The more active particles are inside the seed, the higher the energy of their germination is (Vasiliev, Mashkov & Fatkhutdinov, 2016).

Electrical stimulation of plants can be carried out through the application of electrical, magnetic or electromagnetic fields (Vasiliev, Mashkov & Fatkhutdinov, 2016).

In our opinion, the most promising method is the use of electromagnetic field, as this gives ample opportunities to change its frequency, as well as creates the opportunity to generate amplitude-modulated and frequency-modulated electromagnetic field with the specified waveform.

Electrical Stimulation Method

The use of pulsed electromagnetic field in the area of plant location is suggested for the practical implementation of the method of electromagnetic stimulation of plants. In this case the treatment is carried out as follows: plants are located between two electrodes of different polarity, while the electrode with positive potential is located under the roots of the plants, and the electrode with negative potential is located above the plant. It means that the direction of the externally applied electric field must match the direction of the growth of plants. Therefore, one or more metal electrodes 1 are placed above the plants, and one or more electrodes 2 are placed on the opposite side, which are placed either directly in the soil 7 near the roots of the plants or below the tank with soil, and the plants 6 will be located between the electrodes in a relatively homogeneous electromagnetic field (Figure 1).

Pulsating voltage of certain amplitude and frequency is supplied to the electrodes.

In addition, it is possible to supply to the electrodes not just pulsating (rectified) voltage, but to further modulate it using a specific function. For example, the function of capacitor discharge or meander (U-shaped).

An important issue is the value of voltage, supplied to the electrodes. This voltage is determined by the distance between the electrodes (h), which is roughly equal to the height of the location of the upper electrode relative to the soil, and the required value of electric field intensity (E_{req}), in which the plants are located.

Required field intensity is also determined experimentally. Some researchers, who conducted similar experiments, recommend the intensity to be within the range from 10 to 50 kV/m (Maslobrod, 1981; Medvedev, 1990; Opritov, Pyatygin & Retivin, 1991).

The suggested method can also be used for electromagnetic stimulation of seeds before seeding subject to small adjustments and changes.

Figure 1. Diagram of Electromagnetic Stimulation of Plants: 1 - upper electrode (with a negative potential); 2 - bottom electrode (with a positive potential); 3 - generating power unit with the control unit; 4 - upper electrode support pole; 5 - insulators; 6 - stimulated plants; 7 - soil

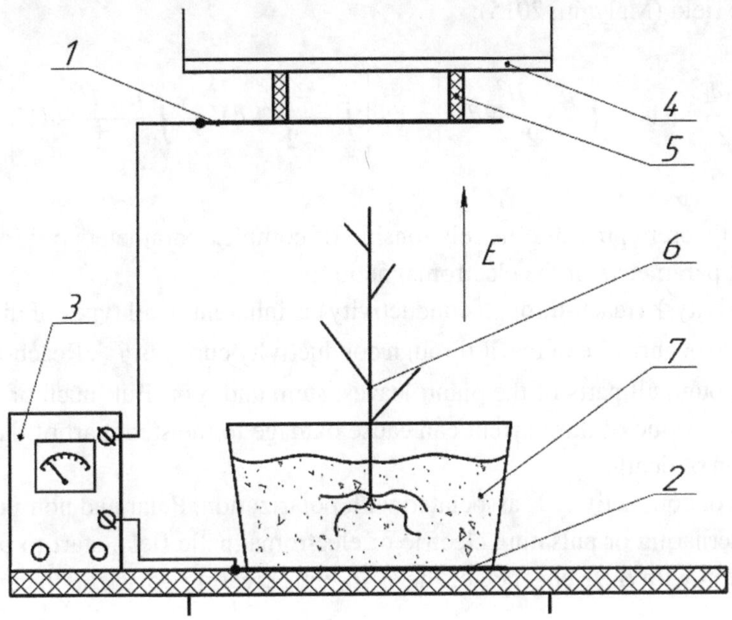

Brief Theoretical Studies

The mechanism of interaction between the electromagnetic field and biological plant is complex due to the fact that even with constant parameters of electric field the biological object itself is uneven by such physical parameters, as specific conductivity $\bar{Y}$, dielectric $\bar{\varepsilon}$ and magnetic $\bar{\mu}$ inductive capacity (Medvedev, 1990; Polunin et al., 2009).

These parameters are complex quantities, depending on the frequency (ω). Depending on the stage of development, humidity and temperature of a biological object, these parameters can apply to conducting ($Y \gg \omega \cdot \varepsilon \cdot \varepsilon_0$), semiconducting ($Y \approx \omega \cdot \varepsilon \cdot \varepsilon_0$) and dielectric ($Y \ll \omega \cdot \varepsilon \cdot \varepsilon_0$) media. Therefore, in order to take into account characteristics of both homogeneous isotropic and nonhomogeneous anisotropic media, complex characteristics should be used:

$$\bar{Y} = Y' + jY'' = Ye^{j\varphi_Y}, \tag{1}$$

$$\bar{\varepsilon} = \varepsilon' + j\varepsilon'' = \varepsilon e^{j\varphi_\varepsilon}, \tag{2}$$

$$\bar{\mu} = \mu' + j\mu'' = \mu e^{j\varphi_\mu}. \tag{3}$$

If the parameters of a medium are expressed in a complex form, the equation of interaction between the electromagnetic field and plant subject to the presence of external forces (i.e. $E_{ef} \neq 0V$) can be conveniently described using the Umov–Poynting theorem, describing the law of conservation of energy of the electromagnetic field (Malygin, 2015):

$$\oint \vec{\Pi} d\vec{S} = \omega \left[\int_V \frac{\varepsilon'' \cdot E_m^2}{2} dV - \int_V \frac{\mu'' \cdot H_m^2}{2} dV \right] - j\omega \left[\int_V \frac{\varepsilon' \cdot E_m^2}{2} dV - \int_V \frac{\mu' \cdot H_m^2}{2} dV \right] - \frac{1}{2} \int \dot{E}_{ef} \ddot{\delta} dV, \quad (4)$$

In this case, the theorem provides the relationship of complex parameters of biological object G, ε and μ with complex parameters of the electromagnetic field.

Specific conductivity Y (reach-through conductivity) is inherent in all types of dielectrics and semiconductors, it leads to occurrence of reach-through conductivity currents $\dot{I}_{rt}$. Reach-through conductivity current flows through all parts of the plant: leaves, stem and root. But much of it runs through the stem - hence the great value of this current can cause damage to the stem part of the plant and, consequently, its inhibition or death.

The second type of conductivity is associated with polarization. Polar and non-polar molecules, being caught in an alternating or pulsating electric or electromagnetic field, start to periodically change their space orientation with the frequency of the field, thus creating polarization current (bias current) $\dot{I}_{bias}$, A:

$$\dot{I}_{bias} = \frac{dD}{dt}, \quad (5)$$

where D is the electric induction, C/m^2 ($D = \varepsilon_0 \varepsilon_r E$) (Malygin, 2015).

Reach-through conductivity current and bias current have the same direction and are cumulated, forming the total current, passing through the plants:

$$\dot{I} = \dot{I}_{rt} + \dot{I}_{bias}, \quad (6)$$

One of the main parameters studied at this stage is the electric field intensity. The most convenient way to evaluate it is to use electrical component (E), kV/m. In this case, the electrical field intensity may be determined as the ratio of the voltage, applied to the electrodes 1 (*Figure* 1), to the distance between the electrodes:

$$\dot{E} = \frac{\dot{U}}{h}, \quad (7)$$

where h is the distance between the electrodes, m.

As a result, the value of total current passing through the plant, will be:

$$\dot{I} = \frac{d}{dt}\varepsilon_0\varepsilon_r\dot{E} + \dot{U}\overline{Y} = \frac{1}{h}\frac{d}{dt}\left(\overline{\varepsilon}_0\dot{U}\right) + \dot{U}\overline{Y}_{rt}, \quad (8)$$

Relative dielectric capacity ε_r of plants (for example, green crops) is 4 ... 5 times more than the dielectric capacity of air. Stimulated plants represent a non-homogeneous medium, consisting of plants and air gaps and form the single extremely inhomogeneous dielectric. Therefore, the electric field intensity in air gaps and plants will not be identical, i.e. the electromagnetic field becomes inhomogeneous and considerably distorted.

The geometric sum of considered currents causes movement of chemicals in plants, accelerating metabolic processes, such as photosynthesis, and in aggregate forming the effect of stimulation of plants. This increases the energy and speed of growth of both the aboveground and root parts of plants.

Experimental Research Technique

The experiment on research of the influence of electromagnetic field on growth and development of the aboveground part of plants of Sanka ultra-early ripening tomatoes, produced by AILITA, was conducted in 2018.

The studied factor was the value of electrical field intensity. It was necessary to identify the degree of impact of the electric field of varying intensity on the plant growth rate.

The electric field has been created between pairs of electrodes, placed on a certain equal distance from each other, to which the voltage of a preset value was supplied from generating plants. The voltage was rectified and pulsating. The frequency of pulses in all variants was 100 s^{-1} (sine-wave voltage with the frequency of 50 Hz rectified by bridge rectifier). Three variants (graduations) of voltages, equal to 40V, 220V and 10kV and of the intensities of the field for variants equal to B1 = 0.15 kV/m, B2 = 0.75 kV/m and B3 = 37.0 kV/m have been studied. Wide scale of values of field intensities can be explained by the need for identification of the effective value of the field intensity by the factor of plant response to stimulation, since its value was unknown before the start of the experiment.

Each variant of the studied factor was set to be done four times, including the control variant, when plants are not subjected to electrical stimulation.

Layout of the replications by variants is presented in Table 1.

Variants B1-B3 are in descending order of electrical field intensity. This is for the reduction of possible impact of the field on control variant K.

Table 1. Layout of plants by replications and variants

Replications	Factor Variants							
	-	B3	-	B2	-	B1	-	K
1	X	Υ	X	Υ	X	Υ	X	Υ
2	X	Υ	X	Υ	X	Υ	X	Υ
3	X	Υ	X	Υ	X	Υ	X	Υ
4	X	Υ	X	Υ	X	Υ	X	Υ

Free space marked with the sign "X" is provided for between neighboring variants in order to reduce the cross impact of electric fields with different intensity.

General view of the experimental laboratory unit is shown in Figure 2. The generator set consists of three voltage sources (PS1, PS2, PS3), generating pulsating (rectified) voltage of various values, three pairs of electrodes and connecting wires, through which each pair of electrodes is connected to appropriate source of voltage (*Figure* 2). Pairs of electrodes are located above and below the stimulated plants: electrode with positive potential is located under the cartridge, inside of which the cups with plants are placed, and electrode with negative potential is suspended on insulators over the plants.

Therefore, the electric field direction coincides with the direction of the growth of plants, that promotes the movement of minerals and nutrients, accelerates photosynthesis and the rate of growth and development of plants. Such approach is particularly relevant in growing green crops.

Supplementary lighting of plants using LED phyto lamps with the capacity of 100W, produced by RosSvet, was carried out during the experiment (Figure 3). Artificial supplementary lighting in research is required to simulate natural sunlight for proper growth of plants, i.e. accurately recreate the natural environment of their growth.

Illumination at the level of sprouts was equal to 7,250.6 lux, which complies with the regulatory requirements.

Parameters of the luminous flux, generated by the lamp, have been studied in order to ensure the reliability of studies. Studies were carried out using TKA-VD-2 spectral colorimeter.

As a result, data on luminous flux was collected, reflecting its color coordinates and chromaticity coordinates, illumination (Table 2) and spectrogram (Figure 4).

Figure 2. General view of the experimental laboratory unit for stimulation of plants: 1 - power supply PS1 (10 kV); 2 - power supply PS2 (220V); 3 - power supply PS3 (40V); 4 - negative polarity electrodes; 5 - insulators; 6 - cartridge with cells of plants and positive polarity electrodes

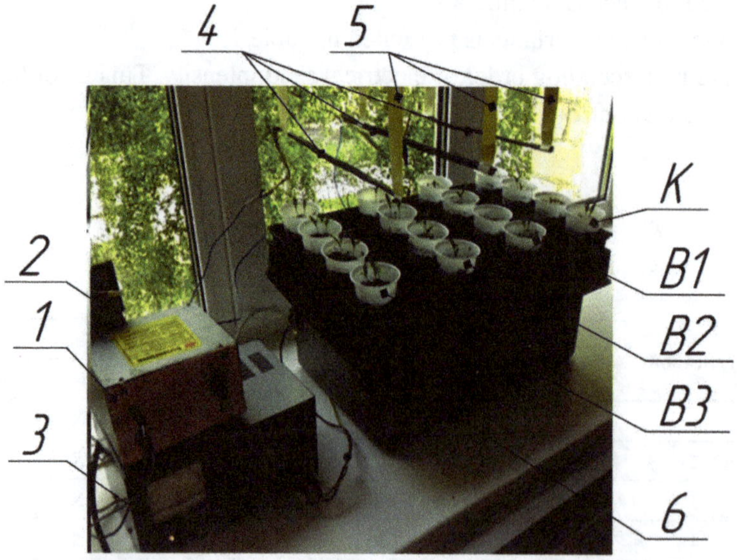

Figure 3. Supplementary lighting of plants during experimental research

Table 2. Characteristics of luminous flux

Program Data	Color Coordinates	Chromaticity Coordinates	Illumination
Date: Time: May 29, 2018, 11:58:37 a.m. CIE 1931 RGB, Ref.White:E, Ad.Metod:None, Gamma: 2.2	X = 125 Y = 100 Z = 147 R = 219 G = 84 B = 148	x = 0.337 y = 0.268 u' = 0.243 v' = 0.435 r = 0.486 g = 0.187	E = 7,250.6 lux E = 673.6 fc

Figure 4. Spectrogram of ROSSVET LED Phyto lamp

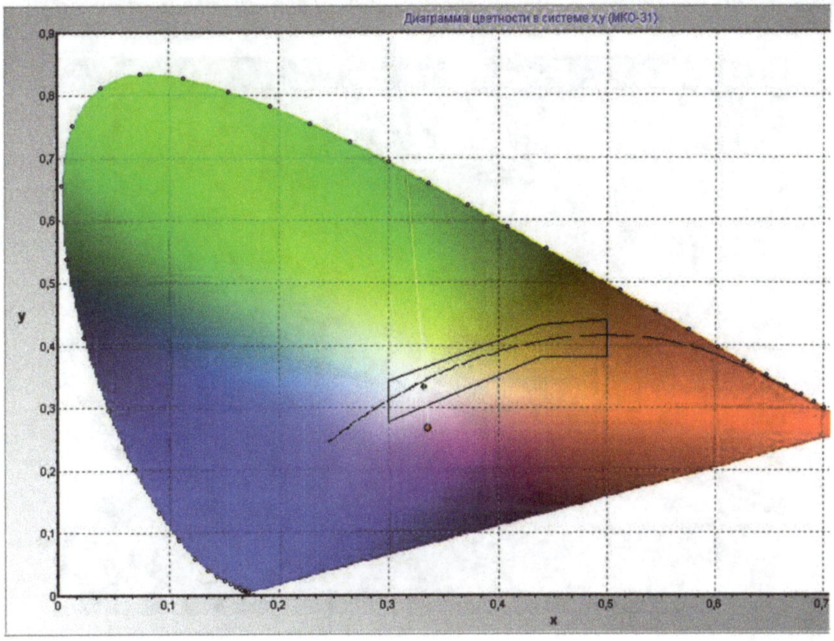

Duration of supplementary lighting was even longer: from 4.00 to 9.00 a.m. and from 6.00 to 11.00 p.m. The obtained color coordinates and spectrogram confirm that the range of the luminous flux of the lamp mostly contains red and blue parts of the spectrum, that is most effective for plants.

Electric stimulation has been carried out in the morning and the evening. The duration of each session was 3 hours (from 6.00 to 9.00 a.m. and from 4.00 to 7.00 p.m. respectively). The total daily duration was 6 hours.

Long duration of supplementary lighting is explained by the peculiarity of the spatial orientation of the window, located at the site of the experiment, as well as a large number of cloudy days.

The total daily duration of supplementary lighting was 10 hours.

Duration of the experiment was 50 days. Shoots began to appear on 9-12th day. Upon completion of the experiment, all plants were cut at the ground level, the length of the aboveground part and weight was measured.

Summary of experiment results is provided in Table 5.

As follows from Figure 5, even with a naked eye one can see that the plants in variant B3 (leftmost row) are higher and have bunchy tops.

The results of measurement of plant parameters, such as height and weight of the aboveground part, are presented in Tables 3, 5 and 7 according to the variant, and Tables 4, 6 and 8 present the results of the analysis of the experiment results.

Replications for each variant are groups of plants, grown in one cup (Figure 5). Accordingly, each variant includes four replications. Each replication includes three plants (some include two or one since some of them failed to germinate). In such cases, the table cells are marked with dashes.

Analysis of data in Table 4 shows that the coefficient of variation for the second and sixth replications exceeds 5%. Variant B1 was carried out with the least amount of field intensity. Plant height is consistent with the control, i.e. the coefficient of variation of non-stimulated plants is much greater (υ_{B1}

Figure 5. Summary of the results of experimental research of tomato plant stimulation

Table 3. Plant height measurement results (Variant B1)

Replication No.	Plant No. in the *i*-th Replication			Average Height by Replication, mm	Average Height for Variant B1, mm
	1	2	3		
1	251	-	-	251.0	256.8
2	236	278	265	259.7	
3	271	277	267	271.7	
4	229	248	258	245.0	

Table 4. Results of the analysis of experimental data for variant B1

Replication No.	Root-Mean-Square Deviation by Replications σ_{rep}, %	Coefficient of Variation by Replications υ_{rep}, %	Root-Mean-Square Deviation for Variant B1 σ_{B1}, %	Coefficient of Variation for Variant B1 υ_{B1}, %
1	11.0	4.38	16.12	6.27
2	21.5	8.28		
3	5.0	1.85		
4	14.7	6.01		

Table 5. Plant height measurement results (Variant B2)

Replication No.	Plant No. in the *i*-th Replication			Average Height by Replication, mm	Average Height for Variant B2, mm
	1	2	3		
1	265	259	251	262.0	264.7
2	272	248	256	260.0	
3	281	277	262	279.0	
4	278	263	261	270.5	

>> υ_K.) It means that stimulation using weak electric field gives the opposite effect - there is no height growth, but the variation increases.

The Table shows that the average height of the leaf part (which is equal to 264.7 mm) insignificantly exceeds the control. It means that simulation using the field intensity of 0.75 kV/m has no significant effect.

Analysis of Table 6 shows that the coefficient of variation for Variant B2 is 4.02%, which is less than that during the control. It means that the plants in this variant are more even along the height.

The average height of the leaf part (which is equal to 315.25 mm) significantly exceeds the control by 58.55 mm. It means that stimulation using the intensity of 37 kV/m has the greatest positive impact on the growth of plants.

Increase of plant growth rate within this experiment was accompanied by a significant reduction in the coefficient of variation (which is equal to 3.89% - approx. 0.5% lower than that during the control).

Table 6. Results of the analysis of experimental data for variant B2

Replication No.	Root-Mean-Square Deviation by Replications σ_{rep}, %	Coefficient of Variation by Replications υ_{rep}, %	Root-Mean-Square Deviation for Variant B2 σ_{B2}, %	Coefficient of Variation for Variant B2 υ_{B2}, %
1	7.02	2.68	10.64	4.02
2	12.22	4.70		
3	10.02	3.59		
4	9.29	3.43		

Table 7. Plant height measurement results (variant B3)

Replication No.	Plant No. in the *i*-th Replication			Average Height by Replication, mm	Average Height for Variant B3, mm
	1	2	3		
1	300	321	309	310	315.25
2	332	320	311	321	
3	299	315	307	307	
4	338	306	325	323	

Table 8. Results of analysis of experimental data for variant B3

Replication No.	Root-Mean-Square Deviation by Replications σ_{rep}, %	Coefficient of Variation by Replications υ_{rep}, %	Root-Mean-Square Deviation for Variant B3 σ_{B3}, %	Coefficient of Variation for Variant B3 υ_{B3}, %
1	10.54	3.40	12.27	3.89
2	10.54	3.28		
3	8.0	2.61		
4	16.09	4.98		

It means that stimulation using the electric (electromagnetic) field with the medium intensity allows to make the plant even along the height and get a significant increase in biomass (*Table 9, 10*).

Results of the control experiment show significant variance in the values of coefficients of variation by replications - from 0.39% to 6.36%. The reason for this is not clear yet, because all control replications were located next to each other.

To visualize the results of the analysis of experimental data, the histogram (diagram) of the coefficients of variation (ν) and root-mean-square deviations (σ) by variants, including control, has been generated (Figure 6).

The studied variants are located on the diagram in the same order, in which they were located during the experiment (Figure 5). I.e. they are located in descending order of electrical field intensity. This location is necessary to reduce the impact of electric field on the control variant.

Table 9. Plant height measurement results (Variant K)

Replication No.	Plant No. in the *i*-th Replication			Average Height by Replication, mm	Average Height for Variant K, mm
	1	2	3		
1	260	246	-	251.3	256.7
2	277	278	248	267.7	
3	252	251	253	252.0	
4	248	254	262	254.7	

Table 10. Results of analysis of experimental data for variant K

Replication No.	Root-Mean-Square Deviation by Replications σ_{rep}, %	Coefficient of Variation by Replications υ_{rep}, %	Root-Mean-Square Deviation for Variant K σ_K, %	Coefficient of Variation for Variant B1 υ_K, %
1	7.77	3.09	10.98	4.28
2	17.04	6.36		
3	1.0	0.39		
4	7.02	2.76		

Conclusions

It has been theoretically established that substances of biological objects (plants) generate two types of current (let-through and polarization currents) when interacting with the pulsed electric field of one polarity. Electric currents and magnetic fields affect the rate of movement of substances (electrolytes) in the plant's body. Ultimately this should lead to an acceleration of photosynthesis and, as a consequence, acceleration of growth of plant biomass.

In Variant B3 the field intensity was the greatest - 37 kV/m. In this case, the stimulation showed the best effect. In this experiment the average height of the leaf part is equal to 315.25 mm, which significantly exceeds the control by 58.55 mm. It means that simulation using the field intensity of 37 kV/m greatly increases the biomass growth rate.

Increase of plant growth rate within this experiment was accompanied by a significant reduction in the coefficient of variation (which is equal to 3.89% - approx. 0,5% lower than that during the control).

It means that stimulation using the electric (electromagnetic) field with the medium intensity allows to make the plant even along the height, which is critically important for production of vegetables for sale, as well as to get a significant increase in biomass compared to the control.

Figure 6. diagram of distribution of coefficients of variation (v) and root-mean-square deviations (σ) by Variants

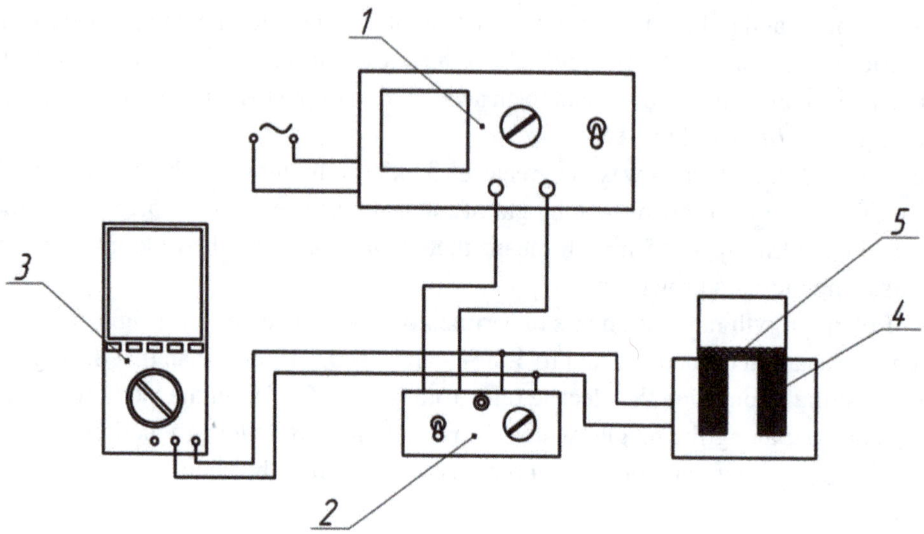

MAGNETIC STIMULATION OF CROP SEEDS

Special attention is paid to production of environmentally clean products with minimal or no use of chemicals. As a result, one of the methods to solve this problem is the electrophysical impact on plants and their seeds (Vasiliev, Mashkov & Fatkhutdinov, 2016; Yudaev, Tibirkov & Azarov, 2012; Kasakova et al., 2018).

Stimulation of seeds using the magnetic field is an effective and promising direction, that eliminates the negative impact on the genotype of seeds and allows to increase crop yields by enhancing the germinating energy and capacity (Syrkin, 2018).

There are studies on stimulation of crop seeds in constant and alternating magnetic fields with different parameters of magnetic field and treatment time (Kuleshov, Ereshko & Khronyuk, 2011; Temerkieva & Plieva, 2016).

Purpose of the research is to increase the efficiency of crop production through stimulation of seeds using the pulsed magnetic field.

- *Objectives of the research*: study of the impact of the magnetic field on various crop seeds and to evaluate laboratory germination and growth rates of plants.

Research Material and Methods

Preliminary laboratory studies of the impact of pulsed magnetic field on lentil and flax seeds, conducted in 2017, showed positive results of germination of seeds and growth rate of plants. Best results were obtained when stimulating seeds in the magnetic field with the frequency of 30 Hz. Studies have shown that pre-seeding exposure of seeds yielded no positive results, and thereafter was used selectively to individual crops.

The laboratory studies of seeds of spring and winter wheat, barley, lentils and millet were conducted in 2018. Experimental laboratory unit, providing the U-shaped pulsed magnetic field, was used for stimulation (Figure 7).

Unit consists of induction coil 4 (Figure 7), installed in a W-shaped core, control unit 2, power supply 1, multimeter 3 and container with seeds 5. Unit allows to create the magnetic field, affecting the seeds with the intensity of up to 180 A/m and the pulse frequency of 10 Hz to 2 kHz. Duration of the stimulation has been monitored using a stopwatch.

During the research, the specified current frequency, which created pulsed magnetic field with the help of induction coil, was installed on control unit 2 using multimeter 3. The next step was to install the container with seed crops on coil 3 inside the W-shaped core, after which the stopwatch was turned on. After a given time, the container with seeds was removed, and the next operating mode was set.

The experiment on the germination of seeds on labmats was basically conducted during the studies of the impact of magnetic fields on the seeds of different crops (Figure 8). Additional experiment on growing plants in soil (Figure 9), aimed at identifying the growth rate and development of plants, was conducted for such crops, as spring wheat and lentils.

The following factors have been set during the research: magnetic field frequency, duration of stimulation and hold up time before seeding. Table 11 shows the research program.

For comparative evaluation, a portion of seeds was not subjected to stimulation and was germinate as the control sample.

Figure 7. Experimental unit of magnetic seed stimulation: a) general view of the unit; b) unit layout: 1 - power supply; 2 - control unit; 3 - multimeter; 4 - induction coil; 5 - container with seeds

During the experiments on germination of seeds on the labmat, the laboratory germination of seeds, which was held on the seventh day after seeding, was taken into account. The length of the seedlings was measured using a ruler.

The experiment on cultivation of plants in the soil took into account the evenness of sprout germination and plant length. Interim measurement was conducted on the tenth day after germination. Control measurement was conducted on the 30th day after germination.

Research Results

Table 11 shows the results of the research of spring wheat germination on the labmats and growing of sprouts in the soil. Seeding was carried out without the hold up time.

Figure 8. Spring wheat seedlings on labmat (a) and preparation of the experiment on stimulation of winter wheat, barley and millet using the magnetic field (b)

Figure 9. Wheat seedlings in the cartridge with the soil

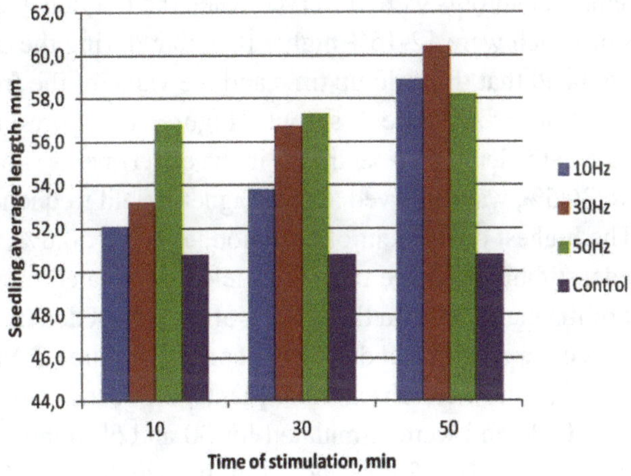

Research of laboratory germination showed the average germination of 93%, which fulfils the requirements of laboratory germination, allowable deviation of the samples has not been exceeded. Therefore, the results of experiments are considered reliable. Average germination of all samples of stimulated seeds turned out to be 94%, which is 5% higher than the controlled average germination, which is equal to 89%. Greatest germination rates were identified in samples, subjected to stimulation for 30 minutes under the frequencies of 10 Hz and 30 Hz, which amounted to 99% in each sample. Therefore, when stimulating seeds with these rates of time and frequency, laboratory germination will be 10% higher compared to the control.

Table 11. Seed magnetic field stimulation research program

Experiment No.	Crop	Magnetic Field Frequency, Hz	Stimulation Duration, min	Hold Up Time Before Seeding, days	Type of Experiment
1	Spring wheat	10-30-50	10-30-50	0-1-2	Germination on the labmat, growing in the soil
2	Lentils	10-30-50	1-5-10	0-1-2	Germination on the labmat, growing in the soil
3	Spring wheat	10-30-50	10-30-50	No hold up	Germination on the labmat
4	Winter wheat	10-30-50	10-30-50	0-1-2	Germination on the labmat
5	Millet	10-30-50	10-30-50	0-1-2	Germination on the labmat
6	Barley	10-30-50	10-30-50	0-1-2	Germination on the labmat

Analysis of research of the length of seedlings showed that the average length of seedlings of seeds, stimulated in the magnetic field, was higher than that during the control.

The greatest value of the average length of seedlings was observed in plants, stimulated at the frequency of 30 Hz for 50 minutes (fig. 7, fig. 8) which was equal to 60.4 mm, that is 19% higher than the control sample equal to 50.8 mm. Seeds, stimulated for 50 minutes at the frequency of 10 Hz and 50 Hz, also showed high rates, equal to 58.9 mm and 58.2 mm respectively.

High rates of average length of sprouts were observed when the frequency of the magnetic field was equal to 50 Hz, the values of which were 12-15% higher than that during the control.

Therefore, it can be concluded that the hold up time and the value of the frequency of the magnetic field have the direct impact on the growth rate of sprouts (Figure 10, Figure 11).

Research of the difference of the lengths of seedlings in the experiments showed that the lowest coefficient of variation, equal to 36.5%, was observed at the magnetic field frequency of 50 Hz and stimulation time of 10 minutes. The highest coefficient of variation, equal to 50.8%, was observed during the control. Therefore, the seeds, stimulated in the magnetic field, germinated more evenly.

Research of the impact of magnetic field on the growth of plants, seeded in soil, has showed positive growth rate in all experiments compared to that during the control (Figure 12, Figure 13). The maximum values of the average length of plants were observed in plants, the seeds of which were exposed to the magnetic field frequency of 50 Hz and were stimulated for 30 and 50 minutes. The average length of sprouts of this experiment amounted to 355 mm and 358 mm, which is 15.6% and 16.6% higher than the average length of the plants during the control, which is 307 mm.

Research of the difference of the lengths of plants in each sample showed that the lowest coefficient of variation, equal to 38.2%, was observed at the magnetic field frequency of 50 Hz and stimulation time of 10 minutes, while the coefficient of variation during the control was equal to 56%. Therefore, the plants, the seeds of which were stimulated in the magnetic field, germinated more evenly.

Table 12 shows part of the results of studies of the impact of magnetic field on seeds of lentils, wheat, barley and millet.

Studies of the impact of magnetic field on lentil seeds also included three variable factors: magnetic field frequency, duration of stimulation and hold up time before seeding. Table 12 presents the results

Figure 10. Histogram of seedling average length dependence on the time of stimulation at the magnetic field frequency of 10 Hz, 30 Hz, 50 Hz

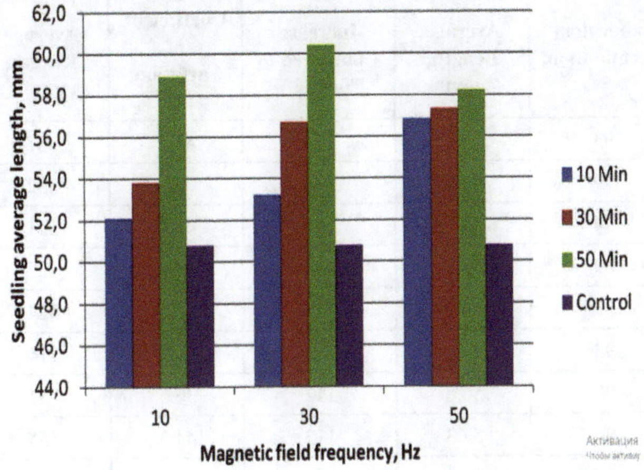

Figure 11. Histogram of sprout average length dependence on the magnetic field frequency at the time of stimulation of 10 min, 30 min and 50 min

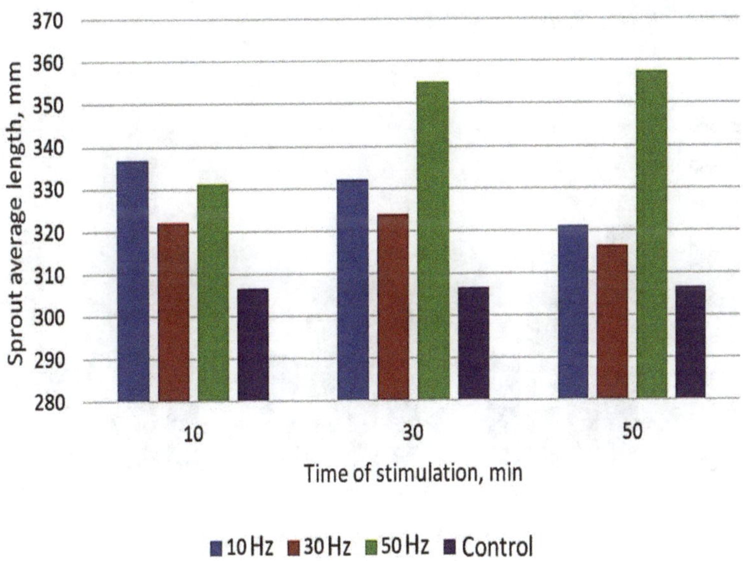

of germination and measurement of average seedling length subject to the treatment time of 1 min, hold up time of 1 day and frequencies of 10 Hz, 30 Hz and 50 Hz.

Research results show that the impact of magnetic field gives positive dynamics of seed germination and seedling growth rates. Lentil seeds, treated with magnetic field at the frequency of 30 Hz, have the greatest average length of sprouts, equal to 42.6 mm, which exceeds the average length during the control

Table 12. Results of measurement of germination, length of wheat seedlings and sprouts

Magnetic Field Frequency, Hz	Hold Up Time, min	Germination on the Labmat				Growing in the Soil		
		Laboratory Germination, %	Average Length, mm	Increase Compared to Control,%	Coefficient of Variation, %	Average Length, mm	Increase Compared to Control,%	Coefficient of Variation, %
10	10	93	52.1	102	45.3	337	109.8	48.9
	30	99	53.8	106	38.1	332	108.1	39.5
	50	94	58.9	116	43.9	321	104.5	47.4
30	10	89	53.2	105	42.0	322	104.9	51.1
	30	99	56.7	112	38.7	324	105.5	41.2
	50	93	60.4	119	40.1	316	102.9	40.8
50	10	90	56.8	112	36.5	331	107.8	38.2
	30	96	57.3	113	44.5	355	115.6	49.6
	50	96	58.2	115	40.8	358	116.6	42.3
Control		89	50.8	100	51.8	307	100.0	56.2

Figure 12. Histogram of sprout average length dependence on the time of stimulation at the magnetic field frequency of 10 Hz, 30 Hz, 50 Hz

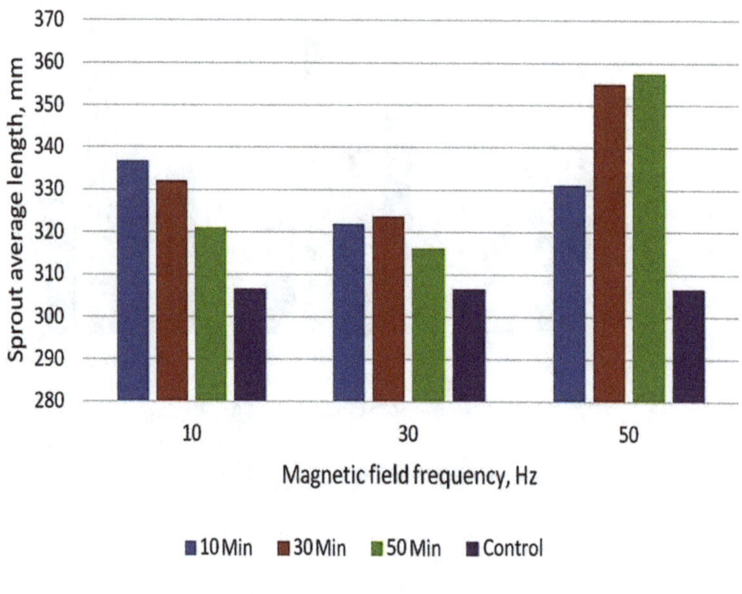

by 17%. The lowest coefficient of variation of the difference of the length of seedlings is observed at the magnetic field frequency of 50 Hz and is equal to 44.3%, while during the control this coefficient of variation is equal to 50.1%.

The average length of sprouts at the hold up time of one day was lower than that of stimulated seedlings immediately prior to seeding. Therefore, it can be concluded that the hold up before seeding of one day reduces the stimulating effect on seeds.

Figure 13. Histogram of sprout average length dependence on the magnetic field frequency at the time of stimulation of 10 min, 30 min and 50 min

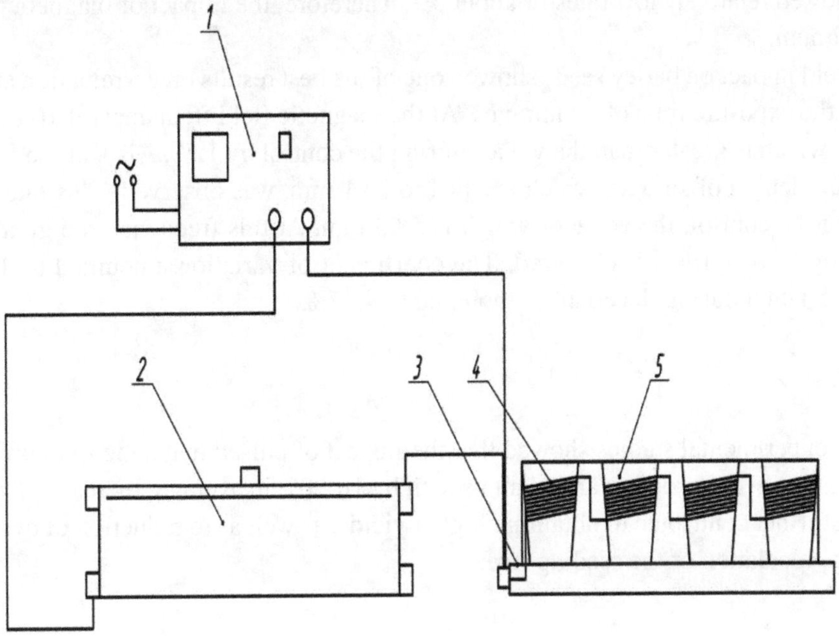

Stimulation of spring wheat seeds within the range of time from 10 to 50 seconds and at frequencies of 10 Hz, 30 Hz and 50 Hz also has a positive effect by such criteria, as germination and the average length of sprouts. Table 3 shows the results of the research of stimulation of seeds for 50 seconds. There was an increase in germination of seeds up to 6.1% at frequency of 30 Hz compared with that during the control. Also, at the preset frequency the maximum average length of sprouts, which is equal to 31 mm, was observed. This rate is 12% higher than that during the control, the result of which is 27.6%. The difference of the length of sprouts proved to be insignificant during the calculation of the coefficient of variation.

As a result of reached conclusions, the pre-condition of seed stimulation before seeding is created.

Research of the magnetic field impact on wheat seeds showed a positive result for germination and growth rate of seedlings. Table 3 presents the results of research of stimulation of winter wheat at exposure time of 10 minutes without hold up before seeding and at the magnetic field frequency of 10 Hz, 30 Hz and 50 Hz. Germination of seeds of winter wheat, exposed to magnetic field, was higher than that during the control by an average of 6.5%. The highest germination was observed in wheat, exposed to the magnetic field frequency of 10 Hz. At this frequency, high average length of sprouts, which is equal to 47.3 mm, was observed. This rate is 20% higher than that during the control, the value of which was 39.2 mm. Analysis of the difference of the length of the seedlings also showed that the coefficient of variation at the frequency of 10 Hz was equal to 42.3%, which is lower than that during the control, the value of which was 51.4%. Therefore, the 10 Hz magnetic field parameter is the most favorable for winter wheat.

Research of the magnetic field impact on millet seeds showed a slight increase in germination, which reached 100% during the experiment at frequencies of 10 Hz and 50 Hz. Research of the average length of seedlings showed relatively low rates of about 5%. Therefore, the impact of magnetic field on millet seeds was minimum.

Magnetic field impact on barley seeds showed one of the best results on germination and growth rate of seedlings at the exposure time of 30 minutes. At the magnetic field frequency of 10 Hz, germination rate was 100%, which is greater than the values during the control by 12%. Also, at the frequency of 10 Hz the maximum length of sprouts, which is equal to 74.1 mm, was observed. This rate is 26% higher than that during the control, the value of which is 58.3 mm. At this frequency the greatest difference between the lengths of sprouts is observed. The coefficient of variation amounted to 45%, while the coefficient of variation during the control amounted to 43.7%.

Conclusion

Analysis of the experimental studies showed that the impact of pulsed magnetic field on crop seeds increases germination, seed sprouting and plant growth by 10-20% in average. Increase of germination of crop seeds will further contribute to obtaining higher yields, as well as to reduction of overconsumption of seeds during calculation of the seeding rate.

MAGNETIC STIMULATION OF VEGETABLE CROPS

The magnetic field is an effective, economical and safe way to stimulate plants (Vasilev et al., 2018). The method of stimulation in the magnetic field is not currently being used. Studies carried out by various scientists have shown that the impact of magnetic fields on plants gives a positive effect and increases the growth rate of plants by 15-20%.

The experimental device for stimulation of plants in a magnetic field has been designed (Figure 14).

Experimental unit has been designed on the basis of cartridges for seedlings, some cells of which have induction coil of copper wire.

The unit includes power supply 1, three resistors 2 and induction coils 4, installed on cartridge 3 with cells 5. Induction coils are divided into three batteries, eight pcs. each. Electric current is supplied to each battery from the power supply unit via 2 resistors. Resistors are required to adjust the current force in each battery. Coil is installed outside the cartridge cell. The coils in the battery are connected in series. To reduce the mutual action of magnetic fields of different batteries, as well as to determine the impact of the downward magnetic field on plants, part of cells, located between the batteries, was without coils.To adjust the exposure time, the power supply is connected to the network via a timer.

Procedure for Experimental Unit Operation. Seeds are seeded into cells for the experiment. The unit is to be turned on after occurrence of first sprouts. The magnetic flux, which is directed upward along the plant height, is generated inside the coils subjected to the electric current. The plants are exposed every day in the specified period. This device is capable of adjusting the magnetic field intensity up to 5,000 A/m.

Radish and tomato seedlings are used as the plants being studied (Figure 15, Figure 16, Table 13).

Part of plants was seeded into open soil in order to determine the impact of seedling magnetic field stimulation. Each experiment was carried out using five bushes (Figure 17).

Table 13. Results of measurement of length of lentil seedlings

No.	Crop	Magnetic Field Frequency, Hz	Stimulation Duration	Experiment on the Germination of Seeds on the Labmat			
				Germination, %	Average Length, mm	Increase Compared to Control, %	Coefficient of Variation, %
1	Lentils	10	1 min	94.3	38.6	106	49.7
		30		96.2	42.6	117	46.0
		50		95.6	40.4	111	44.3
		Control		93.9	36.3	100	50.1
2	Spring wheat	10	50 sec	92.2	28.9	105	52.1
		30		95.4	31.0	112	49.3
		50		94.3	28.3	103	48.6
		Control		89.3	27.6	100	50.7
3	Winter wheat	10	10 min	93.6	47.3	120	42.3
		30		80.2	45.5	116	45.4
		50		83.8	40.7	104	46.7
		Control		79.3	39.2	100	51.4
4	Millet	10	10 min	100.0	56.6	105	48.3
		30		98.0	57.5	107	47.9
		50		100.0	60.4	112	52.1
		Control		95.0	53.9	100	50.6
5	Barley	10	30 min	100.0	74.1	126	45.2
		30		93.0	65.2	118	38.4
		50		96.0	64.1	110	41.9
		Control		88.0	58.3	100	43.7

Table 14. Program of Research of Magnetic Field Impact on Plants

Experiment No.	Crop	Stimulation Duration, h	Magnetic Field Intensity, A/m
1	Radish	1	1000-3000-5000
2	Tomato seedlings	3	1000-3000-5000

Plants were measured after completion of the experiment. Results of the experiment on radish magnetic field stimulation showed that the length of the leaf parts of plants, exposed to stimulation, proved to be higher than those during the control (Figure 18). Maximum dimensions of radish leaf part were observed in plants that were exposed to the magnetic field intensity of 3,000 A/m. The size of the leaf part was 43% more than the size of the control plants, not subjected to treatment.

It was also found that the leaf parts of plants located in close proximity to the cells with coils were greater that those during the control.

Similarly, studies of the magnetic field impact on the growth and development of tomato seedlings were carried out. Plants were measured after completion of the experiment. The experiment showed that the green part of plants, exposed to the magnetic field with the intensity of 3,000 A/m, was greater

Figure 14. Electric Circuit of the Experimental Unit for Magnetic Seed Stimulation: 1 - power supply; 2 - resistor; 3 - cartridge; 4 - induction coil; 5 – cell

Figure 15. Experimental cell-based unit for radish magnetic field stimulation

than that of the control plants by 16% (Figure 19). Plants, stimulated using the magnetic field with the intensity of 1,000 A/m and 5,000 A/m, showed good results as well. Dimensions of the green part of plants were greater than those during the control, by 15% and 10% respectively.

In order to determine tomato yield, tomatoes were collected, weighed and visually inspected after completion of the ripening period (Figure 17).

Studies of the impact of seedling magnetic field stimulation on tomato yield revealed that the yield of plants, stimulated in magnetic field, was higher than that during the control (Figure 20).

The greatest result was observed for plants, stimulated in the magnetic field with the intensity of 3,000 A/m, and amounted to 7.18 kg (collected from five bushes), while the result during the control amounted to 5.8 kg. The yield increased approximately by 24%. The weight of collected tomatoes from

Figure 16. Experimental cell-based unit for tomato seedling magnetic field stimulation

Figure 17. Cultivation of tomatoes in open soil: a) tomato bed; b) harvesting

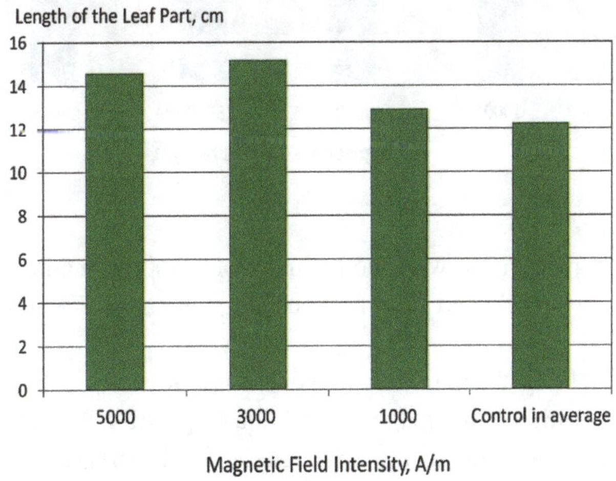

Figure 18. Diagram of dependence of the length of radish leaf part on magnetic field intensity

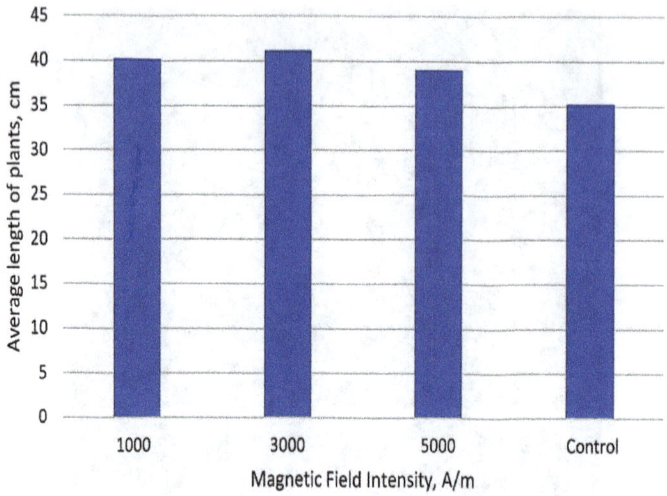

Figure 19. Diagram of dependence of the length of the tomato leaf part on magnetic field intensity

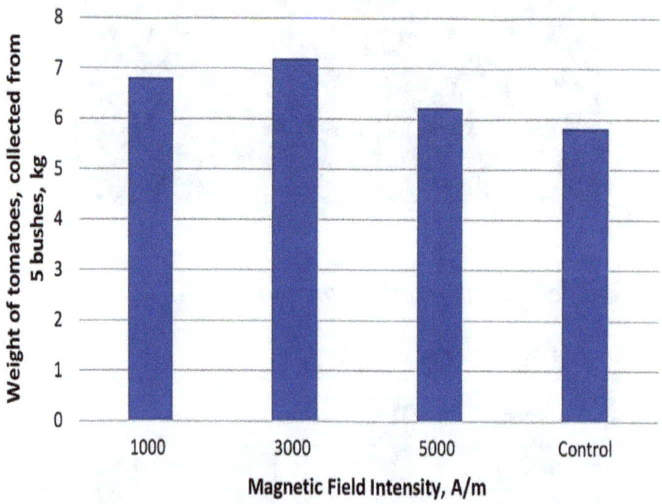

plants, stimulated in the magnetic field with the intensity of 1,000 A/m and 5,000 A/m, was equal and amounted to 6.8 kg and 6.2 kg (collected from five bushes), which is 17% and 7% higher than that during the control.

According to the results of experimental research of stimulation of radish and tomato seedlings using the magnetic field, it was found, that the magnetic field has a positive impact on the growth and development of plants. Stimulation of plants using the magnetic field increases the growth rate by 10-20%. The yield is increased by 10-20% in average.

Figure 20. Diagram of dependence of the tomato weight on magnetic field intensity

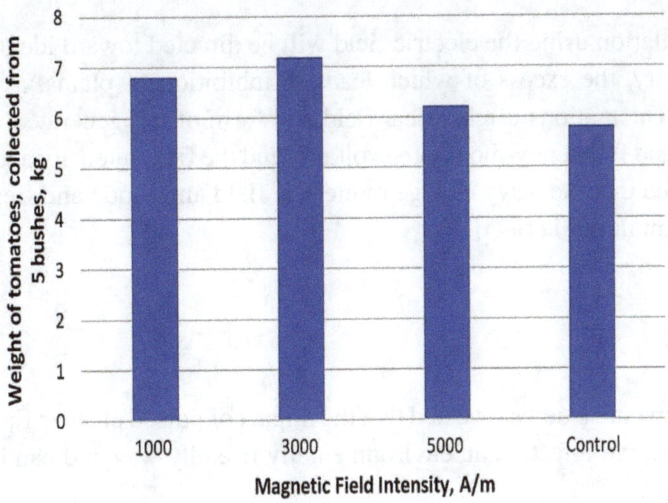

SOLUTIONS AND RECOMMENDATIONS

The studies revealed the objective positive impact of electric and magnetic fields on acceleration of growth and development of plants. When placed into electric field, the plants of green and vegetable crops interact with it. The nature of interaction is determined by the field type (alternating or pulsating), its frequency and intensity.

At the theoretical stage, parameters for interaction between the electric and magnetic fields and substances of plant bodies, affecting the movement of these substances (electrolytes), were established. Ultimately this leads to an acceleration of photosynthesis and, as a consequence, acceleration of growth of plant biomass.

Results of experimental studies confirmed the theoretical assumption.

When the field intensity was the greatest - 37 kV/m, the stimulation showed the best effect. In this experiment the average height of the leaf part is equal to 315.25 mm, which significantly exceeds the control by 58.55 mm. It means that simulation using the field intensity of 37 kV/m greatly increases the biomass growth rate, which is critically important for green vegetable crops.

Significant reduction in the coefficient of variation (which is equal to 3.89% - approx. 0,5% lower than that during the control) is a very important achievement.

Results of studies of stimulation of seeds in a magnetic field have shown that the greatest effect was achieved when the exposure time was equal to 30-50 minutes. Therefore, units of volumetric stimulation, where the seeds are placed into a container, in which the stimulation will be carried out for a specified time, should be used for seed treatment. Stimulation of seeds for 30 and 50 seconds also showed positive results, which allows to use the conveyor method of stimulation. In this case, the flow of seeds, which goes through the area of magnetic field generation during a very short period of time, is generated.

Studies of the magnetic field impact on the growth and development of plants showed that it is possible to use bottom coils for stimulation of several plants in order to reduce the material consumption.

FUTURE RESEARCH DIRECTIONS

Future studies of stimulation using the electric field will be directed toward identifying the upper limit of electric field intensity, the excess of which leads to inhibition of plants, as well as the rationale for optimal frequency fluctuations of electrical field and form of the generated voltage. It means that amplitude-modulated and frequency-modulated voltage (and the field intensity accordingly) are planned to be used. It is expected that the wavy change of electric field amplitude and frequency will cause additional awakening (stimulating) effect.

CONCLUSION

Theoretical and experimental studies revealed that the impact of pulsed electric field with positive direction on plants has a positive effect, is an environmentally friendly way and can be used in production of green vegetable crops.

Studies of the magnetic field impact on crop plants showed positive dynamics of seed germination and seedling growth rates. Magnetic field impact on seeds allows to improve laboratory germination by 10%, thereby reducing the seeding rate. Increase in growth rate by 10-15% would result in more rapid development and increased yields.

The use of magnetic field to stimulate plants during their growing period will ensure 10-20% higher developed and even sprouts, as well as will increase yield by 10-20%.

REFERENCES

Aladjadjiyan, A. (2012). Physical factors for plant growth stimulation improve food quality. In A. Aladjadjiyan (Ed.), *Food production – approaches, challenges and tasks* (pp. 145–168). Rijeka: InTech. doi:10.5772/32039

Batygin, N. F. (1986). *Ontogenez vysshikh rasteniy* [Ontogenesis of higher plants]. Moscow: Agropromizdat.

Bilalis, D. J., Katsenios, N., Efthimiadou, A., Karkanis, A., Khah, E. M., & Mitsis, T. (2013). Magnetic field pre-sowing treatment as an organic friendly technique to promote plant growth and chemical elements accumulation in early stages of cotton. *Australian Journal of Crop Science*, *7*(1), 46–50.

Cwintal, M., Dziwulska-Hunek, A., & Wilczek, M. (2010). Laser stimulation effect of seeds on quality of alfalfa. *International Agrophysics*, *24*(1), 15–19.

Dardeniz, A., Tayyar, S., & Yalcin, S. (2006). Influence of low-frequency electromagnetic field on the vegetative growth of grape cv. Uslu. *Journal of Central European Agriculture*, *7*(3), 389–395.

De Souza, A., Garcнa, D., Sueiro, L., Gilart, F., Porras, E., & Licea, L. (2006). Pre-sowing magnetic treatments of tomato seeds increase the growth and yield of plants. *Bioelectromagnetics*, *27*(4), 247–257. doi:10.1002/bem.20206 PMID:16511881

Flyrez, M., Carbonell, V., & Martínez, E. (2007). Exposure of maize seeds to stationary magnetic fields: Effects of germination and early growth. *Environmental and Experimental Botany*, *59*(1), 68–75. doi:10.1016/j.envexpbot.2005.10.006

Goncharov, A. A., Berezhnaya, I. E., & Gursky, I. G. (1994). *Vliyaniye fizicheskikh metodov predposevnoy obrabotki semyan sel'skokhozyaystvennykh kul'tur na ikh semennyye kachestva* [The influence of physical methods of pre-sowing treatment of seeds of agricultural crops on their seed qualities]. Zelenograd.

Gordeev, A. M., & Shershnev, V. B. (1991). *Elektrichestvo v zhizni rasteniy* [Electricity in plant life]. Moscow: Nauka.

Goussous, S. J., Samarah, N. H., Alqudah, A. M., & Othman, M. O. (2010). Enhancing seed germination of four crop species using an ultrasonic technique. *Experimental Agriculture*, *46*(2), 231–24. doi:10.1017/S0014479709991062

Gut, M. (2007). Impact of alternating electric field on potato tuber growth and cropping. *Inżynieria Rolnicza*, *8*(96), 73–79.

Hernandez, A. C., Dominguez, P. A., Cruz, O. A., Ivanov, R., Carballo, C. A., & Zepeda, B. R. (2010). Laser in agriculture. *International Agrophysics*, *24*(4), 407–422.

Jedlička, J., Paulen, O., & Ailer, Š. (2014). Influence of magnetic field on germination, growth and production of tomato. *Potravinárstvo*, *8*(1), 150–154. doi:10.5219/349

Jiang, J., He, X., Li, L., Li, J., Shao, H., Xu, Q., ... Dong, Y. (2014). Effect of cold plasma treatment on seed germination and growth of wheat. *Plasma Science & Technology*, *16*(1), 54–58. doi:10.1088/1009-0630/16/1/12

Kasakova, A. S., Yudaev, I. V., Fedorishchenko, M. G., Mayboroda, S. Y., Ksenz, N. V., & Voronin, S. M. (2018). New approach to study stimulating effect of the pre-sowing barley seeds treatment in the electromagnetic field. *Online Journal of Biological Sciences*, *18*(2), 197–207. doi:10.3844/ojbsci.2018.197.207

Kuleshov, A.N., Ereshko, A.S., & Khronyuk, V.B. (2011). Primeneniye magnitnykh poley postoyannykh magnitov dlya predposevnoy obrabotki semyan yachmenya [Use of magnetic fields of permanent magnets for presowing treatment of barley seeds]. *Vestnik agrarnoy nauki Dona*, *1*, 95-100.

Maffei, M. E. (2014). Magnetic field effects on plant growth, development, and evolution. *Frontiers in Plant Science*, *5*, 445. doi:10.3389/fpls.2014.00445 PMID:25237317

Mahmood, M., Bee, O. B., Mahmud, T., & Subramaniam, S. (2011). The growth and biochemical responses on in vitro cultures of Oncidium taka orchid to electromagnetic field. *Australian Journal of Crop Science*, *5*(12), 1577–1587.

Malygin, V.M. (2015) Teorema Umova-Poytinga i vektor plotnosti potoka elektromagnitnoy energii: raznyye usloviya, raznyye resheniya [Umov-Poiting theorem and density vector of electromagnetic energy flux: different conditions, different solutions]. *Prostranstvo i vremya*, *3*(21), 103-109.

Marinković, B., Grujić, M., Marinković, D., Crnobarac, J., Marinković, J., Jaćimović, G., & Mircov, D. V. (2008). Use of biophysical methods to improve yields and quality of agricultural products. *Journal of Agricultural Sciences*, *53*(3), 235–242. doi:10.2298/JAS0803235M

Marks, N., & Szecywka, P. S. (2010). Impact of variable magnetic field stimulation on growth of aboveground parts of potato plants. *International Agrophysics*, *24*, 165–170.

Martínez, E., Carbonell, M. V., Flyrez, M., Amaya, J. M., & Maqueda, R. (2009). Germination of tomato seeds (Lycopersicon esculentum, L.) under magnetic field. *International Agrophysics*, *23*, 45–49.

Maslobrod, S.N. (1981). *Elektricheskiy yazyk rasteniy* [Electric plant language]. Kishinov: Shtiitsa.

Medvedev, S.S. (1990). Elektricheskiye polya i rost rasteniy [Electric fields and plant growth]. *Elektronnaya obrabotka materialov*, *3,* 68-74.

Mihai, A. L., Dobrin, D., Magureanu, M., & Popa, M. E. (2014). Positive effects of non-thermal plasma treatment on radish seeds. *Romanian Reports in Physics*, *66*(4), 1110–1117.

Molamofrad, F., Lotfi, M., Khazaei, J., Tavakkol-Afshari, R., & Shaiegani-Akmal, A. (2013). The effect of electric field on seed germination and growth parameters of onion seeds (Allium cepa, L.). *Advanced Crop Science*, *3*(4), 291–298.

Novitsky, Y. I. (1973). Magnitnoye pole v zhizni rasteniy [Magnetic field in plant life]. In. V.N. Chernihiv (Ed.), Problemy kosmicheskoy biologii [Problems of space biology] (pp. 164-188). Moscow: Nauka.

Novitsky, Y. I. (1987). Reaktsiya rasteniy na magnitnyye polya [The reaction of plants to magnetic fields]. In Y. A. Kholodov (Ed.), *Reaktsiya biologicheskikh sistem na magnitnyye polya* [The Reaction of Biological Systems to Magnetic Fields] (pp. 117–130). Moscow: Nauka.

Novitsky, Yu. I., Strekova, V. Yu., & Tarakanova, G. A. (1971). Deystviye postoyannogo magnitnogo polya na rost rasteniy [The effect of a constant magnetic field on plant growth]. In Y. A. Kholodov (Ed.), *Vliyaniye magnitnykh poley na biologicheskiye ob"yekty* [Effect of magnetic fields on biological objects] (pp. 69–88). Moscow: Nauka.

Opritov, V. A., Pyatygin, S. S., & Retivin, V. G. (1991). *Bioelektrogenez u vysshikh rasteniy* [Bioelectrogenesis in higher plants]. Moscow: Nauka.

Pietruszewski, S., Muszynski, S., & Dziwulska, A. (2007). Electromagnetic field and electromagnetic radiation as non-invasive external stimulants for seeds (selected methods and responses). *International Agrophysics*, *21*, 95–100.

Polunin, V.N., Zhidkov, T.V., Bel'tyukov, L.P., & Kuprov, A.V. (2009) Vliyaniye elektromagnitnogo polya na posevnyye rostovyye i produktivnyye svoystva ozimoy pshenitsy [Influence of the electromagnetic field on sown growth and productive properties of winter wheat]. *Vestnik agrarnoy nauki Dona*, *3*, 12-16.

Racuciu, M. (2011). 50 Hz Frequency Magnetic Field Effects on Mitotic Activity in the Maize Root. *Romanian Journal of Biophysics*, *21*(1), 53–62.

Radhakrishnan, R., & Kumari, B. D. R. (2012). Pulsed magnetic field: A contemporary approach offers to enhance plant growth and yield of soybean. *Plant Physiology and Biochemistry*, *51*, 139–144. doi:10.1016/j.plaphy.2011.10.017 PMID:22153250

Ramnek, G.M. (1911). *Vliyaniye elektrichestva na pochvu: Ionizatsiya pochvy i usvoyeniye atmosfernogo azota* [Effect of Electricity on the Soil: Ionization of the Soil and Acquisition of Atmospheric Nitrogen]. Kiyev: Tipografiya universiteta Svyatogo Vladimira.

Sokolov, O. A. (1983). *Kachestvo urozhaya grechikhi* [Buckwheat harvest quality]. Pushchino: ONTI NTSBI AN USSR.

Spirov, G.M., Valueva, Y.V., Merkulova, V.G., Medvedeva, L.N., Lukyanov, N.B., & Zaitsev, A.S. (2008). Eksperimental'noye issledovaniye vliyaniya elektrofizicheskikh faktorov na urozhaynost' ovoshchnykh kul'tur [Experimental study of the influence of electrophysical factors on the yield of vegetable crops]. *Uspekhi sovremennogo yestestvoznaniya*, *6*, 30-38.

Statsyuk, N.V., Takur, K., Smetanina, T.I., & Kuznetsova, M.A. (2016). Reaktsiya rasteniy kartofelya (Solanum tuberosum, L.) raznykh sortov na predposadochnuyu obrabotku klubney impul'snym nizkochastotnym elektricheskim polem [The reaction of potato plants (Solanum tuberosum, L.) Of different varieties to the preplant treatment of tubers with a pulsed low-frequency electric field]. *Sel'skokhozyaystvennaya biologiya*, *51*(3), 360-366.

Strekova, V. Y. (1973). Mitoz i magnitnoye pole [Mitosis and magnetic field]. In V.N. Chernigovsky (Ed.), Problemy kosmicheskoy biologii [Problems of space biology] (pp. 200-204). Moscow: Nauka.

Syrkin, V. A. (2018). Issledovaniya stimulirovaniya semyan v impul'snom magnitnom pole [Studies of seed stimulation in a pulsed magnetic field]. Innovatsionnyye dostizheniya nauki i tekhniki APK [Innovative achievements of science and technology of agriculture] (pp. 346-349). Kinel: RIO SGSKHA.

Temerkieva, Y. M., & Plieva, A. M. (2016). Prorashchivaniye semyan pshenitsy pod vozdeystviyem magnitnogo polya [Germination of wheat seeds under the influence of a magnetic field]. *Aprobatsiya*, *3*(42), 7–9.

Vasilev, S. I., Mashkov, S. V., Syrkin, V. A., Gridneva, T. S., & Yudaev, I. V. (2018). Results of studies of plant stimulation in a magnetic field. *Research Journal of Pharmaceutical, Biological and Chemical Sciences*, *9*(4), 706–710.

Vasiliev, S.I., Mashkov, S.V., & Fatkhutdinov, M.R. (2016). Elektromagnitnoye stimulirovaniye semyan i rasteniy [Electromagnetic stimulation of seeds and plants]. *Sel'skiy mekhanizator*, *7*, 8-9.

Voytova, A.S., Yukin, N.A., & Ubiraylova, V.G. (2016). Slabyy elektricheskiy tok kak faktor stimulyatsii rosta domashnikh rasteniy [Weak electrical current as a factor stimulating the growth of domestic plants]. *Mezhdunarodnyy studencheskiy nauchnyy vestnik*, *4*(3), 364-366.

Yan, D.-L., Guo, Y.-Q., Zai, X.-M., Wan, S.-W., & Qin, P. (2009). Effects of electromagnetic fields exposure on rapid micropropagation of beach plum (Prunus maritima, L.). *Journal of Ecological Engineering*, *35*(4), 597–601. doi:10.1016/j.ecoleng.2008.04.017

Yang, L., & Shen, H. (2011). Effect of electrostatic field on seed germination and seedling growth of Sorbus pohuashanesis. *Journal of Forestry Research*, *22*(1), 27–34. doi:10.100711676-011-0120-9

Yudaev, I. V., Tibirkov, A. P., & Azarov, E. V. (2012) Predposevnaya elektrofizicheskaya obrabotka semyan – perspektivnyy agropriyem resursosberegayushchey tekhnologii vozdelyvaniya ozimoy pshenitsy [Preseeding electrophysical seed treatment is a promising agricultural use of resource-saving winter wheat cultivation technology]. *Izvestiya Nizhnevolzhskogo agrouniversitetskogo kompleksa: Nauka i vyssheye professional'noye obrazovaniye*, *3*(27), 61-66.

Chapter 16
Linear Electromechanical Transducer in the Systems of Welded Joints of Electrodynamic Processing

Andrii Zhyltsov
National University of Life and Environmental Sciences of Ukraine, Ukraine

Igor Kondratenko
National Academy of Sciences of Ukraine, Ukraine

Vyacheslav Vasyuk
National University of Life and Environmental Sciences of Ukraine, Ukraine

ABSTRACT

This chapter is dedicated to establishing characteristics relationships between the induction type impact electromechanical transducer and parameters and quality indicators of electrodynamic effects on the welded joints. The authors developed two-dimensional circle-field mathematical model of transient discharge capacity at the branched electrical circuit with the coil inductance which changes dynamically, allowing by adjusting the parameters of the electromechanical transducer to achieve the necessary technological requirements for the characteristics of electrodynamic processing. Based on mathematical modeling of electrophysical processes in electromechanical transducers, induction type for electrodynamic processing of welded joints, reasonably geometrical parameters massive disk, and the contact area, the necessary conditions are created to reduce residual stresses in the weld joints.

DOI: 10.4018/978-1-5225-9420-8.ch016

INTRODUCTION

Welding is one of the most common methods of metal bonding, which, in comparison with other types of integral joints, has significant advantages. However, the welding process also has drawbacks, one of which is the emergence of residual stresses and deformations, which can adversely affect the quality of the welded structure.

Residual or technological ones are called stresses that exist in the design or in its individual elements in the absence of external power, heat or other influences. In technology, for the purpose of designating residual stresses, the names of technological processes are used, after which they are found: welding stresses, deformation stresses, stresses of hardening, stresses of processing.

The presence of welding residual stresses is one of the reasons for reducing the resource of metal structures by influencing their characteristics of fatigue strength, corrosion resistance and residual molding. The reason for this is the uneven linear or volumetric deformations of adjacent volumes of metal. In the weld and in the zone around it there is a tensile stress, which is close to the threshold of the fluidity of the metal and even exceeds it. In some cases, welding stresses and deformations lead to the destruction of the welded structure, and as a consequence, affect the reliability and durability of the process equipment.

The objective of reducing welding residual stress can be solved in different ways processing: rolling, forging, heat processing, vibration processing, ultrasonic processing, and the shock and explosive load and others. Listed weld processing methods have certain disadvantages, which include the need to create energy and metal-intensive technological equipment, restrictions during processing large structures and significant energy costs.

A promising way to increase the life of welded joints is the method of electrodynamic processing based on the use of effect electroplastic deformation, which is characterized by ease of implementation, reducing process time by several times, the possibility of local action, low energy costs and leads to a decrease in residual stresses by 50-65%. Thus, provided that the current density in the metal reaches a value more than 10^9 A/m^2 and upon application of compressive forces at the level of 20 kN, the effect of electroplasticity is manifested, which consists in the relaxation stress-strain state of metallic materials.

However, despite the widespread use of the phenomenon of electroplastic deformation, the mechanisms that govern this phenomenon are not yet fully known. We can assume that in the process of local elastic discharge, the determining contribution belongs to the intensive dislocation interactions that are caused by the flow of electric current. These interactions depend on the parameters of the current pulse: the shape of the pulse, the amplitude, the duration of the front, and others.

The realization of this pulsed electromagnetic effect can be achieved by electrodynamic processing, which is the simultaneous action of electrodynamic force and current. During the passage of the current through the metal being processed and the action of the pulsed electromagnetic force, deformation processes are initiated, the interaction of which with the welding stress causes residual plastic deformation. The consequence of such interaction is to reduce the level of residual tensile stresses or their transformation into compression stresses, which positively affects the lengthening of the resource of welded joints. The advantages of using this processing are the possibility of a local action, which will enable the processing of objects of any shape and size at low energy costs.

The electrotechnical system for the implementation of electrodynamic processing can be constructively performed as a linear electromechanical converter of induction type of percussion action, which allows you to simultaneously provide both a force effect and transmit a current pulse through a contact electrode to the treatment area. But the combination of two functions in a single device results in simultaneous consideration of both electrodynamics processes and processes in the discharge circuit.

The peculiarity of the proposed electrotechnical device, which has an energy effect on the metal, is the presence of strong scattering fields, so the use of the finite element method and the finite difference method for mathematical modeling leads to a large number of superfluous calculations. Proceeding from this, the method of integration and integro-differential equations is optimal in terms of reducing the search field of the solution, and, consequently, the rational use of computer resources, which is especially relevant when conducting multivariate calculations. The application of this method will make it possible to accurately take into account the scattering fields in an unbounded space in the minimal calculated region: volumes and surfaces of massive bodies.

However, now, the solution of a circuit-field problem, characteristic of the electrodynamic processing of welded joints of metal non-ferromagnetic structures with the use of a linear electromechanical converter of induction type of percussive action, was not considered.

Thus, the development of methods for calculating the interconnected electrophysical processes in the systems of electrodynamic processing of welded joints of metal non-ferromagnetic structures and providing the necessary for technological reasons parameters – effort and current density – actual scientific problem, which is essential for the development of the theory and practice of pulsed electromagnetic systems in the composition of power sources and electromechanical transducers.

Therefore, the purpose of the study was to establish the relationship of structural, electrical and operating characteristics linear electromechanical converter of induction type of shock action with the efficiency of electrodynamic processing of welded joints.

To achieve the purpose solved the following tasks:

- substantiate the necessity of developing new, energy-efficient ones, electrotechnical systems for electrodynamic action on welded joints in non-ferromagnetic metal constructions;
- develop a system of electrodynamic action on welded joints in non-ferromagnetic metal constructions;
- develop on the basis of integro-differential equations a circle-field mathematical model of non-stationary electromagnetic processes in non-uniform electrons for electrodynamic action systems on welded joints in non-ferromagnetic metal constructions;
- apply developed mathematical models for the analysis of currents distribution and electrodynamic forces in the «electrode-welded joint» electromagnetic system, and to substantiate the parameters of the electrodynamic action system for welded joints to provide local relaxation of welding residual stresses in metallic non-ferromagnetic constructions;
- check the validity and adequacy of the developed mathematical model of non-stationary electromagnetic processes in inhomogeneous media by comparing the calculated and experimental data obtained on the physical model of the system of electrodynamic effects on welded joints.

BACKGROUND

Experimentally established (Baranov, Troitskiy, Avraamov, & Shlyapin, 2001; Gromov, 2014; Troitskiy, 2018) that electric current with a density of $10^5 \div 10^6$ A/cm^2 and a duration of ~ 10^{-4} s. It is passed through the metal, causes, along with thermal, also an additional influence on the process of plastic deformation in the metal through non-thermal mechanisms. Studies of the interaction of the electron flow with dislocations involved in the plastic deformation of the metal were carried out. To explain this effect, the hypothesis of electron wind and the effect of a side force effect of a pulsed current, the pinch effect, was proposed (Gromov, 2014).

Therefore, the process of deformation, which takes place during the flow of electric current through the metal with a density of hundreds to thousands of A/cm^2, is called the effect of electroplastic deformation.

Further studies were carried out to study the effect on the metal, which is deformed, by the direct action of an electric current of high density. The purpose of the experiments was to identify the nature of the electroplastic deformation. The influence of side factors – heat, skin effect, and ponderomotive actions of the current was analyzed (Troitskiy, Stashenko, Ryizhkov, Lyashenko, & Kobyilskaya, 2011; Gromov, 2014). A further study of the mechanism of electroplastic deformation was carried out in (Troitskiy, Stashenko, & Ryizhkov, 2011) when studying the effect of current on the stress relaxation process in Zn, Cd, and Pb crystals with a change in the current direction. It was found that relaxation does not occur immediately after changing the direction of the current, but after some time has passed, it is several tens of seconds. Accordingly, the greater the applied voltage and the earlier the direction of current changes from the moment of the beginning of the relaxation, the smaller the time delay.

In studies of the processes of compression, stretching and creep of metal single crystals, it was confirmed that the mechanisms that lead to relaxation are the electron-plastic effect and the ponderomotive electric current (Troitskiy, Baranov, Avraamov, & Shlyapin, 2004). The effect is manifested in a decrease in the deforming force and a change in some structural properties of the metal. The pressure of the ponderomotive forces, in some cases, can reach a value comparable to the yield point.

Research (Conrad, 2000; Troitskiy, Stashenko, Ryizhkov, Lyashenko, & Kobyilskaya, 2011) confirmed the phenomenon of electroplastic deformation. For the experiments, single pulses of similar current density and duration were used. It is established that the efficiency of the effect increases with increasing current density, and the use of single pulses causes a change in the deforming voltage up to 40%.

The appearance of the electroplastic effect is influenced by the current density and the pulse duration, with increasing amplitude current density in the metal, the magnitude of the electroplastic effect increases, starting from the limiting values of the current density of $10^5 \div 10^6$ A/cm^2. The electroplastic effect is also the limiting values for the duration of the pulsed current, arising from values of 10^{-4} s and increasing to values of 10^{-3} s, when thermal effects begin to appear. At smaller values of the pulse duration, the effect is absent. An increase in the pulse repetition rate leads to an increase in the effect, but the magnitude of the change in deformation from one pulse decreases (Baranov, Troitskiy, Avraamov, & Shlyapin, 2001). Studies (Bunget, Salandro, Mears, & Roth, 2010; Salandro, Bunget, & Mears, 2010) on the determination of the dependence of the influence coefficient of the electroplastic effect on the current density.

The study of the phenomenon of electroplastic deformation was performed by many scientists from Russia, Ukraine, America and Israel (Guoyi, 2000; Baranov, Troitskiy, Avraamov, & Shlyapin, 2001; Troitskiy, Kayzer, Layshev, & Gorelik, 2004; Song, Wang, & Gao, 2007; Maltsev, 2008; Troitskiy, Stashenko, & Ryizhkov, 2011; Gromov, 2014).

It was revealed that the electrical effect is fixed in plastic materials under various types of load, including compression and other stress states. It can act along with the joule effect in metal processing with the participation of electric current, such as electrocontact heating and induction heating by eddy currents, where the Joule effect is used.

At present, the use of (Guoyi, 2000; Song, Wang, & Gao, 2007; Troitskiy, 2018) electroplastic deformations in the metalworking industry for the manufacture of wire, powder materials, sheet metal, as well as in stamping, rolling, forging, and rolling has been widely studied and implemented.

Currently there are several theories trying to explain the mechanism of the electron-plastic effect, but each of them does not fully reveal the essence of this phenomenon.

For all the time studying electropulse and electroplastic treatments, scientists have focused on identifying the influence of the components of the electroplastic effect on the plastic properties of the metal. At the same time, the issue of developing an electrical system to reduce residual stresses using an electroplastic effect has not been resolved.

The method of controlling the mechanical characteristics of a metal using an electric current is relatively simple to implement, which explains the interest of scientists in the application of pulsed electromagnetic effects in the direction of its application to reduce welding residual stress.

To carry out a pulsed electromagnetic effect of certain parameters on the metal is possible through the implementation of electrodynamic processing, which consists in the simultaneous action of a pulsed electrodynamic force and a pulsed current.

As a result, deformation processes are initiated in the metal, which interact with welding stresses and cause permanent plastic deformations. The result of the interaction is a reduction in the level of residual tensile stresses or their conversion to compressive stress, which has a positive effect on the extension of the life of welded joints. A device for carrying out such a pulsed electromagnetic effect can be constructively implemented as a linear electromechanical converter of induction type of an impact type of an induction type system; it provides an electrodynamic effect on a welded joint by passing current pulses of specified parameters generated by a source of current pulses.

The main advantages of using electrodynamic processing to reduce the welding residual stresses, compared to other methods, are such (Burkin, Shimov, & Andryukova 2015):

- Equipment is relatively simple, universal, mobile and small, which allows it to use it to handle a wide range of welded objects, the ability to perform a local action;
- Low cost of service and reduction of the technological time in 50-60 times;
- Reducing energy consumption by more than ten times;
- Lack of environmental pollution;
- Electrodynamic treatment of welded joints leads to a reduction of residual stresses by 50-65% (thermal method 30-70%, vibration treatment 40-55%, ultrasonic treatment 16-30%, shock and explosive loading 40-60%, rolling, roasting 10-30%).

Among a number of existing methods for reducing welding residual stresses, electrodynamic processing has significant advantages and requires further research in the direction of developing a mathematical model of the non-stationary electrophysical process of electrodynamic treatment of welded joints and the establishment on its basis of structural, electrical and mode characteristics to provide the parameters of current pulses sufficient for effective regulation of the residual stressed state of metal structures.

Considering the above, it can be concluded that it is necessary to develop methods for calculating interconnected electrophysical processes – discharging a capacitance into an extensive electrical circuit with semiconductor elements and creating the electrodynamic forces necessary for technological reasons in the contact zone of the electrode and the welded joint.

Linear, electromechanical transducers impact action – this class of devices intended for direct transformation of electric energy source into mechanical energy of linear motion of movable elements. The source of energy in these devices is pulsed capacitive energy storage with electronic control system.

By the principle of operation, it is possible to distinguish linear electromechanical transducers of percussion magnetoelectric, electromagnetic, electrodynamic, railgun and induction types (Ismagilov, Afanasev, & Styiskin, 2006). High quality indicators of electromechanical converters caused their widespread use in industry and agriculture.

Linear electromechanical transducer of the induction type is built on the interaction of a magnetic field, which is excited by an alternating current of the inductor winding, with an induced current in the electrically conductive armature. The electromechanical transducer of this type can be coaxial, side disk execution or as a ring accelerator. Usually, an electromechanical converter of this type is made without a magnetic circuit and has a fixed inductor winding and a movable electrically conductive armature, which is a short-circuited multi-turn winding or a massive electrically conducting element in the form of a disk (Bose, 2002; Tyukov, 2006).

A coaxially mounted armature with an inductor winding during electrodynamic interaction with an inductor winding moves in the axial direction. Performing an electric conductive anchor as a massive element in the form of a disk is easier because it has a greater mechanical stiffness and a maximum filling factor. However, the uneven distribution of induced current leads to uneven use of its volume. The principle of operation of the linear electromechanical transducer induction type is as follows. When an inductor winding is excited from a capacitive energy storage in an anchor there is a current, which leads to the emergence of electrodynamic forces along the axis of the anchor.

Perhaps highlight the following advantages of linear electromechanical transducer induction type:

- Not capricious to the parameters of the pulsed energy source;
- Possibility of conducting the conductor windings of the inductor on relatively low currents;
- Possibility of contactless movement of the anchor in relation to the winding of the inductor;
- Acceleration of anchor of any sizes and forms with maintenance of high start accelerations;
- Possibility of distributing electrodynamic forces along the entire length along the anchor;
- The possibility of passing the current through an anchor from both a third-party source and from one source with an excitation winding.

Given the above advantages of linear induction type electromechanical transducers, and especially the opportunity to influence that will include simultaneous action of electrodynamic forces and current, it can be argued that the use of converters of this type is the most appropriate.

Currently working in the development of linear electromechanical transducers inductive type performed mainly towards the development of various design schemes aimed at enhancing their dynamic performance and efficiency. At present, work towards the use of linear induction type electromechanical transducer Impact who has to simultaneous action of electrodynamic forces and power was not fulfilled.

Significant contribution to the theory and practice of constructing electromechanical linear transducer made by such researchers as (Jeon, Katoh, & Iwamoto, 1999; Yampolskiy, 1990; Yamori, Ono, & Kubo 2001; Afonin, & Grebenikov, 2004; Tyukov, 2006; Ismagilov, Afanasev, & Styiskin, 2006; Kondratenko, Bozhko, Zhiltsov, & Vasyuk, 2012; Flora, 2012; Usanov, Moshkin, Kargin, & Volgin, 2015; Oleksenko, 2016) and other.

Given the above advantages of linear electromechanical transducer of induction type, especially the ability to influence, including the simultaneous action of electrodynamic force and current, it can be argued that the use of converters of this type is most appropriate.

The most common approach to solving problems in an electrical system, which includes an electromechanical transducer, is to determine the interaction of bodies that carry current or charge in an electromagnetic field. This approach is based on the solution of the equations of electrodynamics – Maxwell's equations.

Modeling of electromagnetic processes occurring in a given electromechanical system reduces to solving three-dimensional boundary value problems for a system of Maxwell equations in an unbounded domain and supplemented with boundary and initial conditions, formulates an initial-boundary value problem, the solution of which is reduced to modeling electromagnetic processes in electrical devices with massive conductors. In some cases, the solution of such a problem is admissible in a plane-parallel or axisymmetric approximation.

The analysis of the literature indicates that today there are different approaches to the modeling of electrophysical processes taking into account the real electrophysical properties, namely: the finite difference method, the finite element method, the method of integral equations, combined methods and others.

Methods of finite elements and finite differences intensively developed in the works (Sidorov, & Sarapulov, 2010; Neyman, & Neyman, 2015; Shreyner, Polyakov, & Medvedev, 2016) and others.

Based on the finite element method, several numerical-software complexes (Comsol, Ansys, Elcat, Maxwell, FEMM, etc.) has been developed. They require a significant expenditure of computer computing resources on the calculation of one option when considering three-dimensional problems. There are no functional solutions for even more resource-intensive problems of optimal design. In the presence of significant local features of the field, the application of the mentioned packages is calculated gives unreliable results (Astahov, 2004)

One of the most promising is the method of integral and integro-differential equations. This is since in the class of electromechanical devices under consideration, there are usually strong electromagnetic stray fields. The method of integral equations in this case is more efficient, since the desired quantities are distributed only in the volume of massive bodies (eddy currents), on their boundary (electric charges) and on the boundary of ferromagnetic bodies (magnetization currents). Thus, the solution search area is significantly smaller than in the methods of finite differences and finite elements, when it is necessary to search for a solution in the whole, unbounded, domain.

Questions of the application of the integro-integro-differential equation method for mathematical modeling of electromagnetic fields in electrical engineering devices are devoted to the work of well-known scientists (Petrushenko-Kubala, & Petrushenko, 2003; Evdokimov, Kuchaev, Petrushenko, & Kasyan 2010; Zhiltsov, Kondratenko, & Sorokin, 2012) and others.

A feature of the proposed electrical device, which provides energy impact on the metal, is the presence of strong stray fields, so the use of the finite element method and the finite difference method for mathematical modeling leads to a significant number of unnecessary calculations.

On this basis, from the point of view of reducing the search area, the decoupling, and, as a result, the rational use of computer resources, which is especially important during multivariate calculations, is the method of integration and integro-differential equations.

The use of this method will allow you to accurately take into account the field of scattering in an unlimited space with the minimum computational domain: the volumes and surfaces of massive bodies.

Thus, the mathematical modeling of interconnected electrophysical processes – the discharge of capacities on the branched electric circle with semiconductor elements and the creation of the necessary electrodynamic forces in the electrodynamic forces in the contact area of the electrode and the processed sample, it is expedient to do this by using the method of integral and integro-differential equations, which is an actual scientific and applied problem, the solution of which is essential for the reliability and durability of the process equipment.

MAIN FOCUS OF THE CHAPTER

Electrotechnical System for Electrodynamic Processing of Welded Non-Ferromagnetic Joints

The developed of an electrical system (Lobanov, Paschin, Cherkashin, Mihoduy, & Kondaratenko, 2012; Kondratenko, Zhiltsov, & Vasyuk, 2014), designed to reduce welding residual stresses by means of electrodynamic processing. It consists of: an electromechanical transducer of induction type of shock action (Zhiltsov, Kondratenko, & Vasyuk, 2014; Lobanov, Kondratenko, Zhiltsov, Karlov, Paschin, Vasyuk, & Yaschuk, 2016; Bolyuh, Oleksenko, & Schukin, 2015) and a source of current pulses (Kondratenko, Bozhko, Zhiltsov, & Vasyuk, 2012) (Figure. 1).

Figure 1. Electrotechnical system for electrodynamic processing of welded joints

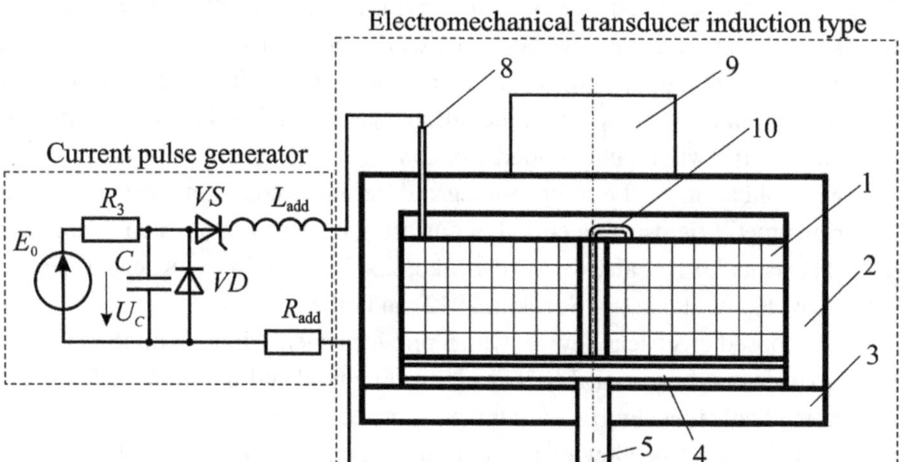

An induction type electromechanical transducer provides, on the one hand, an electrical contact between the electrode through which a current pulse is generated, generated by the source and the welded joint, and on the other hand, a force action.

The constructive elements of the electromechanical transducer of the induction type for electrodynamic processing of metal are the inductor 1 and the massive electrically conductive disk 4 located coaxially with the coil.

In the center of the disk there is a rigidly fixed copper rod 5, at the end of which a tungsten electrode 6 is installed. The inductor 1 and the disk 4 are housed in the housing 2 and are covered with a lid 3 at the bottom. The electromagnetic system is placed at a right angle with respect to the welded joint 7.

To create a reliable initial electrical contact "electrode-welded joint", a load 9 is fixed on the instrument case. To bring current to the electrode device, coil terminals 8, 10 and contact with sample 11 are used. The electromechanical transducer is powered by a current source.

The source of the current pulses is based on charge storage capacitor *C* through a large resistance R_3 from the stabilized source of DC voltage E_0 and subsequent rapid discharge of the capacitor through the inductive electromechanical transducer type, which is characterized by inductance and resistance $R = R_{coil} + R_{add}$, where L_{coil}, R_{coil} – inductance and resistance of the coil 1; L_{add}, R_{add} – additional inductance and resistance, the values of which can be changed. One of the assignments of additional inductance and resistance is the task of the parameters of the pulse of the discharge current.

The coil 1 provides the creation of pressing the efforts of the electrode 6 to the surface of the welded joint 7. This clamping occurs at the time of passage of the current pulse due to the mutual repulsion of the currents in the coil 1 and eddy currents arising in the massive disk 4. As a high-speed switching element, a control thyristor *VS* is used. To record oscillograms of discharge current, an oscilloscope connected to a low inductive shunt is used. R_{shunt}. An electrode device is located in a housing of electrical insulating material 2 with a lid 3 through which the rod 5 passes.

Since during the discharge process, the electrodynamic clamping force subsequently goes into a negative value (disk 4 is attracted to coil 1), there is a threat of breaking the electrical contact of the electrode with the welded joint (Kondratenko, Bozhko, Zhiltsov, & Vasyuk, 2012; Kondratenko, Zhiltsov, & Vasyuk, 2015).

This may lead to electric arc processes with the release of thermal energy in the contact zone, is unacceptable. To eliminate this phenomenon, it is necessary to reduce negative electrodynamic force as much as possible.

The nature of this force is related to the magnitude of the derivative of the magnitude of the current in the circuit and the change in its sign. So, the slower the current in the circuit changes after passing through its maximum value, the less negative force will be. Such a process is characteristic of aperiodic discharge of a capacitor on $R - L$ circuit.

Also one of the possible solutions to this problem is to perform an electrode system based on two independent circuits (Zhiltsov, Kondratenko, & Vasyuk, 2014), one of which performs a dynamic action and presses the electrode to the surface, and the other circuit provides a current pulse through the welded joint.

It is proposed to solve this problem by means of converting equipment. The easiest way to do this is to establish an inverse diode VD parallel to capacitance *C*. At that, a certain time, until the voltage on the capacitor passes through zero, the diode is in the closed state, and then it opens, and the continuation of the transient process occurs already in the circuit, which only includes inductance and active resistance.

The development of an electrical system provides for mathematical modeling of non-stationary electrophysical processes (Kondratenko, Bozhko, Zhiltsov, & Vasyuk, 2012), which will allow to establish constructive, electrical and regime parameters, to ensure the effectiveness of applying an electrodynamic effect on welding residual stresses, namely: electrical circuit parameters C, R, L, U_C which provide, on the one hand, a given electrode clamping force $F(t)$ on prototype and its duration, on the other hand, provide current density $\delta(Q,t)$ in the region of the welded joint to reduce residual stresses.

Let the capacitance C, charged to the voltage U_C, be closed to the system of series connected additional inductance L_{add}, additional resistance R_{add} and resistance R_{coil} inductance coil L_{coil}, massive conductors occupying the volume $D = D_1 \cup D_2 \cup D_3$, which is limited by a smooth surface $S = S_{10} \cup S_{20} \cup S_{30} \cup S_{12} \cup S_{23}$ (Figure 2).

In the general case, the solution of the problem requires solving a three-dimensional boundary value problem for a system of Maxwell equations in an unbounded domain, which reduces to the boundary value problem in terms of the vector magnetic potential and the scalar electric potential. Further, us-

Figure 2. Meridian cross section of an electromagnetic system with an image of an external electric circuit

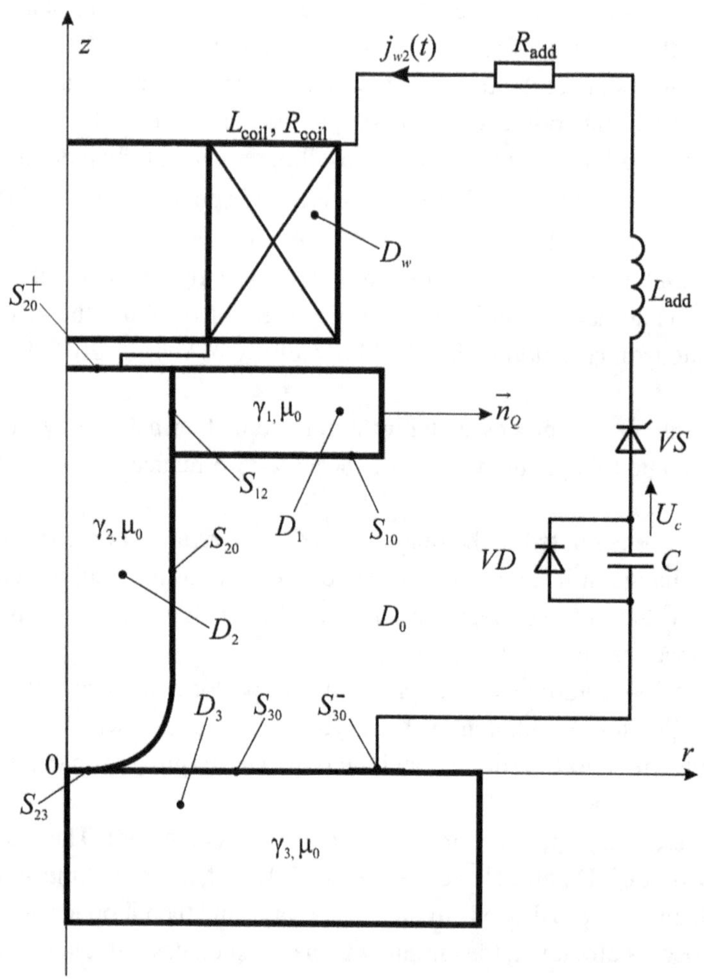

ing the theory of potentials and the method of secondary sources, the problem is reduced to the system of integro-differential equations for the density of vortex currents and the density of a simple layer of electric charges:

$$\frac{\vec{\delta}_q(Q,t)}{\gamma_q\lambda}+\frac{\partial}{\partial t}\int\limits_D\frac{\vec{\delta}(M,t)}{r_{QM}}dV_M-\frac{1}{\varepsilon_0\mu_0}\int\limits_S\sigma(M,t)\frac{\vec{r}_{QM}}{r_{QM}^3}dS_M=-\frac{\partial}{\partial t}\int\limits_{D_W}\frac{\vec{\delta}_W(M,t)}{r_{QM}}dV_M,$$

$Q\in D_q$, $q=1,2,3$;

$$\frac{\chi(Q)\mu_0\varepsilon_0}{2\pi}\frac{\partial}{\partial t}\int\limits_D\frac{\left(\vec{\delta}(M,t),\vec{n}_Q\right)}{r_{QM}}dV_M+\sigma(Q,t)+\frac{\chi(Q)}{2\pi}\int\limits_S\sigma(M,t)\frac{\left(\vec{r}_{QM},\vec{n}_Q\right)}{r_{QM}^3}dS_M=$$

$$=-\frac{\chi(Q)\mu_0\varepsilon_0}{2\pi}\frac{\partial}{\partial t}\int\limits_{D_W}\frac{\left(\vec{\delta}_W(M,t),\vec{n}_Q\right)}{r_{QM}}dV_M-F(Q,t),Q\in S\cup S_{20}^{+}\cup S_{30}^{-},$$

$$\vec{\delta}_W(M,t)=\vec{\delta}_W^{(0)}(M),\ \vec{\delta}(M,t)=\vec{\delta}^{(0)}(M),\ \sigma(M,t)=\sigma^{(0)}(M),$$

$\vec{\delta}_q(Q,t)$ – instantaneous density of eddy currents in the point $Q\in D_q$, $q=1,2,3$; γ_q – conductivity material massive body D_q, $q=1,2,3$; $\lambda_S=\mu_0/(2\pi)$; $\vec{\delta}(M,t)$ – instantaneous current density at the point $M\in D$, $D=D_1\cup D_2\cup D_3$; $\vec{r}_{QM}$ – radius vector directed from the point Q to the point M; $\mu_0=8{,}85\cdot10^{-12}$ F/m – electric constant; $\mu_0=4\pi\cdot10^{-7}$ H/m – magnetic constant; $\sigma(M,t)$ – the instantaneous value of the density of a simple layer of electric charges at the point $M\in S$; $\delta_W(M,t)$ – instantaneous current density in the winding D_w; t – time; $\delta_\alpha^{(0)}(M)$, $\delta_{W\alpha}^{(0)}(M)$ – the density of vortex currents in massive conductors and the coil at the initial moment of time; $\vec{n}_Q$ – outside normal at the point Q to the border S massive body (Fig. 2).

$$\chi(Q)=\begin{cases}\dfrac{\gamma^+(Q)-\gamma^-(Q)}{\gamma^+(Q)+\gamma^-(Q)}, & \text{if}\quad Q\in S;\\ 1, & \text{if}\quad Q\in S_{20}^{+}\cup S_{30}^{-};\end{cases}$$

$$F(Q,t)=\begin{cases}0, & \text{if} \quad Q\in S;\\ \dfrac{2\varepsilon_0}{\gamma_1}\delta_{n_Q}^{+}(t), & \text{if} \quad Q\in S_{20}^{+};\\ \dfrac{2\varepsilon_0}{\gamma_3}\delta_{n_Q}^{-}(t), & \text{if} \quad Q\in S_{30}^{-}.\end{cases}$$

$\gamma^{+}(Q)\left[\gamma^{-}(Q)\right]$ – specific conductivity of the medium at the point $Q\in S$ when approaching it to the border S massive body from the inner (external) side.

Equation system – should be supplemented by equations of the circuit according to the second law of Kirchhoff to the instantaneous values of voltage and current in the circuit:

$$R_{\text{add}}j_w(t)+L_{\text{add}}\frac{dj_w(t)}{dt}+R_{\text{coil}}j_W(t)+\frac{d\Psi(t)}{dt}+\frac{1}{C}\int_0^t j_W(t)dt+u_C(0_+)=0\,,\ 0<t<t_1,$$

$$\Psi(0_+)=\Psi(0_-)=0 \quad \text{or} \quad j_W(0_+)=j_W(0_-)=0\,,\ u_C(0_+)=u_C(0_-)\,;$$

after a moment t_1 triggered diode VD,

$$R_{\text{add}}j_w(t)+L_{\text{add}}\frac{dj_w}{dt}+R_{\text{coil}}j_W(t)+\frac{d\Psi(t)}{dt}=0\,,\ t_1<t,$$

$$\Psi(t_{1+})=\Psi(t_{1-}) \quad \text{or} \quad j_W(t_{1+})=j_W(t_{1-}).$$

$j_W(t)=\delta_W(t)\Delta S_W$ – current in the coil D_W; $\Delta S_W=\Delta r\times\Delta z$ – square spiral; $\Psi(t)$ – instantaneous grappling of the magnetic field, which creates both the current in the coils of the coil, and their eddy currents in the massive disk, with the coil current; $u_C(0_-)$ – initial value of the voltage on the capacitance C; R_{add}, R_{coil}, – additional active resistance of the circle and active resistance of the coil; L_{add} – additional inductance of the electric circuit.

Having solved the system of equations – by numerical methods, we find the current density in each massive conductor, knowing what the density of heat sources in massive conductors, the electromagnetic force acting on the electrode with the disk, etc. can be calculated. Thus, a new two-dimensional field-to-field mathematical model of the transient discharge process of a capacitance on a branched electric circle with a coil, whose inductance is dynamically changing, is developed, which enables to control the geometrical dimensions of the electromechanical system by achieving the electrodynamic characteristics required by the technological requirements.

Mathematical Modeling of Electrophysical Processes in Electrotechnical System for Electrodynamic Treatment of Welded Joints

Assuming that the coil, electrode and plate are cylindrical massive bodies with a common axis of rotation, with which the axis of the cylindrical coordinate system r, α, z (Figure 2) is subsequently combined, then the three-dimensional problem can be reduced to two two-dimensional ones with cross-linked symmetric electromagnetic fields. The resulting electromagnetic field is found by imposing electromagnetic fields defined in each statement of the problem separately.

In the first subtask for the components of the electromagnetic field and vortex currents in massive conductors there is a relation:

$$\vec{E}\left(r,z,t\right)=\vec{e}_{\alpha}E_{\alpha}\left(r,z,t\right),$$

$$\vec{\delta}\left(r,z,t\right)=\delta_{\alpha}\left(r,z,t\right)\vec{e}_{\alpha},$$

$$\vec{A}\left(r,z,t\right)=A_{\alpha}\left(r,z,t\right)\vec{e}_{\alpha},$$

$$\vec{B}\left(r,z,t\right)=B_{r}\left(r,z,t\right)\vec{e}_{r}+B_{z}\left(r,z,t\right)\vec{e}_{z}.$$

In the second subtask:

$$\vec{E}\left(r,z,t\right)=E_{r}\left(r,z,t\right)\vec{e}_{r}+E_{z}\left(r,z,t\right)\vec{e}_{z},$$

$$\vec{\delta}\left(r,z,t\right)=\delta_{r}\left(r,z,t\right)\vec{e}_{r}+\delta_{z}\left(r,z,t\right)\vec{e}_{z},$$

$$\vec{A}\left(r,z,t\right)=A_{r}\left(r,z,t\right)\vec{e}_{r}+A_{z}\left(r,z,t\right)\vec{e}_{z},$$

$$\vec{B}\left(r,z,t\right)=B_{\alpha}\left(r,z,t\right)\vec{e}_{\alpha},$$

$\vec{E}$ – electric field strength; $\vec{\delta}$ – current density, A/m^2; $\vec{A}$ – vector magnetic potential; $\vec{B}$ – magnetic induction; r, z – coordinates of a cylindrical coordinate system; $\vec{e}_z,\vec{e}_\alpha,\vec{e}_r$ – orti cylindrical coordinate system.

For each of these subtask, boundary problems are formulated in terms of vector magnetic potential and scalar electric potential, which, using the potential theory and the concept of secondary sources, are reduced to a system of integro-differential equations with their subsequent numerical solution (Kondratenko, Zhiltsov, & Vasyuk, 2015).

Calculation of the contour parameters to create a compressive force based on the effect of vortex currents in a non-ferromagnetic conductive disk (first subtask). Let's calculate the parameters of the electric circuit from the condition of providing a given force of the pressure of the electrode D_2, which is rigidly connected to the leading disk D_1, and the specified time of its action on the welded joints D_3 (Fig. 2). In the analysis of the transient process in the electric circuit, consider the presence near the disk coil D_1. The effect of vortex currents in other metal conductors is neglected because these bodies are shielded by a disk. D_1. Parameters of the electrical system (Figure 3) for electrodynamic treatment of welded joints are given in (Table 1).

In the subtask under consideration, the current density $\delta_W(t)$ In the coil is unknown, but the known voltage on the capacitance, which is discharged to the coil. In this regard, the equation after integrating at azimuthal angle

$$\frac{\delta_\alpha(Q,t)}{\gamma\lambda_S}+\frac{\partial}{\partial t}\int_S \delta_\alpha(M,t)T(Q,M)dS_M=-\frac{\partial}{\partial t}\int_{S_W}\delta_{W\alpha}(M,t)T(Q,M)dS_M,\ Q\in S,$$

$$\delta_\alpha(M,t)=\delta_\alpha^{(0)}(M),\ \delta_{W\alpha}(M,t)=\delta_{W\alpha}^{(0)}(M),$$

complemented by Kirchoff's second law for instantaneous values of current and voltage in the circuit – .

In the equation:

$$T(Q,M)=\sqrt{\frac{r_M}{r_Q}}f(k).$$

Figure 3. Contour for the creation of a compressive force taking into account the effect of vortex currents in a non-ferromagnetic conductive disk

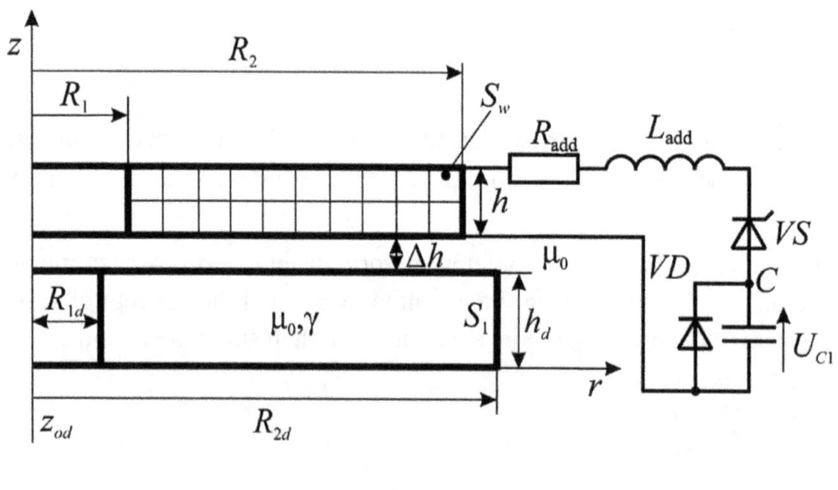

Table 1. Parameters of the electrical system

Current Pulse Generator		
Condenser capacity	C	5140 uF
Voltage on capacitance	U_C	500 V
Additional inductance	L_{add}	9 mH
Resistance to conductors	R_{add}	15 mΩ
Electromechanical Transducer Induction Type		
Inner radius of the coil	$d_{w\,in} = 2R_1$	20 mm
The outer radius of the coil	$d_{w\,ext} = 2R_2$	92 mm
Coil height	h_w	6 mm
Number of spiral in coil winding	w	18
The geometric size of the wire from which the coil winding is made	$S = 0 \times b$	1,5×5,5 mm²
Filling factor coil	K_f	0,7
Spacing between coil and disk	Δh	0,5 mm
Diameter of the disk	$d_{2d} = 2R_{2d}$	97 mm
The thickness of the disc	h_d	8 mm
Diameter of the electrode	d_e	8 mm
The height of the electrode	h_o	40 mm
Diameter of contact pad of electrode with metal	S_{cont}	2 mm²
Coil resistance	R_{coil}	5,46 mΩ

Flow-closure $\Psi(t)$ magnetic field, which creates as a current in the coils of the coil, and they give them eddy currents in a massive disk, with a current i-spiral, $i = 1,2,...w$, the coil is defined as follows:

$$\Psi_i(t) = \int_S \vec{B}(Q,t)\,d\vec{S}_Q = \int_S rot\vec{A}(Q,t)\,d\vec{S}_Q = \oint_l A(Q,t)\,dl_Q = 2\pi A_\alpha(Q_i,t)\,r_{Q_i}, i = 1,2,...w,$$

S – current spiral area; l – its outline; r_{Q_i} – radius i-spiral, $i = 1, 2, ... w$; $A_\alpha(Q_i, t)$ – instantaneous value α- components of the vector magnetic potential at the point Q_i, located in the center i-spiral, coil, $i = 1, 2, ... w$; w – number of spiral in the coil.

Here it is taken into account that the vector potential has only an angular component:

$$A_\alpha(Q,t) = \frac{\mu_0}{2\pi}\int_{S_1}\delta_\alpha(M,t)T(Q,M)dS_M + \frac{\mu_0}{2\pi}\int_{S_W}\delta_{W\alpha}(M,t)T(Q,M)dS_M .$$

Then, flow-through with the current i - th spiral coil equals

$$\Psi_i(t) = 2\pi A_\alpha(Q_i, t)r_{Q_i} .$$

Flow-closure with all reels the reel is like:

$$\Psi(t) = \sum_{i=1}^{w}\Psi_i(t) = 2\pi\sum_{i=1}^{w}A_\alpha(Q_i,t)r_{Q_i} =$$

$$= \mu_0\sum_{i=1}^{w}r_{Q_i}\int_{S_1}\delta_\alpha(M,t)T(Q_i,M)dS_M + \mu_0\sum_{i=1}^{w}r_{Q_i}\int_{S_W}\delta_{W\alpha}(M,t)T(Q_i,M)dS_M .$$

Compatible solution of the system of equations, and – considering allows you to determine the current in the coil inductance, the current density in a massive disk D_1 gives the ability to determine the current in the coil and the strength of its interaction with vortex currents arising under its action in a massive inductive disk, the current density in a massive disk D_1:

$$\vec{f}(Q,t) = \vec{e}_r\delta_\alpha(Q,t)B_z(Q,t) - \vec{e}_z\delta_\alpha(Q,t)B_r(Q,t).$$

The adequacy and reliability of the mathematical modeling of the non-stationary process in the electromagnetic system was confirmed by experimental studies performed on a prototype developed device. In fig. 4, as an example, graphs of current dependence on time for circle parameters are given: U_C=500 V, C=5140 uF, L_{add}=9 mH. Total active resistance of the circle R consists of resistance of conductors R_{add}=15 mΩ and the coil resistance $R_{coil} = 5,46$ mΩ.

In (Figure 4) shows a graph of the dependence of the electrodynamic force transmitted through the contact pad on the test sample. As can be seen from the graphs shown, the experimental and calculated results for determining the value of the current in the transition process of the bit circle are well-coordinated (the mean square error is <3%), and the electrodynamic pressures at the maximum value of 22 kN at the instant that coincides with the maximum value of the current, falls to almost zero.

Figure 4. Results calculation and experimental determination of the current strength in the discharge circuit (A) and calculation of the electromagnetic force of interaction of the current in the winding of the coil with eddy currents in a massive disk (B)

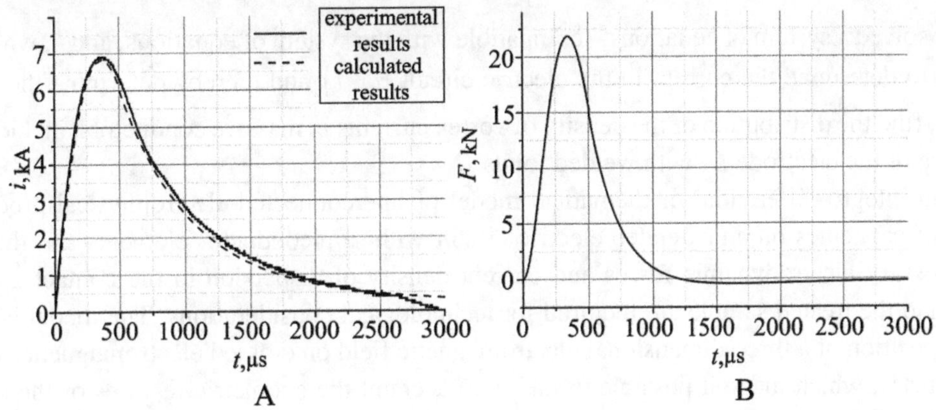

Calculation of the contour parameters to provide the required current density in the contact area of the electrode with a welded connection (second subtask). The system of integro-differential equations (1) – (3) after integrating at azimuthal angle in a cylindrical coordinate system takes the form of:

$$\frac{\delta_r(Q,t)}{\gamma(Q)\lambda_L}+\frac{\partial}{\partial t}\int_S \delta_r(M,t)\sqrt{\frac{r_M}{r_Q}}f(k)dS_M-\frac{1}{\varepsilon_0\mu_0}\int_L \sigma(M,t)\xi_r(Q,M)dL_M=0,$$

$$Q\in S, S=S_1\cup S_2\cup S_3,$$

$$\frac{\delta_z(Q,t)}{\gamma(Q)\lambda_L}+\frac{\partial}{\partial t}\int_S \delta_z(M,t)\sqrt{\frac{r_M}{r_Q}}K(k)kdS_M-\frac{1}{\varepsilon_0\mu_0}\int_L \sigma(M,t)\xi_z(Q,M)dL_M=0,$$

$$Q\in S, S=S_1\cup S_2\cup S_3,$$

$$\frac{\chi(Q)\mu_0\varepsilon_0}{\pi}\frac{\partial}{\partial t}\int_S\left(\vec{\delta}(M,t)\cdot\vec{n}_Q\right)\sqrt{\frac{r_M}{r_Q}}K(k)dS_M+\sigma(Q,t)+\frac{\chi(Q)}{\pi}\int_L\sigma(M,t)\left(\vec{\xi}(Q,M)\cdot\vec{n}_Q\right)dL_M=$$

$$=-\frac{\chi(Q)\mu_0\varepsilon_0}{\pi}\frac{\partial}{\partial t}\int_{S_W}\left(\vec{\delta}_W(M,t)\cdot\vec{n}_Q\right)\sqrt{\frac{r_M}{r_Q}}K(k)dS_M-F(Q,t), Q\in L\cup L_{10}^{+}\cup L_{30}^{-},$$

$$\vec{\delta}_W(M,t)=\vec{\delta}_W^{(0)}(M),\ \vec{\delta}(M,t)=\vec{\delta}^{(0)}(M),\ \sigma(M,t)=\sigma^{(0)}(M),$$

$$\delta^{+}_{n_Q}\left(t\right)=\frac{j_W\left(t\right)}{S^{+}_{10}},\delta^{-}_{n_Q}\left(t\right)=\frac{j_W\left(t\right)}{S^{-}_{30}}.$$

Having solved a system of equations – compatible with the system of equations, and –, which makes it possible to determine the current in the electric circuit $j_W\left(t\right)$ and thereby determine the boundary conditions, find the distribution of the density of vortex currents in massive conductors, including in the contact area of the electrode D_2 with welded joints D_3.

Thus, an integro-differential mathematical model of interconnected electrophysical processes – a discharge of capacities on an extended electric circuit with semiconductor elements and the creation of the necessary electrodynamic forces and current density of dispersion in the contact zone of the electrode and the treated sample are required for technological considerations. The model is based on the decomposition of a three-dimensional electromagnetic field on crossed electromagnetic fields with axial symmetry, which made it possible to take into account the geometric features of the system, to significantly reduce the search areas of the solution of unknown, and as a result, led to the rational use of computer resources.

SOLUTIONS AND RECOMMENDATION

The magnitude of the discharge current and its duration are determined by the electric parameters of the discharge circle: the active resistance, inductance, capacity and voltage on it. The electrodynamic force of pressing the electrode to the metal surface is determined by the magnitude of the discharge current, the value of which is influenced by the ratio of the structural dimensions of the elements of the bit circle - a coil of inductance and a massive disk.

The geometrical parameters of the coil of inductance include: internal diameter, outer diameter, height, number of turns, coefficient of filling of the winding. The geometric parameters of the disk include: internal diameter, outer diameter, thickness. Depending on these parameters, the nature of the transient process in the electric circuit changes, which, in turn, affects the character of the electrodynamic force of pressing the electrode to the plate and the distribution of currents of dispersion. Restrictions on the size of the coil and disk are determined by the overall dimensions of the device.

The task of further research was to: a) determine the geometric parameters of the disk, the distance between the coil and the disk, which provide the above-mentioned strength of the interaction of the current in the coil and the resulting vortex currents in a massive disk; b) determination of the optimal area of the contact of the electrode with the weld seam provided the maximum volume is provided in the welded joint, in which spreading currents with densities greater than 10^9 A/m^2 are created, and processes of reduction of residual stresses occur.

In (Figure 5) the family of curves of equal values of the amplitude of electrodynamic force (in kN), with which a massive electric drive drives off from the coil with current, depending on the thickness of the disk and its diameter, is given. The calculation is performed for such unchanged parameters of the electrotechnical system for electrodynamic processing as shown in (Table 1).

Figure 5. The family of dependences of the amplitude value of the electromagnetic force (kN) acting on the massive non-ferromagnetic disk, on its thickness and diameter: h_d – the thickness of the massive disk; d_{2d} – its outer diameter; A – for voltage at a capacity of 500 V; B – for voltage at a capacity of 750 V

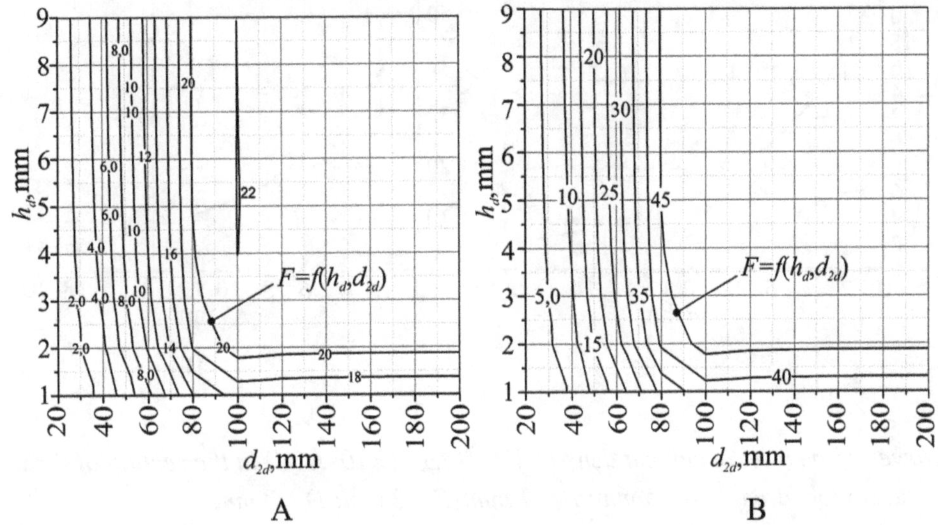

Analysis of the dependencies shown in (Figure 6), indicates that the magnitude of the electrodynamic force remains unchanged after reaching the thickness of the disk more than 4 mm. It is also inappropriate to increase the diameter of the disk to more than 100 mm, that is, it can be argued that the choice of the diameter of the disk, which is larger than the size of the coil, does not lead to an increase in electrodynamic force. It should also be noted that the growth of the magnitude of the amplitude of the electrodynamic force from the voltage value on the capacitance is faster than the quadratic law.

The amplitude of the electrodynamic force largely depends on the size of the gap separating the conductive disk and the coil. In (Figure 6) shows graphs of the amplitude of the amplitude of the z-component of the electromagnetic force acting on a massive disk, depending on the distance between the coil and the disk for two, as above, voltage values on the capacitance of 500 V (Figure 6, A) and 700 V (Figure 6, B). In any case, there is a decrease in the amplitude of the electrodynamic force while simultaneously reducing the amplitude of the pulsed current in the bit circle, which is due to the growth of the inductance of the coil.

Dependence of the current density of dispersion from the contact area of the electrode with the welded connection. One of the most important constructive characteristics of an electromechanical converter of an induction type is the geometric and electrophysical parameters of the electrode.

Considering that the efficiency of electrodynamic processing is determined, first of all, not by the size of the discharge current, but by its density in the contact area of the electrode with the main material, there is a need for the study of these processes.

In (Figure 7), as an example, in the form of curves of equal magnitude ($\times 10^9$ A/m^2), the distribution of the density of the current module in the pilot sample in the contact area for the time point is given, which corresponds to the maximum value of current in the bit circle for the above parameters for a number of values of the area of the contact area . Curves 8, A–D correspond to the contact in the form of a circle,

Figure 6. Chart of the dependence of the amplitude value -component of the electromagnetic force acting on the non-ferromagnetic massive disk, from the gap between the disk and the coil, at the voltage on the container: A – 500 V; B – 750 V

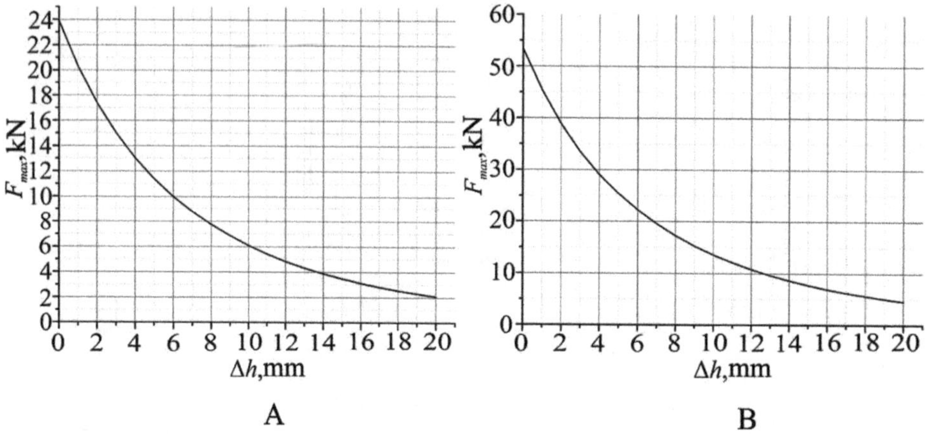

Figure 7. Curves of the level of current density V= $f(z,r)$ (×10⁹ A/м²) in the section of the weld for the diameter of the contact area: A – 0,5 mm; B – 1 mm; C – 2 mm; D – 3 mm

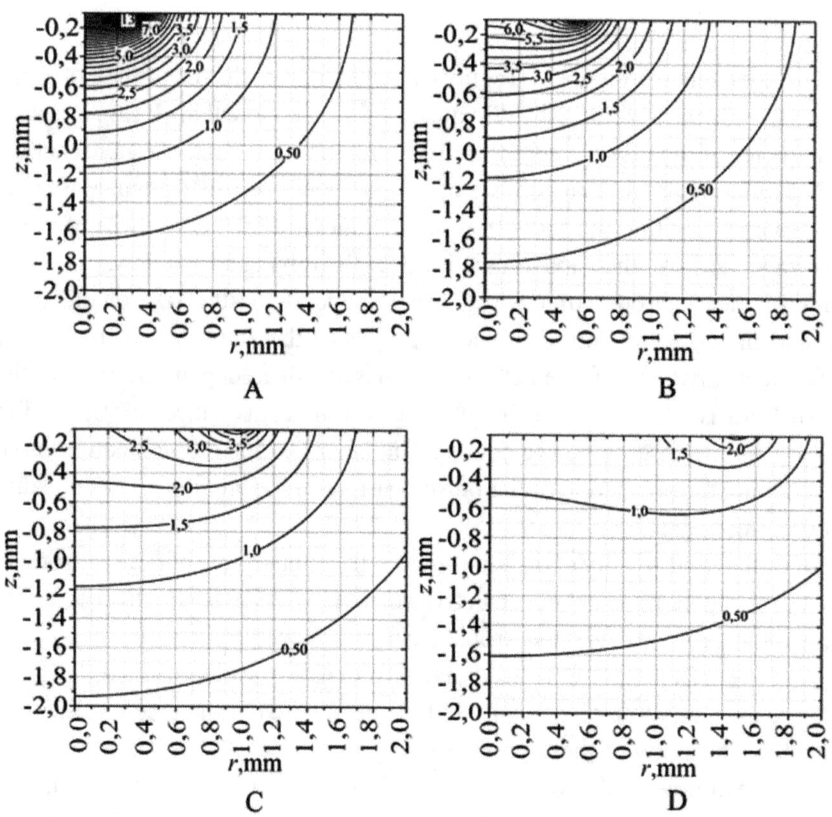

with diameters consecutively 0,5 mm, 1, 2, and 3 mm. The given dependencies allow to determine the volume of areas with a current density exceeding 10^9 A/m^2, which reduces the residual stresses in the weld.

In (Figure 8) shows graphs of the dependence of the volume of the region of the main material, where the current module density exceeds the value of 10^9 A/m^2, from the diameter of the contact area of the electrode with a weld on different voltages on the capacitance (250 V, 500, 750 V). It turns out that the diameter of the contact d_{cont} =2,25 mm, the volume of the area in which the conditions are created for the transformation of the mechanical state of structural materials, has a maximum value. It should also be noted that for a diameter of the contact area more than 4, 4.5 mm, the volume of the region in which the current density module exceeds the value of 10^9 A/m^2, is approaching zero. Based on the mathematical modeling of electrophysical processes in an electromechanical induction type converter for electrodynamic treatment of welded joints, the geometric parameters of a massive disk and contact area, which create the necessary conditions for reducing the residual stresses in a welded joint, are substantiated.

The appearance of the electromechanical transducer Impact induction type shown in (Figure 9).

To ensure the efficiency and reproducibility of electrodynamic processing of welded joints of metal nonferromagnetic structures, it is necessary:

1. Provide a reliable (100%) electrode-metal contact.
2. The electrode at the contact point should have a radius of 2 ... 2,5 mm (Lobanov, Kondratenko, Zhiltsov, Karlov, Paschin, Vasyuk, & Yaschuk, 2016).
3. Optional condition is the contact of the metal electrode with the seam area, the processing may be carried out at a certain distance, provided that the seam enters the area where the current module density will exceed the value A/m^2 which determines the location of the next discharge.

Figure 8. Charts of the dependence of the volume of the weld area where the current module density exceeds the value of 10^9 A/m^2, from the diameter d_{cont} of the contact area of the electrode with the weld seam

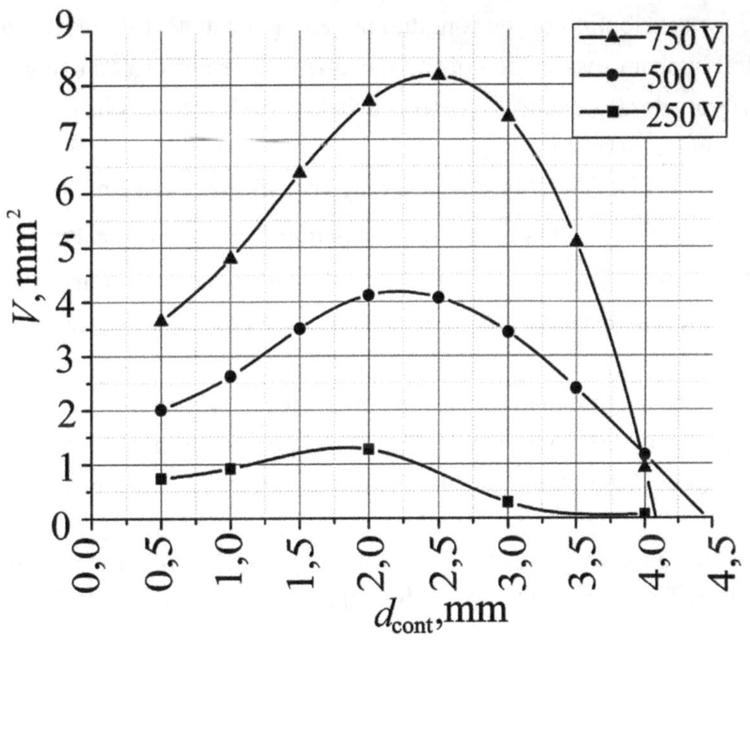

Figure 9. Appearance of an electromechanical converter of an induction type: 1 – a case made of insulating material; 2 – outputs of the coil and electrode; 3 – a reel; 4 – copper layer of a disk; 5 – stainless steel disc;6 – copper rod; 7 – lower cover; 8 – tungsten electrode

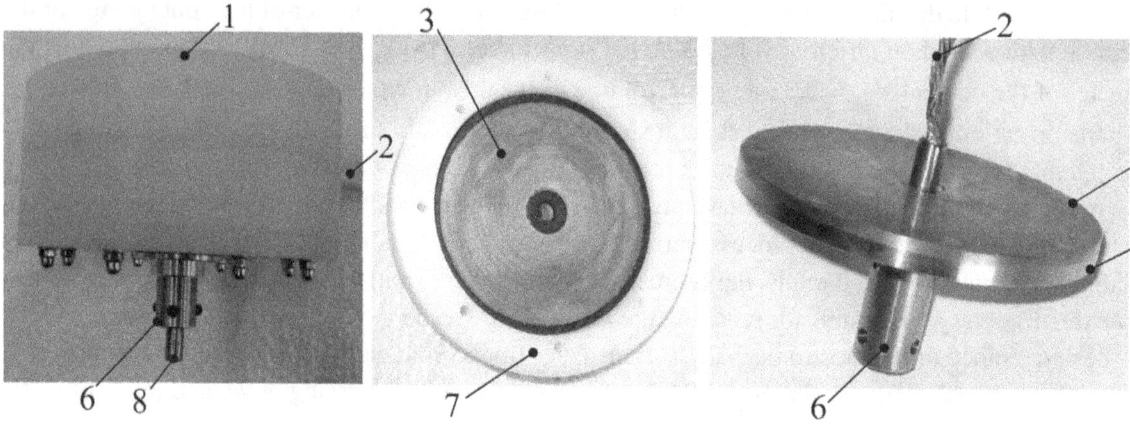

Taking into account the results of the calculation and experiments, the effect of electrodynamic processing in the mode was investigated $U_C = 500$ V, which corresponds to the electrodynamic force F_{max} =22 кН, to reduce the initial level of elastic strain tension σ_x =140 MPa in flat samples of aluminum alloy AMg6.

We tested typical samples of AMg alloy plates in the size of 400x150x2 mm with a central weld. Welded connections were performed by automatic welding of TIG (Ar) in a mode that provides voltage, current and process speed, respectively $U_{зв}$=20 V, $I_{зв}$=170 A and $U_{зв}$=5,5 mm / s. In order to study the effect of the direct passage of the pulsed current through the sample, the treatment of the weld was carried out using the equipment (Lobanov, Paschin, Cherkashin A, Mihoduy, & Kondaratenko, 2012) (Figure 1).

Investigation of the distribution of the longitudinal component σ_x (along the suture line) residual stresses in the central lumbar cross section of the specimens were performed using the method of electronic speckle-interferometry (Lobanov, Pivtorak, Savitskiy, & Tkachuk, 2008). Distribution σ_x before, and also after processing, is shown in (Figure 10).

Yes, the initial level of effort stretching σ_x reaching 165 MPa (curve 1), which is close to the threshold of fluidity $\sigma_{0,2}$ AMG6 alloy. After processing, which implements only the dynamic pressure (current through the electrode does not pass), the level σ_x significantly decreased (curve 2), and in the contact area of the electrode, the tensile stress was transformed into a compression, the level of which reaches $\sigma_x = -30$ MPa.

When processing metal according to the scheme of (Figure 1), which provides the passage of current through the metal sample, the maximum level of compression σ_x amounted to 80 MPa (curve 3), and their area of distribution reached 20 mm, guaranteed to overlap the zone of probable destruction of the welded joint.

Thus, the effect of the direct passage of the current through the metal, in accordance with the mechanism given in (Baranov, Troitskiy, Avraamov, & Shlyapin, 2001; Lobanov, Paschin, & Mihoduy, 2012), stimulates the deformation processes initiated by the electrodynamic pressure of the electrode with the force F_{max}, obtained using the given calculation model not only in the zone of guaranteed electroplastic effect (current density exceeds 10^9 A/m^2), but also beyond its limits.

Listed in (Figure 10) The results of experimental studies confirm the effect of electrodynamic processing on a significant reduction in the residual welding stresses with their associated transformation into compression stress (Lobanov, Kondratenko, Zhiltsov, Karlov, Paschin, Vasyuk, & Yaschuk, 2016).

As is known (Smirnov, 2008), the presence of compression stress fields in the zone of probable destruction of welded joints (seam and periolabial zone) increases their resistance to tiredness over cyclic loads.

In order to confirm the increase of the fatigue strength of the weld plates treated by the scheme (Figure 1), cyclic examinations of samples of welded joints whose geometrical characteristics are indicated in (Figure 11), where the direction and the weld seam treatment area are indicated by a dashed arrow.

Distribution of residual stresses σ_x in the untreated samples was close to that given on curve 1 (Figure 10), and in the treated ones – on curve 3.

The tests were carried out for a period of time before the appearance of a fatigue crack with the current recording of the number of cycles N corresponding to the beginning of the destruction of the samples.

The test machine UPM-02 (Figure 12), which implements the curve of fatigue exams on the bend (Figure 13), was used in a symmetrical cycle of loading at a frequency of 14 Hz.

Here 1 – a moving clutch, 2 – a fixing plate, 3 – a sample, 4 – a counter of cycles, 5 – a fixed clutch. The amplitude of the cycle σ_a were set in the range 80-160 MPa, which corresponds to the range of operational fatigue loads of welded joints AMg6 alloy.

Fatigue curves characterizing cyclic strength at examinations on the bend of the rough (curve 1) and treated samples (curve 2) are shown in (Figure 13).

Figure 10. Distribution of residual stresses σ_x, mPa: curve 1 – output voltage level; curve 2 – tension level after dynamic processing; curve 3 – voltage level after electrodynamic processing

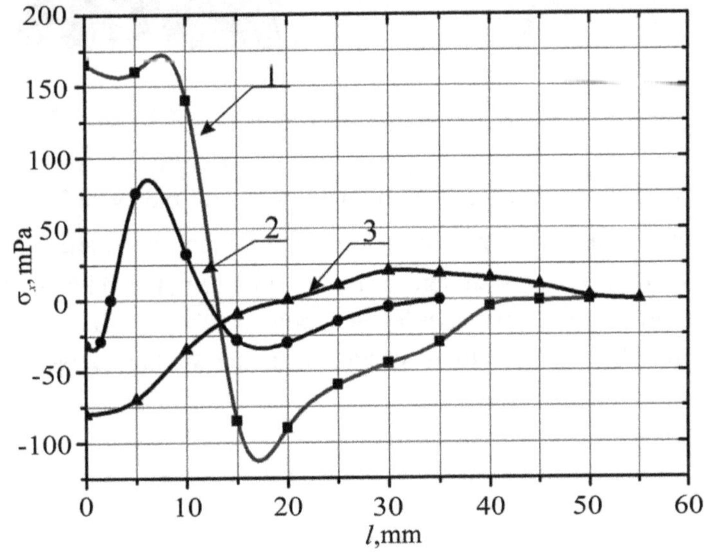

Figure 11. Geometric characteristics of samples of welded joints

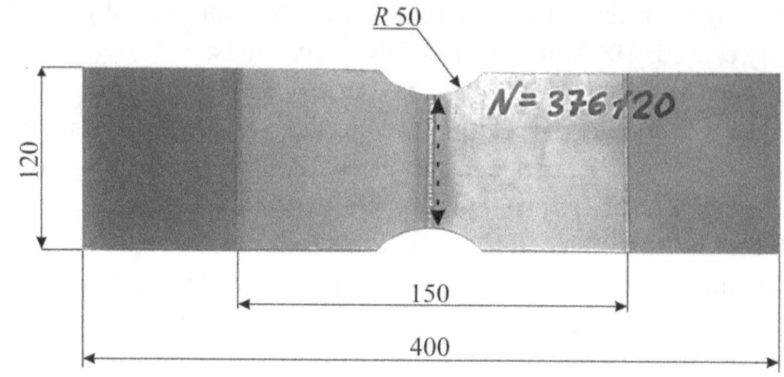

Figure 12. Test machine UPM-02

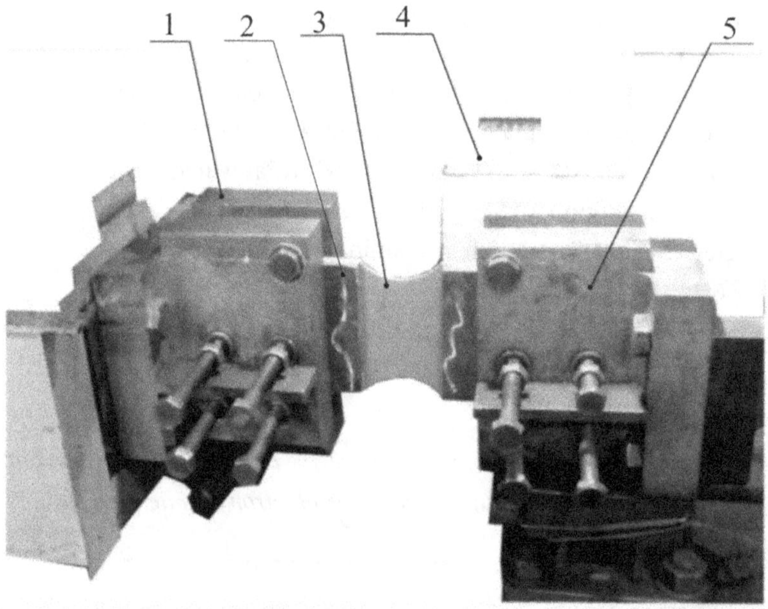

The analysis of the data shows that the maximum amplitude of the cycle $2\sigma_a$=160 MPa durability *N* of processed samples is increased more than twice. For a minimum value $2\sigma_a$=80 МПа, where the effect of residual stresses on the fatigue strength is the most present, the value *N* rise up to three times.

Thus, the guiding of the compression stresses for electrodynamic processing in a mode calculated on the basis of the developed model, contributes to increasing the fatigue strength of welded joints AMg6 alloy.

Figure 13. Fatigue curves characterizing cyclic strength at examinations on bend of rough (curve 1) and processed samples (curve 2).

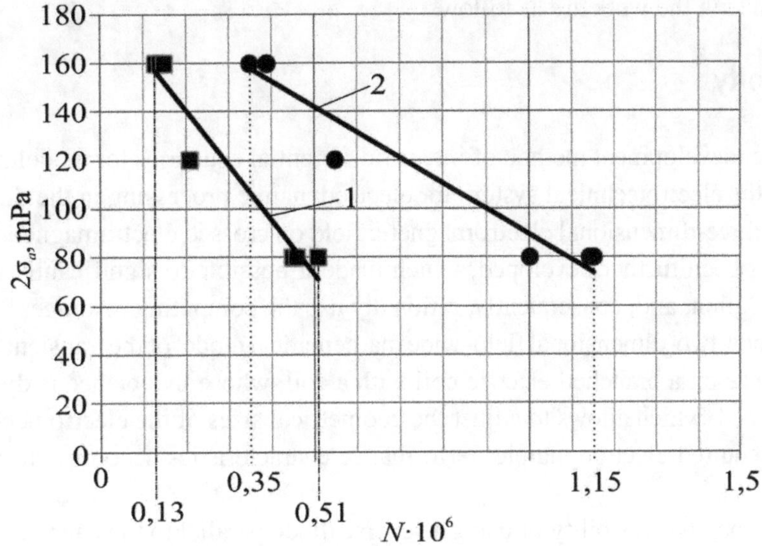

FUTURE RESEARCH DIRECTIONS

The results of the work have been found to be practical in the method of developing an electrotechnical system for electrodynamic treatment of welded joints with adjustable parameters of a bit circle, which became a prerequisite for the creation of an automatic welding system.

The developed methods of calculating electrical parameters of a bit circle can be used to improve the energy parameters of induction type electromechanical converters, which are used in accelerators of massive actuators and shock-force devices.

According to the results of theoretical and experimental research, an algorithm for calculating the characteristics of an electromechanical converter of an induction type was developed, on the basis of which a software product was created in the programming language *Fortran*, which enables, by adjusting the parameters of the electromechanical system, to achieve the required electrodynamic effects according to the technological requirements.

The developed electrotechnical system for electrodynamic effect can be used in the technology of non-destructive determination of residual stresses in nodes and structural elements by the method of electronic speckle-interferometry.

CONCLUSION

The research is devoted to solving the actual scientific problem of developing methods for the calculation of interconnected electrophysical processes in electrodynamic processing systems of welded joints of metallic neferromagnetic constructions and providing necessary technological considerations of parameters - forces and current densities, which is essential for the development of the theory and practice of pulsed electromagnetic systems in the form of power supplies and electromechanical converters. The main results obtained in the work are as follows.

Scientific Novelty

1. Gained further development method of integro-differential equations for modeling electrophysical processes in the electrotechnical system for electrodynamic processing in the direction of decomposition of a three-dimensional electromagnetic field on crossed electromagnetic fields with axial symmetry has been further developed, which made it possible to significantly reduce the search area of the solution, and, consequently, rationally use the computing resources.
2. Developed a new two-dimensional field-wide mathematical model of the transient discharge process of a capacitance on a branched electric coil with a coil, whose inductance is dynamically changing, is developed, which allows to adjust the geometrical sizes of the electromechanical system to achieve the required electrodynamic performance characteristics according to the technological requirements.
3. For the first time, the feasibility of using a reverse diode parallel to the capacitance in the current pulse generator is justified, which allows to reduce the negative electrodynamic force of pressing and, consequently, to reduce the risk of breakage of the electrical contact of the electrode with the welded joint.
4. For the first time, the magnitude and shape of the current in the transient process of discharge of a capacitive drive shunt by a reverse diode into a coil of inductance with allowance for vortex currents in a non-ferromagnetic massive disk located near it is determined, which makes it possible to increase the adequacy of determining the value of the dynamic characteristics of the electromechanical converter of the induction type of shock action.
5. Based on the developed mathematical model of the non-stationary electrophysical process, for the first time, the size of the contact area of the electrodes of an electromechanical converter with a welded joint is justified, which allows to provide the maximum volume of metal, where the conditions for electroplastic deformation are created.

Practical Significance of the Results

1. An electrotechnical system consisting of a source of pulses of currents and an electromechanical converter of an induction type of shock action is developed. The source of current pulses is based on a capacitive energy storage device, which consists of two pulsed capacitors (2x2570 mV, 900 V), charged from an adjustable voltage source (200-900 V). The discharge capacity of the capacitive energy storage device is carried out on a bit circle consisting of a stationary inductance of 11 μH, located in the case of the source of current pulses and inductance, which is installed in the electrode system over a disk of diameter 97 mm which consists of a copper layer of 3 mm, which is fixed to the disk with stainless steel thickness 5 mm. In the bit circle, the conductor supports and the contact resistance between the electrode and the metal structure being processed are included. The diameter of the contact area is assumed to be 2 mm, and the electric resistance of the circle, which includes the electrical resistance of the open thyristor, is 20.46 mΩ. Switching is carried

out by the thyristor of the corresponding class of voltage and current. The thyristor is shunted by a reverse diode, which, on the one hand, serves to prevent the recharge of the capacitive energy storage device by the voltage of the reverse sign, and on the other hand, the transformation of the discharge current type, which is characteristic for the aperiodic discharge of the capacitor.

2. The integro-differential field-field mathematical model of interconnected non-stationary electrophysical processes is developed, which consists of three interconnected blocks: a) the calculation of the transition process in the electric circuit, which includes dynamically changing inductance; b) calculation of components of the electric circle of the "coil-disk" system; c) calculation of the density of current currents at the point of contact of the electrode to the weld joint. The first calculation unit consists of two parts, the first of which is described by inhomogeneous differential equations with zero initial conditions, and the second by homogeneous differential equations with initial conditions corresponding to the state at which the voltage on the capacitive energy storage device reaches zero. The mathematical model makes it possible to calculate the value of the current coupling of the coil currents, the inductance of which is dynamically changing, taking into account the currents of the discharge circle and vortex currents, which are induced in the conductive disk. The feature of the second block is the assumption that there are only azimuthal components of the current, varying in radius and height. And the third block – that the currents of emission have only r and z-component of currents. Such representation allowed to reduce the problem from three-dimensional to two two-dimensional, which allowed to take into account the geometric features of the system, significantly reduce the search area of unknown, and as a result, led to the rational use of computer resources.
3. On the basis of mathematical modeling of electrophysical processes in the electrotechnical system, the current strength (amplitude up to 7 kA) in the transient process of discharge of a capacitive drive shunt by a reverse diode, electrodynamic force of pressure (amplitude greater than 22 kN) of an electrode is determined on the electrodynamic treatment of welded joints welded joints, and the distribution of current density in the section of the weld seam. The boundaries of the region (a zone approximated to half of the ellipsoid of rotation with a large axis of 3, 4 mm and a small half-width of 1, 2 mm) are determined in a welded joint, which correspond to the conditions of the occurrence of the electroplastic effect. This made it possible to substantiate the geometrical parameters of the induction type electromechanical converter, which creates the necessary conditions for reducing the residual stresses in the welds.
4. Reliability and validity of scientific results are ensured by the correct and consistent use of the method of integro-differential equations for the analysis of non-stationary electromagnetic processes in complex electrical engineering systems, the coincidence of the obtained results in the limiting cases with known results, the coordination of mathematical modeling with experimental data (mean square error is <3%).
5. The results of the work have been found to be of practical use in the development of electrotechnical equipment for electrodynamic treatment of welded joints with adjustable parameters of the bit circle, which is a prerequisite for the creation of an automatic welding system.

Economics

1. Analysis of literary sources shows that existing methods of reducing the residual stresses in welded metal compounds (rolling, roasting, heat treatment, vibration processing, ultrasonic treatment, impact and explosive loading, etc.) have significant drawbacks: the need to create energy and metal-based technological equipment; limitations in the processing of large-sized constructions and significant energy costs. Promising way to increase the resource of welded joints is the method of electrodynamic treatment based on the application of the effect of electroplastic deformation, which is characterized by the simplicity of implementation, reducing the time of the process several times, the possibility of local action, low energy consumption and leads to a decrease in residual stresses at 50-65%).

REFERENCES

Afonin, A. A., & Grebenikov, V. V. (2004). Issledovaniya v oblasti razvitiya elektromehanicheskih preobrazovateley energii lineynogo i rotatsionnogo dvizheniya [Research in the development of electromechanical energy converters of linear and rotational motion]. *Proceedings of the Institute of Electrodynamics of NAS of Ukraine*, 2(11), 95–98.

Astahov, V. I. (2004). Matematicheskoe i kompyuternoe modelirovanie elektromagnitnogo polya kak osnova dlya resheniya zadach v elektrotehnike i elektroenergetike [Mathematical and computer simulation of the electromagnetic field as a basis for solving problems in electrical engineering and power engineering]. Izvestiya vuzov. Elektromehanika, 6, 4–6.

Baranov, Y. V., Troitskiy, O. A., Avraamov, Y. S., & Shlyapin, A. D. (2001). Fizicheskie osnovyi elektroimpulsnoy i elektroplasticheskoy obrabotok i novyie materialyi [Physical fundamentals of electropulse and electroplastic treatments and new materials]. Moskow: MGIU.

Bolyuh, V. F., Oleksenko, S. V., & Schukin, I. S. (2015). Sravnitelnyiy analiz lineynyih impulsnyih elektromehanicheskih preobrazovateley elektromagnitnogo i induktsionnogo tipa [Comparative analysis of linear pulse electromechanical converters of electromagnetic and induction type]. Technical Electrodynamics, 4, 20–27.

Bose, B. K. (2002). *Modern power electronics and AC drivers*. Prentice Hall PTR.

Bunget, C., Salandro, W., Mears, L., & Roth, J. T. (2010). *Energy-based modeling of anelectrically-assisted forging process. In Trans. of the North American Manuf* (pp. 38–46). Research Institute of SME.

Burkin, S. P., Shimov, G. V., & Andryukova, E. A. (2015). Ostatochnyie napryazheniya v metalloproduktsii [Residual stresses in metal products]. Ekaterinburg: Ural. un-t.

Conrad, H. (2000). Electroplasticity in metals and ceramics. *Materials Science and Engineering*, *287*(2), 276–287. doi:10.1016/S0921-5093(00)00786-3

Evdokimov, V. F., Kuchaev, A. A., Petrushenko, E. I., & Kasyan, G. I. (2010). Dvumernaya integralnaya model raspredeleniya sinusoidalnyih vihrevyih tokov i elektrodinamicheskih usiliy v kristallizatore s yavnopolyusnyim elektromagnitnyim peremeshivatele [Two-dimensional integral model of the distribution of sinusoidal eddy currents and electrodynamic forces in the mold with an explicit-pole electromagnetic stirrer]. Elektronnoe modelirovanie, 1, 53-75.

Flora, V. D. (2012). *Spetsialnyie elektromagnitnyie i elektromehanicheskie preobrazovateli energii* [Special electromagnetic and electromechanical energy converters]. Zaporizhia.

Gromov, V. E. (2014). Prochnost i plastichnost materialov v usloviyah vneshnih energeticheskih vozdeystviy [Strength and ductility of materials under external energy conditions]. Novokuznetsk: Sibirskiy gosudarstvennyiy industrialnyiy universitet.

Guoyi, T. (2000). Experimental study of electroplastic effect on stainless steel wire 304L. *Materials Science and Engineering*, (281), 263–267.

Ismagilov, F. R., Afanasev, Yu. V., & Styiskin, A. V. (2006). *Vvedenie v konstruirovanie elektromehanicheskih preobrazovateley energii* [Introduction to the design of electromechanical energy transducer]. Moskow: MAI.

Jeon, W. J., Katoh, S., Iwamoto, T., Kamiya, Y., & Onuki, T. (1999). Propulsive characteristics of a novel linear hybrid motor with both induction and synchronous operations. *IEEE Transactions on Magnetics*, *35*(5), 4025–4027. doi:10.1109/20.800743

Kondratenko, I. P., Bozhko, I. V., Zhiltsov, A. V., & Vasyuk, V. V. (2012). Doslidzhennya elektrofizichnih protsesiv u elektrodniy sistemi neruynivnogo viznachennya zalishkovih napruzhen [Investigation of electrophysical processes in the electrode system of non-destructive determination of residual stresses]. Institut elektrozvaryuvannya Im. E. O. Patona NAN UkraYini.

Kondratenko, I. P., Bozhko, I. V., Zhiltsov, A. V., & Vasyuk, V. V. (2012). Matematichne modelyuvannya elektroflzichnih protseslv v sistemah operativnogo neruynlvnogo viznachennya zalishkovih napruzhen [Mathematical modeling of electrophysical processes in systems of operational non-destructive determination of residual stresses]. Technical electrodynamics, 3, 1–22.

Kondratenko, I. P., Zhiltsov, A. V., & Vasyuk, V. V. (2014). Modelirovanie elektrofizicheskih protsessov v elektrotehnicheskom komplekse dlya snizheniya ostatochnyih napryazheniy [Modeling electrophysical processes in electrical complex to reduce residual stresses]. Moscow: Energosberegayuschie tehnologii v zhivotnovodstve i statsionarnoy energetike.

Kondratenko, I. P., Zhiltsov, A. V., & Vasyuk, V. V. (2015). Simulation of discharge capacity axle symmetric systems «coil – non-ferromagnetic massive disk» by the method of integral equations. Computational problems of electrical engineering (CPEE – 2015), 71–73.

Lobanov, L. M., Kondratenko, I. P., Zhiltsov, A. V., Karlov, O. M., Paschin, M. O., Vasyuk, V. V., & Yaschuk, V. Y. (2016). NestatsIonarnI elektroflzichnI protsesi v sistemah znizhennya zalishkovih napruzhen zvarnih z'Ednan [Non-stationary electrophysical processes in systems of reduction of residual stresses of welded joints]. Technical electrodynamics, 6, 10–19.

Lobanov, L. M., Paschin, N. A., Cherkashin, A. V., Mihoduy, O. L., & Kondaratenko, I. P. (2012). Effektivnost elektrodinamicheskoy obrabotki alyuminievogo splava Amg6 i ego svarnyih soedineniy [Efficiency of electrodynamic processing of Amg6 aluminum alloy and its welded joints]. *Avtomaticeskaya Svarka (Kiev)*, (1), 3–7.

Lobanov, L. M., Paschin, N. A., & Mihoduy, O. L. (2012). Vliyanie usloviy nagruzheniya na soprotivlenie deformirovaniyu splava AMg6 pri elektrodinamicheskoy obrabotke [The effect of loading conditions on the resistance to deformation of the AMg6 alloy during electrodynamic processing]. Problemyi prochnosti, 5, 15–26.

Lobanov L. M., Pivtorak V. A., Savitskiy V. V., & Tkachuk G. I., (2008). Diagnostika svarnyih konstruktsiy metodami elektronnoy shirografii i spekl-interferometrii [Diagnostics of welded structures using electronic shear and speckle interferometry]. *Avtomaticheskaya svarka, 11*(667), 195–203.

Maltsev, I. M. (2008). Elektroplasticheskaya prokatka metalla s tokom vyisokoy plotnosti [Electroplastic metal rolling with high density current]. Izvestiya vyisshih uchebnyih zavedeniy, tsvetnaya metallurgiya, 3, 34–38.

Neyman, L. A., & Neyman, V. Y. (2015). Novyie konstruktivnyie resheniya problemyi tochnoy sinhronizatsii vozvratno-postupatelьnogo dvizheniya boyka neupravlyaemoy elektromagnitnoy mashinyi udarnogo deystviya [New constructive solutions of the problem of accurate synchronization of return-transfer movement of the wagon of the controlled electromagnetic machine of shock action]. Aktualьnyie problemyi v mashinostroenii, 2, 280-285.

Oleksenko, S. V. (2016). *Otsenka pokazateley lineynyih elektromehanicheskih preobrazovateley udarnogo deystviya s vyisokoy magnitnoy sovmestimostyu* [Evaluation of linear electromechanical percussion transducers with high magnetic compatibility]. Kharkov.

Petrushenko-Kubala, I. E., & Petrushenko, E. I. (2003). Metod elementarnyih solenoidov modelirovaniya trehmernyih magnitnyih poley v ustroystvah s ferromagnitnyimi serdechnikami i obmotkami s tokom [The method of elementary solenoids modeling three-dimensional magnetic fields in devices with ferromagnetic cores and windings with current]. Elektronnoe modelirovanie, 5, 15–31.

Salandro, W., Bunget, C., & Mears, L. (2010). Modeling and quantification of the electroplastic effect when bending stainless steel sheet metal. *International manufacturing science and engineering conference (MSEC).* 10.1115/MSEC2010-34043

Shreyner R. T., Polyakov V. N., & Medvedev A. V. (2016). Matematicheskoe modelirovanie yavnopolyusnyih sinhronnyih dvigateley s avtomaticheskim podborom parametrov lokalnyih harakteristik namagnichivaniya [Mathematical modeling of polar pole synchronous engines with automatic selection of local parameters magnetization characteristics]. *Elektrichestvo, 2*, 57–65.

Sidorov, O. Y., & Sarapulov, F. N. (2010). Osobennosti issledovaniya lineynogo asinhronnogo dvigatelya metodom konechnyih elementov [Peculiarities of the linear asynchronous motor study by the method of finite elements]. Izvestiya vyisshih uchebnyih zavedeniy. Elektromehanika, 1, 17–20.

Smirnov, A. N. (2008). Metallovedenie svarki i defektyi metalla [Welding metal science and metal defects]. Kemerovo: KuzGTU.

Song, H., Wang, Z., & Gao, T. (2007). Effect of high-density electropulsing treatment on formability of TC4 titanium alloy sheet. *Transactions of Nonferrous Metals Society of China, 17*(1), 87–92. doi:10.1016/S1003-6326(07)60053-3

Troitskiy, O. A., Baranov, Y. V., Avraamov, Y. S., & Shlyapin, A. D. (2004). Fizicheskie osnovyi i tehnologii obrabotki sovremennyih materialov (teoriya, tehnologii, struktura i svoystva) [Physical bases and technologies for processing modern materials (theory, technology, structure and properties)]. Izhevsk: RHD, ANO IKI.

Troitskiy, O. A., Kayzer, A. A., Layshev, R. R., & Gorelik, V. S. (2004). Ispolzovanie elektroimpulsnyih tehnologiy v praktike yuvelirnogo dela [The use of electric pulse technology in the practice of jewelry]. Collection of scientific papers and engineering developments 5-y Rossiyskoy vyistavki «Izdeliya i tehnologii dvoynogo naznacheniya, 2, 364–367.

Troitskiy, O. A., Stashenko, V. I., & Ryizhkov, V. G. (2011). Elektroplasticheskoe volochenie i novyie tehnologii sozdaniya oblegchennyih provodov [Electroplastic drawing and new technologies for creating lightweight wires]. *Voprosyi atomnoy nauki i tehniki (VANT), 4*, 111–117.

Troitskiy, O. A., Stashenko, V. I., Ryizhkov, V. G., Lyashenko, V. P., & Kobyilskaya, E. B. (2011). Elektroplasticheskoe volochenie i novyie tehnologii sozdaniya oblegchennyih provodov [Electroplastic wire drawing and new technologydevelopment light wire]. Voprosyi atomnoy nauki i tehniki, 4, 111–117.

Troitsky, O.A. (2018). Elektroplasticheskiy effekt v metallah [Electroplastic effect in metals. Ferrous Metallurgy]. *Bulletin of Scientific, Technical and Economic Information, 2018*(9), 65-76.

Tyukov, V. A. (2006). *Elektromehanicheskie sistemyi* [Electromechanical systems]. Novosibirsk: NGTU.

Usanov, K. M., Moshkin, V. I., Kargin, V. A., & Volgin, A. V. (2015). *Lineynyie elektromagnitnyie dvigateli i privodyi v impulsnyih protsessah i tehnologiyah* [Linear electromagnetic motors and drives in pulse processes and technologies]. Kurgansk.

Yamori, A., Ono, Y., Kubo, H., Kono, M., & Kawashima, N. (2001). Development of an induction type railgun. *IEEE Transactions on Magnetics, 37*(1), 470–472. doi:10.1109/20.911879

Yampolskiy, Y. G. (1990). O proektirovanii optimalnyih lineynyih impulsnyih elektrodinamicheskih dvigateley vozvratno-postupatelnogo dvizheniya [About designing optimal linear pulse electrodynamic engines of reciprocating motion]. Elektrotehnika, 2, 51–55.

Zhiltsov, A. V., Kondratenko, I. P., & Sorokin, D. S. (2014). The mathematical model of the electromagnetic and mechanical processes is developed in a coaxial-linear engine with massive magnetic conductors. ECONTECHMOD: An International Quarterly Journal on Economics of Technology and Modelling Processes, 1, 69-73.

Zhiltsov, A. V., Kondratenko, I. P., & Vasyuk, V. V. (2014). Rozrahunok parametrIv kontura dlya stvorennya pritiskayuchogo zusillya v elektrotehnIchnomu kompleksI dlya znizhennya zalishkovih napruzhen [Calculation of contour parameters for creation of a compressive force in an electrical complex for reduction of residual stresses]. Energetika I avtomatika, 4, 54–64.

KEY TERMS AND DEFINITIONS

Electric Current Density: A charge value flowing through the unit area per unit time.

Electrical Conductivity: Aт ability of a body to conduct an electrical current, a property of the body or medium that determines the occurrence of an electric current in them under the influence of an electric field.

Electroplastic Effect: Reduction of metal resistance to deformation and increase of its plasticity under the influence of high-density electric current (about 10^5 A/cm^2) or under the influence of intensive electron irradiation.

Filling Factor Coil: A ratio of the cross-sectional area of the winding to the entire area of the coil limb.

Flux Coupling: A physical quantity that represents the total magnetic flux interlocking with all turns of an inductor.

Linear Electromechanical Transducer: A class of devices for the direct conversion of electrical energy into mechanical energy source linear motion rolling elements.

Magnetic Induction: A vector physical value, the main characteristic feature of the magnitude and direction of the magnetic field.

Stresses: A measure of the intensity of internal forces, distributed across the intersections, that is, the force per unit area of intersection of the body.

Chapter 17
Artificial Bee Colony-Based Optimization of Hybrid Wind and Solar Renewable Energy System

Diriba Kajela Geleta
Madda Walabu University, Ethiopia

Mukhdeep Singh Manshahia
Punjabi University Patiala, India

In this chapter, the artificial bee colony (ABC) algorithm was applied to optimize hybrids of wind and solar renewable energy system. The main objective of this research is to minimize the total annual cost of the system by determining appropriate numbers of wind turbine, solar panel, and batteries, so that the desired load can be economically and reliably satisfied based on the given constraints. ABC is a recently proposed meta heuristic algorithm which is inspired by the intelligent behavior of honey bees such as searching for food source and collection and processing of nectar. Instead of gradient and Hessian matrix information, ABC uses stochastic rules to escape local optima and find the global optimal solutions. The proposed methodology was applied to this hybrid system by the help of MATLAB code and the results were discussed. Additionally, it is shown that ABC can be efficiently solve the optimum sizing real-world problems with high convergence rate and reliability. The result was compared with the results of PSO.

INTRODUCTION

Unless there is a difference in consumption from place to place, energy becomes one of the vital elements for human life which comes mainly from fossil fuel. Even though, the conventional energy source has been used as the main energy source for past decades all over the world, due to its number of limitations like the issue of global warming, depletion of its sources and continuous increase in oil prices have forced the world attention for the development and utilization of alternative renewable energy sources (Rubio, Perea, Vazquez, & Os-Moreno., 2012).

DOI: 10.4018/978-1-5225-9420-8.ch017

These Conventional energy sources which include power plants using fossil fuels (natural gas, coal, etc.) are not ever lasting and have a lot of disadvantages like the issue of environmental degradation and continuous fuel supply for operation (Geleta & Manshahia, 2018; Kosmadakis, Sotirios, & and Emmanuel, 2013).

In addition to the above-mentioned drawback of conventional energy source, due to rapid increment of industrialization all over the world, the need for energy was exponentially increases from time to time and also depletion of fossil fuels has been occurred which leads to an initiation for the need of alternative inexhaustible sources of energy to satisfy the continuously increasing energy demand. Another important reason to reduce our consumption of fossil fuels is the growing global warming phenomena (Zong, 2012; Kosmadakis, Sotirios, & and Emmanuel, 2013).

To minimize some of the problems associated with these conventional energy sources, other sources which are environmentally clean, naturally endless, in- exhaustible and renewable energy are getting much attention and growing up nowadays (Geleta & Manshahia, 2017; Luna, Trejo, Vargas, & Os-Moreno, 2012).

Renewable energy sources are environmentally clean, abundant and friendly used power generation sources which has to play an important role for the future power supply by diversifying and maintaining the energy supply market (Kaabeche, Belhamel, & Ibtiouen, 2010). Nowadays, the electrification of rural villages in most of the places have been electricity from extension of main grids and installation of diesel generators as an option. In reality, grid expansion to all the places of such areas is impossible to satisfy the power demand of the society. This is because of either financial constraint or practically not feasible for a lot of reasons like geographical location, low densely populated and a very low power demand. Thus, to increase the power utility coverage, applying standalone renewable energy sources will made such society more beneficiary (Geleta & Manshahia, 2018; Zong, 2012).

In the past decade, energies from wind turbine and solar panel generation technologies have in- creased their use in either wind alone, solar alone or hybrid of wind and solar configurations are com- mon. Moreover, the economic aspects of these renewable energy technologies are gradually increasing at present including the development of their market. Wind and Solar have abundant power which can be exploited as electric energy by the help of wind turbines and solar panels. Based on the daily average data of wind speed, isolation, temperature and power demand, the system capacity is determined to best match the power demand by minimizing the difference between generation and load demand over a 24-hour period (Kaabeche, Belhamel, & Ibtiouen, 2010; Al-Shamma & Khaled, 2012). The capacity of the storage needed to make the system operate independently as a stand-alone system is determined from the hourly information obtained from power demand. The main disadvantage of these technologies is fluctuation of their power output depending on weather condition. To overcome these limitations which can affect the power output, using hybrids of renewable energy technology was important (Geleta & Manshahia, 2018; Luna, Trejo, Vargas, & Os-Moreno, 2012).

The most commonly known renewable energy sources are wind, solar, hydro, biomass, ocean wave, geothermal and tides which are naturally re- placeable. The result of many researches shows that, using hybrid system reduce operation cost through reducing fuel consumption, increasing system efficiency and reduce noise and emission. The only limitation of renewable energy technologies is the fluctuation of their power output which can be managed by applying hybrids renewable energy technology (Geleta & Manshahia, 2017).

Here we use hybrid of wind and solar which applicable at any remote society of the world and relatively cheaper to implement.

This paper is attempts to find the optimal size of Hybrids of wind and solar renewable energy system. The main concern is to determine the size of each components participating in the system means numbers of wind turbine, solar system and batteries, so that the desired load can be satisfied with minimum possible annual cost. Different scholars have been used various methods to optimize the hybrid renewable energy systems.

Some of these methods are the trade-off method by (Gavanidou, Bakirtzis, & Dokopoulos, 1993; Elhadidy & Shaahid, 1999), the least square method used by (Kellog, Nehrir, Venkataramanan, & Gerez, 1998; Borowy & Salameh, 1996). They have presented a methodology for optimization of a PV (Photovoltaics)/wind system based on deficiency of power supply probability (DPSP), relative excess power generated (REPG), unutilized energy probability (UEP). The Iteration method by (Geleta & Manshahia, 2018) was discussed the optimal value of the hybrid system by taking the load balance in to consideration and fix the optimal components through accomplishment of the desired load. The technical approach also called loss of power supply probability (LPSP) by (Diaf, Belhamel, Haddadi, & Louche, 2007) have optimized a hybrid system size based on loss of power supply probability (LPSP) and the levelized cost of energy (LCE). (Belfkira, Zhang, & Barakat, 2011) have used a methodology for sizing optimization of a standalone hybrid wind/PV/diesel energy system. This methodology uses a classical algorithm to determine the optimal unit components by minimizing the total cost of the system and guaranteeing the availability of the energy. (Singh & Kaushik, 2016; Alireza, 2013) have presented artificial bee colony algorithm for Optimal sizing of grid integrated hybrid PV-biomass energy system by the least levelized cost of energy while minimizing annualized cost of the system to find optimum hybrid system configuration. It has been observed from the results that a grid connected hybrid PV-biomass system is cost effective, reliable choice and the proposed algorithm provides better results as compared with other deterministic method and HOMER (Koutroulis, Dionissia, Potirakis, & Kostas, 2006; Ashok, 2007; Wang & Singh, 2009).

In this paper the researchers apply Artificial Bee Colony (ABC) algorithm for solving the eco- nomic aspect meaning minimizing the total annual cost of the defined problem which can satisfy the desired power with high light summery of loss of power supply probability (LPSP), the energy not supplied by the system(ENS), reliability of the system and left further technical aspects of the re- search and comparison of different nature inspired algorithms to solve these hybrids solar and wind renewable energy system for other researchers for further deal. ABC is one of recently developed meta heuristic algorithm which inspire from for- aging behavior of honey bees. Even though it has a number of parameters to be adjusted, it is relatively easy to apply and highly converges to global optimal value.

The remainder of this paper is organized as follows: Section 2 Brief reviews of modelling of the individual components of the hybrid system. Section 3 describes the formulation of fitness problem and possible constraints which has to be satisfied during the optimization process in details. Section 4 reviews Artificial Bee Colony (ABC) optimization with its algorithm. Section 5 presents numerical data in tabular and figurative which are used to apply the algorithm. Section 6 provides the simulation results from MATLAB code and discussion. Finally, the conclusions summarize in Section 7

MODELLING THE HYBRID SYSTEM

One of the important procedures in Optimizing Hybrids of Renewable Energy System (HRES) is modelling its components which can make it suit- able for employment to get the maximum output of the desired load. Here the model of individual components is reviewed. The main components used for configuration in this paper consists of wind turbine, solar panel, battery bank, bi-directional converter and with the main load is supplied primarily from wind turbine and solar panel through the bi-directional converter. Various modeling techniques are developed to model Hybrids of Wind and Solar Renewable Energy System (HWSRES) components in previous studies.

In this study three principal subsystems are included, the solar panel generator, the wind turbine and the battery storage as shown in figure 1 (Eltamaly & Mohamed, 2014).

In the modelling process it is assumed that the power generated from these renewable sources is constant during one-hour duration. The mathematical modelling of different components of the proposed hybrid wind and solar energy with battery system in order to analyze the system performance is discussed in this section.

Modeling Wind Generator

Wind power is one of the most important component and promising source of hybrid renewable energy system. The electric power output from wind turbine at a particular location depends on the type of land surface, wind speed at hub height and the speed characteristics of the turbine.

Since the height of the wind turbine has a valuable effect on output energy from the system, the adjustment of the wind profile for height at the time of installation can be taken into account by using a height adjustment equation. Wind speed at certain height can be calculated by using power law, as shown in equation below (Ilinca, McCarthy, Chaumel, & Retiveau, 2003).

Figure 1. Convergence of ABC algorithm

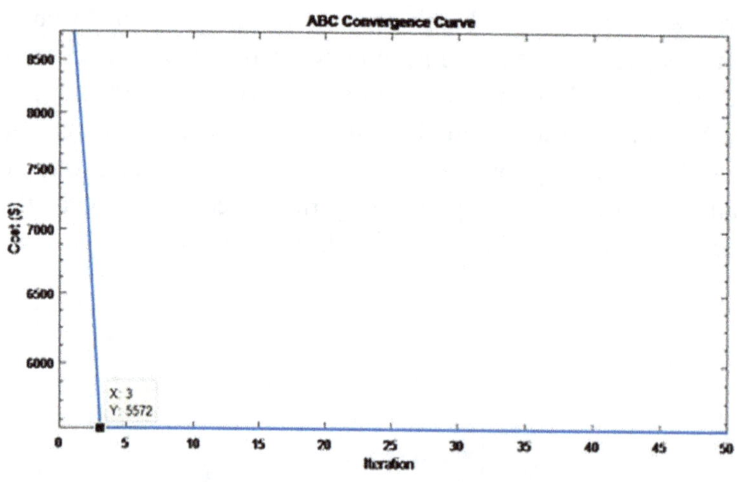

$$V = V_i \left(\frac{h}{h_i} \right)^{\alpha} \tag{1}$$

Where V is the wind speed at hub height h, V_i is the wind speed measured at the reference height h_i and α is the power law exponent who varies with the elevation, the time of day, the season, the wind speed and the temperature. the typical value of $^{1/7}$ corresponding to low roughness surfaces and well exposed sites, is used in this study.

The desired power output information from the component is very important to model the performance of wind generator. Different scholars use different types of models to maximize the desired energy output. Some of them are (Borowy & Salameh, 1996; Borowy & Salameh, 1996) are model the component by taking Weibull parameters in to account (Chedid, Akiki, & Rahman, 1998; Eftichios, Dionissia, Antonis, & Kostas, 2006; Lysen, 1983) assume that the turbine power curve has a linear, quadratic or cubic form. Others like; (Bueno & Carta, 2005) approximate the power curve with a piecewise linear function with a few nodes (Diaf et al.,2007; Borowy & Salameh, 1996).

The produced power of each wind turbine is predicted by using the following equation in terms of the wind speed (Javadi, Mazlumi, & Jalilvand, 2011).

$$P_{WT} = \begin{cases} 0 & V_W \leq V_C, V_W \geq V_F \\ P_R \times \left(\frac{V_W - V_C}{V_R - V_C} \right)^3, & V_C \leq V_W \leq V_R \\ P_R, & V_R \leq V_W \leq V_F \end{cases} \tag{2}$$

Where P_{WT} the Wind turbine output power (Watt), P_R is the Wind turbine rated power (Watt), V_W the wind speed (m/s). V_C, V_F, V_R are Cut-in, Cut-out and rated or nominal speed of the wind turbine (m/s).

Modeling PV Generator

As the activity of modelling for operation and the performance of a PV generator is interested to its maximum power, the PV module's maximum power output behaviors are more practical for PV system assessment. The output power from PV energy subsystem is depend upon the solar radiation on tilted surface, orientation of the PV array against the movement of the sun, the manufacturers data of the PV modules and the ambient temperature at the given time as follows (Etamaly, Mohamed, & Alo-lah, 2015).

$$P_{PV} = R_t A_t \mu_c(t) \tag{3}$$

Where R_t is the radiation on the tilted surface, A_t is the total cell area and $\mu_c(t)$ is the solar cell efficiency at time t.

The instantaneous generating efficiency of the PV modules can be obtained in terms of the cell temperature as shown in the following equation

$$\mu_c(t) = \mu_{cr}\left[1 - \beta_t\left(T_c(t) - T_{cr}\right)\right] \tag{4}$$

Where β_t is the temperature coefficient, ranging from 0.004 to 0.006 per o_C for silicon cells. μ_{cr} and T_{cr} is the theoretical solar cell efficiency and temperature. $T_c(t)$ is the instantaneous solar cell temperature at an ambient temperature T_a.

The instantaneous solar cell temperature $T_c(t)$ can be obtained by the following equation

$$T_c(t) = T_a + 3 * R_t \tag{5}$$

Modeling of Battery Storage

Since the output of the PV cells and the wind turbines are stochastic in nature, proper sizing of battery to maintain the load demand is important. At any time, t the state of charge of battery is de pends up on the previous state of charge and to the energy production and load demand situation of the system during the time t-1. The state of charge of the battery is used as a decision variable for the control of the over charge and discharge. The case of over charge of the battery is occur when excess amount of power is generated by the hybrid system or demand of the load is low at a time. When the state of charge of the battery reaches its maximum value $C_{(batt,\, max)}$, the control system intervenes and stop the charging process where as when it reaches its minimum level $C_{(batt,\, min)}$ the control system disconnects the load to prevent batteries against shortening their life or being distracted (Diaf et al.,2007; Borowy & Salameh, 1996).

When the total output power of the turbine and PV cells is greater than that of the desired load power, the battery is in the state of charging, and the charged quantity of the battery at the moment of t is expressed by (Diaf et al.,2007; Borowy & Salameh, 1996).

$$C_{batt}(t) = C_{batt}(t-1)(1-\sigma) + \left(E_{PV}(t) + E_{WT}(t) - \frac{E_L(t)}{\eta_{inv}}\right) \tag{6}$$

On the other hand, when the load demand is greater than the available energy generated, the battery bank is in discharging state. Therefore, the available battery bank capacity at hour t can be expressed as

$$C_{batt}(t) = C_{batt}(t-1)(1-\sigma) + \left(\frac{E_L(t)}{\eta_{inv}} - E_{PV}(t) + E_{WT}(t)\right) \tag{7}$$

where C_{batt} (t) and C_{batt} $(t$ -1) are the available battery bank capacity in watt hour (Wh) at hour t and t-1, respectively; η_{batt} is the battery efficiency (during discharging process, the battery discharging efficiency was set equal to 1 and during charging, the efficiency is 0.65 - 0.85 depending on the charging current) (Bin et al., 2003). σ is the self-discharge rate of the battery bank. $E_{PV}(t)$ and $E_{WT}(t)$ are the energy generated by PV and wind generators, respectively; $E_L(t)$ is the load demand at hour t and η_{inv} is the inverter efficiency which may considered as constant, 92At any hour, the storage capacity is subject to the following constraints:

$$C_{batt,\min} \leq C_{batt} \leq C_{batt,\max} \tag{8}$$

where $C_{batt,\ min}$ and $C_{batt,\ max}$ is the maximum and minimum allowable storage capacity, respectively. To safe battery or prolong its life, the battery should not be over discharged or overcharged at any time t.

Modeling of System Reliability

In addition to economic aspects of hybrid system, various technical analysis approaches are used in literature to optimizing hybrid renewable energy systems. Among these employed techniques, trade-off method by (Elhadidy & Shaahid, 1999), the least square method used by (Kellog et al., 1998; Borowy & Salameh, 1996) loss of power supply probability (LPSP) by (Lu, Yang, & Burnett, 2002; Kellog et al., 1998). Here, the technical sizing model for the hybrid of wind and solar renewable energy system is developed according to the concept of LPSP to evaluate the reliability of the system. The methodology used can be summarized in the following steps (Diaf et al., 2007).

The total power, $P_{tot}(t)$, generated by the wind turbine and PV generator at hour t is calculated as follows.

$$P_{tot}(t) = P_{PV}(t) + P_{WT}(t) \tag{9}$$

Then, the inverter total input power, $P_{inv}(t)$, is calculated using the corresponding load power requirement as follows:

$$P_{inv}(t) = \frac{P_{load}(t)}{\eta_{inv}} \tag{10}$$

where $P_{load}(t)$ is the power consumed by the load at hour t, η_{inv} is the inverter efficiency. It is to be known that, the desired load demand at any time t $P_{load}(t)$, may or may not be satisfied according to the corresponding values of the total generated power $P_{tot}(t)$ and $SOC(t)$ at that hour. The pro- posed energy management of the PV/Wind hybrid system can be summarized as follows:

1. If [$P_{tot}(t) \geq P_{inv}(t)$] and [$SOC(t-1) \leq SOC_{max}$], then satisfy the load, charge the battery with surplus power $P_{Batt} = (P_{tot}(t) - P_{inv}(t)) * \eta_{Batt}$. Next, check if $SOC(t) > SOC_{max}$. Then, stop the battery charging.
2. If [$P_{tot}(t) > P_{inv}(t)$] and [$SOC(t-1) > SOC_{max}$], then satisfy the load, Stop charging the battery and Dump the surplus power $P_{dump} = (P_{tot}(t) - P_{inv}(t))$
3. If [$P_{tot}(t) = P_{inv}(t)$], then satisfy the load only. No extra process of charging /discharging or dumping power is here.

4. [$P_{tot}(t) < P_{inv}(t)$] and $DOP\ (t-1) < DOP_{max}$, then satisfy the load by dis- charging the battery to overcome the load deficit and check if $DOP\ (t) > DOP_{max}$, then stop battery discharge and set $DOP\ (t) = DOP_{max}\ (t)$.Here may be less amount of subordinate power generating is needed.
5.if [$P_{tot}(t) < P_{inv}(t)$] and $DOP\ (t-1) > DOP_{max}$, then stop the battery discharge and Set deficit $P_{Def} = P_{tot}(t) < P_{inv}(t)$. Here more power generating support will be needed.

The wasted energy (WE), defined as the energy produced and not used by the system, for hour t is calculated as follows (Diaf et al., 2007).

$$WE(t) = P_{tot}(t)\sigma t - \left(\frac{P_{Load}(t)}{\eta_{inv}}\Delta t + \left(\frac{C_{batt,\max} - C_{batt}(t-1)}{\eta_{cha}}\right)\right) \tag{11}$$

The loss of power supply probability, LPSP, for a considered period T, can be defined as the ratio of all the (LPS(t)) values over the total load required during that period. The LPSP technique is considered as technical implemented criteria for sizing a hybrid PV/wind system employing a battery bank. Here, some reliability indices are applied to evaluate capacity of the hybrid system sizing developed to supply the load demand. These are the LPSP technique, which is defined as the probability that an insufficient energy results when the hybrid system is unable to supply the load, meaning the system is reliable when it is able to supply enough power to the electrical load during a certain period, the energy not supplied (ENS) which is considered when generated power is less than the demanded power and the reliability of power supply (RPS). Each of them can be obtained by using the following formulas (Diaf et al, 2007; Bin et al., 2003).

$$LPSP = \frac{\sum_{t=1}^{T} LPS(t)}{\sum_{t=1}^{T} P_{Load}(t)\Delta t} \tag{12}$$

where T is the operation time (in this study, T= 1 year).

The Loss of power supply of the system is the energy deficit to satisfy the desired load and can be calculated as:

$$LPS(t) = P_{Dem}(t) - P_{Gen}(t) \tag{13}$$

The energy not supplied (ENS) by the hybrid system is also given by:

$$ENS = \sum \left(P_{Dem} - P_{Gen}\right) \tag{14}$$

Reliability of power supply (RPS) can be found in terms of LPSP as follows

$$RPS = 1 - LPSP \tag{15}$$

OPTIMIZATION FORMULATION

In order to optimal sizing the hybrid system considered here, it is important to design the proper fitness function which includes all the parameters and possible conditions which affect the desired result directly or indirectly. The main Objective of the sizing Optimization problem is to minimize the total annual cost (f_{TAC}) of the system. For this problem the total annual cost is taken as the sum of initial capital cost (C_{ICC}) and annual maintenance cost. Thus, the problem to be minimized will be stated as follows: (C_{Mnt}) (Geleta & Manshahia, 2018; Zong, 2012; Yang, Zhou, & Lou, 2008).

$$\min f_{TAC} = C_{ICC} + C_{Mnt} \tag{16}$$

Maintenance cost C_{Mnt} of the system occurs during the project life time while capital cost C_{ICC} occurs at the beginning of the project. In order to compare these costs, the initial capital cost has to converted annual capital cost by the capital recovery factor (CRF) can be defined as (Hadidian, Arabi, & Bigdeli, 2016).

$$CRF = \frac{i(1+i)^n}{(1+i)^n - 1} \tag{17}$$

Where i the interest rate and n denote the life span of the system.

Now the initial capital cost of the system can be broken in the annual costs of the wind turbine, solar panel, batteries and backup generator will be given as follows;

$$C_{ICC} = \frac{i(1+i)^n}{(1+i)^n - 1}\left[N_{PV}C_{PV} + N_{WT}C_{WT} + \left(\frac{n}{LS_{Batt}}\right)N_{Batt}C_{Batt} + C_{Backup}\right] \tag{18}$$

Where LS_{Batt} is batteries life span, N_{PV}, N_{WT} and N_{Batt} are numbers of PV panels, wind tur- bine and batteries respectively, C_{PV}, C_{WT}, N_{Batt} and C_{Backup} are unit costs of PV panels, wind tur- bine, batteries and backup generator respectively The unit cost of solar panel C_{PV} is consists of unit cost of PV panel and its installation fee and that of wind turbine C_{WT} is also consists of unit cost of wind turbine and its installation fee as shown on equation (4) next.

$$\begin{aligned} C_{PV} &= C_{PV,unit} + C_{inst,unit} \\ C_{WT} &= C_{WT,unit} + C_{inst,unit} \end{aligned} \tag{19}$$

The number of batteries N_{Batt} which, depends on the number of photovoltaic panel and number of wind turbines are decision variables and determined by the following function

$$N_{Batt}\left(N_{PV}, N_{WT}\right) = Roundup\left(\frac{S_{Req}}{\eta S_{Batt}}\right) \tag{20}$$

where Roundup (.) is a function which returns a number rounded up to an integer number; S_{Req} is required storage capacity; η is usage \% of rated capacity which guarantees battery's life span; and S_{Batt} is rated capacity of each battery.

Similar to the number of batteries, the required storage capacity S_{Req} which is defined as the number of solar panels and wind turbines in the hybrid system can be obtained by using energy curve ΔW defined as:

$$\Delta W = W_{Gen} - W_{Dem} = \int \Delta P dt = \int \left(P_{Gen} - P_{Dem}\right) dt \tag{21}$$

Where W_{Gen} and P_{Gen} are the total energy and power generated respectively and W_{Dem} and P_{Dem} are their respective demand values. Thus, the required storage capacity S_{Req} defined as the number of solar panels and wind turbines is given by:

$$S_{Req}\left(N_{PV}, N_{WT}\right) = \sum_{t=1}^{Maxt}\left(P_{PV}^{t} + P_{WT}^{t} - P_{Dem}^{t}\right)dt - \sum_{t=1}^{Mint}\left(P_{PV}^{t} + P_{WT}^{t} - P_{Dem}^{t}\right)dt \tag{22}$$

Where Max t is the time when total energy (kwh) is highest; Min t is the time when total energy (kwh) is lowest. Δt is unit time under consideration (1hr) here. P_{PV}^{t} and P_{WT}^{t} are the powers generated by solar panel and wind turbine at time t respectively; and P_{Dem}^{t} the total power demand at time t. The total power generated by the components at time t is given by:

$$\begin{aligned} P_{PV}^{t} &= N_{PV} \times P_{PV,Each\,unit}^{t} \\ P_{WT}^{t} &= N_{WT} \times P_{WT,Each\,unit}^{t} \end{aligned} \tag{23}$$

Where P_{PV}^{t} and P_{WT}^{t} are the total powers generated by the wind turbines and solar panels, whereas $P_{PV,Each\,unit}^{t}$ and $P_{WT,Each\,unit}^{t}$ are the power generated from each respective components at a time t.

The annual maintenance cost of the system was calculated by the following equation.

$$C_{Maint} = \left(C_{PV,maint} \times \sum_{t=1}^{24} P_{PV}^{t} \Delta t + C_{WT,maint} \times \sum_{t=1}^{24} P_{WT}^{t} \Delta t\right) \times 365 \tag{24}$$

Constraints

The Fitness function defined by equations (17-25) for the Minimizing Hybrid Wind and Solar Renewable energy system by unit sizing the system components in order to meet the desired load which can minimize the annual cost. Here the following conditions has to considered through the whole process of optimization.

Decision Variables Constraint

$$N_{PV} \in \mathbb{Z}, N_{PV} \geq 0 \, and \, N_{PV} \leq N_{PV,\max}$$
$$N_{WT} \in \mathbb{Z}, N_{WT} \geq 0 \, and \, N_{WT} \leq N_{WT,\max} \quad (25)$$

Power Generated Constraint

The total transferred power from PV and WT to the battery bank is calculated using the following Equation

$$P^{k}_{Total}(t) = N_{PV} P^{k}_{PV}(t) + N_{WG} P^{k}_{WG}(t) \quad (26)$$
$$1 \leq k \leq 365, 1 \leq t \leq 24$$

The power generated from each source $P_{Gen}(i)$ must be less than or equal to the maximum capacity of the source as

$$P_{Gen}(i) \leq P_{Gen.\max}(i) \quad (27)$$

Where i is Number of sources

Battery Constraint

When the activity of operation and performance of HRES is made, proper attention must be given to the life span of battery to satisfy the energy demand and supply of the whole system. The battery should not be over discharged or overcharged. This means that the battery SOC at any hour t must be subject to the following constraint:

$$\left(1 - POD_{\max}\right) \leq SOC \leq SOC_{\max} \quad (28)$$

Where $POD_{\max}$ and $SOC_{\max}$ are the battery maximum permissible depth of discharge and SOC, respectively.

System Reliability constraint

In order to consider the reliability of the system, the loss of power supply probability (LPSP) should be less than the maximum settled LPSP by the user.

$$LPSP \leq LPSP_{\max} \tag{29}$$

ARTIFICIAL BEE COLONY

The artificial bee colony (ABC) algorithm is one of the recently developed population based, stochastic meta-heuristic search technique defined by Karaboga in 2005 (Kefayat, Lashkar, & Nabavi, 2015). It is inspired from foraging and waggle dancing behaviour of honey bees. Colonies of honey bees have a highly structured social organization and division of labour.

In their colony, the honey bees are divided into three different groups: employed bees, onlooker bees and scout bees. An employed bee is the one exploits a specific food source, and then returns to the hive to share information about that food source with the other bees through specialized dances. The other kind of bee waiting on the dance area for making a decision to choose a food source is called onlooker. A bee that carries out random search for discovering new food sources is scout bee (Binitha & Sathya, 2012). The latter two kinds of bees are also known as unemployed bees.

The special dance called a waggle dance made by employed bee is proportional to the quality of the food source and help onlooker bee to make decisions about which food source to choose. The probability of selecting good food source is proportional to its quality meaning, the food source that has higher quality will have a greater chance to be selected by the onlooker bees than the one of lower quality. At the beginning, a set of food source positions are discovered randomly by employed bees and their nectar amount will be measured. Then after getting information about the source from employed bees, selection of food source will made by onlooker bees (Singh & Kaushik, 2016). After this, the nectar of food sources is exploited by employed bees and onlooker bees, and this continual exploitation will ultimately cause them to become exhausted. An employed bee changes the food source position in her memory depending on the local information called visual information and tests the nectar amount to compare the fitness value of the new source with the previous. When the nectar amount of the new position is better than that of the previous one, the bee memorizes the new position and forgets the old one. Otherwise the old is retained.

If the number of visiting bees does not improve the food source, then the source is abandoned and the employed bee becomes a scout. Every bee colony has scout bees to avoid local optima whenever a source is exploited fully. Scout bees carry out a random search to explore the environment in order to find new food source locations (solution).

In artificial honey bees (ABC), the first half of the colony consists of employed bees and the other half is onlooker bees (Singh & Kaushik, 2016). In this algorithm, a possible solution to the problem is represented by the position of a food source and the quality (fitness) of the associated solution was represented by the nectar amount of a food source corresponds to it. The number of food source (number of solutions) in the swarm is equal to the number of employed bees since each employed bee is associated with one and only one food source (Karaboga, Gorkemli, Ozturk, & Karaboga, 2014)

Self-organization is a key feature of a swarm system which results collective behaviour by means of local interactions among simple agents

- **Positive Feedback:** As the nectar amount of food sources increases, the number of onlookers visiting them increases, too.
- **Negative Feedback:** The exploitation process of poor food sources is stopped by bees.
- **Fluctuations:** The scouts carry out a random search process for discovering new food sources.
- **Multiple Interactions:** Bees share their information about food sources with their nest mates on the dance area.

Three essential components of forage selection that leads to the emergence of collective intelligence of honey bee swarms consists of three essential components: food sources, employed foragers and unemployed foragers, and the model defines two leading modes of the behaviour: the recruitment to a rich nectar source and the abandonment of a poor source.

1. **Food Sources:** The value of a food source depends on many factors such as its proximity to the nest, its richness or concentration of its energy, and the ease of extracting this energy.
2. **Employed Foragers:** They are associated with a particular food source which they are currently exploiting or are employed at. They carry with them information about this particular source, its distance and direction from the nest, the profitability of the source and share this information with a certain probability.
3. **Unemployed Foragers:** They are continually at look out for a food source to exploit. There are two types of unemployed foragers: scouts, searching the environment surrounding the nest for new food sources and onlookers waiting in the nest and establishing a food source through the information shared by employed foragers.

The ABC generates a randomly distributed initial population of SN solutions (food source positions), where SN denotes the swarm size. Let $X_{i,j} = \left(x_{i,1}, x_{i,2}, \cdots x_{i,D}\right)$, where D is dimension size or optimization parameters here, be represent the ith solution in the swarm. Each employed bee $X_{i,j}$ generates a new candidate solution $V_{i,j}$ in the neighbourhood of its present position as follows (Karaboga & Basturk, 2007).

$$V_{i,j} = X_{i,j} + \phi_{i,j}\left(x_{i,j} - x_{k,j}\right) \tag{30}$$

Where x_k is randomly selected candidate solution for $\left(i \neq j\right)$ from all employed bees, j is the random dimension index selected from the set $\left\{1, 2, \ldots, D\right\}$ and $\phi_{i,j}$ is the random number in $\left[-1, 1\right]$.

After each candidate source position $v_{i,j}$ is calculated and then evaluated by the artificial bee, its performance is compared with that of $x_{i,j}$. If the new food source has equal or better nectar than the old source, it is replaced with the old one in the memory. Otherwise, the old one is taken as it is. In other words, a greedy selection mechanism is employed as the selection operation between the old and the current food sources.

In general, the algorithmic structure of ABC optimization approach is given as follows:

Initialization Phase

Here a set of food source positions are randomly selected by the bees and their nectar amounts are determined.

Repeat

Employed Bees Phase

Artificial employed bees search for new food sources having more nectar within the neighbourhood of the food source in their memory and evaluate its fitness. Then it shares the nectar information with onlooker bees waiting in the hive by dancing on the dancing area.

Onlooker Bees Phase

Artificial onlooker bees probabilistically choose their food sources depending on the information provided by the employed bees. For this purpose, a fitness-based selection technique can be used, such as the roulette wheel selection method. After a food source for an onlooker bee is probabilistically chosen, a neighbourhood source is determined, and its fitness value is computed. As in the employed bee phase, a greedy selection is applied between two sources.

Scout Bees Phase

When the nectar of a food source is abandoned by the bees, a new food source is randomly determined by a scout bee and replaced with the abandoned one.

Memorize the best solution achieved so far.

Until Cycle = Maximum Cycle Number or a Maximum CPU time.

In this model, it is supposed that at each cycle one scout goes outside for searching a new food source and the number of employed and onlooker bees were equal. The probability P_i for selecting a new food source i is given by:

$$P_i = \frac{fit_i}{\sum_{n=1}^{N} fit_n} \tag{31}$$

Once all onlookers have fixed their food sources, each of them determines a food source in the neighbourhood of his chosen food source and computes its fitness. The best food i among all the sources will be the new location of the food source i. If this source didn't improve the previous one, then that food source is abandoned by its associated employed bee and it becomes a scout for searching a new food source (Javadi et al., 2011). After the new location of each food source is determined, another iteration of ABC algorithm begins. The whole process is repeated again and again till the termination condition is satisfied.

ALGORITHM OF ARTIFICIAL BEE COLONY IN OPTIMAL OF HYBRID RENEWABLE ENERGY SYSTEMS

The main steps of the implementation of the ABC algorithm to solve optimisation problem for the mentioned hybrid system are described as follows (Karaboga & Basturk, 2007)

1. Input annual data of wind speed, the solar radiation and load demand and also initialise the control parameters like maximum cycle number, colony size, population of food sources, dimension of the problem and limit.
2. Generate a randomly distributed population within the range of boundaries of the parameters by using equation (16) above.
3. Set trial counters (to store the number of solution trials) to zero.
4. According to initial guess solutions of numbers of solar PV panels, wind turbines and batteries perform the following steps.
    a. Obtain solar PV panels and wind turbines outputs by the equations given under fitness function.
    b. Determine the number of batteries needed to maintain the load balance.
5. The fitness function as described in (xx) is evaluated for initial food source.
6. Cycle = 1
7. Repeat.
8. Produced a new modified food location for the employed bees by using the following equation

$$v_{i,j}^{new} = x_{i,j} + \phi_{i,j}\left(x_{i,j} - x_{k,j}\right) \tag{32}$$

Where x_k is randomly selected candidate solution for $\left(i \neq j\right)$ from all employed bees, j is the random dimension index selected from the set $\left\{1,2,\ldots,D\right\}$ and $\phi_{i,j}$ is the random number in $\left[-1,1\right]$.

9. If a parameter generated exceeds its predetermined limits, it can be set to an acceptable boundary.
10. Evaluate the fitness function described in (17) using new solutions by following the procedure mentioned in step 5.
11. Apply the greedy selection process for the employed bees.
12. Calculate the probability value, p_i for the solutions using fitness value by equation (16)

13. Produce the new solutions $v_{i,j}^{new}$ by using step (8) for the onlooker bees from the solutions selected depending upon the value of p_i
14. Evaluate the fitness function by using new solutions by following the procedure mentioned in step 5.
15. Apply the greedy selection process for the onlooker bees.
16. Determine the abandoned solution for the scout, if exists and replace it with a new randomly produced solution.
17. Memorise the best solution obtained as of now.
18. Cycle = cycle + 1.
19. Until (cycle = maximum cycle number).

NUMERICAL DATASETS

A hybrid solar and wind renewable energy system, which is designed based on the developed Optimization formulas given (1–13) above is considered here to be evaluated. The numerical examples used in this paper are similar to those data proposed by (Kellog et al., 1998). They were used an installed anemometer and pyranometer located in remote area of South-Central Montana and record the data information for every 30 seconds. They summarized these data as annual average hourly load profile and use an iterative method for component sizing of standalone hybrid (Wind/Pv) system to supply the electrical power demand. Also, (Zong, 2012) improved these data by including all necessary data to make it whole dataset for other researchers to tackle it with different methods.

More over (Geleta & Manshahia, 2018) updated these data by current factors of inflation rate and energy demand due to certain time gap. In their adjustment, those values used as decision variables in Table 1 are multiplied by 3.31%, which is global inflation rate of 2018 and the power demand data given in Table 2 are multiplied by 1.3%, global average energy demand increment value. So, all the numerical data in this paper is similar to the undated data on (Geleta & Manshahia, 2018).

Table 1 show the values of the updated decision variables for the test of the system and initialize the employed nature inspired algorithm (ABC). Here the values mentioned on (Zong, 2012) which related to purchase, for instance solar panel price, wind turbine price, cost of battery and cost of backup generator are updated by the inflation rate and energy demand rate factors mentioned above. Because of life span of each battery is taken as 4 years, 5 times installations are needed during the whole life span of the system.

Table 2. Taken from (Geleta & Manshahia, 2018),which is the updated valued of annual average hourly demand P_{Dem}^{t}, generated power by each components $P_{PV,unit}^{t}$ and $P_{WT,unit}^{t}$ and the difference in power in kilo watt including the total power demand versus total power generated for each unit time of the day. The data of this table originally taken experimentally by Pyrometer and Thermometer by (Kellog et al., 1998). To illustrate this table more to the readers, the following graphs are plotted by the help of MATLAB.

Table 1. Design variables used for Solar and Wind Hybrid System

Variables	Values
Annual interest (i)	6%
Life span of the system(n)	20 years
Solar panel price	$350/panel
Solar panel installation fee	50% of the price
Wind turbine price	$20000/Turbine
Wind turbine installation fee	25% of the price
Unit cost of the battery	$170
Cost of backup generator	$2000
Usage % of battery rated capacity (η)	80%
Batteries rated capacity	2.1Kwh
Batteries life span	4 years
Unit time (Δt)	1hr
Maintenance cost of PV array	0.5cents/Kwh
Maintenance cost of wind turbine	2cents/Kwh

RESULTS AND DISCUSSION

In this paper Artificial Bee Colony algorithm has been applied to solve the problem, Optimization of Hybrid Wind and Solar Renewable Energy System. The obtained results are shown in tables 3-5 below.

Optimization of the proposed system is done with the objective of minimizing the annual cost of the system which can balance the desired load by satisfying all the constraint from equations 26 -31. As shown in tables 3,4 and 5 above, the results were compared with the PSO and result in Ref. [3]. Here we use the result of PSO as a standard result to compare the result of ABC algorithm. AS the result shows in the tables ABC algorithm is promising result when compared with results of PSO and iteration Algorithms

The value of LPSP is almost zero in all the tables. Which indicates that, the load demand of the system was always satisfied in all three systems means wind-battery alone, solar-battery alone and hybrids of wind, solar and battery system.

When results of tables 3-5 are compared, the ABC algorithm was selected the wind turbine only for the optimal value because the cost of wind turbine units for significant power generation is cheaper than the cost of solar panel units for that power generation. The results show that the optimal solution using ABC leads to cost 5572.01$ which is almost equal to the result obtained by PSO. While the optimal cost for implementing iteration method is 5753.09$.

Table 4 shows the Optimization result when only solar panel was power generating system. For this combination the ABC algorithm with optimal value $ 8936.44 was also good result when com-pared with PSO $ 8936.64 and that of iteration method $ 9242.79.

The Optimization results given in table 5 above are the results when hybrids of wind and solar system with battery are used as power generating unit. As shown in the table the result obtained by ABC $ 6827.91 was equal to the result obtained by PSO with equal value of LPSP and better result when compared with that of iteration algorithm $7085.97.

Table 2. Updated daily power data for wind and solar system

Time(t)	P_Dema (KW)	P_WT(KW)	P_PV(W)	ΔP (KW)
1	1.39	0.58	0	-0.81
2	1.25	0.49	0	-0.76
3	1.19	0.48	0	-0.71
4	1.22	0.53	0	-0.69
5	1.34	0.47	0	-0.87
6	1.8	0.51	0	-1.29
7	2.66	0.46	1.6	-2.198
8	2.9	0.46	3.4	-2.437
9	2.52	0.61	10.3	-1.899
10	2.21	0.76	24.6	-1.425
11	2.05	1.1	31.7	-0.918
12	1.94	1.53	35.3	-0.375
13	1.82	1.67	36.6	-0.113
14	1.71	1.89	37.4	0.217
15	1.62	2.43	36.8	0.847
16	1.65	2.45	33.5	0.833
17	1.87	1.91	24.2	0.064
18	2.29	1.76	13.4	-0.517
19	2.58	1.57	5.6	-1.004
20	2.6	1.16	1.5	-1.438
21	2.54	0.87	0	-1.67
22	2.49	0.76	0	-1.73
23	2.28	0.74	0	-1.54
24	1.79	0.7	0	-1.09
25	47.72	25.89		-21.534

Table 3. Optimization Results of wind-battery system

Methodology	No. of Panels	No. of Turbines	No. of Batteries	LPSP	Total Annual Cost ($)
ABC	0	2	9	-0.0853	5572.01
PSO	0	2	9	0.0853	5572.04
(Geleta & Manshahia, 2018)	0	2	9		5753.09

Table 4. Optimization results of solar-battery system

Methodology	No. of Panels	No. of Turbines	No. of Batteries	LPSP	Total Annual Cost
ABC	161	0	16	0.0015	8936.44
PSO	161	0	16	0.0014	8936.64
(Geleta & Manshahia, 2018)	162	0	17		9242.79

Table 5. Solar-Wind hybrid-battery system optimization results

Methodology	No of Panels	No of Turbines	No of Batteries	LPSP	Total Annual Cost ($)
ABC	74	1	11	-0.0016	6827.91
PSO	74	1	11	-0.0016	6827.91
(Geleta & Manshahia, 2018)	74	1	12		7085.97

The desired load of the system was checked by LPSP which almost equals to zero in all the results. Reliability of power supply (RPS) be found from LPSP whose value becomes 1. From these two concepts it is possible to understand that, the system was reliable. Moreover, the energy not supplied (ENS) which is considered when generated power is less than the demanded power was also calculated by the code. The ENS value is -4.070 is only considerable from table 3 Which is there is a need for battery to charge.

The optimal solution of the hybrid system is found when $N_{PV} = 0, N_{WT} = 2\, and\, N_{Batt} = 9$ with the optimal cost of $5572.01 meaning when only wind panel was used as power generating mechanism. But due to stochastic nature of wind, using wind turbine alone may not satisfy the desired load at any time t. Therefore, it is important to use the hybrid of wind along with solar to overcome the drawback comes by weather change and in order to make the power output stable.

As seen from the simulation result of the system, the optimal value by using ABC algorithm was obtained with good convergence rate as shown on g 7 and g 8 when compared with that of PSO algorithm.

Here to maintain the power output of the system, it is a solution to use certain number of solar panels along with wind turbines. To balance the desired energy of the system the possible number of turbines are as follows:

$$N_{WT} = 1 \tag{35}$$

$$N_{WT} = 0 \tag{36}$$

The simulation results based on Equation (21) was in table (5). According to the obtained results, the solar panels, wind turbines, and batteries sup-plied the load demand. The optimal solution is determined as 74 for solar panels, 1 for wind tur-bine, and 11 for batteries with LPSP equals -0.0016 which shows the load is satisfied by these components. The optimal cost value for this combination is 6827.91$.

Table 6. Results obtained by different scholars

Reference	Methodology	Components	NW T	NP V	NBatt	Total Annual Cost ($)
(Zong, 2012)	B&B	Wind only	3	0	11	7924.89
		Solar only	0	160	17	8843.63
		Hybrids	1	72	11	6692.28
(Kellogg et al.,1998)	Iteration	Wind only	2	0	9	5574.00
		Solar only	0	158	16	8677.00
		Hybrids	1	72	11	6691.00
(Alireza, 2013)	DHR	Wind only	2	0	11	5652.66
		Solar only	0	160	17	8844.09
		Hybrids	1	98	13	7988.24

On the other hand, when Equation (22) applied to simulation process as the results shown from table (4), the number of solar panels and batteries in proposed method are 161 and 16 respectively with the optimal value $ 8936.44. When this result was compared with table 4, the result obtained by different scholars on the same project project with different methods [3, 6, 18, 20] the result by ABC is optimal values for all the three possible combinations. The power difference between the generated and demanded powers is illustrated in Fig.5 shows, the Hybrid Solar panel and Wind turbine system without applying the batteries is not capable of supplying the system continuously.

Here, DHR stands for Discrete Harmony Search and B&B stands for Branch and Bound methods of solving Optimization problems.

Hence, when the generated powers through the renewable energy systems are more than the required demand, it can be stored in the batteries. In other words, when the solar and wind powers can't satisfy the demand, the battery energy storage systems compensate the load de cit and improve the supply reliability of the system. According to Fig.5, the power difference must be managed by batteries in charge (positive values) and discharge (negative values) processes.

Additionally, the result obtained in this paper was compared with the results obtained by different scholars in literature which use the same numerical data to optimize the hybrid system by their own different methods as shown in table 6. It was shown that ABC algorithm was superior over all the conventional methods used by the scholars in the three possible combinations (wind alone, solar alone and hybrids of wind and solar). The optimal solution obtained $5572.01 for wind alone is the least value when compared with the values mentioned in table 6. While ABC requires new fitness tests on the new algorithm parameters and population of solutions increases the computational cost.

CONCLUSION AND FUTURE RESEARCH DIRECTIONS

This paper presents an Artificial Bee Colony (ABC) algorithm to solve the optimal sizing of a hybrid Wind and Solar renewable energy system standalone daily average of about 47.72 KW/day power generating by considering certain pre-designed constraints. The main objective of this research is minimizing the system total annual cost by sizing the number of individual components while the load demand is balanced.

For simulation MATLAB code was designed for the fitness function, for ABC algorithm and for PSO algorithm. PSO algorithm was defined here to compare the obtained result with most common meta heuristic algorithm. The optimal number of wind turbine, PV panels, and batteries is determined using Both ABC and PSO to sup-ply the desired load. The proposed method ABC was highly converged to optimal as shown on Figure 1 when compared to that of PSO on Figure 2 and selected the wind alone with battery system in optimal sizing because the cost of wind turbine units for significant power generation is cheaper than the cost of solar panel units for that power generation. The obtained results showed the superiority of ABC than other methods in system cost minimizing. The results mentioned in table 6 by different their own methods were compared with ABC. The Artificial Bee Colony (ABC) gives better result than any of the methods used here in all the possible combinations.

Hopefully, the way of organizing hybrid wind and solar renewable energy system and the updated numerical data, by current factors of inflation rate and energy demand, taken to optimize the system would be very beneficial and become a good benchmark for us and other researchers of the field to apply their our methods and other nature inspired meta heuristic algorithms and also compare this result with other nature inspired algorithms for the future.

ACKNOWLEDGMENT

Authors would like to thank the researchers or academicians whose works have been cited in this review paper. Authors are also grateful to Punjabi University Patiala for offering sufficient library and internet facility

Figure 2. Convergence of PSO Algorithm

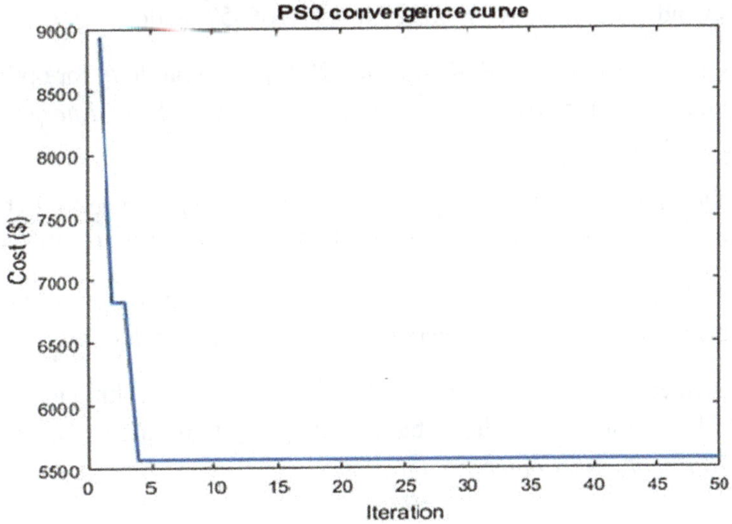

REFERENCES

Al-Shamma, A. A., & Khaled, E. (2012). Optimum sizing of hybrid PV/wind/battery/diesel system consider-ing wind turbine parameters using Genetic Algorithm. *Power and Energy (PECon), IEEE International Conference*, 121-126.

Alireza, A. (2013). Developing a discrete harmony search algorithm for size opti- mization of wind–photovoltaic hybrid energy system. *Solar Energy*, *98*, 190–195. doi:10.1016/j.solener.2013.10.008

Ashok, S. (2007). Optimised model for community-based hybrid energy system. *Renewable Energy*, *32*(7), 1155–1164. doi:10.1016/j.renene.2006.04.008

Belfkira, R., Zhang, L., & Barakat, G. (2011). Optimal sizing study of hybrid wind/PV/diesel power generation unit. *Solar Energy*, *85*(1), 100–110. doi:10.1016/j.solener.2010.10.018

Binitha, S., & Sathya, S. S. (2012). A survey of bio inspired optimization algo-rithms. *International Journal of Soft Comput-ing and Engineering*, *2*(2), 137–151.

Borowy, B. S., & Salameh, Z. M. (1996). Methodology for optimally sizing the combination of a battery bank and PV array in a wind/PV hybrid system. *IEEE Transactions on Energy Conversion*, *11*(2), 367–375. doi:10.1109/60.507648

Borowy, B. S., & Salameh, Z. M. (1996). Methodology for optimally sizing the combi-nation of a battery bank and PV array in a wind/PV hybrid system. *IEEE Transactions on Energy Conversion*, *11*(2), 367–375. doi:10.1109/60.507648

Bueno, C., & Carta, J. A. (2005). Technical–economic analysis of wind-powered pumped hydrostorage systems. Part II: Model application to the island of El Hierro. *Solar Energy*, *78*(3), 396–405. doi:10.1016/j.solener.2004.08.007

Chedid, R., Akiki, H., & Rahman, S. (1998). A decision support technique for the design of hybrid solar-wind power systems. *IEEE Transactions on Energy Conversion*, *13*(1), 76–83. doi:10.1109/60.658207

Diaf, S., Belhamel, M., Haddadi, M., & Louche, A. (2007). A methodology for op-timal sizing of autonomous hybrid PV/wind system. *Energy Policy*, *35*(11), 5708–5718. doi:10.1016/j.enpol.2007.06.020

Eftichios, K., Dionissia, K., Antonis, P., & Kostas, K. (2006). Methodology for optimal sizing of stand alone photovoltaic/wind generator systems using genetic algorithms. *Solar Energy*, *80*(9), 1072–1088. doi:10.1016/j.solener.2005.11.002

Elhadidy, M. A., & Shaahid, S. M. (1999). Optimal sizing of battery storage for hybrid (wind+ diesel) power systems. *Renewable Energy*, *18*(1), 77–86. doi:10.1016/S0960-1481(98)00796-4

Eltamaly, A. M., & Mohamed, M. A. (2014). A novel design and optimization software for autonomous PV/wind/battery hybrid power systems. *Mathematical Problems in Engineering*.

Etamaly, A. M., Mohamed, M. A., & Alo-lah, A. I. (2015). A smart technique for op-timization and simulation of hybrid photo-voltaic/wind/diesel/battery energy systems. *Smart Energy Grid Engineering (SEGE), 2015 IEEE International Conference on*, 1-6.

Fister, I. Y., Jr. (2013). *A brief review of nature-inspired algorithms for optimization.* arXiv preprint arXiv:1307.4186

Gavanidou, E. S., Bakirtzis, A. G., & Dokopoulos, P. S. (1993). A probabilistic method for the evaluation of the performance and the reliability of wind-diesel energy systems. *IEEE Transactions on Energy Conversion*, *8*(2), 197–206. doi:10.1109/60.222705

Gavrilas, M. (2010). Heuristic and metaheuristic optimization techniques with application to power systems. *Proceedings of the 12th WSEAS international conference on Mathematical methods and computational techniques in electrical engineering.*

Geleta, D.K., & Manshahia, M.S. (2017). Optimization of Renewable Energy Systems: A Review. *International Journal of Scientific Research in Science and Technology, 8*(3), 769-795.

Geleta, D.K., & Manshahia, M. S. (2018). Opti-mization of Hybrid Wind and Solar Renewable Energy System by Iteration Method. *International Conference on Intelligent Computing and Optimization*, 98-107.

Geleta, D. K., & Manshahia, M.S. (2017). Nature Inspired Computational Intelligence: A Survey. *International Journal of Engineering. Science and Mathematics*, *6*(7), 769–795.

George Kosmadakis, S. K. (2013). *Renewable and Conventional Electricity Generation Systems.* Technologies and Diversity of Energy Systems.

Hadidian, M. J., & Arabi, N. S., & Bigdeli. (2016). Optimal sizing of a stand-alone hybrid photovoltaic/wind system using new grey wolf optimizer considering re-liability. *Journal of Renewable and Sustainable Energy*, *8*(3).

Ilinca, A., McCarthy, E., Chaumel, J. L., & Retiveau, J. L. (2003). Wind potential assessment of Quebec Province. *Renewable Energy*, *28*(12), 1881–1897. doi:10.1016/S0960-1481(03)00072-7

Javadi, M. R., Mazlumi, K., & Jalilvand, A. (2011). Application of GA, PSO and ABC in optimal design of a stand-alone hybrid system for north-west of Iran. In *Electrical and Elec-tronics Engineering* (pp. 1–203). ELECO.

Kaabeche, A., Belhamel, M., & Ibtiouen, R. (2010). Optimal sizing method for stand-alone hybrid PV/wind power genera-tion system. *Revue des Energies Renouvelables (SMEE'10)*, 205-213.

Karaboga, D., & Basturk, B. (2007). A powerful and efficient algorithm for numerical function optimization: Artificial bee colony (ABC) algorithm. *Journal of Global Optimization*, *39*(3), 459–471. doi:10.100710898-007-9149-x

Karaboga, D., Gorkemli, B., Ozturk, C., & Karaboga, N. (2014). A comprehensive survey: Artificial bee colony (ABC) algorithm and ap-plications. *Artificial Intelligence Review*, *42*(1), 21–57. doi:10.100710462-012-9328-0

Kefayat, M. A., Lashkar, A., & Nabavi, N. (2015). A hybrid of ant colony optimization and artificial bee colony algorithm for probabilistic optimal placement and sizing of distributed energy resources. *Energy Conversion and Management*, *92*, 149–161. doi:10.1016/j.enconman.2014.12.037

Kellog, W., Nehrir, M., Venkataramanan, G., & Gerez, V. (1998). Generation Unit Sizing and Cost Analysis for Stand -alone Wind, Phovoltaic, and Hybrid Wind/PV Systems. *IEEE Transactions on Energy Conversion, 13*(1), 70–75. doi:10.1109/60.658206

Kosmadakis, G., Sotirios, K., & Emmanuel, K. (2013). Renewable and conventional elec-tricity generation systems. *Technologies and diversity of energy systems*. In *Renewable Energy Governance* (pp. 9–3). Springer London. doi:10.1007/978-1-4471-5595-9_2

Koutroulis, E., Dionissia, K., Potirakis, A., & Kostas, K. (2006). Methodol-ogy for optimal sizing of stand-aloMethodol-ogy for optimal sizing of stand-alone photovoltaic/wind-generator systems us-ing genetic algorithms. *Solar Energy, 80*(9), 1072–1088. doi:10.1016/j.solener.2005.11.002

Leijnse, A., & Hassanizadeh, S. M. (1994). Short communication: Model definition and model validation. *Advances in Water Resources, 17*(3), 197–200. doi:10.1016/0309-1708(94)90041-8

Lu, L., Yang, H., & Burnett, J. (2002). Investigation on wind power potential on Hong Kong islandsan analysis of wind power and wind turbine characteristics. *Renewable Energy, 27*(1), 1–12. doi:10.1016/S0960-1481(01)00164-1

Luna, R. R., Trejo, P. M., Vargas, D., & Os-Moreno, G. (2012). Op-timal sizing of renewable hybrids energy sys-tems: A review of methodologies. *Solar Energy, 88*(4), 1077–1088. doi:10.1016/j.solener.2011.10.016

Lysen, H. (1983). Introduction to wind energy. Consultancy Services. *Wind Energy (Chichester, England)*, 82–91.

Rubio, R., Perea, M. T., Vazquez, D. V., & Os-Moreno, G. J. (2012). Op-timal sizing of renewable hybrids energy systems: A review of methodologies. *Solar Energy, 86*(4), 1077–1088. doi:10.1016/j.solener.2011.10.016

Singh, S., & Kaushik, S. C. (2016). Optimal sizing of grid integrated hybrid PV-biomass energy system using artificial bee colony algorithm. *IET Renewable Power Generation, 10*(5), 642–650. doi:10.1049/iet-rpg.2015.0298

Wang, L., & Singh, C. (2009). Multicriteria design of hybrid power generation systems based on a modi ed particle swarm optimization algorithm. *IEEE Transactions on Energy Conversion, 24*(1), 163–172. doi:10.1109/TEC.2008.2005280

Yang, H., Zhou, W., & Lou, C. (2008). Optimal sizing method for stand-alone hybrid solar wind system with LPSP technology by using genetic algorithm. *Solar Energy, 82*(4), 354–367. doi:10.1016/j.solener.2007.08.005

Zong, W. G. (2012). Size optimization for a hybrid photovoltaic wind energy system. *Electrical Power and Energy Systems, 42*(1), 448–451. doi:10.1016/j.ijepes.2012.04.051

KEY TERMS AND DEFINITIONS

Constraints: Are the conditions which can affect solutions of the given optimization problem.

Fitness Function: It is a kind of objective function developed for a particular function in order to find the desired solution.

Hybrid Energy: The application of two or more energy technologies to maximize the efficiency of the system is known as hybrid energy.

Chapter 18
The Study of Luminescence Spectra of Seeds of Crop Species for Diagnostic Quality

Alexey Bashilov
Moscow Aviation Institute, Russia

Mikhail Belyakov
https://orcid.org/0000-0002-4371-8042
Branch of the National Research University "MPEI" in Smolensk, Russia

ABSTRACT

In this chapter, optical luminescent biological objects diagnostics methods and biotissues are considered. According to the previously developed method, excitation and photoluminescence spectra agricultural plants seeds, including cereals, legumes, fodder, technical, and vegetable, were measured. The typical excitation spectrum lies in the range of 355-500 nm and has two maxima: the main one at 424 nm and the side one at 485 nm. The luminescence spectrum lies in the range of 420-650 nm and has a maximum in the region of 500-520 nm. The maximum luminescence is less pronounced than in the excitation spectrum. The measured spectral luminescence characteristics forage plants seeds by scarification. Due to the scarification forage plants seeds spectral characteristics increase. In Galega seeds with multiple scarification, observed qualitative changes in the excitation spectrum was associated with the appearance of a new maximum at a wavelength of 423 nm. Similarly, for clover seeds, the obtained results can be used to create seed diagnostics devices.

INTRODUCTION

Due to the growing demand of the world's population for food, there is a growing need to intensify the production of high-quality agricultural products. One of the directions of such intensive development is the development and implementation of modern high-performance methods and devices for the diagnosis of seed.

DOI: 10.4018/978-1-5225-9420-8.ch018

Optical methods and means of diagnostics and control are highly accurate, selective, Express, as well as remote and non-destructive. They have stable parameters of probing radiation, and the received signals can be received by radiation receivers with a wide range of characteristics, amplified and processed with the help of modern computer programs with the issuance of complex results. Other advantages of optical and optoelectronic diagnostic devices are simplicity and safety of their operation, a minimum of subjective factors and the possibility of integration into existing modern agricultural machines and devices.

There are works considering the use of luminescent analysis (Gaevskj, 2002), polarization-reflective and fluorescent spectroscopy (Ovchinnikova et al, 2005), including laser-induced (Ryabova et al, 2006) for the study and diagnosis of biological tissues. To date, methods and installations of optical diagnostics of biological medical facilities have been developed (Monich et al, 1994). Analysis of luminescence spectra is used to determine the protein content, vitamins, starch, the potato tubers landscaping degree, detection of rot and diseases of citrus fruits, onions, potatoes and grapes (Bashilov, 2005; Sventickji et al, 1990; Baek et al, 2013; Li et al, 2016; Leemans et al, 2017; Belzile et al, 2004; Liu et al, 2016; Noh et al, 2006). Optical methods are used to diagnose fruits of fruit trees.

In plant breeding and seed production due to optical density thin ismel received seeds in the near infrared region can be used to diagnose the level of Gib-radnoti with the use of a calibration series, the samples with known level of hybridity.

A common method of in vivo optical diagnostics of biotissues is fluorescent spectroscopy of biotissues. Despite significant achievements in the field of in vivo reflective and fluorescent spectroscopy of biotissues, the possibilities of methods are far from exhausted. The main difficulty facing the development of methods is the limited number of experimentally measured parameters and sufficient statistical data.

The aim of this work is to study the spectral characteristics and parameters of excitation and luminescence of seeds of agricultural plants in order to diagnose their quality. The influence of multiple mechanical scarification on the luminescence of seeds of forage plants was also studied.

SPECTRAL LUMINESCENT CHARACTERISTICS INVESTIGATION OF AGRICULTURAL PLANTS SEEDS

Optical spectral excitation characteristics measurement and luminescence of agricultural crops seeds were carried out on the basis of hardware and software complex, consisting of a multifunctional spectrofluorometer "Fluorate-02-Panorama", a computer with installed software "PanoramaPro" and an external camera for the investigated samples ("Lumex", 2018).

Mathematical processing of measurement results was carried out by means of the supplied software or other software products, for which it is provided to export the measurement results in ASCII and MS Excel formats. The method of measurement of excitation and agricultural seeds luminescence spectra developed By M. V. Belyakov is described in more detail in (Belyakov et al, 2016).

To study the luminescence spectra seeds of crops were used: cereals (wheat, rye, triticale, barley, oats, corn, millet), legumes (peas, soybeans, white beans), herbaceous (clover, buckwheat, amaranth, rape, Vika, sunflower, lentils, Galega Oriental, flax), vegetables (pumpkin, cucumber, radish, tomatoes, zucchini, cayenne).

The spectral measurements results of wheat seeds are shown in Figure 1.

The excitation spectrum of wheat seeds lies in the range of 355...500 nm and has two peaks: the main at 424 nm and the side at 485 nm. The luminescence spectrum lies in the range of 420 ... 650 nm and has a maximum in the region of 500...520 nm. The luminescence maximum is less shown than in the excitation spectrum.

The spectral characteristics of excitation and seeds luminescence of other crops were studied by the same method.

All investigated seeds contain in their spectrum the excitation of maximum 424 and 485 nm, at least 455, 530 and others. Luminescence spectra are single-modal, located in the visible region and have maxima in the range 480...520 nm. The level of luminescence signal strongly depends on the seeds color. Due to the fuzzy maximum of the spectra (especially luminescence), it is difficult to determine the exact Stokes shift. It is approximately 68 ... 96 nm between the highest peaks. For all studied spectra the Stokes-Lommel law is used. The mirror symmetry rule (left-hand rule) is not done, which indicates the glow of simple molecules. The half-width of the excitation spectrum is less than the half-width of the luminescence spectrum (for example, for wheat seeds 85...105 nm and 155 nm, respectively).

At the same time it was stated that the seeds of crops such as amaranth, sunflower, canola, vetch almost do not luminesce, owing to their dark color. The spectra of excitation and luminescence of seeds in most grains (wheat, rye, triticale, buckwheat, oats), legumes (peas, beans) and vegetables (pepper, tomato) are similar. Optical spectral properties of plant seeds are described in more detail in the monograph.

ANALYSIS OF SPECTRA STATISTICAL PARAMETERS

It is advisable to analyze the spectra by determining their parameters, by means of which you can get an idea of the most significant features of the photoluminescence spectra. The analysis identified the

Figure 1. Spectral characteristics of excitation (1) wheat seeds and luminescence (2)

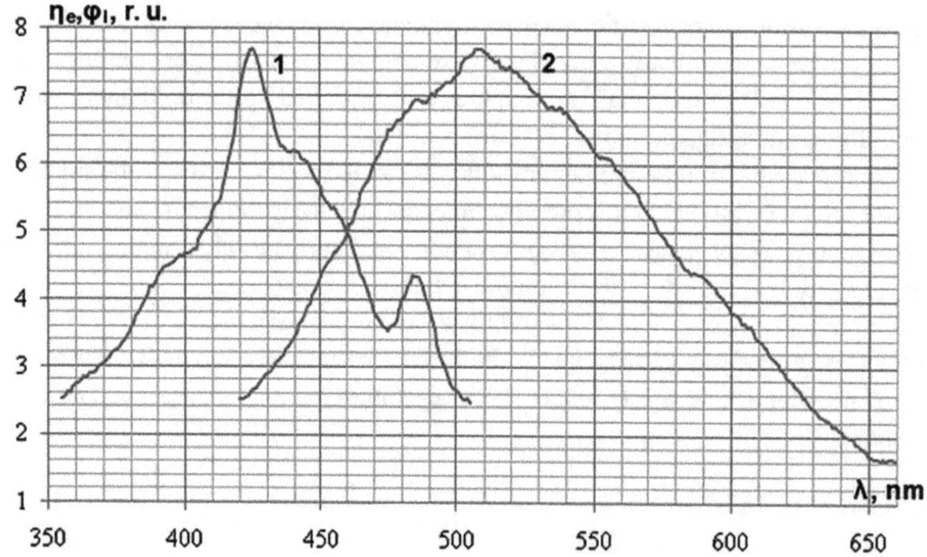

following parameters: mathematical expectation M_λ, variance σ^2, the spectrum asymmetry degree *As*, the excess E_λ, the luminescence energy *E*.

For the spectra treatment and parameters calculation the program MicrocalOrigin7 was used. The calculations results for seeds of some plants are given in table 1.

From table 1 it follows that the expectation of most studied crops is within 472...550 nm, except millet and zucchini in which it is shifted to the long-wave region of about 600 nm and flax in which 413 nm. Flax has the highest dispersion due to the relatively weak luminescence signal. The smallest dispersion is of seeds and vegetable plants. All spectra have right-hand asymmetry (*As*>0). The most asymmetric spectra of zucchini seeds (*As*=0.80), beans (*As*=0.67), clover (*As*=0.64) and soybean (*As*=0.60). Range of millet is almost symmetrical. Also all spectra are flat-vertex (E_λ<0). The value of the kurtosis takes values from -0.16 in zucchini to -1.04 in cayenne and tomato. The energy of the photo-luminescence spectrum is within 2.08 ... 3.05 eV.

THE SPECTRAL SEEDS LUMINESCENT CHARACTERISTICS CHANGE OF FORAGE PLANTS BY SCARIFICATION

A comparison of the luminescence noscarification and scarification seeds galegov various Eastern germination was made. The experiment consisted of mechanical scarification of 400 Galega seeds using

Table 1. Average value of photoluminescence spectra parameters of plant seeds

Culture, Variety	M_λ, nm	σ^2	*As*	E_λ	*E*, eV
Wheat, Skipetr	508	1372	0.52	-0.54	2.45
Rye, Saratovskaya	546	1560	0.22	-0.92	2.28
Triticale, TI-17	528	2436	0.45	-0.64	2.37
Barley, Gonar	528	1810	0.21	-0.82	2.36
Corn, Amatus	473	1160	0.24	-0.78	2.63
Buckwheat, Sapfir	543	2494	0.23	-0.72	2.31
Millet, Saratovskoe	601	2653	0.01	-0.51	2.08
Pea, Faraon	521	1303	0.28	-0.87	2.39
Bean, Nehvi	510	1533	0.67	-0.20	2.44
Soy, Renta	529	1866	0.60	-0.30	2.36
Galega, Magistr	549	1336	0.23	-0.77	2.45
Clover, Dymkovskij	545	2332	0.64	-0.28	2.29
Flax, Soyuz	413	3058	0.56	0.45	3.05
Squash, Gribovskij	600	1239	0.80	-0.16	2.80
Cucumber, Nezhinskij	550	756	0.10	-0.98	2.48
Pepper, Mar'ya	505	825	0.01	-1.04	2.45
Radish, Saksa	529	520	0.07	-1.03	2.34
Tomato, Sibirskij	501	721	0.06	-1.04	2.48

sandpaper and subsequent measurement of the spectral characteristics of excitation and luminescence with germination of initial and scarified seeds control.

Figure 2 shows a comparison of the averaged excitation and luminescence spectra of two species of four hundred seeds of scarified and non-scarified goat. Operating range of excitation spectrum: is 420-505 nm; main maximum: 461 nm; side maximum: 485 nm. Luminescence spectrum range: 480-630 nm. The luminescence spectrum graph reaches a maximum in the range from 527 nm to 535 nm. The value of the excitation and luminescence signal is presented in relative units (r. u.).

In addition, the graph shows that due to the scarification of seeds, the maximum value of the η_e of the two - modal excitation spectrum increased relative to the excitation spectrum of non-scarified seeds by 28%, the position of the maxima did not change but the ratio of their intensity changed: the non-scarified first maximum has an intensity of 0.9 r. u. greater than the second maximum and the scarified one-by 2 r. u. The change in the intensity of the luminescence spectrum is similar to the change in the excitation spectrum: the intensity increased by 28%.

For mathematical processing "PanoramaPro" determined the integrals under the curves of the luminescence and excitation. In this case the first under the curves of the luminescence spectrum has a physical meaning of the luminescence flux. Analyzing the results, we noticed that the luminescence can increase by 185% and fall to 42%. It should also be noted that in 7% luminescence from scarified seeds measurements becomes less than in non-scarified ones.

Integrals under excitation curves were determined for two spectral regions: 420-470 nm and 470-505 nm to determine the ratio between the maxima. To do this, we found the average value of all calculated intervals (table 2).

Table 2 shows that the average value of the integral in the range of 420-470 nm is greater than the average value of the integral in the range of 470-505 nm by 1.29 times for the spectra of non-scarified seeds. And for the spectra of scarified seeds-1.54 times. Scarification increased this ratio by 20%.

In addition, an analysis of photoluminescence spectra galegov East with the help of the software package Microcal Origin7.0 was conducted, with the aim of gaining an understanding of the increased

Figure 2. Comparison of the averaged excitation spectra and luminescence of two types of seeds Galega: 2,4 – spectra of excitation and luminescence notscarifed; 1,3 – spectra of excitation and luminescence scsrifed

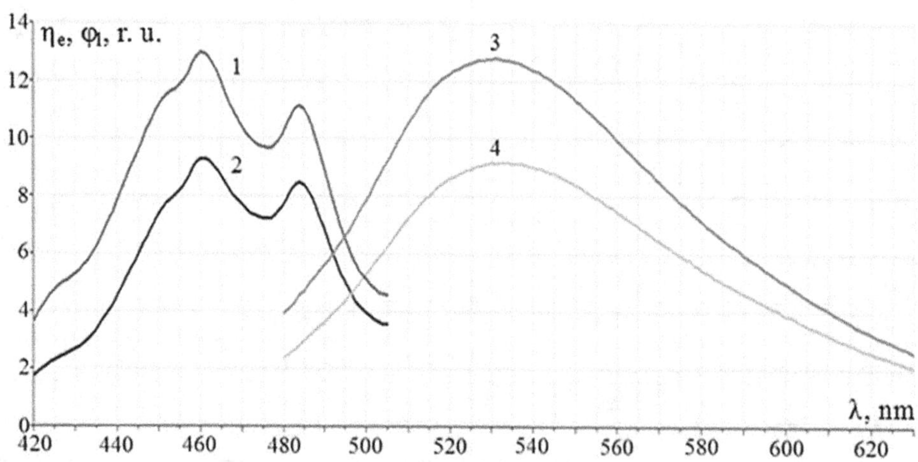

Table 2. Averaged parameters of excitation and luminescence spectra of notscarified and scarified Galega seeds

Seeds	Excitation					Luminescence	
	420-470		470-505		420-505	$\varphi_{l,max}$, r.u.	$\Phi_{aver.}$, r.u.
	$H_{aver.}$, r.u.	$\eta_{e,max}$, r.u.	$H_{aver.}$, r.u.	$\eta_{e,max}$, r.u.	$H_{aver.}$, r.u.		
Notscarified	307	9.3	238	8.4	545	9.2	872
Scarified	503	13.0	325	11.0	828	12.8	1197

bandwidth supports spectrum. For this purpose, the parameters of the averaged excitation and luminescence spectra of two types of four hundred Galega seeds (Figure 2) were determined and recorded in table 3.

Analyzing the data obtained from the table, it can be noted that such parameters as the expectation, dispersion, static moment 3 and 4 of magnitude orders, the energy of the spectrum, the effective width of the spectrum, the frequency of the maximum spectrum and the center of gravity, the frequency at the level of 0.5 scarification has virtually no effect. In addition, it can be seen that the asymmetry for excitation has negative values and for luminescence – positive. At the same time, the energy of the spectrum is positive for both excitation and luminescence.

A similar experience was done with clover seeds. 50 seeds of tetraploid clover were used. The operating range of the excitation spectrum is 380-500 nm, the operating range of the luminescence spectrum is

Table 3. Statistical and frequency parameters of excitation and luminescence Galega seeds

Parameter	Excitation		Luminescence	
	Notscarified	Scarified	Notscarified	Scarified
M_{λ}, nm	466	464	548	546
σ^2	411	441	1281	1277
Static moment of the 3rd order μ_3	-1189	-531	13330	15248
Static moment of the 4th order μ_4	$0.374 \bullet 10^6$	$0.413 \bullet 10^6$	$3.75 \bullet 10^6$	$3.79 \bullet 10^6$
As	-0.14	-0.057	0.29	0.33
E_{λ}	-0.79	-0.88	-0.71	-0.67
E, eV	2.27	2.68	2.27	2.28
Effective spectrum width Δf, Hz	0.314	0.330	0.340	0.395
Maximum spectrum frequency f_0, Hz	0.234	0.240	0.363	0.367
The frequency of the gravity centre f_c, Hz	0.219	0.235	0.323	0.332
Frequency f_1, Hz at 0.5	0.043	0.050	0.132	0.147
Frequency f_2, Hz at 0.5	0.360	0.389	0.530	0.542
The increased bandwidth supports μ_1	1.432	1.407	1.231	1.190
The increased bandwidth supports μ_2	1.341	1.373	1.097	1.075
The increased bandwidth supports μ_3	1.354	1.417	1.099	1.074

470-680 nm. Then averaged all the spectra obtained (Figure 3) and processed the results in the software package MicrocalOrigin.

The peaks of the luminescence spectra are in the range 504-519 nm, the peaks of the absorption curves lie in the following ranges: the main peak at 445-451 nm, the side peak at 483-488 nm.

Then the integrals under the excitation and luminescence curves were calculated. After the scarification the integral under the excitation curves increases by 1.62 times and under the luminescence curves - by 1.45 times. The peaks of the curves are the excitation at 448 nm and 486 nm home side for noscarification seeds, 448 nm and 485 nm home side for scarification seeds; the peaks of the curves of the luminescence are located at 515 nm for noscarification seeds and 513 for scarification.

To determine the ratio between the maximum the integrals under the curve of excitation for the two parts of the spectrum were determined: 380-475 nm, 475-500 nm. It was found that the average value of the integral in the range 380-475 nm for the spectra of non - scarified clover is greater than the average value of the integral in the range 475-500 nm by 3.76 times and for the scarified-by 3. 78 times.

Then the mathematical processing for the averaged clover spectra was carried out and the results were shown in table 4.

From the analysis of the data presented in the table it follows that for the luminescence and excitation spectra the following changes in spectral characteristics are observed: a significant increase in the area; a slight change in the total energy of the spectrum and dispersion; a shift in the center of gravity to the short-wave region (about 3nm). During scarification the coefficient of kurtosis for luminescence spectra increases and for excitation spectra – decreases. A distinctive feature of the excitation spectra is symmetry with respect to the gravity of the curve center.

Figure 3. Averaged excitation and luminescence spectra of tetraploid clover seeds: notscarified (1,2) and scarified (3,4)

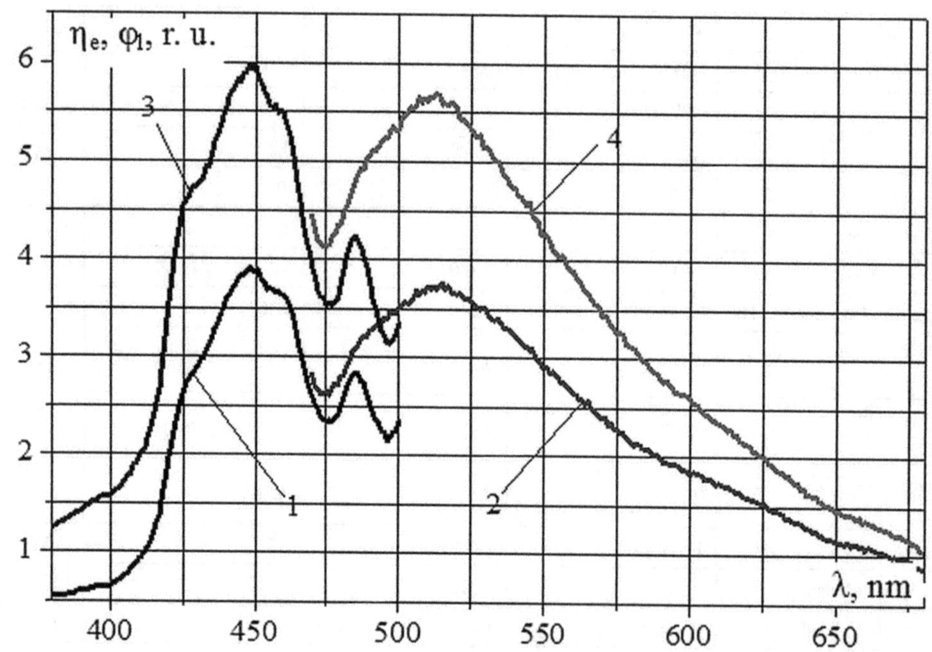

Table 4. Statistical and frequency parameters of excitation and luminescence clover seeds

Parameter	Excitation		Luminescence	
	Notscarified	Scarified	Notscarified	Scarified
M_{λ}, nm	452	449	553	549
σ^2	736	819	2961	2832
Static moment of the 3rd order μ_3	0.00	0.00	81443	86141
Static moment of the 4th order μ_4	$1.37 \cdot 10^6$	$1.60 \cdot 10^6$	$1.99 \cdot 10^7$	$1.93 \cdot 10^7$
As	0.00	0.00	0.51	0.57
E_{λ}	-0.48	-0.61	-0.72	-0.60
E, eV	2.76	2.78	2.26	2.28
Area	271	438	487	707

A similar experiment was conducted for lupine seeds. It should be noted that the shell of lupine seeds is dense and conceived of dark color, so that the signal intensity is very low. With weak scarification, when the shell is partially damaged, but the seed body is not visually visible, there are practically no differences in the spectra removed. Scarification was performed so that the shell was completely damaged, exposing the area of the seed body.

The operating range of the excitation spectrum is 300-460 nm, the operating range of the luminescence spectrum is 450-600 nm. For scarified seeds the main peak of the excitation spectrum lies at 424-428 nm, the side peak at 442-448 nm, the peak of the luminescence spectrum goes to the left edge of the measurement range, and for non – scarified seeds, the peak of the excitation spectrum shifts to the right edge of the measurement range, the peak of the luminescence spectrum is at 515-532 nm. Then averaged all obtained spectra (Figure 4).

From figure 4 it is seen that the luminescence spectra doodley, and side maximum appears only after scarification of the seed. Also note that the ratio of the amplitudes of the excitation and luminescence spectra increases with increasing signal intensity. To quantify this, we find the ratio of integrals under these curves: for non – scarified lupine seeds, the ratio of integrals of the luminescence and absorption spectra was 1.24, and for scarified seeds-2.01.

Then the spectra were processed, the results of which are shown in table 5.

Analyzing the data given in the table, it can be concluded that the following changes in spectral characteristics are observed for the luminescence and excitation spectra during scarification: a significant increase in the area and a decrease in the dispersion; a slight change in the total energy of the spectrum; a shift in the center of gravity to the long-wave region (more than 10 nm). In the case of scarification, the coefficient of kurtosis for excitation spectra increases and changes to the opposite sign, while for luminescence spectra it increases not so much and remains negative. A distinctive feature of Lupin spectra is symmetry with respect to the center of gravity of the curve.

The first experience in the study of the degree scarification the influence of Galega seeds on its luminescent characteristics was the mechanical scarification of 200 Galega East seeds using sandpaper. The scarification was of different intensity: 100 seeds subscripition (single scarification), 100 – salescareersonline (scarification again).

Figure 4. Averaged excitation and luminescence spectra of lupine seeds nesterilizovanny (1,2) and scarification (3,4)

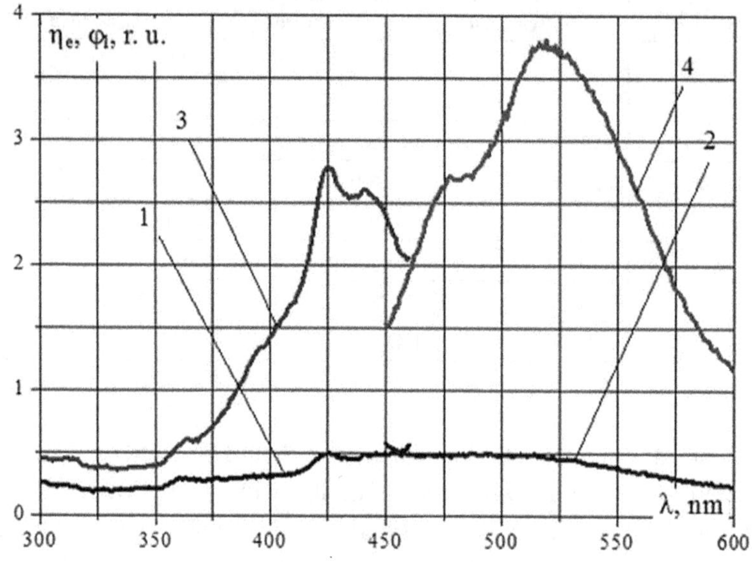

Table 5. Statistical and frequency parameters of excitation and luminescence of lupine seeds

Parameter	Excitation		Luminescence	
	Notscarified	Scarified		Notscarified
M_{λ}, nm	394	410	517	521
σ^2	2117	1502	1698	1399
Static moment of the 3rd order μ_3	0.00	0.00	15317	4233
Static moment of the 4th order μ_4	$8.91 \cdot 10^6$	$7.69 \cdot 10^6$	$5.67 \cdot 10^6$	$4.20 \cdot 10^6$
As	0.00	0.00	0.22	0.08
E_{λ}	-1.01	0.41	-1.03	-0.85
E, eV	3.19	3.05	2.42	2.39
Area	51	195	63	392

Averaging of excitation and luminescence spectra was clone (figure 5). The results of germination of Galega seeds were obtained and recorded in table 6. In addition, for mathematical processing, PanoramaPro determined integrals under excitation and luminescence curves and found their average value. Integrals under excitation curves were determined for two spectral regions: 380-475 nm and 475-505 nm to determine the ratio between the maxima.

The operating range of the excitation spectrum: 380-505 nm; the main maximum in non-scarified and weakly scarified seeds: 461 nm, and in strongly scarified such main maxima two: 450 and 460 nm; the side maximum in all three: 485 nm. Also, strongly scarified seeds have an additional side maximum

Figure 5. Excitation and luminescence spectra of Galega seeds: 3,6 –notscarified; 2,5 – subscarified; 1,4-strongly scarified

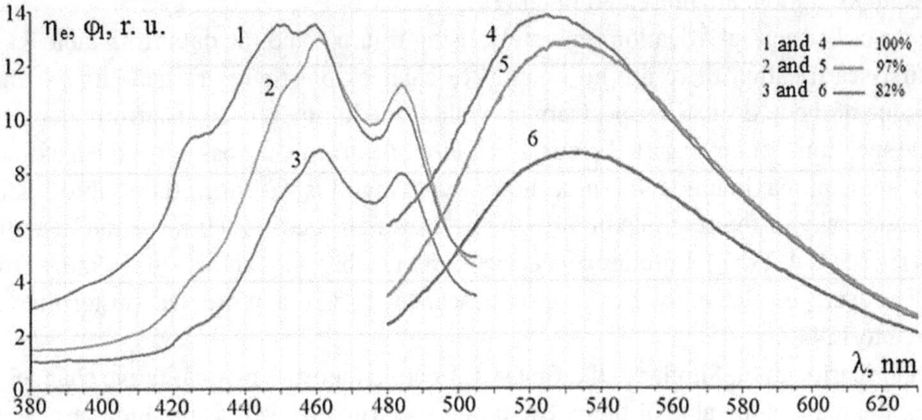

Table 6. Luminescence parameters of Galega seeds of different eastern germination under strong and weak scarification

Scarification	B, %	Excitation		Luminescence
		$H_{aver.}$, r.u.		$\Phi_{aver.}$, r.u.
		380-475	475-505	
Strong	100	1202	212	1282
Weak	97	680	292	1194
No scarification	82	340	165	833

at a wavelength of 425 nm. Luminescence spectrum range: 480-630 nm. The luminescence spectrum graph reaches the maximum in the range from 527 nm to 535 nm. Moreover, for strongly scarified seeds this range is shifted to the left: 520-530 nm.

In addition, the graph shows that due to scarification, the maximum value of the η_e of the two – modal excitation spectrum increased relative to the excitation spectrum of non – scarified seeds: for weakly scarified seeds-by 32%, and for strongly scarified seeds-by 34%. The position of maxima in non-scarified and weakly - scarified seeds has not changed but the ratio of their intensity has changed: in non-infected seeds the first maximum has an intensity of 0.9 r. u. greater than the second maximum, and in weakly-scarified seeds-2 r. u. If two main maxima of strongly scarified seeds are taken as one, the ratio of their intensity maxima differs by 3.1 r. u. The change in the intensity of the luminescence spectrum is similar to the change in the excitation spectrum: the intensity increased by 32 and 34%, respectively.

The second experience – the measurement of luminous characteristics of ten individual seeds galegov with increasing degree of scarification.

The experiment was done with the aim of identifying similar behaviors of the spectra of the seed galegov with increasing degree of scarification. Ten seeds were subject to scarification, each of which was scarified five times, and the degree of scarification increased each time. For measurements, the operating range of the excitation spectrum was: 420-505 nm; and the luminescence spectrum: 480-630 nm. For mathematical processing "PanoramaPro" determined the integrals under the curves of the luminescence and excitation, are presented in table 7.

Next we describe each scarification separately, using figure 6 and the data from table 7.

At the first scarification there are no qualitative changes but there are quantitative changes: the maximum value of the η_e excitation spectrum, as well as the luminescence spectrum increased by 30% relative to the spectrum of non-scarified seeds. Excitation spectrum: the position of the maxima remained unchanged (the main maximum is 460 nm; the side maximum is 484 nm); the ratio of their intensity changed (the intensity of the first maximum in non - scarified seeds is 0.9 r. u. more than the second, and in scarified seeds-1.2 r. u.). Luminescence spectrum: reaches a maximum in the range from 528 nm to 536 nm. The average value of the flow of luminescence at 350 r. u. more than nesterilizovanny seeds and amounted to 1244 r. u.

The second scarification. Similar to the first scarification there are no qualitative changes. Quantitative changes: the maximum value of the η_e excitation spectrum, as well as the luminescence spectrum increased by 36% relative to the spectrum of non-scarified seeds. Excitation spectrum: the position of the maxima remained unchanged (the main maximum is 460 nm; the side maximum is 484 nm); the ratio of their intensity changed (the intensity of the first maximum in twice scarified seeds is 1.9 r. u.). more than the second, in comparison with the same change in the intensity of non – scarified seeds-0.9 r. u.). Luminescence spectrum: maximum in the same range as the first scarification. The average value of the luminescence flux increased compared with the previous scarification, it is equal to 1353 r. u.

The third scarification. There are qualitative and quantitative changes. The maximum value of the η_e excitation spectrum, as well as the luminescence spectrum increased by 28% relative to the spectrum of non-scarified seeds, and relative to the spectrum of seeds, scarified twice, decreased by 13%. Excitation spectrum: the share of the spectrum in the short-wave excitation region increased; the position of the maxima remained unchanged (the main maximum is 460 nm; side maximum-484 nm), however, it became most pronounced to show two additional side maximum at a wavelength of 451 and in the range

Figure 6. Excitation and luminescence spectra of Galega seeds: 1,2 – nesterilizovanny; skalpirovaniya: 3,4 – once, 5,6-two times, 7,8-three times, 9,10-four times, 11,12-five times

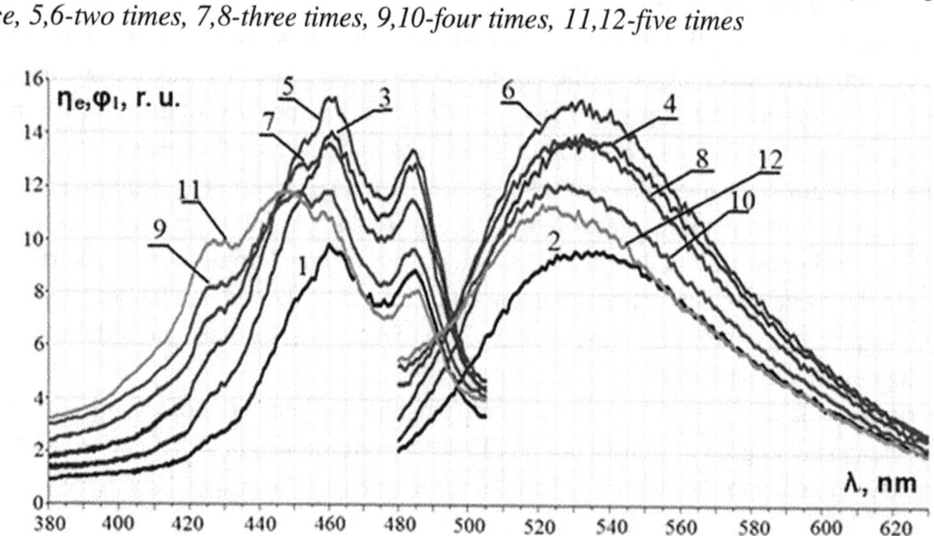

Table 7. Luminescence parameters of Galega seeds of different scarification degrees

Scarification	Excitation		Luminescence
	$H_{aver.}$, r.u.		$\Phi_{aver.}$, r.u.
	380-475	475-505	
No scarification	395	192	894
Initial	609	278	1244
Second	719	298	1353
Third	703	251	1269
Fourth	810	216	1145
Fifth	774	193	1054

of 425-432 nm; changed the ratio of the intensity of the main and side maxima (the intensity of the first at 2.1 r. u. more than the second). Luminescence spectrum: the maximum area is shifted to a shorter wavelength range-from 526 nm to 534 nm. The value of the flow of luminescence decreased to 1269 r. u.

The fourth scarification. There are also qualitative and quantitative changes. The maximum value of η_e excitation spectrum and luminescence spectrum increased by 17.6% relative to the spectrum of non-scarified seeds but decreased by 15% relative to the spectrum of seeds, scarified three times. Excitation spectrum: the share of the spectrum in the short-wave excitation region increased; the presence of two main maxima at the wavelength of 450 and 460 nm, the side maximum is the same-484 nm; an additional side maximum in the range of 425-432 nm; the ratio of the main intensity (take two for one) and the side peaks did not change (2.1 r. u.). The luminescence spectra: the region of the maximum is shifted further to shorter wavelengths from 523 nm to 530 nm. The flux of luminescence decreases (equal to 1145 r. u.).

The fifth scarification. Again, there are both qualitative and quantitative changes. Excitation spectrum: the maximum value of η_e increased by 16.9% relative to the spectrum of non-scarified seeds, relative to the spectrum of seeds scarified four times has not changed, only slightly decreased (0.8%); the main peak in the range of 445-451 nm becomes predominant over the former main peak (460 nm) and shifts to a shorter wavelength region; the side maximum is the same – 484 nm; additional side maximum in the range of 425-432 nm; the intensity ratio of the main and side maxima is 3.7 r. u. and is expressed brighter than in the previous spectra. Luminescence spectrum: the maximum area is further shifted to the short-wave region - from 520 nm to 527 nm; the maximum value of η_e increased by 14% relative to the spectrum of non-scarified seeds and decreased by 6% relative to the spectrum of seeds, scarified four times. The value of the luminescence flux was 1054 r. u., it continues to decrease.

Analyzing the results of the calculation of integrals, we noticed that luminescence can increase: for seeds, scarified 1 time by 87%, 2 times-109%, 3 times-120%, 4 times-148%, 5 times-25%; and fall: for seeds, scarified 1 time to 12%, 2 times-to -2%, 3 times - to -17%, 4 times - to -25%, 5 times-to -70%.

Integrals under excitation curves were determined for two spectral regions: 420-470 nm and 470-505 nm to determine the ratio between the maxima. To do this, we found the average value of all calculated integrals.

For spectra of non-scarified seeds, the average value of the integral in the range 380-475 nm is greater than the average value of the integral in the range 475-505 nm for seeds scarified: once in 2.01 times, twice – in 2.19 times, three times – in 2.41 times, 4 times – in 3.75 times, 5 times – in 4.01 times.

Similarly, the excitation spectra of $\eta_e(\lambda)$ and luminescence of $\varphi_l(\lambda)$ tetraploid clover seeds were measured; the operating range of the absorption spectrum was 380-500 nm, the operating range of the luminescence spectrum was 470-680 nm.

Spectra were measured nesterilizovanny of clover seeds and also after first and second clarificati. Measurements were carried out for 50 seeds. The curves are averaged, the result is shown in figure 7. The main peaks of the averaged excitation curves lie in the range of 448-450 nm, side-484 nm; peaks of the averaged luminescence curves lie in the range of 506-516 nm.

As can be seen from the figure, the signal intensity increases after the first and second scarifications. Let us calculate the integrals under the curve of the excitation and luminescence spectra, and the integral under the curve of luminescence is a relative capacity (flow) of luminescence. It was found that the area under the excitation curves after the first scarification increased by 2.38 times and after the second - 3.88 times relative to the spectrum of non-scarified seeds. The flow after the first scarification increases 1.98 times and after the second - 3 times relative to the original.

To determine the ratio between the maxima of the absorption curves we calculate the integrals under the curves in the ranges 380-475 nm and 475-500 nm. Initially the maxima had a ratio of 3.30 after the first scarification their ratio was 4.27, and after the second – 4.62. This suggests that the degree of scarification decreases the severity of the side maximum.

Then we carried out mathematical processing. The results are given in table 8.

Based on the data obtained it can be concluded that for the excitation and luminescence spectra, during the scarification process, the center of gravity shifts to the short-wave region and the total energy and area

Figure 7. Spectra of excitation and luminescence of clover seeds tetraploid th: 1,2 – not scarifed, 3,4 - scarifed once (weak), 5,6 - scarifed twice (badly), respectively

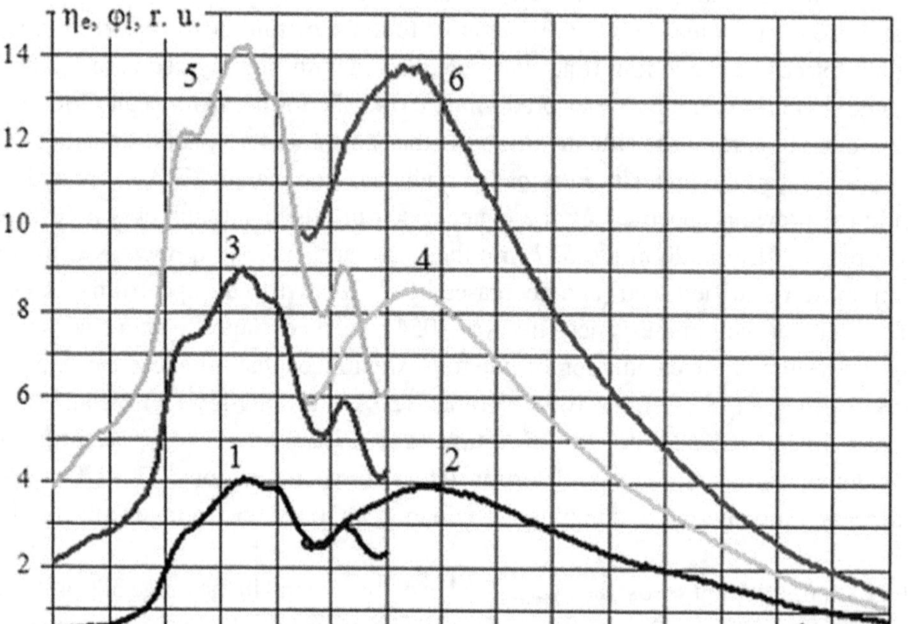

Table 8. Statistical parameters of clover seeds tetraploid in double scarification

Excitation								
Scarification	M_λ, nm	σ^2	μ_3	$\mu_4 \bullet 10^6$	As	E_λ	E, eV	H, r.u.
No	453	715	0.00	1.31	0.00	-0.44	2.75	279.6
1 time	447	834	0.00	1.63	0.00	-0.66	2.79	666.7
2 times	445	857	0.00	1.68	0.00	-0.71	2.80	1086
Luminescence								
Scarification	M_λ, nm	σ^2	$\mu_3 \bullet 10^4$	$\mu_4 \bullet 10^7$	As	E_λ	E, eV	Φ, r.u.
No	552	2813	7.59	1.85	0.51	-0.67	2.26	505.2
1 time	545	2585	8.70	1.75	0.66	-0.38	2.30	988.7
2 times	542	2482	9.02	1.70	0.73	-0.24	2.31	1516

under the spectrum curve increase. The variance of the excitation spectra increases, and the spectrum of luminescence decreases. The asymmetry coefficient for excitation spectra does not change and is equal to zero and for luminescence spectra it increases. The kurtosis of the spectra for luminescence is increased and for excitation spectra decreases. Stokes shift (by mathematical expectation) is: for non – scarified seeds – 99.86 nm; for the first scarification – 98.27 nm; for the second scarification – 96.72 nm.

For the experiment on multiple scarification 20 clover seeds were used. Initially the excitation and luminescence spectra of non-scarified seeds were removed. Then 8 scarifications were carried out, after each of which the spectra were repeatedly removed. The averaged absorption and photoluminescence spectra of non-scarified seeds and with varying degrees of scarification are shown in figure 8.

Figure 8. Averaged spectra of excitation and luminescence of seeds clover tetrap-rodnogo: not-scarifed (1,2), scarifed one time (3,4), twice (5,6), three times (7,8), four times (9,10), five times (11,12), six times (13,14), seven times (15,16) and eight times (17,18)

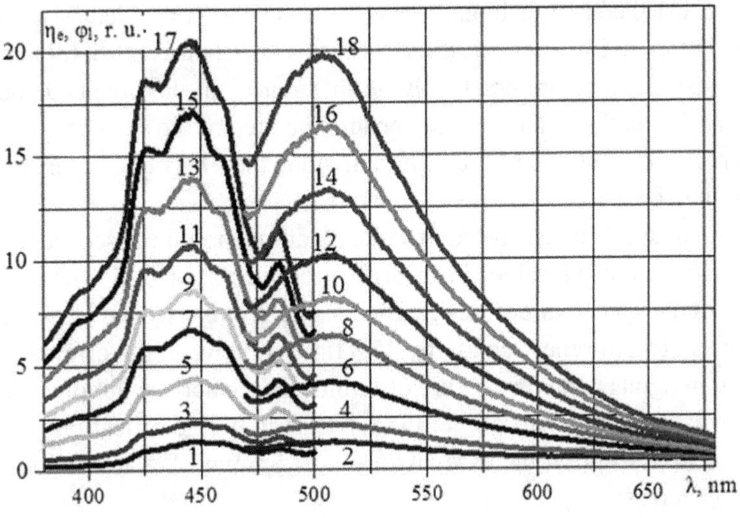

It can be seen from figure 8 that with an increase in the degree of scarification, two additional side peaks appear: the first in the range of 425-435 nm with peaks at 426-428 nm, and the second, less pronounced, at 455-465 nm.

In the framework of mathematical analysis let us calculate the integrals under the curves of the luminescence. After the first scarification, the flow increased 1.57 times, after the second – 2.89 times, after the third-4.24 times, the fourth-5.32 times, the fifth-6.53 times, the sixth-8.36 times, the seventh-10.2 times and after the eighth-12.1 times relative to the flow of non-scarified seeds.

We find two ratios of integrals under the absorption curves at intervals: 380-435 nm, 435-500 nm and 380-475 nm, 475-500 nm. For spectra nscarevocationurl clover the average value of the integral in the range 380-475 nm more than the average value of the integral in the range 475-500 nm 3.44 times; for the spectra of the seeds after the first scarification – in 3.97 times, after the second – 4.44%, the third at 4.72 times, the fourth – 4.94 times, the fifth is at 5.10%, the sixth – 5.25%, seventh – 5.32 times and after the eighth – 5.52 times. This suggests that with an increase in the degree of scarification side maximum, located at 475-500 nm becomes less pronounced in relation to the main. In the range of 380-435 nm is greater than the average value of the integral in the range of 435-500 nm in 2.68 times; for seed spectra, scarified once-2.00 times, twice-1.70 times, three times-1.62 times, four times-1.55 times, five times-1.52 times, six times-1.49 times, seven times-1.49 times and eight times-1.47 times. This means that with increasing degree of scarification side high, located on 380-435 nm grows in relation to the principal, that is, becomes more pronounced.

The results of spectra processing in Microcal Origin package are given in table 9. Analyzing these data, we can say that for the luminescence and excitation spectra each subsequent scarification is characterized by: an increase in the area, the asymmetry coefficients (for luminescence) and the kurtosis, the total energy of the spectrum; a decrease in the dispersion value; a shift in the center of gravity to the short-wave region. For luminescence spectra by the seventh scarification, the kurtosis changes its sign to the opposite one (the shape of the curve becomes more peaked than the normal distribution curve). The excitation spectra are characterized by an asymmetry coefficient equal to zero which indicates the symmetry of the curve with respect to the center of gravity.

The excitation spectra of $\eta_e(\lambda)$ and luminescence of $\varphi_l(\lambda)$ 50 lupine seeds were measured; the operating range of the excitation spectrum was 300-460 nm, the operating range of the luminescence spectrum was 450-600 nm. Spectra were measured nesterilizovanny lupine seeds and every time after two clarificati.

It should be noted that the shell of lupine seeds is dense and conceived of dark color, so that the signal intensity is very low. With weak scarification, when the shell is partially damaged but the seed body is not visually visible, there are practically no differences in the obtained spectra. Scarification was performed so that the shell was damaged, exposing the area of the seed body.

The averaged curves are shown in figure 9. It is obvious that the signal intensity after the first and second scarification increases.

Let us calculate the integrals under the curves of excitation and luminescence. It was found that the area under the excitation curves after the first scarification increased 11.5 times, and after the second - 32.1 times relative to the spectrum of non-scarified seeds. Luminescence after the first scarification increases by 16.1 times and after the second - by 45.0 times relative to the original.

The ratio between integrals under the curves of luminescence and excitation spectra was found. For notscarifed lupine seeds it amounts to 1.56, for the subsequent two scarification – 2.18 and 2.19. Thus, the flow with respect to absorption increases with the beginning of scarification and with a greater degree of it begins to decrease.

Table 9. Average parameters of excitation and luminescence notscarification and scarification seeds, cleverer multiple scarification

Luminescence								
	M_λ, nm	σ^2	μ_3	μ_4	As	E_λ	E, eV	Φ, r. u.
not	552	3211	100907	$2.30 \cdot 10^7$	0.55	-0.77	2.27	165
1	548	2957	100042	$2.11 \cdot 10^7$	0.62	-0.59	2.28	260
2	544	2727	99330	$1.95 \cdot 10^7$	0.70	-0.38	2.30	477
3	542	2577	98195	$1.83 \cdot 10^7$	0.75	-0.24	2.31	701
4	540	2493	96615	$1.77 \cdot 10^7$	0.78	-0.16	2.31	880
5	539	2439	96028	$1.72 \cdot 10^7$	0.80	-0.10	2.32	1081
6	538	2365	94074	$1.66 \cdot 10^7$	0.82	-0.03	2.32	1383
7	537	2328	93289	$1.63 \cdot 10^7$	0.83	0.01	2.33	1680
8	536	2285	92902	$1.60 \cdot 10^7$	0.85	0.07	2.33	2000
Excitation								
	M_λ, nm	σ^2	μ_3	μ_4	As	E_λ	E, eV	H, r. u.
not	452	739	0.00	$1.38 \cdot 10^6$	0.00	-0.47	2.76	94
1	447	833	0.00	$1.63 \cdot 10^6$	0.00	-0.65	2.78	168
2	445	875	0.00	$1.73 \cdot 10^6$	0.00	-0.74	2.80	337
3	444	876	0.00	$1.73 \cdot 10^6$	0,00	-0.75	2.81	517
4	443	877	0.00	$1.73 \cdot 10^6$	0.00	-0.76	2.81	663
5	443	877	0.00	$1.72 \cdot 10^6$	0.00	-0.76	2.81	831
6	442	869	0.00	$1.70 \cdot 10^6$	0.00	-0.75	2.81	1076
7	442	861	0.00	$1.68 \cdot 10^6$	0.00	-0.74	2.81	1312
8	442	851	0.00	$1.65 \cdot 10^6$	0.00	-0.73	2.82	1563

It is also seen from the figure that the luminescence spectrum is two-modal, and the excitation spectra during scarification additional maxima appear. To determine the ratio between the maxima of the luminescence curves we calculate the integrals under the curves in two ranges: 450-485 nm and 485-600 nm. Initially, the highs had a ratio of 3.55 after the first scarification their ratio was 4.18, and after the second – 4.33. This suggests that the degree of scarification decreases the severity of the side maximum.

Calculate the ratio of integrals under the absorption curves at intervals of 300-435nm and 385-460 nm. For the spectra of non-scarified lupine seeds, this ratio is 2.58, for the spectra of scarified seeds – 1.99, and after the second scarification – 2.80, i.e. after the first scarification, the side maximum becomes more pronounced, and after the second scarification it undergoes a slight decrease with respect to the main maximum with respect to the initial value.

The maxima of averaged excitation and luminescence curves for scarified seeds are as follows: absorption - 424-426 nm main, 439-440 nm side, luminescence - 484 nm side, 517-520 nm main.

Then we performed mathematical processing, the results of which are presented in table 10.

After analyzing the data from the table, we obtained that for the luminescence spectra there is no change in the total energy of the spectrum and for the excitation spectra the changes are of an oscillatory nature. For both excitation and luminescence spectra an oscillatory shift of the center of gravity is

Figure 9. Spectra of excitation and luminescence of lupine seeds: 1,2 - notscarifed, 3,4 - scarifed once (weak), 5,6 - scarifed (greatly), respectively

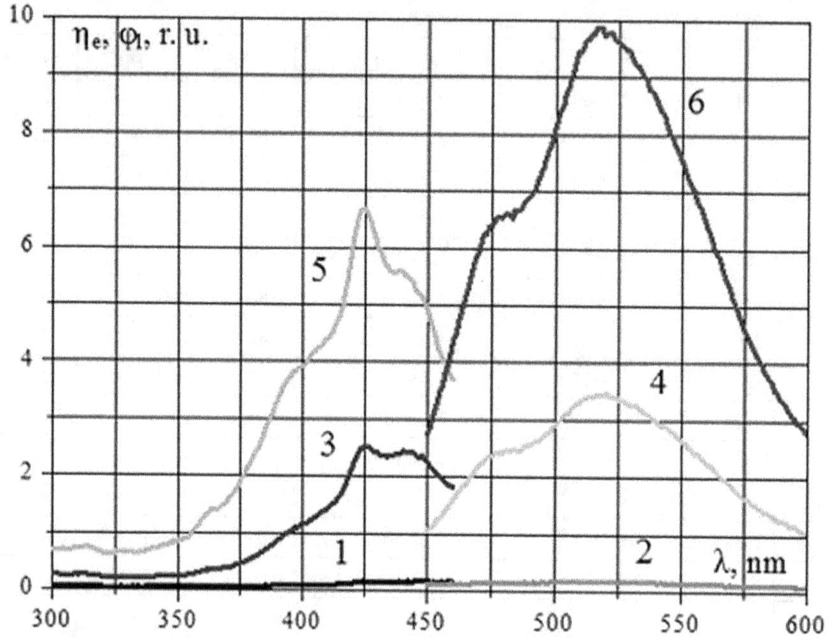

Table 10. Statistical parameters of lupine seeds with double scarification

				Excitation				
Scarification	**M_λ, nm**	**σ^2**	**μ_3**	**$\mu_4 \cdot 10^6$**	**As**	**E_λ**	**E, eV**	**H, r.u.**
No	395	2291	0.00	10.1	0.00	-1.08	3.19	14.05
1 time	412	1500	0.00	8.30	0.00	0.69	3.04	161.86
2 times	407	1486	0.00	7.06	0.00	0.20	3.08	451.23
				Luminescence				
Scarification	**M_λ, nm**	**σ^2**	**$\mu_3 \cdot 10^3$**	**$\mu_4 \cdot 10^6$**	**As**	**E_λ**	**E, eV**	**Φ, r.u.**
No	522	1622	4.52	5.21	0.07	-1.02	2.39	21.96
1 time	522	1370	4.472	4.06	0.09	-0.84	2.39	353.44
2 times	522	1343	3.53	3.94	0.07	-0.81	2.39	987.25

observed in the long-wave and short-wave regions. The area under the curve increases for all spectra during the scarification process. The kurtosis for luminescence does not change, for excitation - fluctuates. For the excitation spectrum at the first scarification, a change in the sign of the kurtosis to the opposite is observed. The variance of all the spectra is reduced at each scarification. Lupin spectra are characterized by an asymmetry coefficient equal to zero which indicates the symmetry of the curve relative to the center of gravity. Stokes shift (by mathematical expectation) is: for non – scarified seeds – 127.03 nm; for the first scarification – 109.29 nm; for the second scarification-114.95 nm.

Also, the laying of seeds for germination was carried out, the data obtained are given in table 11.

To conduct the experiment with lupine 20 seeds were used. Initially, the excitation and luminescence spectra of non-scarified seeds were removed. Then 8 scarifications were carried out, after each of which the spectra were repeatedly removed. The averaged absorption and photoluminescence spectra of non-scarified seeds and with varying degrees of scarification are shown in figure 10.

From figure 10 it can be seen that with an increase in the degree of scarification, the excitation spectrum has additional maxima: more pronounced at 355-365 nm and less bright at 390-400 nm, and a side maximum with peaks at 442-448 nm loses its severity.

Let us calculate the integrals under the curves of the luminescence. After the first scarification, the flow increased 4.84 times, after the second – 13.3 times, after the third – 22.9 times, the fourth-31.3 times, the fifth – 36.7 times, the sixth – 45.2 times, the seventh – 49.4 times and after the eighth-56.9 times relative to the flow of non-scarified seeds. The ratio between integrals under the curves of luminescence and excitation spectra was found. For notscarified lupine seeds it is 1.25, for subsequent scarification–

Table 11. Germination and luminescent parameters of lupine seeds

Scarification	Germination Energy, %	B, %	φ_p, r.u.	Ф, r.u.
No	4	8	0.19	21.96
1 time	32	96	3.46	353.4
2 times	82	98	9.89	987.3

Figure 10. Averaged excitation and luminescence spectra of lupine seeds: notscarfied (1,2), scarified once (3,4), twice (5,6), three times (7.8), four times (9.10), five times (11.12), six times (13.14), seven times (15,16) and eight times (17,18)

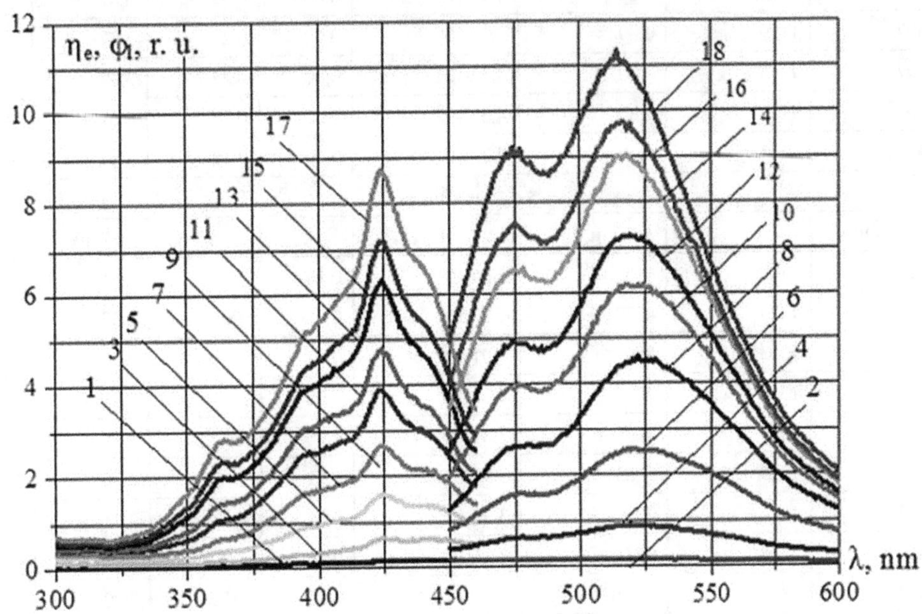

1.94, 2.24, 2.34, 2.19, 2.11, 2.00, 1.92 and 1.84. Thus, the flow with respect to absorption increases with the beginning of scarification, and with its large degree begins to decrease.

We calculate two ratios of integrals under the absorption curves: at intervals of 300-370 nm and 370-460 nm. The average value of the integral in the range of 300-370 nm is greater than the average value of the integral in the range of 370-460 nm by 2.08 times; for the spectra of seeds after the first scarification – 4.46 times, after the second – 5.34 times, the third – 5.76 times, the fourth – 5.63 times, the fifth – 5.61 times, the sixth – 5.34 times, the seventh – 5.45 times and after the eighth – 5.48 times. I. e. when after the first three scarifications, the side maximum becomes more pronounced, and after the scarification of a strong degree, it undergoes a slight decrease in relation to the main maximum. The results of spectra processing in The MicrocalOrigin package are given in table 12.

Analyzing the data from table 12, it was found that for the luminescence and excitation spectra, the change in the total energy of the spectrum and the center of gravity (for excitation spectra) was observed only for the first seed scarification. There are no significant changes in subsequent scarifications. For luminescence spectra, a shift of the center of gravity to the long-wave region before the third scarification is observed, followed by a shift to the short-wave region. The kurtosis coefficient and the area under the curve (for luminescence spectra) increases for all spectra with each scarification. For the excitation

Table 12. Average parameters of excitation and luminescence nesterilizovanny and skalpirovaniya lupine seeds by repeated scarification

Excitation								
	M_λ, nm	σ^2	μ_3	μ_4	As	E_λ	*E*, eV	H, r. u.
no	393	2147	0.00	$8.98 \cdot 10^6$	0.00	-1.05	3.20	15
1	408	1658	0.00	$8.32 \cdot 10^6$	0.00	0.03	3.07	48
2	408	1423	0.00	$6.74 \cdot 10^6$	0.00	0.33	3.07	113
3	408	1303	0.00	$5.79 \cdot 10^6$	0.00	0.41	3.06	187
4	407	1244	0.00	$5.15 \cdot 10^6$	0.00	0.33	3.07	272
5	407	1199	0.00	$4.70 \cdot 10^6$	0.00	0.27	3.07	332
6	407	1166	0.00	$4.37 \cdot 10^6$	0.00	0.22	3.07	432
7	407	1174	0.00	$4.37 \cdot 10^6$	0.00	0.17	3.07	491
8	408	1163	0.00	$4.29 \cdot 10^6$	0.00	0.17	3.07	590
Luminescence								
	M_λ, nm	σ^2	μ_3	μ_4	As	E_λ	*E*, eV	Φ, r. u.
no	519	1708	11528	$5.66 \cdot 10^6$	0.16	-1.06	2.41	19
1	521	1423	4377	$4.33 \cdot 10^6$	0.08	-0.86	2.39	92
2	523	1353	989	$4.01 \cdot 10^6$	0.02	-0.81	2.38	253
3	524	1321	30	$3.88 \cdot 10^6$	0.00	-0.78	2.38	436
4	521	1324	3096	$3.89 \cdot 10^6$	0.06	-0.78	2.39	596
5	519	1325	5083	$3.90 \cdot 10^6$	0.11	-0.78	2.40	700
6	517	1320	7859	$3.90 \cdot 10^6$	0.16	-0.76	2.41	863
7	516	1318	9721	$3.91 \cdot 10^6$	0.20	-0.75	2.42	943
8	514	1314	12150	$3.93 \cdot 10^6$	0.26	-0.73	2.42	1085

spectrum at the first scarification, there is a change in the sign of the kurtosis to the opposite; the coefficient of kurtosis increases to the third scarification, then this value decreases. The variance of all the spectra is reduced at each scarification. Lupin spectra are characterized by an asymmetry coefficient equal to zero which indicates the symmetry of the curve relative to the center of gravity.

A study of the time dependence of the luminescent properties of 30 Eastern Galega seeds during scarification was carried out. The excitation and luminescence spectra of non-scarified seeds, the spectra measured immediately after scarification and the spectra measured one day after scarification and 7 days after scarification were measured. The excitation and luminescence spectra are plotted and averaged (figure 11).

Havinf got the results of germination, we listed in them table 13. The process of increasing seed germination after their scarification has inertia: immediately after scarification increases from 82% to 90%, and after 7 days - up to 100%. Determined the integrals under the curves of excitation and luminescence, and found their average.

The operating range of the excitation spectrum: 420-505 nm; the main maximum in non-scarified, measured immediately after scarification and a day after scarification – 460 nm, and measured in 7 days – 461 nm. Luminescence spectrum range: 480-630 nm. The luminescence spectrum graph reaches the maximum in the range from 527 nm to 535 nm. Moreover, for seeds measured after 7 days this range is slightly shifted to the left.

Figure 11 and table 13 show that after the scarification the relative luminescence flux continues to increase: immediately after the scarification – by 31%, a day – by 23% and after 7 days – by 16.7% from the previous measurement. It follows that it is recommended to carry out scarification a week before the expected sowing.

To carry out a similar experiment, excitation and luminescence spectra were measured for 20 clover seeds of different terms that have passed since the scarification (on the day of scarification, the next day, every other day, etc.). After the curves were averaged, the results are presented in figure 12.

The maxima of the averaged curves of the excitation spectra are in the range 447-450 nm-the main maximum and 485-486 nm – the side maximum, the maxima of the averaged curves of the luminescence spectra are in the range 511-520 nm.

Figure 11. Excitation and luminescence Spectra Galega seeds: 4,8-notscarified; 3,7-measured immediately after scarification; 2,6-measured one day after scarification; 1,5-measured 7 days after scarification

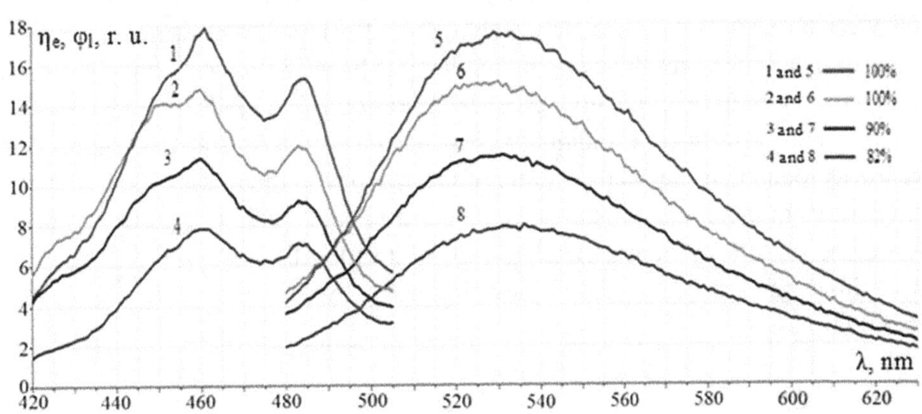

Table 13. Luminescence parameters of Galega eastern seeds depending on time after scarification

Seeds	B, %	Excitation	Luminescence
		$H_{aver.}$, r.u.	$\Phi_{aver.}$, r.u.
Noscarification	82	362	724
Measured immediately after scarification	90	556	1056
Measured one day after scarification	100	808	1362
Measured 7 days after scarification	100	896	1590

Figure 12. Averaged excitation and luminescence spectra of clover seeds: notscarificed (1,2), scarified and measured on the first day (3,4), the second day (5,6), the third day(7,8), the fourth day (9,10), the fifth day (11,12), eighth day (13,14) and eleventh day (15,16)

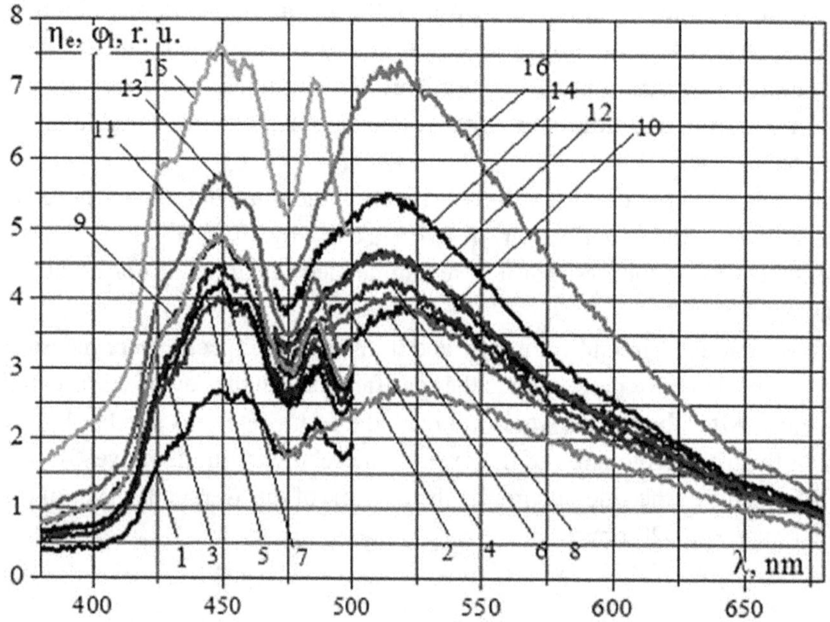

For the mathematical treatment we calculated the integrals under the Cree-ment of excitation and luminescence. It can be noted that immediately after the scarification luminescence can increase by 35%, 41% – on the second day, 43% –the third day, 54% –the fourth day, 55% –the fifth, 79% –the eighth and 135% –the eleventh day relative to the luminescence of non-scarified seeds.

The integrals under the curve of excitation were determined for two spectral regions: 380-475 nm, 475-500 nm. It was found that the average value of the integral in the range of 380-475 nm is greater than the average value of the integral in the range of 475-500 nm by 2.93 times; for the spectra of scarified seeds measured on the day of scarification – by 2.99 times, the second day – by 3.38 times, the third – by 3.19 times, the fourth – by 3.45 times, the fifth – by 3. 33 times, the eighth-in 3.45 times, eleventh – in 3.06 times.

At the same time, clover seeds were laid for germination. For this purpose 700 seeds were scarified. Laying of 100 seeds was carried out in stages for different time passed from the moment of scarification (on the day of scarification, the next day, every other day, etc.). Germination energy and germination (*B*) for different periods following scarification are given in table 14.

Laying of seeds on the second day after scarification was carried out unsuccessfully, the degree of damage to samples by fungi is strong, the results of the experiment are not applicable for further processing and statistics.

As it can be seen from the data presented in table 14, the germination, as expected from previous studies with other crops, increases markedly over time. A sharp increase in germination from 10% for non - scarified and 13% for scarified first term (bookmark on the day of scarification) to 88% for the seventh term (after 10 days after scarification can be explained by the fact that it is not known in advance which seeds are germinated and which are not. During the calculation of germinated seeds part of the seeds laid fo the germination, had no signs of swelling, which suggests that they were initially dissimilar and non-viable, it is also possible the presence of random factors that influenced the germination of seeds.

Table 15 presents the results of excitation treatment and luminescence spectra in MicrocalOrigin.

From the analysis of the data in table 15, it follows that for all clover spectra the area under the curve and the total energy of the spectrum increases. For excitation spectra, the spectrum curve is symmetric with respect to the center of gravity, the dispersion increases and the value of the kurtosis decreases; the center of gravity of the spectra slightly shifts to the short-wave region. For luminescence spectra, the excess and the asymmetry coefficient increase, the dispersion decreases; the center of gravity of the spectra slightly shifts to the short-wave region.

A similar experiment was conducted with lupine seeds. After the curves were averaged the results are shown in figure 13.

The picture shows that the spectra of scarified seeds at different times are randomly arranged and superimposed on each other.

Let us calculate the integrals under the curves of the luminescence. After scarification, the flow increased by 7.78 times, on the second day – by 7.53 times, on the third-by 7.39 times, the fourth-by 7.35 times, the fifth – by 7.96 times and on the eighth – by 7.76 times relative to the flow of non-scarified seeds. From this it can be concluded that the use of lupine changes after a certain period of time after

Table 14. Clover germination and luminescent parameters

Time After Scarification	*B*, %	φ_l, r.u.	Φ, r.u.
Not scarification	10	2.84	385.3
0 days	13	4.02	542.1
1 day	41	4.23	520.6
2 days	-	4.50	552.3
3 days	54	4.89	592.5
4 days	67	4.91	597.4
7 days	73	5.77	688.7
10 days	88	7.65	905.6

Table 15. Average excitation and luminescence parameters notscarification and scarification of clover seeds depending on the time after scarification

Excitation								
Day	**M_λ, nm**	**σ^2**	**μ_3**	**μ_4**	**As**	**E_λ**	**E, eV**	**H, r. u.**
no	453	751	0.00	$1.45{\cdot}10^6$	0.00	-0.43	2.74	191
1 day	453	772	0.00	$1.49{\cdot}10^6$	0.00	-0.50	2.75	289
Day 2	451	757	0.00	$1.45{\cdot}10^6$	0.00	-0.47	2.76	291
Day 3	453	714	0.00	$1.32{\cdot}10^6$	0.00	-0.42	2.75	302
Day 4	451	775	0.00	$1.48{\cdot}10^6$	0.00	-0.53	2.76	346
Day 5	451	779	0.00	$1.50{\cdot}10^6$	0.00	-0.53	2.76	349
Day 8	450	785	0.00	$1.51{\cdot}10^6$	0.00	-0.55	2.76	412
Day 11	450	865	0.00	$1.73{\cdot}10^6$	0.00	-0.69	2.77	606
Luminescence								
Day	**M_λ, nm**	**σ^2**	**μ_3**	**μ_4**	**As**	**E_λ**	**E, eV**	**Φ, r. u.**
no	558	2931	58988	$1.87{\cdot}10^7$	0.37	-0.82	2.24	385
1 day	552	2950	82723	$1.99{\cdot}10^7$	0.52	-0.71	2.27	521
Day 2	557	2912	61997	$1.86{\cdot}10^7$	0.39	-0.80	2.25	542
Day 3	552	2939	82249	$1.98{\cdot}10^7$	0.52	-0.70	2.27	552
Day 4	551	2851	82815	$1.92{\cdot}10^7$	0.54	-0.64	2.27	593
Day 5	551	2848	80758	$1.91{\cdot}10^7$	0.53	-0.65	2.27	598
Day 8	550	2788	82468	$1.87{\cdot}10^7$	0.56	-0.59	2.28	689
Day 11	551	2639	72688	$1.70{\cdot}10^7$	0.54	-0.55	2.27	906

Figure 13. Averaged excitation and luminescence spectra of lupine seeds: notscarified (1,2), scarified and measured on the first day (3,4), the second day (5,6), the third day(7,8), the fourth day (9,10), the fifth day (11,12) and on the eighth day (13,14)

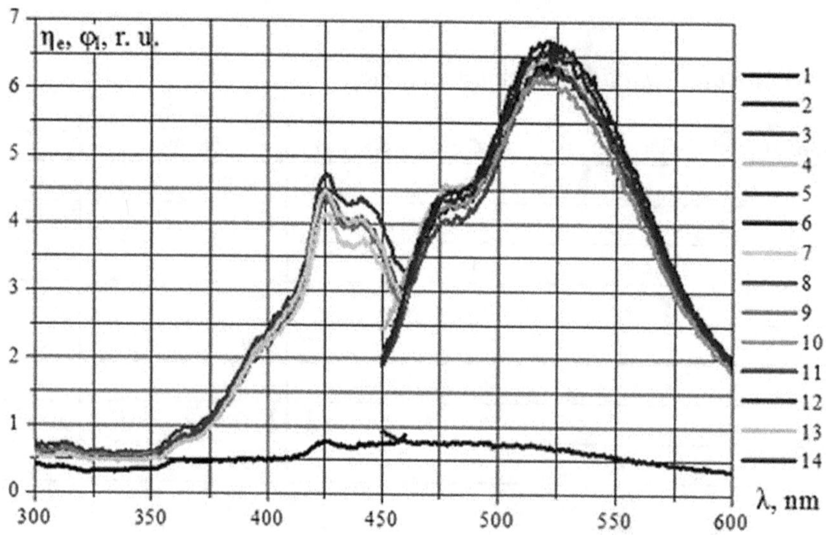

scarification is impractical, since significant changes in the characteristics of the excitation and luminescence spectra do not occur.

The results of data processing spectra in Microcal Origin are given in table 16.

From the analysis of the data given in table 16 it follows that the change in the total energy of the spectrum, dispersion, center of gravity and kurtosis is observed only for scarified seeds on the first day of observation. There are no significant changes in the following days. Lupine seeds are characterized by symmetry of spectral excitation and luminescence curves relative to the center of gravity. The coefficient of kurtosis for the excitation spectrum during scarification changed the sign to the opposite (the shape of the spectrum curve became more peaked than the normal distribution curve).

The results of this study can justify the development of technology for the diagnosis of plant seeds. The further direction of research is the development of an optical luminescent analyzer of germination of seeds of forage plants during scarification.

CONCLUSION

1. The studied plant seeds have excitation spectra with maxima of 424 and 485 nm, less often 455, 530 and others. Luminescence spectra are single-modal, located in the visible region and have maxima in the range 480...520 nm.

Table 16. Average excitation and luminescence parameters notscarification and scarification lupine seeds depending on the time factor

Excitation								
Day	**M_λ, nm**	**σ^2**	**μ_3**	**μ_4**	**As**	**E_λ**	**E, eV**	**H, r. u.**
no	393	2109	0.00	$8.75 \cdot 10^6$	0.00	-1.03	3.20	81
1 day	411	1453	0.00	$7.46 \cdot 10^6$	0.00	0.53	3.05	325
Day 2	411	1432	0.00	$7.44 \cdot 10^6$	0.00	0.63	3.04	297
Day 3	411	1393	0.00	$7.01 \cdot 10^6$	0.00	0.61	3.04	282
Day 4	411	1401	0.00	$7.12 \cdot 10^6$	0.00	0.62	3.05	299
Day 5	412	1343	0.00	$6.74 \cdot 10^6$	0.00	0.73	3.04	303
Day 8	413	1333	0.00	$6.86 \cdot 10^6$	0.00	0.86	3.03	296
Luminescence								
Day	**M_λ, nm**	**σ^2**	**μ_3**	**μ_4**	**As**	**E_λ**	**E, eV**	**Φ, r. u.**
no	511	1481	1522	$4.35 \cdot 10^6$	0.27	-1.02	2.44	86
1 day	522	1395	3664	$4.18 \cdot 10^6$	0.07	-0.85	2.39	669
Day 2	523	1371	2514	$4.07 \cdot 10^6$	0.05	-0.84	2.39	647
Day 3	523	1361	1649	$4.03 \cdot 10^6$	0.03	-0.83	2.38	635
Day 4	522	1377	3687	$4.09 \cdot 10^6$	0.07	-0.84	2.39	632
Day 5	523	1358	2262	$4.00 \cdot 10^6$	0.05	-0.83	2.38	684
Day 8	523	1360	2373	$4.02 \cdot 10^6$	0.05	-0.83	2.38	667

2. The spectra of excitation and luminescence of seeds of most grain, leguminous and vegetable crops are similar. Such spectra can be considered typical in the design of devices for diagnosing the quality of seeds.
3. The mathematical expectation of the studied cultures majority is within 472 ... 550 nm. All spectra have right-hand asymmetry. All spectra are flat-topped. The energy spectrum of the photoluminescence is in the range of 2.08...3.05 eV.
4. When the seeds of forage plants are scarified, the spectral characteristics increase. In Galega seeds with multiple scarification observed qualitative changes in the excitation spectrum associated with the appearance of a new maximum at a wavelength of 423 nm. Similarly for clover seeds.
5. For clover seeds each subsequent scarification is characterized by an increase in the area, the asymmetry and kurtosis coefficients, the total energy of the spectrum and the shift of gravity center to the short-wave region. For luminescence spectra by the seventh scarification, the kurtosis changes its sign to the opposite. Spectra of lupine seeds with multiple scarification show only quantitative changes. Lupin spectra are characterized by an asymmetry coefficient equal to zero.

REFERENCES

Afanas'ev, V. N., Afanas'eva, G. V., Biryukov, S. V., & Beleckij I. P. (2009). *Mnogofunkcional'noe ustrojstvo dlya diagnostiki i sposob testirovaniya biologicheskih ob"ektov* [Multifunctional device for diagnostics and testing of biological objects]. Patent RF. no 2363948.

Baek, I., Cho, B., Kim, M., & Kim,Y. (2013) Determination of optimal excitation and emission wavebands for detection of defect cherry tomato by using fluorescence emission and excitation matrix. *Proc. SPIE 8721, Sensing for Agriculture and Food Quality and Safety V.*, 872108

Bashilov, A. M. (2005). *EHlektronno-opticheskoe zrenie v agrarnom proizvodstve [Electro-optical sight in the agricultural production]*. Moscow, Russia: GNU VIEHSKH.

Bashilov, A. M., & Belyakov, M. V. (2015). *Spektral'nye harakteristiki lyuminescencii i otrazheniya semyan agrokul'tur [Spectral characteristics of luminescence and reflection of agricultural seeds]*. Moscow, Russia: FBGNU VIEHSKH.

Belyakov, M. V. (2015). Tipovye spektral'nye harakteristiki semyan rastenij [Typical spectral characteristics of plant seeds]. *Estestvennye i tekhnicheskie nauki, 11*, 521-525.

Belyakov, M. V. (2016). Metodika issledovaniya lyuminescentnyh svojstv semyan rastenij na spektrofluorimetre «Flyuorat-02-Panorama» [Methods of investigation of luminescent properties of plant seeds on the spectrofluorometer "Fluorat-02-Panorama»]. *Nauchnaya zhizn, 3*, 18-26.

Belyakov, M. V., & Kharitonova, D. A. (2016). Research of galega orientalis seeds with luminescent methods during scarification. In *Proceedings of Conference XII International scientific and practical conference*. Sheffield, UK: Science and Education LTD.

Belzile, C., Belanger, M., Viau, A., Chamberland, M., & Roy, S. (2004). An operational fluorescence system for crop assessment [Monitoring Food Safety, Agriculture, and Plant Health]. *Proceedings of the Society for Photo-Instrumentation Engineers*, 5271.

Bergmann, A. (2009). *Diagnostic method for disorders using copeptin.* Patent US. no 2009221009.

Gaevskij, N. A. (2002). *Kriterii i metodologiya ocenki strukturno-funkcional'nogo sostoyaniya al'gocenoza na osnove fluorescentnogo analiza* [Criteria and methodology for assessing the structural and functional state of algocenosis based on fluorescent analysis] (Unpublished doctoral dissertation). Institut biofiziki SO RAN, Krasnoyarsk, Russia.

Leemans, V., Marlier, G., Destain, M., Dumont, B., & Mercatoris, B. (2017). Estimation of leaf nitrogen concentration on winter wheat by multispectral imaging. *Proc. SPIE 10213. Hyperspectral Imaging Sensors: Innovative Applications and Sensor Standards, 2017.*

Li, C., Wang, X., & Meng, Z. (2016). Tomato seeds maturity detection system based on chlorophyll fluorescence. *Proc. SPIE 10021. Optical Design and Testing, 7.*

Lu, Y., Li, R., & Lu, R. (2016). Detection of fresh bruises in apples by structured-illumination reflectance imaging. *Proc. SPIE 9864. Sensing for Agriculture and Food Quality and Safety, 8*, 986406.

Monich, V. A., Monich, E. A., Golikov, V. M., Zhirkov, A. R., & Malinovskaya S. L. (1994). *Sposob izmereniya intensivnosti lyuminescencii v ob"eme sredy, preimushchestvenno biologicheskih ob"ektov* [The method of measuring the intensity of luminescence of the total environment, mainly of biological objects]. Patent RF. no 2012213.

Noh, H., & Lu, R. (2006) Integrating fluorescence and interactance measurements to improve apple maturity assessment. *Proc. SPIE 6381, Optics for Natural Resources, Agriculture, and Foods.*

Ovchinnikova, I. A. (2005). *Monitoring sostoyaniya biotkanej metodami polyarizacionno-otrazhatel'noj i fluorescentnoj spektroskopii* [Monitoring of biological tissues by the methods of polarization-reflective and fluorescent spectroscopy] (Unpublished doctoral dissertation). Saratovskij gosudarstvennyj universitet Saratov, Russia.

Ryabova, A. V. (2006) *Kombinirovannyj spektroskopicheskij metod analiza ehffektivnosti sensibilizatorov v biologicheskih ob"ektah* [A combined spectroscopic method of analysis of the effectiveness of sensitizers in biological objects] (Unpublished doctoral dissertation). Institut obshchej fiziki RAN, Moscow, Russia.

Sventickij, I.I., Timchenko, S.D., & Usmanov, S.A. (1990). *Opredelenie kachestva semyan po opticheskim spektral'nym harakteristikam* [Determination of seed quality by optical spectral characteristics]. Pushchino, Russia: AN SSSR, Nauch. centr biol. issled., In-t pochvovedeniya i fotosinteza, In-t radiotekhniki i ehlektron.

The Company Lumex. (2018). Technical characteristics of the spectrofluorimeter Fluorat-02-Panorama [Data file]. Retrieved from: http://www.lumex.ru/catalog/flyuorat-02-panorama.php#specification

Chapter 19
Researches of Technology Electrohydraulic Effect:
Impact on Water

Jorge Vinna Sabrejos
The Antenor Orrego Private University, Peru

Alexey Nikolaevich Vasilyev
https://orcid.org/0000-0002-7988-2338
Federal Scientific Agroengineering Center VIM, Russia

Alexander Anatolievich Belov
https://orcid.org/0000-0002-9216-9852
Federal Scientific Agroengineering Center VIM, Russia

Viktor Nikolaevich Toporkov
Federal Scientific Agroengineering Center VIM, Russia

Andrey Anatolievich Musenko
Federal Scientific Agroengineering Center VIM, Russia

ABSTRACT

The purpose of the chapter is to study the technology and technical means of electrohydraulic action on water. The authors justify the relevance of the research. The design of the original negative electrode tip is being developed to increase the density of the electromagnetic field and reduce power loss. The design parameters of the electrohydraulic installation are shown. Modeling of factors influencing the process of electrohydraulic treatment of water according to the Plackett-Berman plan and the random balance method is carried out; significant and insignificant factors are identified. The operation modes of the electrohydraulic installation are determined and optimized experimentally. The substantiation of the economic feasibility of using electrohydraulic water treatment technology in farms is being conducted. The prospects and scope of electrohydraulic technology are determined.

DOI: 10.4018/978-1-5225-9420-8.ch019

INTRODUCTION

According to statistics from the National Fruit and Vegetable Union, the gross yield of greenhouse vegetables in Russia as of October 2018 was 803 thousand tons. These figures were 25% ahead of the same period in 2017, including cucumbers – 490,6 thousand tons (+ 14,6%), tomatoes – 299,9 thousand tons (+ 46,8%), other vegetable crops – 12,5 thousand tons (+ 26,2%) (Belov, 2018). In this regard, the authors note the promise of exploratory research on improving the processes of vegetable growing in protected soil in order to increase the efficiency of agricultural production. The purpose of the chapter is the study of the technology and technical means of electrohydraulic action on water.

BACKGROUND

Plants need to be provided with nutrients that are in large quantities in water solutions of the soil, but in a form that is not digestible. To translate useful organics and minerals into a form that is easily accessible to plants, it is necessary to degrade the cellulose and lignin shell of membranes, inside which are located useful substances for plants. The most suitable for these purposes, in terms of cheapness, simplicity of execution and efficiency, is the electrohydraulic method, the essence of which is that during high-voltage pulsed discharges within a liquid, high and ultrahigh pressures reach to hundreds of thousands of atmospheres (Locke, Sato, Sunka, Hoffmann & Chang, 2006). These pressures can perform useful mechanical work when a complex of physical and chemical phenomena occurs. These are both resonant phenomena and high-intensity ultrasonic vibrations, as a result of which chemical bonds in molecules are destroyed, and then combined again, forming new substances that improve soil fertility. In nature, it takes tens and hundreds of years, and with electrohydraulic (EH) processing - less than a second (Müller, Gelfond, Eisert, 2014). Electrohydraulic effect (EHE) is developed by Yutkin L.A. for long ago (Yutkin, 1986). The fundamental and applied values of discovery are proved by the author. EH-effect is widely used in various branches of science and technology, including industry, medicine and agriculture (Kutter, 1969). The electrohydraulic effect is increasingly being used in various technological processes of plant growing. The authors propose to use this phenomenon for the treatment of water or solutions for irrigation in plant cultivation for agrotechnical purposes. Due to the cheapness and simplicity of devices for producing fertilizers by this method, this technology can be widely used in greenhouses, personal subsidiary and peasant (farmer) farms. The advantage of EH-technology is the absence of a negative impact on the environment (Mamutov, Golovashchenko, Mamutov & John Bonnenc, 2015). In addition, these same devices can also be used to treat water taken from any well. Studies have shown that in electrohydraulically treated water from any source, the amount of nitrogen compounds dissolved in water can increase hundreds of times even with a small expenditure of energy. Mostly this is characterized by the supply of air through the water under low pressure. After EH exposure, activated water promotes active leaching from the soil and passes into the soluble state of the chemical elements needed by plants (Toporkov, 2017). Thus, the authors have identified a proposal to use water for watering plants as a substitute for nitrogen fertilizer, which is considered more preferable than the introduction of chemical fertilizers into the soil or dirt, the impact of which is negative from an environmental point of view. In addition, is valuable the fact that the soil in the case of watering with water, rather than introducing by chemistry, can be reused seasonally throughout the year depending on the type of crop. This will have a positive effect on the economic component in terms of eliminating downtime of the soil or protected ground. Nutrient solution is a key factor in growing vegetable and flower crops. The main sources of water supply for irrigating crops are surface waters, water from artesian wells and wells. Industrial-scale tap water is practically not used for these purposes, due to its high cost. The ideal option for irrigation would be rainwater, since there are almost no salts and bicarbonates in it, however, of course, it cannot be a constancy source for which it can be counted on. Growing plants need good water quality, but very often they suffer due to the fact that irrigation water is heavily polluted by pathogenic microorganisms. Kila, blackleg, white rot, gall nematode, strick, late blight, etc. are transmitted through soil moisture. All this adversely affects growth and plant development. There are several methods of water disinfection: ultraviolet radiation, reverse osmosis, water ozonation (Tijani, Fatoba, Madzivire & Petrik, 2014). The authors propose to use the EH-treatment of aqueous nutrient solutions as a disinfection and sterilization a certain specie. The electrohydraulic effect is a powerful source of ultrasound, as well as of ultraviolet and x-ray radiation, and is capable of forming atomic oxygen (Novikova & Mikhailova, 2004). All this can destroy pathogens in water. The authors note the prospect of using nutrient solutions treated with the help of the EH-effect in hydroponics, that is, growing plants without soil, on artificial nutrient solutions of a water base, which are intended to obtain high and stable yields at low production costs.

TECHNICAL TOOL FOR EH-EXPOSURE

The developed installation allows you to vary the parameters of exposure to the object, that is, aqueous solutions, to achieve the intended result. Technical novelty is the working body of the installation. The device is a cylinder, which is a container for a set of technological fluid - an aqueous solution. The material of the body of the working fluid is titanium, the choice of which is due to the characteristic properties of lightness, strength and corrosion resistance in accordance with figure 1.

Figure 1. Model of the device for electrohydraulic water treatment: a) spatial model; b) top view without covers with a combination of electrodes "needle" - "needle": 1 – branch pipes, 2 – tank, 3 – high potential electrode, 4 – negative electrode

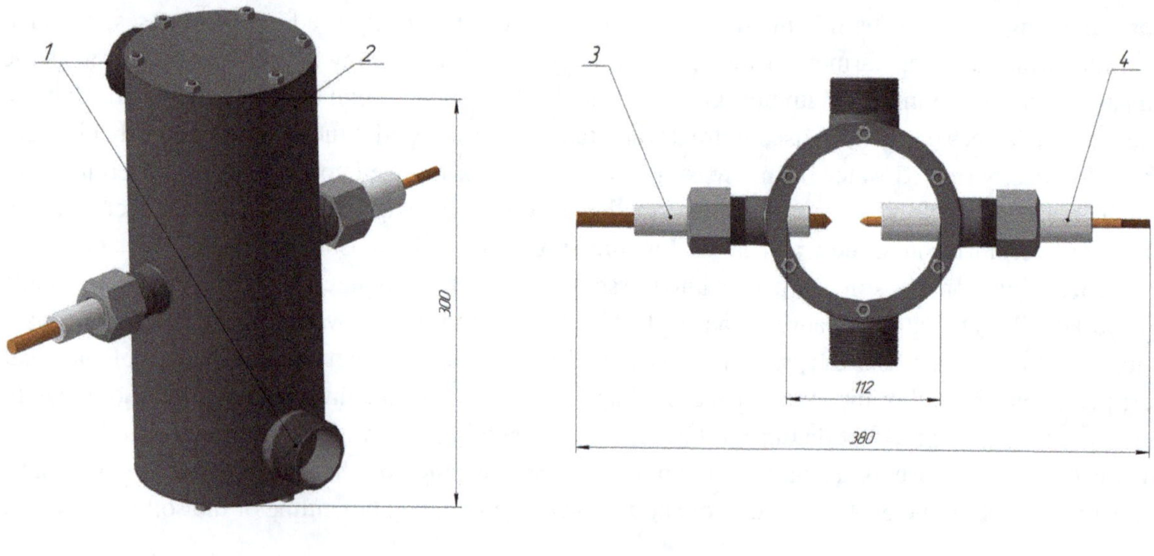

The above tank 2 contains inlet and outlet branch pipes 1 for the supply and removal of the process fluid, respectively. Functionally, the device is a flowing type. At the same time, the device has technological holes for fixing the electrodes of high voltage pulses (Liu, Lin, Pan, Zhang, Li, Li, Liu, 2018). The difference from the existing equipment analogues is in the design features of the negative electrode. Researchers are proposing to modernize the shape of the electrodes, namely the working tip. Instead of the disk, which is negative electrode in one of works of Yutkin L.A., the authors used a cup-shaped hemispherical disk in the shape of a semisphere. This allows you to purposefully concentrate and compact the electromagnetic field, collect the antinodes of the modes, that is, increase the intensity of the electromagnetic field per unit area. Therefore, whis is reduce the dissipated energy in the plane of the electrode. In relation to the analogue studies were conducted with the same area of the working plane of the electrodes. Revealed an increase in the energy of the spark discharge relative to the analog.

The high potential electrode 4 is structurally made of a copper core with an insulator, the material of which must be a dielectric. It is advisable to use fluorinated polymers in accordance with the level of high voltage. The negative electrode 3 has a similar design with some functional distinctive features. This is the adjustment of the air gap between the electrodes in which the spark occurs (Cook, Gleeson, Roberts & Rogers, 1997). A discharge that forms in the aquatic environment is considered working. At the same time, it should be noted that the formative and working gapes should be considered interdependent. Along with this, the cross-sectional area of the rods and tips of the electrodes is chosen. As a result of the formation of spark discharges between the electrodes, the aqueous solution is subjected to EH-exposure.

At researches the various form of tips of electrodes was chosen. The following combinations of types of positive and negative electrode were used, respectively: “needle - needle”, “needle - disk” and “needle - hemispherical disk”. The characteristic features of the EH-effect, namely, the formation of a discharge, were obtained for all combinations of electrode shapes, but for different source voltages. The authors noted the insignificance of voltage differences. It is more preferable to use the hemispherical tip of the negative electrode developed by the authors in view of the required lower voltages when forming a discharge of the same length.

The combination of “needle - needle” electrodes was structurally obtained without the use of tips; the surface area of the electrodes is 1 mm^2. The voltage on the charging capacity was 36 kV at a working and formative intervals of 10 mm. Deviations of the distances of the worker and the formative gaps insignificantly affect the parameters of the spark discharge. At a voltage of less than 36 kV, only a coronation was observed on the working gap in tap water with a capacitor of 0,1 μF capacitor. A corona discharge of indigo appeared near the tip of both positive and negative electrodes (Huang, Yan, Wang, Zhang, Liu & Yan, 2014). The volume of water treated by the EG effect in the device varied from 2.5 to 5 liters. The volume of water loaded in the working body does not affect the conditions for the occurrence of the EH-effect, however, the water level must be above the plane of the electrode. The surface area of the tip of the negative electrode “disk” was 1520 mm^2, hemispherical - 1700 mm^2. The maximum gaps of the working and formative intervals were 20 mm, while the operating voltage was up to 50 kV in tap water with a discharge capacity of 0,2 μF.

THE ASSESSMENT OF FACTORS INFLUENCING THE PROCESS ELECTRO-HYDRAULIC WATER TREATMENT

The undoubted advantage of the EH-technology is that the regulation of the discharge parameters allows to control these processes and selectively influence their passage. An analytical review of the collected theoretical information on the topic under consideration and experiments carried out taking into account the theoretical foundations of the generation and propagation of electromagnetic waves in the high-voltage range in media shows that the process of electrohydraulic treatment of solutions in combination with a large number of factors, but only a small part of them has a significant effect on result (Budnikov, 2018). Therefore, it is important to identify the most significant factors affecting the yield of digestible forms of fertilizers during EG processing. Along with this, have the opportunity to vary the modes of operation of the installation. The parameters of exposure and operating modes of the facility are proposed to be considered as factors, the size and level of which mainly determine and change the output results of experiments (Vasiliev, 2016). Modeling of factors influencing the process of EH-processing is carried out in two ways. The first method is based on the plan of Plackett-Berman and is intended to sifting non-essential factors. The second method is based on applying the random balance method to isolate the most significant factors influencing the process of EH-exposure in aqueous solutions with ionic conductivity (Hartman, Letsky & Shefer, 1977).

A SIFTING EXPERIMENT BASED ON THE PLACKETT-BERMAN PLAN

In this regard, a two-tier experiment was planned. The first level includes the choice of an impressive number of variable factors of the main most significant initial factors on the basis of which this process takes place (Storchevoy, 2012). To solve the above, a separate experiment is conducted, called sifting. At the second level, the degree of influence on the object of study of the most important factors is determined. This is achieved by constructing a mathematical model due to the solution of the problem of planning a two-level experimental formulation based on the use of the Plackett-Berman plan.

The planning of a two-level experiment is based on the operation of choosing the number of experiments and the requirements for conducting them. Determined by their number, sufficient to resolve the question posed. Then the required degree of accuracy is determined. The main properties and characteristics of the object of research vary according to special functional dependencies. Next, a mathematical model characterized by specific statistical properties is developed. It displays, compiles and structures the results of the experiment.

At the initial stage of planning an experiment, the authors need to solve the following tasks:

- Identify the factors that influence in varying degrees on the system;
- Analyze the factors;
- Systematize them according to the form and method of management, denoting their levels;
- Develop a matrix of sifting experiment.

Authors construct the mathematical model according to the principle - from simple to complex. The object of study is represented in the form of an n-pole or a black box system, where n - is the number of input parameters or, in our case, factors.

Input and output parameters that determine the state of the object of study, classify into groups with individual features:

1. The group $X = (x_1, ..., x_k)$ - the initial independent factors that influence the behavior of the system. These parameters appear to be controlled, with the help of which the given technological mode is realized.
2. The group $Z = (z_1, ..., z_m)$ corresponds to factors that influence the behavior of the system and do not allow their purposeful change, including the initial information of the previous links in the technological chain.
3. The group $Q = (q_1, ..., q_n)$ consists of uncontrollable factors that influence changes in a given system. They show perturbations that can be measured quantitatively which is difficult or impossible (not given in the tasks).
4. The group $Y = (y_1, ..., y_p)$ - output variables, target values with certain optimization parameters. This group can be correlated system response to the effects or dependencies between input and output factors.

Theoretical analysis along with a priori information showed that during EH-treatment of solutions, the following 12 factors can affect the yield of nitrate nitrogen in water.

Sifting experiment will perform according to plan of Plackett-Berman. The choice of this method is due to the following considerations:

- The selection of significant factors can be made at the lowest possible expenses (the minimum Number of experiments);
- The orthogonality of these plans allows you to reduce the resource processing results.

To use the plan of Plackett-Berman to highlight significant variables, add three fictitious factors to the above. Fictitious factors are located under a random index number of 8, 13, 15. In this case, it will be necessary to conduct $N = 16$ experiments for 15 factors. The selected factors vary on two levels: the lower level corresponds to the smaller value of the *Hmin* factor (-1), the upper one - to the greater *Hmax* (+1). Choose so that they are on the boundaries of the planning area.

As a result, obtain the values of the variation levels, which have been in table 1.

Due to certain design features of the installation, it is not possible to vary the temperature of the solution and the environment at this stage of the experiments. In this regard, the factors *Z1* and *Z2* in the first approximation are neglected. Based on the data in table 1, the authors compose a matrix of the Plackett-Berman plan with factors in coded units and with objective function in physical units in accordance with table 2.

1. The calculation of the effects of individual factors.

The estimate of the effect of B_i is equal to the difference between the sums of the values of the objective function for factor x_i at levels +1 and -1 divided by *N/2*:

Table 1. Factors affecting the system

No. p/p	Classification of Factors and Objective Function	Decipherment of Factors and Objective Function	Levels of Variation	
			-1	+1
1	*X1*	The magnitude of the applied voltage, kV	10	70
2	*X2*	The duration of the experiment, s	60	600
3	*X3*	Capacity of storage capacitors, uF	0,025	0,2
4	*X4*	Volume of test chamber, l	1	5
5	*X5*	Pulse frequency, Hz	1	50
6	*X6*	Pulse energy, J	1,25	490
7	*X7*	Electrode area, mm²	1	2500
8	*X8*	Fictitious factor	-	-
9	*X9*	Number of pulses, pcs.	10	300
10	*X10*	The inductance of the discharge circuit, mH	50	200
11	*X11*	Water source (lake, sea)	lake	sea
12	*X12*	Electrode material (copper, titanium)	copper	titanium
13	*X13*	Fictitious factor	-	-
14	*X14*	Electrode shape, (needle, circle)	needle	circle
15	*X15*	Fictitious factor	-	-
16	*Z1*	Solution temperature, °C	0	50
17	*Z2*	Ambient temperature, °C	-20	+20
18	*Y*	Nitrogen yield, mg / l	10	1500

$$B_i = \frac{\sum_{j=1}^{N} y^j x_i^j}{N/2}. \quad (1)$$

In accordance with (1) are:

$B_{i1} = 392{,}25$; $B_{i2} = 33$; $B_{i3} = 299{,}75$; $B_{i4} = -28{,}75$; $B_{i5} = 44$; $B_{i6} = 197$; $B_{i7} = -250{,}25$;

$B_{i8} = 32{,}5$; $B_{i9} = 593{,}5$; $B_{i10} = -247{,}5$; $B_{i11} = -215{,}75$; $B_{i12} = -29{,}25$; $B_{i13} = -40{,}75$;

$B_{i14} = -40{,}25$; $B_{i15} = 44{,}5$.

The values of a_i are equal to half of the respective effect estimates.

Table 3 shows the levels of factors and the effects of B_i (or the coefficients a_i), which are determined in accordance with expression 1.

2. Check the significancy of the parameters.

To identify significant factors, *t*-criterion is used and the condition is checked:

Table 2. The matrix of the Plackett-Berman plan

No. p/p	Factor Levels															Response
	X1	*X2*	*X3*	*X4*	*X5*	*X6*	*X7*	*X8*	*X9*	*X10*	*X11*	*X12*	*X13*	*X14*	*X15*	*Y*
1	+	+	+	+	-	+	-	+	+	-	-	+	-	-	-	1500
2	+	+	+	-	+	-	+	+	-	-	+	-	-	-	+	390
3	+	+	-	+	-	+	+	-	-	+	-	-	-	+	+	110
4	+	-	+	-	+	+	-	-	+	-	-	-	+	+	+	1500
5	-	+	-	+	+	-	-	+	-	-	-	+	+	+	+	25
6	+	-	+	+	-	-	+	-	-	-	+	+	+	+	-	97
7	-	+	+	-	-	+	-	-	-	+	+	+	+	-	+	51
8	+	+	-	-	+	-	-	-	+	+	+	+	-	+	-	540
9	+	-	-	+	-	-	-	+	+	+	+	-	+	-	+	540
10	-	-	+	-	-	-	+	+	+	+	-	+	-	+	+	413
11	-	+	-	-	-	+	+	+	+	-	+	-	+	+	-	319
12	+	-	-	-	+	+	+	+	-	+	-	+	+	-	-	108
13	-	-	-	+	+	+	+	-	+	-	+	+	-	-	+	365
14	-	-	+	+	+	+	-	+	-	+	+	-	-	+	-	51
15	-	+	+	+	+	-	+	-	+	+	-	-	+	-	-	413
16	-	-	-	-	-	-	-	-	-	-	-	-	-	-	-	10

Processing of experimental results (Hartman, Letsky & Shefer, 1977).

$$\left|a_i\right| \geq t_{cr} \cdot S_i, \tag{2}$$

where t_{cr} – critical value of the t-distribution for the significance level α and φ degrees of freedom; S_i^2 –estimate of the dispersion of the coefficient a_i.

The variance of observation errors is estimated using special experiments, introducing into the plan fictitious factors from x_{l+1} to x_{N-1} in accordance with the expression:

$$S_l^2 = \frac{4k\left(a_{l+1}^2 + a_{l+2}^2 + \ldots + a_{N-1}^2\right)}{4k - l - 1}. \tag{3}$$

Regarding the structuring of fictitious factors and in accordance with (3), for calculating the variance of observation errors, the expression is obtained:

$$S_l^2 = \frac{4k\left(a_8^2 + a_{13}^2 + a_{15}^2\right)}{4k - l - 1}, \tag{4}$$

where $k = 4$, $l = 12$.

In accordance with (4) determine

Table 3. The selection of significant factors

Classification of Factors	Decipherment of Factors	Levels of Variation		α_i	Significancy
		-1	+1		
X1	The magnitude of the applied voltage, kV	10	70	196,125	yes
X2	The duration of the experiment, s	60	600	-123,75	yes
X3	Capacity of storage capacitors, uF	0,025	0,2	149,875	yes
X4	Volume of test chamber, l	1	5	98,500	yes
X5	Pulse frequency, Hz	1	50	22,000	not
X6	Pulse energy, J	1,25	490	-14,375	not
X7	Electrode area, mm^2	1	2500	-125,125	yes
X8	Fictitious factor	-	-	16,250	-
X9	Number of pulses, pcs.	10	300	296,750	yes
X10	The inductance of the discharge circuit, mH	50	200	16,500	not
X11	Water source (lake, sea)	lake	sea	-107,875	yes
X12	Electrode material (copper, titanium)	copper	titanium	-14,625	not
X13	Fictitious factor	-	-	20,375	-
X14	Electrode shape, (needle, circle)	needle	circle	-20,125	not
X15	Fictitious factor	-	-	22,250	-

$$S_l^2 = 16 \cdot 391,421875.$$

The variance of the estimates of the coefficient a_i is determined from the expression:

$$S_i^2 = \frac{S_l^2}{4 \cdot k}. \quad (5)$$

Thus, in accordance with (5) receive

$$S_i \approx 19,78.$$

Taking into account $\alpha = 0,05$ and $\varphi = 3$ from the table of t-distribution (distribution according to the Student's t-test) find $t_{\kappa p} = 3,18$.

The significancy of the indexes is checked in accordance with (2) and is shown in table 3

$$\left|a_i\right| \geq 3,18 \cdot 19,78 \approx 62,9.$$

As a result of the significance test, the edge factors, that is, significant and insignificant factors that have a certain impact on the target output function in accordance with the initial indicators, were identified. In order to determine the estimates of the influence of each of the essential factors and for their more descriptive analysis, their distribution is compiled according to the degree of influence on the target value in accordance with the diagram in figure 2.

As a result of the sifting experiment based on the Plackett-Berman plan, it was revealed that the significant factors affecting the yield of nitrate forms of nitrogen are:

- X1 the magnitude of the applied voltage;
- X3 capacity of storage capacitors;
- X6 pulse energy;
- X7 electrode area;
- X9 number of pulses;
- X10 the inductance of the discharge circuit;
- X11 water source.

Of the above, the most significant factors are:

- X1 the magnitude of the applied voltage;
- X3 capacity of storage capacitors;
- X9 number of pulses.

SIMULATION OF SIGNIFICANT FACTORS BY RANDOM BALANCE METHOD

In addition to the sifting experiment according to the Plackett-Berman plan, the distribution of significant factors was performed using the random balance method (Hartman, Letsky & Shefer, 1977).

Figure 2. Distribution of factors according to the degree of their influence on the target value

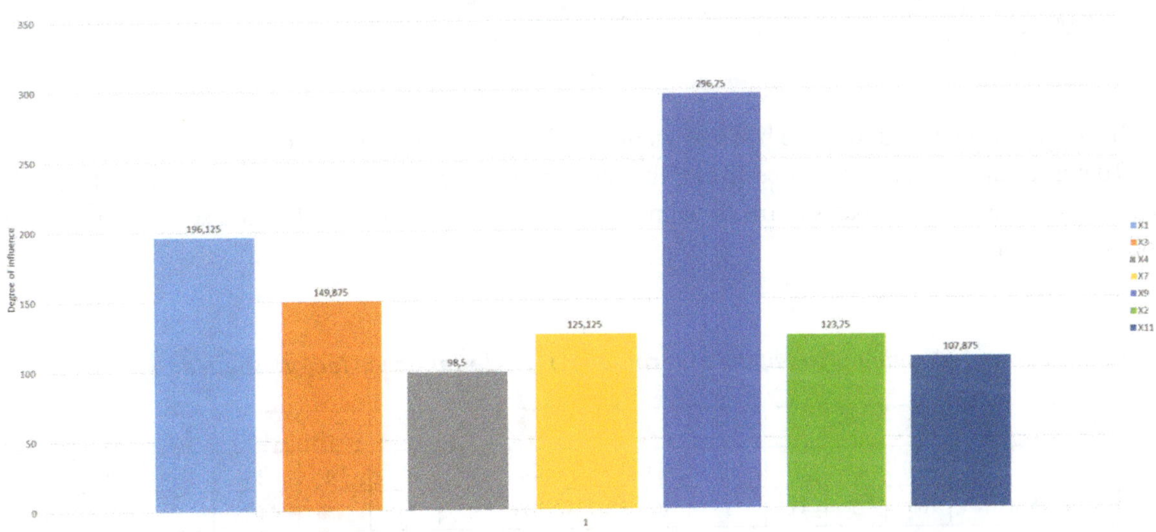

Estimation of the number of the above factors, their number - 15, allows the authors to perform the procedure for selecting the planning of the experiment and its processing. In this context, it is advisable to use supersaturated plans. This makes sense in order to structure the number of experiments within the limits of which statistical processing will be carried out with minimal funds and costs. It is proposed to apply the method of random balance. The developer of this method is Saterzuite.

According to this method, significant factors are singled out from a large set of variables. They also take into account their paired interactions. It is assumed that from the entire flow of factors only their insignificant or significant part, what will be established as a result of the experiment, will probably have a significant impact on the output dependent magnitude. Impact parameters and operating modes of the installation with a minor impact can be considered the so-called "noise".

The mathematical model includes 15 linear effects and 105 paired interactions. The planning edge areas of the experiment were developed for 15 process-influencing factors, where *x1* – is the level of applied voltage; *x2* – the duration of the experiment; *x3* – inductance; *x4* – material test chamber; *x5* – the area of the electrodes; *x6* – is the number of pulses; *x7* – is the volume of the test chamber; *x8* – electrode material; *x9* – is the pulse frequency; *x10* – the shape of the electrodes; *x11* – is the pulse energy; *x12* – capacity of storage capacitors; *x13* – type of aqueous solution; *x14* – is the temperature of the aqueous solution; *x15* – ambient temperature. Factors vary according to two levels. The lower level is the minimum value of the factor (x_{min}). The upper level corresponds to the maximum value of the factor (x_{max}). When selecting them, it is necessary to take into account the condition of respecting the greatest possible difference between them, which reflects their location at the boundaries of the planning area. Rationing factors is determined by the following expression:

$$x_i = \frac{x_i^* - x_{i0}^*}{L_i}, \tag{6}$$

where x_i^* – is the natural form of the factor; x_{i0}^* – is the natural form of the zero factor; L_i – is the variation interval.

$$L_i = \frac{x_{i<0:A}^* - x_{i<8=}^*}{2}. \tag{7}$$

In accordance with table 4, the boundaries of the planning area are defined.

To construct the plan of the experiment, the mixing of randomly formed samples, that is, the set of test cases obtained on the basis of fractional plans, is used. The resulting factors are structured into the following groups.

Table 4. Regional planning areas of the experiment for the 15 influence factors studied

Factors		x_1*	x_2*	x_3*	x_4*	x_5*	x_6*	x_7*	x_8*	x_9*	x_{10}*	x_{11}*	x_{12}*	x_{13}*	x_{14}*	x_{15}*
Levels	+1	70	600	20	Ti	2500	300	5	Cu	50	C	490	0,2	L	50	20
	-1	10	60	5	Fe	1	100	1	Ti	1	N	1,25	0,025	S	0	0

1. *x1, x2, x3, x4.*
2. *x5, x6, x7, x8.*
3. *x9, x10, x11, x12.*
4. *x13, x14, x15.*

An experiment plan of type 2^4 is chosen. Through a random sampling of numerical values from the matrix of the plan 2^4, rows are marked by a random method. This is done individually for each of the composed groups of factors. For 16 experiments, 16 lines are selected. For this sample, the following sequence of lines are obtained:

- **Group 1:** 10,6, 3,16, 4,15, 14,1, 5,9, 7,13, 2,8, 12,11;
- **Group 2:** 13,3, 7,16, 2,9, 2,16, 6,12, 13,7, 9,12, 3,6;
- **Group a 3:** 5,10, 15,16, 3,4, 15,4, 6,9, 9,3, 16,5, 6,10;
- **Group 4:** 7,15, 11,8, 16,3, 5,13, 4,12, 6,2, 2,1, 14,9.

The developed matrix of the plan along with the results of the experiment are presented in accordance with table 5.

The next stage is the construction of a scatterplot. On the dependency graph, the values of the resulting function are plotted over each of the 15 factors considered in accordance with the scatter diagram in figure 3. Both levels of varying factors +1 and -1 are noted and medians are found, that is, the mean lines of the factors studied. The criterion of the degree of influence of the corresponding factor is deter-

Table 5. The plan and the results of the experiment according to the method of random balance

No. p/p	x_1	x_2	x_3	x_4	x_5	x_6	x_7	x_8	x_9	x_{10}	x_{11}	x_{12}	x_{13}	x_{14}	x_{15}	y^1
1	+	-	-	+	-	-	+	+	-	-	+	-	-	+	+	811
2	+	-	+	-	-	+	-	-	+	-	-	+	-	+	+	144
3	-	+	-	-	-	+	+	-	-	+	+	+	-	+	-	193
4	+	+	+	+	+	+	+	+	+	+	+	+	+	+	+	1245
5	+	+	-	-	+	-	-	-	-	+	-	-	+	+	+	163
6	-	+	+	+	-	-	-	+	+	+	-	-	-	+	-	907
7	+	-	+	+	+	-	-	-	-	+	+	+	-	-	+	223
8	-	-	-	-	+	+	+	+	+	+	-	-	-	-	+	110
9	-	-	+	-	+	-	+	-	+	-	+	-	+	+	-	56
10	-	-	-	+	+	+	-	+	-	-	-	+	+	+	-	860
11	-	+	+	-	-	-	+	+	-	-	-	+	+	-	+	107
12	-	-	+	+	-	+	+	-	-	+	-	-	+	-	-	1489
13	+	-	-	-	-	-	-	+	+	+	+	+	+	-	-	180
14	+	+	+	-	+	+	-	+	-	-	+	-	-	-	-	220
15	+	+	-	+	-	+	-	-	+	-	+	-	+	-	+	983
16	-	+	-	+	+	-	+	-	+	-	-	+	-	-	-	1022

Figure 3. The scatter diagram for 15 factors

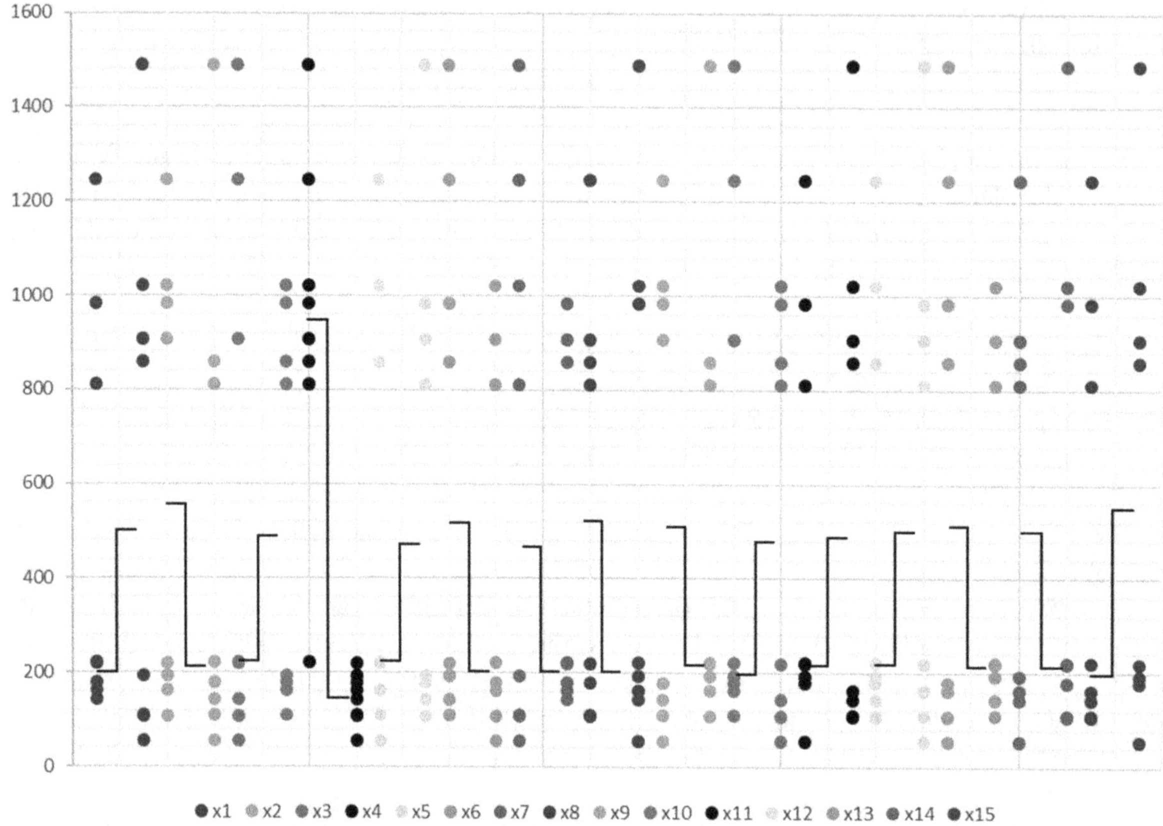

mined by the difference between the medians. The most significant impact has factor *x4*. This follows from the analysis of chart. The visual identification of dominant factors by the magnitude of the medians is allowed to be replaced by a method based on the determination of the number of "peculiar points".

"Peculiar points" are the definitions of the output function for the upper maximum level of the factor $(x_i = +1)$, which are less than the smallest or greater than the largest value corresponding to another level $(x_i = -1)$. The combination of such graphic points is allowed to characterize the importance of factors.

The estimation of contributions is determined by the average value of the output function, for which the selected factor x_4 is located at a high level of +1. Then, the average value of the output function at a low level of -1 is subtracted from it. The coefficients of the experiment plan are determined similarly. Thus, the estimation of the factor x_4 is equal to $B_4^1 = 795{,}875$. The value B_4^1 allows you to get a parameter estimate $a_4^1 = B_4^1/2 = 397{,}9375 \approx 398$.

To find other significant factors, the influence of factor x_4 on the output function is eliminated. For this purpose, the value of $2 \cdot a_4^1 = 796$ is subtracted from all output functions for which x4 was at the level of +1 in accordance with table 6. Thus, the effect of the factor x_4 on these quantities is terminated. Following this action, a newly calculated vector of experimental results is formed. The scatter diagram is constructed, according to which factor x_4 already interrupts having an effect. Studies of the scatter diagram show that the factors x_2, x_{13}, x_{14} have the greatest influence.

Table 6. The results of the experiment according to the method of random balance

Experiment Number	y^2	y^3	y^4	y^5	y^6
1	15	15	25	279	238
2	144	144	167	310	158
3	193	72	105	216	105
4	449	146	179	562	410
5	163	-140	-140	131	90
6	111	-10	-10	-10	-10
7	-573	-573	-573	-190	-342
8	110	110	143	414	414
9	56	-126	-116	124	13
10	64	-118	-95	33	-78
11	107	-196	-186	-43	-154
12	693	511	544	544	544
13	180	-2	-2	109	-43
14	220	99	122	361	320
15	187	-116	-93	288	247
16	226	105	115	243	132

The next step is to assess the contributions of a_2, a_{13}, a_{14} for factors x_2, x_{13}, x_{14}. For each group of numerical values is the average value of the output function. Estimates of the contributions of factors are calculated by the differences between the sums of average values for high and low levels according to the following expression:

$$B_2^2 = \frac{y_3^m + y_4^m + y_5^m + y_6^m + y_{11}^m + y_{14}^m + y_{15}^m + y_{16}^m}{8} - \frac{y_1^m + y_2^m + y_7^m + y_8^m + y_9^m + y_{10}^m + y_{12}^m + y_{13}^m}{8}, \quad (8)$$

where $m = 2$.

B_{13}^2 and B_{14}^2 are defined similarly. Estimates of the coefficients are calculated by the above expression. The values of the estimates of contributions and ratios thus more clearly allow us to characterize the degree of materiality of factors than the difference in medians. For the estimates of the coefficients a_2, a_{13}, a_{14}, the following numerical values were obtained:

$a_2^2 = 60{,}43$; $a_{13}^2 = 90{,}81$; $a_{14}^2 = 2{,}81$.

The numerical value of the coefficient estimate $a_{14}^2 = 2{,}81$ compared with the other two. Therefore, further calculation will be carried out in case of removal only from the influence of factors x_2 and x_{13}. This is achieved by subtracting from those output functions y^2, for which $x_2 = +1$ and $x_{13} = +1$, the values of B_2^2 and B_{13}^2 are correspondingly. Then a vector of results y^3 is obtained and a new scatter diagram is developed, which shows that factors x_6 and x_7 have the greatest influence. Calculation of parameter estimates is carried out according to the above method.

The result is the following:

$a_6^3 = -11{,}5$; $a_7^3 = -4{,}9$.

To isolate significant factors from the remaining ones, it is necessary to remove the vector of the y^3 results from the influence of factors x_6 and x_7. This is achieved by subtracting from the output function the values of B_6^3 and B_7^3 at x_6 and x_7, which assume a high level of values, respectively. The following scatterplot is constructed for the newly obtained vector y^4. Processing is carried out by visualization and calculation of medians. From which we can conclude the remaining factors that predominate or dominate. These are x_5, x_{11} and x_{15}. Next, new factors are calculated. They turn out to be equal:

$a_5^4 = -64{,}38$; $a_{11}^4 = -55{,}69$; $a_{15}^4 = -71{,}31$.

By the already well-founded law, the influence of the selected factors on the response of the function is excluded. As a result, the vector y5 is found. For this vector once again it is necessary to construct the fifth scatter diagram.

Visually are determined by the dominant factors x1 and x12. Further, the following coefficients are calculated:

$a_1^5 = 20{,}5$; $a_{12}^5 = 55{,}7$.

After eliminating the influence of factors x1 and x12 on the output function, the vector y6 is determined. The constructed scatter diagram of factors clearly demonstrates that the influence of the remaining factors is negligible and is defined approximately as one order of magnitude.

Ultimately, the following estimates are determined:

$a_1^5 = 20{,}5$; $a_2^2 = 60{,}43$; $a_4^1 = 398$; $a_5^4 = -64{,}38$; $a_6^3 = -11{,}5$; $a_7^3 = -4{,}9$; $a_{11}^4 = -55{,}69$; a_{12}^5

$= 55{,}7$; $a_{13}^2 = 90{,}81$; $a_{14}^2 = 2{,}81$; $a_{15}^4 = -71{,}31$.

The final stage is to test the significancy of the coefficients on the basis of the t-criterion. This condition is written in accordance with the expressions (2), (3), (4), (5).

This validation can only be performed if each cell of the table contains at least two values of the output function.

For example, we calculate the verification of the significance of the coefficient a_1:

$$S_1^2 = \frac{(310-197)^2 + (562-197)^2 + (-190-197)^2 + (109-197)^2}{3} = 101169,$$

$$S_2^2 = 58059,\ S_3^2 = 36088,\ S_4^2 = 19875,$$

$$S_F^2 = 1560,$$

$$S^2 = \frac{1}{16} \cdot 1560 \cdot \frac{4}{4} = 97,5\,,$$

S = 9,9, φ = 12.

When α = 0,05, the critical value $t_{кр}$ is 2,18. In this connection, the a_1 estimate is insignificantly different from zero.

This completes the definition of linear effects. Next, the most basic of pairwise interactions is highlighted. Their total number is equal to the number of combinations of 15 to 2, that is, C_{15}^{2} = 15. There is no need to build the scatter diagram for such a large number of paired interactions; moreover, this process will appear cumbersome and complex. A heuristic visual selection method will be used to find the most important pairwise interactions. The basis of this method is the following rule. The emergence of prominent points on high and low levels $x_i x_j = +1$ and $x_i x_j = -1$ is necessary. Then the pairwise interaction of the factors x_i and x_j can be ranked as significant. In the first case, x_i and x_j must be the same characters, and in the second - different. It follows from this that the pair interaction of such factors will be significant, in which the number of distinguished points will be large both at the same and at different levels. In fact, for these factors, the lower level of the scattering diagram should be similar to the mirror image of the upper one in accordance with figure 4.

According to this statement, pairwise interactions are visually selected from the scatterplot, for which a new diagram is constructed. Corresponding tables are being developed for calculating the parameters. Due to the correlation between some rows, part of the cells in the tables remain unfilled. In this regard, the quantitative assessment of pair interactions is performed only on the basis of medians.

The following pairwise interactions are significant:

X_1X_{11}, X_3X_{10}, X_5X_{12}, X_6X_{13}, X_8X_{14}, $X_{11}X_{15}$.

Figure 4. The scatter diagram for paired interactions

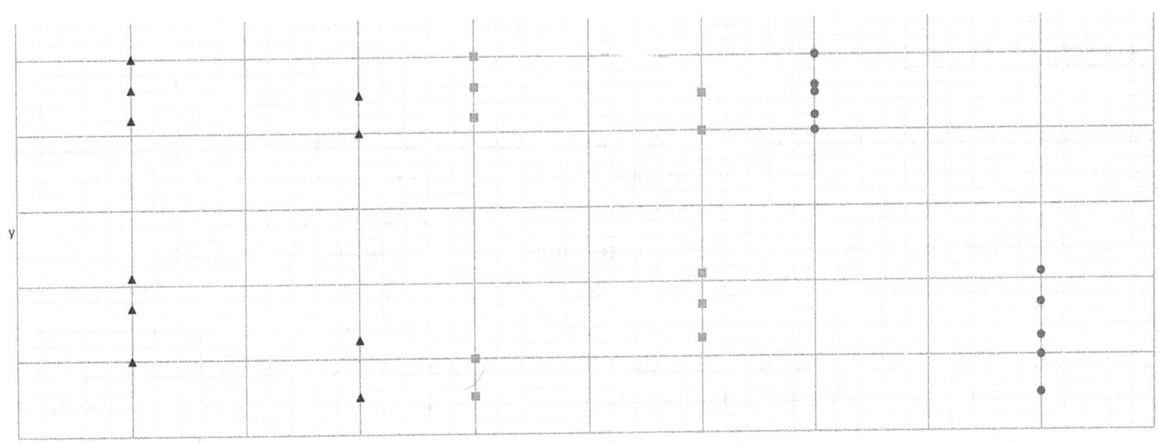

This process of separating the materiality of couples can continue indefinitely. Therefore, at this stage a certain stopping criterion is required. It is possible to use the F - criterion (Hartman, Letsky & Shefer, 1977).

$$F = \frac{S_m^2}{S^2}, \tag{9}$$

where $S_m^{\ 2}$ – is the estimate of the variance of the experimental results relative to their arithmetic average value at the r-step of the operation; S^2 – is the estimate of the variance of observation errors calculated on the basis of the results of several parallel observations. The change in the dispersions $S_m^{\ 2}(r)$ depending on the step number r of the factor extraction procedure can be analyzed using the scatter plot. This dependence characterizes the change in the variances of the results of experiments. Following this diagram, it can be revealed that after five steps this dispersion is quite small. By repeating the experiments at one point of the plan, the estimate $s^2 = 32$ is found $(\varphi = 7)$. Along with this, for $S_m^{\ 2}(5) = 74{,}4$ (φ_m = 12) it turns out

$$F = \frac{74,4}{32} = 2,32 < 3,5 = F_{:@}\alpha = 0,05\,.$$

All significant factors are highlighted. This indicates that the settlement operations for this procedure can be terminated.

The random balance method used in this topic due to the simplicity of processing and the possibility of varying factors at several levels revealed significant and insignificant factors influencing the process of electrohydraulic water treatment without recourse to technical tools and computer programs for processing experiments. In the end, inconsequential factors are reasonably defined: the duration of the experiment; test chamber material; the volume of the test chamber; pulse frequency.

Thus, the results of the constructed optimization models for identifying significant and insignificant factors on the basis of two methods coincide and are combined.

EXPERIMENTAL STUDIES

The EH-effect can be considered as the conversion of electrical energy into mechanical or kinetic. The essence of this effect is that when a high-voltage pulsed discharge in a liquid, or in an open or closed vessel, ultra-high hydraulic pressures occur around the discharge zone (Mikiporis, 2003).

Vibrations and movement detected of the aquatic environment as a result of exposure to the EH in the developed installation. The fluid, having received acceleration from the discharge channel that expands with high speed, moves from it in all directions, forming in the place where there was a discharge, a significant cavitation cavity, causing the first (main) hydraulic hit. Then the cavity closes at high speed, creating a second cavitation hydraulic hit. At this point, the single cycle of the electrohydraulic effect ends, and it can be repeated, in accordance with a given repetition rate, an unlimited number of times.

Only a few stages of discharge appear: the occurrence of a conductive channel between the electrodes, the energy in the discharge channel is separated, the final stage, all electrical processes end. The electrohydraulic effect is characterized by the mode of energy release on the active resistance of the circuit, which is close to critical (Zhu, He, Gao, Tan, Yue & Chang, 2014).

For a sharp increase in the hydraulic and hydrodynamic effects and the amplitude of the shock effect, when a pulsed electric discharge is carried out in an ion-conducting liquid, the impulse should have the shortest possible duration, the maximum steepness of the front and his shape close to aperiodic. The steepness of the front of the current pulse determines the rate of expansion of the discharge channel, and the instantaneous power of the current pulse, reaching hundreds of thousands of kilowatts when voltage is applied to the discharge electrodes of several tens of kilovolts, causes a sharp and significant pressure in the liquid, which causes a strong mechanical effect of the pulse discharge (Zhang, Zhu, Huang, Liu & Yan, 2017).

The main factors affecting the electrohydraulic effect are high and ultra-high impulse hydraulic pressures, leading to the appearance of shock waves with sonic and supersonic speeds; emerging cavitation processes capable of covering relatively large volumes of fluid; electromagnetic fields; intense pulsed light, heat, ultraviolet radiation; multiple ionization of compounds and elements contained in a liquid (Liu, Ding, Han, Wu, Jing, Zhou, Qiu & Zhang Y., 2017). All these factors make it possible to exert a variety of physical and chemical effects on the fluid and objects placed in it (Jones & Kunhardt, 1994). The high efficiency of the electrohydraulic effect (up to 80% due to the direct conversion of electrical energy into mechanical energy without intermediate links), as well as the wide possibilities of electrohydraulic action are the basis for its use by the authors (Mackersie, Timoshkin & MacGregor, 2005).

To implement the EH-technology, the authors apply a well-known electric principal circuit based on a high-voltage pulsed energy source formed on the use of capacitive storage (Pijaudier-Cabot, Christian, Reess, Wen, Maurel, Rey-Berbeder & De Ferron, 2016). The discharge circuit of the RC-generator consists from the series-connected charged capacitance C, inductance L, and active resistance R. Inductance L includes the internal inductance of the drive, the inductance of the load, and the connecting wires (Woo, Noh, An, Kang, Kim & Song, 2017). The same elements are included in the active resistance R. For linear parameters of the discharge circuit, the transient process is described by known equations. The type of discharge depends on the coefficient γ, equal to $\gamma = 0{,}5R$. The discharge can be oscillatory $(\gamma < 1)$, aperiodic $(\gamma > 1)$ and critical $(\gamma = 1)$. For electrohydraulic technologies, the discharge type is close to aperiodic. Capacitive storage provides a rapid release of stored energy in a resistive load with a damped oscillatory current or with an aperiodic, close to critical (Zia, Fazli & Soltanpour, 2017).

It was established experimentally that for the formation of a discharge, a current pulse with a steep front up to $2 \cdot 10^{11}$ A/s and with an absolute value of current up to 250 kA is required. During electrohydraulic shock, a plasma channel appears between the electrodes with a temperature of 15 to 30 °C and the pressure rises to thousands of atmospheres. The voltage of the secondary winding of the transformer ranged from 30 to 40 kV. The frequency of the pulses of spark discharges ranged from 1 to 50 Hz. Energy costs for the process of electrohydraulic water treatment in personal subsidiary farms will be 120 kWh / year, in peasant (farmer) farms - 1800 kWh / year. The increase in income due to the use of an electrohydraulic installation in personal subsidiary farms will amount to 9100 rubles per year, when used in peasant (farmer) farms, 600,000 rubles per year.

The experiments were conducted on tap, lake, pond, saline, artesian and distilled water. Research results are changes in the chemical, physical, and microbiological properties of water and aqueous solutions (Liu, Ding, Han, Wu, Jing, Zhou, Qiu & Zhang, 2017). The total mineralization increases, the hydrogen index increases, salt ions are formed and converted and precipitated. Turbidity and increase in water temperature as a result of changes in density and electrophysical properties. Disinfection of water was detected, a decrease in colony-forming microorganisms was revealed, which indicates a bactericidal effect.

When conducting research, the following technical measuring instruments were used:

- ITAN pH-meter / ionomer to determine the pH, redox potential, the concentration of nitrate compounds in water;
- fluid analyzer conductometric multichannel ATLANT 1200 to determine the conductivity and temperature;
- EnSURE luminometer for measuring the microbial number of bacteria in water.

Based on the materials of the study, there is a promising and applied value, which has already been proven by the original source of the discovery of the EH-effect, but it is being developed and complemented by us both because of its former value and for increasing global environmental problems of water use and resource management of water resources.

FUTURE RESEARCH DIRECTIONS

The authors plan EH-tillage in stationary and field conditions in order to obtain digestible forms of fertilizers, disinfecting waste of livestock farms and wastewater, mathematical and computer modeling of the EH-impact process, justification of the economic feasibility of the proposed technological and technical solutions.

CONCLUSION

The main conclusions by the authors are formulated as follows.

1. As a result of EH-exposure to fluids with ionic conductivity, the pathogenic microflora in it decreases, the digestibility of nutrients in the aquatic environment increases, the nitrogen content increases, thus the technology is universal, that is, a single device performs several operations.
2. Negative electrode design was developed - a hemispheric cup-shaped hemispherical disk for concentrating the electromagnetic field in order to increase the length of the spark discharge at a constant source voltage by reducing the dissipated power loss.
3. As a result of the sifting experiment, it was revealed that the significant factors influencing the nitrogen output of the nitrogen form in aqueous solutions are the magnitude of the applied voltage (30-40 KV), the capacitance of storage capacitors (0.025-0.2 μF), the number of pulses (100-300).

4. The energy costs of the EH-effect on water and the economic efficiency of the EH-exposure are shown. Shows the economic efficiency of the EH-exposure; Energy costs for the process of electro-hydraulic water treatment in personal subsidiary farms will be 120 kWh / year, in peasant (farmer) farms - 1800 kWh / year. The increase in income due to the use of an electrohydraulic exposure in personal subsidiary farms will amount to 9100 rubles per year, when used in peasant (farmer) farms, 600,000 rubles per year.
5. The scope of the EH-treated water is irrigation and irrigation in crop production in personal subsidiary and peasant (farmer) farms.

REFERENCES

Belov, A. A. (2018). Modeling the estimation of factors of influence on the process electrohydraulic treatment of water. *Vestnik NGIEI, 11*(90), 28–39.

Budnikov, D. A. (2018). Laboratory installation and planning experiment on wheat microwave heating of various densities. *Innovations in Agriculture, 2*(27), 36-40.

Cook, J. A., Gleeson, A. M., Roberts, R. M., & Rogers, R. L. (1997). A spark-generated bubble model with semi-empirical mass transport. *The Journal of the Acoustical Society of America, 101*(4), 1908–1920. doi:10.1121/1.418236

Hartman, K., Letsky, E., & Shefer, V. (1977). *Planning an experiment in the study of technological processes*. Moscow: Mir.

Huang, Y., Yan, H., Wang, B., Zhang, X., Liu, Z., & Yan, K. (2014). The electro-acoustic transition process of pulsed corona discharge in conductive water. *Journal of Physics. D, Applied Physics, 47*(25), 169–175. doi:10.1088/0022-3727/47/25/255204

Jones, H. M., & Kunhardt, E. E. (1994). The influence of pressure and conductivity on the pulsed breakdown of water. *IEEE Transactions on Dielectrics and Electrical Insulation, 1*(6), 1016–1025. doi:10.1109/94.368641

Kutter, H. K. (1969). *The electrohydraulic effect: potential application in rock fragmentation*. University of Michigan Library.

Liu, Q., Ding, W., Han, R., Wu, J., Jing, Y., Zhou, H., Qiu, A., & Zhang, Y. (2017). Fracturing effect of electrohydraulic shock waves generated by plasma-ignited energetic materials explosion. *IEEE Transactions on Plasma Science, 45*(3), 423-431.

Liu, Y., Li, Z., Li, X., Lin, F., & Pan, Y. (2016). Experiments on the fracture of simulated stratum by underwater pulsed discharge shock waves. Diangong Jishu Xuebao: China Machine Press, 31(24), 71-78.

Liu, Y., Lin, F., Pan, Y., Zhang, Q., Li, H., Li, Z., & Liu, S. (2018). *Pipeline scale removing and rock stratum fracturing device based on electrohydraulic pulse shockwaves*. Patent WO2018058401 CN, МПК E21B 43/26. Huazhong University of Science and Technology.

Locke, B. R., Sato, M., Sunka, P., Hoffmann, M.R., & Chang, J.-S. (2006). Electrohydraulic discharge and nonthermal plasma for water treatment. *Industrial and Engineering Chemistry Research, 45*(3), 882-905.

Mackersie, J. W., Timoshkin, I. V., & MacGregor, S. J. (2005). Generation of high-power ultrasound by spark discharges in water. *IEEE Transactions on Plasma Science*, *33*(5), 1715–1724. doi:10.1109/TPS.2005.856411

Mamutov, A. V., Golovashchenko, S. F., Mamutov, V. S., & John Bonnenc, J. F. (2015). Modeling of electrohydraulic forming of sheet metal parts. *Journal of Materials Processing Technology*, *219*, 84–100. doi:10.1016/j.jmatprotec.2014.11.045

Mikiporis, Yu. A. (2003). Influence of electrohydraulic effect on the properties of liquids (emulsions and solutions). *Electronic processing of materials. Institute of Applied Physics of the Academy of Sciences of Moldova (Chisinau)*, *6*, 34–37.

Müller, T., Gelfond, L., & Eisert, S. (2014). *Method and device for the disintegration of a recyclable item*. Patent 2771120 DE, B02C 19/18. IMPULSTEC GMBH.

Novikova, G. V., & Mihailova, O. V. (2004). The use of electromagnetic radiation in the poultry industry. Mechanization and Electrification of Agriculture, 9.

Pijaudier-Cabot, G., Christian, L.B., Reess, T., Wen, C., Maurel, O., Rey-Berbeder, F., & De Ferron, A. (2016). *Electrohydraulic Fracturing of Rocks*. Wiley-ISTE.

Storchevoy, V. F. (2012). Mathematical modeling of stationary processes of the ionizer-ozonizer. *K.A. Timiryazeva*, *2*, 78–82.

Tijani, J. O., Fatoba, O. O., Madzivire, G., & Petrik, L. F. (2014). A review of combined advanced oxidation technologies for the removal of organic pollutants from water. Water, Air, & Soil Pollution, 225(9).

Toporkov, V. N. (2017). Electrotechnological method of obtaining fertilizer from the soil and water for greenhouses, LPH and small-scale farms. *Vestnik VIESH*, *3*(28), 49–55.

Vasiliev, A. N. (2016). Research methodology for microwave convective grain processing. *Innovations in Agriculture, 3*(18), 143-153.

Woo, M. A., Noh, H. G., An, W. J., Kang, B. S., Kim, J., & Song W. J. (2017). Numerical study on electrohydraulic forming process to reduce the bouncing effect in electromagnetic forming. *The International Journal of Advanced Manufacturing Technology, 89*(5), 1813-1825.

Yutkin, L. A. (1986). Electrohydraulic effect and its application in industry. *Engineering*.

Zhang, L., Zhu, X., Huang, Y., Liu, Z., & Yan, K. (2017). Effects of water pressure on plasma sparker's acoustic characteristics. *International Journal of Plasma Environmental Science & Technology*, *11*(1), 60–63.

Zhu, L, He, Z. H., Gao, Z. W., Tan, F. L., Yue, X. G., & Chang, J. S. (2014). Research on the influence of conductivity to pulsed arc electrohydraulic discharge in water. *Journal of Electrostatics, 72*(1), 1-6.

Zia, M., Fazli, A., & Soltanpour, M. (2017). Warm Electrohydraulic Forming: A Novel High-Speed Forming Process. *Procedia Engineering*, *207*, 323–328. doi:10.1016/j.proeng.2017.10.782

Compilation of References

13. Adomavicius, V., Kharchenko, V., Valickas, J., & Gusarov, V. (2013). Res-based microgrids for environmentally friendly energy supply in agriculture. In *Conference Proceeding - 5th International Conference, TAE 2013: Trends in Agricultural Engineering 2013*, Prague, Czech Republic, September 2-3 (pp. 51-55).

Abdalla, G. (2015). Performance Characteristics of Automotive Air Conditioning System with Refrigerant R134a and its alternatives. *International Journal of Energy and Power Engineering*, *4*(3), 168–177.

Adamavichus, V.B., Gusarov, V.A., Valitskas, I.Y., & Kharchenko V.V. (2016). Arguments and new opportunities to increase the share of energy supply of agriculture through renewable energy. *Innovations in agriculture*, *5*, 391-396.

Adams, E. P., Slake, G. R., Martin, W. P., & Boelter, D. H. (1960). Influence of soil compaction on crop growth and development. *Trans. 7-th Intern. Congr. Soil Science*, *1*, 171–178.

Adler, Yu. P., Markova, E. V., & Granovskiy, Yu. V. (1976). Planirovaniye eksperimenta pri poiske optimal'nykh usloviy [Planning an experiment when searching for optimal conditions]. Moskow: Nauka.

Adomavicius, V. B., Kharchenko, V. V., Gusarov, V. A., & Valitskas, I. Yu. (2013). Sources of controlled power in micro networks. *Alternative Energy and Ecology*, *7*, 54–59.

Afanas'ev, V. N., Afanas'eva, G. V., Biryukov, S. V., & Beleckij I. P. (2009). *Mnogofunkcional'noe ustrojstvo dlya diagnostiki i sposob testirovaniya biologicheskih ob"ektov* [Multifunctional device for diagnostics and testing of biological objects]. Patent RF. no 2363948.

Afonin, A. A., & Grebenikov, V. V. (2004). Issledovaniya v oblasti razvitiya elektromehanicheskih preobrazovateley energii lineynogo i rotatsionnogo dvizheniya [Research in the development of electromechanical energy converters of linear and rotational motion]. *Proceedings of the Institute of Electrodynamics of NAS of Ukraine*, 2(11), 95–98.

Ageev, L. E. (1978). *Osnovy rascheta optimal'nykh i dopustimykh rezhimov raboty mashinno-traktornykh agregatov* [Basics of calculating the optimal and allowable modes of operation of machine-tractor units]. Leningrad, Russia: Kolos.

Ahmed, A. A. H., Moustafa, M. K., & Marth, E. H. (2016). Incidence of Bacillus cereus in milk and some milk products. *Journal of Food Protection*, *46*(2), 126–128. doi:10.4315/0362-028X-46.2.126 PMID:30913599

Ahmed, A. M., & Shimamoto, T. (2014). Isolation and molecular characterization of Salmonella enterica, Escherichia coli O157: H7 and Shigella spp. from meat and dairy products in Egypt. *International Journal of Food Microbiology*, *168–169*, 57–62. doi:10.1016/j.ijfoodmicro.2013.10.014 PMID:24239976

Airlight Energy. (n.d.). DSolar SunFlower. Retrieved from http://www.airlightenergy.com/high-concentration-photovoltaic-thermal/

Ajzenberg, Y. U. B. (Ed.). (2006). *Spravochnaya kniga po svetotekhnike [The reference book on light engineering]*. Moscow, Russia: Znak.

Aladjadjiyan, A. (2012). Physical factors for plant growth stimulation improve food quality. In A. Aladjadjiyan (Ed.), *Food production – approaches, challenges and tasks* (pp. 145–168). Rijeka: InTech. doi:10.5772/32039

Alanod Corporation. (n.d.). Miro-Silver products. Retrieved from https://www.alanod.com/products/

Aleksandrov, G. (2018). *Rezhimy raboty vozdushnyh linij ehlektroperedachi* [Modes of operation of overhead power lines]. Retrieved from http://www.cpk-energo.ru/metod/AlexandrovLEP.pdf

Alferova L.K., Yuferev L.Y., & Yufereva A.A. (2015). K metodike rascheta svetodiodnogo osveshcheniya [More about methodic procedure of calculation of light emitting diode-based illumination]. *Innovatsii v sel'skom khozyaystve [Innovations in agrarian sector], 2*(12), 31-34. (in Russian)

Al-Holy, M., Wang, Y., Tang, J., & Rasco, B. (2005). Dielectric properties of salmon (Oncorhynchus keta) and sturgeon (Acipenser transmontanus) caviar at radio frequency (RF) and microwave (MW) pasteurization frequencies. *Journal of Food Engineering*, *70*(4), 564–570. doi:10.1016/j.jfoodeng.2004.08.046

Alireza, A. (2013). Developing a discrete harmony search algorithm for size opti- mization of wind–photovoltaic hybrid energy system. *Solar Energy*, *98*, 190–195. doi:10.1016/j.solener.2013.10.008

Almeco-Tinox Solar. (n.d.). Tinox Energy. Retrieved from http://www.almecogroup.com/en/pagina/53-tinox-energy-cu

Al-Shamma, A. A., & Khaled, E. (2012). Optimum sizing of hybrid PV/wind/battery/diesel system consider-ing wind turbine parameters using Genetic Algorithm. *Power and Energy (PECon), IEEE International Conference*, 121-126.

Amaya, J. (1996). Effects of stationary magnetic fields on germination and growth of seeds. *Hortic. Abst*, *68*, 1363.

Amerhanov, R. A. (2008). Jenergeticheskij potencial solnechnoj radiacii i jekonomicheskaja celesoobraznost' primenenija gelioustanovok v Krasnodarskom krae i Jakutii [Energy potential of solar radiation and economic feasibility of solar installations in the Krasnodar Territory and Yakutia]. *Works of the Kuban State Agrarian University*, *1*, 26–32.

Anderson, W., O'Reilly, C. E., Grant, I. R., Donaghy, J., Rowe, M., O'Mahony, P., & Harvey, P. (2004). Surveillance of bulk raw and commercially pasteurized cows' milk from approved Irish liquid-milk pasteurization plants to determine the incidence of mycobacterium paratuberculosis. *Applied and Environmental Microbiology*, *70*(9), 5138–5144. doi:10.1128/AEM.70.9.5138-5144.2004 PMID:15345392

Annual Energy Outlook. (2017). Retrieved from http://www.eia.gov/outlooks/aeo/

Antoshchenko, V. S., & Nesterenkov, A. G. (2010). Calculation of a combined thermal photovoltaic system with concentrated solar radiation fluxes. *Energy and Fuel Resources of Kazakhstan*, *2*, 27–31.

Aronova, E. S., & Grilihes, V. A. (2006). Metodika rascheta real'noj plotnosti solnechnogo izluchenija pri proektirovanii fotojelektricheskih ustanovok [Method for calculating the real density of solar radiation in the design of photovoltaic installations]. *SPbSU Scientific and Technical Statements*, *6*(1), 62–66.

Arrigoni, N., Cammi, G., Savi, R., De Cicco, C., Ricchi, M., Garbarino, C. A., & Pongolini, S. (2016). Estimation of Mycobacterium avium subsp. paratuberculosis load in raw bulk tank milk in Emilia-Romagna Region (Italy) by qPCR. *MicrobiologyOpen*, *5*(4), 551–559. doi:10.1002/mbo3.350 PMID:26991108

Ashok, S. (2007). Optimised model for community-based hybrid energy system. *Renewable Energy*, *32*(7), 1155–1164. doi:10.1016/j.renene.2006.04.008

Astahov, V. I. (2004). Matematicheskoe i kompyuternoe modelirovanie elektromagnitnogo polya kak osnova dlya resheniya zadach v elektrotehnike i elektroenergetike [Mathematical and computer simulation of the electromagnetic field as a basis for solving problems in electrical engineering and power engineering]. Izvestiya vuzov. Elektromehanika, 6, 4–6.

Astana Solar. (n.d.). Approved tariffs for renewable energy sources. Retrieved from http://astanasolar.kz/ru/news/utverzhdeny-tarify-na-vozobnovlyaemye-istochniki-energii

Atkinson, L. K. (2000). *Single pressure absorption heat pump analysis* [dissertation]. Georgia Institute of Technology.

Atwater, M., & Ball, J. T. (1978). A numerical solar radiation model based on standard meteorogical observation. *Solar Energy*, *21*(3), 163–170. doi:10.1016/0038-092X(78)90018-X

Aung, H. M. (2015). Renewable Energy Development in Myanmar. In The Energy Dialogue between Russia and ASEAN in the Field of Renewable Energy and Clean Energy Technologies.

Aung, Ko., & Shestopalova, T. A. (2016). Problems of creating a comprehensive system of electricity and heat supply of the off-line consumer of Myanmar. In *Proceedings of the Twenty-second international scientific and technical conference of students and graduate students "Radio electronics, Electronics and Energy"* (p. 316). Moscow: Nauka.

Aung, T.P. (2013, June). Introduction of MES and Current status of Energy Efficiency in Myanmar. Energy and Renewable Energy Committee. University of Technology Thonburi Campus, Bangkok, Thailand.

Aung, Ko., Malinin, N. K., & Shestopalova, T. A. (2016). *Statement of the problem of optimizing the composition of the energy complex on the basis of renewable energy sources for the integrated power supply to the rural consumer of Myanmar.*

Aygun, O., & Pehlivanlar, S. (2006). Listeria spp. in the raw milk and dairy products in Antakya, Turkey. *Food Control*, *17*(8), 676–679. doi:10.1016/j.foodcont.2005.09.014

Bacci, C., Lanzoni, E., Alpigiani, I., Boni, E., Vismarra, A., Bonardi, S., & Brindani, F. (2014). Listeria monocytogenes in raw milk and artificially contaminated aliquots. *Large Animal Review*, *20*(4), 175–180.

Baehr, R. (2001). Transformer technology state-of-the-art and trends of future development. *ELECTRA-CIGRE*, 13-19.

Baek, I., Cho, B., Kim, M., & Kim,Y. (2013) Determination of optimal excitation and emission wavebands for detection of defect cherry tomato by using fluorescence emission and excitation matrix. *Proc. SPIE 8721, Sensing for Agriculture and Food Quality and Safety V.*, 872108

Bajla, J. (1998). *Penetrometrické merania pôdnych vlastností* [Penetrometrical Measurements of the Soil Properties]. Nitra: SUA in Nitra.

Balakshin, OB, Kukharenko, B.G. (2015). The emergence and development of flutter rotor blades axial turbocharger. *Problems of mechanical engineering and reliability of machines*, *3*, 24-29.

Balbuen, R. H., Terminiello, A. M., Claverie, J. A., Casado, J. P., & Marlats, R. (2000). Soil compaction by forestry harvester operation, evolution of physical properties. *Revista Brasileira de Engenharia Agrícola e Ambiental*, *4*, 453–459.

Bali, O. S., Lajnef, R., Felfoul, I., Attia, H., & Ayadi, M. A. (2013). Original article detection of escherichia coli in unpasteurized raw Milk. *International Journal of Agricultural and Food Science*, *3*(2), 53–55.

Barakat, A. M. A., Sobhy, M. M., El Fadaly, H. A. A., Rabie, N. S., Khalifa, N. O., Hassan, E. R., & Zaki, M. S. (2015). Zoonotic hazards of campylobacteriosis in some areas in Egypt. *Life Science Journal*, *12*(7), 9–14.

Baranov, Y. V., Troitskiy, O. A., Avraamov, Y. S., & Shlyapin, A. D. (2001). Fizicheskie osnovyi elektroimpulsnoy i elektroplasticheskoy obrabotok i novyie materialyi [Physical fundamentals of electropulse and electroplastic treatments and new materials]. Moskow: MGIU.

Bartolomeoli, I., Maifreni, M., Frigo, F., Urli, G., & Marino, M. (2009a). Occurrence and characterization of Staphylococcus aureus isolated from raw milk for cheesemaking. *International Journal of Dairy Technology*, *62*(3), 366–371. doi:10.1111/j.1471-0307.2009.00498.x

Bashilov, A. M., Gordeev, Y. A., Belyakov, M. V., & Shirokih, T. V. (2011). Svetodiod uvelichivaet vskhozhest [The led light increases the germination rate]. *Sel'skij mekhanizator*, *10*, 32-33.

Bashilov, A. M. (2005). *EHlektronno-opticheskoe zrenie v agrarnom proizvodstve [Electro-optical sight in the agricultural production]*. Moscow, Russia: GNU VIEHSKH.

Bashilov, A. M., & Belyakov, M. V. (2008). Metodika monitoringa chuvstvitel'nosti semyan k vozdejstviyu opticheskogo izlucheniya [Method of monitoring the sensitivity of seeds to the effects of optical radiation]. *Vestnik FGOU VPO MGAU Seriya. Agroinzheneriya*, *2*(27), 22–24.

Bashilov, A. M., & Belyakov, M. V. (2011). Optiko-ehlektronnaya aktivizaciya semyan izlucheniem svetodiodov [Optoelectronic activation of the seed radiation of the LEDs]. *Vestnik FGOU VPO MGAU Seriya. Agroinzheneriya*, *1*(46), 7–9.

Bashilov, A. M., & Belyakov, M. V. (2015). *Spektral'nye harakteristiki lyuminescencii i otrazheniya semyan agrokul'tur [Spectral characteristics of luminescence and reflection of agricultural seeds]*. Moscow, Russia: FBGNU VIEHSKH.

Batygin, N.F. (1980). Biologicheskiye osnovy predposevnoy obrabotki semyan i zony yeye effektivnosti [Biological basis of presowing treatment of seeds and the zone of its effectiveness]. *Sel'skokhozyaystvennaya biologiya [Agricultural Biology]*, *15*(4), 504-509.

Batygin, N. F. (1986). *Ontogenez vysshikh rasteniy* [Ontogenesis of higher plants]. Moscow: Agropromizdat.

Beckley, P. (1999). Modern steels for transformers and machines. *Power Engineering Journal*, *13*(4), 190–200. doi:10.1049/pe:19990403

Belenov, A. T. Kharchenko, V.V., Rakitov, S.A., Daus, Y.V., Yudaev, I.V. (2016). The Experience of Operation of the Solar Power Plant on the Roof of the Administrative Building in the Town of Kamyshin, Volgograd Oblast. *Applied Solar Energy, 52*(2), 105-108.

Belfkira, R., Zhang, L., & Barakat, G. (2011). Optimal sizing study of hybrid wind/PV/diesel power generation unit. *Solar Energy*, *85*(1), 100–110. doi:10.1016/j.solener.2010.10.018

Belitskaya, M. N., & Gribust, I. R. (2009). Stabilizatsiya i produktivnost' zernovykh agrotsenozov pri ispol'zovanii nanostrukturirovannogo rastvora bishofita [Stabilization and productivity of grain agrocenoses using nanostructured bischofite solution]. In Nanotekhnologii i nanomaterialy: sovremennoye sostoyaniye i perspektivy razvitiya v usloviyakh Volgogradskoy oblasti [Nanotechnologies and nanomaterials: current state and development prospects in the conditions of the Volgograd region] (pp. 62-65). Volgograd: Izdatel'stvo VolgGU [VolgGU Publishing].

Belitskaya, M. N., & Gribust, I. R. (2011). Izucheniye vozmozhnostey ispol'zovaniya nanostrukturirovannogo rastvora bishofita v rasteniyevodstve [Study of the possibilities of using nanostructured bischofite solution in plant growing]. In Nanotekhnologii i nanomaterialy: sovremennoye sostoyaniye i perspektivy razvitiya v usloviyakh Volgogradskoy oblasti [Nanotechnologies and nanomaterials: current state and development prospects in the conditions of the Volgograd region] (pp. 42-46). Volgograd: Izdatel'stvo VolgGU [VolgGU Publishing].

Belitskaya, M. N., Gribust, I. R., & Nefedieva, E. E. (2012). Ustanovleniye reglamentov primeneniya EKHA vody v zernovykh agrotsenozakh [Establishment of regulations for the use of ECA water in grain agrocenoses]. *Izvestiya Nizhnevolzhskogo agrouniversitetskogo kompleksa: Nauka i vyssheye professional'noye obrazovaniye* [Proceedings of the Nizhnevolzhsky agrouniversity complex: Science and higher professional education], 2 (26), 3-8.

Belov, A. A. (2018). Modeling the assessment of factors influencing the process of electro-hydraulic water treatment. *Vestnik Ngiei, 11*, 103–112.

Belov, A. A. (2018). Modeling the estimation of factors of influence on the process electrohydraulic treatment of water. *Vestnik NGIEI, 11*(90), 28–39.

Belyakov, M. V. (2008). *Optiko-ehlektronnaya tekhnologiya i sredstva upravleniya biologicheskoj aktivnost'yu semyan* [Optoelectronic technology and means of controlling biological activity of seeds] (Unpublished doctoral dissertation). FGOU VPO MGAU, Moscow, Russia.

Belyakov, M. V. (2015). Tipovye spektral'nye harakteristiki semyan rastenij [Typical spectral characteristics of plant seeds]. *Estestvennye i tekhnicheskie nauki, 11*, 521-525.

Belyakov, M. V. (2016). Metodika issledovaniya lyuminescentnyh svojstv semyan rastenij na spektrofluorimetre «Flyuorat-02-Panorama» [Methods of investigation of luminescent properties of plant seeds on the spectrofluorometer "Fluorat-02-Panorama»]. *Nauchnaya zhizn, 3*, 18-26.

Belyakov, M. V., & Kharitonova, D. A. (2016). Research of galega orientalis seeds with luminescent methods during scarification. In *Proceedings of Conference XII International scientific and practical conference*. Sheffield, UK: Science and Education LTD.

Belyakov, M. V. (2009). Primenenie opticheskoj stimulyacii dlya semyan soi [Application of optical stimulation for soybean seeds]. *Proceedings of Conference Activation of the role of young scientists-the way to the formation of innovative potential of agriculture. Collection of materials of the international scientific-practical conference devoted to the 70th anniversary of Professor M. Gordeev.*

Belyakov, M. V., & Rybkina, S. V. (2011). *Vozdejstvie opticheskogo izlucheniya na semena drevesnyh rastenij dlya stimulyacii rostovyh processov [The effect of optical radiation on the seeds of woody plants to stimulate growth processes].* Smolensk, Russia: Universum.

Belzile, C., Belanger, M., Viau, A., Chamberland, M., & Roy, S. (2004). An operational fluorescence system for crop assessment [Monitoring Food Safety, Agriculture, and Plant Health]. *Proceedings of the Society for Photo-Instrumentation Engineers*, 5271.

Ben Kahla, I., Boschiroli, M. L., Souissi, F., Cherif, N., Benzarti, M., Boukadida, J., & Hammami, S. (2011). Isolation and molecular characterisation of Mycobacterium bovis from raw milk in Tunisia. *African Health Sciences, 11*(3 Suppl. 1), S2–S5. doi:10.4314/ahs.v11i3.70032 PMID:22135638

Bergmann, A. (2009). *Diagnostic method for disorders using copeptin.* Patent US. no 2009221009.

Berzan, V. P. (2011). Aspects of the problem of stimulating the introduction of heat pumps. Problems of regional energy, 1, 91–94.

Besekersky, V. A., & Popov, E. P. (2004). *Theory of automatic control systems. Profession, 4*, 752.

Bespamyatnova, N. M. (2002). Nauchno-metodicheskiye osnovy adaptatsii pochvoobrabatyvayushchikh i posevnykh mashin [Scientific and methodical bases of adaptation of tillage and sowing machines]. – Rostov n/D: OOO .

Bespamyatnova, N. M. (2010). Vibratsii v tekhnologicheskikh protsessakh [Vibrations in technological processes]. Zernograd: VNIPTIMESKh.

Bezrukih, P. P. (2008). *Resursy i jeffektivnost' ispol'zovanija vozobnovljaemyh istochnikov jenergii v Rossii: uchebnoe posobie* [Resources and Efficiency of Renewable Energy Sources in Russia: Tutorial]. Moscow: Kniga–Renta.

Bezrukikh, P. P. (2010). Wind power.

Bilalis, D. J., Katsenios, N., Efthimiadou, A., Karkanis, A., Khah, E. M., & Mitsis, T. (2013). Magnetic field pre-sowing treatment as an organic friendly technique to promote plant growth and chemical elements accumulation in early stages of cotton. *Australian Journal of Crop Science*, *7*(1), 46–50.

Binitha, S., & Sathya, S. S. (2012). A survey of bio inspired optimization algo-rithms. *International Journal of Soft Comput-ing and Engineering*, *2*(2), 137–151.

Blaszczak, U., Aziz, D., & Gryko, L. (2017). Influence of the spectral composition of LED lighting system on plants cultivation in a darkroom. *Proc. SPIE 10445, Photonics Applications in Astronomy, Communications, Industry, and High Energy Physics Experiments.*

Blokh, A. G., Zhuravlev, Y. A., & Ryzhkov, L. N. (1991). *Radiation heat transfer: a Handbook*. Moscow, Russia: Energoatomizdat.

Bobryshev, F. I., Starodubtseva, G. P., & Popov, V. F. (2000). Effektivnyye sposoby predposevnoy obrabotki semyan [Effective methods of pre-sowing seed treatment]. *Zemledeliye*, *3*, 45.

Bogdal, D. (2005). Interaction of Microwaves with Different Materials. In D. Bogdal (Ed.), Microwave-Assisted Organic Synthesis: One Hundred Reaction Procedures. The Netherlands: Elsevier. doi:10.1016/S1460-1567(05)80014-5

Bolyuh, V. F., Oleksenko, S. V., & Schukin, I. S. (2015). Sravnitelnyiy analiz lineynyih impulsnyih elektromehanicheskih preobrazovateley elektromagnitnogo i induktsionnogo tipa [Comparative analysis of linear pulse electromechanical converters of electromagnetic and induction type]. Technical Electrodynamics, 4, 20–27.

Booth, A. I. (1992). *Elektronno-ionnyye protsessy vodnykh struktur zhivykh organizmov i produktov ikh pererabotki* [Electron-ion processes of aquatic structures of living organisms and their products]. Moscow.

Borodin, I. F. (Ed.). (1996). Ispol'zovanie kogerentnogo ehlektromagnitnogo izlucheniya v proizvodstve produkcii rastenievodstva [The use of coherent electromagnetic radiation in crop production]. *Doklady RASKHN*, *6*, 41–44.

Borowy, B. S., & Salameh, Z. M. (1996). Methodology for optimally sizing the combination of a battery bank and PV array in a wind/PV hybrid system. *IEEE Transactions on Energy Conversion*, *11*(2), 367–375. doi:10.1109/60.507648

Bose, B. K. (2002). *Modern power electronics and AC drivers*. Prentice Hall PTR.

Boxler, C., Augustin, W., & Scholl, S. (2013). Fouling of milk components on DLC coated surfaces at pasteurization and UHT temperatures. *Food and Bioproducts Processing*, *91*(4), 336–347. doi:10.1016/j.fbp.2012.11.012

Briganti, A. A. (2001). *Residential Heat Pumps*. Moscow. *ABOK*, *5*, 24–30.

Budnikov, D. A. (2018). Laboratory installation and planning experiment on wheat microwave heating of various densities. *Innovations in Agriculture, 2*(27), 36-40.

Bueno, C., & Carta, J. A. (2005). Technical–economic analysis of wind-powered pumped hydrostorage systems. Part II: Model application to the island of El Hierro. *Solar Energy*, *78*(3), 396–405. doi:10.1016/j.solener.2004.08.007

Bulatov, U. N., Krukov, A. V., & Hung, C.Z. (2014). Automatic controllers for distributed generation plants. Bratsk: Bratsk State University.

Bunea, M., Johnston, K., Bonner, C., Cousins, P., Smith, D., & Rose, D. (2010). Simulation and characterization of high efficiency back contact cells for low-concentration photovoltaics. In *Proceedings of the 35th IEEE Photovoltaic Specialists Conference PVSC* (pp. 823-826). 10.1109/PVSC.2010.5617188

Bunget, C., Salandro, W., Mears, L., & Roth, J. T. (2010). *Energy-based modeling of anelectrically-assisted forging process. In Trans. of the North American Manuf* (pp. 38–46). Research Institute of SME.

Buonomano, A., Calise, F., & Vicidimini, M. (2016). Design, Simulation and Experimental Investigation of a Solar System Based on PV Panels and PVT Collectors. *Energies*, *9*(7), 497. doi:10.3390/en9070497

Burkin, S. P., Shimov, G. V., & Andryukova, E. A. (2015). Ostatochnyie napryazheniya v metalloproduktsii [Residual stresses in metal products]. Ekaterinburg: Ural. un-t.

Burmistrov, A.A., Vissarionov, V.I., & Deryugin, G.V. (2007). Methods for calculating renewable energy resources: a tutorial. MPEI.

Burmistrov, A. A., Vissarionov, V. I., Deryugina, G. V., Kuznetsova, V. A., Kunakin, D. N., & Malinin, N. K. (2009). *Methods for calculating the resources of renewable energy sources.* MPEI.

Capelli, E., Sacchi, R., Ghitti, M., Feligini, M., Bonacina, C., Panelli, S., & Brambati, E. (2014). One-year investigation of Clostridium spp. occurrence in raw milk and curd of Grana Padano cheese by the automated ribosomal intergenic spacer analysis. *Food Control*, *42*, 71–77. doi:10.1016/j.foodcont.2014.02.002

Caudron, C., Héchard, C., Natorp, J. C., Berri, M., Souriau, A., Rodolakis, A., & Camuset, P. (2007). Comparison of Coxiella burnetii shedding in milk of dairy Bovine, Caprine, and Ovine Herds. *Journal of Dairy Science*, *90*(12), 5352–5360. doi:10.3168/jds.2006-815 PMID:18024725

Chandrasekaran, S., Ramanathan, S., & Basak, T. (2012). Microwave material processing-a review. *AIChE Journal. American Institute of Chemical Engineers*, *58*(2), 330–363. doi:10.1002/aic.12766

Chea, L. C., Hakansson, H., & Karlsson, B. (2013). Performance evaluation of new two axes tracking pv-thermal concentrator. *Journal of Civil Engineering and Architecture*, *7*(12), 1485–1493.

Chedid, R., Akiki, H., & Rahman, S. (1998). A decision support technique for the design of hybrid solar-wind power systems. *IEEE Transactions on Energy Conversion*, *13*(1), 76–83. doi:10.1109/60.658207

Chong, C. M., & Datta, A. K. (1999). Thawing of foods in a microwave oven: II. Effect of load geometry and dielectric properties. *The Journal of Microwave Power and Electromagnetic Energy*, *34*(1), 22–32. doi:10.1080/08327823.1999.11688385 PMID:10355128

Chow, T. T. (2010). A review on photovoltaic/thermal hybrid solar technology. *Applied Energy*, *87*(2), 365–379. doi:10.1016/j.apenergy.2009.06.037

Christiansson, A., Bertilsson, J., & Svensson, B. (1999). Bacillus cereus spores in raw milk: Factors affecting the contamination of milk during the grazing period. *Journal of Dairy Science*, *82*(2), 305–314. doi:10.3168/jds.S0022-0302(99)75237-9 PMID:10068952

Chugayeva, V. I. (2008). Patent 2187888. Russian Federation.

Chupackina, G. N., Maslennikov, P. V., & Skrypnik, L. N. (2011). Prirodnyye antioksidanty (ekologicheskiy aspect) [Natural anti-oxidants (ecological aspect)]. (in Russian)

Cicorin, A. (2012). *Ocenka poter' holostogo hoda v sel'skih silovyh transformatorah pri ih ehkspluatacii* [Estimation of no-load losses in rural power transformers during their operation]. Unpublished doctoral dissertation.

Clare, D. A., Bang, W. S., Cartwright, G., Drake, M. A., Coronel, P., & Simunovic, J. (2005). Comparison of sensory, microbiological, and biochemical parameters of microwave versus indirect UHT fluid skim milk during storage. *Journal of Dairy Science*, *88*(12), 4172–4182. doi:10.3168/jds.S0022-0302(05)73103-9 PMID:16291608

Conrad, H. (2000). Electroplasticity in metals and ceramics. *Materials Science and Engineering*, *287*(2), 276–287. doi:10.1016/S0921-5093(00)00786-3

Constellation. (2016). Which is more energy efficient? Microwave vs toaster oven vs oven. Retrieved from https://blog.constellation.com/2016/05/20/toaster-oven-vs-microwave-energy-efficiency/

Cook, J. A., Gleeson, A. M., Roberts, R. M., & Rogers, R. L. (1997). A spark-generated bubble model with semi-empirical mass transport. *The Journal of the Acoustical Society of America*, *101*(4), 1908–1920. doi:10.1121/1.418236

Coronel, P., Simunovic, J., & Sandeep, K. P. (2003). Temperature profiles within milk after heating in a continuous-flow tubular microwave system operating at 915 MHz. *Journal of Food Science*, *68*(6), 1976–1981. doi:10.1111/j.1365-2621.2003.tb07004.x

Covenry, J. S. (2005). Performance of a concentrating photovoltaic/thermal solar collector. *Solar Energy*, *78*(2), 211–222. doi:10.1016/j.solener.2004.03.014

Cruz, J. A., Salbilla, B. A., Kanazava, A., & Kramer, D. M. (2001). Inhibition of plastocyanin to P_{700} electron transfer in Chlamydomonas reinhardtii by hyperosmotic stress. *Plant Physiology*, *127*(3), 1167–1179. doi:10.1104/pp.010328 PMID:11706196

Cwintal, M., Dziwulska-Hunek, A., & Wilczek, M. (2010). Laser stimulation effect of seeds on quality of alfalfa. *International Agrophysics*, *24*(1), 15–19.

D'agostino, M., Wagner, M., Vazquez-boland, J. A., Kuchta, T., Karpiskova, R., Hoorfar, J., & Cook, N. (2016). A validated PCR-based method to detect Listeria monocytogenes using raw milk as a food model—Towards an international standard. *Journal of Food Protection*, *67*(8), 1646–1655. doi:10.4315/0362-028X-67.8.1646 PMID:15330529

D'amico, D. J., & Donnelly, C. W. (2011). Characterization of Staphylococcus aureus strains isolated from raw milk utilized in small-scale Artisan cheese production. *Journal of Food Protection*, *74*(8), 1353–1358. doi:10.4315/0362-028X.JFP-10-533 PMID:21819666

Daminelli, P., Bertasi, B., Dalzini, E., Losio, M. N., Bernini, V., & Varisco, G. (2015). Survey of prevalence and seasonal variability of Listeria monocytogenes in raw cow milk from Northern Italy. *Food Control*, *60*, 466–470.

Dapice, D. (n.d.). *Electricity Demand and Supply in Myanmar.*

Dardeniz, A., Tayyar, S., & Yalcin, S. (2006). Influence of low-frequency electromagnetic field on the vegetative growth of grape cv. Uslu. *Journal of Central European Agriculture*, *7*(3), 389–395.

Daus, Y. V., & Yudaev, I. V. (2016). Designing of Software for Determining Optimal Tilt Angle of Photovoltaic Panels. In Proceedings of International Conference on Education, Management and Applied Social Science (EMASS2016). Beijing: DEStech Publications, Inc.

Daus, Yu. V., & Yudaev, I. V. (2016). Designing of Software for Determining Optimal Tilt Angle of Photovoltaic Panels. In *Proceedings of International Conference on Education, Management and Applied Social Science (EMASS2016)*. Beijing: DEStech Publications, Inc.

Daus, Y., Kharchenko, V., & Yudaev, I. V. (2018). Solar Radiation Intensity Data as Basis for Predicting Functioning Modes of Solar Power Plants. In *Handbook of Research on Renewable Energy and Electric Resources for Sustainable Rural Development* (pp. 283–309). Hershey, PA: IGI Global.

Daus, Yu. V., & Kharchenko, V. V. (2018). Evaluating the applicability of data on total solar-radiation intensity derived from various sources of actinometric information. *Applied Solar Energy*, *54*(1), 71–76. doi:10.3103/S0003701X1801005X

Daus, Yu. V., Kharchenko, V. V., & Yudaev, I. V. (2016a). Evaluation of Solar Radiation Intensity for the Territory of the Southern Federal District of Russia when Designing Microgrids Based on Renewable Energy Sources. *Applied Solar*, *52*(2), 124–129.

Daus, Yu. V., Yudaev, I. V., & Stepanchuk, G. V. (2018). Reducing the costs of paying for consumed electric energy by utilizing solar energy. *Applied Solar Energy*, *54*(2), 137–141. doi:10.3103/S0003701X18020056

de Bruin, A., van der Plaats, R. Q. J., de Heer, L., Paauwe, R., Schimmer, B., Vellema, P., & van Duynhoven, Y. T. H. P. (2012). Detection of Coxiella burnetii DNA on small-ruminant farms during a Q fever outbreak in the Netherlands. *Applied and Environmental Microbiology*, *78*(6), 1652–1657. doi:10.1128/AEM.07323-11 PMID:22247143

De Souza, A., Garcнa, D., Sueiro, L., Gilart, F., Porras, E., & Licea, L. (2006). Pre-sowing magnetic treatments of tomato seeds increase the growth and yield of plants. *Bioelectromagnetics*, *27*(4), 247–257. doi:10.1002/bem.20206 PMID:16511881

de Souza, V., Ribeiro, L. F., Oliveira, M. C., Borges, L. A., de Medeiros, M. I. M., Maluta, R. P., & Barbosa, M. M. C. (2016). Antimicrobial resistance and virulence factors of Escherichia coli in cheese made from unpasteurized milk in three cities in Brazil. *Foodborne Pathogens and Disease*, *13*(9), 469–476. doi:10.1089/fpd.2015.2106 PMID:27258947

Deboli, R., Calvo, A., & Preti, C. (2017). Whole-body vibration: Measurement of horizontal and vertical transmissibility of an agricultural tractor seat. *International Journal of Industrial Ergonomics*, *58*, 69–78. doi:10.1016/j.ergon.2017.02.002

Deriugina, G.V., Zay, Y.L., & Tyagunov, M.G. (2017). Data verification for use in the regional renewable energy sources geographic information system. Energetik, 5, 36-40.

Diaf, S., Belhamel, M., Haddadi, M., & Louche, A. (2007). A methodology for op-timal sizing of autonomous hybrid PV/wind system. *Energy Policy*, *35*(11), 5708–5718. doi:10.1016/j.enpol.2007.06.020

Distributed Generation, D. (2013). An Overview of Recent Policy and Market Developments. Retrieved from https://www.publicpower.org/files/PDFs/Distributed%20Generation–Nov2013.pdf

Dmitriev, A. M., & Stratskevich, L. K. (1986). *Stimulyatsiya rosta rasteniy* [Stimulation of plant growth]. Minsk: Uradzhay.

Dobermann, T. (2016, March). Energy in Myanmar. International Growth Centre.

Donovan, K. O. (1959). The occurrence of Bacillus Cereus in milk and on dairy equipment. *The Journal of Applied Bacteriology*, *22*(1), 131–137. doi:10.1111/j.1365-2672.1959.tb04617.x

Dubey, S., & Tay, A. A. O. (2012). Experimental Study of the Performance of Two Different Types of Photovoltaic Thermal (PVT) Modules under Singapore Climatic Conditions. *Journal of Fundamentals of Renewable Energy and Applications*, *2*, 1–6. doi:10.4303/jfrea/R120313

Duffie, J. A., & Beckman, W. A. (2006). *Solar Engineering of Thermal Processes* (3rd ed.). Hoboken, New Jersey: John Wiley and Son Inc.

Dzhurabaev, M. (1986). Primeneniye elektroaktivirovannoy vody v sel'skom khozyaystve [Use of electro-activated water in agriculture]. *Mekhanizatsiya i elektrifikatsiya sel'skogo khozyaystva [Agricultural mechanization and electrification], 11*, 51-53.

Ebel, F., Idler, S., Prede, G., & Scholz, D. (2008). *Fundamentals of Automation Technology, technical book*. Germany: Denkendorf.

Ebisawa, M., Shoji, K., Kato, M., Shimomura, K., Goto, F., & Yoshihara, T. (2008). Supplementary ultraviolet radiation B together with blue light at night increased quercetin content and flavonol synthase gene expression in leaf lettuce (Lactuca sativa L.). *Environment Control in Biology, 46*(1), 1–11. doi:10.2525/ecb.46.1

Eftekhari, M., & Mosavari, N. (2016). Isolation and molecular identification of Mycobacterium from commercially available pasteurized milk and raw milk samples collected from two infected cattle farms in Alborz Province, Iran. *International Journal of Mycobacteriology, 5*, S222–S223. doi:10.1016/j.ijmyco.2016.10.005 PMID:28043568

Eftichios, K., Dionissia, K., Antonis, P., & Kostas, K. (2006). Methodology for optimal sizing of stand alone photovoltaic/wind generator systems using genetic algorithms. *Solar Energy, 80*(9), 1072–1088. doi:10.1016/j.solener.2005.11.002

Ekadewi, A., & Djatmiko, I. (2013). The optimal tilt angle of a solar collector. *Energy Procedia, 32*, 166–175. doi:10.1016/j.egypro.2013.05.022

Elhadidy, M. A., & Shaahid, S. M. (1999). Optimal sizing of battery storage for hybrid (wind+ diesel) power systems. *Renewable Energy, 18*(1), 77–86. doi:10.1016/S0960-1481(98)00796-4

Ellison, W. J. (2007). Permittivity of pure water at standard atmospheric pressure, over the frequency range 0-25 THz and the temperature range 0-100 °C. *Journal of Physical and Chemical Reference Data, 36*(1), 1–18. doi:10.1063/1.2360986

El-Sharoud, W. M. (2009). Prevalence and survival of Campylobacter in Egyptian dairy products. *Food Research International, 42*(5–6), 622–626. doi:10.1016/j.foodres.2009.01.009

Eltamaly, A. M., & Mohamed, M. A. (2014). A novel design and optimization software for autonomous PV/wind/battery hybrid power systems. *Mathematical Problems in Engineering*.

Engineering Simulation & 3D Design Software – ANSYS. (n.d.). Retrieved from http://www.ansys.com/

Ershova, I.G., Ershov M.A., Poruchikov D.V. (2017). Justification of the regulation of the transmission of a low-potential energy source in a system based on a heat pump. *Journal Fundamental foundations of mechanics, 2*, 32-33.

Escher, W., Paredes, S., Zimmermann, S., & Chin, L. O. (2012). Thermal management and overall performance of a high concentration pv. In *Proceeding of the 8th International Conference on Concentrator Photovoltaics (CPV-8)* (pp. 239-243). 10.1063/1.4753877

Etamaly, A. M., Mohamed, M. A., & Alo-lah, A. I. (2015). A smart technique for op-timization and simulation of hybrid photo-voltaic/wind/diesel/battery energy systems. *Smart Energy Grid Engineering (SEGE), 2015 IEEE International Conference on*, 1-6.

Evdokimov, V. F., Kuchaev, A. A., Petrushenko, E. I., & Kasyan, G. I. (2010). Dvumernaya integralnaya model raspredeleniya sinusoidalnyih vihrevyih tokov i elektrodinamicheskih usiliy v kristallizatore s yavnopolyusnyim elektromagnitnyim peremeshivatele [Two-dimensional integral model of the distribution of sinusoidal eddy currents and electrodynamic forces in the mold with an explicit-pole electromagnetic stirrer]. Elektronnoe modelirovanie, 1, 53-75.

Fagundes, H., Barchesi, L., Filho, A. N., Ferreira, L. M., & Oliveira, C. A. F. (2010). Occurrence of staphylococcus aureus in raw milk produced in dairy farms in São Paulo State, Brazil. *Brazilian Journal of Microbiology, 41*(2), 376–380. doi:10.1590/S1517-83822010000200018 PMID:24031507

Fang, Y., Cheng, J., Roy, R., Roy, D. M., & Agrawal, D. K. (1997). Enhancing densification of zirconia-containing ceramic-matrix composites by microwave processing. *Journal of Materials Science, 32*(18), 4925–4930. doi:10.1023/A:1018624223909

Fayzullin, R., & Starikov, V. (2016). *Propusknaya sposobnost' lehp i meropriyatiya po eyo povysheniyu* [Capacity of power lines and measures for its improvement]. Ural State Mining University. Retrieved from http://science.ursmu.ru/upload/doc/2016/06/15/11_elektrotehnicheskie_kopleksy_i_sistemy.pdf

Fedio, W. M., & Jackson, H. (2013). Incidence of Listeria monocytogenes in raw bulk milk in Alberta. *Canadian Institute of Food Science and Technology Journal, 23*(4–5), 236–238.

Fellmeth, T., Ebert, M., & Efinger, R. (2014). *Industrially feasible all-purpose metal-wrap- through concentrator solar cells.* In *Proceedings of the 40th IEEE PVSC* (pp. 2106-2110). 10.1109/PVSC.2014.6925340

Feodorova, O. A. (2010). *Supramolekulyarnaya khimiya* [Supramolecular chemistry]. Moscow: RHTU. (in Russian)

Feuerhuber, K., Lindert, S. O., & Schlacher, K. (2013). Vibration attenuation by semi-active dampers. *Paper presented at 14th International Conference on Computer Aided Systems Theory (EUROCAST)*, Las Palmas, Spain.

Filippov, A. K., Bityuckij, N. P., & Fedorov M. A. (1997). *Ustrojstvo dlya plazmennoj obrabotki semyan rastenij* [Device for plasma treatment of plant seeds]. Patent RF. no 2076555.

Fister, I. Y., Jr. (2013). *A brief review of nature-inspired algorithms for optimization.* arXiv preprint arXiv:1307.4186

Flora, V. D. (2012). *Spetsialnyie elektromagnitnyie i elektromehanicheskie preobrazovateli energii* [Special electromagnetic and electromechanical energy converters]. Zaporizhia.

Flórez, M., Carbonell, M. V., & Martínez, E. (2007). Exposure of maize seeds to stationary magnetic fields: Effects on germination and early growth. *Environmental and Experimental Botany, 59*(1), 68–75.

Flyrez, M., Carbonell, V., & Martínez, E. (2007). Exposure of maize seeds to stationary magnetic fields: Effects of germination and early growth. *Environmental and Experimental Botany, 59*(1), 68–75. doi:10.1016/j.envexpbot.2005.10.006

Forbes Solar. (n.d.). Solar Cogeneration System. Retrieved from https://www.forbesmarshall.com/Download/Solar%20Cogeneration%20System.pdf

Gaevskij, N. A. (2002). *Kriterii i metodologiya ocenki strukturno-funkcional'nogo sostoyaniya al'gocenoza na osnove fluorescentnogo analiza* [Criteria and methodology for assessing the structural and functional state of algocenosis based on fluorescent analysis] (Unpublished doctoral dissertation). Institut biofiziki SO RAN, Krasnoyarsk, Russia.

Ganiev, R.F., Balakshin, OB, Kukharenko, B.G. (2015). Turbulence flow and flutter rotor blades in an axial turbocharger. *Problems of mechanical engineering and reliability of machines, 6*, 3-10.

Gaska, R., & Hang Dzh, Z. (2007). Ul'trafioletovo izuchayushchie diody (UV-emitting diodes). *Svetotekhnika, 6*, 38–39.

Gavanidou, E. S., Bakirtzis, A. G., & Dokopoulos, P. S. (1993). A probabilistic method for the evaluation of the performance and the reliability of wind-diesel energy systems. *IEEE Transactions on Energy Conversion, 8*(2), 197–206. doi:10.1109/60.222705

Gavrilas, M. (2010). Heuristic and metaheuristic optimization techniques with application to power systems. *Proceedings of the 12th WSEAS international conference on Mathematical methods and computational techniques in electrical engineering.*

Geleta, D.K., & Manshahia, M.S. (2017). Optimization of Renewable Energy Systems: A Review. *International Journal of Scientific Research in Science and Technology, 8*(3), 769-795.

Geleta, D. K., & Manshahia, M.S. (2017). Nature Inspired Computational Intelligence: A Survey. *International Journal of Engineering. Science and Mathematics*, *6*(7), 769–795.

Geleta, D.K., & Manshahia, M. S. (2018). Opti-mization of Hybrid Wind and Solar Renewable Energy System by Iteration Method. *International Conference on Intelligent Computing and Optimization*, 98-107.

George Kosmadakis, S. K. (2013). *Renewable and Conventional Electricity Generation Systems*. Technologies and Diversity of Energy Systems.

Giezendanner, N., Meyer, B., Gort, M., Müller, P., & Zweifel, C. (2009). Rohmikh-Assoziierte Staphylococcus Aureus Intoxikation Bei Kindern. *Schweizer Archiv fur Tierheilkunde*, *151*(7), 329–331. doi:10.1024/0036-7281.151.7.329 PMID:19565455

Giliberto, L., Perrotta, G., Pallara, P., Weller, J. L., Fraser, P. D., Bramley, P. M., ... Tavazza, M. (2005). Giuliano Manipulation of the blue light photoreceptor cryptochrome 2 in tomato affects vegetative development, flowering time, and fruit antioxidant content. *Plant Physiol.*, 199–208. doi:10.1104/pp.104.051987

Godzhaev, Z.A., Shevtsov, V.G., Lavrov, A.V., & Rusanov, A.V. (2015). Metodika rascheta maksimalnogo kontaktnogo davleniya kolesnogo dvizhetelya na pochvu s ispolzovaniem universalnoj kharakteristiki shiny [The method of calculation of maximum contact pressure wheel on soil mover with the use of universal bus characteristics]. *Alternative Energy Sources in the Transport and Technological Complex: Problems and Prospects for Rational introduction*, *2*(1), 83-89.

Golikov, I., & Vinogradov, A. (2017). *Adaptivnoe avtomaticheskoe regulirovanie napryazheniya v sel'skih ehlektricheskih setyah 0,38 kV* [Adaptive automatic voltage regulation in rural power networks of 0.38 kV]. Orel, Russia: FGBOU VO Orel GAU.

Golovatskaya, I. F. (2005). Rol' kriptokhroma 1 i fitokhromov v regulyatsii fotomorfogeneticheskikh reaktsiy rasteniy na zelenom svetu [Role of cryptochrome 1 and phytochromes in regulation of plants photomorphogenetic reactions on green light]. Fiziologiya rasteniy [Plants physiology], 6, 822-829. (in Russian)

Golovatskaya, I. F. (2016). Morfogenez rasteniy i yego regulyatsiya. Chast' 1: Fotoregulyatsiya morfogeneza rasteniy [Morphogenesis of plants and its regulation. Part 1: Photoregulation of plants morphogenesis: Educational manual]. Tomsk: Publishing house of Tomsk State University. (in Russian)

Goncharov, A. A., Berezhnaya, I. E., & Gursky, I. G. (1994). *Vliyaniye fizicheskikh metodov predposevnoy obrabotki semyan sel'skokhozyaystvennykh kul'tur na ikh semennyye kachestva* [The influence of physical methods of pre-sowing treatment of seeds of agricultural crops on their seed qualities]. Zelenograd.

Gordeev, A. M., & Shershnev, V. B. (1991). *Elektrichestvo v zhizni rasteniy* [Electricity in plant life]. Moscow: Nauka.

Gorshkov, V. G. (2004). Heat pumps. Analytical review. In Handbook of industrial equipment.

Goryachkin, V. P. (1965) Zemledel'cheskaya mekhanika. Sobraniye sochineniy [Agricultural mechanics. Moscow, Russia: Kolos.

Goryagina, E. B. (2011). Predobrabotka nezrelykh zarodyshey gibridov rapsa yarovogo i gorchitsy beloy na etape vvedeniya v kul'turu in vitro [Pre-treatment of immature embryos of spring rape and white mustard hybrids at the stage of introduction into culture in vitro]. In *Materialy VI mezhdunarodnoy konferentsii molodykh uchenykh i spetsialistov [Proceedings of the VI International Conference of Young Scientists and Specialists]* (pp. 66-69). VNIIMK.

Goto, E., Hayashi, K., Furuyama, S., Hikosaka, S., & Ishigami, Y. (2016). Effect of UV light on phytochemical accumulation and expression of anthocyanin biosynthesis genes in red leaf lettuce. *Acta Horticulturae*, (1134), 179–186. doi:10.17660/ActaHortic.2016.1134.24

Goussous, S. J., Samarah, N. H., Alqudah, A. M., & Othman, M. O. (2010). Enhancing seed germination of four crop species using an ultrasonic technique. *Experimental Agriculture*, *46*(2), 231–24. doi:10.1017/S0014479709991062

Grant, E. H., Buchanan, T. J., & Cook, H. F. (1957). Dielectric behavior of water at microwave frequencies. *The Journal of Chemical Physics*, *26*(1), 156–161. doi:10.1063/1.1743242

Grigor'yeva, O. (2014). Sposoby podgotovki semyan k posevu [Methods for preparing of seeds for sowing]. *LesProm*, *6*(104), 176–177.

Groisman, Y., & Gedanken, A. (2008). Continuous flow, circulating microwave system and its application in nanoparticle fabrication and biodiesel synthesis. *The Journal of Physical Chemistry C*, *112*(24), 8802–8808. doi:10.1021/jp801409t

Gromov, V. E. (2014). Prochnost i plastichnost materialov v usloviyah vneshnih energeticheskih vozdeystviy [Strength and ductility of materials under external energy conditions]. Novokuznetsk: Sibirskiy gosudarstvennyiy industrialnyiy universitet.

Groover, M. (2007). *Automation, production systems, and computer-integrated manufacturing*. Prentice Hall Press.

Gude, V. G. (2017). Microwave–Mediated Biofuel Production. In Microwave–Mediated Biofuel Production. doi:10.1201/9781315151892

Gundogan, N., & Avci, E. (2014). Occurrence and antibiotic resistance of Escherichia coli, Staphylococcus aureus and Bacillus cereus in raw milk and dairy products in Turkey. *International Journal of Dairy Technology*, *67*(4), 562–569. doi:10.1111/1471-0307.12149

Guoyi, T. (2000). Experimental study of electroplastic effect on stainless steel wire 304L. *Materials Science and Engineering*, (281), 263–267.

Gusarov, V.A. (2014). Improving the efficiency of electronic devices for autonomous power supply of rural consumers. *Innovations in agriculture*, *2*, 39-48.

Gusarov, V.A. (2017). Microgas turbine power plants for autonomous power supply. *Gas turbine technologies*, *6*, 8 - 12.

Gusarov, V. A., Gusarova, O. F., & Kulagin, Ya. V. (2013). The use of gas turbine technologies in agricultural production. *Alternative Energy and Ecology*, *2*, 76–79.

Gusarov, V. A., Zadde, V. V., Nikitin, B. A., & Kargiev, V. M. (2004). Automatic complex uninterrupted power supply. In *Proceedings of the International Scientific and Technical Conference* (Vol. 4, pp. 263 – 268).

Gusavor, V. A. (2016). Prospects for distributed energy. *Bulletin of the VIESH*, *3*, 77–83.

Gut, M. (2007). Impact of alternating electric field on potato tuber growth and cropping. *Inżynieria Rolnicza*, *8*(96), 73–79.

Gwida, M. M., & AL-Ashmawy, M. A. M. (2014). AL-Ashmawy, M. A. M. (2014). Culture versus PCR for Salmonella species identification in some dairy products and dairy handlers with special concern to its Zoonotic importance. *Veterinary Medicine International*, 1–5. doi:10.1155/2014/502370

Hadden, R. L. (2008). The geology of Burma (Myanmar): An annotated bibliography of Burma's geology, geography and earth science. Geospatial Information Library (GIL).

Hadidian, M. J., & Arabi, N. S., & Bigdeli. (2016). Optimal sizing of a stand-alone hybrid photovoltaic/wind system using new grey wolf optimizer considering re-liability. *Journal of Renewable and Sustainable Energy*, *8*(3).

Haitsma, O., De Jonge, R., Besselse, M., Mulder, B., Van Pelt, W., Van Steenbergen, J., & Van Duynhoven, Y. T. H. P. (2009). A prolonged outbreak of Salmonella Typhimurium infection related to an uncommon vehicle: Hard cheese made from raw milk. *Epidemiology and Infection, 137*(11), 1548–1557. doi:10.1017/S0950268809002337 PMID:19296867

Hakansson, I. (1990). A method for characterizing the state of compactness of the plough layer. *Soil & Tillage Research, 16*(1-2), 105–120. doi:10.1016/0167-1987(90)90024-8

Hall, C. W., & Trout, G. M. (1968). *Milk Pasteurization*. Westport, Connecticut: AVI Publishing Company, Inc.

Hamelin, J., Mehl, J. B., & Moldover, M. R. (1998). The static dielectric constant of liquid water between 274 and 418 K near the saturated vapor pressure. *International Journal of Thermophysics, 19*(5), 1359–1380. doi:10.1023/A:1021979401680

Hanoi, M. (2012, November). Renewable Energy and Rural Development in Myanmar.

Hao, X., Little, C., Zheng, J. M., & Cao, R. (2016). Far-red LEDs improve fruit production in green house tomato grown under high-pressure sodium lighting. *Acta Horticulturae*, (1134): 95–102. doi:10.17660/ActaHortic.2016.1134.13

Hartman, K., Letsky, E., & Shefer, V. (1977). *Planning an experiment in the study of technological processes*. Moscow: Mir.

Hernandez, A. C., Dominguez, P. A., Cruz, O. A., Ivanov, R., Carballo, C. A., & Zepeda, B. R. (2010). Laser in agriculture. *International Agrophysics, 24*(4), 407–422.

Herve, A. G., Tang, J., Luedecke, L., & Feng, H. (1998). Dielectric properties of cottage cheese and surface treatment using microwaves. *Journal of Food Engineering, 37*(4), 389–410. doi:10.1016/S0260-8774(98)00102-2

Hetz, E. J. (2001). Soil compaction potential of tractors and other heavy agricultural machines used in Chile. *Agricultural Mechanization in Asia. Africa and Latin America, 32*, 38–42.

Hosseini, R., Hosseini, N., & Khorasanizadeh, H. (2011). An Experimental study of combining a Photovoltaic System with a heating System. In *Proceedings of the World Renewable Energy Congress*, Linkoping, Sweden (pp. 2993-3000). 10.3384/ecp110572993

Ho, T., Htwe, K. K., Yamasaki, N., Zhang, G. Q., Ogawa, M., Yamaguchi, T., & Hirai, K. (1995). Isolation of Coxiella burnetii from dairy cattle and ticks, and some characteristics of the isolates in Japan. *Microbiology and Immunology, 39*(9), 663–671. doi:10.1111/j.1348-0421.1995.tb03254.x PMID:8577279

Huang, Y., Yan, H., Wang, B., Zhang, X., Liu, Z., & Yan, K. (2014). The electro-acoustic transition process of pulsed corona discharge in conductive water. *Journal of Physics. D, Applied Physics, 47*(25), 169–175. doi:10.1088/0022-3727/47/25/255204

Hunt, K., Schelin, J., Rådström, P., Butler, F., & Jordan, K. (2012). Classical enterotoxins of coagulase-positive Staphylococcus aureus isolates from raw milk and products for raw milk cheese production in Ireland. *Dairy Science & Technology, 92*(5), 487–499. doi:10.100713594-012-0067-4

Husu, J. R., Seppänen, J. T., Sivelä, S. K., & Rauramaa, A. L. (1990). Contamination of Raw Milk by Listeria monocytogenes on Dairy Farms. *Journal of Veterinary Medicine, Series B, 37*(1–10), 268–275.

Hu, Y., Li, P., & Jiang, J. (2007). Developing a new supplemental lighting device with ultra-bright white LED for vegetables. *Proc. SPIE 6486, Light-Emitting Diodes, Research, Manufacturing, and Applications, 11*, 64861A.

Hyland, G. J. (2003). Bio-Electromagnetism. In F.A. Popp, & L.V. Beloussov (Eds.), Integrative Biophysics-Biophotonics (pp. 117-148). Dordrecht: Kluwer Academic Publisher.

Iacoboni, M., Chow, J., & Wheless, E. (2012). Calabasas Landfill Microturbine Power Generation Project: Lessons Learned after One-Year of Operation. Retrieved from https://www.capstoneturbine.com/

Ibrahim, A., Jin, G. L., Daghigh, R., Salleh, M. H. M., Othman, M. Y., & Ruslan, M. H.. (2009). Hybrid photovoltaic thermal (PV/T) air and water based solar collectors suitable for building integrated applications. *American Journal of Environmental Sciences*, 618–624.

Ibrahim, A., Othman, M. Y., Ruslan, M. H., Mat, S., & Sopian, K. (2011). Recent advances in flat plate photovoltaic/ thermal (PV/T) solar collectors. *Renewable & Sustainable Energy Reviews*, *15*(1), 352–365. doi:10.1016/j.rser.2010.09.024

Ilinca, A., McCarthy, E., Chaumel, J. L., & Retiveau, J. L. (2003). Wind potential assessment of Quebec Province. *Renewable Energy*, *28*(12), 1881–1897. doi:10.1016/S0960-1481(03)00072-7

Inada, K. (1978). Spectral dependence of photosynthesis in crop plants. *Acta Horticulturae*, *87*(87), 177–184. doi:10.17660/ ActaHortic.1978.87.18

Informatsionnyy obzor. (2016). Operativnyye dannyye [Informational review "United Energy System of Russia: Interim Results]. Retrieved from http://so–ups.ru/fileadmin/files/company/reports/ups–review/2016/ups_review_june16_1.pdf

IRENA. (2016). *REmap: Roadmap for a Renewable Energy Future, 2016 Edition. International Renewable Energy Agency*. Abu Dhabi: IRENA.

IRENA. (2016). *Renewable Energy Outlook for ASEAN: a REmap Analysis. International Renewable Energy Agency (IRENA). Abu Dhabi and ASEAN Centre for Energy*. Jakarta: ACE.

Isachenko, V. P. (1975). *Heat transfer processes. Textbook for high schools*. Moscow, Russian: Energy.

Ismagilov, F. R., Afanasev, Yu. V., & Styiskin, A. V. (2006). *Vvedenie v konstruirovanie elektromehanicheskih preobrazovateley energii* [Introduction to the design of electromechanical energy transducer]. Moskow: MAI.

Izenberg, Yu. B. (1983). *Reference book on light-techniques*. Moscow: Energopromizdat. (in Russian)

Izmailov, A., Shevtsov, V., Lavrov, A., Godzhaev, Z., & Pryadkin, V. (2015). Application of the Universal Tire Characteristic for Estimating the Maximum Pressure of a Pneumatic Tractor Wheel on the Ground. *SAE*. doi:10.4271/2015-01-2760

JA SOLAR. (n.d.). Retrieved from http://jasolar.com/site/solar_Mono/564#level-2

Janardhan, V. & Galloway, D. (2001). Thinking about the box: Ten ways to mitigate distribution transformer losses. *ElectricLight& Power, 79*(10).

Javadi, M. R., Mazlumi, K., & Jalilvand, A. (2011). Application of GA, PSO and ABC in optimal design of a stand-alone hybrid system for north-west of Iran. In *Electrical and Elec-tronics Engineering* (pp. 1–203). ELECO.

Jedlička, J., Paulen, O., & Ailer, Š. (2014). Influence of magnetic field on germination, growth and production of tomato. *Potravinárstvo*, *8*(1), 150–154. doi:10.5219/349

Jeon, W. J., Katoh, S., Iwamoto, T., Kamiya, Y., & Onuki, T. (1999). Propulsive characteristics of a novel linear hybrid motor with both induction and synchronous operations. *IEEE Transactions on Magnetics*, *35*(5), 4025–4027. doi:10.1109/20.800743

Jiang, J., He, X., Li, L., Li, J., Shao, H., Xu, Q., ... Dong, Y. (2014). Effect of cold plasma treatment on seed germination and growth of wheat. *Plasma Science & Technology*, *16*(1), 54–58. doi:10.1088/1009-0630/16/1/12

Ji, X., Li, M., Lin, W., Wang, W., Wang, L., & Xi, J. (2012). Modeling and characteristic parameters analysis of a trough concentrating photovoltaic/thermal system with gas and super cell arrays. *International Journal of Photoenergy*.

Jones, H. M., & Kunhardt, E. E. (1994). The influence of pressure and conductivity on the pulsed breakdown of water. *IEEE Transactions on Dielectrics and Electrical Insulation*, *1*(6), 1016–1025. doi:10.1109/94.368641

Kaabeche, A., Belhamel, M., & Ibtiouen, R. (2010). Optimal sizing method for stand-alone hybrid PV/wind power genera-tion system. *Revue des Energies Renouvelables (SMEE'10)*, 205-213.

Kakinoki, Y., Kato, Y., Ogawa, K., Nakao, A., Okai, Z., & Katsuyama, T. (2013). Efficient plant growth using automatic position-feedback laser light irradiation. *Proc. SPIE 8881. Sensing Technologies for Biomaterial, Food, and Agriculture, 2013*. doi:10.1117/12.2030575

Kalinin, M. I., & Kudryavtsev, E. P. (2006). Patent 2292000. Russian Federation.

Kalnin I.M., Savitsky I.K. (2000). Heat pumps: yesterday, today, tomorrow. *Refrigeration equipment*, *10*, 2-6.

Kamilov, A. G., Muminov, R. A., & Tursunov, M. N. (2008). Ocenka jeffektivnosti sistemy solnechnogo jelementa i kollektora v uslovih zharkogo klimata (Evaluation of system effectiveness solar cell and collector in a hot climate). *Geliotehnika*, *2*, 32–35. (in Russian)

Karaboga, D., & Basturk, B. (2007). A powerful and efficient algorithm for numerical function optimization: Artificial bee colony (ABC) algorithm. *Journal of Global Optimization*, *39*(3), 459–471. doi:10.100710898-007-9149-x

Karaboga, D., Gorkemli, B., Ozturk, C., & Karaboga, N. (2014). A comprehensive survey: Artificial bee colony (ABC) algorithm and ap-plications. *Artificial Intelligence Review*, *42*(1), 21–57. doi:10.100710462-012-9328-0

Karns, J. S., Gorski, L., McCluskey, B. J., Perdue, M. L., & Van Kessel, J. S. (2010). Prevalence of Salmonellae, Listeria monocytogenes, and Fecal Coliforms in bulk tank milk on US dairies. *Journal of Dairy Science*, *87*(9), 2822–2830. PMID:15375040

Kartashev, I. (2001). *Upravlenie kachestvom ehlektroehnergii* [Power Quality in Power Supply Systems]. Moscow, Russia: Izdatel'skij dom MEHI.

Kasakova, A. S., Yudaev, I. V., Fedorishchenko, M. G., Mayboroda, S. Y., Ksenz, N. V., & Voronin, S. M. (2018). New approach to study stimulating effect of the pre-sowing barley seeds treatment in the electromagnetic field. *Online Journal of Biological Sciences*, *18*(2), 197–207. doi:10.3844/ojbsci.2018.197.207

Kataria, S., Baghel, L., & Guruprasad, K. N. (2017). Pre-treatment of seeds with static magnetic field improves germination and early growth characteristics under salt stress in maize and soybean. *Biocatalysis and Agricultural Biotechnology*, *10*, 83–90. doi:10.1016/j.bcab.2017.02.010

Kaynak-Onurdag, F., Okten, S., & Sen, B. (2016). Screening Brucella spp. in bovine raw milk by real-time quantitative PCR and conventional methods in a pilot region of vaccination, Edirne, Turkey. *Journal of Dairy Science*, *99*(5), 3351–3357. doi:10.3168/jds.2015-10637 PMID:26971148

Kazwala, R. R., Daborn, C. J., Kusiluka, L. J. M., Jiwa, S. F. H., Sharp, J. M., & Kambarage, D. M. (1998). Isolation of Mycobacterium species from raw milk of pastoral cattle of the Southern Highlands of Tanzania. *Tropical Animal Health and Production*, *30*(4), 233–239. doi:10.1023/A:1005075112393 PMID:9760715

Kefayat, M. A., Lashkar, A., & Nabavi, N. (2015). A hybrid of ant colony optimization and artificial bee colony algorithm for probabilistic optimal placement and sizing of distributed energy resources. *Energy Conversion and Management*, *92*, 149–161. doi:10.1016/j.enconman.2014.12.037

Kefeli, V. I., & Tom, S. I. (1985). *Rost rasteniy i yego regulyatsiya (geneticheskiye i fiziologicheskiye aspekty)* [Plant growth and its regulation (genetic and physiological aspects)]. Chisinau: Shtiintsa. [Shtiints]

Keller, R.B. (2009). Flavonoids: Biosynthesis, Biological Effects and Dietary Sources. Nova Science Publishers.

Kellog, W., Nehrir, M., Venkataramanan, G., & Gerez, V. (1998). Generation Unit Sizing and Cost Analysis for Stand -alone Wind, Phovoltaic, and Hybrid Wind/PV Systems. *IEEE Transactions on Energy Conversion, 13*(1), 70–75. doi:10.1109/60.658206

Kessy, B. M., John, J., Rajashekara, G., Kassem, I. I., Gebreyes, W., Kazwala, R. R., & Kashoma, I. P. (2015). Prevalence and antimicrobial resistance of Campylobacter isolated from dressed beef Carcasses and raw milk in Tanzania. *Microbial Drug Resistance, 22*(1), 40–52. PMID:26153978

Kevenk, T. O., & Terzi Gulel, G. (2016). Prevalence, antimicrobial resistance and Serotype distribution of Listeria monocytogenes isolated from raw milk and dairy products. *Journal of Food Safety, 36*(1), 11–18. doi:10.1111/jfs.12208

Kharchenko, V., Gusarov, V., & Bolshev, V. (2019). Reliable Electricity Generation in RES-Based Microgrids. In Handbook of Research on Smart Power System Operation and Control (pp. 162-187).

Kharchenko, V., Gusarov, V., & Bolshev, V. (2019). Reliable Electricity Generation in RES-Based Microgrids. In Handbook of Research on Smart Power System Operation and Control (pp. 162–187). Hershey, PA: IGI Global. doi:10.4018/978-1-5225-8030-0.ch006

Kharchenko, V., Panchenko, V., Tikhonov, P. V., & Vasant, P. (2018). Cogenerative PV thermal modules of different design for autonomous heat and electricity supply. In Handbook of Research on Renewable Energy and Electric Resources for Sustainable Rural Development (pp. 86-119). Hershey, PA: IGI Global. doi:10.4018/978-1-5225-3867-7.ch004

Kharchenko, V. V. (2014). PVThermal modules. In *Proceedings of XI International Annual Conference "Renewable and small energy"* (pp. 248-257).

Kharchenko, V. V., & Gusarov, V. A. (2015). The principles of the formation of a generating complex of microgrids based on renewable energy sources. *Bulletin of Agrarian Science Don, 32*, 71–83.

Kharchenko, V. V., Nikitin, B. A., & Tikhonov, P. V. (2010). Estimation and forecasting of PV cells and modules parameters on the basis of the analysis of interaction of a sunlight with a solar cell material. In *Proceedings of 4th International Conference TAE 2010, Trends in Agricultural Engineering* (pp. 307-310).

Kharchenko, V., Nikitin, B., Tikhonov, P., & Gusarov, V. (2013). Investigation of experimental flat PV thermal module parametres in natural conditions. In *Conference Proceeding - 5th International Conference, TAE 2013: Trends in Agricultural Engineering*, Prague, Czech Republic, September 2-3 (pp. 51-55).

Kholmanskiy, A. (2015). Activation energy of water structural transitions. *Journal of Molecular Structure, 1089*, 124–126. doi:10.1016/j.molstruc.2015.02.049

Kholmanskiy, A. (2016). Chirality anomalies of water solutions of saccharides. *Journal of Molecular Liquids, 216*, 683–687. doi:10.1016/j.molliq.2016.02.006

Kholmanskiy, A., Smirnov, A., & Parmon, V. (2018). Bleaching of solutions of chlorophyll a and b with blue and red light. *High Energy Chemistry, 52*(1), 6–13. doi:10.1134/S001814391801006X

Kireev, V. V., Lazeev, N. A., & Stepanenko, P. P. (2003). Saving energy resources through the use of natural cold. Storage and processing of agricultural resources, 10, 10-13.

Kislovskiy, L. D. (1982). Rol' vody v labil'nosti poverkhnostnykh struktur [The role of water in the lability of surface structures]. Moskow: VINITI.

Kiva, O. V., Khodurskyi, V. Y. (2010). Doslidzhennia ta rozrobka prystroiu dlia peredposivnoi obrobky nasinnia tsukrovoho buriaku [Research and development of the device for pre-sowing processing of seeds of sugar beet]. *Visnyk Poltavskoi derzhavnoi ahrarnoi akademii*, 4, 176 – 178.

Klassen, V. I. (1982). Omagnichivaniye vodnykh system [Magnetization of water systems]. Moskow: Khimiya.

Kolchin, N. N., & Fomin, S. L. (2006). *Potato storage: State and development prospects.* Potatoes and vegetables. *KARTO and OV, 1*, 28–31.

Kolin, A. R., Sergeyev, V. V., & Gorbatsevich, N. A. (2008). Vozdeystviye gradiyentnym magnitnym polem na posadochnyy material i vegetiruyushchiye kartofel'nyye rasteniya [Effect of a gradient magnetic field on planting material and vegetative potato plants]. Russkiye vysokiye tekhnologii. Retrieved from http://skutis.ucoz.ru/publ/26-1-0-13

KOMPAS-3D v17. (n.d.). Retrieved from http://kompas.ru/

Kondrat'eva, N. P., Krasnolutskaya, M. G., Dukhtanova, N. V., & Obolensky, N. V. (2019). Effect of ultraviolet radiation the germination rate of tree seeds. *IOP Conf. Series: Earth and Environmental Science, 226.* 10.1088/1755-1315/226/1/012049

Kondratenko, I. P., Bozhko, I. V., Zhiltsov, A. V., & Vasyuk, V. V. (2012). Doslidzhennya elektrofizichnih protsesiv u elektrodniy sistemi neruynivnogo viznachennya zalishkovih napruzhen [Investigation of electrophysical processes in the electrode system of non-destructive determination of residual stresses]. Institut elektrozvaryuvannya Im. E. O. Patona NAN UkraYini.

Kondratenko, I. P., Bozhko, I. V., Zhiltsov, A. V., & Vasyuk, V. V. (2012). Matematichne modelyuvannya elektroflzichnih protseslv v sistemah operativnogo neruynlvnogo viznachennya zalishkovih napruzhen [Mathematical modeling of electrophysical processes in systems of operational non-destructive determination of residual stresses]. Technical electrodynamics, 3, 1–22.

Kondratenko, I. P., Zhiltsov, A. V., & Vasyuk, V. V. (2014). Modelirovanie elektrofizicheskih protsessov v elektrotehnicheskom komplekse dlya snizheniya ostatochnyih napryazheniy [Modeling electrophysical processes in electrical complex to reduce residual stresses]. Moscow: Energosberegayuschie tehnologii v zhivotnovodstve i statsionarnoy energetike.

Kondratenko, I. P., Zhiltsov, A. V., & Vasyuk, V. V. (2015). Simulation of discharge capacity axle symmetric systems «coil – non-ferromagnetic massive disk» by the method of integral equations. Computational problems of electrical engineering (CPEE – 2015), 71–73.

Kondrat'eva, N. P. (2001). Vliyanie predposevnoj obrabotki semyan yarovoj pshenicy na urozhajnost' [Influence of presowing treatment of spring wheat seeds on yield]. *Mekhanizaciya i ehlektrifikaciya sel'skogo hozyajstva, 12*, 17.

Konstantinov, I., Zakonov, I., Vakatov, A., Korovin, S., & Abdrahmanov, A. (2007). Vnedrenie metodiki rascheta tekhnicheskih poter' ehlektroehnergii vo vnutristancionnyh ehlektricheskih setyah generiruyushchih predpriyatij OAO «Tatehnergo» [Implementation of the methodology for calculating technical losses of electricity in the internal electrical networks of generating enterprises of JSC "Tatenergo"]. In Sbornik dokladov 5-go nauchno-tekhnicheskogo seminara-vystavki «Normirovanie i snizhenie poter' ehlektricheskoj ehnergii v ehlektricheskih setyah». Moscow, Russia: DialogEHlektro.

Kosmadakis, G., Sotirios, K., & Emmanuel, K. (2013). Renewable and conventional elec-tricity generation systems. *Technologies and diversity of energy systems.* In *Renewable Energy Governance* (pp. 9–3). Springer London. doi:10.1007/978-1-4471-5595-9_2

Kozyrskyi, V., Savchenko, V., & Sinyavsky, O. (2018). Presowing Processing of Seeds in Magnetic Field. In Handbook of Research on Renewable Energy and Electric Resources for Sustainable Rural Development (pp. 576–620). Hershey, PA: IGI Global. doi:10.4018/978-1-5225-3867-7.ch024

Kravchenko, V. A. (2010). *Povishenie dinamicheskih i ekspluatatsionnyh pokazateley selskohozyastvennyh mashinno-traktornyh agregatov* [Increase of dynamic and operational parameters of agricultural machine and tractor units] [PhD dissertation]. FSEE HE ABSSAEA, Zernograd Russia.

Kravchenko, V. A., Senkevich, S. E., Senkevich, A. A., Goncharov, D. A., & Duryagina, V. V. (2008). RU Patent 2398147 F16H47/04.

Kravchenko, V. A., Senkevich, S. E., Senkevich, A. A., Yarovoy, V. G., & Tolstoukhov, Yu. S. (2004). RU Patent 2222440, B60K 17/10

Kribus, A., Kaftori, D., Mittelman, G., Hirshfeld, A., Flitsanov, Y., & Dayan, A. (2006). A Miniature Concentrating Photovoltaic and Thermal System. *Energy Conversion and Management*, *47*(20), 3582–3590. doi:10.1016/j.enconman.2006.01.013

Kryajkov, V. M., Godzhaev, Z. A., Shevtcov, V. G., & Lavrov, A. V. (2015). Park traktorov: Sostoyanie i napravlenie razvitiya [Tractor fleet: the status and direction of development]. *Rural Mechanic*, *9*, 3–5.

Kudra, T., & Van, F. R. (1991). Heating characteristics of milk constituents in a microwave pasteurization system. *Journal of Food Science*, *56*(4), 931–934. doi:10.1111/j.1365-2621.1991.tb14608.x

Kuleshov, A.N., Ereshko, A.S., & Khronyuk, V.B. (2011). Primeneniye magnitnykh poley postoyannykh magnitov dlya predposevnoy obrabotki semyan yachmenya [Use of magnetic fields of permanent magnets for presowing treatment of barley seeds]. *Vestnik agrarnoy nauki Dona*, *1*, 95-100.

Kurgan, V. P., & Pankin, A. A. (2016). Optimal speed control electromechanical positioning system. Bulletin of the Samara State Technical University, 1(49), 116-121.

Kutter, H. K. (1969). *The electrohydraulic effect: potential application in rock fragmentation*. University of Michigan Library.

Kuznetsov, A. I., & Spiridonov, V. T. (2009). The state and prospects for the development of potato in Chuvashia. In *Proceedings of the scientific-practical conference "Prospects for the innovative development of potato."* Cheboksary: PMC CR "Agro-innovations.".

Kuznetsov, N. G. (2006). *Stabilizatsiya rezhimov raboty skorostnyh mashinno-traktornyh agregatov* [Stabilization of the operating modes for high-speed machine-tractor units]. Volgograd, Russia: Volgograd State Agricultural Academy.

Kyaw, S.S.N. (2012, March). Myanmar Electricity Outlook with reference to Demand Scenario.

Labutina, E. V. (1988). *Vozdeystviye sochetaniya magnitoaktivirovannoy vody, vodno-go i pitatel'nogo rezhimov pochvy na formirovaniye urozhaya tomatov pri dozhdeva-nii v usloviyakh Volgo-Akhtubinskoy poymy* [The impact of a combination of magnetically activated water, water and nutrient regimes of the soil on the formation of a tomato crop during rainfall under the conditions of the Volga-Akhtuba floodplain] [dissertation]. Volgogradskiy gosudar-stvennyy universitet [Volgograd State University], Volgograd.

Larionova, A. P. (1993). Biologicheskoye obosnovaniye obrabotki semyan zlakovykh kul'tur i donnika elektricheskim polem koronnogo razryada [Biological substantiation of the treatment of seeds of cereals and clover electric field of the corona discharge]. In Sovershenstvovaniye nauchno-teoreticheskogo i metodicheskogo urovnya prepodavaniya fiziologii rasteniy [Improving the scientific-theoretical and methodological level of teaching plant physiology] (pp. 63). Smolensk.

Lavrov, A. V. (2018). K voprosu razrabotki intellektualnoj tekhnologii vyseva semennogo materiala [To the question of the development of intellectual technology of seeding seed material]. *Engineering Bulletin of the Don*, *4*. Retrieved from http://ivdon.ru/ru/magazine/archive/n4y2018/5329

Lavrov, A., Smirnov, I., & Litvinov, M. (2018). Justification of the construction of a self-propelled selection seeder with an intelligent seeding system. In *MATEC Web of Conferences*. 10.1051/matecconf/201822405011

Lavrov, A. V., Kryukovskaya, N. S., & Petrischev, N. A. (2018). Otsenka vozdejstviya na pochvu kolesnykh dvizhitelej samokhodnoj selektsionnoj seyalki [Evaluation of impact on soil wheel drivers of self-propelled selection seeder]. *Agrimachinery and Energy*, *4*(21), 95–106.

Leemans, V., Marlier, G., Destain, M., Dumont, B., & Mercatoris, B. (2017). Estimation of leaf nitrogen concentration on winter wheat by multispectral imaging. *Proc. SPIE 10213. Hyperspectral Imaging Sensors: Innovative Applications and Sensor Standards, 2017.*

Leijnse, A., & Hassanizadeh, S. M. (1994). Short communication: Model definition and model validation. *Advances in Water Resources*, *17*(3), 197–200. doi:10.1016/0309-1708(94)90041-8

Levin, M. I. (1999). Current state of the problem of diesel automatics in foreign practice and domestic experience. *Dvigatelestroyeniye*, *4*, 28–31.

Li, B., Sun, D., Hu, M., & Liu, J. (2019). Automatic starting control of tractor with a novel power-shift transmission. *Mechanism and Machine Theory*, *131*, 75–91. doi:10.1016/j.mechmachtheory.2018.09.012

Li, C., Wang, X., & Meng, Z. (2016). Tomato seeds maturity detection system based on chlorophyll fluorescence. *Proc. SPIE 10021. Optical Design and Testing*, *7*.

Lichtenthaler H. K., Buschmann C. (2001). Chlorophylls and Carotenoids: Measurement and Characterization by UV-VIS Spectroscopy. *Current Protocols in Food Analytical Chemistry.*

Lin, U. (2011, June). ASEAN Countries' Presentation on Renewable Energy Projects and Business Opportunities (Myanmar), Myanmar Engineering Society (MES). In ASEAN Countries Presentation on Renewable Energy, BITEC, Bangkok.

Liu, Q., Ding, W., Han, R., Wu, J., Jing, Y., Zhou, H., Qiu, A., & Zhang, Y. (2017). Fracturing effect of electrohydraulic shock waves generated by plasma-ignited energetic materials explosion. *IEEE Transactions on Plasma Science, 45*(3), 423-431.

Liu, Y., Li, Z., Li, X., Lin, F., & Pan, Y. (2016). Experiments on the fracture of simulated stratum by underwater pulsed discharge shock waves. Diangong Jishu Xuebao: China Machine Press, 31(24), 71-78.

Liu, Y., Lin, F., Pan, Y., Zhang, Q., Li, H., Li, Z., & Liu, S. (2018). *Pipeline scale removing and rock stratum fracturing device based on electrohydraulic pulse shockwaves.* Patent WO2018058401 CN, МПК E21B 43/26. Huazhong University of Science and Technology.

Liu, B. Y. H., & Jordan, R. C. (1961). Daily insolation on surfaces tilted towards the equator. *ASHRAE Journal*, *3*(10), 53–63.

Li, Z., Ji, J., & Xu, M. (2013) Designation of rapid detection system for chlorophyll fluorescence parameters based on LED irradiation. *Proc. SPIE 8768, International Conference on Graphic and Image Processing (ICGIP 2012).* 10.1117/12.2010885

Lobanov L. M., Pivtorak V. A., Savitskiy V. V., & Tkachuk G. I., (2008). Diagnostika svarnyih konstruktsiy metodami elektronnoy shirografii i spekl-interferometrii [Diagnostics of welded structures using electronic shear and speckle interferometry]. *Avtomaticheskaya svarka, 11*(667), 195–203.

Lobanov, L. M., Kondratenko, I. P., Zhiltsov, A. V., Karlov, O. M., Paschin, M. O., Vasyuk, V. V., & Yaschuk, V. Y. (2016). NestatsIonarnI elektroflzichnI protsesi v sistemah znizhennya zalishkovih napruzhen zvarnih z'Ednan [Non-stationary electrophysical processes in systems of reduction of residual stresses of welded joints]. Technical electrodynamics, 6, 10–19.

Lobanov, L. M., Paschin, N. A., & Mihoduy, O. L. (2012). Vliyanie usloviy nagruzheniya na soprotivlenie deformirovaniyu splava AMg6 pri elektrodinamicheskoy obrabotke [The effect of loading conditions on the resistance to deformation of the AMg6 alloy during electrodynamic processing]. Problemyi prochnosti, 5, 15–26.

Lobanov, L. M., Paschin, N. A., Cherkashin, A. V., Mihoduy, O. L., & Kondaratenko, I. P. (2012). Effektivnost elektrodinamicheskoy obrabotki alyuminievogo splava Amg6 i ego svarnyih soedineniy [Efficiency of electrodynamic processing of Amg6 aluminum alloy and its welded joints]. *Avtomaticeskaya Svarka (Kiev)*, (1), 3–7.

Lobyshev, V. I. (2005). Water is a sensor to weak forces including electromagnetic fields of low intensity. *Electromagnetic Biology and Medicine*, *24*(3), 449–461. doi:10.1080/15368370500382248

Locke, B. R., Sato, M., Sunka, P., Hoffmann, M.R., & Chang, J.-S. (2006). Electrohydraulic discharge and nonthermal plasma for water treatment. *Industrial and Engineering Chemistry Research, 45*(3), 882-905.

Loftis, A. D., Priestley, R. A., & Massung, R. F. (2010). Detection of Coxiella burnetii in commercially available raw milk from the United States. *Foodborne Pathogens and Disease*, *7*(12), 1453–1456. doi:10.1089/fpd.2010.0579 PMID:20704507

Loginov, M. I. (1986). Predposevnaya obrabotka semyan ul'trafioletovymi luchami [Pre-sowing seed treatment with ultraviolet rays]. *Lyon i konoplya*, *2*, 28-29.

Loncarevic, S., Jørgensen, H. J., Løvseth, A., Mathisen, T., & Rørvik, L. M. (2005). Diversity of Staphylococcus aureus enterotoxin types within single samples of raw milk and raw milk products. *Journal of Applied Microbiology*, *98*(2), 344–350. doi:10.1111/j.1365-2672.2004.02467.x PMID:15659189

Loschenko, A. V. (2016). *Snizhenie uplotneniya pochvy i povrezhdeniya kornevoj sistemy dvizhitelyami mobilnykh ehnergeticheskikh sredstv* [Reducing of soil compaction and root damage by movers of mobile energy tools]. In V.I. Orobinskiy & V.G. Kozlova (Eds.), *Proceedings of the international scientific-practical conference: Modern scientific and practical solutions of the XXI century* (pp. 231-234). Voronezh, Russia: Voronezh State Agrarian University.

Lubote, R., Shahada, F., & Matemu, A. (2014). Prevalence of Salmonella spp. and Escherichia coli in raw milk value chain in Arusha, Tanzania. *American Journal of Research Communication*, *2*(9), 1–13.

Lu, L., Yang, H., & Burnett, J. (2002). Investigation on wind power potential on Hong Kong islandsan analysis of wind power and wind turbine characteristics. *Renewable Energy*, *27*(1), 1–12. doi:10.1016/S0960-1481(01)00164-1

Luna, R. R., Trejo, P. M., Vargas, D., & Os-Moreno, G. (2012). Op-timal sizing of renewable hybrids energy sys-tems: A review of methodologies. *Solar Energy*, *88*(4), 1077–1088. doi:10.1016/j.solener.2011.10.016

Lurie, A. B. (1969). *Dinamika regulirovaniya navesnykh selskokhozyaystvennykh agregatov* [The dynamics of regulation of mounted agricultural units]. Leningrad, Russia: Mechanical Engineering.

Lurie, A. B. (1981). *Statisticheskaya dinamika sel'skokhozyaystvennykh agregatov* [Statistical dynamics of agricultural units]. Moscow, Russia: Kolos.

Lu, Y., Li, R., & Lu, R. (2016). Detection of fresh bruises in apples by structured-illumination reflectance imaging. *Proc. SPIE 9864. Sensing for Agriculture and Food Quality and Safety*, *8*, 986406.

Luzzana, M., Morandi, S., Cremonesi, P., Pisoni, G., Moroni, P., Brasca, M., & Castiglioni, B. (2007). Detection of enterotoxigenic Staphylococcus aureus isolates in raw milk cheese. *Letters in Applied Microbiology*, *45*(6), 586–591. doi:10.1111/j.1472-765X.2007.02231.x PMID:17916131

Lysakov, A. A., Ivanov, R. V. (2014). Vliyaniye magnitnogo polya na sokhrannost' kartofelya [The influence of the magnetic field on the conservation of potatoes]. *Uspekhi sovremennogo estestvoznaniya*, *8*, 103-106.

Lysen, H. (1983). Introduction to wind energy. Consultancy Services. *Wind Energy (Chichester, England)*, 82–91.

Mackersie, J. W., Timoshkin, I. V., & MacGregor, S. J. (2005). Generation of high-power ultrasound by spark discharges in water. *IEEE Transactions on Plasma Science*, *33*(5), 1715–1724. doi:10.1109/TPS.2005.856411

Maffei, M. E. (2014). Magnetic field effects on plant growth, development, and evolution. *Frontiers in Plant Science*, *5*, 445. doi:10.3389/fpls.2014.00445 PMID:25237317

Magnani, M., Brandi, G., Casiere, A., Amagliani, G., Omiccioli, E., & Bruce, I. (2004). Direct detection of Listeria monocytogenes from milk by magnetic based DNA isolation and PCR. *Food Microbiology*, *21*(5), 597–603. doi:10.1016/j.fm.2003.10.008

Magnino, S., Vicari, N., Boldini, M., Rosignoli, C., Nigrelli, A., Andreoli, G., & Fabbi, M. (2009). Rilevamento di Coxiella burnetii nel latte di massa di alcune aziende bovine Lombarde. *Large Animal Review*, *15*(1), 3–6.

Mago, P., & Luck, R. (2012). Evaluation of the potential use of a combined micro-turbine organic Rankine cycle for different geographic locations. *Applied Energy*, *102*, 1324–1333. doi:10.1016/j.apenergy.2012.07.002

Mahmoodi, M. M. (2010). Occurrence of Listeria monocytogenes in raw milk and dairy products in Noorabad, Iran. *Journal of Animal and Veterinary Advances*, *9*(1), 16–19. doi:10.3923/javaa.2010.16.19

Mahmood, M., Bee, O. B., Mahmud, T., & Subramaniam, S. (2011). The growth and biochemical responses on in vitro cultures of Oncidium taka orchid to electromagnetic field. *Australian Journal of Crop Science*, *5*(12), 1577–1587.

Mahmood, S., & Usman, M. (2014). Consequences of Magnetized Water Application on Maize Seed Emergence in Sand Culture. *Journal of Agricultural Science and Technology*, *16*(1), 47–55.

Malkin, Y.S., Furtat, I.Ye., Zhuravska, N.Y., Usachov, V.P. (2014). Perspektyvy stvorennia resursozberihaiuchykh tekhnolohii shliakhom mahnitnoi obrobky vody ta vodnykh rozchyniv [Prospects for resource-saving technologies through magnetic treatment of water and water solutions]. *Ventyliatsiia, osvitlennia ta vodopostachannia*, 17, 120-127.

Malkin, Y.S., Zhuravska, N.Y., Kovalenko, N.O. (2015). Protses obrobky vody v mahnitnykh poliakh [The process of water treatment in magnetic fields]. *Ventyliatsiia, osvitlennia ta vodopostachannia*, *18*, 70-74.

Maltsev, I. M. (2008). Elektroplasticheskaya prokatka metalla s tokom vyisokoy plotnosti [Electroplastic metal rolling with high density current]. Izvestiya vyisshih uchebnyih zavedeniy, tsvetnaya metallurgiya, 3, 34–38.

Malygin, V.M. (2015) Teorema Umova-Poytinga i vektor plotnosti potoka elektromagnitnoy energii: raznyye usloviya, raznyye resheniya [Umov-Poiting theorem and density vector of electromagnetic energy flux: different conditions, different solutions]. *Prostranstvo i vremya*, *3*(21), 103-109.

Malý, V., & Kučera, M. (2014). Determination of mechanical properties of soil under laboratory conditions. *Research in Agricultural Engineering*, *60*(Special Issue), 66–69. doi:10.17221/37/2013-RAE

Mamutov, A. V., Golovashchenko, S. F., Mamutov, V. S., & John Bonnenc, J. F. (2015). Modeling of electrohydraulic forming of sheet metal parts. *Journal of Materials Processing Technology*, *219*, 84–100. doi:10.1016/j.jmatprotec.2014.11.045

Mansouri-Najand, L., & Khalili, M. (2007). Detection of shiga-like toxigenic Escherichia coli from raw milk cheeses produced in Kerman-Iran. *Veterinarski Arhiv*, *77*(6), 515–522.

Marinković, B., Grujić, M., Marinković, D., Crnobarac, J., Marinković, J., Jaćimović, G., & Mircov, D. V. (2008). Use of biophysical methods to improve yields and quality of agricultural products. *Journal of Agricultural Sciences*, *53*(3), 235–242. doi:10.2298/JAS0803235M

Marks, N., & Szecywka, P. S. (2010). Impact of variable magnetic field stimulation on growth of aboveground parts of potato plants. *International Agrophysics*, *24*, 165–170.

Marnissi, E. (2014). Presence of Listeria monocytogenes in raw milk and traditional dairy products marketed in the north-central region of Morocco. *African Journal of Food Science*, *7*(5), 87–91. doi:10.5897/AJFS2013.0992

Marshall, J. C., Soboleva, T. K., Jamieson, P., & French, N. P. (2016). Estimating bacterial pathogen levels in New Zealand bulk tank milk. *Journal of Food Protection*, *79*(5), 771–780. doi:10.4315/0362-028X.JFP-15-230 PMID:27296424

Martin, S., & Katja, W. (2019). BWP Market Figures 2018: Sustainable growth with upward air, a clear signal for politics. Retrieved from https://www.waermepumpe.de/fileadmin/user_upload/2019-01-28_BWP_Absatzzahlen_2018_fin.pdf

Martínez, E., Carbonell, M. V., Flyrez, M., Amaya, J. M., & Maqueda, R. (2009). Germination of tomato seeds (Lycopersicon esculentum, L.) under magnetic field. *International Agrophysics*, *23*, 45–49.

Martinez, E., Florez, M., & Carbonell, M. V. (2017). Stimulatory Effect of the Magnetic Treatment on the Germination of Cereal Seeds. *International Journal of Environment Agricultural Biotechnology*, *2*(1), 375–381.

Martinez, F. R., Pacheco, A. D., Aguilar, C. H., Pardo, G. P., & Ortiz, E. M. (2014). Effects of magnetic field irradiation on broccoli seed with accelerated aging. *Acta Agrophysica*, *21*(1), 63–73.

Masateru, N., & Masato, M. (2013). Single-mode microwave reactor used for continuous flow reactions under elevated pressure. *Industrial & Engineering Chemistry Research*, *52*(12), 4683–4687. doi:10.1021/ie400199r

Maslobrod, S.N. (1981). *Elektricheskiy yazyk rasteniy* [Electric plant language]. Kishinov: Shtiitsa.

Medvedev, S. S. (2013). Plants physiology. St. Petersburg: BHV Petersburg. (in Russian)

Medvedev, S.S. (1990). Elektricheskiye polya i rost rasteniy [Electric fields and plant growth]. *Elektronnaya obrabotka materialov, 3,* 68-74.

Meunier-Goddik L., Sandra, S. (2016). Liquid Milk Products: Pasteurized Milk. *Reference Module in Food Science.*

Mihai, A. L., Dobrin, D., Magureanu, M., & Popa, M. E. (2014). Positive effects of non-thermal plasma treatment on radish seeds. *Romanian Reports in Physics*, *66*(4), 1110–1117.

Mikheev, M. A., & Mikheeva, I. M. (1977). *Basics of heat transfer*. Moscow, Russia: Energy.

Mikiporis, Yu. A. (2003). Influence of electrohydraulic effect on the properties of liquids (emulsions and solutions). *Electronic processing of materials. Institute of Applied Physics of the Academy of Sciences of Moldova (Chisinau)*, *6*, 34–37.

Mikitinets, Z. G., & Kanibolotsky, V. G. (1984) Vliyaniye omagnichennoy vody na fi-ziologicheskiye protsessy rasteniy [The effect of magnetized water on the physiological processes of plants]. In Vsesoyuznoye nauchno-proizvodstvennoye sove-shchaniye po primeneniyu opticheskogo izlucheniya v sel'skokhozyaystvennom proizvodstve pri vypolnenii Prodovol'stvennoy programmy [All-Union Scientific and Production Meeting on the use of optical radiation in agricultural production during the implementation of the Food Program] (pp. 53). L'vov [Lviv].

Milichko, V.A., Shalin, A.S., Mukhin, I.S., Kovrov, A.E., Krasilin, A.A., Vinogradov, A.V., … Simovsky, K.R. (2016). *Solar photovoltaics: current state and development trends. Success in the physical sciences*, *186*(8), 802-852.

Ministry of agriculture of the Russian Federation (2017). *Federalnaya nauchno-tekhnicheskaya programma razvitiya selskogo khozyajstva na 2017 - 2025 gody [Federal scientific and technical program of agricultural development for 2017-2025].* Moscow, Russia: Printing office of the Ministry of agriculture of the Russian Federation.

Ministry of agriculture of the Russian Federation. (2018). *Itogi raboty otrasli rastenievodstva v 2017 godu i zadachi na 2018 god [The results of the work of the crop industry in 2017 and objectives for 2018]*. Moscow: Printing office of the Ministry of agriculture of the Russian Federation.

Mioni, R., Giacometti, F., Cascone, G., Bianchi, M., Piva, S., Serraino, A., ... Merialdi, G. (2015). Human campylobacteriosis related to the consumption of raw milk sold by vending machines in Italy: Quantitative risk assessment based on official controls over four years. *Preventive Veterinary Medicine*, *121*(1–2), 151–158. PMID:26142145

Molamofrad, F., Lotfi, M., Khazaei, J., Tavakkol-Afshari, R., & Shaiegani-Akmal, A. (2013). The effect of electric field on seed germination and growth parameters of onion seeds (Allium cepa, L.). *Advanced Crop Science*, *3*(4), 291–298.

Monich, V. A., Monich, E. A., Golikov, V. M., Zhirkov, A. R., & Malinovskaya S. L. (1994). *Sposob izmereniya intensivnosti lyuminescencii v ob"eme sredy, preimushchestvenno biologicheskih ob"ektov* [The method of measuring the intensity of luminescence of the total environment, mainly of biological objects]. Patent RF. no 2012213.

Morton, J. (2002). *AVR: an introductory course*. Elsevier.

Moussa-Boudjemaa, B., Kihal, M., Lopez, M., & Gonzalez, J. (2004). The incidence of Bacillus cereus spores in Algerian raw milk: A study of the chief sources of contamination. *Archiv fur Lebensmittelhygiene*, *55*(4), 94–96.

Müller, T., Gelfond, L., & Eisert, S. (2014). *Method and device for the disintegration of a recyclable item.* Patent 2771120 DE, B02C 19/18. IMPULSTEC GMBH.

Naer, V. A., & Garachuk, V. K. (1982). Theoretical bases of thermoelectric cooling. *Energy*, 120.

Naguib, K., & Shouman, M. T. (1972). Identification and typing of Clostridia in raw milk in Egypt. *The Journal of Applied Bacteriology*, *35*(4), 525–530. doi:10.1111/j.1365-2672.1972.tb03733.x PMID:4346755

Nakoryakov, V. E., & Elistratov, S. L. (2007). Ecological aspects of the use of vapor compression heat pumps. *Energy*, *4*, 76–83.

NASA Atmospheric Science Data Center. (n.d.). Retrieved from https://eosweb.larc.nasa.gov/sse

Natale, A., Busani, L., Comin, A., De Rui, S., Buffon, L., Nardelli, S., & Ceglie, L. (2009). First report of bovine Q-fever in north-eastern Italy: Preliminary results. *Clinical Microbiology and Infection*, *15*, 144–145. doi:10.1111/j.1469-0691.2008.02154.x PMID:19438637

Nazari, R., Godarzi, H., Rahimi Baghi, F., & Moeinrad, M. (2014). Enterotoxin gene profiles among Staphylococcus aureus isolated from raw milk. *Majallah-i Tahqiqat-i Dampizishki-i Iran*, *15*(4), 409–412. PMID:27175141

Nekhoroshev, D. A. (2014). *Stabilizatsiya rezhimov MTA izpolzovaniem pnevmogidravlicheskoy mufty stsepleniya* [Stabilization of MTU modes using pneumatic hydraulic clutch] [PhD dissertation]. Volgograd State Agrarian University, Volgograd, Russia.

Nekhoroshev, D. D., & Nekhoroshev, D. A. (2015). Energeticheskoe sredstvo v transmissii kolesnogo traktora [Power means in the transmission of the wheeled tractor]. *Unique Research of the XXI Century*, *12*(12), 40–41.

Nesterenkov, A. G., Nesterenkov, P. A., & Nesterenkova, L. A. (2015). Patent No. 30003 of 04.13.2015. Republic of Kazakhstan, Astana.

Nesterenkov, P. A., Abdullaev, K. A., & Nesterenkova, L. A. (2017). Cogeneration systems with radiation concentration - a new type of "equipment for solar energy. In *Materials of the World Congress of Engineers and Scientists "Energy of the future: innovative scenarios and methods for their implementation" WSEC-2017* (pp. 264-270).

Nesterenkov, P. A., Nesterenkov, A. G., & Nesterenkova, L. A. (2016). *Fundamentals of designing hybrid concentrator solar systems*. In *Proceeding of the 12th International Conference on Concentrator Photovoltaics (CPV-12)*. 10.1063/1.4962073

Nesterenkov, P. A., Nesterenkov, A. G., & Nesterenkova, L. A. (2018). *Cogeneration solar systems with concentrators of solar radiation. In Handbook of Research on Renewable Energy and Electric Resources for Sustainable Rural Development*. Hershey, PA: IGI Global.

Next Generation Wind and SolarPower. (2016). From cost to value. Retrieved from https://www.iea.org/publications/freepublications/publication/NextGenerationWindandSolarPower.pdf

Neyman, L. A., & Neyman, V. Y. (2015). Novyie konstruktivnyie resheniya problemyi tochnoy sinhronizatsii vozvratno-postupatelьnogo dvizheniya boyka neupravlyaemoy elektromagnitnoy mashinyi udarnogo deystviya [New constructive solutions of the problem of accurate synchronization of return-transfer movement of the wagon of the controlled electromagnetic machine of shock action]. Aktualьnyie problemyi v mashinostroenii, 2, 280-285.

Nicole, C. C. S., Charalambous, F., Martinakos, S., van de Voort, S., Li, Z., Verhoog, M., & Krijn, M. (2016). Lettuce growth and quality optimization in a plant factory. *Acta Horticulturae*, (1134), 231–238. doi:10.17660/ActaHortic.2016.1134.31

Nizharadze, T. S. (2013). Vliyaniye ekologicheskikh priyemov predposevnykh obrabotok semyan yachmenya na porazhennost' listosteblevymi boleznyami [The impact of environmental practices of presowing treatment of barley seeds on the incidence of leaf stem diseases]. *Izvestiya Orenburgskogo gosudarstvennogo agrarnogo universiteta*, *6*(44), 56-58.

Noh, H., & Lu, R. (2006) Integrating fluorescence and interactance measurements to improve apple maturity assessment. *Proc. SPIE 6381, Optics for Natural Resources, Agriculture, and Foods.*

Novikova, G. V., & Mihailova, O. V. (2004). The use of electromagnetic radiation in the poultry industry. Mechanization and Electrification of Agriculture, 9.

Novitsky, Y. I. (1973). Magnitnoye pole v zhizni rasteniy [Magnetic field in plant life]. In. V.N. Chernihiv (Ed.), Problemy kosmicheskoy biologii [Problems of space biology] (pp. 164-188). Moscow: Nauka.

Novitsky, Yu. I. (1973). Magnitnoye pole v zhizni rasteniy [Magnetic field in plant life]. In. V.N. Chernihiv (Ed.), Problemy kosmicheskoy biologii [Problems of space biology] (pp. 164-188). Moscow: Nauka.

Novitsky, Yu. I. (1987). Reaktsiya rasteniy na magnitnyye polya. [The reaction of plants to magnetic fields] In Yu. A. Kholodov (Ed.), *Reaktsiya biologicheskikh sistem na magnitnyye polya* [The Reaction of Biological Systems to Magnetic Fields] (pp. 117–130). Moscow: Nauka.

Novitsky, Yu. I., Strekova, V. Yu., & Tarakanova, G. A. (1971). Deystviye postoyannogo magnitnogo polya na rost rasteniy. [The effect of a constant magnetic field on plant growth] In Yu. A. Kholodov (Ed.), *Vliyaniye magnitnykh poley na biologicheskiye ob"yekty* [Effect of magnetic fields on biological objects] (pp. 69–88). Moscow: Nauka.

O'Donnell, E. T. (1995). The incidence of Salmonella and Listeria in raw milk from farm bulk tanks in England and Wales. *International Journal of Dairy Technology*, *48*(1), 25–29. doi:10.1111/j.1471-0307.1995.tb02433.x

Oleksenko, S. V. (2016). *Otsenka pokazateley lineynyih elektromehanicheskih preobrazovateley udarnogo deystviya s vyisokoy magnitnoy sovmestimostyu* [Evaluation of linear electromechanical percussion transducers with high magnetic compatibility]. Kharkov.

Opritov, V. A., Pyatygin, S. S., & Retivin, V. G. (1991). *Bioelektrogenez u vysshikh rasteniy* [Bioelectrogenesis in higher plants]. Moscow: Nauka.

Ortolani, M. B. T., Yamazi, A. K., Moraes, P. M., Viçosa, G. N., & Nero, L. A. (2009). Microbiological quality and safety of raw milk and soft cheese and detection of Autochthonous Lactic acid bacteria with Antagonistic activity against Listeria monocytogenes, Salmonella Spp., and Staphylococcus aureus. *Foodborne Pathogens and Disease*, *7*(2), 175–180. doi:10.1089/fpd.2009.0390 PMID:19839761

Osadchenko, I. M., Kharchenko, O. V., & Churzin, V. N. (2009). Povysheniye posevnykh kachestv semyan arbuza, dyni i kabachka s primeneniyem biologicheski aktivnykh veshchestv [Increase of sowing qualities of seeds of watermelon, melon and zucchini with the use of biologically active substances]. In *Izvestiya Nizhnevolzhskogo agrouniversitetskogo kompleksa: Nauka i vyssheye professional'noye obrazovaniye* [Proceedings of the Nizhnevolzhsky agrouniversity complex: Science and higher professional education] (pp. 49-53).

Osepchuk, J. M. (1984). A history of microwave heating applications. *IEEE Transactions on Microwave Theory and Techniques*, *32*(9), 1200–1224. doi:10.1109/TMTT.1984.1132831

Ospina, A., & Quijano, N. (2016). Distributed control of small–scale power systems using noncooperative games. *International Journal of Electrical Power & Energy Systems*, *82*, 535–544. doi:10.1016/j.ijepes.2016.03.065

Otchet o funktsionirovanii YeES Rossii v 2015 godu [Report on the functioning of the UES of Russia in 2015]. (n.d.). Retrieved from http://www.rcit.su/inform–rf–ees2015.html#inf–ees2015–21

Otchet o razvitii VIE i predlozheniya v energeticheskuyu strategiyu Rossii [Report on the development of renewable energy sources and proposals for the energy strategy of Russia]. (2014). Retrieved from http://ac.gov.ru/files/content/1578/11–02–14–energostrategy–2035–pdf.pdf

Othman, M. Y. H., Ruslan, H., Sopian, K., & Jin, G. L. (2009). Performance study of photovoltaicthermal (PV/T) solar collector with V-grooved absorber plate. *Sains Malaysiana*, 537–541.

Ovchinnikova, I. A. (2005). *Monitoring sostoyaniya biotkanej metodami polyarizacionno-otrazhatel'noj i fluorescentnoj spektroskopii* [Monitoring of biological tissues by the methods of polarization-reflective and fluorescent spectroscopy] (Unpublished doctoral dissertation). Saratovskij gosudarstvennyj universitet Saratov, Russia.

Pakhomova, V. M. (1995). Osnovnyye polozheniya sovremennoy teorii stressa i nespe-tsificheskiy adaptatsionnyy sindrom u rasteniy [The main provisions of the modern theory of stress and non-specific adaptation syndrome in plants]. *Tsitologiya*, *37*(1), 66–91.

Panchenko, V.A., Strebkov, D.S., Polyakov, V.I. & Arbuzov, Yu.D. (2015). Vysokovol'tnye solnechnye moduli s naprjazheniem 1000 V (High-voltage solar modules with a voltage of 1000 V). *Alternative energy and ecology*, *19*(183), 76-81. (in Russian)

Paneto, B. R., Schocken-Iturrino, R. P., Macedo, C., Santo, E., & Marin, J. M. (2007). Occurrence of toxigenic Escherichia coli in raw milk cheese in Brazil. *Arquivo Brasileiro de Medicina Veterinária e Zootecnia*, *59*(2), 508–512. doi:10.1590/S0102-09352007000200035

Patent of the Russian Federation for invention N° 2505755. Mayorov, V.A., Panchenko, V.A. & Strebkov, D.S. Solnechnyj fotojelektricheskij modul' s parabolotoricheskim koncentratorom (Solar photovoltaic thermal module with a parabolic concentrator). Application: 2011153585/28, 28.12.2011. Published: 01.27.2014. Bul. No. 3. [in Russian].

Petrushenko-Kubala, I. E., & Petrushenko, E. I. (2003). Metod elementarnyih solenoidov modelirovaniya trehmernyih magnitnyih poley v ustroystvah s ferromagnitnyimi serdechnikami i obmotkami s tokom [The method of elementary solenoids modeling three-dimensional magnetic fields in devices with ferromagnetic cores and windings with current]. Elektronnoe modelirovanie, 5, 15–31.

Petukhov, V. S. (1967). *Heat transfer and resistance in laminar flow in pipes*. Moscow, Russia: Energy.

Pietruszewski, S., Muszynski, S., & Dziwulska, A. (2007). Electromagnetic field and electromagnetic radiation as non-invasive external stimulants for seeds (selected methods and responses). *International Agrophysics*, *21*, 95–100.

Pijaudier-Cabot, G., Christian, L.B., Reess, T., Wen, C., Maurel, O., Rey-Berbeder, F., & De Ferron, A. (2016). *Electrohydraulic Fracturing of Rocks*. Wiley-ISTE.

Pilavachi, P. A. (2000). Power generation with gas turbine systems and combined heat and power. *Applied Thermal Engineering*, *20*(15-16), 1421–1429. doi:10.1016/S1359-4311(00)00016-8

Pilavachi, P. A. (2002). Mini - and micro-gas turbines for combined heat and power. *Applied Thermal Engineering*, *22*(18), 2003–2014. doi:10.1016/S1359-4311(02)00132-1

Podleoeny, J., Pietruszewski, V., & Podleoena, A. (2004). Efficiency of the magnetic treatment of broad bean seeds cultivated under experimental plot conditions. *International Agrophysics*, *18*, 65–71.

Pogodin, N. N., Kuchko, V. V., Barsukevich, F. A., & Shatilo, S. V. (2008). Uplotnenie pochv selskokhozyajstvennoj tekhnikoj [Soil compaction by agricultural machinery]. *Land Improvement*, *1*(59), 70–74.

Polevoy, V. V. (2001). Fiziologiya tselostnosti rastitel'nogo organizma [Physiology of the integrity of the plant organism]. *Fiziologiya rasteniy*, *48*(4), 631–643.

Polifke, G. (2016). Hydrodamp - hydraulic torsional vibration damper for tractors. *Paper presented at Conference on Agricultural Engineering*, Koln, Germany.

Polifke, G. (2017). Hydrodamp - hydraulic torsional vibration damper for tractors and construction machines design for a race-sensitive drive train. Paper presented at *VDI Conference on Couplings and Clutch Systems in Drives 2017 / 2nd VDI Conference on Vibration Reduction in Mobile Systems*, Ettlingen, Germany.

Polivaev, O. I., Gorban, L. K., Vorohobin, A. V., & Vedrinsky, O. S. (2018). Decrease of dynamic loads in mobile energy means. *Paper presented at Conference Series: Materials Science and Engineering (IOP)*, Voronezh, Russia. 10.1088/1757-899X/327/4/042083

Polivaev, O. I., Zhidenko, K. S., Maksimov, I. I., & Levkin, I. G. (2017). Uprugo-dempfiruyushchiy privod vedushchih koles traktora [Elastic-damping drive of the tractor's driving wheels]. *Paper presented at Modern Trends in the Development of Technology and Technical Means in Agriculture*, Voronezh, Russia.

Polivaev, O. I., & Vedrinsky, O. S. (2013). *Vliyanie uprugodempfiruyushchego privoda vedushchikh koles traktora na dinamicheskie nagruzki v transmissii pri razgone MTA* [Influence of the elastic-damping drive of the tractor's driving wheels on the dynamic loads in the transmission when accelerating the MTU]. (pp. 80–83). Innovative Technologies and Technical Means for Agroindustrial Complex.

Polivaev, O. I., & Vedrinsky, O. S. (2014). Analiz vliyaniya uprugodempfiruyushchego privoda koles na dinamicheskie nagruzki v transmissii traktora pri razgone [Analysis of the effect of the elastic-damping drive of the wheels on the dynamic loads in the tractor's transmission during acceleration]. *The Bulletin of the Don Agrarian Science*, *28*, 5–9.

Polivaev, O. I., Vedrinsky, O. S., & Derkanosova, N. M. (2016). Povishenie dolgovechnosti stsepleniya traktorov za schet uprugofriktsionnogo dempfera [Increasing the durability of tractor traction due to elastic-friction damper]. *Science and Education in Modern Conditions*, *12*, 226–230.

Polivaev, O. I., & Voicshev, V. S. (2013). Snizhenie uplotneniya pochvy dvizhitelyami mobilnykh ehnergeticheskikh sredstv [Reduction of soil compaction by movers of mobile energy resources]. *Bulletin of Voronezh State Agrarian University*, *1*(36), 57–59.

Polunin, V.N., Zhidkov, T.V., Bel'tyukov, L.P., & Kuprov, A.V. (2009) Vliyaniye elektromagnitnogo polya na posevnyye rostovyye i produktivnyye svoystva ozimoy pshenitsy [Influence of the electromagnetic field on sown growth and productive properties of winter wheat]. *Vestnik agrarnoy nauki Dona, 3*, 12-16.

Popel, O. S. (2010). *Atlas resursov solnechnoj jenergii na territorii Rossii* [Atlas of solar energy resources in Russia]. Moscow: OIVT RAN.

Porto, A. F., Sadicoff, B. L., Amorim, M. C. V., & De Mattos, M. C. S. (2002). Microwave-assisted free radical bulk-polyaddition reactions in a domestic microwave oven. *Polymer Testing, 21*(2), 145–148. doi:10.1016/S0142-9418(01)00061-7

Poulek, V., Strebkov, D. S., Persic, I. S., & Libra, M. (2012). Towards 50 years lifetime of PV panels laminated with silicone gel technology. *Solar Energy, 86*(10), 3103–3108. doi:10.1016/j.solener.2012.07.013

Prisnyi, A. A. (2008). *Biophysics: lectures course: manual.* Belgorod. (in Russian)

Pshechenkov, K. A., & Stockings, B. A. (2008). *Tuber ozonation reduces storage loss, improves product quality and increases yield. KARTO and OV, 3*, 7.

Purdum, J. J., & Levy, B. (2012). *Beginning C for Arduino.* Apress. doi:10.1007/978-1-4302-4777-7

Purygin, P. P., Vasil'yeva, T. I., Purygin, V. A., Sovetkin, D. A., Tsaplev, D. A. (2015). Vliyaniye predposevnoy obrabotki semyan l'na na rost i biokhimicheskiye pokazateli prorostkov [Effect of presowing treatment of flax seeds on the growth and biochemical parameters of seedlings]. *Estestvennonauchnaya seriya, 10*(132), 166–173.

Putintsev, A. F. (1997). Obrabotka semyan elektromagnitnym polem [Seed treatment with electromagnetic field]. *Zemledeliye, 4*, 45.

Pyndak, V. I., Lagutin, V. V., & Yushkin, A. V. (2001). Perspektivy primeneniya ekologicheski chistykh aktivirovannykh vodnykh rastvorov v rasteniyevodstve [Prospects for the use of environmentally friendly activated aqueous solutions in crop production]. Povolzhskiy ekologicheskiy vestnik [Volga ecological journal], 8, 119-122). Volgograd: Izdatel'stvo VolgGU [VolgGU Publishing].

Qian, F., Sun, J., Cao, D., Tuo, Y., Jiang, S., & Mu, G. (2017). Experimental and modelling study of the denaturation of milk protein by heat treatment. *Han-gug Chugsan Sigpum Hag-hoeji, 37*(1), 44–51. doi:10.5851/kosfa.2017.37.1.44 PMID:28316470

Racuciu, M. (2011). 50 Hz Frequency Magnetic Field Effects on Mitotic Activity in the Maize Root. *Romanian Journal of Biophysics, 21*(1), 53–62.

Radhakrishnan, R., & Kumari, B. D. R. (2012). Pulsed magnetic field: A contemporary approach offers to enhance plant growth and yield of soybean. *Plant Physiology and Biochemistry, 51*, 139–144. doi:10.1016/j.plaphy.2011.10.017 PMID:22153250

Rahman, T., Akon, T., Sheuli, I. N., & Hoque, N. (2015). Microbiological analysis of raw milk, pasteurized milk and yogurt samples collected from different areas of Dhaka city, Bangladesh. *Journal of Bangladesh Academy of Sciences, 39*(1), 31–36. doi:10.3329/jbas.v39i1.23655

Rall, V. L. M., Vieira, F. P., Rall, R., Vieitis, R. L., Fernandes, A., Candeias, J. M. G., & Araújo, J. P. (2008). PCR detection of staphylococcal enterotoxin genes in Staphylococcus aureus strains isolated from raw and pasteurized milk. *Veterinary Microbiology, 132*(3–4), 408–413. doi:10.1016/j.vetmic.2008.05.011 PMID:18572331

Ramalingam, R. (2018). Seed pretreatment with magnetic field alters the storage proteins and lipid profiles in harvested soybean seeds. *Physiology and Molecular Biology of Plants, 24*(2), 343–347. doi:10.100712298-018-0505-8 PMID:29515328

Ramnek, G.M. (1911). *Vliyaniye elektrichestva na pochvu: Ionizatsiya pochvy i usvoyeniye atmosfernogo azota* [Effect of Electricity on the Soil: Ionization of the Soil and Acquisition of Atmospheric Nitrogen]. Kiyev: Tipografiya universiteta Svyatogo Vladimira.

Rana, K. K., & Rana, S. (2014). Microwave reactors: A brief review on its fundamental aspects and applications. *Open Access Library Journal*, *1*, 1–20.

Ratanadecho, P., Aoki, K., & Akahori, M. (2002). The characteristics of microwave melting of frozen packed beds using a rectangular waveguide. *IEEE Transactions on Microwave Theory and Techniques*, *50*(6), 1495–1502. doi:10.1109/TMTT.2002.1006410

Rawat, P., Debbarma, M., Mehrotra, S., & Sudhakar, K. (2014). Design, development and experimental investigation of solar photovoltaic/thermal (PV/T) water collector system. *International Journal of Science. Environmental Technology*, *3*(3), 1173–1183.

Ray D., McMichael D. (1982). Heat pumps. *Energoizdat, 224.*

Razzaq, A. (2016). Occurrence of Shiga toxin producing E. coli from raw milk. *Pure and Applied Biology*, *5*(2), 270–276. doi:10.19045/bspab.2016.50035

Remund, J., Kunz, S., & Lang, R. (1999). METEONORM: Global meteorological database for solar energy and applied climatology. In *Solar Engineering Handbook (version 4.0)*. Bern, Germany: Meteotest.

Rezgo, G. Y., & Nikolaeva, M. A. (2010). *Introduction of innovative storage technologies as a way to solve the problem of ensuring food security. Food Industries*, *4*, 35–37.

Rock, K. T., Mugizi, D. R., Ståhl, K., Magnusson, U., & Boqvist, S. (2016). The milk delivery chain and presence of Brucella spp. antibodies in bulk milk in Uganda. *Tropical Animal Health and Production*, *48*(5), 985–994. doi:10.100711250-016-1052-3 PMID:27026231

Rodionova, A., V., Borovkov M., S., Ershov, M., A. (2012). Justification of the selected frequency of electromagnetic radiation in physical prophylaxis of harbols. *Niva Povolz'A*, *1*, 108-110.

Rola, J. G., Czubkowska, A., Korpysa-Dzirba, W., & Osek, J. (2016). Occurrence of Staphylococcus aureus on farms with small scale production of raw milk cheeses in Poland. *Toxins*, *8*(3), 62. doi:10.3390/toxins8030062 PMID:26950152

Rossitto, P. V., Cullor, J. S., Crook, J., Parko, J., Sechi, P., & Cenci-Goga, B. T. (2012). Effects of UV Irradiation in a continuous turbulent flow UV reactor on microbiological and sensory characteristics of cow's milk. *Journal of Food Protection*, *75*(12), 2197–2207. doi:10.4315/0362-028X.JFP-12-036 PMID:23212017

Rusanov, V.A (1998*). Problema pereuplotneniya pochv dvizhitelyami i ehffektivnye puti ee resheniya [The problem of soil compaction movers and effective ways to solve it]*. Moscow, Russia: all-Russian Institute of mechanization.

Ryabova, A. V. (2006) *Kombinirovannyj spektroskopicheskij metod analiza ehffektivnosti sensibilizatorov v biologicheskih ob"ektah* [A combined spectroscopic method of analysis of the effectiveness of sensitizers in biological objects] (Unpublished doctoral dissertation). Institut obshchej fiziki RAN, Moscow, Russia.

Rymkevich, A. A., Kostyrya, A. M., & Kachkin, A. A. (2009). Patent 2350850.

Sage E.P., & White C.S. (1982). Optimal control systems. *Radio and communication, 392.*

Salandro, W., Bunget, C., & Mears, L. (2010). Modeling and quantification of the electroplastic effect when bending stainless steel sheet metal. *International manufacturing science and engineering conference (MSEC)*. 10.1115/MSEC2010-34043

Savin, V. N. (1981). *Deystviye ioniziruyushchego izlucheniya na tselostnyy rastitel'nyy organizm* [The effect of ionizing radiation on an integral plant organism]. Moscow: Energoizdat.

Selivanov, M. I., Timofeev, V. N., & Chekmarev, G. E. (2006). Patent 52468. Russian Federation.

Senkevich, A. A. (2008). Povysheniye effektivnosti funktsionirovaniya posevnogo mashinno-traktornogo agregata putem ustanovki v transmissiyu traktora klassa 1,4, uprugodempfiruyushchego mekhanizma [Increase the efficiency of the sowing machine-tractor unit by installing a class 1.4 tractor, an elastic-damping mechanism in the transmission] [PhD dissertation]. FSEE HE ABSSAEA, Zernograd Russia. 142 p.

Senkevich, S., Kravchenko, V., Duriagina, V., Senkevich, A., & Vasilev, E. (2019). Optimization of the Parameters of the Elastic Damping Mechanism in Class 1.4 Tractor Transmission for Work in the Main Agricultural Operations. In P. Vasant, I. Zelinka, & G. W. Weber (Eds.), *Intelligent Computing & Optimization. ICO 2018*. Cham: Springer. doi:10.1007/978-3-030-00979-3_17

Sen, Z. (2008). *Solar Energy Fundamentals and Modeling Techniques, Atmosphere, Environment, Climate Change and Renewable Energy* (1st ed.). Switzerland: Springer.

Sergeev, A. G. (2017). *Ehkologicheskaya problema – uplotnenie pochvy* [Ecological problem – soil compaction]. In *Materials of the IV all-Russian scientific-practical conference in Saratov state agrarian University named after N.I. Vavilova: Technogenic and natural safety* (pp. 338-340). Saratov, Russia: Amirit.

Sevela P. & Olesen, B.W. (2013). Development and Benefits of Using PVT Compared to PV. *Sustainable Building Technologies,* 90-97.

Seyoum, E. T., Woldetsadik, D. A., Mekonen, T. K., Gezahegn, H. A., & Gebreyes, W. A. (2015). Prevalence of listeria monocytogenes in raw bovine milk and milk products from central highlands of Ethiopia. *Journal of Infection in Developing Countries*, *9*(11), 1204–1209. doi:10.3855/jidc.6211 PMID:26623629

Shekhovtsov, V. V., Sokolov-Dobrev, N. S., & Potapov, P. V. (2016). Decreasing of the Dynamic Loading of Tractor Transmission by means of Change of the Reactive Element Torsional Stiffness. *Procedia Engineering*, *150*, 1239–1244. doi:10.1016/j.proeng.2016.07.129

Shevtsov, V. G., Soloveychik, A. A., Rusanov, A. V., & Lavrov, A. V. (2014). The use of universal characteristics of a tire in determining maximum pressure of a wheel running on soil. In *Proceedings of the International Research & Practice Video Conference: Topical research trends in the twenty-first century: Theory and practice* (pp. 169-173). Voronezh, Russia: Voronezh Public Forest Engineering Academy.

Short, A. L. (1955). The temperature coefficient of expansion of raw milk. *The Journal of Dairy Research*, *22*(1), 69–73. doi:10.1017/S0022029900007561

Shreyner R. T., Polyakov V. N., & Medvedev A. V. (2016). Matematicheskoe modelirovanie yavnopolyusnyih sinhronnyih dvigateley s avtomaticheskim podborom parametrov lokalnyih harakteristik namagnichivaniya [Mathematical modeling of polar pole synchronous engines with automatic selection of local parameters magnetization characteristics]. *Elektrichestvo, 2*, 57–65.

Shubert, F. (2008). *Svetodiody (LEDs)*. Moscow, Russia: Fizmatlit.

Shyam, S., & Rajeev, K. (2011). Estimation of Hourly Solar Radiation on Horizontal and Inclined Surfaces in Western Himalayas. *Smart Grid and Renewable Energy*, *2*(1), 45–52. doi:10.4236gre.2011.21006

Sidorov, O. Y., & Sarapulov, F. N. (2010). Osobennosti issledovaniya lineynogo asinhronnogo dvigatelya metodom konechnyih elementov [Peculiarities of the linear asynchronous motor study by the method of finite elements]. Izvestiya vyisshih uchebnyih zavedeniy. Elektromehanika, 1, 17–20.

Sidortsov, I. G. (2007). Ustanovka dlya predposevnoy obrabotki semyan [Plant for presowing seed treatment]. *Tekhnika v sel'skom khozyaystve*, *3*, 61-62.

Silaev, A. A. (1963). *Spektral'naya teoriya podressorivaniya transportnykh mashin* [Spectral theory of the suspension of transport vehicles]. Moscow, Russia: Mashgiz.

Singh, D., Basu, C., Meinhardt-Wollweber, M., & Roth, B. (2015). LEDs for Energy Efficient Green house Lighting. Hannover Centre for Optical Technologies.

Singh, S., Gupta, D., Jain, V., & Sharma, A. K. (2015). Microwave processing of materials and applications in manufacturing industries: A review. *Materials and Manufacturing Processes*, *30*(1), 1–29. doi:10.1080/10426914.2014.952028

Singh, S., & Kaushik, S. C. (2016). Optimal sizing of grid integrated hybrid PV-biomass energy system using artificial bee colony algorithm. *IET Renewable Power Generation*, *10*(5), 642–650. doi:10.1049/iet-rpg.2015.0298

Sircov, D. V. (2009). Chuvashia, the prospects for the development of potato. *Agrotrade*, *1*, 8–11.

Slade, A., & Garboushian, V. (2005). 27.6% efficient silicon concentrator cell for mass production. In *Proceeding of the 20th European Photovoltaic Solar Energy Conference*, 701.

Smirnov A., Kholmanskiy A., Ukhanova V. (2018). Optimization of lighting spectrum of greenhouse vegetables by using light emitting diodes. *International Journal of Research in Pharmacy and Biosciences*, *5*(4), 11-17.

Smirnov A.A. (2017). Vliyaniye UF-A radiatsii na biosintez antotsianov krasnolistnogo salata [Influence of UV-A radiation on anthocyans biosynthesis of red-leafed lettuce]. *Innovatsii v sel'skom khozyaystve [Innovations in agrarian sector]*, *1*, 6-11. (in Russian)

Smirnov A.A. (2018). Zavisimost' biosinteza pigmentov i produktivnosti tomata ot spektral'nogo sostava izlucheniya [Dependence of pigments biosynthesis and productivity of tomato on spectral composition of irradiation]. *Innovatsii v sel'skom khozyaystve [Innovations in agrarian sector]*, *3*(28), 78-86. (in Russian)

Smirnov A.A., Kholmansky A.S. (2017). Zavisimost' fotosinteza pigmentov i produktivnosti tomata ot spektral'nogo sostava obluchatelya [Dependence of pigments photosynthesis and productivity of tomato on spectral composition of irradiator]. *Nauchnaya zhizn' [Scientific life]*, *10*, 14-19. (in Russian)

Smirnov, A. N. (2008). Metallovedenie svarki i defektyi metalla [Welding metal science and metal defects]. Kemerovo: KuzGTU.

Smirnov, A. V. (2010). *Improving the efficiency of solar power plants concentrators with high-voltage photovoltaic cells* [Doctoral dissertation].

Smirnov, A.V., Kosmynin, A.V., Khvostikov, A.S., Shchetinin, B.C., Ivanova, N.A. (2016). Problems of operation of turbochargers ICE and ways to improve their reliability. *Problems of mechanical engineering and machine reliability*, *2*, 67 - 71.

Smith, C. A., & Corripio, A. B. (2012). *Principles and Practice of Automatic Process Control*. Editorial Félix Varela.

Smith, W. L., Ruzante, J. M., Cullor, J. S., Gardner, I. A., & Thornton, C. G. (2007). Modified culture protocol for isolation of Mycobacterium avium subsp. paratuberculosis from raw milk. *Foodborne Pathogens and Disease*, *3*(4), 457–460. PMID:17199528

Snowden, M., Cope, K., & Bugbee, B. (2016). Sensitivity of Seven Diverse Species to Blue and Green Light: Interactions with Photon Flux. *PLoS One*, *11*(10), e0163121. doi:10.1371/journal.pone.0163121 PMID:27706176

Soane, B. D. (1970). The ground pressure of wheels and traks. *Power Farm*, *44*(4), 40–44.

Sokolov A.V., Yuferev L.Yu. (2014). Modelirovaniye spektrov svetodiodnykh matrichnykh svetil'nikov [Modeling of spectra of light emitting diode-based matrix light fixtures]. *Innovatsii v sel'skom khozyaystve [Innovations in agrarian sector]*, *2*(7), 65-72. (in Russian)

Sokolov A.V., Yuferev L.Yu. (2015). Modelirovaniye spektra osveshcheniya svetodiodnym obluchatelem [Modeling of spectra of illumination by light emitting diode irradiator]. *Mechanization and electrification of agriculture*, *8*, 22-24. (in Russian)

Sokolov, O. A. (1983). *Kachestvo urozhaya grechikhi* [Buckwheat harvest quality]. Pushchino: ONTI NTSBI AN USSR.

Solimpeks. Volther Hybrid PV-T Panels. (n.d.). Retrieved from http://www.solimpeks.com

Song, H., Wang, Z., & Gao, T. (2007). Effect of high-density electropulsing treatment on formability of TC4 titanium alloy sheet. *Transactions of Nonferrous Metals Society of China*, *17*(1), 87–92. doi:10.1016/S1003-6326(07)60053-3

Soomro, A. H., Arain, M. A., Khaskheli, M., & Bhutto, B. (2009). Isolation of Escherichia Coli from raw milk and milk products in relation to public health sold under market conditions at Tandojam, Pakistan. *Pakistan Journal of Nutrition*, *1*(3), 151–152.

Spiridonov, A. A. (1981). *Planirovanie eksperimenta pri issledovanii tehnologicheskikh protsessov* [Planning an experiment in the study of technological processes]. Moscow, Russia: Mechanical Engineering.

Spiridonov, A. A. (1981). *Planirovanie eksperimenta pri issledovanii tekhnologicheskikh protcessov* [Planning an Experiment in the Study of Technological Processes]. Moscow, Russia: Mashinostroenie.

Spirov, G.M., Valueva, Y.V., Merkulova, V.G., Medvedeva, L.N., Lukyanov, N.B., & Zaitsev, A.S. (2008). Eksperimental'noye issledovaniye vliyaniya elektrofizicheskikh faktorov na urozhaynost' ovoshchnykh kul'tur [Experimental study of the influence of electrophysical factors on the yield of vegetable crops]. *Uspekhi sovremennogo yestestvoznaniya*, *6*, 30-38.

Standards PJSC Rosseti. (2017). STO 34.01-21.1-001-2017. Raspredelitel'nye ehlektricheskie seti napryazheniem 0,4-110 kV. Trebovaniya k tekhnologicheskomu proektirovaniyu [STO 34.01-21.1-001-2017. Distributive electric networks of 0.4-110 kV. Requirements for technological design]. Russia: PJSC Rosseti

Standards Russia. (1988). *GOST 3484.1-88. Transformatory silovye. Metody ehlektromagnitnyh ispytanij [GOST 3484.1-88. Power transformers. Electromagnetic test methods].*

Statsyuk, N.V., Takur, K., Smetanina, T.I., & Kuznetsova, M.A. (2016). Reaktsiya rasteniy kartofelya (Solanum tuberosum, L.) raznykh sortov na predposadochnuyu obrabotku klubney impul'snym nizkochastotnym elektricheskim polem [The reaction of potato plants (Solanum tuberosum, L.) Of different varieties to the preplant treatment of tubers with a pulsed low-frequency electric field]. *Sel'skokhozyaystvennaya biologiya*, *51*(3), 360-366.

Stephan, R., Schumacher, S., Tasara, T., & Grant, I. R. (2007). Prevalence of Mycobacterium avium subspecies paratuberculosis in Swiss raw milk cheeses collected at the retail level. *Journal of Dairy Science*, *90*(8), 3590–3595. doi:10.3168/jds.2007-0015 PMID:17638968

Storchevoy, V. F. (2012). Mathematical modeling of stationary processes of the ionizer-ozonizer. *K.A. Timiryazeva*, *2*, 78–82.

Strebkov, D. S., Kirsanov, A. I., Irodionov, A. E., Panchenko, V. A., & Mayorov, V. A. (2015). Patent 2557272. Russian Federation.

Strebkov, D.S., Mayorov, V.A. & Panchenko, V.A. (2013). Solnechnyj teplofotojelektricheskij modul' s parabolotoricheskim koncentratorom [Solar photovoltaic thermal module with a parabolic concentrator]. *Alternative energy and ecology*, 1/2, 35-39.

Strebkov, D.S., Persits, I.S. & Panchenko, V.A. (2014). Solnechnye moduli s uvelichennym srokom sluzhby [Solar modules with extended service life]. *Innovations in agriculture. Theoretical and scientific and practical journal. Innovations in renewable energy, 3*(8), 154-158. (in Russian)

Strebkov, D. S., Bobovnikov, N. Yu., Irodionov, A. E., Kirsanov, A. I., Panchenko, V. A., & Filippchenkova, N. S. (2016). Programma "Odin million solnechnyh krysh" v Rossii [Program "One Million Sun Roofs" in Russia]. *Vestnik VIESH, 3*(24), 80–83. (in Russian)

Strebkov, D. S., Mayorov, V. A., Panchenko, V. A., Osmakov, M. I., & Plokhikh, S. A. (2013). Solnechnaja ustanovka s matrichnymi fotojelementami i koncentratorom (A solar installation with matrix photocells and a concentrator). *Elektro, 2*, 50–52. (in Russian)

Strebkov, D. S., Panchenko, V. A., Irodionov, A. E., & Kirsanov, A. I. (2015). Razrabotka krovel'noj solnechnoj paneli (Development of a roofing solar panel). *Vestnik VIESH, 4*(21), 107–111. (in Russian)

Strebkov, D. S., Polyakov, V. I., & Panchenko, V. A. (2013). Issledovanie vysokovol'tnyh solnechnyh kremnievyh modulej (Investigation of high-voltage solar silicon modules). *Alternative Energy and Ecology, 6*(2), 36–42. (in Russian)

Strekova, V. Y. (1973). Mitoz i magnitnoye pole [Mitosis and magnetic field]. In V.N. Chernigovsky (Ed.), Problemy kosmicheskoy biologii [Problems of space biology] (pp. 200-204). Moscow: Nauka.

Sudnova, V. (2000). *Kachestvo ehlektricheskoj ehnergii [Power quality]*. Moscow, Russia: EHnergoservis.

Sudnova, V., Prigoda, V., & Hakimov R. (2007). Principy postroeniya AIIS monitoringa PKEH i upravlenie kachestvom ehlektroehnergii [Principles of building monitoring PQ indexes and quality management of electric energy]. *Promyshlennaya ehnergetika, 3*, 37-42.

Sun, J., Wang, W. L., & Yue, Q. Y. (2016). Review on microwave-matter interaction fundamentals and efficient microwave-associated heating strategies. *Materials (Basel), 9*(4), 1–25. doi:10.3390/ma9040231 PMID:28773355

SunPower. (n.d.). Retrieved from https://us.sunpower.com/buy-solar-cells/

Sunsystems. (n.d.). Retrieved from http://www.sunsystem.bg/en/fotovoltaika/PV-T/

Sventickij, I.I., Timchenko, S.D., & Usmanov, S.A. (1990). *Opredelenie kachestva semyan po opticheskim spektral'nym harakteristikam* [Determination of seed quality by optical spectral characteristics]. Pushchino, Russia: AN SSSR, Nauch. centr biol. issled., In-t pochvovedeniya i fotosinteza, In-t radiotekhniki i ehlektron.

Syrkin, V. A. (2018). Issledovaniya stimulirovaniya semyan v impul'snom magnitnom pole [Studies of seed stimulation in a pulsed magnetic field]. Innovatsionnyye dostizheniya nauki i tekhniki APK [Innovative achievements of science and technology of agriculture] (pp. 346-349). Kinel: RIO SGSKHA.

Syromyatnikov, V.F. (1989). Adjustment of automation of ship power plants. *Shipbuilding, 352*.

Tarchevsky, I. A. (2002). *Signal'nyye sistemy kletok* [Cell signaling systems]. Moscow: Nauka.

Teixeira-da-Silva, J., & Dobránszki, J. (2014). Impact of magnetic water on plant growth. *Environmental and Experimental Biology, 12*, 137–142.

Temerkieva, Y. M., & Plieva, A. M. (2016). Prorashchivaniye semyan pshenitsy pod vozdeystviyem magnitnogo polya [Germination of wheat seeds under the influence of a magnetic field]. *Aprobatsiya, 3*(42), 7–9.

The Company Lumex. (2018). Technical characteristics of the spectrofluorimeter Fluorat-02-Panorama [Data file]. Retrieved from: http://www.lumex.ru/catalog/flyuorat-02-panorama.php#specification

Tijani, J. O., Fatoba, O. O., Madzivire, G., & Petrik, L. F. (2014). A review of combined advanced oxidation technologies for the removal of organic pollutants from water. Water, Air, & Soil Pollution, 225(9).

Tikhomirov, A. A., Sharupich, V. P., & Lisovsky, G. M. (2000). Svetokul'tura rasteniy: biofizicheskiye i biotekhnologicheskiye osnovy [Photoculture of plants: biophysical and biotechnological basics]. Novosibirsk: Siberian department of Russian Academy of Sciences. (in Russian)

Timofeev, N., Timofeev, A. V., & Timofeev, D. V. (2006). Patent 118406. Russian Federation.

Timofeev, V. N., Chekmarev, G. E., & Galkina, N. A. (2002). Patent 31637. Russian Federation.

Toporkov, V. N. (2017). Electrotechnological method of obtaining fertilizer from the soil and water for greenhouses, LPH and small-scale farms. *Vestnik VIESH, 3*(28), 49–55.

Troitskiy, O. A., Baranov, Y. V., Avraamov, Y. S., & Shlyapin, A. D. (2004). Fizicheskie osnovyi i tehnologii obrabotki sovremennyih materialov (teoriya, tehnologii, struktura i svoystva) [Physical bases and technologies for processing modern materials (theory, technology, structure and properties)]. Izhevsk: RHD, ANO IKI.

Troitskiy, O. A., Kayzer, A. A., Layshev, R. R., & Gorelik, V. S. (2004). Ispolzovanie elektroimpulsnyih tehnologiy v praktike yuvelirnogo dela [The use of electric pulse technology in the practice of jewelry]. Collection of scientific papers and engineering developments 5-y Rossiyskoy vyistavki «Izdeliya i tehnologii dvoynogo naznacheniya, 2, 364–367.

Troitskiy, O. A., Stashenko, V. I., & Ryizhkov, V. G. (2011). Elektroplasticheskoe volochenie i novyie tehnologii sozdaniya oblegchennyih provodov [Electroplastic drawing and new technologies for creating lightweight wires]. *Voprosyi atomnoy nauki i tehniki (VANT), 4*, 111–117.

Troitskiy, O. A., Stashenko, V. I., Ryizhkov, V. G., Lyashenko, V. P., & Kobyilskaya, E. B. (2011). Elektroplasticheskoe volochenie i novyie tehnologii sozdaniya oblegchennyih provodov [Electroplastic wire drawing and new technologydevelopment light wire]. Voprosyi atomnoy nauki i tehniki, 4, 111–117.

Troitsky, O.A. (2018). Elektroplasticheskiy effekt v metallah [Electroplastic effect in metals. Ferrous Metallurgy]. *Bulletin of Scientific, Technical and Economic Information, 2018*(9), 65-76.

Tsai, C., Huang, C., Chen, C., & Yue, C. (2017). The study of LED light source illumination conditions for ideal algae cultivation. *Proc. SPIE 10107. Smart Photonic and Optoelectronic Integrated Circuits, 19.*

Tullberg, J. N., Hunter, M. N., Paull, C. J., & Smith, G. D. (1990). Why control field traffic. In *Proceedings of Queensland. Department of Primary Industries Soil Compaction Workshop*, Toowoomba, Australia (pp. 13–25).

Tyukov, V. A. (2006). *Elektromehanicheskie sistemyi* [Electromechanical systems]. Novosibirsk: NGTU.

Tyutereva E.V., Ivanova A.N., Voitsekhovskaya O. V. (2014). K voprosu o roli khlorofilla b v ontogeneticheskikh adaptatsiyakh rasteniy [More about role of chlorophyll-b in plants ontogenetic adaptations]. *Successes of modern biology, 134*(3), 249-256. (in Russian)

Untila, G. G., Kost, T. N., Chebotareva, A. B., Zaks, L. B., Sitnikov, A. M., Soldodukh, O. I., & Schwartz, M. Z. (2012). Solar cell of n-type, two-sided, concentrator. *Physics and Technology of Semiconductors, 46*, 1217–1223.

Urmantsev, Yu. A., & Gudskov, N. L. (1986). Problema spetsifichnosti i nespetsifichno-sti otvetnykh reaktsiy rasteniy na povrezhdayushcheye vozdeystviye [The problem of specificity and nonspecificity of plant responses to damaging effects]. *Zhurnal Obshchei Biologii [Journal of General Biology, 48*(3), 337–349.

Usanov, K. M., Moshkin, V. I., Kargin, V. A., & Volgin, A. V. (2015). *Lineynyie elektromagnitnyie dvigateli i privodyi v impulsnyih protsessah i tehnologiyah* [Linear electromagnetic motors and drives in pulse processes and technologies]. Kurgansk.

Usman, A. M., Kwagga, J., Junaid, K., & Abdulkadir, I. (2016). Detection of mycobacteria in raw milk and assessment of risk factors among Fulani herdsmen in Bwari Area Council, Abuja, Nigeria. *International Journal of Infectious Diseases, 45*, 248. doi:10.1016/j.ijid.2016.02.554

Usman, M., & Mukhtar, N. (2014). Prevalence of Listeria Monocytogenese in raw milk in Iran. *Ncbi.Nlm.Nih. Gov, 7*(12), 8–11.

Vartapetian, B. B. (1985). *Anaerobioz i strukturno-funktsional'nyye perestroyki rastitel'noy kletki* [Anaerobiosis and structural and functional rearrangements of the plant cell]. (pp. 175–198). Moscow: Nauka.

Vasant, P., Weber, G. V., & Dieu, V. N. (2016). Handbook of research on modern optimization algorithms and applications in engineering and economics. Hershey, PA: IGI Global. doi:10.4018/978-1-4666-9644-0

Vasbinder, A. J., & De Kruif, C. G. (2003). Casein-whey protein interactions in heated milk: The influence of pH. *International Dairy Journal, 13*(8), 669–677. doi:10.1016/S0958-6946(03)00120-1

Vasilev, S. I., Mashkov, S. V., Syrkin, V. A., Gridneva, T. S., & Yudaev, I. V. (2018). Results of studies of plant stimulation in a magnetic field. *Research Journal of Pharmaceutical, Biological and Chemical Sciences, 9*(4), 706–710.

Vasiliev, A. N. (2016). Research methodology for microwave convective grain processing. *Innovations in Agriculture, 3*(18), 143-153.

Vasiliev, A. N., Budnikov, D. A., Krausp, V. R., Dubrovin, V. A., Toporkov, V. N., Nurgaliyev, I. S., . . . Kazakova, V. A. (2017). Metody energeticheskogo vozdeystviya na semena prioritetnykh zernovykh i ovoshchnykh kul'tur razlichnykh sortov, rasteniya i sel'skokhozyaystvennyye materialy. kontseptsiya ispol'zovaniya elektrotekhnologiy dlya obrabotki kormov, udobreniy, otkhodov rasteniyevodstva. Nauchno obosnovannyye parametry energosberegayushchikh kombinirovannykh ustanovok dlya obezzarazhivaniya vozdukha i poverkhnostey. [Methods of energetic action on seeds of high- priority grain and vegetable crops of various sorts; plants and agrarian materials. Concept of use of electric technologies for processing of fodder, fertilizers, wastage of plants cultivation. Scientifically substantiated parameters of energy-saving combined installations for detoxication of air and surfaces]. Federal scientific agro-engineering center of VIM. (in Russian)

Vasiliev, G.P. (2002). Economically reasonable level of thermal protection of buildings. *Energy saving, 5*, 54-57.

Vasiliev, S.I., Mashkov, S.V., & Fatkhutdinov, M.R. (2016). Elektromagnitnoye stimulirovaniye semyan i rasteniy [Electromagnetic stimulation of seeds and plants]. *Sel'skiy mekhanizator, 7*, 8-9.

Vasilieva, I. G., & Timofeev, V. N. (2006). Patent 100873. Russian Federation.

Vasilieva, I. G., & Timofeev, V. N. (2013). Patent 123909. Russian Federation.

Vasiliev, A., Ershova, I., Belov, A., Timofeev, V., Uhanova, V., Sokolov, A., & Smirnov, A. (2018). Energy-saving system development based on heat pump. *Amazonia Investiga, 17*, 219–227.

Vasiliev, G. P., & Krundyshev, N. S. (2002). Energy-efficient rural school in the Yaroslavl region. *ABOK, 5*, 34–38.

Vasilyev, G. P. (2006). *Heat and cool supply of buildings and structures using low-potential thermal energy of the surface layers of the Earth*. Moscow: Red Star.

Vasilyeva, I. G. (2006). Patent 109507. Russian Federation.

Vasilyeva, I. G. (2006). Patent 117256. Russian Federation.

Vasilyeva, I. G. (2012). Analysis of ground heat exchangers to maintain the temperature regime in vegetable storehouses. In *Proceedings of the international scientific-practical conference "Actual issues of improving the technology of production and processing of agricultural products: Mosolovskie readings"* (pp. 137-138).

Vasilyeva, I. G., Novikova, G. V., & Timofeev, V. N. (2012). Improvement of the electronic three-way valve in the heat pump installation of the potato storage. In *Proceedings of the VIII All-Russian Scientific and Practical Conference of Young Scientists, Postgraduates and Students "Youth and Innovation"* (pp. 205-208).

Vasilyeva, I.G., Novikova G.V., & Timofeev V.N. (2011). Mechanization of potato storage processes. *Agrarian Science for Agriculture, 2*, 90-94.

Vasilyeva, I.G., Novikova G.V., Timofeev V.N. (2011). Improving the efficiency of storage of potatoes at catering facilities. *Bulletin of the International Academy of Refrigeration, 4*, 27- 29.

Vasilyeva, I. G. (2010). Improving the efficiency of storage of agricultural products at catering facilities. *Food Industries, 8*, 19–21.

Velicheva, Z. M., & Velichev, P. G. (1993). Vliyaniye oblucheniya semyan lyupina svetom He-Ne lazera na nekotoryye fiziologo-biokhimicheskiye osobennosti semyan i rasteniy, ikh produktivnost' i kachestvo urozhaya [The effect of irradiation of lupine seeds with He-Ne laser light on some physiological and biochemical characteristics of seeds and plants, their productivity and crop quality]. In Sovershenstvovaniye nauchno-teoreticheskogo i metodicheskogo urovnya prepodavaniya fiziologii rasteniy [Improving the scientific-theoretical and methodological level of teaching plant physiology] (pp. 65).

Veselovsky, V. A., Veselova, T. V., & Chernavsky, D. S. (1993). Stress rasteniy. Biofizicheskiy podkhod [Stress plants. Biophysical approach]. *Fiziologiya rasteniy, 40*(3), 553–557.

Viessmann. *Vitosol Solar collector 100-F*. Retrieved 15.01.2019, from https://www.viessmann-us.com/en/residential/solar/flatplate-collectors/vitosol_100-f.html

Villamiel, M., López-Fandiño, R., Corzo, N., Martínez-Castro, I., & Olano, A. (1996). Effects of continuous flow microwave treatment on chemical and microbiological characteristics of milk. *European Food Research and Technology, 202*(1), 15–18. PMID:8717091

Villamiel, M., Munoz, M. M., Hernandez, A., & Corzo, N. (1998). β-lactoglobulin denaturation and furosine formation during continuous microwave treatment of milk at temperatures of 90-120 °C. *Milchwissenschaft. Milk Science International, 53*(8), 434–436.

Vinogradov A., Kopylov, R., & Matveev, A. (2015). Obosnovanie sozdaniya mobil'nogo izmeritel'nogo kompleksa po ocenke poter' ehlektroehnergii v silovyh transformatorah [The rationale for the creation of a mobile measuring complex for the assessment of electricity losses in power transformers]. *Agrotekhnika i ehnergoobespechenie, 2*(6), 36-43.

Vinogradov, A., Vasiliev, A., Bolshev, V., Semenov, A., & Borodin, M. (2018). Time Factor for Determination of Power Supply System Efficiency of Rural Consumers. In Handbook of Research on Renewable Energy and Electric Resources for Sustainable Rural Development (pp. 394–420). Hershey, PA: IGI Global. doi:10.4018/978-1-5225-3867-7.ch017

Vinogradov, A., Bolshev, V., Vinogradova, A., Kudinova, T., Borodin, M., Selesneva, A., & Sorokin, N. (2019a). A System for Monitoring the Number and Duration of Power Outages and Power Quality in 0.38 kV Electrical Networks. In P. Vasant, I. Zelinka, & G. W. Weber (Eds.), *Intelligent Computing & Optimization. ICO 2018*. Cham: Springer.

Vinogradov, A., Borodin, M., Bolshev, V., Makhiyanova, N., & Hruntovich, N. (2019b). Improving the Power Quality of Rural Consumers by Means of Electricity Cost Adjustment. In *Advanced Agro-Engineering Technologies for Rural Business Development*. Hershey, PA: IGI Global.

Vissarionov, V.I. Burmistrov, A.A., Deryugina, G.V., Kuznetsova, V.A., Kunakin, D.N., Malinin, N.K., & Pugachev, R.V. (2008). *Solnechnaja jenergetika: Uchebnoe posobie dlja vuzov [Solar energy: A textbook for high schools]*. Moscow: Izdatel'skij dom MJeI.

Voytova, A.S., Yukin, N.A., & Ubiraylova, V.G. (2016). Slabyy elektricheskiy tok kak faktor stimulyatsii rosta domashnikh rasteniy [Weak electrical current as a factor stimulating the growth of domestic plants]. *Mezhdunarodnyy studencheskiy nauchnyy vestnik, 4*(3), 364-366.

Waak, E., Tham, W., & Danielsson-Tham, M. L. (2002). Prevalence and fingerprinting of Listeria monocytogenes strains isolated from raw whole milk in farm bulk tanks and in dairy plant receiving tanks. *Applied and Environmental Microbiology, 68*(7), 3366–3370. doi:10.1128/AEM.68.7.3366-3370.2002 PMID:12089016

Wacker, R., Augustin, J. C., Boulais, C., Ben Cheikh, M. H., & Peladan, F. (2011). Modeling the occurrence of Mycobacterium avium subsp. paratuberculosis in bulk raw milk and the impact of management options for exposure mitigation. *Journal of Food Protection, 74*(7), 1126–1136. doi:10.4315/0362-028X.JFP-11-005 PMID:21740715

Wang, L., & Singh, C. (2009). Multicriteria design of hybrid power generation systems based on a modi ed particle swarm optimization algorithm. *IEEE Transactions on Energy Conversion, 24*(1), 163–172. doi:10.1109/TEC.2008.2005280

Westhoff, D. C. (1978). Heating milk for microbial destruction: A historical outline and update. *Journal of Food Protection, 41*(2), 122–130. doi:10.4315/0362-028X-41.2.122 PMID:30795181

Winter, C.-J. (1991) Solar Power Plants. Springer. doi:10.1007/978-3-642-61245-9

Woo, M. A., Noh, H. G., An, W. J., Kang, B. S., Kim, J., & Song W. J. (2017). Numerical study on electrohydraulic forming process to reduce the bouncing effect in electromagnetic forming. *The International Journal of Advanced Manufacturing Technology, 89*(5), 1813-1825.

Wu, S., Zhu, L., Zhao, F., Yang, B., Chen, Z., Cai, R., & Chen, J. (2013). Effect of LED lamping on the chlorophylls of leaf mustard. *Proc. SPIE 8761, PIAGENG 2013: Image Processing and Photonics for Agricultural Engineering*.

Yagodin, B. A., Zhukov, Y. P., & Kobzarenko, V. I. (2002). Agrokhimiya [Agrochemistry]. Moskow: Kolos.

Yamasaki, S., Awasthi, S. P., Hinenoya, A., Shima, A., Elbagory, A. R. M., Iguchi, A., & Ombarak, R. A. (2016). Prevalence and pathogenic potential of Escherichia coli isolates from raw milk and raw milk cheese in Egypt. *International Journal of Food Microbiology, 221*, 69–76. doi:10.1016/j.ijfoodmicro.2016.01.009 PMID:26824810

Yamori, A., Ono, Y., Kubo, H., Kono, M., & Kawashima, N. (2001). Development of an induction type railgun. *IEEE Transactions on Magnetics, 37*(1), 470–472. doi:10.1109/20.911879

Yampolskiy, Y. G. (1990). O proektirovanii optimalnyih lineynyih impulsnyih elektrodinamicheskih dvigateley vozvratno-postupatelnogo dvizheniya [About designing optimal linear pulse electrodynamic engines of reciprocating motion]. Elektrotehnika, 2, 51–55.

Yan, D.-L., Guo, Y.-Q., Zai, X.-M., Wan, S.-W., & Qin, P. (2009). Effects of electromagnetic fields exposure on rapid micropropagation of beach plum (Prunus maritima, L.). *Journal of Ecological Engineering*, *35*(4), 597–601. doi:10.1016/j.ecoleng.2008.04.017

Yang, H., Zhou, W., & Lou, C. (2008). Optimal sizing method for stand-alone hybrid solar wind system with LPSP technology by using genetic algorithm. *Solar Energy*, *82*(4), 354–367. doi:10.1016/j.solener.2007.08.005

Yang, L., & Shen, H. (2011). Effect of electrostatic field on seed germination and seedling growth of Sorbus pohuashanesis. *Journal of Forestry Research*, *22*(1), 27–34. doi:10.100711676-011-0120-9

Yobouet, B. A., Kouamé-Sina, S. M., Dadié, A., Makita, K., Grace, D., Djè, K. M., & Bonfoh, B. (2014). Contamination of raw milk with Bacillus cereus from farm to retail in Abidjan, Côte d'Ivoire and possible health implications. *Dairy Science & Technology*, *94*(1), 51–60. doi:10.100713594-013-0140-7

You, K. Y., You, L. L., Yue, C. S., Mun, H. K., & Lee, C. Y. (2017). Physical and Chemical Characterization of Rice Using Microwave and Laboratory Methods. In Rice - Technology and Production (pp. 81 - 99). U.S.: InTech. doi:10.5772/66001

You, K. Y. (2015). *RF Coaxial Slot Radiators: Modeling, Measurements, and Applications*. U.S.: Artech House.

You, K. Y. (2017). Materials Characterization Using Microwave Waveguide System. In S. K. Goudos (Ed.), *Microwave Systems and Applications* (pp. 341–358). U.S.: InTech. doi:10.5772/66230

Yudaev, I. V., Tibirkov, A. P., & Azarov, E. V. (2012) Predposevnaya elektrofizicheskaya obrabotka semyan – perspektivnyy agropriyem resursosberegayushchey tekhnologii vozdelyvaniya ozimoy pshenitsy [Preseeding electrophysical seed treatment is a promising agricultural use of resource-saving winter wheat cultivation technology]. *Izvestiya Nizhnevolzhskogo agrouniversitetskogo kompleksa: Nauka i vyssheye professional'noye obrazovaniye*, *3*(27), 61-66.

Yudaev, I. V. (2015). Opyt ispol'zovaniya VIE na sel'skikh territoriyakh i v rekreatsionnykh zonakh v regionakh YuFO [Experience of using renewable energy sources in rural areas and in recreational areas in the regions of the Southern Federal District]. *Don Agrarian Science Bulletin*, *1*, 82–92.

Yuferev L.Yu, Sokolov A.V. (2014). Energosberegayushchaya sistema osveshcheniya dlya teplichnykh rasteniy [Energy-saving illumination system for greenhouse plants]. *Innovatsii v sel'skom khozyaystve [Innovations in agrarian sector]*, *4*(9), 78-81. (in Russian)

Yuferev, L. (2018). The Resonant Power Transmission System. In Handbook of Research on Renewable Energy and Electric Resources for Sustainable Rural Development (pp. 534-560). doi:10.4018/978-1-5225-3867-7.ch022

Yuferev, L., & Sokolov, A. (2018). Energy-Efficient Lighting System for Greenhouse Plants. In V. Kharchenko & P. Vasant (Eds.), Handbook of Research on Renewable Energy and Electric Resources for Sustainable Rural Development (pp. 204–229). Hershey, PA: IGI Global. doi:10.4018/978-1-5225-3867-7.ch009

Yutkin, L. A. (1986). Electrohydraulic effect and its application in industry. *Engineering*.

Zay, L. & Tyagunov, M.G. (2015a). Prototype of the pilot GIS Regional Renewable Energy Sources. In *International Congress REENCON-XXI Renewable Energy XXI Century: Energy and Economic Efficiency*, Moscow, (pp. 169-172). Moscow: Nauka [Science].

Zay, L. & Tyagunov, M.G. (2016). Evaluation of the technical potential of solar, wind, tidal and small electric power industry of Myanmar. In *Proceedings of the Twenty-second International Scientific and Technical Conference of Students and Postgraduates "Radio Electronics, Electrical Engineering and Energy,"* Moscow, MPEI, (pp. 321). Moscow: Nauka [Science].

Zay, L. & Tyagunov, M.G. (2016, July).Creating regional geographic information system to determining optimal placements of power generation based on renewable energy resources. *Papers of the SGEM Conference*, Republic of Bulgaria.

Zay, Y.L. & Tyagunov, M.G. (2017).Research on the efficiency of the use of solar-wind-hydraulic energy complex in the republic of Myanmar union.

Zay, Y.M. & Vissarionov, V.I. (2013). Study of the energy characteristics of regional wind energy in the Republic of the Union of Myanmar [Ph.D. thesis]. Moscow Energy Institute.

Zhang, L., Zhu, X., Huang, Y., Liu, Z., & Yan, K. (2017). Effects of water pressure on plasma sparker's acoustic characteristics. *International Journal of Plasma Environmental Science & Technology, 11*(1), 60–63.

Zharkov, S. V. (2014). Assessment and enhancement of the energy supply system efficiency with emphasis on the cogeneration and renewable as main direction for fuel saving. *International Journal of Energy Optimization and Engineering, 3*(4), 1–20. doi:10.4018/ijeoe.2014100101

ZHelezko, YU. (2002). O normativnyh dokumentah v oblasti kachestva ehlektroehnergii i uslovij potrebleniya reaktivnoj moshchnosti [About normative documents in the field of power quality and consumption conditions of jet capacity]. *EHlektricheskie stancii, 6*, 18–24.

Zheng, E. L., Fan, Y. D., Zhu, R., Zhu, Y., & Xian, J. Y. (2016). Prediction of the vibration characteristics for wheeled tractor with suspended driver seat including air spring and MR damper. *Journal of Mechanical Science and Technology, 30*(9), 4143–4156. doi:10.100712206-016-0826-x

Zhiltsov, A. V., Kondratenko, I. P., & Sorokin, D. S. (2014). The mathematical model of the electromagnetic and mechanical processes is developed in a coaxial-linear engine with massive magnetic conductors. ECONTECHMOD: An International Quarterly Journal on Economics of Technology and Modelling Processes, 1, 69-73.

Zhiltsov, A. V., Kondratenko, I. P., & Vasyuk, V. V. (2014). Rozrahunok parametrIv kontura dlya stvorennya pritiskayuchogo zusillya v elektrotehnIchnomu kompleksI dlya znizhennya zalishkovih napruzhen [Calculation of contour parameters for creation of a compressive force in an electrical complex for reduction of residual stresses]. Energetika I avtomatika, 4, 54–64.

Zhu, L, He, Z. H., Gao, Z. W., Tan, F. L., Yue, X. G., & Chang, J. S. (2014). Research on the influence of conductivity to pulsed arc electrohydraulic discharge in water. *Journal of Electrostatics, 72*(1), 1-6.

Zhu, X., Guo, W., Jia, Y., & Kang, F. (2015). Dielectric properties of raw milk as functions of protein content and temperature. *Food and Bioprocess Technology, 8*(3), 670–680. doi:10.100711947-014-1440-5

Zia, M., Fazli, A., & Soltanpour, M. (2017). Warm Electrohydraulic Forming: A Novel High-Speed Forming Process. *Procedia Engineering, 207*, 323–328. doi:10.1016/j.proeng.2017.10.782

Zimmermann, S., Helmers, H., Tiwari, M. K., Escher, W., Paredes, S., & Neves, P. (1556). … Michel, B. (2013). *Advanced liquid cooling in hcpvt systems to achieve higher energy efficiencies. AIP Conference Proceedings*, 248–251.

Ziyaee, A., & Roshani, M. R. (2012). A survey study on Soil compaction problems for new methods in agriculture. *International Research Journal of Applied and Basic Sciences, 3*(9), 1787–1801.

Zong, W. G. (2012). Size optimization for a hybrid photovoltaic wind energy system. *Electrical Power and Energy Systems, 42*(1), 448–451. doi:10.1016/j.ijepes.2012.04.051

Zoz, F., & Grisso, R. (2003). Traction and Tractor Performance. *Paper presented at Agricultural Equipment Technology Conference*, Louisville, KT.

About the Contributors

Valeriy Kharchenko is a researcher and consultant in a wide range of energy saving fields, including energy policy at national and international level, energy RTD strategy, RUE and RES. His education includes: 1994 - 1995 Tacis Training Programme (Manager in energy); 1994 Tashkent State Technical University, Uzbekistan (professor); 1987 State Higher Training Courses on Patenting and Invention (Certificate with distinction); 1987 Institute of Electronics, Moscow (Doctor of Engineering Sciences Degree); 1969 Institute of Inorganic Chemistry, Novosibirsk (Candidate of Chemistry Degree); 1956 - 1961 Tashkent Polytechnic Institute (Technical University) (Technology engineer). His international activity experience: 1994-1996 - counterpart expert in the Project of EU "Monitoring and evaluation of the energy (non nuclear) projects implementation in Uzbekistan" (carried out by European companies); 1998-2000 - the expert for evaluation of submitted proposals INTAS (ID 716); 1996-2000 - senior expert of the Central Asia Energy Advisory Group (a project under the EC DG XVII's SYNERGY Programme) with obligations included duties for formation of expert groups, peering and estimation of their reports and preparing general Interim and Final annual reports; 2000-2001 years -UNDP Program, Global Environmental Facility, Government of Uzbekistan (Joinnt Project UZB/98/G42/A/1G/99), expert at the premises. Among his obligations are evaluation and reception of reports of local expert groups on various aspects of the organization of heating and hot water supply system of Tashkent city Area of Scientific activity; development of new technological approaches for superpure silicon fabrication; improvement of silicon epitaxial layers technology; application of silicon layers in various devices, including PVcells; application of installations on the basis of solar cells in practice; development of new approaches for production of silicon for solar cells; application of a solar energy for electricity and heat producrion; participation in a number of international projects of the Eurocommission and UNDP on an energy policy, energy saving (RUE), and renewable energy sources (RES); works on increase of efficiency of use of energy in systems of heating and hot water supply. He currently holds the position of Chief Scientific Officer of the Federal Scientific Agroengineering Center. He is the official reviewer of a number of international journals, including Solat Energy, Deputy Editor-in-Chief of European Journal of Renewable Energy, member of the editorial boards of several international journals, honorary professor of the Institute of Hydropower and Renewable Energy of the National Research University MEI. Under his editorship, a number of books have been published by IGI Global (USA).

Pandian Vasant is a senior lecturer at Department of Fundamental and Applied Sciences, Faculty of Science and Information Technology, Universiti Teknologi PETRONAS in Malaysia. He holds PhD (UNEM, Costa Rica) in Computational Intelligence, MSc (UMS, Malaysia, Engineering Mathematics) and BSc (2nd Class Upper-Hons in Mathematics, UM, Malaysia) in Mathematics. He has co-authored research papers and articles in national journals, international journals, conference proceedings, conference paper presentation, and special issues lead guest editor, lead guest editor for book chapters, edited books, and keynote lecture. (248 publications indexed in Web of Science). In the year 2009, Dr. Pandian Vasant was awarded top reviewer for the journal Applied Soft Computing (Elsevier), awarded outstanding reviewer in the year 2015 for ASOC (Elsevier) journal and Top reviewer for Sentinels of Science: Computer Science (Oct. 2015 - Sept. 2016). He has 28 years of working experience at the various universities from 1989-2019. Currently he is Editor-in-Chief of IJEOE, Member of American Mathematical Society, MERLIN (TDTU, Vietnam), and NAVY (TUO, Czech Republic) Research Groups.

* * *

Alexey Bashilov graduated from the Moscow Institute of agricultural engineers (MGAU. Goryachkin V. P.), faculty of electrification and automation of agriculture in 1973, Doctor of technical Sciences (2001), Professor (2006).

Maria Belitskaya graduated from Volgograd state pedagogical Institute named after S. Serafimovich in 1974. In 1986 she defended her Candidate thesis (equivalent to PhD) at Voronezh Institute of forestry. In 2004 she defended her Doctoral thesis at the Kuban state agrarian University. Maria Belitskaya is the author of 109 scientific publications. Research areas: assessment of the conditions and problems of biodiversity conservation; organization, dynamics and mechanisms of sustainability of agro-forestry and urban ecosystems; problems of ecological traits and adaptation process of entomofauna to the environment. Invasion of alien species of insects. Biological and physical means of plant protection; influence of environmental factors on biological systems; scientific basis of rational use and reproduction of biological resources.

Alexander Belov is a doctor of technical sciences, senior researcher of Federal State Budgetary Institution "Federal Scientific Agro-Engineering Center VIM".

Mikhail Belyakov is the head of the Department of "Optoelectronic systems" of Smolensk branch Of the national research University "MPEI". He has more than 150 publications in the field of spectroscopy, application of energy-efficient radiation sources, optical diagnostics of biological organisms. He is a member of the Federal educational Association for Photonics, instrumentation, optical and biotechnical systems.

Vadim Bolshev is a researcher with Federal Scientific Agroengineering Centre VIM and specializes in research in the field of power supply for rural consumers. Vadim studied at the Orel Technical College as an electrical technician from 2004 to 2008, at the Orel State Agrarian University as an electrical engineer from 2008 to 2012. In 2016 he graduated from the postgraduate course at the Orel State Agrarian University. Vadim has a great professional experience in the field of working with electrical equipment. So he worked in the water and sewer utility "Orelvodokanal" for 3.5 years as an electrician (2010 - 2013), in the scientific and clinical multidisciplinary center named after Z.I. Krugloi for 3 years as an engineer (2013-2016) and 1.5 years as chief engineer (2016-2018). To date, he works in the Federal Scientific Agro-Engineering Center of VIM as a research fellow. The field of scientific activity is to develop methods and tools aimed at improving power supply efficiency . This includes, first of all, the development of methods and devices for monitoring power quality and the technical state of power supply system elements as well as the automation of these systems.

Maxim Borodin is an Associate Professor of the Power Supply Department of the Orel State Agrarian University named after N.V. Parakhin and specializes in research into improving the performance of rural consumers. Maxim studied at the Orel State Agrarian University at the Faculty of Agrotechnics and Energy Supply, specializing in electrification and automation of agriculture, received a diploma with honors, and also has a second higher education in the specialty "Public and municipal finance." In 2014 he defended his thesis for the degree of Candidate of Technical Sciences with a degree in Electrotechnical Complexes and Systems. Maxim is an employee of the electrical laboratory the OrelSAU. The field of scientific activity is the development of methods and tools aimed at improving the efficiency of power supply systems. This includes, first and foremost, the development of methods and devices for monitoring the quality of electrical energy, the development of methods and means to ensure the quality of electrical energy. Author and co-author of more than 40 publications, 2 inventions, 2 monographs and 4 teaching aids.

Alexey Bukreev is a graduate student with Orel State Agrarian University and specializes in research in the field of electrical technologies, electrical equipment and power supply of rural consumers. Alexey studied at the Orel State Agrarian University for a bachelor's degree from 2011 to 2015, for Master's degree from 2015 to 2017. Alexey has extensive professional experience in the field of electrical equipment: he worked for 2 years as an electrician with Montazh Orel LLC (2013–2015), 1 year as an engineer (2015–2016) and 1 year as a chief engineer (2016 - 2017). The field of scientific activity is the development of methods and tools aimed at improving the efficiency of energy surveys of agricultural facilities. This, first of all, includes the development of methods and devices for identifying conductors and portable devices for studying the time and energy characteristics of electrical equipment operation modes.

Vasyl Bunko graduated from the Faculty of Energy of Agricultural Production of the Berezhansk Agro-Industrial Institute of the National Agrarian University and received a qualification as an energy engineer. In 2013 he defended his Ph.D. thesis. Since 2014, he has been working as an assistant professor of the Department of Energy and Automation. In 2018 he was appointed Dean of the Faculty of Power Engineering and Electrical Engineering. Has more than 100 scientific and teaching works.

Cheong Yew Chong is currently senior lecturer at the Tunku Abdul Rahman University College, Malaysia. He received his Bachelor of Food Science and Technology and his PhD in Food Safety from the Universiti Putra Malaysia. His interests focus on adaptive response of pathogen, antimicrobial coating and dynamics of microbial biofilm formation.

Soumana Datta got B.Sc Botany (Hons) Degree, in 1982 she got M.Sc in Botany Degree, and in 1988 she got Ph.D in Plant Nematology. She is Invited member, Rajasthan state biodiversity board Fellow, Indian Botanical society Fellow, Society for Applied Biotechnologists Life member, Indian science congress association Life member, Indian women scientist association Life member, Nematological society of India Life Member, Blue Planet society Member, Indian association for angiosperm taxonomy

Former Secretary, Rajasthan Univ Womens Association(RUWA) Member, Research action group(RAG), Forest dept, Rajasthan. Research areas Plant nematology, Biochemistry, Stress biology, Biosystematics and Ethnobotany Eight research scholars registered for Ph.D presently.

Yuliia Daus received the diploma of specialist in Energy Management in Zaporizhzhya National Technical University. In the period 2008 – 2014 worked at the same university as an assistant of the Power supply of industrial enterprises department. From 2012 to 2015 studied at post graduate school on Electrotechnical complexes and systems specialty. Now work as an editor of the scientific journal " Don Agricultural Science Bulletin ". In 2017 got my master's degree in Electrical Engineering and power.

Galina M. Deriugina is Assistant Professor of the Department of Hydro power and Renewable Energy Sources, Moscow Power Engineering Institute.

Valerij Drevin graduated from the Natural-geographical faculty at Volgograd State Pedagogical University named after A. S. Serafimovich. 2010-2012 –head of the Chemistry faculty of Volgograd State Agrarian University. Since January 2013 – Dean of the "Processing technologies and commodity science" faculty of Volgograd State Agrarian University. In 1999 defended thesis on "Synthesis and properties of peroxides based on poly-and perfluorinated carbonyl compounds". Main scientific field: chemistry of organometallic compounds; chemistry and study of properties of natural minerals; chemistry and study of properties of food products. He has 113 publications, including 1 scientific work in foreign countries.

Vera Dyachenko received the diploma of specialist in Power Engineering in Zaporizhzhya National Technical University. Works at the same university as associated professor of the Power supply of industrial enterprises department. In 2013 got degree of Candidate of Technical Sciences.

Mikchail Ershov is a candidate of chemical sciences, patent specialist in JSC "Pharmasintez", Chuvash State Pedagogical University I.Y.Yakovleva and ChSAA. He studied Inorganic Chemistry at ChGPU. I. Y. Yakovlev.

Irina Ershova is a Candidate of Technical Sciences. Research experience - more than 8 years, scholar of the President of the Russian Federation. Eurodoc member - European Council for Doctoral Candidates and Junior Researchers (2012-2017); Member of the Moscow Chuvash National Cultural Autonomy; I take an active part in the public life of the republic and country, in conferences, competitions, forums, including abroad; I am the organizer of the first international conference "Achievements of Young Scientists in the Field of Energy Saving" (2013).

Giedrius Ge is a Council Member (Union of Young Scientists of Lithuania), Researcher (Institute of Environmental Engineering) and Dean of the Faculty of Technology (Kauno kollegija / University of Applied Sciences).Studies wind energy at Lithuanian Energy Institute Worked at Vytautas Magnus University and KTU Kaunas University of Technology / Kaunas University of Technology. Lithuanian Energy Institute Kaunas, Energetika. Let's separate Year of release: 2007 Vytautas Magnus University Environmental science · Kaunas Kaunas University of Technology, Kaunas University of Technology Environment Engineering.

Diriba Kajela Geleta is a PhD candidate at Department of Mathematics, Punjabi University, Patiala, Punjab state, India Home Country: Ethiopia Adress: Madda Walabu University, Bale Robe, Oromia state, Ethiopia.

Igor Golikov graduated from "Oryol State Agrarian University" in 2012. Enrolled in full-time graduate school at the department of "Power". Studying in graduate school (2012-2015), successfully passed candidate exams, collected statistical data on voltage deviation levels in rural electrical networks of 0.38 kV and time of output of these indicators beyond standard values, developed an experimental model of an adaptive automatic voltage control system in rural 0.38kV electrical networks, co-authored 12 scientific papers. In 2016 he defended his thesis on the topic "Adaptive automatic voltage regulation in rural electrical networks of 0.38 kV", received the degree of Candidate of Technical Sciences. He currently works as a senior lecturer at the Department of Electrical Supply of Orel State Agrarian University, and is engaged in research work aimed at developing methods and devices for improving the quality of electricity supplied to rural consumers.

Irina Gribust graduated from Volgograd agricultural Academy. She defended her Candidate thesis (equivalent to PhD) at the end of post-graduate study at all – Russian research Institute of agroforestry in 2009. The author of 49 publications. Research areas: Assessment of the conditions and problems of biodiversity conservation. Organization, dynamics and mechanisms of sustainability of agroforestry and urban ecosystems. Problems of ecological traits and adaptation process of entomofauna to the environment. Invasion of alien species of insects. Biological and physical means of plant protection. Influence of environmental factors on biological systems. Scientific basis of rational use and reproduction of biological resources.

Pavel Ivanov from 1998-2003, was a student of Azov-Black Sea state Agroengineering Academies. 2003-2006 post-graduate student of the Department "tractors and cars". 2006-2010 high school teacher at the Department "theoretical mechanics". 2010-2012high school teacher at the Department "quality management in of Agroengineering". Since 2012 teacher of higher school at the Department "technical service in the agro-industrial complex". In 2012 defended thesis on competition for the degree of candidate of technical Sciences. Dissertation topic -Substantiation of parameters of the working body of the sub- soil sowing of cereal grains in the conditions of pneumatic transportation of seeds in to coulters.

Igor Kondratenko is a famous scientist in the field of electrical and power engineering. Doctor of Engineering Sciences (2006), corresponding member of the NAS of Ukraine (2015). Head of the Department of Electromagnetic Systems of the Institute of Electrodynamics of the NAS of Ukraine (since 2005). Laureate of Khruschev V.M. (2010) and Proskura G.F. (2014) awards of the NAS of Ukraine. Author of over 180 scientific publications including a monograph. Research interests of I.P. Kondratenko are directed at addressing one of the major problems of modern energetics – the problems of improving the efficiency of energy consumption in different economic sectors, especially when there are nonlinear, asymmetric and non-stationary energy consumers such as the most energy-intensive industries. His research activity is focused on the development of the methods for analyzing electromagnetic fields in nonlinear media with transient conditions, development of theoretical foundations for creating effective electric power systems and complexes.

Volodymyr Kozyrskyi graduated department of the electrification of Ukrainian Agricultural Academy (now The National University of Life and Environmental Sciences of Ukraine) in 1978. He was appointed director of the research institute of engineering and technology in September 2005. He appointed director of the Education and Research Institute of Energetics, Automation and Energy Efficiency in 2009.

Natalya Kryukovskaya graduated from Moscow State Technical University. N.E. Bauman specialty: "Renovation of funds and objects of material production." Awarded the qualification "Engineer". Passed refresher courses in the direction: "Design". From 2010 to 2016 she worked at the Federal Scientific Agroengineering Center VIM as a design engineer in the laboratory "Technical service in the agricultural sector." Since 2016, she has been transferred to the laboratory "Mobile energy and running systems" to the post of research officer. Research interests: modernization of existing and development of mobile power equipment models, mobile and stationary diagnostic tools for agricultural equipment and its components, tools for the maintenance and repair of agricultural equipment, development of equipment for the restoration of worn parts.

Alexandr Lavrov graduated from Moscow State Technical University named N.E. Bauman on the specialty Engineer of Agriculture. In 2013 he graduated as candidate of technical science on theme "Optimization of the quantitative-age composition of the tractor amount of an agricultural organization in conditions of limited resources" in the specialty "Technologies and means of agricultural mechanization". He works in the Federal Scientific Agroengineering Center VIM in the position of head of the laboratory "Mobile energy and drives systems." Research interests: monitoring the state of resource support for machine technologies, operation of mobile power equipment and operating modes, optimization of the design of mobile power equipment and the composition of the tractor fleet.

Pavel Lavrukhin from 1976 -1981, was a student of Azov-Black sea Institute of agricultural mechanization. 1981 -1985 engineer of the North-Caucasus machinery testing station. 1985 – 1994 research associate, Department of mechanization field crops 1994 - 2001 senior researcher of the scientific-production company "Agrotehnic» in 2000 defended thesis for the degree of candidate of technical Sciences. 2001 - 2005 associate Professor of the Department "Theoretical mechanics» 2005-2018 associate Professor of " Tractors and cars»

Mukhdeep Singh Manshahia is working as Assistant Professor in the Department of Mathematics, Punjabi University Patiala. He has published more than 30 research paper in journal of national and international repute and edited one book. He has presented papers in more than 20 national and international seminars and conferences. He has also participated in many international workshops on wireless communication and sensor networks. His areas of interest are Modeling and optimization of RES, Internet of Things (IoT), Wireless Sensor Networks, Pervasive computing, Artificial intelligence and ICT.

Sergey Mashkov graduated from the Engineering faculty of Samara State Agricultural Academy. In 2009 defended the master's thesis on a subject "Economic assessment of agricultural machinery in the production technology of crop production (on materials of the Samara region)". 2009-2016 – the deputy dean for study at the Engineering faculty of Samara State Agricultural Academy. Since July, 2016 – the

Dean of the Engineering faculty of Samara State Agricultural Academy. Main scientific areas: "Electrotechnologies and electric equipment in agriculture", Agroinzheneriya. Has 90 publications, including: 2 publications in foreign countries.

Thu Yein Min is a postgraduate student of the Department of Hydro power and Renewable Energy Sources, Moscow Power Engineering Institute.

Maksim Moskovskiy is a doctor of technical science (2017), professor of Russian Academy of Science, member of CIGR (International Commission of Agricultural and Biosystems Engineering), representative of Russia in CIGR, Head of department of technologies and equipment for selection operations Federal agroengeneering center VIM. Scientific interests - development of intellectual systems for selection technologies and equipment - optimization of seed and grain production cycles - conducts research on post-harvest processing of seeds and grain - application of polymeric materials in agriculture.

Andrey Musenko is a graduate student, engineer of Federal State Budgetary Institution "Federal Scientific Agro-Engineering Center VIM".

Elena Nefed'eva graduated from Penza state pedagogical Institute. In 1997 she defended her thesis at the Department of plant physiology of the Moscow state agricultural Academy named after K. A. Timiryazev. After completing her doctoral studies, she defended her doctoral dissertation at the same institution in 2011. The author of 130 publications. Research areas: The role of pressure in the regulation of plant growth and development. Hormesis and stress in plants after abiotic factors. The conditions of the biopolymers of the seeds during stress and aging. Study of the effectiveness and phytotoxicity of modern fungicides. Development of methods to reduce phytotoxicity of complex protectants. The conditions of urban green spaces. Design of urban green spaces taking into account the influence of environmental factors.

Alexander Nesterenkov since 2008 has been the Head of the Laboratory "Plasma Technologies and Renewable Energy Sources" at JSC "Kazakh Research Institute of Power Engineering named academician Sh.Ch. Chokin". Honored Power Engineer of the Republic of Kazakhstan (certificate N°376 dated 11.29.2011). Author of more than 30 inventions in the field of plasma technology, physics of electric arc discharge, solar energy, nanotechnology. Author of more than 55 scientific articles in various fields of physics, plasma coating technology, technology for the production of thermal insulation materials, etc.

Peter Nesterenkov from 2017 to the present works in Al-Farabi Kazakh National University, Faculty of Mechanics and Mathematics, Department of Computer Science as a lecturer. In 2015–2018, 10 scientific articles were published, a chapter in a collection of cited monographs; 4 innovative patents of the Republic of Kazakhstan for an invention were received. Presented a report "Cogeneration plants with radiation concentration - a new type of equipment for solar energy" at the International Congress of Engineers and Scientists WSEC-2017. Co-author of the project, which passed the National Competition and presented at the International Exhibition "Astana EXPO-2017". Winner of the international competition of start-up projects "Russian Startup Tour 2016".

Vladimir Panchenko from 2003 to 2009 studied at "Bauman Moscow State Technical University" at the faculty "Power Engineering " with qualification of engineer. From 2009 to 2012 studied at the graduate school of the ''All-Russian Scientific Research Institute of Agriculture Electrification'', specializing in "Renewable Energy Installations". Passed an internship program on the International Energy Program in the Solar Energy Center (SEC), the Certificate under the Indian Technical and Economic Cooperation (ITEC), the Program of the Ministry of External Affairs (MEA), the Government of India - India, Delhi, November-December 2011. Participated in the International Conference-Exhibition Intersolar Europe 2012, Germany, Munich, June 11 - 15, 2012 as part of the delegation of the Ministry of Energy of the Russian Federation Ph.D. thesis "Development and research of solar photovoltaic thermal module with a paraboloid-type concentrator" was protected on July 2, 2013. From 2013 to 2017 head of the laboratory ''Solar photovoltaic modules with an experimental - technological site''. The Council of Young Scientists of the VIESH was elected the Chairman of the Council of Young Scientists and was appointed a member of the Academic Council of the VIESH in 2014 - 2017. In 2014 was appointed as a contact person and a member of the International Solar Energy Society of the Russian section of the community. Since September 2015 associate professor at the ''Moscow State University of Railway Engineering''. Since 2017, a senior researcher at the ''Solar Energy Laboratory'' FSBSI "Federal Scientific Agroengineering Center VIM". Passed the advanced training in 2016 in ''Russian Engineering Academy of Management and Agribusiness '' for an additional professional program '' Automated Information Processing and Management Systems''. Passed the advanced training in 2017 in ''Moscow State University of Railway Engineering' the program "Using e-learning and distance educational technologies in the implementation of educational programs on distance learning". In 2016 awarded the grant ''Young lecturer of Moscow State University of Railway Engineering''. In 2018 awarded the Scholarship of the President of the Russian Federation for 2018 - 2020 for young scientists and graduate students engaged in advanced research and development in priority areas of the modernization of the Russian economy. Has 3 medals of exhibitions, 3 diplomas of the winner of competitions, participated in more than 80 conferences and exhibitions of various levels, has more than 120 publications, including patents, monographs and educational publications. Area of interest includes three-dimensional modeling, use of solar energy, electric transport and distance education.

Sergey Rakitov graduated from FSBEI HE Volgograd State Agrarian University. Now Director of the Suburban MPG branch of PJSC "Volgogradoblelectro."

Jorge Sabrejos, Ph.D. (engineering), is a researcher of the Antenor Orrego Private University, Department of Engineering. Was born in Trujillo, Peru.

Vitaliy Savchenko graduated with honors in 2008 Department of Energy and Automation, National University of Life and Environmental Sciences of Ukraine in specialty "Automated process control." In 2012 he defended his thesis for the degree of candidate of technical sciences, specialty 05.09.03 - Electrotechnical complexes and systems on "electro-technological complex for potato pre-treatment in a magnetic field." Has over 150 scientific and educational works.

Anna Senkevich from 1997-2002, was a student of the Agroengineering University. 2002-2005 postgraduate-course student of the Department of Tractors and Cars. 2008 defended thesis for the degree of candidate of technical sciences. 2018 awarded the title of associate Professor.

Sergey Senkevich from 1997-2002, was a student of Agroengineering University. 2002-2005 post-graduate-course student of the Department of Tractors and Cars. 2006 defended thesis for the degree of candidate of technical sciences. 2005-2009 worked as an assistant at the Department of the Tractors and Cars. 2009 awarded the title of associate Professor of the Department of Tractors and Cars. 2009-2013 he worked as an assistant professor of the Department of Tractors and Cars. From 2013 to January 2019, Senior Researcher at the Laboratory of Mobile Energy Means. From January 2019 to the present, Senior Researcher of the Laboratory of Automated Drive of Mobile Energy Means in the Federal Scientific Agroengineering Center VIM (Moscow, Russia).

Nikolay Sergeev graduated from the Azov-Black Sea Institute of Agricultural Mechanization, specializing in mechanization of agriculture. In 2009 defended a thesis on the topic "Reduction of energy costs in the operation of a tilled aggregate due to changes in the structure of tires of a class 1.4 tractor". In 2012 was awarded with the academic rank of associate professor at the Tractors and automobiles department. Currently works as an assistant professor of the Tractors and automobiles department in the Don State Agrarian University.

Sergey Shevchenko is a corresponding member of RAS, Doctor of Agricultural Sciences, Director of Samara Research Scientific Institute of Agriculture, Bezenchuk, Samara Region. Born in 1960 in the city of Chapayevsk, Kuibyshev region. He studied at the Kubyshev Agricultural Institute, after which he was qualified as the scientist agronomist. For the creation of varieties of spring wheat in 1998 he was awarded the Provincial Prize in Science. He is the author of more than 150 scientific papers and 2 teaching tutors, 4 monographs, guidelines, catalog of varieties, 14 patents.

Oleksandr Sinyavsky graduated in 1985 Faculty of Agricultural automation of Ukrainian Agricultural Academy. He worked at the department of electricdrive and electrotechnologies from 1992. In 1992 he defended his thesis on "Automated electrical preparation of nutrient solution in hydroponic greenhouses." Has more than 230 scientific and educational works.

Kiril Sirakov is Assistant Professor, Department of Electrical Power Supply and Equipment, Ph.D., University of Ruse, Bulgaria. Birth – 1967 Ruse, Bulgaria. In 1993 graduated from the Electrical Power Supply and Equipment department, University of Ruse. Research interests: electrical power supply and engineering; energy efficiency in industry and agriculture; reactive power controlling; electric power networks and systems; automation in agricultural and industrial production. Author & co-author of 123 scientific publications.

Alexander Smirnov, CSc. (Engineering), is a Senior Researcher FSAC VIM (2009-present). He finished the postgraduate school of All-Russian Research Institute of Electrification of Agriculture in 2009. He has authored 26 articles in the best Russian newspapers. His scientific works are devoted to the study of the effects of electrophysical effects (microwave, ozone, optical radiation) on agricultural objects.

Dmitry Tikhomirov is doctor of technical sciences, professor. He is a graduate of the Faculty of Electrification and Automation for Industry and transport of the Moscow Power Engineering Institute (MPEI). Since 1991 he is working in the All-Russian Institute of Agriculture Electrification, which entered into the Federal Scientific Agroengineering Centre VIM in 2016. Head of the Department of

Energy supply of agriculture, chief researcher. Direction of scientific work: "electrical and heat supply of technological processes in agriculture." Scientific research is aimed at reducing the energy intensity of agricultural production, energy saving and improving the reliability of energy supply of the agro-industrial complex. He is the author of more than 160 scientific works.

Viktor Toporkov, Ph.D. (engineering), is a senior specialist of Federal State Budgetary Institution "Federal Scientific Agro-Engineering Center VIM".

Michael G. Tyagunov is a Doctor of Technical Sciences, Professor of the Department of Hydro power and Renewable Energy Sources, Moscow Power Engineering Institute. Academician of the Academy of Electrical Engineering.

Aleksey Vasiliev is a doctor of technical sciences, professor, Department Head of Federal State Budgetary Institution "Federal Scientific Agro-Engineering Center VIM". A graduate of the Faculty of electrification of the Azov-Black Sea institute of agricultural mechanization.

Sergey Vasilyev graduated from the Engineering faculty at the Samara State Agricultural Academy. In 2007 defended the master's thesis on the subject "Improvement of Method and Technical Means for Horizontal Measurement of Hardness of the Soil at Introduction of Technology of Coordinate Agriculture". Since 2009 –associate professor of "Electrification and Automation of Agrarian and Industrial Complex" department of the Engineering faculty of the Samara State Agricultural Academy. Main scientific areas: "Electrotechnologies and electric equipment in agriculture". Has 65 publications.

Alexander Vinogradov is a leading researcher at the Federal Scientific Agroengineering Center VIM and specializes in research into improving power supply system efficiency for rural consumers. Alexander studied at the Kostroma State Agricultural Academy from 1994 to 1999, majoring in Electrification and Automation of Agriculture. From 1999 to 2018 he worked as an assistant, senior lecturer, head of the department, associate professor of the Power Supply Department of the Orel State Agrarian University. Since 2018 he has been a leading researcher at the Federal Scientific Agroengineering Center VIM. Alexander has been the General Director of the LLC "Information and Energy Center AVPS-Innovation" since 2007 Alexander has experience in the preparation, organization and implementation of educational services, the development of teaching technologies, the development and creation of technical training facilitiesas well as experience in conducting research, developing theoretical and practical aspects of various kinds of interactions including social ones. He also has practical experience in conducting energy surveys, designing power supply systems, installing electrical equipment, managing the design organization and designing. He has 4 dan karate-do and 3 dan kobudo. Research interests: research is being carried out to develop new approaches to the creation of power supply systems, automation of electric networks, the development of energy accounting systems, the provision of quality of electricity supplied to consumers, energy saving, electrical modeling of human and society, educational technologies, use of renewable energy sources. He is an author and a co-author of more than 200 scientific and methodical works, including 33 patents, 9 monographs, 6 essays. He is also an author and a co-author of 4 collections of literary works. He is the author and the editor-in-chief of the newspaper "High Voltage Truth" and the editor-in-chief of the journal "Agrotechnics and energy supply".

Alina Vinogradova is a researcher at the Federal Scientific Agroengineering Center VIM and specializes in research in the field of power supply for rural consumers. Alina used to study at the Oryol State Agrarian University from 2003 to 2008. In 2015 she defended her thesis. Alina has great professional experience in the field of designing electrical networks. So she worked as a design engineer (electrician) in JSC Orelenergoremont in the period from 2007 to 2008; as an engineer for project work at JSC Orelenergoremont from 2008 to 2009; as an engineer of the design department of VL in JSC "Voronezhenergoproekt" from 2009 to 2010; as an assistant of the Department of "Power Supply" in OrelSAU from 2010 to 2011; as a senior lecturer of the department "Power supply" in OrelSAU from 2011 to 2016; as an associate professor of the Department of "Power Supply" in OrelSAU from 2016 to 2018. At the moment she works in the federal scientific agroengineering center VIM as a research assistant. The field of scientific activity is to develop methods and tools aimed at improving the efficiency of power supply. This includes, first of all, the development of methods and devices for monitoring the quality of electrical energy and the technical state of the elements of power supply systems as well as its automation.

Kok Yeow You was born in 1977. He obtained his B.Sc. Physics (Honours) degree in Universiti Kebangsaan Malaysia (UKM) in 2001. He pursued his M.Sc. in Microwave at the Faculty of Science in 2003 and his Ph.D. in Wave Propagation at the Institute for Mathematical Research in 2006 in Universiti Putra Malaysia (UPM), Serdang, Selangor, Malaysia. Recently, he is a senior lecturer at School of Electrical Engineering, Faculty of Engineering, Universiti Teknologi Malaysia (UTM), Skudai, Johor, Malaysia. His main personal research interest is in the theory, simulation, and instrumentation of electromagnetic wave propagation at microwave frequencies focusing on the development of microwave passive devices and sensors for medical and agricultural applications. He has authored or coauthored over a hundred publications.

Andrii Zhyltsov graduated from the Faculty of Physics of M.V. Frunze Simferopol State University (now V.I. Vernadsky Taurida National University) in the specialty "Physics" and received a qualification of a specialist "Physicist. Teacher". From 1998 to 2000 he studied at the post-graduate school of M.V. Frunze Simferopol State University on the specialty "Theoretical Electrical Engineering". From 2000 to 2005 he was a specialist of the Department of Applied Electrodynamics of V.I. Vernadsky Taurida National University. In 2005 he defended a dissertation for the degree of a candidate of technical sciences on the topic "Improving the efficiency of calculating magnetic fields in piecewise homogeneous linear and nonlinear environments by the method of integral equations" in the specialty theoretical electrical engineering. From 2005 to 2008 he was Ph.D. student of G. E. Puhov Institute of Modeling Problems in Energy of NAS of Ukraine. From 2008 to 2009 he was Associate Professor of the Department of Power Supply of Education and Research of Energy and Automation of the National University of Life and Environmental Sciences of Ukraine. In 2009 he defended his dissertation for the degree of Doctor of Technical Sciences in specialty mathematical modeling and computational methods. The subject of the doctoral dissertation is "Mathematical modeling of non-stationary processes of electromagnetic influence on liquid metal". From 2009 to 2011 he was the head of the Department of Power Supply of Education and Research Institute of Power Engineering and Automation of National University of Life and Environmental Sciences of Ukraine. Since February 2011 to September 2018 he was the head of the department of electrical machines and the operation of electrical equipment, and from September

2018 is the head of the department of electrical engineering of electromechanics and electrotechnology. Research interests: analysis and synthesis of electromagnetic fields, calculation and optimization of electrical equipment and structures, synthesis of electric and magnetic systems, analytical and numerical methods of calculating electromagnetic, thermal and other physical fields and processes, experimental methods of investigations of electromagnetic fields and processes.

Index

F

G

H

K

L

M

N

O

P

Q

R

S

T

V

W

Y

www.ingramcontent.com/pod-product-compliance
Lightning Source LLC
LaVergne TN
LVHW080530300926
849162LV00007B/1575